Springer Series in Statistics

Advisors:
P. Bickel, P. Diggle, S. Fienberg, U. Gather,
I. Olkin, S. Zeger

Springer Series in Statistics

Alho/Spencer: Statistical Demography and Forecasting
Andersen/Borgan/Gill/Keiding: Statistical Models Based on Counting Processes
Atkinson/Riani: Robust Diagnostic Regression Analysis
Atkinson/Riani/Ceriloi: Exploring Multivariate Data with the Forward Search
Berger: Statistical Decision Theory and Bayesian Analysis, 2nd edition
Borg/Groenen: Modern Multidimensional Scaling: Theory and Applications, 2nd edition
Brockwell/Davis: Time Series: Theory and Methods, 2nd edition
Bucklew: Introduction to Rare Event Simulation
Cappé/Moulines/Rydén: Inference in Hidden Markov Models
Chan/Tong: Chaos: A Statistical Perspective
Chen/Shao/Ibrahim: Monte Carlo Methods in Bayesian Computation
Coles: An Introduction to Statistical Modeling of Extreme Values
Devroye/Lugosi: Combinatorial Methods in Density Estimation
Diggle/Ribeiro: Model-based Geostatistics
Dudoit/Van der Laan: Multiple Testing Procedures with Applications to Genomics
Efromovich: Nonparametric Curve Estimation: Methods, Theory, and Applications
Eggermont/LaRiccia: Maximum Penalized Likelihood Estimation, Volume I: Density Estimation
Fahrmeir/Tutz: Multivariate Statistical Modeling Based on Generalized Linear Models, 2nd edition
Fan/Yao: Nonlinear Time Series: Nonparametric and Parametric Methods
Ferraty/Vieu: Nonparametric Functional Data Analysis: Theory and Practice
Ferreira/Lee: Multiscale Modeling: A Bayesian Perspective
Fienberg/Hoaglin: Selected Papers of Frederick Mosteller
Frühwirth-Schnatter: Finite Mixture and Markov Switching Models
Ghosh/Ramamoorthi: Bayesian Nonparametrics
Glaz/Naus/Wallenstein: Scan Statistics
Good: Permutation Tests: Parametric and Bootstrap Tests of Hypotheses, 3rd edition
Gouriéroux: ARCH Models and Financial Applications
Gu: Smoothing Spline ANOVA Models
Györfi/Kohler/Krzyżak/Walk: A Distribution-Free Theory of Nonparametric Regression
Haberman: Advanced Statistics, Volume I: Description of Populations
Hall: The Bootstrap and Edgeworth Expansion
Härdle: Smoothing Techniques: With Implementation in S
Harrell: Regression Modeling Strategies: With Applications to Linear Models, Logistic Regression, and Survival Analysis
Hart: Nonparametric Smoothing and Lack-of-Fit Tests
Hastie/Tibshirani/Friedman: The Elements of Statistical Learning: Data Mining, Inference, and Prediction
Hedayat/Sloane/Stufken: Orthogonal Arrays: Theory and Applications
Heyde: Quasi-Likelihood and its Application: A General Approach to Optimal Parameter Estimation
Huet/Bouvier/Poursat/Jolivet: Statistical Tools for Nonlinear Regression: A Practical Guide with S-PLUS and R Examples, 2nd edition
Ibrahim/Chen/Sinha: Bayesian Survival Analysis
Jiang: Linear and Generalized Linear Mixed Models and Their Applications
Jolliffe: Principal Component Analysis, 2nd edition

(continued after index)

Albert W. Marshall
Ingram Olkin

Life Distributions

Structure of Nonparametric, Semiparametric, and Parametric Families

With 138 Illustrations

Springer

Albert W. Marshall
Department of Statistics
University of British Columbia
Vancouver, BC Canada
and mailing address
2781 W. Shore Drive
Lummi Island, WA 98262
almarshall@earthlink.net

Ingram Olkin
Department of Statistics
Stanford University
Stanford, CA 94305
USA
iolkin@stat.stanford.edu

Library of Congress Control Number: 2007925439

ISBN-13: 978-0-387-20333-1   e-ISBN-13: 978-0-387-68477-2

Printed on acid-free paper.

© 2007 Springer Science+Business Media, LLC
All rights reserved. This work may not be translated or copied in whole or in part without the written permission of the publisher (Springer Science+Business Media, LLC, 233 Spring Street, New York, NY 10013, USA), except for brief excerpts in connection with reviews or scholarly analysis. Use in connection with any form of information storage and retrieval, electronic adaptation, computer software, or by similar or dissimilar methodology now known or hereafter developed is forbidden.
The use in this publication of trade names, trademarks, service marks, and similar terms, even if they are not identified as such, is not to be taken as an expression of opinion as to whether or not they are subject to proprietary rights.

9 8 7 6 5 4 3 2 1

springer.com

To our grandchildren for the pleasure, joy and affection they give us:

AWM	IO
Nathalie and Laurent	Leah and Jared
Sam and Ben	Noah and Sophia
	Rachel and Jeremy

To the memory of our colleagues; their friendship enriched our lives:

Robert E. Bechhofer, Z.W. (Bill) Birnbaum, James D. Esary, Jack Kiefer, Gerald J. Lieberman, Frank Proschan, Ron Pyke, Milton Sobel, Herbert Solomon

Preface

For many years the authors have been interested in developing methods for generating multivariate distributions, especially for positive data. Part of the motivation was to find models that would be useful in reliability and survival analysis. This led us to the idea of writing a book on multivariate nonnormal distributions. We thought that an introductory or reference chapter on the univariate case would be necessary. A preliminary effort to write that chapter was the genesis of the present book. We soon recognized what should have been obvious, that our original idea was overly ambitious. Even in just the univariate case, we found that to make the writing a manageable project we needed to impose some boundaries on the topics covered. This led to the decision to limit the book to probabilistic aspects of the subject and not to include statistical topics, which itself would make for another book.

Initially, we had in mind an audience having a background typical of someone with a master's degree in statistics, mathematics, or engineering. But the desire to be inclusive in order to make the book a more valuable reference led us to include a number of proofs that are somewhat more advanced than we would have liked. However, we believe that most of the results can be understood by our originally intended audience, and reading of the proofs may not be essential. Indeed, we have tried to provide motivations and insights to help the reader focus on the implications, rather than the proofs. Nevertheless, we have paid close attention to the proofs and have omitted only a few. Some proofs are similar to those in print, but many are new, and we hope they will provide further insights into the theory.

The reader should not hesitate to begin reading this book from almost any place; references to required earlier material have been made as needed. In fact, some of the topics of earlier chapters might best be appreciated if this kind of reading is followed.

One aspect of this book not present elsewhere is an effort to classify and understand various categories of parameters. Scale and location parameters are familiar in discussions of the normal distribution, but other kinds of parameters, often referred to as "shape parameters," also play a fundamental role in statistics. A number of such parameters are named and studied in this book. Here one of the important questions addressed is how parameters relate to orderings of distributions.

For engineering applications, a number of books on reliability theory are in print, and a number of texts are available on survival analysis aimed at medical applications. This book should not be viewed as a competitor to any such books, but we believe that it can stand alone or be profitably used in conjunction with other books to provide an increased depth of understanding.

Because a number of different distributions are surveyed in this book, comparisons with the many books on specific distributions or compendia on general distributions, as, for example, the volumes of Johnson, Kotz and Balakrishnan (1994, 1995), are inevitable. We find those volumes to be indispensable references, but they do not emphasize connections between various distributions to the extent done here.

The authors recognize that this book is by no means a complete treatment of its subject. Although the bibliography is extensive, we know that it is not complete. A few papers and their contents were intentionally omitted in a failed attempt to control the size of the book. But more serious are the papers that we inadvertently missed—papers with results that belong here. For these omissions, we offer our sincere apologies both to their authors and to our readers. For us, time was a factor; a manuscript was promised the publisher several years ago, and we both wanted to avoid posthumous publication.

We have made a special effort to give attribution to the originator of ideas and results, and we would appreciate readers bringing to our attention cases where we have not been accurate.

Suggestions for Using this Book

Convexity and total positivity are important powerful methods that are used throughout this book. Consequently, the reader should become familiar with the contents of Chapter 21. Readers well-versed in probability may skip Chapter 20, but for others this chapter may serve as a refresher of some of the needed concepts. Similarly, Chapter 24 provides a quick summary of some of the needed results from analysis.

Chapters 2 to 4 and 7 on ordering of distributions, mixtures, nonparametric families, and semiparametric families constitute the central theoretical portions of the book. These chapters may be referred to for readers wishing to find specific theoretical results in the theory of reliability and survival analysis.

Chapters 8 to 15 deal with specific parametric families, and here the focus is on connecting for each family the basic concepts in Chapters 2 to 4 and 7 to the parametric family. Thus, for example, there is a discussion on the family of inverse Gaussian distributions and the behavior of the hazard rate.

A discussion of coincidence of semiparametric families and stability of semiparametric families constitute the contents of Chapters 18 and 19, and are not as basic as the the earlier chapters.

Because in applications, parameters are often functions of covariates, Chapters 16 and 17 provide a review of their use.

A key feature in this book is an attempt to create a calculus of distributions. By this we show how different distributions may arise from a common origin, and how the hierarchy of distributions can be created. In so doing, we also provide a warehouse of tools that are used to provide new proofs to many of the results in this field.

Albert W. Marshall
Ingram Olkin
March 2007

Acknowledgements

The writing of this book did not occur in a vacuum. Numerous individuals have contributed over the years with their research, their ideas freely offered, and their encouragement. Our interest in distributions for nonnegative data started at the Boeing Scientific Research Laboratories where one of us (AWM) was a staff member and the other (IO) was a consultant. There, staff members and consultants included Jim Esary, George Marsaglia, Frank Proschan, Sam Saunders, Richard Barlow, Z.W.(Bill) Birnbaum and Ron Pyke. Burt Colvin, the Head of the Mathematics Group provided a great working ambiance. To all, we offer our thanks.

More recently, some of our colleagues have reviewed drafts of at least some chapters. Reviewing someone else's book is a labor of friendship, and we are most thankful to Barry Arnold, Henry Block, Mark Brown, Moshe Shaked, Tom Savits and Nozer Singpurwalla for their thoughtful comments.

Over the long gestation period of this book, we occasionally worked together at our own Universities. In addition, we spent time at the University of Augsburg through the generosity of Professor Friedrich Pukelsheim, and we worked together at the University of Neuchatel with the support of Professor Yadolah Dodge.

Our thanks to The Institute of Mathematical Statistics for permission to reprint some of the graphs. Initial versions of other graphs were made by Sarah Childress Emerson, Guillaume Christian Horel and Gill Ward. Final versions sometimes required more accurate numerical methods, and these were skillfully done by Julia Olkin. The graphs are an important part of this book, and we offer our sincere thanks.

We credit the National Science Foundation and the National Research Council of Canada for providing some support during part of the time that we were writing this book.

Finally but most importantly, we thank our spouses for tolerating our long periods of hibernation during the many years that this project was underway.

Contents

Preface vii
Acknowledgements xi
Basic Notation and Terminology xix

Part I. Basics

1. Preliminaries 3
   A. Introduction 3
   B. Probabilistic Descriptions 7
   C. Moments and Other Expectations 22
   D. Families of Distributions 25
   E. Mixtures of Distributions: Introduction 26
   F. Parametric Families: Basic Examples 28
   G. Nonparametric Families: Basic Examples 30
   H. Functions of Random Variables 32
   I. Inverse Distributions: The Lorenz Curve and the
      Total Time on Test Transform 35

2. Ordering Distributions: Descriptive Statistics 47
   A. Magnitude 49
   B. Dispersion 61
   C. Shape 67
   D. Cone Orders 76

3. Mixtures 79
   A. Basic Ideas 80
   B. The Conditional Mixing Distribution 83
   C. Limiting Hazard Rates 86

D. Hazard Transforms of Mixtures 88
E. Mixtures and Minima 92
F. Preservation of Orders Under Mixtures 94

Part II. Nonparametric Families

4. **Nonparametric Families: Densities and Hazard Rates** 97
   A. Introduction 97
   B. Log-Concave and Log-Convex Densities 98
   C. Monotone Hazard Rates 103
   D. Bathtub Hazard Rates 120
   E. Determination of Hazard Rate Shape 133

5. **Nonparametric Families: Origins in Reliability Theory** 137
   A. Coherent Systems 137
   B. Monotone Hazard Rate Averages 151
   C. New Better (Worse) Than Used Distributions 161
   D. Decreasing Mean Residual Life Distributions 169
   E. New Better (Worse) Than Used in Expectation Distributions 173
   F. Additional Nonparametric Families of Distributions 177
   G. Summary of Relationships and Closure Properties 180
   H. Shock Models 182
   I. Replacement Policies: Renewal Theory 187
   J. Some Additional Families 192

6. **Nonparametric Families: Inequalities for Moments and Survival Functions** 195
   A. Results Concerning Moments 195
   B. Bounds for Survival Functions 198

Part III. Semiparametric Families

7. **Semiparametric Families** 217
   A. Introduction 217
   B. Location Parameters 220
   C. Scale Parameters 224
   D. Power Parameters 228

E.	Frailty and Resilience Parameters: Proportional Hazards and Reverse Hazards	232
F.	Tilt Parameters: Proportional Odds Ratios, Extreme Stable Families	242
G.	Hazard Power Parameters	256
H.	Moment Parameters	258
I.	Laplace Transform Parameters	260
J.	Convolution Parameters	261
K.	Age Parameters: Residual Life Families	264
L.	Successive Additions of Parameters	265
M.	Mixing Semiparametric Families	267
N.	Summary of Order Properties	283
O.	Additional Semiparametric Families	284
P.	Distributions not Admitting Parameters	285

Part IV. Parametric Families

8. The Exponential Distribution **291**
    A. Defining Functions 292
    B. Characterizations of the Exponential Distribution 296
    C. Some Basic Properties of Exponential Distributions 302

9. Parametric Extensions of the Exponential Distribution **309**
    A. The Gamma Distribution 310
    B. The Weibull Distribution 321
    C. Exponential Distributions with a Resilience Parameter 333
    D. Exponential Distributions with a Tilt Parameter 338
    E. Generalized Gamma (Gamma–Weibull) Distribution 348
    F. Weibull Distribution with a Resilience Parameter 353
    G. Residual Life of the Weibull Distribution 355
    H. Weibull Distribution with a Tilt Parameter 355
    I. Generalized Gamma Convolutions 359
    J. Summary of Distributions and Hazard Rates 360

10. Gompertz and Gompertz–Makeham Distributions **363**
    A. The Gompertz Distribution 364
    B. The Extensions of Makeham 375
    C. Further Extensions of the Gompertz Distribution 390
    D. Summary of Distributions and Hazard Rates 396

11. Pareto and F Distributions and Their Parametric Extensions ... 399
   A. Introduction ... 399
   B. Pareto Distributions ... 400
   C. Generalized F Distribution ... 411
   D. The F Distribution ... 418
   E. Ordering Pareto and F Distributions ... 423
   F. Another Generalization of the Pareto Distribution ... 424

12. Logarithmic Distributions ... 427
   A. Introduction ... 427
   B. The Lognormal Distribution ... 431
   C. Log Logistic Distributions ... 441
   D. Log Extreme Value Distributions ... 442
   E. The Log Cauchy Distribution ... 443
   F. The Log Student's t Distribution ... 445
   G. Alternatives for the Logarithm Function ... 445

13. The Inverse Gaussian Distribution ... 451
   A. The Inverse Gaussian Distribution ... 452
   B. The Generalized Inverse Gaussian Distribution ... 459
   C. The Birnbaum–Saunders Distribution ... 466

14. Distributions with Bounded Support ... 473
   A. Introduction ... 473
   B. The Uniform Distribution and One-Parameter Extensions ... 475
   C. The Beta Distribution ... 479
   D. Additional Two-Parameter Extensions of the Uniform Distribution ... 489
   E. Introduction of a Scale Parameter ... 493
   F. Algebraic Structure of the Distributions on $[0, 1]$ ... 494

15. Additional Parametric Families ... 497
   A. Noncentral Chi-Square Distributions ... 497
   B. Noncentral F Distributions ... 501
   C. A Noncentral Beta Distribution and the Noncentral Squared Multiple Correlation Distribution ... 504
   D. Log Distributions from Nonnegative Random Variables ... 509
   E. Another Extension of the Exponential Distribution ... 518
   F. Weibull–Pareto–Beta Distribution ... 521

	Contents	xvii

G. Composite Distributions 523
H. Stable Distributions 529

Part V. Models Involving Several Variables

16. Covariate Models **533**
   A. Introduction 533
   B. Some Regression Models 536
   C. Regression Models for Other Parameters 540

17. Several Types of Failure: Competing Risks **541**
   A. Definitions and Notation 542
   B. The Problem of Identifiability 547
   C. Assumption of Independence 549
   D. Verifiability of Independence 554
   E. Known Copula 555
   F. Positively Dependent Latent Variables 557

Part VI. More About Semi-parametric Families

18. Characterizations Through Coincidences of Semiparametric Families **563**
   A. Introduction 564
   B. Coincidences Leading to Continuous Distributions 568
   C. Coincidences Leading to Discrete Distributions 596
   D. Unresolved Coincidences 607

19. More About Semiparametric Families **611**
   A. Introduction: Stability Criteria 611
   B. Classification of Parameters 612
   C. Derivation of Families 619
   D. Orderings Generated by Semiparametric Families 626
   E. Related Stronger Orders 630

Part VII. Complementary Topics

20. Some Topics from Probability Theory **635**
   A. Foundations 635
   B. Moments 644
   C. Convergence 650

D.	Laplace Transforms and Infinite Divisibility	653
E.	Some Discrete Distributions	658
F.	Poisson and Pólya Processes: Renewal Theory	663
G.	Extreme-Value Distributions	669
H.	Chebyshev's Covariance Inequality	673
I.	Multivariate Basics	674

21. Convexity and Total Positivity 687
   A. Convex Functions 687
   B. Total Positivity 694

22. Some Functional Equations 701
   A. Cauchy's Equations 701
   B. Variants of Cauchy's Equations 704
   C. Some Additional Functional Equations 712

23. Gamma and Beta Functions 717
   A. The Gamma Function 717
   B. The Beta Function 722

24. Some Topics from Analysis 729
   A. Basic Results from Calculus 729
   B. Some Results Concerning Lebesgue Integrals 731

References 733

Author Index 763
Subject Index 771

Basic Notation and Terminology

Throughout this book, the terms "increasing" and "decreasing" are used, respectively, to mean "nondecreasing" and "nonincreasing." Thus, the statement that ϕ is *increasing* means that

$$\phi(x) \leq \phi(y) \text{ whenever } x < y.$$

If the stronger condition

$$\phi(x) < \phi(y) \text{ whenever } x < y$$

holds, then ϕ is said to be *strictly increasing*. Similar use is made of the terms "decreasing" and "strictly decreasing."

Notation

(a) For any real number a, the notation $\bar{a} = 1 - a$ is often used. This same notation is used for real-valued functions ϕ, that is, $\bar{\phi}(x) = 1 - \phi(x)$.

(b) log is always a natural log, that is, the base is e. log 0 is taken to be $-\infty$, and $-\infty + x = -\infty$ for all real x.

(c) $\phi(t-) = \lim_{x \uparrow t} \phi(x)$, $\phi(t+) = \lim_{x \downarrow t} \phi(x)$.

(d) If a real-valued function of a real variable changes sign twice, first from + to − and then from − to + as its argument increases (0 values discarded), it is said that the function changes sign twice, "in the order $+, -, +$". Similar notations are used for other sign change patterns.

Section and Equation Numbering

Throughout this book, chapters are numbered, and sections within chapters are labeled with capital letters, whereas subsections are labeled with lower case letters, starting with "a" in each section. Equations are numbered, restarting at the beginning of each section.

Reference to equations by number only means that this equation is in the same section as the reference. Equations referenced by letter and number, say in the form B (7), refer to equation (7) of Section B in the chapter containing the reference. Equations referenced from another chapter are given the complete designation such as 9.B(3); this refers to the third equation of Section B in Chapter 9.

Part I

Basics

1

Preliminaries

Probability is a mathematical discipline with aims akin to those, for example, of geometry or analytical mechanics. In each field we must carefully distinguish three aspects of the theory: (a) the formal logical content, (b) the intuitive background, and (c) the applications. The character, and the charm, of the whole structure cannot be appreciated without considering all three aspects in their proper relation.
William Feller, *Introduction to Probability and Its Application*, Vol. 1, p. 1.

A. Introduction

Although the title of this book refers to reliability and survival analysis, nonnegative random variables arise in a wide variety of applications. Life-lengths of man-made devices or of biological organisms are, respectively, the focus of reliability and survival analysis. But other types of waiting times also arise in applications; these can be waiting times for delays in traffic, intervals between earthquakes or floods, or time periods required for learning a task. Nonnegative random variables also arise as magnitudes related to physical objects; these may be anthropomorphic measurements, crack lengths, tree diameters or heights, wind speeds, material strengths, stream flows, rainfall, tire wear, or chemical composition. Economics is another area of applications in which nonnegative random variables arise; income, firm size, prices, and actuarial losses are by their nature nonnegative.

By contrast, the normal distribution, which has long played a central role in statistics, allows for the corresponding random variables to take on all real values—both negative and positive. This is the case for measurement errors, the context in which the normal distribution first arose. For nonnegative random variables with standard deviations that are small compared to their means, the normal distribution has been widely used and can often provide excellent approximations. In other cases, the normal distribution may be an inappropriate model, and alternatives must be considered.

For nonnegative random variables, there is no distribution as pervasive as the normal distribution, with its foundation in the central limit theorem. This means that a wide variety of distributions share relative importance. The purpose of this book is to investigate the origins and properties of the various distributions for nonnegative random variables.

a. Statistical Motivation

For the analysis of data, several statistical approaches are in common use. These approaches form the following hierarchy:

Distribution-free (nonparametric) methods. Statistical methods that do not depend on any assumptions about the underlying distribution are attractive because there are no assumptions to question. Although attractive from this point of view, conclusions may be weaker than what might be possible with some plausible assumptions.

Qualitatively conditioned methods. Familiarity with the origins of the data may make qualitative assumptions reasonable. For example, a practitioner may know that data comes from a distribution with a decreasing density, a unimodal density, or that the density has a heavy right hand tail. They may suspect from physical considerations that the hazard rate is monotone or that it is initially decreasing and eventually increasing. A number of statistical procedures are known to test the validity of such assumptions. Others are based upon such assumptions, and they can lead to insights not easily obtainable with distribution-free methods.

Semiparametric methods. There are a number of possible hybrid methods that involve an unspecified distribution and a specific parametric structure. Perhaps the best known such model is the proportional hazards model, in which the survival function of the unspecified distribution is raised to a positive power; this power is then a parameter. Models such as this are called semiparametric models.

Parametric methods. Finally, one may be willing to assume that the data comes from a specified parametric family. This is perhaps the best known approach to statistical problems, and of course, the assumption of normality is the most familiar.

b. Use of Models

Except for the distribution-free methods, all of the statistical approaches described above depend upon what are sometimes called "models." Because the assumption of an inappropriate model can lead to erroneous conclusions, why are models ever used?

In some applications, the assumption of a model involves little or no risk. For example, the central limit theorem may make the normal distribution a clear choice. Or, with count data, it may be that the binomial or Poisson distribution clearly applies. Unfortunately, in many applications the "appropriate" model is not at all clear, but what constitutes an appropriate model clearly depends upon how the model is used and what is expected of it.

Parametric models were introduced early in studies of human life lengths. In this context, life tables or mortality tables are fundamental. Ideally, a mortality table starts with a fixed group of individuals, all born at the same time, and records the number living at the end of each successive year until all have passed away. Such a table represents an empirical record with data grouped by years. Due to both random fluctuations and errors in statements of age at death, the empirical "rates of mortality" calculated from such mortality tables are irregular. Consequently, actuaries concerned with the pricing of life insurance and annuities adopted various methods of *graduation*, described by Spurgeon (1932) as "the endeavor to arrive at the true law of mortality underlying the rough results disclosed by the data."

It was often assumed that the deaths in any year occurred uniformly over the year; for purposes of graduation, De Moivre (1724) made the assumption of uniformity over even longer periods of time, though he recognized that this approximation is not strictly true. Again for the purposes of graduation, Gompertz (1825) introduced the parametric model now known as the Gompertz distribution as an approximation to the "true law of mortality." Gompertz worked closely with data, and then developed a theoretical basis for his distribution, but to obtain a good fit to the data he found it necessary to divide ages into three groups, using different parameter values in each age group.

In a number of other applications, parametric models have played a prominent historical role in statistical theory, and were long studied under the older rubric of "curve fitting." Some of the best known early work is that of Pearson (1895), who created a set of "curves" or "frequency distributions" that might be suitable for fitting to data arising in a variety of contexts. Other sets of distributions were constructed by Bruns, Charlier, Edgeworth, Kapteyn, Thiele, and others. Elderton and Johnson (1969, p. 2) indicate that "The advantages of any system of curves depend on the simplicity of the formulae and the number of classes of observations that can be dealt with satisfactorily,...." See also Elderton (1906, 1934), Elderton and Johnson (1969), Särndal (1971), Cramér (1972). Thus, there was a focus upon both the richness of applications and mathematical simplicity.

What has been the motivation for selecting models? Fisher (1958, p. 41) writes that "From a limited experience, for example, of individuals of a species, or of the weather of a locality, we may obtain some idea of the infinite hypothetical population from which our sample is drawn, and so of the probable nature of future samples to which our conclusions are to be applied." Thus, prognosis or prediction is one purpose underlying this description, but other reasons are also stated in the literature. Models introduce parameters, the interpretation and behavior of which leads to insights. Models also permit the automation of statistical analysis and provide a sense of objectivity.

Historically, there was an underlying assumption that knowing that a distribution provides a reasonable fit to data is sufficient. But what is reasonable for one purpose may not be reasonable for another. This concern was captured by Kingman (1978), who noted that "Although it is often possible to justify the use of a distribution empirically, simply because it appears to fit the data, it is more satisfactory if the structure of the distribution reflects plausible features of the underlying mechanism."

Many of the models discussed in this book can be derived from assumptions that sometimes may be plausible on physical grounds. Properties and consequences of various qualitative assumptions are discussed, particularly as they relate to the structure of parametric and semiparametric models. The aim is to help the practitioner utilize knowledge of the physical origins of data when making a model choice.

Choice of a parametric model may also be based upon the data, particularly with the utilization of likelihood methods. However, even with a considerable amount of data it is often the case that several parametric families will appear to provide reasonable fits. Consequently, an understanding of the structure of these families and their origins can be important.

A substantial body of literature exists concerning model choice. This book does not include discussions of data-based model choice methods or the actual statistical methods to be used once a model is selected.

B. Probabilistic Descriptions

To mathematically describe the distribution of a random variable, various alternative functions are in common use. These functions include distribution functions, survival functions, densities, hazard rates, mean residual lives, and total time on test transforms. When they exist, any of these functions can be obtained, at least theoretically, from any other. But there are good reasons to be interested in all of these functions; none is uniformly best. Sometimes, one has a particularly simple form whereas others are awkward to work with. Perhaps more important is the fact that certain aspects of a distribution are revealed more clearly by one function than any other. Different people may have different preferences, depending upon the intuition they have developed. Also, some of these functions may be easier to estimate than others.

a. Distribution Functions and Survival Functions

B.1.a. Definition. The function F defined on the interval $(-\infty, \infty)$ by

$$F(x) = P\{X \leq x\}$$

is called the *distribution function* of X.

Distribution functions are sometimes called "cumulative distribution functions." When more than one random variable is being discussed, the distribution function of X is sometimes denoted by F_X. At other times, distributions are distinguished by a numerical subscript.

A distribution function F is nondecreasing and right continuous (i.e., $\lim_{z \downarrow x} F(z) = F(x)$). Moreover, $\lim_{z \to -\infty} F(z) = 0$ and $\lim_{z \to \infty} F(z) = 1$. Any function with these properties is a distribution for some random variable.

Some authors require distribution functions to be *left continuous*, and this is an equally acceptable convention, but not the one adopted here, and not in common use.

B.1.b. Definition. The function $\bar{F}$ defined on the interval $(-\infty, \infty)$ by

$$\bar{F}(x) = P\{X > x\}$$

is called the *survival function* of X.

Of course, $\bar{F} = 1 - F$ and so it might seem superfluous to introduce the survival function. But it is often the case that for nonnegative random variables, the survival function is more meaningful and takes a more convenient form than the better known distribution function.

The survival function is sometimes called the "survivor function" or the "reliability." Various notations for this function have been used in the literature; the "bar" notation was introduced by Frank Proschan and was used by Barlow and Proschan (1965). Since that time this notation has become widely used and is used in this book as well.

b. Probability Mass Functions and Density Functions

For any random variable, the distribution function and survival function always exist. This advantage is not enjoyed by probability mass functions or density functions, but these functions sometimes have other advantages.

Suppose first that the random variable X can take on only a finite or countable number of values. For example, X might be the number of trials required to obtain "heads" in repeated tosses of a coin. Then X (and F) is said to be *discrete*. Discrete distribution functions are step functions.

B.2.a. Definition. If $x_1, x_2, x_3, \ldots$ is the set of possible values of X and $p(x_i) = P\{X = x_i\}, i = 1, 2, \ldots$, then

$$F(x) = \sum_{x_i \leq x} p(x_i),$$

and p is called the *probability mass function* of X.

When X takes on all values in some (possibly infinite) interval of the real line, it is often possible to write F as an integral.

B.2.b. Definition. If f is a nonnegative function for which

$$F(x) = \int_{-\infty}^{x} f(z)\, dz, \text{ for all real } x,$$

then f is called a *probability density* of X (or F).

When a density exists, X (or F) is said to be *absolutely continuous*. When densities exist, they are not unique because they can be altered arbitrarily at isolated points without changing the integral. More specifically, they can be altered on a set of Lebesgue measure 0. In most examples, F is differentiable (except possibly at a few isolated points) and the derivative $f = F'$ is the usual form of the probability density.

B.2.c. Cautionary note. In a number of instances in this book, the shape of a density is discussed; for example, densities may be decreasing on $[0, \infty)$ or they may be unimodal. Such properties hold only for the "right" version of the density. Often, this is the derivative of the distribution function. But in what follows, it is often tacitly assumed that the "right" version of the density is under consideration.

Because densities are nonnegative and integrate to one, they cannot be increasing or decreasing on the entire real line. The densities f encountered in this book have the property that $\lim_{x \to \infty} f(x) = \lim_{x \to -\infty} f(x) = 0$. However, it is not difficult to construct examples where these limits fail.

If a density exists, then it has often been the preferred description of a distribution. The most studied and used absolutely continuous distribution, the normal distribution, has a density with a convenient mathematical expression whereas the distribution function cannot be written in closed form. Statisticians and researchers in various fields are trained to look at histograms, empirical approximations of densities, and they develop a feeling for their behavior.

B.3. Definition. A distribution F is said to be *concentrated* on the closed interval $[a, b]$ if $F(x) = 0$ for all $x < a$, and $F(x) = 1$ for all $x > b$. The *support* of the distribution F is the set of all points x such that $F(x + \varepsilon) - F(x - \varepsilon) > 0$, for all $\varepsilon > 0$.

It can be shown that the support of a distribution is a closed set, and in most examples discussed in this book, it is an interval. If F is concentrated on the interval $[a, b]$, then the support of F is a subset of that interval. When F is absolutely continuous and is concentrated on the closed interval $[a, b]$, then there is a natural version f of the corresponding density that satisfies $f(x) = 0$, for all $x \notin [a, b]$. Sometimes, the support of F is defined as the closure of the set of all points x such that $f(x) > 0$, but this presumes that an appropriate version of the density has been chosen.

Clearly, to say that F is said to be concentrated on the closed interval $[a, b]$ is less precise than to say that F has the support $[a, b]$, but it is easier to check and is sufficient for most purposes.

c. Unimodality

The idea of unimodality is best understood in terms of a density, but is perhaps best defined in terms of the distribution function.

B.4. Definition. A distribution function F is said to be *unimodal* with mode at m if $F(x)$ is convex in $x < m$ and concave in $x > m$.

When F is unimodal and has a continuous density f, then $f(x)$ is increasing in $x < m$ and decreasing in $x > m$ so that $f(m)$ is a maximum of f. If the density looks like the profile of a flat-topped mountain, then F is linear over some interval and the mode m may not be unique even though F is still said to be "unimodal." When the mode of a unimodal distribution is not unique, the set of modes form an interval.

The concept of a mode is sometimes extended to allow multiple modes (which do not together form an interval); these "modes" are local maxima of the density.

For more about unimodality see Section 20.A.e.

d. Hazard Functions and Hazard Rates

Some aspects of an absolutely continuous distribution, important or even essential in certain contexts, can be seen more clearly from the hazard rate than from either the distribution function or density function. Moreover, hazard rates have a distinct intuitive appeal, as described below.

B.5. Definition. The function R defined on $(-\infty, \infty)$ by

$$R(x) = -\log \bar{F}(x) \tag{1}$$

is called the *hazard function* of F, or of X.

For a nonnegative random variable, $R(0-) = 0$, R is increasing, and $\lim_{x \to \infty} R(x) = \infty$; any function with these properties is a hazard function. Hazard functions are sometimes convenient mathematical tools, but in contrast to hazard rates, they have little apparent intuitive appeal.

B.6. Definition. If F is an absolutely continuous distribution function with density f, then the function r defined on $(-\infty, \infty)$ by

$$\begin{aligned} r(x) &= \frac{f(x)}{\bar{F}(x)}, \quad \text{if } \bar{F}(x) > 0 \\ &= \infty, \quad \text{if } \bar{F}(x) = 0 \end{aligned} \tag{2}$$

is called a *hazard rate* of F, or of X.

For x such that $\bar{F}(x) = 0$, the ratio $f(x)/\bar{F}(x)$ is indeterminate; the value ∞ is arbitrary but sometimes convenient. Because densities are not unique, hazard rates are not unique, but fortunately, there is usually a natural version, identified except possibly on one or two points.

B.7. Notes on terminology. Particularly in the biostatistics literature, hazard rates are often called "hazard functions," and then hazard functions are called "cumulative hazard functions." This is a serious source of potential confusion; the term "hazard function" is used in the literature with two distinct meanings, and authors do not always clearly indicate their usage.

There is another potential source of confusion: because the survival function is sometimes called the "reliability," it is sometimes denoted by "R." The function R is called the hazard function in this book, but it has also been called the "hazard potential" by Singpurwalla (2003), who employs an innovative interpretation of it.

In the reliability literature, the hazard rate is very often called the "failure rate," though some authors object to this term because of its potential for confusion with another concept that arises in renewal theory (see Section 20.F.b and Sherwin (1997)). The term "mortality rate" has also been used. In actuarial work, the hazard rate has long been called the "force of mortality," but the terms "age specific force of mortality" and "intensity of mortality" have also been used. For the normal distribution, the reciprocal of the hazard rate is known as "Mills' ratio." It is unfortunate that such a confusing array of terminology is in use.

Although hazard rates are not unique, more often than not reference is made to "the hazard rate." When F is absolutely continuous and the hazard function R is differentiable, then its derivative is a hazard rate.

To more fully understand hazard rates, it is helpful to note that

$$r(x) = \lim_{\Delta \downarrow 0} \frac{P\{x < X \leq x + \Delta \mid X > x\}}{\Delta}.$$

Thus,

$$\Delta r(x) \approx P\{x < X \leq x + \Delta \mid X > x\}.$$

Consequently, $\Delta r(x)$ can be thought of as the conditional probability, given survival up to time x, of death or failure in the next small increment Δ of time. It is this interpretation that makes the concept of a hazard rate so useful, both in theory and in applications. It is interesting to note that actuaries long used the hazard rate, i.e., the "force of mortality," and did not really focus on the density ("curves of death" in their terminology) until the time of Karl Pearson.

From (1) and (2), it can be seen that if $F(0) = 0$, then

$$\bar{F}(x) = \exp\{-R(x)\} = \exp\left\{-\int_0^x r(z)\,dz\right\}. \tag{3}$$

This key formula shows how to retrieve the survival function from the hazard rate. But the second equality of (3) is valid only if F is absolutely continuous. In particular, it fails when F has a discrete part, even though F has a density apart from isolated points of discontinuity. See Singpurwalla and Wilson (1993) for a discussion of the validity of (3).

When F is absolutely continuous, it can be seen from the second equality of (3) that

$$r(x) = \frac{dR(x)}{dx}.$$

From (2) and (3) it follows that a function r is the hazard rate of some distribution on $(0, \infty)$ if and only if

(i) $r(x) \geq 0,$ for all $x > 0$,

(ii) $\int_0^x r(t)\,dt < \infty,$ for some $x > 0$,

(iii) $\int_0^\infty r(t)\,dt = \infty$,

(iv) $\int_0^x r(t)\,dt = \infty$ implies $r(z) = \infty,$ for all $z > x$.

Condition (ii) requires some explanation. If the distribution F corresponding to r satisfies $F(x) < 1$ for all x, then the integral of (ii) is finite for all $x < \infty$. But if there is a number $a < \infty$ such that $F(x) < 1$ for $x < a$ and $F(a) = 1$, then the integral (ii) is finite only for $x < a$. Condition (iv) results from the way the hazard rate is defined for $x > a$.

B.8. Proposition. If F is concentrated on $[0, a]$ and has the hazard rate r, then $\limsup_{x \uparrow a} r(x) = \infty$.

Proof. This is a consequence of (3) which shows that r is integrable on $[0, x]$ for $x < a$, but the integral must diverge on $[0, a]$. □

B.9. Observation. If the hazard rate r is decreasing at $x = x_0$, then the corresponding density f is also decreasing at x_0. This follows directly from (2).

It is shown in Section H that if $Z = \min[X, Y]$, then the hazard rate of Z is the sum of the hazard rates of X and Y. This mechanism produces a variety of hazard rate shapes.

e. Reverse Hazard Functions and Reverse Hazard Rates

The reverse hazard function is defined in a manner similar to the hazard function $R(x) = -\log \bar{F}(x)$, but with the distribution function F replacing the survival function $\bar{F}$. In addition, the minus sign is omitted to make it, like the hazard function, increasing.

B.10. Definition. The function S defined on $(-\infty, \infty)$ by

$$S(x) = \log F(x) \qquad (4)$$

is called the *reverse hazard function* of F, or of X. If F is an absolutely continuous distribution function with density f, then a function s defined on $(-\infty, \infty)$ by

$$s(x) = f(x)/F(x) \qquad (5)$$

is called a *reverse hazard rate* of F, or of X.

Note that $S(x) = \log F(x)$, whereas $R(x) = -\log \bar{F}(x)$; with these definitions, both functions are increasing.

The reverse hazard function and reverse hazard rate of F have not played a prominent role in the literature. The reverse hazard rate was introduced by von Mises (1936) and was discussed briefly by Barlow, Marshall and Proschan (1963), who note that $s(-x)$ is the hazard rate of $-X$, thus the terminology here. More recently it has been discussed in some detail by Block, Savits and Singh (1998); see also Shaked and Shanthikumar (1994, p. 24).

From (4) and (5), it can be seen that if $F(0) = 0$,

$$F(x) = \exp\{S(x)\} = \exp\left\{\int_0^x s(z)\, dz\right\}. \qquad (6)$$

The reverse hazard rate has largely been ignored in the literature primarily because it does not have the strong intuitive content of the hazard rate. These functions do not play a big role in this book, but they arise, e.g., in the study of domains of attraction for extreme value distributions in Section 20.G.

Note that if the reverse hazard rate is increasing at a point, say at x_0, then the density f is increasing at x_0. This means that a distribution F with increasing reverse hazard rate must, at some finite point m, satisfy $F(m) = 1$.

B.11. Proposition. The hazard rate r and reverse hazard rate s have the following monotonicity properties:

$$s \text{ is increasing} \Rightarrow r \text{ is increasing},$$
$$r \text{ is decreasing} \Rightarrow s \text{ is decreasing}.$$

These results are weak, and the implications do not reverse. They can be easily obtained from the relationship $r(x) = s(x)[F(x)/\bar{F}(x)]$. For a further result, see Proposition B.17.

f. The Residual Life Distribution

The distribution of remaining life for an unfailed item of age t is often of interest and plays a recurring role in what follows.

B.12. Definition. Let F be a distribution function such that $F(0) = 0$. The *residual life distribution* F_t of F at t is defined for all $t \geq 0$ such that $\bar{F}(t) > 0$ by

$$\bar{F}_t(x) = \frac{\bar{F}(x+t)}{\bar{F}(t)}, \quad x \geq 0. \tag{7}$$

If F has a density f, then F_t has density f_t and hazard rate r_t given by

$$f_t(x) = \frac{f(x+t)}{\bar{F}(t)}, \quad x \geq 0, \tag{8}$$

$$r_t(x) = \frac{f(x+t)}{\bar{F}(x+t)}$$
$$= r(x+t), \quad x \geq 0. \tag{9}$$

Clearly, the residual life distribution F_t is a conditional distribution of the remaining life given survival up to time t. This distribution is of considerable practical interest because the remaining life of devices (used cars, etc.) or of biological entities (people, for example) is often of interest.

B.13. Proposition. If the hazard rate r of F has a finite positive limit, $\lim_{t \to \infty} r(t) = \lambda$, then F_t converges in distribution to an exponential distribution (defined in Section F.a) with parameter λ as $t \to \infty$.

Proof. From (3), it follows that

$$-\log \bar{F}_t(x) = -\log \bar{F}(x+t) + \log \bar{F}(t) = \int_t^{t+x} r(z)\, dz \to \lambda x \text{ as } t \to \infty.$$

□

Proposition 20.G.5 gives a related but more general result. Limits of residual life distributions have been used by Rojo (1996) to categorize distributions according to "tail length."

g. The Mean Residual Life Function

In order to introduce the concept of a mean residual life, it is necessary to anticipate Section C of this chapter and define the "mean" or "expectation" of a random variable.

B.14. Definition. Suppose that the random variable X has the distribution function F and that the integral

$$\int_{-\infty}^{\infty} |x|\, dF(x)$$

exists (is finite). Then, the *expected value* EX of X exists and is given by the integral

$$EX = \int_{-\infty}^{\infty} x\, dF(x).$$

The expected value of X is also called the *mean* of X, or the *expectation* of X, and is often denoted by μ.

B.14.a. Proposition.

$$EX = \int_0^{\infty} \bar{F}(x)\, dx - \int_{-\infty}^0 F(x)\, dx; \tag{10a}$$

for nonnegative random variables, that is, for distributions such that

$F(x) = 0$ for $x < 0$,

$$EX = \int_0^\infty \bar{F}(x)\,dx. \tag{10b}$$

Proof. To obtain (10a), make use of Fubini's theorem 24.B.1 to compute

$$EX = \int_{-\infty}^\infty x\,dF(x) = \int_{x=0}^\infty \int_{z=0}^x dz\,dF(x) - \int_{x=-\infty}^0 \int_{z=x}^0 dz\,dF(x)$$

$$= \int_{z=0}^\infty \int_{x=z}^\infty dF(x)\,dz - \int_{z=-\infty}^0 \int_{x=-\infty}^z dF(x)\,dz$$

$$= \int_0^\infty \bar{F}(x)\,dx - \int_{-\infty}^0 F(x)\,dx.$$

See Figure B.1. □

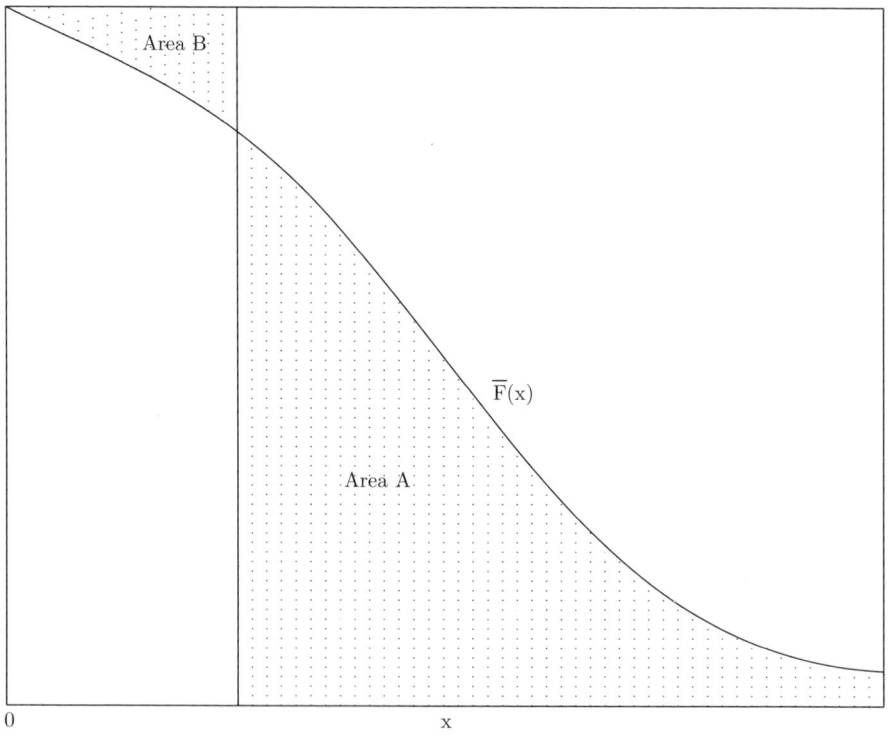

Fig. B.1. The expectation in terms of area: $EX =$ Area $A -$ Area B

B.15. Definition. The *mean residual life function* $m(t)$ is the mean of the residual life distribution F_t as a function of t. More explicitly, when F has a finite mean μ and $F(x) = 0$, for $x < 0$, the mean residual life function is given by

$$m(t) = \int_0^\infty \frac{\bar{F}(x+t)}{\bar{F}(t)}\, dx = \int_t^\infty \frac{\bar{F}(z)}{\bar{F}(t)}\, dz = \int_t^\infty \frac{(t-z)}{\bar{F}(t)}\, dF(z) \quad (11)$$

for t such that $\bar{F}(t) > 0$,

$$= 0, \quad \text{if } \bar{F}(t) = 0.$$

Formula (11) explains the terminology of this definition. Other terms have been used for this function; in the context of actuarial science, it has been called the *average excess claim* or the *mean excess function*.

The mean residual life function provides yet another way to describe a distribution. To see that it determines the survival function, first compute directly that for t such that $\bar{F}(t) > 0$,

$$\frac{d}{dt} \log \int_t^\infty \bar{F}(z)\, dz = -\frac{1}{m(t)}.$$

Now, integrate both sides of this equation from 0 to x and make use of (10) to obtain

$$\int_0^x \frac{dt}{m(t)} = \log \int_0^\infty \bar{F}(t)\, dt - \log \int_x^\infty \bar{F}(t)\, dt$$

$$= \log \mu - \log m(x) - \log \bar{F}(x).$$

This yields, for t such that $\bar{F}(t) > 0$,

$$\bar{F}(t) = \frac{\mu}{m(t)} \exp\left\{-\int_0^t \frac{dz}{m(z)}\right\}. \quad (12)$$

Here is the survival function in terms of the mean residual life function; this result is given by Cox (1962), Muth (1977), and Gupta (1979). Equation (12) is somewhat reminiscent of the much better known equation (3), which shows how to retrieve the survival function from the hazard rate.

As noted by Muth (1977), the hazard rate can also be directly obtained from the mean residual life function through the equation

$$r(t) = \frac{m'(t) + 1}{m(t)}. \tag{13}$$

This result can be verified directly by differentiating $m(t)$ using the second form given in (11). Because $r(t) \geq 0$, it follows from (13) that the derivative of the mean residual life has a lower bound that holds for all F;

$$m'(t) \geq -1. \tag{14}$$

This result was also noted by Muth (1977).

An excellent review of the theory and applications of the mean residual life function is provided by Guess and Proschan (1988). See also Hall and Wellner (1981), Kupka and Loo (1989), and Ghai and Mi (1999) for further discussions of the mean residual life.

h. Equilibrium Distributions

Let F be a distribution function with finite mean μ such that $F(x) = 0$ for $x < 0$, and let

$$f_{(1)}(x) = \frac{\bar{F}(x)}{\mu} = \frac{\bar{F}(x)}{\int_0^\infty \bar{F}(z)\,dz}, \quad x \geq 0, \tag{15}$$

$$= 0, \quad x < 0.$$

The density function $f_{(1)}$ arises in later chapters and in the context of renewal theory (see Section 20.F.b) where the corresponding distribution is called the *equilibrium distribution* or the *stationary renewal distribution*.

For any distribution F such that $F(0) = 0$, $f_{(1)}(0) = 1/\mu$, and with this, F can be retrieved from $f_{(1)}$. Clearly, every equilibrium density $f_{(1)}$ is decreasing. A density g is an equilibrium density if and only if (i) $g(x) = 0, x < 0$, (ii) g is decreasing on $[0, \infty)$, and (iii) $g(0) < \infty$.

It is straightforward to show that the hazard rate $r_{(1)}$ of the equilibrium distribution is the reciprocal of the mean residual life, that is,

$$r_{(1)}(t) = \frac{1}{m(t)}. \tag{16}$$

For a further discussion of equilibrium distributions and generalizations, see Section 20.B.c.

i. The Odds Ratio

If A and B are two mutually exclusive events, it is common especially in the content of gambling to speak of the "odds of A against B." This quantity is the ratio $P\{A\}/P\{B\}$ of the probabilities of the two events. Here, the two events are "survival beyond time x" and "failure by time x." It is in this context that odds ratios are often used in the medical literature, where comparisons are sometimes made between the odds ratio for a treatment group and the odds ratio for a control group.

Notation for odds ratios has not been standardized. Here, the odds of surviving and of not surviving both are considered, for which the notations $\emptyset^+$ and $\emptyset^-$ are used. Usually these odds ratios are not considered simultaneously, in which case there is no need to distinguish them. Then, notations such as "OR," "O," and "Odds" have all been used. In the medical literature, the notation "ω" for the odds ratio is particularly common.

B.16. Definition. The function $\emptyset^+$, defined for x such that $F(x) > 0$ by

$$\emptyset^+(x) = \bar{F}(x)/F(x), \tag{17a}$$

is called the *odds ratio of surviving beyond time x*. The function $\emptyset^-$ defined for x such that $\bar{F}(x) > 0$ by

$$\emptyset^-(x) = F(x)/\bar{F}(x) \tag{17b}$$

is called the *odds ratio of failure by time x*.

Sometimes, the term "odds" is used in place of "odds ratio." If both $F(x) > 0$ and $\bar{F}(x) > 0$, then $\emptyset^+(x) = 1/\emptyset^-(x)$. These odds ratios can be defined for all x with the convention that they take the value ∞ when a denominator is 0.

When the odds ratios exist,

$$\bar{F}(x) = \emptyset^+(x)/[1 + \emptyset^+(x)] = 1/[1 + \emptyset^-(x)]. \tag{18}$$

When both the odds ratios exist and F has a density,

$$\emptyset^+(x)r(x) = s(x) \quad \text{and} \quad \emptyset^-(x)s(x) = r(x), \tag{19}$$

where r and s are the hazard rate and reverse hazard rate, respectively.

Note that the odds ratio $\emptyset^+(x)$ is decreasing in x and $\emptyset^-(x)$ is increasing in x. It can be checked that the odds ratio $\emptyset_t^+$ of the residual life distribution at t is given by

$$\emptyset_t^+ = [\bar{F}(t)/\bar{F}(x+t)] - 1. \tag{20}$$

B.17. Proposition. Monotonicity of the hazard rate r, reverse hazard rate s, and convexity of odds ratios have the following relationships:

$$s \text{ is increasing} \Rightarrow r \text{ is increasing} \Rightarrow \emptyset^- \text{ is convex},$$
$$r \text{ is decreasing} \Rightarrow s \text{ is decreasing} \Rightarrow \emptyset^+ \text{ is convex}.$$

Proof. The implications relating r and s are given in Proposition B.11. Assuming that derivatives exist, the other implications can be verified by showing that the derivatives of $\emptyset^-$ and $\emptyset^+$ are monotone under the given conditions. □

The fact that r is decreasing implies $\emptyset^+$ is convex is given by Kirmani and Gupta (2001).

The odds ratio $\emptyset^-(x)$ for a distribution function F has all of the properties of the hazard function of another distribution, say H_-. That is,

$$\bar{H}_-(x) = e^{-F(x)/\bar{F}(x)} \tag{21}$$

defines a survival function. Similarly, $-\emptyset^+(x)$ has all of the properties of a reverse hazard rate such that

$$H_+(x) = e^{-\bar{F}(x)/F(x)} \tag{22}$$

is a distribution function. Equations (21) and (22) yield some possibly unexpected connections between familiar pairs of distributions. See, for example, Sections 10.A.e and 11.B.o.

Equations (21) and (22) can be solved for $\bar{F}$ to yield

$$\bar{F}(x) = \frac{1}{1 - \log \bar{H}_-(x)} = \frac{-\log H_+(x)}{1 - \log H_+(x)}.$$

Direct calculations show that if F has the hazard rate r, then H_- has the hazard rate $r(\cdot)/\bar{F}(\cdot)$. Thus, if r is increasing, then the hazard rate of H_- is increasing.

j. Inverse Distribution Functions

Inverse distribution functions also characterize a distribution, and as such could well find a place in this section. Because inverse distribution functions involve more technicalities and are perhaps somewhat less important, their consideration is delayed until Section I.

k. Summary

Table B.1. exhibits various functions that define a survival function. For the inverse distribution function, see Section I later in this chapter.

Table B.1. Alternatives for determination of a survival function

Function	Survival function
Density $f(x) = F'(x)$	$\bar{F}(x) = \int_x^\infty f(z)\,dz$
Hazard function $R(x) = -\log \bar{F}(x)$	$\bar{F}(x) = \exp\{-R(x)\}$
Hazard rate $r(x) = R'(x) = f(x)/\bar{F}(x)$	$\bar{F}(x) = \exp\left\{-\int_0^x r(z)\,dz\right\}$
Reverse hazard function $S(x) = \log F(x)$	$\bar{F}(x) = 1 - \exp\{S(x)\}$
Reverse hazard rate $s(x) = S'(x) = f(x)/F(x)$	$\bar{F}(x) = 1 - \exp\left\{\int_0^x s(z)\,dz\right\}$
Residual life distribution $\bar{F}_t(x) = \bar{F}(x+t)/\bar{F}(t)$	$\bar{F}(x) = \bar{F}_0(x)$
Mean residual life function $m(t) = \int_0^\infty \bar{F}(x+t)/\bar{F}(t)\,dx$	$\bar{F}(t) = \dfrac{\mu}{m(t)} \exp\left\{-\int_0^t \dfrac{dz}{m(z)}\right\}$
Odds ratios $\emptyset^+(x) = \bar{F}(x)/F(x)$ $\emptyset^-(x) = F(x)/\bar{F}(x)$	$\bar{F}(x) = \emptyset^+(x)/[1+\emptyset^+(x)]$ $\bar{F}(x) = 1/[1+\emptyset^-(x)]$
Equilibrium distribution $f_{(1)}(x) = \bar{F}(x)/\mu, \quad x \geq 0$	$\bar{F}(x) = [f_{(1)}(x)]/[f_{(1)}(0)],$ when $F(0) = 0$
Inverse distribution function $F^{-1}(p) = \sup\{z\,:\,F(z) \leq p\},$ $\quad 0 \leq p < 1$ $\quad = \sup\{z\,:\,F(z) < 1\},$ $\quad p = 1$	$\bar{F}(x) = 1 - \inf\{p\,:\,F^{-1}(p) \geq x\}$

C. Moments and Other Expectations

The expected value of a random variable is defined in Definition B.13. For discrete random variables, the expected value of X can be written in the notation of Definition B.2.a as

$$EX = \sum_{n=1}^{\infty} x_i p(x_i);$$

for absolutely continuous random variables with density f,

$$EX = \int_{-\infty}^{\infty} x f(x)\, dx.$$

Of course, for nonnegative random variables, the lower limit of the integral can be replaced by 0.

For a random variable with distribution function F such that $F(t) = 0$ for $t < 0$, it is well known (Proposition 20.B.1) that if $Y = \psi(X)$ and if the expectation EY exists, then

$$EY = E\psi(X) = \int_0^{\infty} \psi(x)\, dF(x). \tag{1}$$

A case of particular interest is $\psi(x) = x^r$, and then

$$\mu_r = EX^r = \int_0^{\infty} x^r\, dF(x) \tag{2}$$

is called the rth *moment of* X. The rth moment may or may not exist, i.e., the integral of (2) may or may not converge. According to Proposition 20.B.4, if $\mu_r < \infty$ for some $r > 0$, then $\mu_s < \infty, 0 \le s \le r$. Thus, the existence of the rth moment gurarantees the existence of all smaller positive moments. Of course, the first moment is just the expectation of X; this moment is also called the *mean of* X and is customarily denoted by μ, with the subscript 1 being omitted.

a. Moment Generating Functions and Laplace Transforms

Various generating functions can be used to find the moments μ_r when r is a positive integer.

C.1. Definition. The function

$$\mathrm{mgf}(s) = E\,e^{sX} = \sum_{j=0}^{\infty} \frac{s^j EX^j}{j!}$$

is called the *moment generating function* of X.

The moment generating function is finite for all s in some interval of the form $(-\infty, a)$ where $a \geq 0$. In case $a > 0$ and r is a positive integer, the rth derivative of the moment generating function evaluated at $s = 0$ yields the rth moment:

$$EX^r = \frac{d^r}{ds^r}\mathrm{mgf}(s)|_{s=0}.$$

In some contexts of this book, the less well-known normalized moments are more convenient than the moments themselves. The rth *normalized moment* λ_r of X is defined as

$$\lambda_r = \mu_r/\Gamma(r+1), \qquad (3)$$

where Γ is the well-known gamma function discussed in Chapter 23.

For purposes of computation, formulas (1) and (2) are not always the most convenient, especially when neither a probability mass function nor a density exists, or when there is a simple expression for $\bar{F}$. In this case, and for certain theoretical purposes, it is useful to note the alternative expression

$$EX^r = r \int_0^{\infty} \bar{F}(x) x^{r-1}\, dx. \qquad (4)$$

This formula, repeated in Proposition 20.B.3, can be established from (2) through an integration by parts (see 20.A.1); alternatively, some readers may prefer to write $x^r = \int_0^x r z^{r-1}\, dz$ in (2) and make a change in the order of integration.

Yet another form of the rth moment that is sometimes used in computations is given in Proposition 20.B.4, namely,

$$EX^r = \int_0^1 [F^{-1}(p)]^r\, dp. \qquad (5)$$

The formulas (2), (4), and (5) all provide ways to compute EX^r. For any given distribution F, the formulas can vary widely in their ease of use.

The existence or nonexistence of moments can sometimes be determined quite easily from the hazard rate, as indicated by Proposition 20.B.6.

C.2. Measures of location. The first moment EX of X is often used as a measure of the "center" or "location" of the distribution of X. Indeed, if the density of X were used as a profile and cut from a sheet of metal, then the cut-out would balance at EX. Another measure sometimes used to locate the "center" of the distribution of X is the *median* med X. A median is a point m such that $P\{X > m\} \leq 1/2$ and $P\{X < m\} \leq 1/2$. For strictly increasing distribution functions, the median is unique, and can be defined in terms of the inverse distribution (see Section I).

Another concept related to location is the mode (Definition B.4). If a unimodal density is symmetric about some point and has a finite expectation, that point is simultaneously a mode, a median, and an expectation (first moment).

C.3. Measures of spread. When EX^2 is finite, the *variance of X* (or F) exists and is defined by

$$\sigma^2 = Var(X) = E(X - EX)^2 = EX^2 - (EX)^2; \qquad (6)$$

The variance is often used as a measure of spread or dispersion. The variance and the *standard deviation* σ of X are perhaps less important for random variables with support $[0, \infty)$ than they are for distributions with support $(-\infty, \infty)$, but they are still the most commonly used measures of dispersion.

The *coefficient of variation of X* (or F) is defined as the ratio

$$CV(X) = \sigma/\mu; \qquad (7)$$

because $CV(aX) = CV(X)$, for all $a > 0$, the coefficient of variation is used as a measure of scale invariant dispersion.

A measure of concentration, the opposite of dispersion or spread, is the Gini index of Definition I.14 below.

C.4. Transforms. Several kinds of transforms are defined in terms of expectations. In particular, the *Laplace transform* ϕ of X is defined

as
$$\phi(s) = E\, e^{-sX}.$$

For nonnegative random variables, the Laplace transform exists for all $s \geq 0$ and may exist for some or all values of $s < 0$. Laplace transforms and several of their important properties are discussed in Section 20.D.

The moment generating function mgf(s) of X is related to the Laplace transform through the equation

$$\mathrm{mgf}(s) = \phi(-s).$$

Finally, the *Mellin transform* mel(s) of X is defined by

$$\mathrm{mel}(s) = EX^s$$

for all values of s such that the expectation is finite.

Unlike the Laplace transform, the Mellin transform does not necessarily determine the distribution of X.

D. Families of Distributions

Families of distributions indexed by a real number or by several real numbers are called *parametric families*, and the indexing variables are called *parameters*. For a distribution F having a parameter θ, the notations $F(\cdot\,|\,\theta)$ and F_θ are used interchangeably in what follows.

The most familiar parameters are location and scale parameters, which are best introduced by way of the following definition.

D.1. Definition. Distributions F and G are said to be *of the same type* if, for some real number b and some $a > 0$,

$$F(x) = G(ax + b), \quad \text{for all } x.$$

There is a symmetry in this definition relating F and G, and it can equivalently be said that

$$F\left(\frac{x-b}{a}\right) = G(x).$$

Whether in the form $ax + b$ or $(x-b)/a$, a is said to be a "scale parameter" and b is called a "location parameter." To avoid confusion in this book, the following terminology is adopted.

D.2. Definition. A parametric family $\{F(\cdot \mid \lambda), \lambda > 0\}$ of the form $F(x \mid \lambda) = F(\lambda x \mid 1)$ is said to be a *scale parameter family* and λ is called a *scale parameter*.

As noted above, the alternative definition that would call $1/\lambda$ a scale parameter rather than λ itself could just as well have been adopted, and would be more natural in some contexts. In particular, such an alternative definition would be necessary to make the standard deviation of the normal distribution a scale parameter. But in the context of nonnegative random variables where the exponential distribution plays a central role, Definition D.2 is more convenient and it simplifies typography.

Location parameters do not play a central role in the study of life distributions because these distributions are concerned with nonnegative random variables that have a natural location. However, a number of other kinds of parameters are important and are discussed in detail in Chapter 7.

E. Mixtures of Distributions: Introduction

If F_1 and F_2 are distribution functions and $0 < \pi < 1$,

$$F = \pi F_1 + (1 - \pi) F_2, \tag{1a}$$

or equivalently,

$$\bar{F} = \pi \bar{F}_1 + (1 - \pi) \bar{F}_2, \tag{1b}$$

then F is said to be a *mixture* of F_1 and F_2. Unless F is the distribution function of a constant random variable, it can arise as a mixture of two different distributions in infinitely many ways so that a decomposition or mixture representation of the form (1) is not unique. But in practical applications, there is often one mixture that is natural. For example, data on humans can naturally be separated according to ethnic origins or by gender. It is often the case that data cannot be fully understood without recognizing the mixture aspect, so it is important to see what can be learned from mixture representations. Sometimes, such representations are also helpful in theoretical studies, especially when the components of the mixture are relatively simple to understand and work with.

E.1. Definition. Let $\mathcal{F} = \{F_\theta \mid \theta \in \Theta\}$ be a family of distributions and let G be a distribution on Θ. Then,

$$F(x) = \int_\Theta F_\theta(x)\, dG(\theta) \tag{2}$$

is the *mixture of $\mathcal{F}$ with respect to G*.

Mixtures have sometimes been called *compound distributions*.

It is easy to see that the densities (if they exist) and survival functions of mixtures are mixtures of the corresponding densities and survival functions. That is,

$$f(x) = \int_\Theta f_\theta(x)\, dG(\theta),$$

$$\bar{F}(x) = \int_\Theta \bar{F}_\theta(x)\, dG(\theta).$$

But the corresponding formulas are not true for hazard functions or hazard rates. When the distributions F_θ of (2) have densities f_θ, F has the hazard rate

$$r(x) = \frac{\displaystyle\int_\Theta f_\theta(x)\, dG(\theta)}{\displaystyle\int_\Theta \bar{F}_\theta(x)\, dG(\theta)}. \tag{3}$$

An interesting special case is the mixture of but two distributions, say F_1 and F_2, as in (1). Let $\bar{\pi} = 1 - \pi$. Then, when densities exist, (2) and (3) take the form

$$F(x) = \pi F_1(x) + \bar{\pi} F_2(x) \tag{4}$$

and

$$r(x) = \frac{\pi f_1(x) + \bar{\pi} f_2(x)}{\pi \bar{F}_1(x) + \bar{\pi} \bar{F}_2(x)}. \tag{5}$$

In this special case, $\bar{F}(x) = \pi \bar{F}_1(x) + \bar{\pi} \bar{F}_2(x)$.

A more complete discussion of mixtures is given in Chapter 3.

F. Parametric Families: Basic Examples

In this section, the exponential, Weibull, gamma, and lognormal distributions are briefly introduced to make them available for illustrative purposes. These distributions are discussed in detail in Chapters 8, 9, and 12.

a. The Exponential Distribution

The one parameter family of exponential distributions, often referred to simply as "the exponential distribution," is without competition for the position of the most fundamental, basic family of life distributions. For this distribution, the parameter $\lambda > 0$ is a scale parameter and

$$\bar{F}(x) = e^{-\lambda x}, \quad x \geq 0, \tag{1}$$

$$f(x) = \lambda e^{-\lambda x}, \quad x \geq 0, \text{ and} \tag{2}$$

$$r(x) = \lambda, \quad x \geq 0. \tag{3}$$

For the exponential distribution, it is easy to see that

$$\frac{\bar{F}(x+t)}{\bar{F}(t)} = \bar{F}(x), \tag{4}$$

that is, the conditional probability of surviving an additional period of time x, given survival up to time t, is the same as the unconditional probability of survival to time x. In fact, this property characterizes the exponential distribution (see Proposition 8.B.1). Both (3) and (4) can be interpreted as saying that an item with an exponential distribution is not affected by wear or ageing, and it is this property that provides the basis for the importance of the distribution.

If X has an exponential distribution with parameter λ, then for $r > -1$,

$$\mu_r = EX^r = \int_0^\infty x^r \lambda e^{-\lambda x} \, dx = \Gamma(r+1)/\lambda^r, \tag{5}$$

where Γ is the usual gamma function, discussed in Chapter 23. Thus, exponential distributions have finite moments of all orders greater than -1 and they have a simple form. The normalized moments defined by (3) of Section C have an even simpler form and they play a special role in Section 6.A.

The terms "exponential distribution" and "exponential family" should not be confused. The former is a specific parametric family, whereas the latter is a broad family often of use in statistical analysis because it has a convenient form and includes a wide range of distributions.

b. The Gamma Distribution

The two parameter family of gamma distributions includes the exponential distribution as a special case. Whereas the density of the gamma distribution has a nice form, the survival function and hazard rate can be written in closed form only for certain parameter values. Again, the scale parameter $\lambda > 0$; additionally, there is a shape parameter $\nu > 0$ and

$$f(x \mid \lambda, \nu) = \lambda^\nu x^{\nu-1} e^{-\lambda x}/\Gamma(\nu), \quad x \geq 0. \tag{6}$$

With the shape parameter $\nu = 1$, this is just an exponential distribution. The fact that this density integrates to unity is a direct consequence of the usual definition of the gamma function as an integral (Definition 23.A.1).

When the shape parameter ν is an integer,

$$\bar{F}(x \mid \lambda, \nu) = \sum_{k=0}^{\nu-1} e^{-\lambda x} (\lambda x)^k / k!, \quad x \geq 0. \tag{7}$$

Of course, the hazard rate can be easily obtained using (6) and (7), but the resulting expression is awkward and needs to be evaluated numerically to be understood. See Chapter 11 for further details.

c. The Weibull Distribution

The Weibull distribution is another two parameter family that includes the exponential distribution. This family has a scale parameter λ and a shape parameter α, both positive. Unlike the gamma distribution, the survival function here has a simple form, specifically

$$\bar{F}(x) = \exp\{-(\lambda x)^\alpha\}, \quad x \geq 0. \tag{8}$$

Differentiation of this survival function yields the density function

$$f(x) = \alpha\lambda(\lambda x)^{\alpha-1} \exp\{-(\lambda x)^\alpha\}, \quad x \geq 0. \tag{9}$$

Clearly,
$$r(x) = \alpha\lambda(\lambda x)^{\alpha-1}, \quad x \geq 0. \tag{10}$$

For a more complete discussion of the Weibull distribution see Chapter 9.

d. The Lognormal Distribution

If Y is a random variable having a normal distribution and $X = e^Y$, then X is said to have a *lognormal distribution*. Because the normal distribution function does not have a closed form, neither the distribution function nor the hazard rate of the lognormal distribution can be expressed in closed form. But the density can be obtained from the density of the normal distribution.

The lognormal distribution can be usefully parameterized in several ways, three of which are noted here. Suppose that $-\infty < \mu, \beta < \infty, 0 < \lambda, \sigma, \alpha$, and let

$$\mu = -\log\lambda, \quad \sigma = 1/\alpha, \quad \text{and} \quad \beta = \alpha^2\mu = \mu/\sigma^2. \tag{11}$$

For $x > 0$, the density is given by

$$f(x) = \frac{1}{\sqrt{2\pi}\sigma x} \exp\left\{\frac{-(\log x - \mu)^2}{2\sigma^2}\right\} \tag{12a}$$

$$= \frac{\alpha}{\sqrt{2\pi}x} \exp\left\{\frac{-[\log(\lambda x)^\alpha]^2}{2}\right\} \tag{12b}$$

$$= x^\beta \exp\{-\beta^2/2\alpha^2\} \frac{\alpha}{\sqrt{2\pi}x} \exp\left\{\frac{-[\log x^\alpha]^2}{2}\right\}. \tag{12c}$$

Thus, the density can be expressed in terms of the parameters $(\mu, \sigma), (\lambda, \alpha)$, or (α, β). Further discussion of the lognormal distribution can be found in Chapter 12.

G. Nonparametric Families: Basic Examples

A number of nonparametric families of life distributions are discussed in Chapter 4; here, some basic examples are briefly introduced.

a. Log-Concave and Log-Convex Densities

The logarithm of a number of standard densities is either convex or concave. (A discussion of convexity and log convexity is given in Section 21.A.)

(i) The normal density is a well-known example of a log-concave density.
(ii) It can be seen from (6) and the concavity of the logarithm function that the gamma density is log concave for $\nu \geq 1$ and log convex for $\nu \leq 1$.
(iii) The Weibull density (9) is log concave for $\alpha \geq 1$ and log convex for $\alpha \leq 1$.

These facts have implications, both probabilistic and statistical, that make them worth noting. Some conditions for log convexity or log concavity and various probabilistic consequences are given in Chapters 3 and 4.

b. Monotone Hazard Rates

The notion of a monotone hazard rate has played an important role in reliability theory since the early 1960s. It cannot be escaped in any serious study of life distributions. A distribution F is said to have an *increasing hazard rate* (IHR) if it has a density f for which $r = f/\bar{F}$ is increasing. (Definition 4.C.1 of this concept is slightly more general in that it does not require the existence of a density.) A random variable with a distribution having an increasing hazard rate is also said to be IHR.

As already mentioned in Section B, the importance of the hazard rate stems from the interpretation of $r(t)\,dt$ as the conditional probability of failure in the interval $[t, t+dt]$ given survival up to time t. With an increasing hazard rate, the probability of failure in the next instant of time increases as the device or organism ages, a property intuitively appealing as a mathematical description of "wearout."

The similarly defined notion of a *decreasing hazard rate* (DHR) is of less obvious interest because it appears to be a mathematical description of what might be called "wearin." Nevertheless, wines, cheeses, and violins provide examples of items that may improve with age, so there are some direct applications. But perhaps more importantly, the property arises in mixtures, as discussed in Chapter 3.

If a density is log concave [convex], then the corresponding hazard rate is increasing [decreasing]. These facts are the content of Proposition 4.B.8.a, and an indirect proof is obtained in Chapter 2. See Remark 2.A.15.a.

H. Functions of Random Variables

The minimum, maximum, and a sum of two independent random variables are briefly discussed here; these functions are encountered in several places in this book. One function, the reciprocal, of a single random variable is also mentioned.

H.1. Reciprocals. If X is a positive random variable with survival function $\bar{F}$, then $Y = 1/X$ is also a positive random variable. The distribution function G of Y is given by

$$G(y) = \bar{F}(1/y), \quad y > 0.$$

When F is a gamma or Weibull distribution, then sometimes the distribution G is called the "inverse" gamma or "inverse" Weibull distribution. There is some possible confusion in this terminology, because the name of the inverse Gaussian distribution is well entrenched and that distribution is not the distribution of the reciprocal of a Gaussian variate; it arises in a quite different way. The term "inverse" is also used in another way in Section I. In this book, the term *reciprocal distribution* is used for the distribution of the reciprocal of a random variable.

H.2. Minima. The minimum $Z = \min(X, Y)$ of two random life lengths X and Y arises when there are two distinct possible causes of failure. Think of X as the waiting time for "death" due to one cause and Y as the waiting time for "death" due to the other cause. Ordinarily, it is not possible to observe both X and Y, but only their minimum. In this setup, the two causes of death are called "competing risks," and they are the subject of further study in Chapter 17. This theory was developed primarily with medical applications in mind. But there is another point of view in which minima arise in engineering applications.

If a device has two components, both of which are essential for the device to function, then the life length of the device is the minimum of the two component life lengths. Devices of this kind are called "series systems"; they are another reason why the distribution of the minimum of two random variables is interesting. More general systems are considered in Chapter 5.

Let X and Y be independent random variables and let $Z = \min(X, Y)$. If the respective distributions of X, Y, and Z are F, G, and H, it follows that

$$\bar{H}(x) = \bar{F}(x)\,\bar{G}(x). \tag{1}$$

If F and G have densities f and g, then H has the density

$$h(x) = f(x)\,\bar{G}(x) + \bar{F}(x)\,g(x) \tag{2}$$

and hazard rate

$$r_H(x) = \frac{h(x)}{\bar{H}(x)} = \frac{f(x)}{\bar{F}(x)} + \frac{g(x)}{\bar{G}(x)}. \tag{3}$$

Thus, the hazard rate of the minimum is the sum of the component hazard rates. It follows from (3) that if X and Y are independent and have increasing hazard rates, then Z has an increasing hazard rate. Other consequences of (3) are noted later in this book. Of course, (1)–(3) fail to hold when X and Y are not independent.

In the context of competing risks, improper random variables (those that need not be finite valued) may be of some practical interest because there may be causes of death that affect some, but not all, individuals. If $Z = \min[X, Y]$, then Z is a proper random variable as long as at least one of the variables X and Y is proper. In this book, distribution functions are assumed to be proper, but improper distribution functions can be constructed by multiplying a proper one by a factor less than one.

H.3. Maxima. When high reliability is important, systems are often constructed with redundancy. In a simple case, the system has two components and it functions as long as at least one of the components functions. Such a system is called a "parallel" system, and its life length is the maximum of the component life lengths. The renal system in humans is an example of a parallel system because it consists of two kidneys, and one alone is sufficient.

Let X and Y be independent random variables and let $Z = \max[X, Y]$. If the respective distributions of X, Y, and Z are F, G, and K, it follows that

$$K(x) = F(x)G(x). \tag{4}$$

If F and G have densities f and g, then K has the density

$$k(x) = f(x)G(x) + F(x)g(x) \tag{5}$$

and hazard rate

$$r_K(x) = \frac{k(x)}{\bar{K}(x)} = \frac{f(x)G(x) + g(x)F(x)}{1 - F(x)G(x)}. \tag{6}$$

This is a contrast to (3), because unlike the minimum, the hazard rate of the maximum does not have a simple expression in terms of the component hazard rates. On the other hand, the reverse hazard rate for the maximum Z is given in terms of the reverse hazard functions of X and Y by

$$s_K(x) = \frac{k(x)}{K(x)} = \frac{f(x)}{F(x)} + \frac{g(x)}{G(x)},$$

and this is a counterpart to (3).

H.4. Sums. Sometimes, a spare part is available to be placed in service when the original part fails. This terminology is standard in industry, but organ transplantation can also be considered as the utilization of a "spare part." Together, the original part and the spare part act as a system, which has a life length that is the sum of the two component life lengths. This system is similar to the parallel system in that it has built-in redundancy.

Again, let X and Y be independent nonnegative random variables, but now, let $Z = X + Y$. If the respective distributions of X, Y, and Z are F, G, and H, then

$$H(x) = \int_0^\infty F(x-z)\,dG(z) = \int_0^x F(x-z)\,dG(z); \tag{7}$$

H is called the *convolution* of F and G and is denoted by $H = F * G$. Because H is the distribution of $Z = X + Y$, it is apparent that convolutions are commutative and distributive, that is,

$$F * G = G * F \quad \text{and} \quad F_1 * (F_2 * F_3) = (F_1 * F_2) * F_3.$$

When F and G have densities f and g, then $H = F * G$ has the density

$$h(x) = \int_0^x f(x-z)\, dG(z) = \int_0^x f(x-z)g(z)\, dz, \tag{8a}$$

and h is said to be the *convolution* of f and g. In case the random variables are not nonnegative, (8a) is replaced by

$$h(x) = \int_{-\infty}^{\infty} f(x-z)\, dG(z) = \int_{-\infty}^{\infty} f(x-z)g(z)\, dz. \tag{8b}$$

Note that (7) has the form of E(2) with $F_\theta(x) = F(x - \theta)$, so the convolution can be viewed as a mixture.

I. Inverse Distributions: The Lorenz Curve and the Total Time on Test Transform

Inverse distribution functions sometimes play an important role and must be handled with some care because they have no generally accepted definition. The problem is that the definition involves a degree of arbitrariness. Both the total time on test transform and the Lorenz curve, discussed in this section, are often defined in terms of inverse distributions.

a. Inverse Distribution Functions

Inverse distributions are essentially the same as quantiles and are sometimes called "quantile functions"; the important statistical role that these functions play is surveyed by Parzen (2004). For a distribution function F, a number q satisfying $F(x) \leq p$ for all $x < q$ and $F(q) \geq p$ is called a pth *quantile* of F. The apparent asymmetry of this definition is due to the fact that distribution functions have been assumed to be right continuous. If F is strictly increasing on its support, then for $0 < p < 1$, the pth quantile is unique. But if F is "flat" at level p, then there is a closed interval of values all of which qualify as pth quantiles. Defining an inverse distribution function is essentially the same as defining pth quantiles in such a way that they are unique, and this is where the arbitrariness appears. Here, to avoid technicalities, the basic ideas are introduced under the assumption that F is strictly increasing. For a discussion of the general case, see Section 20.A.f.

I.1. Definition. For a strictly increasing distribution function F, the *inverse* F^{-1} of F is the function defined by

$$F^{-1}(p) = \sup\{z : F(z) \leq p\} = \inf\{z : F(z) \geq p\}, \quad 0 < p < 1. \quad (1)$$

Similarly, the *inverse* $\bar{F}^{-1}$ of the survival function $\bar{F}$ is defined by

$$\bar{F}^{-1}(p) = \sup\{z : \bar{F}(z) \geq p\} = \inf\{z : \bar{F}(z) \leq p\}, \quad 0 < p \leq 1,$$
$$= \sup\{z : \bar{F}(z) > 0\} = \inf\{z : \bar{F}(z) = 0\}, \quad p = 0.$$

With these definitions,

$$F^{-1}(1-p) = \bar{F}^{-1}(p) \quad (2)$$

and $F^{-1}(1) = \infty$.

From the inverse F^{-1} of F, F can be recovered via the formula

$$F(z) = \inf\{p : F^{-1}(p) > z\}. \quad (3)$$

b. The Total Time on Test Transform

Suppose that several items are placed on test for a fixed period of time to determine their life lengths. Some of the items may fail during the test period, but others may still be functioning at the termination of the test. The *total time on test statistic* is the sum of all observed complete and incomplete lifetimes. This statistic has a theoretical limit as the number of items placed on test goes to infinity, a limit known as the *total time on test transform*. The total time on test transform was introduced by Barlow, Bartholomew, Bremner and Brunk (1972) for its usefulness in certain estimation problems. It was further studied by Barlow and Campo (1975) who discuss its use as a tool for selecting a model for data analysis. Barlow (1979) discusses the total time on test transform distribution and its properties.

I.2. Definition. The function H_F^{-1} defined on the interval $[0, 1]$ by

$$H_F^{-1}(p) = \int_0^{F^{-1}(p)} \bar{F}(x)\,dx \quad (4)$$

is called the *total time on test transform* of the distribution function F. See Figure I.1a,b.

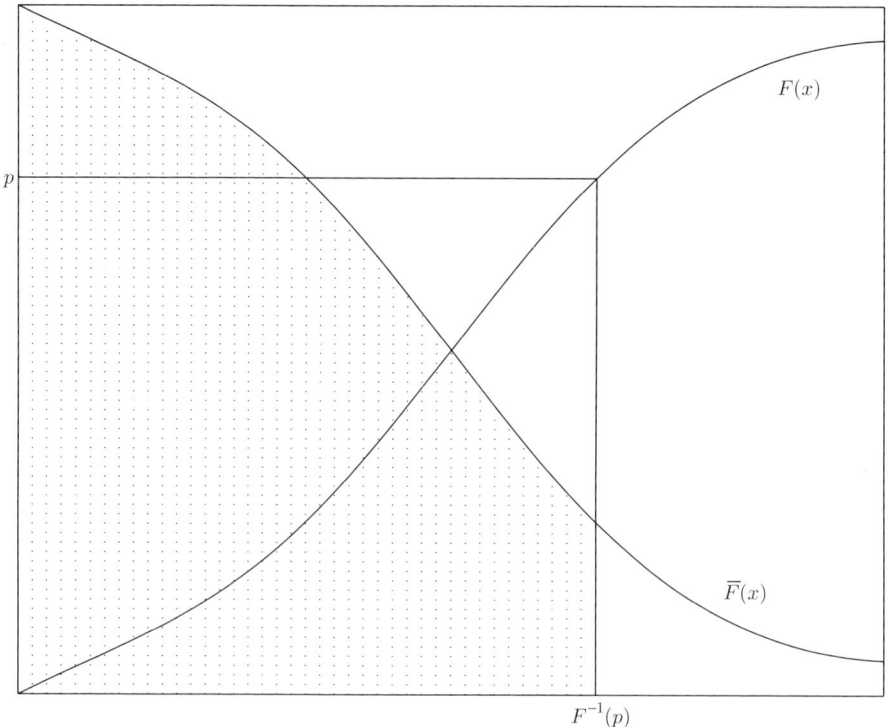

Fig. I.1a. The total time on test transform (shaded area)

When the distribution F is clear from the context, the notation

$$\psi(p) = H_F^{-1}(p)$$

is sometimes used for typographical simplicity. The function H_F is called the *total time on test transform distribution*.

Note that $H_F^{-1}(0) = 0$, and if F has mean $\mu \leq \infty$, then because $\bar{F}^{-1}(1) = \infty$, it follows from C(4) that $H_F^{-1}(1) = \mu$. Because F is an increasing function, so is F^{-1}, and this means H_F^{-1} is an increasing function. Consequently, H_F is increasing, which together with $H_F(0) = 0$, and $H_F(\mu) = 1$, shows that H_F is a distribution function concentrated on $[0, \mu]$.

I.3.a. Example. Suppose that F is an exponential distribution (discussed in Section F.a). Then,

$$\bar{F}^{-1}(p) = -\frac{1}{\lambda} \log(1-p), \quad 0 \leq p \leq 1,$$

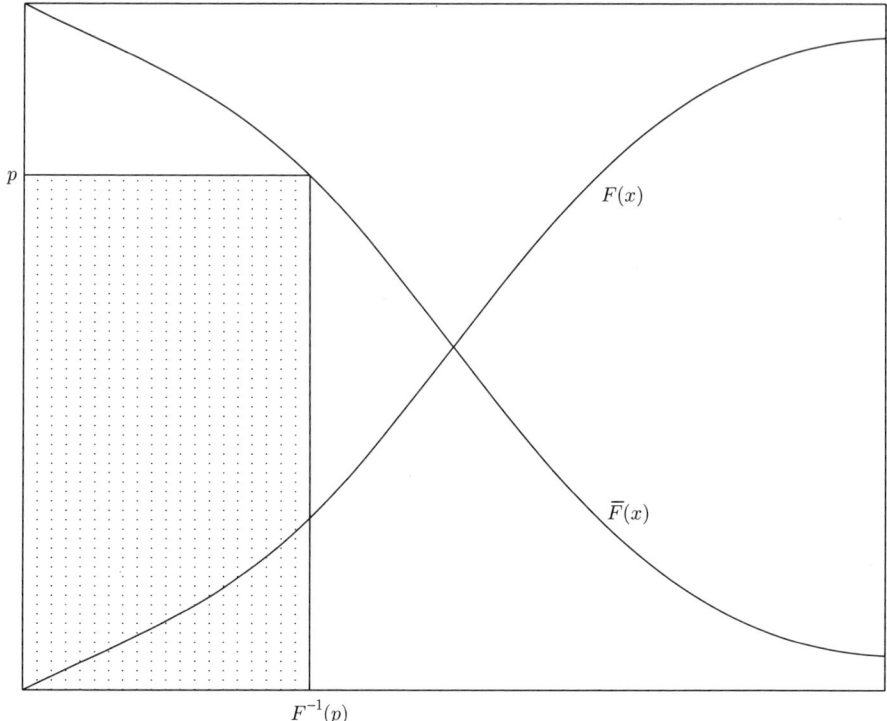

Fig. I.1b. The total time on test transform (shaded area)

and straightforward computations using (4) show that

$$H_F^{-1}(p) = \frac{p}{\lambda}, \quad 0 \le p \le 1.$$

Thus,

$$H_F(x) = \lambda x, \quad 0 \le x \le \lambda.$$

For this distribution, $\mu = 1/\lambda$ so that as expected, $H_F(\mu) = 1$.

I.3.b. Example. Consider the discrete distribution that places probability p_i at the point $x_i, i = 1, 2, \ldots, n$, where $x_1 < x_2 < \cdots < x_n$. Let $P_i = \sum_{j=1}^{i} p_j, i = 1, 2, \ldots, n$, so that $P_n = \sum_{j=1}^{n} p_j = 1$. For this distribution, it can be seen from Figure I.2 that

$$H_F^{-1}(P_r) = x_1 + (1 - P_1)(x_2 - x_1) + (1 - P_2)(x_3 - x_2) \\ + \cdots + (1 - P_{r-1})(x_r - x_{r-1}). \tag{5a}$$

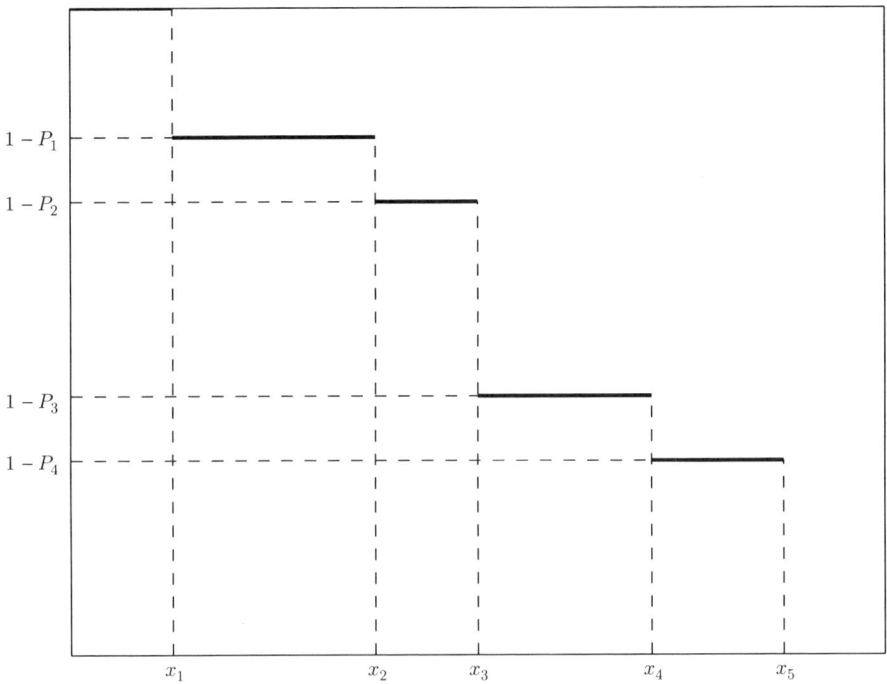

Fig. I.2. The survival function of a discrete distribution (Example I.3.b)

I.3.c. Empirical Distributions. If F is the empirical distribution function based upon the ordered observations $X_1 < X_2 < \cdots < X_n$, then F has the structure of Example I.3.b with $P_i = i/n$ and $x_i = X_i, i = 0, 1, \ldots, n$. Suppose n items are placed on test and the observations are their successive failure times. There are n items on test until the first failure occurs at time X_1 and $n - 1$ items on test from time X_1 to time X_2; the number of items on test continues to diminish one by one until all items have failed. The total time of exposure up to the rth failure is given by

$$T(X_r) = nX_1 + (n-1)(X_2 - X_1) + \cdots + (n-r+1)(X_r - X_{r-1}); \tag{5b}$$

this statistic is known as the *total time on test statistic*. By comparing (5a) and (5b), it can be seen that

$$H_F^{-1}(r/n) = T(x_r)/n. \tag{6}$$

I.4. Proposition. As $n \to \infty$ while $r/n \to p$, the total time on test statistic converges uniformly in p to the total time on test transform.

The proof of this proposition is omitted. The result is similar to the well-known *Glivenko–Cantelli theorem* (Billingsley, 1995, p. 269), which states that empirical distributions converge uniformly to the parent distribution F as the sample size goes to infinity.

Clearly, the distribution function F determines H_F^{-1}; that the reverse is also true is not quite so easy to verify.

Consider first the case that F is absolutely continuous and strictly increasing on its support. Make the change of variables $\zeta = F(x)$ and rewrite

$$H_F^{-1}(p) = \int_0^p \frac{1-\zeta}{f(F^{-1}(\zeta))}\, d\zeta. \tag{7}$$

Let

$$u(\zeta) = 1/f(F^{-1}(\zeta))$$

and differentiate both sides of (7) with respect to p to obtain

$$\frac{dH_F^{-1}(p)}{dp} = (1-p)u(p). \tag{8}$$

By solving for u and integrating, it follows, with the notation $U(p) = \int_0^p u(\zeta)\, d\zeta$, that

$$U(x) = F^{-1}(x) = \int_0^x \frac{1}{1-p}\, dH_F^{-1}(p),$$

which yields $F(x) = U^{-1}(x)$ in terms of H_F^{-1}.

It is possible to show that H_F^{-1} determines F in the discrete case by using (5a) with various values of r to solve for the x_i and p_i.

I.5. Proposition (Barlow, Bartholomew, Bremner and Brunk, 1972). If F has the hazard rate r, then H_F^{-1} is differentiable and

$$\frac{d}{dp} H_F^{-1}(p)\bigg|_{p=F(x)} = \frac{1}{r(x)}.$$

Proof. This is a direct consequence of (8). □

Remark. Proposition I.5 and B(16) together reveal an interesting connection; the mean residual life of a distribution coincides with the derivative of the total time on test transform of the equilibrium distribution.

I.6. Proposition. If $G(x) = F(\lambda x)$ for some $\lambda > 0$ and all $x \geq 0$, then

$$H_F^{-1}(p) = \lambda H_G^{-1}(p), \quad 0 \leq p \leq 1.$$

Proof. Use Proposition 20.A.8 or make direct use of (4) to obtain

$$H_F^{-1}(p) = \int_0^{\lambda G^{-1}(p)} \bar{G}(u/\lambda)\, du = \lambda H_G^{-1}(p). \qquad \square$$

c. Normalized Total Time on Test Transform

Note that if F has a finite first moment μ_F, then

$$\lim_{p \to 1} H_F^{-1}(p) = \int_0^\infty \bar{F}(u)\, du = \mu_F. \tag{9}$$

Because of (9) and Proposition I.6, it is possible and convenient for many purposes to consider the *normalized total time on test transform*, defined for distributions F with finite expectation μ_F by

$$K_F^{-1}(p) = H_F^{-1}(p)/\mu_F. \tag{10}$$

The notation

$$\tilde{\psi}(p) = K_F^{-1}(p)$$

is sometimes used in later chapters where the distribution F is clear from the context.

Some shape characteristics of the normalized total time on test transform are related in simple ways to the behavior of the hazard rate (when it exists). Several results of this kind are given in Chapters 4 and 5. With the aid of Proposition I.6, the normalized total time on test transform determines the distribution apart from a scale parameter.

d. The Lorenz Curve and Gini Index

The normalized total time on test transform is closely related to another function, the Lorenz curve. This curve was first defined by Lorenz (1905) for empirical distributions as follows: For ordered observations $x_{(1)} \leq \cdots \leq x_{(n)}$, let $L(i/n) = (\sum_{j=1}^{i} x_{(j)})/(\sum_{j=1}^{n} x_{(j)})$. The Lorenz curve is obtained by linearly interpolating between the points $(i/n, L(i/n))$. The Lorenz curve has proven to be of great interest in economics, particularly with reference to inequality of incomes. See, for example, Kleiber and Kotz (2003) and the references therein.

The purpose of the following definition is to extend this idea to arbitrary distributions.

I.7. Definition. The *Lorenz curve* L of a distribution function F with finite expectation is defined by

$$L(p) = \frac{\int_0^p F^{-1}(u)\,du}{\int_0^1 F^{-1}(u)\,du} = \frac{\int_0^{F^{-1}(p)} x\,dF(x)}{\int_0^\infty x\,dF(x)}, \quad 0 \leq p \leq 1.$$

The function

$$L^*(p) = \int_0^p F^{-1}(u)\,du = \int_0^{F^{-1}(p)} x\,dF(x), \quad 0 \leq p \leq 1,$$

without the normalization, is also sometimes called the Lorenz curve. For a comparison of the total time on test transform and the Lorenz curve see Figure I.3. The above definition is due to Gastwirth (1971), but see also Gastwirth (1972); alternative equivalent conditions go back to Lorenz (1905). The Lorenz curve is extensively discussed by Arnold (1983, Sections 4.2.1 and 4.2.6) and also by Arnold (1987). Because inverse distributions are increasing, it follows from the first form given for $L(p)$ that Lorenz curves are convex. For a brief discussion of convexity, see Chapter 21.

I.8. Proposition (Chandra and Singpurwalla, 1981).

$$L(p) = K_F^{-1}(p) - F^{-1}(p)[1 - FF^{-1}(p)]/\mu_F.$$

Proof. Rewrite L as

$$L(p) = \frac{\int_0^{F^{-1}(p)} z\,dF(z)}{\int_0^\infty z\,dF(z)} = \int_0^{F^{-1}(p)} \int_0^z dt\,dF(z)/\mu_F.$$

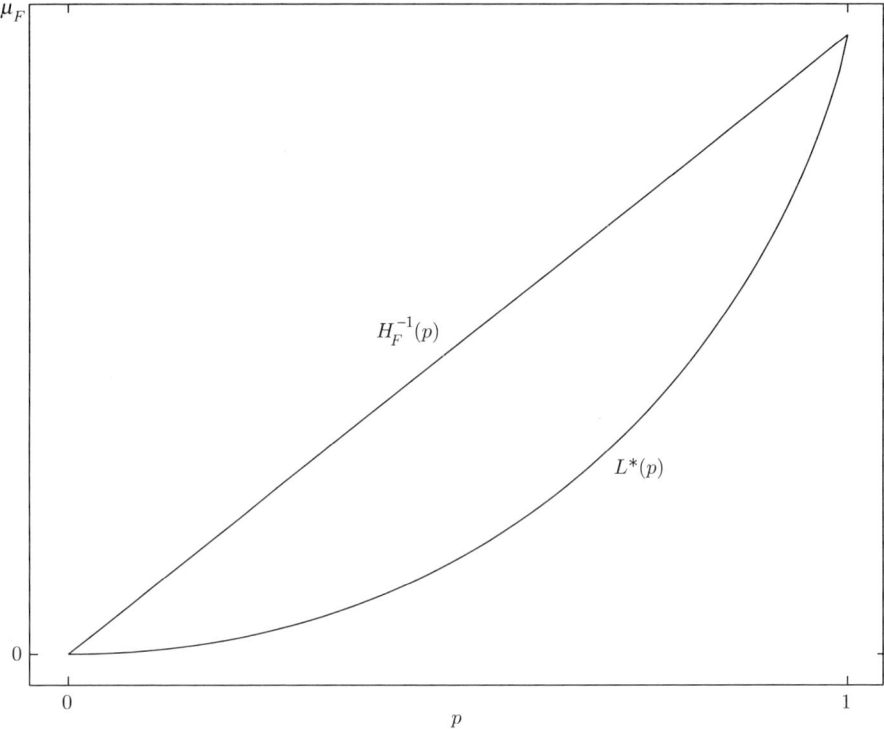

Fig. I.3. The total time on test transform and the Lorenz curve without normalization (for the exponential distribution)

With $F^{-1}(u) = z$, this becomes

$$L(p) = \frac{\int_0^{F^{-1}(p)} \int_t^{F^{-1}(p)} dF(z)\, dt}{\mu_F} = \frac{\int_0^{F^{-1}(p)} [\bar{F}(t) - F^{-1}\bar{F}(p)]\, dt}{\mu_F}$$

$$= \frac{\int_0^{F^{-1}(p)} \bar{F}(t)\, dt}{\mu_F} - \frac{\int_0^{F^{-1}(p)} F^{-1}\bar{F}(p)\, dt}{\mu_F}$$

$$= K_F^{-1}(p) - [F^{-1}(p)\bar{F}F^{-1}(p)][1/\mu_F]$$

$$= K_F^{-1}(p) - F^{-1}(p)[1 - FF^{-1}(p)][1/\mu_F]. \qquad \square$$

This expression can be simplified if F is strictly increasing on its support. In that case, $FF^{-1}(p) = p$ and

$$L(p) = K_F^{-1}(p) - F^{-1}(p)\frac{1-p}{\mu_F}. \qquad (11)$$

I.9. Example. If $\bar{F}(x) = e^{-\lambda x}, x > 0$, then $F^{-1}(p) = -[\log(1-p)]/\lambda$. Thus, $K_F^{-1}(p) = p$ and $L(p) = p + (1-p)\log(1-p)$.

The Lorenz curve is scale invariant, as evident in Example I.9. As noted above, the Lorenz curve is convex, and clearly, $L(0) = 0, L(1) = 1$. Thus, the Lorenz curve falls below the line $l(p) = p, 0 \le p \le 1$. Twice the area between this line and the Lorenz curve is called the Gini index of F, a quantity formally defined below. Gini (1912) proposed this index as a measure of concentration, inequality, or diversity as a competitor to the variance. See also Yitzhaki (2003).

Because the Lorenz curve is scale invariant (or because of Proposition 20.A.8), it is not possible to recover the underlying distribution F from L. On the other hand, F^{-1} can be obtained by differentiating L^*, and then F can be recovered using (3).

I.10. Definition. The integral

$$G_F = 2\int_0^1 [p - L(p)]\,dp$$

is called the *Gini index* or the *Gini index of concentration* of F.

From the area interpretation of the Gini index, it can be seen that $0 \le G_F \le 1$. See Figure I.4.

Various alternative formulations of the Gini index are known. A particularly interesting one is the following:

$$G_F = 1 - \frac{\int_0^\infty [\bar{F}(x)]^2\,dx}{\mu_F} = \frac{\int_0^\infty F(x)\bar{F}(x)\,dx}{\mu_F} = \frac{2\,Cov(X, F(X))}{\mu_F}$$

$$= \frac{\int_{-\infty}^\infty \int_{-\infty}^\infty |x-y|\,dF(x)\,dF(y)}{2\mu_F}. \tag{12}$$

Here, it is clear that the Gini index is defined only if $\mu_F < \infty$, a condition imposed above for defining the Lorenz curve. The equality of the first two expressions for G_F is a consequence of C(4) with $r = 1$. The last expression for G_F involving the covariance can be obtained using Definition 20.I(6) and an integration by parts.

For a survey of results concerning the Gini index and generalizations of it, see Kleiber and Kotz (2002). For an excellent survey that

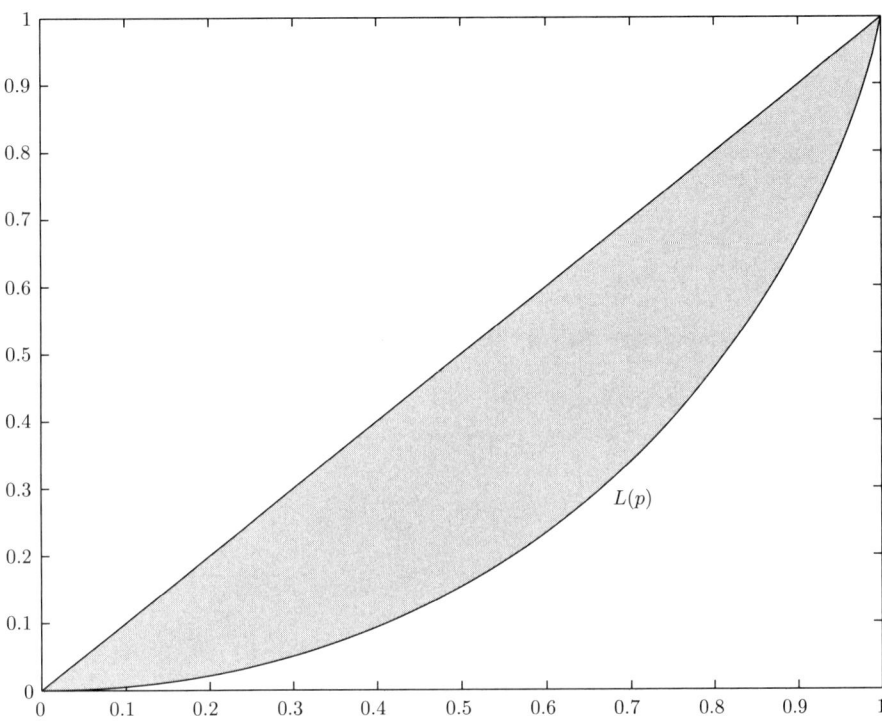

Fig. I.4. Half the Gini index as a shaded area

includes much of the results given in this section as well as additional related results, see Pham and Turkkan (1994). For a discussion of the Lorenz curve and Gini index in the context of an application, see Losinger (1997); estimation considerations are studied by Gastwirth (1972).

2

Ordering Distributions: Descriptive Statistics

> Some are and must be greater than the rest.
> Alexander Pope, *Essay on Man* (ep. IV, I. 49)

Characteristics of distributions or densities such as location, dispersion, skewness, and kurtosis have long been used for descriptive purposes. Early on, measures of such characteristics were proposed, though precise definitions of the characteristics may even now remain elusive; the characteristics were often defined as "that which the measure measures." The standard deviation as a measure of dispersion or spread is a familiar example, but measures of skewness and kurtosis based on moments were also proposed by Pearson (1895).

A second approach to distributional characteristics is followed in this chapter; for a given characteristic of interest, an ordering $F \leq G$ is introduced to make precise the idea that F has less of the characteristic than does G. This approach was used by Mann and Whitney (1947) who introduced what is now called "stochastic order"; it was used by Birnbaum (1948) in a study of "peakedness." Other important orders were introduced by van Zwet (1964) and the notion of ordering distributions was brought into clear focus by Lehmann (1955) and by Bickel and Lehmann (1975). The introduction of an ordering to represent the idea that one distribution has more of some characteristic than another requires careful consideration of the characteristic's nature.

Once an ordering appropriate for a given characteristic has been found, proposed measures of the characteristic can be subjected to the test that they be *order-preserving*. That is, if $F \leq G$ in the ordering, then a measure m of that characteristic should satisfy $m(F) \leq m(G)$. Additional properties may also be required of the measure.

The orders considered in this chapter do not all naturally relate to a standard characteristic; still, these orders are geometrically meaningful, especially when restricted to the comparison of distributions of nonnegative random variables.

Location and dispersion are especially familiar as they relate to the normal distribution, whereas skewness and kurtosis are concepts especially useful in the consideration of departures from normality. Kurtosis is a concept most often applied to symmetric distributions, although MacGillivray and Blanda (1988) and Blanda and MacGillivray (1988, 1990) extend the idea to the nonsymmetric case. It is even more difficult to see how these ideas might extend to distributions of nonnegative random variables, where the origin prescribes the location, and the mean, a location parameter for the normal distribution, becomes what here will be called a *magnitude parameter*. For distributions on $[0, \infty)$, magnitude and dispersion are often related. Note that for the exponential distribution, the mean and standard deviation are both equal to $1/\lambda$; the parameter λ affects both magnitude and spread.

Historically, descriptive statistics indicating location, spread, and other characteristics were oriented toward describing density shapes. But some important characteristics of distributions are not readily apparent from the density. For example, tail behavior and the existence of moments is usually not clear from a graph of the density, but they are sometimes quite obvious from a graph of the hazard rate. Sometimes, such characteristics can also be seen clearly from properties of the total time on test transform. So to more fully understand the properties of a distribution, it is often advantageous to look at more than one function describing the distribution.

This chapter provides only a brief and limited exposure to the topic of ordering distributions; for more complete treatments, see the books devoted to the subject written by Shaked and Shanthikumar (1994, 2007), Szekli (1995), or Müller and Stoyan (2002) (see also Stoyan, 1977, 1983). Sections 19.D and 19.E also discuss orderings, particularly as they relate to semiparametric families; the treatment there is somewhat more abstract than that offered in this chapter.

The orderings $\leq$ of distributions considered in this chapter have two properties; they are both reflexive and transitive. That is,

$$F \leq F, \quad \text{for all distribution functions } F \text{ (reflexivity)}$$

and

$$F \leq G \quad \text{and} \quad G \leq H \quad \text{implies } F \leq H \text{ (transitivity)}.$$

Orderings with these two properties are called *preorders*; if in addition they satisfy the condition that

$$F \leq G \quad \text{and} \quad G \leq F \quad \text{implies } F = G,$$

then the orders are called *partial orders*. Preorders and partial orders, unlike numerical measures of a characteristic, can recognize when two distributions are too disparate to be compared: $F \leq G$ and $G \leq F$ may both be false.

In what follows, it is sometimes convenient to write $X \leq Y$ to mean that the distribution F of X and the distribution G of Y satisfy $F \leq G$.

In general, the orders considered here can be classified as either related to magnitude, dispersion, or some other aspect of "shape." Many of the comparisons involving shape are scale invariant, which is not a usual property of dispersion orders, but still the shape orderings are related to dispersion and are called "variability" orders by Shaked and Shanthikumar (1994, 2007). Whatever the name, degenerate distributions form the smallest class of distributions in these orders, but there is no largest distribution.

For parametric families with two parameters, it may be desirable that one parameter orders the family according to magnitude and the other parameter orders the family according to dispersion or shape. Many such relations are established in later chapters, although for some reason they are mostly not well known. Understanding these orders and their occurrences in parametric families will help illuminate the role that various parameters play in parametric families.

A. Magnitude

The usual stochastic order, defined below, has a number of properties that make it what might be called a "magnitude order." Other orders are included in this section because they imply stochastic order, but their properties may differ.

a. The Usual Stochastic Order

Suppose that $\widetilde{X}$ and $\widetilde{Y}$ are random variables such that always $\widetilde{X} \leq \widetilde{Y}$. Even though $\widetilde{X}$ and X have the same distribution, $\widetilde{Y}$ and Y have the same distribution, it need not be that $X \leq Y$. But still, it seems reasonable to say that in some sense, X is less than Y. This idea leads to the "usual" concept of stochastic order, an order in which X is less than Y if and only if the survival function of X is everywhere less than the survival function of Y. It is surprising that such a simple and fundamental idea is not very old; apparently, it was first introduced by Mann and Whitney (1947). However, a method for comparing two estimators of a parameter, due to Pitman (1937), could be regarded as a forerunner of stochastic order.

A.1. Definition. If X and Y are random variables such that $P\{X > z\} \leq P\{Y > z\}$ for all z, then X is said to be *stochastically smaller* than Y. This relationship is often notated by $X \leq_{st} Y$, or by $F \leq_{st} G$, where X has distribution F and Y has distribution G.

A note of caution: $F \leq_{st} G$ means that $\bar{F}(z) \leq \bar{G}(z)$ for all z, and consequently, $F(z) \geq G(z)$ for all z. This is a potential source of confusion.

The condition $\bar{F}(z) \leq \bar{G}(z)$, for all z, that one survival function dominates another is often easily checked, suggestive both of examples and potential applications, and it arises in a number of contexts. It is useful, for example, in comparing treatments in a medical experiment in which X may be either the convalescent or survival time associated with one treatment and Y may be the corresponding time for another treatment. Or X and Y might be the earnings resulting from different business strategies. In economics and utility theory, stochastic order is called "first order stochastic dominance." See D.4, where second-order stochastic dominance is defined.

If $P\{X \leq Y\} = 1$, then clearly $X \leq_{st} Y$. From this fact, a number of examples follow. If there exists a random variable Z and functions g and h such that $X = g(Z), Y = h(Z)$, and $g(z) \leq h(z)$ for all z, then $X \leq_{st} Y$. If there exists a number a such that $P\{X \leq a\} = 1$ and $P\{Y \geq a\} = 1$, then again $X \leq_{st} Y$.

Several equivalent conditions for stochastic ordering are worth noting.

A.2. Proposition. The following conditions are equivalent:

$$X \leq_{st} Y. \tag{1}$$

$$E\phi(X) \leq E\phi(Y) \quad \text{for all increasing functions } \phi \text{ such that the expectations exist.} \tag{2}$$

$$\phi(X) \leq_{st} \phi(Y) \quad \text{for all increasing functions } \phi. \tag{3}$$

There exist random variables $\tilde{X}$ and $\tilde{Y}$ such that X and $\tilde{X}$ have the same distribution, Y and $\tilde{Y}$ have the same distribution, and

$$P\{\tilde{X} \leq \tilde{Y}\} = 1, \tag{4}$$
$$GF^{-1}(u) \leq u, \quad 0 < u < 1, \tag{5}$$

where X has distribution function F and Y has distribution function G.

The equivalence of (1) to (4) is straightforward to prove. The concept of stochastic ordering for random variables is unrelated to issues of their dependence or independence. The fundamental fact that (4) and (1) are equivalent is due to Lehmann (1955); (4) introduces a joint distribution and makes a "probability one" statement; and apart from its intuitive content, which was used to introduce this section, (4) can sometimes be quite useful in theoretical computations. Condition (5) has been given by Lehmann and Rojo (1992).

Certain consequences of a stochastic ordering are quite conveniently obtained from (2). In particular, with $\phi(x) = x^r$, it follows directly from (2) that $X \leq_{\text{st}} Y$ implies $EX^r \leq EY^r$, for $r \geq 0$, and $EX^r \geq EY^r$, for $r \leq 0$. With $\phi(x) = e^{sx}$, it follows from (2) that $Ee^{sX} \leq Ee^{sY}$ for $s \geq 0$, and $Ee^{sX} \geq Ee^{sY}$ for $s \leq 0$.

It follows directly from (3) that if $X \leq_{\text{st}} Y$, then $\phi(X) \leq_{\text{st}} \phi(Y)$ for all increasing functions ϕ. Thus, $X \leq_{\text{st}} Y$ implies $X^r \leq_{\text{st}} Y^r$ for all $r > 0$, and other examples may come to mind.

A.3. Proposition. If $X \leq_{\text{st}} Y$ and $U \leq_{\text{st}} V$, where X and U and Y and V are independent, then

$$X + U \leq_{\text{st}} Y + V.$$

If these random variables are nonnegative with probability 1, then also

$$XU \leq_{\text{st}} YV.$$

A.4.a. Example. If X has distribution F (where $F(0) = 0$), and if $Z = aX$, where $a \geq 1$, then $X \leq_{\text{st}} Z$.

Here, the distribution of Z is obtained from that of X by "stretching the axis out" away from the origin; whatever the value of X, Z is always bigger than X because $a \geq 1$. In this example, a can be replaced by any random variable that, with probability 1, is at least 1.

A.4.b. Example. If X and Y are random variables, and $Z = \min(X, Y)$, then always $Z \leq X$ and $Z \leq Y$. Consequently, $Z \leq_{\text{st}} X$ and $Z \leq_{\text{st}} Y$, with no requirement that X and Y be independent. On the other hand, if $Z = \max(X, Y)$, or X and Y are nonnegative random variables and $Z = X + Y$, then $Z \geq_{\text{st}} X$ and $Z \geq_{\text{st}} Y$. These observations are further expanded in Example A.8.

Somewhat more detailed discussions of stochastic ordering is given by Marshall and Olkin (1979, Chapter 17), Shaked and Shantikumar (1994, 2007), and Müller and Stoyan (2002).

b. Hazard Rate Order

Suppose that X and Y are life lengths of two devices or organisms and that $X \leq_{\text{st}} Y$. If the organisms are both observed to be alive at time $t > 0$, one might conjecture that the residual lives would also be stochastically ordered. The following definition is motivated by the fact that this conjecture is false. A counterexample can be obtained from Example A.6 below by choosing a hazard rate that fails to satisfy (8). Another counterexample is given below in Example A.8.

A.5. Definition. Let X and Y be random variables with corresponding distribution functions F and G. Then X is said to be *smaller in the hazard rate ordering* than Y, denoted by $X \leq_{\text{hr}} Y$ or by $F \leq_{\text{hr}} G$ if

$$\bar{F}_t(x) = P\{X > x + t \mid X > t\} \leq P\{Y > x + t \mid Y > t\} = \bar{G}_t(x)$$
for all x and $t \geq 0$ such that $\bar{F}(t) > 0, \bar{G}(t) > 0.$ \hfill (6a)

It follows from the definition of conditional probability that (6a) is equivalent to

$$\det \begin{vmatrix} \bar{F}(t) & \bar{F}(x+t) \\ \bar{G}(t) & \bar{G}(x+t) \end{vmatrix} \geq 0$$
for all x and $t \geq 0$ such that $\bar{F}(t) > 0, \bar{G}(t) > 0,$

and this in turn can be written as

$$\det \begin{vmatrix} \bar{F}(u) & \bar{F}(v) \\ \bar{G}(u) & \bar{G}(v) \end{vmatrix} \geq 0 \quad \text{for all } u \leq v \text{ such that } \bar{F}(u) > 0, \bar{G}(u) > 0,$$
\hfill (6b)

that is,

$$\frac{\bar{F}(z)}{\bar{G}(z)} \text{ is decreasing in } z \text{ such that } \bar{G}(z) > 0. \tag{6c}$$

Condition (6c) was introduced and utilized by Brown (1984) in a study of the distance between two distributions. For its convenience, (6c) is taken as the definition of the hazard rate ordering by Nanda and Shaked (2001). As noted by Lehmann and Rojo (1992), with the substitution $u = \bar{G}(z)$, condition (6c) can be rewritten as

$$\frac{\bar{F}\bar{G}^{-1}(u)}{u} \text{ is increasing in } u, \quad 0 < u < 1. \tag{6d}$$

Condition (6b) relates the hazard rate ordering to total positivity (see Chapter 21).

In case X and Y are absolutely continuous random variables with corresponding hazard rates r_X and r_Y obtained by differentiating the hazard functions, then (6c) is equivalent to

$$r_X(z) \geq r_Y(z), \quad \text{for all } z \geq 0. \tag{6e}$$

This explains the terminology of Definition A.5, but take care to note that the larger hazard rate belongs to the random variable smaller in the hazard rate order. From (6a) it follows that

$$X \leq_{\text{hr}} Y \text{ implies } X \leq_{\text{st}} Y. \tag{6f}$$

If in (6b) the survival functions are replaced by distribution functions, then another order, called the *reverse hazard rate order*, is obtained. Thus, X is less than Y in the reverse hazard rate order if

$$\det \begin{vmatrix} F(x) & F(y) \\ G(x) & G(y) \end{vmatrix} \geq 0 \quad \text{for all } x \leq y. \tag{7a}$$

In terms of conditional probabilities, (7a) can be rewritten as

$$P\{X \leq x \mid X \leq y\} \geq P\{Y \leq x \mid Y \leq y\}, \quad x \leq y. \tag{7b}$$

When densities exist, the condition counterpart to (6e) is

$$s_X(z) \leq s_Y(z) \quad \text{for all } z \geq 0, \tag{7c}$$

where s is the reverse hazard rate of 1.B(5). It can be seen from 1.B(6) that if X is less than Y in the reverse hazard rate order, then also, X is less than Y in the sense of stochastic order. But note the comparison of (6e) and (7c).

The hazard rate and reverse hazard rate orderings were introduced and studied by Keilson and Sumita (1982).

A.6. Example. Suppose that X has distribution F (where $F(0) = 0$), and $Y = aX$ where $a \geq 1$. If X has a hazard rate r such that

$$x\, r(x) \text{ is increasing in } x > 0, \tag{8}$$

then $X \leq_{\text{hr}} Y$.

Condition (8) is obtained by computing the hazard rate of Y. The condition is satisfied if r is increasing, but it may fail when r is decreasing. For a more detailed discussion of this example, see Propositions 7.C.6 and 7.C.6.a.

A.7. Example. For all positive constants C, $X \leq_{\text{hr}} X + C$ if and only if the hazard rate of X is increasing.

Proof. This result follows from the fact that if X has hazard rate r, then $X + C$ has hazard rate r_C given by

$$r_C(x) = r(x + C), \quad x \geq C,$$
$$= 0, \quad 0 \leq x \leq C.$$
□

A.8. Example. If X and Y are independent positive random variables, and $Z = \min(X, Y)$, then it is clear from 1.H(3) that $Z \leq_{\text{hr}} X$ and $Z \leq_{\text{hr}} Y$. On the other hand, if $W = \max(X, Y)$, it might be expected that $W \geq_{\text{hr}} X$ and $W \geq_{\text{hr}} Y$, but this turns out to be false even though the weaker stochastic order holds. To show that W need not be greater than X or Y in the hazard rate order, let F, G, and H respectively, be the distributions of X, Y, and Z. Then, from 1.H(6) and some elementary algebra, it follows that at x, the hazard rate of W is less than that of X if and only if

$$\frac{g(x)}{\bar{G}(x)} \leq \frac{f(x)}{F(x)\bar{F}(x)}. \tag{9}$$

The left-hand side of (9) is a hazard rate, and thus is finitely integrable at least in some interval $[0, \varepsilon]$, but because $F(0) = 0$, the right-hand side of (9) is not finitely integrable in any such interval. Thus, (9) holds for sufficiently small values of x. But examples can be constructed for which (9) fails for large values of x, because then the comparison (9) is essentially a comparison of the hazard rates of X and Y.

Equation (9) is the condition that the hazard rate of W is less than that of X; the condition that the hazard rate of W is less than that of Y is obtained by interchanging F and G. The assumption that both inequalities are violated leads to a contradiction, so it follows that

$$\frac{h(x)}{\bar{H}(x)} \leq \max\left(\frac{g(x)}{\bar{G}(x)}, \frac{f(x)}{\bar{F}(x)}\right).$$

Because the right-hand side of this inequality is not the hazard rate of a function of X and Y, the result cannot be translated as a natural hazard rate ordering.

In view of the above discussion, it is rather curious that

$$Z = \min(X, Y) \leq_{\text{hr}} \max(X, Y) = W.$$

This result can be generalized to other order statistics; if the independent nonnegative random variables $X_1, \ldots, X_n$ are ordered to obtain $X_{(1)} \leq \cdots \leq X_{(n)}$, then $X_{(k)} \leq_{\text{hr}} X_{(k+1)}, k = 1, \ldots, n-1$. For this result, see Shaked and Shanthikumar (1994, p. 22; 2007, p. 31).

If $Z = X + Y$, then as noted in Example A.4.b, $Z \geq_{\text{st}} X$ and $Z \geq_{\text{st}} Y$. But again, stochastic ordering cannot be replaced by hazard rate ordering, as can be seen from Example A.7.

In view of the fact that the hazard rate can be interpreted as the probability of failure in the next instant of time given survival up to a specified time, it is clear that the hazard rate ordering would be appropriate for comparing the life lengths of identical devices, one operating in a more hazardous environment than the other.

c. Likelihood Ratio Order

A further strengthening of (6a) of Definition A.5 leads to a still stronger condition that turns out to be quite useful.

A.9. Definition. The random variable X is said to be *smaller in the likelihood ratio ordering* than the random variable Y if for all u,

$$P\{X > u \mid a < X \le b\} \le P\{Y > u \mid a < Y \le b\}, \tag{10}$$

whenever $a < b$ and the conditional probabilities are defined.

This relationship is denoted by $X \le_{\text{lr}} Y$ or by $F \le_{\text{lr}} G$, where F is the distribution of X and G is the distribution of Y.

The following proposition provides a useful way of verifying a likelihood ratio order when densities exist.

A.10. Proposition. If X and Y are absolutely continuous random variables, then $X \le_{\text{lr}} Y$ if and only if there are versions f and g of the corresponding densities such that

$$f(u)g(v) \ge f(v)g(u) \quad \text{for all } u \le v. \tag{11}$$

Proof. First, suppose that (10) holds and that $a \le u \le b$. Rewrite (10) in the form

$$\frac{P\{u < X \le b\}}{P\{a < X \le b\}} \le \frac{P\{u < Y \le b\}}{P\{a < Y \le b\}}. \tag{12a}$$

Subtracting both sides of this inequality from 1 yields

$$\frac{P\{a < X \le u\}}{P\{a < X \le b\}} \ge \frac{P\{a < Y \le u\}}{P\{a < Y \le b\}}; \tag{12b}$$

multiplication of (12a) and (12b) yields

$$P\{a < X \le u\}P\{u < Y \le b\} \ge P\{a < Y \le u\}P\{u < X \le b\},$$
$$a \le u \le b. \tag{13}$$

It follows from a change of variables in (13) that when $u \le b \le v$,

$$P\{u < X \le b\}P\{b < Y \le v\} \ge P\{u < Y \le b\}P\{b < X \le v\}. \tag{14}$$

Multiplication of (13) and (14) yields

$$P\{a < X \le u\}P\{b < Y \le v\} \ge P\{a < Y \le u\}P\{b < X \le v\}. \quad (15)$$

Now, let $a \to u$ and $b \to v$ so as to obtain (11).

Next, suppose that (11) holds. Then, $f(y)g(x) \le f(x)g(y)$, for $x \le y$; integration on y from u to b and on x from a to u gives $P\{u < X \le b\}P\{a < Y \le u\} \le P\{u < Y \le b\}P\{a < X \le u\}$. Addition of $P\{u < X \le b\}P\{u < Y \le b\}$ to both sides yields (10). □

Clearly, the terminology of Definition A.9 comes from the fact that when denominators are positive, (11) can be rewritten as the comparison

$$\frac{f(u)}{g(u)} \ge \frac{f(v)}{g(v)} \quad \text{for all } u \le v, \quad (16a)$$

of likelihood ratios.

A.10.a. Proposition (Lehmann and Rojo, 1992). The condition $F \le_{\mathrm{lr}} G$ is equivalent to the condition that

$$\bar{F}\bar{G}^{-1}(p) \text{ is convex in } p, \quad 0 \le p \le 1. \quad (16b)$$

Proof. Make use of the fact 21.A.3 (iv) that a differentiable function is convex if and only if its derivative is increasing. Assume that densities exist and note that from 24.A.4.b with F and G interchanged, it follows that the derivative of $\bar{F}\bar{G}^{-1}(p)$ is given by

$$\frac{f\bar{G}^{-1}}{g\bar{G}^{-1}}. \quad (16c)$$

Because $\bar{G}^{-1}$ is a decreasing function, the quantity (16c) is increasing if and only if (16a) holds. □

A.11. Proposition. Suppose that X and Y are absolutely continuous random variables with respective densities $f(\cdot \mid \theta_1)$ and $f(\cdot \mid \theta_2)$ in the same parametric family. Then, $X \le_{\mathrm{lr}} Y$ whenever $\theta_1 < \theta_2$ if and only if there is a version $f(\cdot \mid \theta)$ of the density such that $f(x \mid \theta)$ is totally positive of order 2 in x and θ.

Proof. According to Definition 21.B.1, $f(x \mid \theta)$ is totally positivity of order 2 in x and θ if and only if

$$\det \begin{bmatrix} f(x_1 \mid \theta_1) & f(x_1 \mid \theta_2) \\ f(x_2 \mid \theta_1) & f(x_2 \mid \theta_2) \end{bmatrix} \geq 0 \quad \text{for all } x_1 < x_2, \theta_1 < \theta_2.$$

But this now is just a restatement of (11), where $f(\cdot \mid \theta_1) = f(\cdot)$ and $f(\cdot \mid \theta_2) = g(\cdot)$. □

A.12. Proposition. If $X \leq_{\text{lr}} Y$, then the density f of X crosses the density g of Y exactly once, and at this crossing, f crosses g from above.

Proof. It follows directly from (16a) that the ratio $f(u)/g(u)$ is decreasing. Consequently, if $f(u_0)/g(u_0) = 1$, then $f(u)/g(u) \geq 1$, for $u \leq u_0$, and $f(u)/g(u) \leq 1$, for $u \geq u_0$. It follows that the density f of X can cross the density g of Y at most once, and if there is a crossing, f must cross g from above. Because both densities integrate to 1, there must be at least one crossing. □

Directly from (10), it follows that

$$X \leq_{\text{lr}} Y \text{ implies } X \leq_{\text{hr}} Y \tag{17}$$

and similarly, $X \leq_{\text{lr}} Y$ implies that X is less than Y in the reverse hazard rate order.

The likelihood ratio and hazard rate orders are connected through the equilibrium distributions, which are defined and discussed in Sections 1.B.h and 20.B.c. For distributions F and G with corresponding equilibrium distributions $F_{(1)}$ and $G_{(1)}$, it is a direct consequence of their definition, (6c), and (11) that

$$F_{(1)} \leq_{\text{lr}} G_{(1)} \text{ if and only if } F \leq_{\text{hr}} G.$$

This fact is noted by Navarro, Belzunce and Ruiz (1997).

It is often the case that densities have a nice form but survival functions and hazard rates do not. Then, the likelihood ratio order can be much easier to verify than the hazard rate order.

A.13. Example. Consider the gamma distribution with density given by 1.F(6). The survival function of this distribution cannot be given in closed form unless the shape parameter ν is an integer and even then, the hazard rate is awkward to study. In the shape parameter ν of this distribution, is the family stochastically ordered or hazard rate ordered? Although stochastic order can easily be shown using the derivation (iii) of Section 9.A.b, the question of hazard rate order is unpleasant to attack directly. But consider the ratio of densities $f(x \mid \lambda, \theta)/f(x \mid \lambda, \nu) = Cx^{\theta-\nu}$, where C is a positive constant. This ratio is decreasing in x whenever $\theta \leq \nu$, so that in the likelihood ratio ordering, the distribution is increasing in the shape parameter. This means that in both the hazard rate ordering and in the stochastic ordering, the distribution also is increasing in the shape parameter.

A.14. Example. Suppose that X has distribution F (where $F(0) = 0$), and that $Y = aX$ where $a \geq 1$. If X has a density f satisfying

$$f(x)/f(x/a) \text{ is decreasing in } x \geq 0 \quad \text{for all } a > 1, \tag{18}$$

then $X \leq_{\text{lr}} Y$.

For a proof of this and related results, see Propositions 7.C.8 and 7.C.8.a.

A.15. Example. For all positive constants C, $X \leq_{\text{lr}} X + C$ if and only if the logarithm of the density of X is concave.

Proof. Because the density of $X + C$ is given by $f(x - C), x \geq C$, it follows from (10) that $X \leq_{\text{lr}} X + C$ if and only if for all $u < v$, $f(u)f(v - C) \geq f(v)f(u - C)$. From Definition 21.B.7 and Proposition 21.B.8, it follows that this condition for all $C > 0$ is equivalent to log concavity of f. □

A.15.a. Remark. Because $X \leq_{\text{lr}} X + C$ implies $X \leq_{\text{hr}} X + C$, as noted in (17), it follows from Example A.15 that if the density of X is log concave, then $X \leq_{\text{hr}} X + C$ for all positive constants C. But according to Example A.7, this is equivalent to the condition that X has an increasing hazard rate.

Thus, if X has a log-concave density, then X has an increasing hazard rate. The importance of this fact is noted in Proposition 4.B.8, and is proved by more standard methods in Lemma 21.B.15.

There is a clear connection between the likelihood ratio ordering and the theory of total positivity, which is briefly reviewed in Chapter 22. In fact, some of the most interesting examples of the likelihood ratio ordering come from parametric families of densities that are totally positive of order two in the argument and a parameter. The prime example of this is the family of exponential distributions, but the property also holds for the family of gamma distributions with fixed shape parameter.

Applications of the hazard rate ordering in reliability have been discussed by Boland, El-Neweihi and Proschan (1994).

d. Preservation Properties of Magnitude Orders

A.16. Monotone transformations. The condition

$$X \leq Y \text{ if and only if } \phi(X) \leq \phi(Y)$$

for all increasing functions ϕ, holds when the order $\leq$ is stochastic order $\leq_{\text{st}}$, hazard rate order $\leq_{\text{hr}}$, or likelihood ratio order $\leq_{\text{lr}}$.

For the hazard rate order, this result is given by Nanda and Shaked (2001).

A.17. Residual life distribution. According to Definition 1.B.12, the residual life distribution F_t of F at t is given for all t such that $\bar{F}(t) > 0$ by

$$\bar{F}_t(x) = \bar{F}(x+t)/\bar{F}(t), \quad x \geq 0. \tag{19}$$

An ordering $\leq$ of distribution functions is said to be *closed under the formation of residual life distributions* if $F \leq G$ implies that the corresponding residual life distributions F_t and G_t satisfy $F_t \leq G_t$.

As has already been noted, the stochastic order $\leq_{\text{st}}$ is not preserved under the formation of residual life distributions. Indeed, this fact was used to motivate the introduction of the hazard rate order.

A.18. Proposition. Both the hazard rate order $\leq_{\text{hr}}$ and the likelihood ratio order $\leq_{\text{lr}}$ are preserved under the formation of residual life distributions.

Proof. Preservation of the hazard rate order follows from the fact that the residual life distribution of F_t at u is just the residual life distribution of F at $t+u$. In case the hazard rate r of F exists, the result is even more transparent because the hazard rate r_t of F_t satisfies $r_t(u) = r(u+t)$.

Preservation of the likelihood ratio order follows from the fact that for $a < u < b$,

$$\frac{\bar{F}_t(u) - \bar{F}_t(b)}{\bar{F}_t(a) - \bar{F}_t(b)} = \frac{\bar{F}(u+t) - \bar{F}(b+t)}{\bar{F}(a+t) - \bar{F}(b+t)}. \tag{20}$$

Divide both sides of (20) by $b - u$ and take the limit as $b \to u$ to conclude that

$$\frac{f_t(u)}{\bar{F}_t(a) - \bar{F}_t(b)} = \frac{f(u+t)}{\bar{F}(a+t) - \bar{F}(b+t)}. \tag{21}$$

If (16a) holds, then

$$\frac{f(u+t)}{g(u+t)} \geq \frac{f(v+t)}{g(v+t)} \quad \text{for all } u \leq v;$$

use (21) and the corresponding relation for the distribution G to conclude that

$$\frac{f_t(u)}{g_t(u)} \geq \frac{f_t(v)}{g_t(v)} \quad \text{for all } u \leq v. \qquad \square$$

A.19. Mixtures. Suppose that $F = \pi_1 F_1 + \pi_2 F_2$, where π_1 and π_2 are nonnegative weights that add to 1. Similarly, suppose that $G = \pi_1 G_1 + \pi_2 G_2$. The ordering $\leq$ is said to be *closed under the formation of mixtures* if $F_1 \leq G_1$ and $F_2 \leq G_2$ implies $F \leq G$, for all $\pi_1 = 1 - \pi_2$ in $[0, 1]$.

Stochastic order is clearly closed under the formation of mixtures. On the other hand, neither the hazard rate order nor the likelihood ratio order is preserved under mixtures, as is shown by the following example.

A.20. Example. Suppose that $\bar{F}_i(x) = \exp\{-\lambda_i x\}$ and that $\bar{G}_i(x) = \exp\{-\theta_i x\}$, $i = 1, 2$. It is straightforward to check that if $\lambda_i \geq \theta_i$, $i = 1, 2$, then $F_1 \leq_{\text{lr}} G_1$ and $F_2 \leq_{\text{lr}} G_2$. But if $\lambda_2 > \theta_2 > \lambda_1 = \theta_1$ and λ_2 is large, then even the hazard rate order $F \leq_{\text{hr}} G$ fails to hold. To verify this, considerable calculation is required.

A number of other magnitude orderings can be found in the literature. Of special note is the work of Lehmann and Rojo (1992), who discuss modifications of the three orders introduced in this section. Further related results are provided by Navarro, Belzunce and Ruiz (1997).

B. Dispersion

It is noted at the beginning of this chapter that for distributions of nonnegative random variables, magnitude and dispersion are closely related. However, it is possible to separate the two concepts to a substantial degree.

a. Convex Order

Recall that $X \leq_{\text{st}} Y$ if and only if $E\phi(X) \leq E\phi(Y)$ for all increasing functions ϕ for which the expectations exist. A number of comparisons can be made that are similar in form to this, but with the class of increasing functions ϕ replaced by some other class; such orders are discussed in Section D, but the following important example belongs in this section.

B.1. Definition. If

$$E\phi(X) \leq E\phi(Y) \text{ for all convex functions } \phi \text{ such that the expectations exist,} \tag{1}$$

then X is said to be *smaller in the convex order* than Y. In this case, the notation $X \leq_{\text{cx}} Y$ or $F \leq_{\text{cx}} G$ is used, where F and G, respectively, are the distribution functions of X and Y. The convex order is sometimes called "balayage order" (see, e.g., Meyer, 1966, p. 239). The French word "balayage" means "sweeping," and is used in the sense of spreading out.

Another important order, closely related to the convex order, is defined with concave functions replacing the convex functions of (1). This replacement is equivalent to changing the sign in the inequality of (1), and thus to reverse the order: X is less than Y in the convex order if and only if X is greater than Y in the concave order, sometimes denoted by $X \geq_{\text{cv}} Y$. Because $X \leq_{\text{cx}} Y$ if and only if $X \geq_{\text{cv}} Y$, a result concerning one of these orders can be easily translated into a result concerning the other order.

By considering the cases $\phi(x) = x$ and $\phi(x) = -x$ in (1), it is easy to see that $X \leq_{\text{cx}} Y$ implies $EX = EY$. This means that comparisons are made only between distributions that are equal in magnitude as measured by the expectation.

The choice $\phi(x) = x^2$ leads to the conclusion that $X \leq_{\text{cx}} Y$ implies Var $X \leq$ Var Y. Thus, the convex order is stronger than the more familiar variance comparison, but it is restricted by the condition $EX = EY$.

As noted above, $X \leq_{\text{cx}} Y$ implies Var $X \leq$ Var Y, so the corresponding standard deviations are also ordered. Other measures d that preserve the convex order (i.e., satisfy the condition $X \leq_{\text{cx}} Y$ implies $d(X) \leq d(Y)$) can be used to compare dispersions. Because convexity is all that is needed, such examples as $E|X - EX|$, the mean deviation from the mean, come to mind.

The following proposition follows immediately from a somewhat more general theorem due to Karamata (1932); see, for example, Marshall and Olkin (1979, p. 449).

B.2. Proposition. The ordering $X \leq_{cx} Y$ is equivalent to the conditions

(i) $EX = EY$,

(ii) $\int_t^\infty \bar{F}(x)\, dx \leq \int_t^\infty \bar{G}(x)\, dx$ for all t.

Proof. Suppose that $X \leq_{cx} Y$. Because the function $(x-a)^+ = \max[x-a, 0]$ is convex, it follows that $E(X-t)^+ \leq E(Y-t)^+$, that is, (ii) holds. It has already been noted that $X \leq_{cx} Y$ implies (i).

Next, suppose that (i) and (ii) hold. By Proposition 21.A.19, any convex function can be approximated by linear combinations of the functions $-x, x$, and $(x-a)^+$, from which it follows that $X \leq_{cx} Y$. □

B.3. Proposition. Suppose that X and Y are nonnegative random variables. If $EX = EY$ and if $\bar{F}$ crosses $\bar{G}$ only once, and from above, then $X \leq_{cx} Y$.

Proof. According to 1.B(10b) the condition $EX = EY$ can be rewritten as

$$\int_0^\infty \bar{F}(x)\, dx = \int_0^\infty \bar{G}(x)\, dx;$$

This means that $\bar{F}$ and $\bar{G}$ must cross at least once. Because $\bar{F}$ crosses $\bar{G}$ from above, (ii) of Proposition B.2 holds for all t above the crossing point. The single crossing property insures that

$$\int_t^\infty \bar{G}(x)\, dx - \int_t^\infty \bar{F}(x)\, dx$$

is decreasing; because this difference is negative for large t and 0 for $t = 0$, it must be less than or equal to 0 for all t.

A more formal proof can be obtained using total positivity. Suppose that $F(x) - G(x)$ has a single crossing, from + to - as x ranges from 0 to ∞. Let $K(x, y)$ be defined as in 21.B(5), that is, $K(x, y) = 0$, if $x < y$, and $K(x, y) = 1$, if $x \geq y$, $-\infty < x, y < \infty$, then K is totally positive and its variation diminishing property Definition 21.B.12 insures that

$$\int_0^\infty [\bar{F}(x) - \bar{G}(x)]K(x,t)\, dx = \int_t^\infty [\bar{F}(x) - \bar{G}(x)]\, dx$$

has at most one sign change, from + to − if one occurs. Because equality holds at $t = 0$, it follows that this integral is nonpositive, and (ii) of Proposition B.2 holds. □

The fact that Proposition B.3 also holds without the condition that the random variables are nonnegative is proved by Shaked and Shanthikumar (1994, p. 65, 66) by using different methods.

B.4. Example. A distribution F such that $F(0-) = 0$ is said to have an increasing hazard rate average (IHRA) if its hazard function R satisfies the condition that $R(x)/x$ is increasing. According to Example 5.B.3, an IHRA survival function (defined in Section 5.B) can cross an exponential survival function only from above. Thus, if Y has an exponential distribution G, X has the IHRA distribution F, and $EX = EY$, then according to Proposition B.3, $X \leq_{cx} Y$.

To more fully understand why the convex order deserves to be called a "dispersion order," first recall that by Jensen's inequality,

$$\phi(EY) \leq E\phi(Y)$$

for all convex functions ϕ such that the expectations exist. (see Proposition 21.A.12.) This is a comparison like (1) but where X is a random variable degenerate at EY, and so has no dispersion at all.

b. Convex Order and Majorization

It is well known that the convex order arises in the context of empirical distributions. If X takes on the values $x_1, \ldots, x_n$ each with probability $1/n$, and Y takes on the values $y_1, \ldots, y_n$ each with probability $1/n$, then (1) holds if and only if $(x_1, \ldots, x_n)$ is *majorized* by $(y_1, \ldots, y_n)$. For this result and an introduction to the concept of majorization, see Marshall and Olkin (1979, Section A, Chapter 1) or Ando (1989). Majorization was carefully constructed to represent the idea that the x_i's are "more nearly equal" than are the y_i's, and this would mean that the random variable X is "less dispersed" than is the random variable Y.

To generalize the idea of majorization, suppose that $M_x(y)$ is a distribution function in the real variable y, for each fixed x, and is measurable in the real variable x, for each fixed y. Suppose further that

$$F_Y(y) = \int M_x(y)\, dF_X(x), \tag{2}$$

$$x = \int_y dM_x(y). \tag{3}$$

Although relative dispersion relates to the distributions of X and Y and is unrelated to their joint distribution, to better understand the ideas here it may be helpful to think of $M_x(y)$ as the conditional probability

$$M_x(y) = P\{Y \le y \mid X = x\}.$$

Then (2) is just the usual relation of unconditioning and (3) says that

$$E(Y \mid X = x) = x.$$

If (3) holds, then when $X = x$, either $Y = X$ or Y takes on several values, the mean of which is x. This makes Y "more dispersive" than X. Now, make use of Jensen's inequality for conditional expectations (Proposition 21.A.13 and Example 21.A.13.a) and compute that

$$E\phi(Y) = E[E(\phi(Y) \mid X)] \ge E(\phi(Y) \mid X) = E\phi(X).$$

Thus, (2) and (3) together lead to the conclusion that $E\phi(X) \le E\phi(Y)$ for all convex functions ϕ, that is, $X \le_{\text{cx}} Y$. There is a converse to this result: If $X \le_{\text{cx}} Y$, then there exists a function $M_x(y)$ that satisfies (2) and (3). This fact is a special case of Choquet's theorem; see Meyer (1966, p. 277) or Phelps (1966). A further discussion of this topic is beyond the scope of this book.

c. Dispersive Order

Random variables are comparable in the convex order only if they have equal expectations. The dispersive order discussed here avoids this limitation.

B.5. Definition (Doksum, 1969). Let X be a random variable with distribution function F and Y be a random variable with distribution function G. If

$$F^{-1}(q) - F^{-1}(p) \le G^{-1}(q) - G^{-1}(p), \quad 0 \le p \le q \le 1, \tag{4}$$

then X is said to be *smaller than Y in the dispersive order*, written $X \leq_{\text{disp}} Y$.

The following conditions are equivalent to (4):

(i) $G^{-1}F(x) - x$ is increasing in x such that $0 < F(x) < 1$.

(ii) As x ranges over the interval for which $0 < F(x) < 1$, for all real c, $G^{-1}F(x) - x - c$ has at most one sign change, from $-$ to $+$ if one occurs.

(iii) As x ranges over the interval for which $0 < F(x) < 1$, for all real c, $F(x) - G(x + c)$ has at most one sign change, from $-$ to $+$ if one occurs.

(iv) As x ranges over the interval for which $0 < F(x) < 1$, for all real c, $\bar{F}(x) - \bar{G}(x + c)$ has at most one sign change, from $+$ to $-$ if one occurs.

Condition (iii) is to be compared with C.11(iv), where scale takes the place of location.

B.6. Proposition.

$X \leq_{\text{disp}} Y$ if and only if $X + c \leq_{\text{disp}} Y$ for all real c,
$X \leq_{\text{disp}} aX$ for all $a \geq 1$.

B.7. Proposition. If the supports of F and G have the same finite left hand endpoint, that is, if $F^{-1}(0) = G^{-1}(0) > -\infty$, then $X \leq_{\text{disp}} Y$ implies $X \leq_{\text{st}} Y$.

Proof. This follows directly from the definition by taking $p = 0$. □

Shaked and Shanthikumar (1994, p. 75; 2007, p. 154) obtain conditions under which the conclusion of stochastic order in Proposition B.7 can be strengthened to hazard rate order, and they obtain other results connecting these two orders. These results show the close connection between orders relating to magnitude and dispersion for nonnegative random variables.

Proposition B.7 exhibits a character of the dispersive order $\leq_{\text{disp}}$ that is quite different from the convex order $\leq_{\text{cx}}$ because stochastic order precludes the equality of expectations, a requirement of the convex order. Nevertheless, there is a connection between the orders, as indicated by the following proposition.

B.8. Proposition (Shaked and Shanthikumar, 1994, 2007). If X and Y are random variables with finite means and $X \leq_{\text{disp}} Y$, then

$$(X - EX) \leq_{\text{cx}} (Y - EY).$$

Considerably more is known about the dispersive order (see Shaked and Shanthikumar (1994, 2007) and the references contained there). For connections with the total time on test transform, see Bartoszewicz (1986, 1995).

B.9. Example. Suppose that Y has an exponential distribution, that is, $\bar{G}(x) = e^{-\lambda x}, x \geq 0$. To consider condition (i) of B.5, first compute that

$$G^{-1}(z) = -\frac{1}{\lambda} \log(1-z).$$

Thus, $G^{-1}F(x) - x = [-\frac{1}{\lambda} \log \bar{F}(x)] - x$ is increasing in x such that $0 < F(x) < 1$ if and only if

$$\exp\{G^{-1}F(x) - x\} = \exp\left\{\left[-\frac{1}{\lambda} \log \bar{F}(x)\right] - x\right\} = \frac{e^{-x}}{[\bar{F}(x)]^{1/\lambda}}$$

is increasing in x such that $0 < F(x) < 1$. Consequently, $X \leq_{\text{disp}} Y$ if and only if $e^{-\lambda x}/\bar{F}(x)$ is increasing in x such that $0 < F(x) < 1$.

If F has a density, by differentiating, this condition can be recast as the condition that $r(x) \geq \lambda$, for all x such that $0 < F(x) < 1$, where the hazard rate r is the derivative of the hazard function $R = -\log \bar{F}$.

C. Shape

MacGillivray and Blanda (1988) define distributions F and G to have the same shape if for some a and b, $F(x) = G(ax + b)$ for all x, i.e., F and G are of the same type. Because this book is devoted to nonnegative random variables, a slightly different definition is adopted.

C.1. Definition. Distribution functions F and G have the *same shape* if either

(i) for some a and b, $F(x) = G(ax + b)$ for all x,

or

(ii) $F(x) = G(x) = 0$ for all $x < 0$, and for some $a > 0$, $F(x) = G(ax)$ for all x.

The orderings of this section all have the property that if F and G have the same shape, then $F \leq G$ and $G \leq F$. This means that shape orders are preorders but not partial orders.

a. The Lorenz Order

Because magnitude, as reflected in a scale change, is intimately related to dispersion, the convex order compares distributions only if their expectations are equal. A modification of the definition eliminates this severe restriction by rescaling the two distributions before the conditions of the convex order are applied.

C.2. Definition. Let X and Y be nonnegative random variables with finite expectations. Then X is said to be smaller in the *Lorenz order* than Y, written $X \leq_{\text{Lorenz}} Y$, if

$$X/EX \leq_{\text{cx}} Y/EY, \tag{1}$$

that is,

$$E\phi(X/EX) \leq_{\text{cx}} E\phi(Y/EY)$$

for all convex functions ϕ such that the expectations exist.

Clearly, the Lorenz order is a shape order that applies to nonnegative random variables and compares dispersion *after* scale normalizations to achieve unit expectations.

C.3. Proposition. Suppose that X and Y are nonnegative random variables with equal finite expectations. Then, $X \leq_{\text{Lorenz}} Y$ if and only if $L_X(p) \geq L_Y(p), 0 \leq p \leq 1$, where L_X and L_Y are, respectively, the Lorenz curves of the distributions of X and Y. See Definition 1.I.7.

For more complete discussions of the Lorenz order, see Arnold (1987, Chapter 3), Kakwani (1980), Marshall and Olkin (1979), and Shaked and Shanthikumar (1994, 2007). As with the convex order, the Lorenz order has been used in economics to compare income distributions and the risks associated with different prospects.

b. The Gini Index and the Coefficient of Variation

Measures of shape that preserve the Lorenz order are sometimes regarded as "scale-free measures of dispersion" even though this may be a contradiction of terms. In economics, a commonly used measure of this kind is the Gini index (Definition 1.I.10).

C.4. Proposition. If X has the distribution function F and Y has the distribution function G, and $X \leq_{\text{Lorenz}} Y$, then $Gini(F) \leq Gini(G)$.

Proof. This fact follows directly from Proposition C.3 and Definition 1.I.10. □

C.5. Proposition. The Gini index is independent of scale. To state this more precisely, suppose that the random variable X has the distribution function F. Then X/λ, has a distribution function $F(\cdot \mid \lambda)$ given by $F(x \mid \lambda) = F(\lambda x)$, and the Gini index of $F(\cdot \mid \lambda)$ is independent of λ.

Proof. According to 1.I.(12),

$$Gini(F(\cdot \mid \lambda)) = \frac{\int_0^\infty F(\lambda x) \bar{F}(\lambda x)\, dx}{\int_0^\infty \bar{F}(\lambda x)\, dx}. \qquad (2)$$

With the change of variables $y = \lambda x$ in both numerator and denominator, the result follows. □

Another commonly used measure of shape in the sense of the Lorenz order is the coefficient of variation

$$CV(X) = \sigma/\mu \qquad (3)$$

defined in 1.C.3 for a random variable X with expectation μ and standard deviation σ. This measure is independent of scale. Because $(x-\mu)^2$ is convex, it follows that

$$X \leq_{\text{Lorenz}} Y \Rightarrow CV(X) \leq CV(Y).$$

Because $X \leq_{\text{cx}} Y$ implies that $EX = EY$ and again because $(x-\mu)^2$ is convex, it also follows that

$$X \leq_{\text{cx}} Y \Rightarrow CV(X) \leq CV(Y).$$

On the other hand, the stochastic, hazard rate, or likelihood ratio orderings of Section A do not imply that the coefficients of variation are ordered. Simple counterexamples are obtained by letting $Y = X + c$,

$c > 0$. Then, $CV(X) > CV(Y)$ but $X \leq_{\text{st}} Y$; $X \leq_{\text{hr}} Y$ when X has an increasing hazard rate and $X \leq_{\text{lr}} Y$ when X has a log-concave density.

c. Skewness and the Convex Transform Order

The notion of skewness is intended to represent departure of a density from symmetry (or sometimes even departure from normality), where one tail of the density is "stretched out" more than the other. Because symmetry is most natural for distributions with support $(-\infty, \infty)$, the notions of asymmetry may be most natural in that case; but they also have a place in describing distributions with support $[0, \infty)$.

The notion of skewness has been applied particularly to unimodal densities. If the mode of such a density is to the left of "center" and the right-hand tail is relatively long, then the density is said to be "skewed to the right." This is the kind of skewness most often encountered in distributions with support $[0, \infty)$.

"Departure from symmetry" is a rather vague phrase, and the historical approach to making it more precise was to define measures of skewness; some of the various proposed measures are given in Section C.e.

An alternative to measuring skewness is to find an ordering $\leq$ for which "$F \leq G$" captures the essence of what "F is less skewed than G" should mean. Perhaps surprisingly, there is a relatively simple and reasonable way to arrive at such an order.

If $Y = \psi(X)$, what properties should ψ have in order that Y have a distribution "more skewed" than X? To answer this question, imagine that the density f of X is graphed on a sheet of rubber, a sheet that becomes thinner and thinner toward the right, and thus more and more easily stretched toward the right. Now, grasp the right-hand edge of the rubber sheet, stretch it out, and watch the density change shape. If f was symmetric and unimodal before stretching, then after stretching f has become a new density g which is also unimodal, but which has a relatively long right-hand tail; g is *skewed to the right*. The flexibility requirement of the rubber sheet simply means that the horizontal axis has been transformed by an increasing function ψ with increments increasing as one moves to the right, i.e., $\psi(x + \Delta) - \psi(x)$ is increasing in x. This means that ψ is convex. The following proposition is needed to formalize these ideas.

C.6. Proposition. Suppose that $Y = \psi(X)$, where ψ is strictly increasing. Suppose further that X has distribution function F and Y has distribution function G. Then, $\psi = G^{-1}F$.

Proof. Compute that

$$G(y) = P\{\psi(X) \le y\} = P\{X \le \psi^{-1}(y)\} = F(\psi^{-1}(y)),$$

so that

$$y = G^{-1}F(\psi^{-1}(y)).$$

This means that $\psi = G^{-1}F$. □

Although the definition of ψ in Proposition C.6 is in terms of $G^{-1}F$, it should be noted that according to 20.A(6),

$$G^{-1}F = \bar{G}^{-1}\bar{F}. \tag{4}$$

The above discussion together with Proposition C.6 motivates the following ordering for skewness due to van Zwet (1964).

C.7. Definition. Suppose that X has distribution function F and Y has distribution function G. Then, X is said to be *smaller in the convex transform order* than Y, written $X \le_c Y$, if $G^{-1}F(x)$ is convex in x on the support of X. The notation $F \le_c G$ is also used to mean $X \le_c Y$.

The convex transform order is not to be confused with the convex order discussed in Section B.

d. Geometric Interpretation of the Convex Transform Order

For consideration of the $\le_c$ order, various equivalencies are sometimes convenient. The following are given without proof because each step should be clear.

The notation concerning crossings used in the following proposition is described in Section 21.B.g.

C.8. Proposition. The following conditions are equivalent:

(i) $\bar{G}^{-1}\bar{F}$ is convex.

(ii) For all real a, b, $\bar{G}^{-1}\bar{F}(x) - (ax + b)$ has at most two changes of sign; if there are two sign changes, they are in the order $+, -, +$.

(iii) For all real a, b, $\bar{F}(x) - \bar{G}(ax + b)$ has at most two changes of sign; if there are two sign changes, they are in the order $-, +, -$.

(iv) For all real a, b, $\bar{F}(ax + b) - \bar{G}(x)$ has at most two changes of sign; if there are two sign changes, they are in the order $-, +, -$.

(v) For all real a, b, $\bar{G}(x) - \bar{F}(ax+b)$ has at most two changes of sign; if there are two sign changes, they are in the order $+, -, +$.

(vi) For all real a, b, $\bar{F}^{-1}\bar{G}(x) - (ax+b)$ has at most two changes of sign; if there are two sign changes, they are in the order $-, +, -$.

Note that the last condition states that $\bar{F}^{-1}\bar{G}(x)$ is concave; it is sometimes convenient to check concavity of $\bar{F}^{-1}\bar{G}$ rather than convexity of $\bar{G}^{-1}\bar{F}$ (see Proposition 21.A.7).

Note that above-mentioned consideration may as well be limited to $a > 0$, for otherwise there can be only one crossing. So the convex transform order is equivalent to the condition that G, or any distribution of the same type as G, has a survival function that crosses $\bar{F}$ at most twice, and if there are two crossings, $\bar{F}$ has the smaller values in the upper tail.

C.9. Proposition. If $X \leq_c Y$, then $aX + b \leq_c Y$ and $X \leq_c aY + b$ for all $a > 0$ and real b.

This proposition can be verified either directly from the definition or from the geometric interpretation of the ordering. It indicates that the order is a proper shape order under either condition of Definition C.1, so it is useful for all random variables, nonnegative or not.

e. Measures of Skewness

It has been generally accepted (see, e.g., Arnold and Groeneveld, 1995) that any measure γ of skewness should satisfy certain conditions. To state these conditions, it is convenient to write $\gamma(X)$ and $\gamma(F)$ interchangeably, it being understood that X has the distribution F.

$$\gamma(X) = \gamma(aX + b) \quad \text{for all } a > 0 \text{ and all } b. \tag{5}$$
$$\gamma(X) = -\gamma(-X). \tag{6}$$
$$\text{If } F \leq_c G, \text{ then } \gamma(F) \leq \gamma(G). \tag{7}$$

Note that (7) implies

$$\gamma(F) = 0, \quad F \text{ is symmetric.} \tag{8}$$

Several well-known measures of skewness have been proposed. Early on Sir Francis Galton proposed the measure

$$\frac{\text{lower quartile} + \text{upper quartile} - 2(\text{median})}{\text{upper quartile} - \text{lower quartile}}. \tag{9}$$

A measure of skewness proposed by Pearson (1895) for unimodal distributions is

$$[\text{mean} - \text{mode}]/[\text{standard deviation}]. \tag{10}$$

Of course, this is 0 for symmetric unimodal distributions and invariant under scale change. For a distribution of the random variable X, a more commonly used measure of skewness, proposed by Edgeworth (1904) and Charlier (1906), is

$$\frac{E(X-\mu)^3}{\sigma^3} = \frac{E[X-E(X)]^3}{\{E[X-EX]^2\}^{3/2}}, \tag{11}$$

which requires the existence of a third moment. This measure of skewness is also zero for symmetric distributions and invariant under scale change. It is not hard to show by example that either of these measures can also be zero for distributions with nonsymmetric densities. The measure (10) satisfies (5) to (8), but van Zwet (1964, p. 16) shows by a counterexample that (9) fails to satisfy (7). See Oja (1981) and Arnold and Groeneveld (1995) for these results and other measures of skewness.

f. The Star Order

The star order is defined in a fashion very similar to the convex transform order, but replaces the convexity condition by the weaker star-shaped condition (see Definition 21.A.8). Note the lack of symmetry in the terminology that is entrenched in the literature; the star order is closely related to the convex transform order, not the convex order as the terminology might suggest.

Again, let X and Y be nonnegative random variables with respective distribution functions F and G. Assume that the support of X is a possibly infinite interval.

C.10. Definition. Suppose that X has distribution function F and Y has distribution function G. Then X is said to be *smaller in the star order* than Y, written $X \leq_* Y$, if $G^{-1}F$ is starshaped in x, i.e., $G^{-1}F(x)/x$ is increasing in $x \geq 0$. Similarly, write $F \leq_* G$ to mean that $X \leq_* Y$.

g. Geometric Interpretation of the Star Order

The order $\leq_*$ has an interpretation similar to that of the convex transform order; the following proposition is to be compared with Proposition C.8.

C.11. Proposition. The following conditions are equivalent:

(i) $G^{-1}F = \bar{G}^{-1}\bar{F}$ is starshaped.
(ii) $\bar{G}^{-1}\bar{F}(x)/x$ is increasing in $x \geq 0$.
(iii) For all real a, $[\bar{G}^{-1}\bar{F}(x)/x] - a$ has at most one sign change; if there is a sign change, it is in the order $-, +$.
(iv) For all real a, $\bar{F}(x) - \bar{G}(ax)$ has at most one sign change; if there is a sign change, it is in the order $+, -$.
(v) For all real a, $\bar{F}(ax) - \bar{G}(x)$ has at most one sign change; if there is a sign change, it is in the order $+, -$.
(vi) For all real a, $\bar{G}(x) - \bar{F}(ax)$ has at most one sign change; if there is a sign change, it is in the order $-, +$.
(vii) For all real a, $\bar{F}^{-1}\bar{G}(x) - ax$ has at most one sign change; if there is a sign change, it is in the order $+, -$.
(viii) $F^{-1}G(x)/x$ is decreasing in $x > 0$.

Condition (iv) says that $X \leq_* Y$ if and only if $\bar{F}$ crosses $\bar{G}$ at most once, and only from above, no matter how Y is scaled; this is to be compared with B.5(iii) where the change is in location rather than scale. One can think of a collection of grid lines obtained by graphing $\bar{G}$ with all possible scalings; then $X \leq_* Y$ if $\bar{F}$ passes through the grid by crossing grid lines only from above. That this is a shape order is the content of the following proposition.

C.12. Proposition. If $X \leq_* Y$, then $aX \leq_* Y$, for all $a > 0$.

Proof. This result follows from the equivalence of (iv) and (vii). □

C.13. Proposition. Suppose that $F \leq_* G$ and that for a fixed $r > 0$,

$$\int_0^\infty x^r \, dF(x) = \int_0^\infty x^r \, dG(x).$$

If ψ is an increasing function, then

$$\int_0^\infty \psi(x) x^{r-1} \bar{F}(x) \, dx \leq \int_0^\infty \psi(x) x^{r-1} \bar{G}(x) \, dx.$$

Proof. If $F \leq_* G$, then $\bar{F}$ crosses $\bar{G}$ at most once, and only from above. Because the rth moments coincide, they must cross at least once (see Section 21.B.f). Hence, $\bar{F}$ must cross $\bar{G}$ exactly once and from above. Let x_0 be a solution of $\bar{F}(x) = \bar{G}(x)$. Then $x^{r-1}\bar{F}(x)$ crosses $x^{r-1}\bar{G}(x)$

at x_0, and from above. It follows with the aid of 1.C(4) that

$$\int_0^\infty \psi(x) x^{r-1} \bar{F}(x) \, dx - \int_0^\infty \psi(x) x^{r-1} \bar{G}(x) \, dx$$
$$= \int_0^\infty [\psi(x) - \psi(x_0)][x^{r-1}\bar{F}(x) - x^{r-1}\bar{G}(x)] \, dx.$$

Because the two factors of the integrand have opposite signs for all x, the result follows. □

h. The Superadditive Order

C.14. Definition. Suppose that X has distribution function F and Y has distribution function G. Then, X is said to be *smaller in the superadditive order* than Y, written $X \leq_{\text{su}} Y$, if $G^{-1}F = \bar{G}^{-1}\bar{F}$ is superadditive in x, i.e., $G^{-1}F(x+y) \geq G^{-1}F(x) + G^{-1}F(y), x, y \geq 0$. Similarly, $F \leq_{\text{su}} G$ means $X \leq_{\text{su}} Y$.

Note that the superadditive order is related to the convex transform order, not the convex order as its name might suggest.

The superadditive order is perhaps less interesting than the star order; in this book, there are few instances where the superadditive order is encountered. However, the convex transform order, the star order, and the superadditive order, are all used in reliability theory, with G taken to be an exponential distribution (see Chapter 4).

i. Some Order Relationships

It follows from Definition 21.A.10 that

$$X \leq_c Y \Rightarrow X \leq_* Y \Rightarrow X \leq_{\text{su}} Y. \tag{12}$$

Also it can be shown (Shaked and Shanthikumar, 1994, p. 107, 2007, p. 231) that if X and Y are nonnegative random variables with finite strictly positive expectations, then the star order implies the Lorenz order.

j. Summary

$$X \leq_c Y \Rightarrow X \leq_* Y \quad \begin{array}{l} \Rightarrow \ X \leq_{\text{su}} Y \\ \Rightarrow \ X \leq_{\text{Lorenz}} Y \end{array}$$

D. Cone Orders

The literature includes definitions and discussions of a bewildering array of distribution orders, and this chapter cannot begin to review the entire subject. Cone orders constitute one class of orders that needs to be mentioned, if only briefly.

Both stochastic order $\leq_{st}$ and convex order $\leq_{cx}$ can be defined in terms of expectations of functions in a specified class: $X \leq Y$ if $E\phi(X) \leq E\phi(Y)$ for all ϕ in the class. For stochastic order, the class consists of all increasing functions for which the expectations exist; for convex order, the class consists of all convex functions for which the expectations exist. Both classes of functions form convex cones; a class of functions C forms a convex cone if $\phi_1, \phi_2 \in C$ implies $a_1\phi_1 + a_2\phi_2 \in C$, for all $a_1, a_2 > 0$.

D.1. Definition. Let C be a convex cone of measurable functions. Write $X \leq_C^{st} Y$ to mean that

$$E\phi(X) \leq E\phi(Y), \text{ for all } \phi \text{ in } C \text{ such that the expectations exist.}$$

The order $\leq_C^{st}$ is called a *cone order*.

In addition to stochastic and convex order, a number of examples of cone orders can be found in the literature.

D.2. Laplace transform order. For nonnegative random variables X and Y, write $X \leq_{Lt} Y$ to mean that $E \exp\{-sX\} \geq E \exp\{-sY\}$, for all $s > 0$, that is, $-E \exp\{-sX\} \leq -E \exp\{-sY\}$. Here, the convex cone consists of all functions the negative of which is completely monotone, that is, of all mixtures of functions of the form $\phi(x) = -\exp\{-sx\}, s > 0$ (see Definition 20.D.4 and Proposition 20.D.5). The Laplace transform order is discussed, e.g., by Shaked and Shanthikumar (1994, p. 95; 2007, Chapter 5). This order is weaker than stochastic order because the condition A(2) is required for only a limited class of increasing functions.

D.3. Increasing convex order. Let C be the class of convex increasing functions, and denote the resulting cone order by $\leq_{icx}$. The cone here is a proper subset of the class of all convex functions, and consequently $X \leq_{cx} Y$ implies $X \leq_{icx} Y$. But the converse is false (see Shaked and Shanthikumar, 2007, p. 184).

D.4. Increasing concave order: Second-order stochastic dominance. Let C be the class of concave increasing functions, and denote

D. Cone Orders

the resulting cone order by $\leq_{ssd}$. This order, called *second-order stochastic dominance*, arises in economics and utility theory. In this context, the usual stochastic order is called *first-order stochastic dominance* (see Hong and Herk, 1996) for this application and for further references). It can be shown with the aid of Proposition 21.A.19 that $X \leq_{ssd} Y$ if and only if the corresponding distribution functions F and G satisfy the condition

$$\int_{-\infty}^{x} F(z)\, dz \geq \int_{-\infty}^{x} G(z)\, dz, \quad \text{for all } x.$$

This inequality is often used as a definition of second-order stochastic dominance.

A number of other examples, extensions to higher dimensions, and theoretical results about cone orders are given by Marshall (1991).

The following display reviews the relationships between the orders discussed in this section.

$$X \leq_{cx} Y \Rightarrow X \leq_{icx} Y$$
$$\Uparrow$$
$$X \leq_{st} Y \Rightarrow X \leq_{Lt} Y$$

3
Mixtures

It often happens that data from several populations is mixed and information about which subpopulation gave rise to individual data points is unavailable. For example, measurements of life lengths of a device may be gathered without regard to the manufacturer, or data may be gathered on humans without regard, say, to blood type. If the ignored variable (manufacturer or blood type) has a bearing on the characteristic being measured, then the data are said to come from a mixture. In actuality, it is hard to find data that are not some kind of a mixture, because there is almost always some relevant covariate that is not observed. Mixture models arise in a number of applications and statistical settings, many of which are discussed by Titterington, Smith and Makov (1985). Important early work on mixtures was done by Teicher (1960, 1962). For a general discussion of mixtures, see Lindsay (1995) or Ord (1972).

Mixtures also play a central role in Bayesian statistics, not from a physical mixing of several populations but from a lack of precise knowledge of the exact distribution from which data is obtained. In either case, the mathematics is much the same, although interpretations of results may differ.

From a purely mathematical point of view, functions such as sums, products, or ratios of independent random variables have distributions that take on the form of a mixture.

Mixtures are defined and briefly discussed in Section 1.E, and can be found throughout much of this book. Here, some basic general results are given.

A. Basic Ideas

Let $\mathcal{F} = \{F(\cdot \mid \theta) : \theta \in \Theta\}$ be a family of distributions indexed by a parameter θ which takes values in a set Θ. When θ can be regarded as a random variable with a distribution function G, then

$$F(x) = \int_\Theta F(x \mid \theta) \, dG(\theta) \qquad (1)$$

is the *mixture of $\mathcal{F}$ with respect to G*, and G is called the *mixing distribution*. Mixtures have sometimes been called "compound distributions."

Bayesian terminology is somewhat different. In this context, the distribution $F(\cdot \mid \theta)$ is sometimes called the *model distribution*, F is called the *predictive distribution*, and G is the *prior distribution*. Both the model distribution and the prior distribution may be conditioned by additional information, which is suppressed here because statistical considerations are not discussed in this book. For further discussions of Bayesian analysis, see, for example, Barlow (1985), Harris and Singpurwalla (1968), or Press (2003).

With the exception of degenerate distributions (distribution functions taking only the values 0 and 1), all distributions have nontrivial mixture representations, and such representations are not unique. In applications, there is often at least one natural mixture representation, the recognition of which is sometimes important or perhaps even essential for an understanding of the underlying random variable of interest.

It may be worth noting that the mixture (1) need not be proper even though $F(x \mid \theta)$ is proper for all θ. For example, if $F(x \mid \theta)$ is an exponential distribution with parameter θ and if G is a Poisson distribution (defined in Section 20.E.c), then $\lim_{x \to \infty} F(x) < 1$.

Finite mixtures and, in particular, mixtures of but two distributions are of special interest, and are relatively easy to understand. Such mixtures are discussed, e.g., by Titterington, Smith and Makov (1985).

Mixtures are used in later chapters to derive new parametric families of distributions from old ones; this is done by using a mixing distribution G, which has a parameter; the mixture retains that parameter so that it may yield a new parametric family. Sometimes, a mixture representation is also helpful in theoretical studies, especially when the components of the mixture are relatively simple to understand and work with.

If F is given by (1), then the corresponding survival function is given by

$$\bar{F}(x) = \int_\Theta \bar{F}(x \mid \theta) \, dG(\theta). \qquad (2)$$

If a density $f(\cdot \mid \theta)$ of $F(\cdot \mid \theta)$ exists for all θ and F is given by (1), then F has the density f given by

$$f(x) = \int_\Theta f(x \mid \theta) \, dG(\theta). \qquad (3)$$

But similar representations are not true for hazard functions, hazard rates, or residual life distributions. If F has the representation (1), then its residual life survival function $\bar{F}_t$ is given by

$$\bar{F}_t(x) = \frac{\int_\Theta \bar{F}(x+t \mid \theta) \, dG(\theta)}{\int_\Theta \bar{F}(t \mid \theta) \, dG(\theta)}, \qquad (4)$$

and when densities exist, the distribution F of (1) has hazard rate

$$r(x) = \frac{\int_\Theta f(x \mid \theta) \, dG(\theta)}{\int_\Theta \bar{F}(x \mid \theta) \, dG(\theta)}. \qquad (5)$$

Thus, both the residual life distribution and the hazard rate are ratios of mixtures. However, these ratios can be written as actual mixtures, but with a mixing distribution that depends upon t in the case of (4) or x in the case of (5); this observation, discussed further in Section B, turns out to provide important insights into the behavior of the mixture, most of which are well known in Bayesian analysis.

A.1. Proposition. Let $r(x \mid \theta) = f(x \mid \theta)/\bar{F}(x \mid \theta)$; if $r_l \leq r(x \mid \theta) \leq r_u$ for all θ in the support of G, then $r_l \leq r(x) \leq r_u$, where $r(x)$ is given by (5).

The proof of this straightforward consequence of (5) is omitted. Results essentially equivalent to Proposition A.1 have been obtained by Block and Joe (1997) and Badia, Berrade, Campos and Navascués (2001).

The special case in which G puts mass at but two points is sufficiently important to examine in detail. Then F is the mixture of but two distributions, say F_1 and F_2, and (1) takes the form

$$F(x) = \pi F_1(x) + \bar{\pi} F_2(x), \tag{6}$$

where $\bar{\pi} = 1 - \pi$.

Suppose that X_i has distribution F_i, $i = 1, 2$, and suppose that these random variables are independent. Let I be a random variable independent of the X_i, taking only the values 0 and 1, for which

$$P\{I = 1\} = \pi, \quad P\{I = 0\} = \bar{\pi}.$$

Then, the mixture

$$X = IX_1 + (1 - I)X_2 \tag{7}$$

has the distribution function F of (6). Equation (7) shows explicitly that X is sometimes X_1 and sometimes X_2.

When (6) holds, the hazard rate (5) becomes

$$r(x) = \frac{\pi f_1(x) + \bar{\pi} f_2(x)}{\pi \bar{F}_1(x) + \bar{\pi} \bar{F}_2(x)} = p(x)r_1(x) + \bar{p}(x)r_2(x), \tag{8}$$

where

$$p(x) = \frac{\pi \bar{F}_1(x)}{\pi \bar{F}_1(x) + \bar{\pi} \bar{F}_2(x)}, \quad \bar{p}(x) = 1 - p(x), \tag{9}$$

and r_i is the hazard rate of $F_i, i = 1, 2$. Thus, (8) exhibits r in the form of a mixture or average of $r_1(x)$ and $r_2(x)$, with weights that are functions of the argument x. It follows that

$$\min\left[r_1(x), r_2(x)\right] \le r(x) \le \max\left[r_1(x), r_2(x)\right]. \tag{10}$$

It should be intuitively clear that the weights $p(x)$ or $\bar{p}(x)$ placed on the smallest hazard rate are increasing, because as a device ages without failure, the likelihood that the device came from a relatively strong component of the mixture (small hazard rate) should increase. This is the content of the following proposition.

A.2. Proposition. Suppose that r is given by (8) and p is given by (9). If $r_1(x) < r_2(x)$, then $\frac{d}{dz}p(z)|_{z=x} \geq 0$, with strict inequality if $\bar{F}_1(x)\bar{F}_2(x) > 0$.

First Proof. A direct differentiation of $p(z)$ yields

$$\frac{d}{dz}p(z) = \frac{\pi\bar{\pi}\bar{F}_1(z)\bar{F}_2(z)[r_2(z) - r_1(z)]}{[\pi\bar{F}_1(z) + \bar{\pi}\bar{F}_2(z)]^2} \geq 0,$$

and clearly strict inequality holds if $\bar{F}_1(x)\bar{F}_2(x) > 0$. □

Second Proof. Suppose that $\bar{F}_1(x)\bar{F}_2(x) > 0$; otherwise (9) simplifies and the result is immediate. It follows from (9) that

$$p(z) = \frac{\pi}{\pi + \bar{\pi}[\bar{F}_2(z)/\bar{F}_1(z)]}, \tag{11}$$

so it is only necessary to show that $\bar{F}_2(z)/\bar{F}_1(z)$ is strictly decreasing in z at $z = x$. But

$$\bar{F}_2(z)/\bar{F}_1(z) = \exp\left\{-\int_0^z [r_2(u) - r_1(u)]\, du\right\} \tag{12}$$

and the result follows. □

A.3. Example. Suppose that $\bar{F}_i(x) = \exp\{-\lambda_i x\}, i = 1, 2$ are exponential distributions, where $\lambda_1 < \lambda_2$, and let $F(x) = \pi F_1(x) + \bar{\pi}F_2(x)$. Then (11) becomes

$$p(x) = \frac{\pi}{\pi + \bar{\pi}e^{-(\lambda_2 - \lambda_1)x}}.$$

Because $\lambda_1 < \lambda_2$, $p(x)$ is increasing in x. This case of Proposition A.2 is given by Mi (1998).

Results similar to Proposition A.2 are true more generally, and these are the topic of Section B.

B. The Conditional Mixing Distribution

Equations A(4) and A(5) exhibit the residual life survival function and the hazard rate of the distribution function F given by A(1). These quantities are not exhibited as mixtures, but instead are written as ratios of mixtures. It is useful to write the ratios as true mixtures, but with a mixing distribution that depends upon the age t. Note that if

$$dH(\theta \mid t) = \frac{\bar{F}(t \mid \theta) \, dG(\theta)}{\int_\Theta \bar{F}(t \mid \xi) \, dG(\xi)}, \qquad (1)$$

then

$$\bar{F}_t(x) = \frac{\int_\Theta \bar{F}(x + t \mid \theta) \, dG(\theta)}{\int_\Theta \bar{F}(t \mid \xi) \, dG(\xi)} = \frac{\int_\Theta \bar{F}_t(x \mid \theta) \bar{F}(t \mid \theta) \, dG(\theta)}{\int_\Theta \bar{F}(t \mid \xi) \, dG(\xi)}$$

$$= \int_\Theta \bar{F}_t(x \mid \theta) \, dH(\theta \mid t) \qquad (2)$$

and

$$r(t) = \frac{\int_\Theta f(t \mid \theta) \, dG(\theta)}{\int_\Theta \bar{F}(t \mid \xi) \, dG(\xi)} = \frac{\int_\Theta r(t \mid \theta) \bar{F}(t \mid \theta) \, dG(\theta)}{\int_\Theta \bar{F}(t \mid \xi) \, dG(\xi)} = \int_\Theta r(t \mid \theta) \, dH(\theta \mid t). \qquad (3)$$

When $\Theta = [0, \infty)$, (1) can be replaced by the notationally simpler

$$H(\theta \mid t) = \frac{\int_0^\theta \bar{F}(t \mid \xi) \, dG(\xi)}{\int_0^\infty \bar{F}(t \mid \xi) \, dG(\xi)}. \qquad (4)$$

The distribution H is called the *conditional mixing distribution given* $X > t$. It is natural to expect that H depends upon t because survival to time t indicates that the "luck of the draw" has yielded an item from a relatively robust part of the mixture. Or, in the Bayesian context, "survival to time t" is information which, when incorporated into the prior distribution, shifts the prior toward more robust distributions. As time passes, one may expect that the conditional mixing distribution puts more and more of its weight on parts of the mixture with higher survival probabilities (lower hazard rates). Recall that if $r_X(z) \geq r_Y(z)$, then X is said to be smaller in the hazard rate order than Y; this is a possible source of confusion that comes about because higher hazard rates mean lower survival probabilities.

B. The Conditional Mixing Distribution

B.1. Proposition. Suppose that $\mathcal{F} = \{F(\cdot \mid \theta) : \theta \geq 0\}$ is a family of distributions with the property that

$$r(x \mid \theta) \text{ is decreasing in } \theta \quad \text{for all } x \text{ in an interval } I. \tag{5}$$

Then, the family $\{H(\cdot \mid x) : x \geq 0\}$ of conditional mixing distributions is hazard rate increasing in $x \in I$.

Proof. The hazard rate order of Definition 2.A.5 is equivalent to 2.A(6e) when hazard rates exist, as is the case here. If G has the density g, then H has the hazard rate r_H given by

$$r_H(\theta \mid x) = \frac{\bar{F}(x \mid \theta)g(\theta)}{\int_\theta^\infty \bar{F}(x \mid \xi) \, dG(\xi)} = \frac{g(\theta)}{\int_\theta^\infty [\bar{F}(x \mid \xi)/\bar{F}(x \mid \theta)] \, dG(\xi)}. \tag{6}$$

In (6), $\xi \geq \theta$; it follows from (5) that in this case,

$$\bar{F}(x \mid \xi)/\bar{F}(x \mid \theta) = \exp\left\{-\int_0^x [r(u \mid \xi) - r(u \mid \theta)] \, du\right\}$$

is increasing in x, and from the second form of $r_H(\theta \mid x)$ in (6) it is apparent that $r_H(\theta \mid x)$ is decreasing in x. □

The hypothesis (5) is stronger than necessary, and a more general result can be obtained without the use of densities and hazard rates. To this end, let H be given by (4) and let

$$\bar{H}_\eta(\theta \mid x) = \bar{H}(\theta + \eta \mid x)/\bar{H}(\eta \mid x), \, \theta, \eta, \quad x \geq 0$$

be the residual survival function of H at η.

B.2. Proposition. If, for some fixed x and y,

$$\bar{F}(x \mid \alpha)\bar{F}(y \mid \beta) \geq \bar{F}(y \mid \alpha)\bar{F}(x \mid \beta), \quad 0 \leq \alpha < \beta, \tag{7}$$

then

$$\bar{H}_\eta(\theta \mid x) \leq \bar{H}_\eta(\theta \mid y) \quad \text{for all } \eta, \theta > 0, \tag{8}$$

i.e., in the notation of Definition 2.A.5, $\bar{H}(\cdot \mid x) \leq_{\text{hr}} \bar{H}(\cdot \mid y)$.

Proof. It follows from (7) that

$$\int_\eta^{\theta+\eta} \bar{F}(x \mid \alpha) \, dG(\alpha) \int_{\theta+\eta}^\infty \bar{F}(y \mid \beta) \, dG(\beta)$$
$$\geq \int_\eta^{\theta+\eta} \bar{F}(y \mid \alpha) \, dG(\alpha) \int_{\theta+\eta}^\infty \bar{F}(x \mid \beta) \, dG(\beta). \qquad (9)$$

Clearly,

$$\int_{\theta+\eta}^\infty \bar{F}(x \mid \alpha) \, dG(\alpha) \int_{\theta+\eta}^\infty \bar{F}(y \mid \beta) \, dG(\beta)$$
$$= \int_{\theta+\eta}^\infty \bar{F}(y \mid \alpha) \, dG(\alpha) \int_{\theta+\eta}^\infty \bar{F}(x \mid \beta) \, dG(\beta); \qquad (10)$$

addition of (9) and (10) gives

$$\int_\eta^\infty \bar{F}(x \mid \alpha) \, dG(\alpha) \int_{\theta+\eta}^\infty \bar{F}(y \mid \beta) \, dG(\beta)$$
$$\geq \int_\eta^\infty \bar{F}(y \mid \alpha) \, dG(\alpha) \int_{\theta+\eta}^\infty \bar{F}(x \mid \beta) \, dG(\beta),$$

which is just a way of rewriting (8). □

Condition (7) would be a total positivity condition (see Section 21.B) if it were required to hold for all $x \leq y$.

B.3. Corollary. If $\bar{F}(z \mid \theta)$ is totally positive of order 2 in θ and $z \geq 0$, then $\bar{H}(\cdot \mid x)$ is hazard rate increasing in $x \geq 0$.

Proof. This is immediate from Proposition B.2 because the total positivity condition is that (7) holds for all $x < y$. □

B.4. Alternative proof of Proposition B.1. When densities exist, condition (5) is equivalent to the condition that $\bar{F}(x \mid \alpha)/\bar{F}(x \mid \beta)$ is decreasing in $x \in I$ whenever $\alpha < \beta$. Thus, (7) holds for all $x < y$ in I, so the conclusion of Proposition B.1 follows from Proposition B.2. □

C. Limiting Hazard Rates

C.1. Proposition. Recall that the mixture $F(x) = \pi F_1(x) + \bar{\pi} F_2(x)$ has the hazard rate $r(x) = p(x)r_1(x) + \bar{p}(x)r_2(x)$ given by A(8) and

A(9). If the limits of $r_1(t), r_2(t)$, and $r(t)$ as $t \to \infty$ all exist, then

$$\lim_{t \to \infty} r(t) = \min\{\lim_{t \to \infty} r_1(t), \lim_{t \to \infty} r_2(t)\}.$$

Proof. Under the hypotheses of this proposition, it follows from A(11) and A(12) that with p as defined in A(9),

$$\lim_{x \to \infty} p(x) = 1,$$

and so the result is immediate. □

C.1.a. Example. As in Example A.3, suppose that $\bar{F}_i(x) = \exp\{-\lambda_i x\}, i = 1, 2$ are exponential distributions and $\lambda_1 < \lambda_2$. Then according to A(8), $F = \pi F_1 + \bar{\pi} F_2$ has the hazard rate

$$r(x) = \frac{\lambda_1 \pi + \lambda_2 \bar{\pi}\, e^{-(\lambda_2 - \lambda_1)x}}{\pi + \bar{\pi}\, e^{-(\lambda_2 - \lambda_1)x}}.$$

This hazard rate is decreasing in x and $\lim_{t \to \infty} r(t) = \lambda_1$.

Proposition C.1 is stated without proof by Vaupel and Yashin (1985). It has the following intuitive content, recognizable as a Bayesian interpretation. Consider an item with the mixture distribution $F = \pi F_1 + \bar{\pi} F_2$. As the item ages without failure, the conditional probability that the item came from the stronger of the two populations increases. This stronger population is the one with the lower hazard rate.

In Proposition C.1, the limits $\lim_{t \to \infty} r_1(t)$ and $\lim_{t \to \infty} r_2(t)$ are assumed to exist, but of course this is not always the case. More refined results are obtained by Block and Joe (1997) that consider limiting values of the ratio $r(t)/r_1(t)$.

Block, Li and Savits (2001) give an example in which $\lim_{t \to \infty} r_1(t)$ and $\lim_{t \to \infty} r_2(t)$ both exist, but $\lim_{t \to \infty} r(t)$ fails to exist; an essential feature of this example is the property that $\lim_{t \to \infty} r_2(t) = \infty$ sufficiently fast that $\lim_{t \to \infty} \bar{p}(x) r_2(x) > 0$ (whereas $\lim_{t \to \infty} \bar{p}(x) = 0$ because $\lim_{x \to \infty} p(x) = 1$) (see A(8)).

The following proposition is a variant of Theorem 4.1 of Block, Mi, and Savits (1993). It can be viewed as an extension of Proposition C.1, with the discrete mixing distribution replaced by a general distribution. Proposition C.2 is more technically involved than most results in this book. Additional results regarding limiting values of hazard rates of mixtures are given by Block, Li and Savits (2003b).

C.2. Proposition. Let $\bar{F}(x) = \int_\Theta \bar{F}(x \mid \theta) \, dG(\theta)$. Suppose that $\lim_{t \to \infty} r(t)$ exists, where r given by A(5) is the hazard rate of F. Suppose that $r(x \mid \theta) = f(x \mid \theta)/\bar{F}(x \mid \theta)$ is such that $\lim_{x \to \infty} r(x \mid \theta) = \rho_\theta$ exists uniformly, for all θ, and let $\rho = \inf\{\rho_\theta : \theta \in \Theta\}$. Suppose further that $r(x \mid \theta)$ is measurable in θ, for all x, so that $A_\varepsilon = \{\theta : \rho \le \rho_\theta \le \rho + \varepsilon\}$ and their complements A_ε^c are measurable sets. If for every $\varepsilon > 0, P\{\theta : \rho \le \rho_\theta \le \rho + \varepsilon\} > 0$, then $\lim_{x \to \infty} r(x) = \rho$.

Proof. Let $\pi = P\{A_\varepsilon\}$. Then F can be written in the form A(6) where

$$F_1(x) = \frac{\int_{A_\varepsilon} F(x \mid \theta) \, dG(\theta)}{\pi} \quad \text{and} \quad F_2(x) = \frac{\int_{A_\varepsilon^c} F(x \mid \theta) \, dG(\theta)}{\bar{\pi}}.$$

By Proposition A.1 and the uniform convergence of $r(x \mid \theta)$ to ρ_θ, the hazard rate of F_1 has lim inf and lim sup in the interval $[\rho, \rho + \varepsilon]$. Because ε is arbitrary, this means that the limiting value of the hazard rate of F_1 exists and is ρ. By Proposition C.1 the limiting hazard rate of F is also ρ. □

The following example shows that the uniform convergence required in Proposition C.2 is essential. It is a rather striking example where all components of a mixture have limiting hazard rates of infinity, whereas the limiting hazard rate of the mixture is zero.

C.3. Example (Gupta and Gupta, 1996; Block, Li and Savits, 2001). Let $r(x \mid \theta) = \theta \gamma x^{\gamma-1}, \gamma > 1$, be the hazard rate of a Weibull distribution, and let G have the gamma density

$$g(\theta) = \lambda^\nu \theta^{\nu-1} e^{-\lambda \theta}, \quad \lambda, \nu, \theta > 0.$$

Then, the mixture $\bar{F}(x) = \int_0^\infty \bar{F}(x \mid \theta) \, dG(\theta)$ has the hazard rate $r(x) = (\nu \gamma x^{\gamma-1})/(\lambda + x^\gamma)$. Here, $\lim_{x \to \infty} r(x) = 0$ even though $\lim_{x \to \infty} r(x \mid \theta) = \infty$. But the convergence is not uniform.

For another example where the conclusions of Proposition C.2. fail because convergence is not uniform, see Example 4.C.7.b and Section 10.A.h.

D. Hazard Transforms of Mixtures

The hazard function of a mixture, determined by the hazard functions ρ_θ of the components $F(\cdot \mid \theta)$ together with the mixing distribution G, is given explicitly by means of the hazard transform.

D. Hazard Transforms of Mixtures

It may be helpful to consider first the case that G is a finitely discrete distribution with mass p_i at the point $\theta_i, i = 1, \ldots, n$, where $\sum_{i=1}^{n} p_i = 1$. Let $\rho_i(x) = -\log \bar{F}(x \mid \theta_i)$ be the hazard function of the ith component of the mixture, $i = 1, \ldots, n$. The hazard transform of this mixture is given by

$$\eta(\rho_1, \ldots, \rho_n) = -\log \sum_{i=1}^{n} p_i\, e^{-\rho_i}. \tag{1a}$$

This function gives the hazard function $-\log \sum_{i=1}^{n} p_i \bar{F}(x \mid \theta_i) = -\log \sum_{i=1}^{n} p_i\, e^{-\rho_i(x)}$ of the mixture $\sum_{i=1}^{n} p_i \bar{F}(x \mid \theta_i)$ in terms of the hazard functions ρ_i of the components of the mixture. In (1a), the dependency of η on the argument x and the p_i are suppressed to simplify notation.

Note that the hazard transform has n arguments because the mixture has n components. This means that when the mixture has an infinite number of components, the hazard transform has an infinite number of arguments. For a mixture of the form

$$\bar{F}(x) = \int_\Theta \bar{F}(x \mid \theta)\, dG(\theta),$$

it is necessary to replace the vector argument $(\rho_1, \ldots, \rho_n)$ of (1a) by a more general version, which is denoted here by the expression $\langle \rho_\theta, \theta \in \Theta \rangle$. This reduces to a familiar vector only when Θ has a finite number of points.

D.1. Definition. The function η defined by

$$\eta(\langle \rho_\theta, \theta \in \Theta \rangle) = -\log \int_\Theta e^{-\rho_\theta}\, dG(\theta), \quad 0 \leq \rho_\theta \leq \infty \tag{1b}$$

is called the *hazard transform* η *of the mixture* A(1).

Note that the dependency of η on the distribution G has been suppressed in this notation just as the dependency on the p_i is suppressed in (1a) to simplify the notation.

The usual set notation $\{\rho_\theta, \theta \in \Theta\}$ can be found in the literature in place of $\langle \rho_\theta, \theta \in \Theta \rangle$, but this is not quite correct because elements of a set have no specified order as do the components of a vector. Clearly in (1b), the subscript θ is not arbitrarily assigned, but must properly match the argument of G.

90   3. Mixtures

In what follows, the notation $\rho = \langle \rho_\theta, \theta \in \Theta \rangle$ is sometimes used for simplicity. Just as with vectors

$$\alpha \rho + \bar{\alpha} \rho^* = \langle \alpha \rho_\theta + \bar{\alpha} \rho_\theta^*, \theta \in \Theta \rangle,$$

where as usual, $\bar{\alpha} = 1 - \alpha$.

The hazard transform has the property that the hazard function R of the mixture A(1) is given by $R(t) = \eta \langle R(t \mid \theta), \theta \in \Theta \rangle$.

D.2. Proposition (Esary, Marshall and Proschan, 1970). The hazard transform of a mixture is concave.

Proof. Let Z be a random variable taking values in Θ with distribution G, and write

$$\int_\Theta e^{-\rho_\theta} \, dG(\theta) = E \, e^{-\rho_Z}.$$

By Hölder's inequality 24.B.5,

$$E \, e^{-\alpha \rho_Z} e^{-(1-\alpha)\rho_Z^*} \leq (E \, e^{-\rho_Z})^\alpha (E \, e^{-\rho_Z^*})^{1-\alpha}, \quad 0 \leq \alpha \leq 1,$$

which together with (1) yields

$$\eta(\alpha \rho + \bar{\alpha} \rho^*) \geq \alpha \eta(\rho) + \bar{\alpha} \eta(\rho^*),$$

where $\bar{\alpha} = 1 - \alpha$. □

D.3. Proposition. Suppose the hazard rate of $F = \pi F_1 + \bar{\pi} F_2$ is given by A(8). If r_1 and r_2 are differentiable at x, then

$$r'(x) \leq \max \left[r'_1(x), r'_2(x) \right]. \tag{2}$$

First Proof. Write r in the form

$$r(x) = \frac{\pi r_1(x) \bar{F}_1(x) + \bar{\pi} r_2(x) \bar{F}_2(x)}{\pi \bar{F}_1(x) + \bar{\pi} \bar{F}_2(x)}$$

and differentiate to conclude, with tedious but elementary algebra, that

$$r'(x) = p(x) r_1'(x) + \bar{p}(x) r_2'(x) - p(x) \bar{p}(x) [r_1(x) - r_2(x)]^2,$$

where $p(x)$ is as in A(9). This means that

$$r'(x) \leq p(x) r_1'(x) + \bar{p}(x) r_2'(x) \leq \max [r_1'(x), r_2'(x)]. \qquad \square$$

Second Proof. The algebraic tedium of the first proof can be avoided by using the notion of a hazard transform. Write $R(x) = \eta(R_1(x), R_2(x))$, where $\eta(u, v) = -\log(\pi e^{-u} + \bar{\pi} e^{-v})$ is the hazard transform of the mixture A(6). Use subscripts on η to indicate partial derivatives and omit the argument x for notational simplicity. Also for notational simplicity, write $\eta_i = \eta_i(R_1, R_2)$ and $\eta_{ij} = \eta_{ij}(R_1, R_2)$, $i, j = 1, 2$. Then according to 24.A.6,

$$R' = \eta_1 R_1' + \eta_2 R_2'$$

and

$$R'' = \eta_{11}(R_1')^2 + \eta_{22}(R_2')^2 + 2\eta_{12} R_1' R_2' + \eta_1 R_1'' + \eta_2 R_2''.$$

Direct calculation shows that

$$\eta_{11}(u, v) = \eta_{12}(u, v) = \eta_{22}(u, v) = \frac{-\pi \bar{\pi} e^{-(u+v)}}{(\pi e^{-u} + \bar{\pi} e^{-v})^2} \leq 0.$$

Thus,

$$R'' \leq \eta_1 r_1' + \eta_2 r_2' \leq \max [r_1', r_2'].$$

Here, the second inequality follows from the easily verified fact that $\eta_1 + \eta_2 = 1$. $\quad\square$

Note that a counterpart lower bound for the derivative of r cannot be given; for example, when both F_1 and F_2 are exponential distributions, the component hazard rates have zero derivatives everywhere, but r has a strictly negative derivative.

D.4. Corollary. If all component distributions of a mixture have hazard rates decreasing and differentiable at the point x, then the mixture has a hazard rate decreasing at x.

Proof. For a mixture of two distributions, the result follows immediately from Proposition D.3. From this fact, an induction argument shows that it holds for any mixture of a finite number of distributions, and the proof is completed by a limiting argument. □

Corollary D.4 refers to monotonicity at a point, but that monotonicity may extend to all points, in which case the hazard rates are decreasing.

D.4.a. Corollary. If all component distributions of a mixture have decreasing hazard rates, the mixture has a decreasing hazard rate.

D.4.b. Remark. This result appears again as Proposition 4.C.16. It has important consequences in reliability theory and in other applications because it sometimes explains why decreasing hazard rates are encountered unexpectedly by those who are not aware of this mixture result. Corollary D.4.a shows, in particular, that a mixture of exponential distributions has a decreasing hazard rate. Distributions with decreasing hazard rate are further discussed in Section 4.C.

D.5. Proposition. If η is the hazard transform of a mixture and $\langle \rho_\theta, \theta \in \Theta \rangle, \langle \rho_\theta^*, \theta \in \Theta \rangle$ are similarly ordered, i.e.,

$$(\rho_\alpha - \rho_\beta)(\rho_\alpha^* - \rho_\beta^*) \geq 0 \quad \text{for all } \alpha, \beta \in \Theta,$$

then

$$\eta(\boldsymbol{\rho} + \boldsymbol{\rho}^*) \leq \eta(\boldsymbol{\rho}) + \eta(\boldsymbol{\rho}^*). \tag{3}$$

Proof. The proof uses Chebyshev's "other" inequality 20.H.1; according to this inequality,

$$E e^{-\rho_\Theta} e^{-\rho_\Theta^*} \geq E e^{-\rho_\Theta} E e^{-\rho_\Theta^*},$$

which upon taking logarithms yields the claim of the proposition. □

E. Mixtures and Minima

Suppose that $X = \min[X_1, X_2]$, where X_i has distribution $G_i, i = 1, 2$. If X_1 and X_2 are independent, then the survival function $\bar{F}$ of X is given by

$$\bar{F}(x) = P\{\min[X_1, X_2] > x\} = P\{X_1 > x, X_2 > x\} = \bar{G}_1(x)\bar{G}_2(x). \tag{1}$$

This is a simple version of the competing risk model discussed more fully in Section 17.A. Here, the X_i can be regarded as potential waiting times for failures due to two different causes, and the actual failure time is their minimum. Because there are two different possible causes of failure, the distribution can be thought of as a mixture. In fact, the equation

$$\bar{F}(x) = \bar{G}_1(x)\bar{G}_2(x) = \pi \bar{F}_1(x) + \bar{\pi} \bar{F}_2(x) \tag{2}$$

can be solved using ordinary algebra to obtain the solution

$$\bar{F}_1(x) = \frac{\int_x^\infty \bar{G}_2(z)\, dG_1(z)}{\int_0^\infty \bar{G}_2(z)\, dG_1(z)}, \qquad \bar{F}_2(x) = \frac{\int_x^\infty \bar{G}_1(z)\, dG_2(z)}{\int_0^\infty \bar{G}_1(z)\, dG_2(z)}, \tag{3}$$

and

$$\pi = \int_0^\infty \bar{G}_2(z)\, dG_1(z), \qquad \bar{\pi} = 1 - \pi = \int_0^\infty \bar{G}_1(z)\, dG_2(z).$$

There is a partial converse proposition; if $\bar{F}(x) = \pi \bar{F}_1(x) + \bar{\pi} \bar{F}_2(x)$, then (1) holds with

$$\bar{G}_1(x) = \exp\left\{-\int_0^x \frac{\pi}{\pi \bar{F}_1(z) + \bar{\pi} \bar{F}_2(z)}\, dF_1(z)\right\},$$

$$\bar{G}_2(x) = \exp\left\{-\int_0^x \frac{\bar{\pi}}{\pi \bar{F}_1(z) + \bar{\pi} \bar{F}_2(z)}\, dF_2(z)\right\}.$$

However, it may be that one of these survival functions is not proper; i.e., it may be that either $\bar{G}_1(\infty) > 0$ or $\bar{G}_2(\infty) > 0$. But at least one of these survival functions must be proper.

The above explanation shows that the models (1) and (2) are mathematically equivalent provided only that one of the variables X_1 or X_2 in (1) may take the value ∞ with positive probability. But of course, motivations for (1) and (2) can be quite different.

E.1. Example. If F_1 and F_2 are exponential distributions with respective parameters λ_1 and λ_2, $\lambda_1 \neq \lambda_2$, then (3) holds where

$$\bar{G}_1(x) = [\bar{\pi} + \pi\, e^{(\lambda_2 - \lambda_1)x}]^{\lambda_1/(\lambda_1 - \lambda_2)}, \qquad x \geq 0,$$

and G_2 is obtained from G_1 by interchanging λ_1 and λ_2 as well as π and $\bar{\pi}$. If $\lambda_1 < \lambda_2$, then G_1 is proper and G_2 is improper. These distributions are related to the distributions of Section 7.D.

F. Preservation of Orders Under Mixtures

Suppose that $X \leq Y$ and $U \leq V$ for some order $\leq$. With the mixture representation A(7) in mind, let I be a random variable, independent of X, Y, U, and V such that $P\{I = 1\} = 1 - P\{I = 0\}$ and let $W = IX + (1-I)U, Z = IY + (1-I)V$. Is $W \leq Z$ in the same order?

F.1. Proposition. Suppose that $(X, Y), (U, V)$, and I are mutually independent. If $X \leq_{st} Y$ and $U \leq_{st} V$, then $W \leq_{st} Z$.

Proof. By using (4) of Proposition 2.A.2, it is possible to assume that the stochastic orders are orderings that hold with probability 1. Then the result is immediate. □

Proposition F.1 fails to hold if stochastic order is replaced by the hazard rate order.

F.2. Example. Let r_X, r_Y, r_U, and r_V be the hazard rates of X, Y, U, and V, and let r_W and r_Z be the hazard rates of $W = IX + (1-I)U$ and $Z = IY + (1-I)V$. Retain the assumption of independence made in Proposition F.1. If

$$r_X(x) = 0, \quad x < 1,$$
$$= e^x, \quad x \leq 1,$$
$$r_Y(x) = e^x, \quad x \geq 0,$$
$$r_U(x) = r_V(x) = \lambda, \quad x \geq 0,$$

then $r_W > r_Z$ whenever $\lambda < e^e(e^x - 1)/(e - 1)$.

Part II

Nonparametric Families

4
Nonparametric Families: Densities and Hazard Rates

> Chance, too, which seems to rush along with slack reins, is bridled and governed by law.
>
> Boethius (Anicius Manlius Severinus Boethius, 480–525),
> *The Consolation of Philosophy*

A. Introduction

A number of nonparametric families of life distributions have been studied, particularly in the context of reliability theory; the most important of these families are discussed in this chapter. Because the families are defined by properties that have physically meaningful interpretations, an assumption that a distribution lies in a particular family sometimes can be justified by a physical understanding of the failure mechanism. For these nonparametric families, many statistical procedures are available that can be regarded as standing between parametric and standard distribution-free analyses.

Nonparametric families have received little attention in medical contexts, where analyses have tended to be based either on parametric models or have used standard nonparametric methods. In fact, most of the nonparametric families discussed in this chapter have at times been rejected in medical contexts because it is thought that long-term survival usually has a bathtub-shaped hazard rate. But the same criticism can be leveled against the commonly used parametric families, none of which exhibit such hazard rates (see Miller, 1981, p. 15).

98   4. Nonparametric Families: Densities and Hazard Rates

Bathtub-shaped hazard rates form the class of distributions discussed in Section D of this chapter.

Apart from the usefulness of the nonparametric families for data analyses, an understanding of their properties is quite useful when studying particular parametric families because the members of these families usually belong to at least one of the nonparametric families. An understanding of the properties that define the nonparametric families can play an important role in the choice of parametric family.

In the study of parametric families of distributions, only a few nonparametric families are encountered repeatedly. Basic results for these commonly encountered families are the subject of this chapter; additional details are given in Chapter 6.

A useful outline of the results of this chapter and of Chapter 5 has been given by Hollander and Proschan (1984); these authors also discuss statistical procedures for the various nonparametric classes introduced. A review of several classes has been given by Johnson, Kotz and Balakrishnan (1995, pp. 663–680). Additional nonparametric classes continue to appear in the literature, not always with clear motivation; in this book, no attempt has been made for a complete discussion of the topic.

B. Log-Concave and Log-Convex Densities

Log-concave and log-convex densities are important partly because they are often encountered and partly because they have interesting properties. Results regarding unimodality, closure under convolutions or mixtures, and hazard rate behavior are given below. For the purposes of this book, log concavity of the density is important also because it implies other properties with clear physical meanings, and because it is often easier to verify than the sometimes more interesting but weaker properties.

The importance of log-concave densities in statistics was perhaps first recognized by Karlin and Rubin (1956). More recently, the concept has played a role in economic theory (see An, 1998 and the references therein).

B.1. Definition. If F is an absolutely continuous distribution function, with some version f of the density having the property that $\log f$ is concave, then F is said to have a *log-concave* density. If $f(x) = 0$, $x < 0$, and $\log f$ is convex on $[0, \infty)$, then F is said to have a *log-convex* density.

The reason for the lack of symmetry here with the requirement that $f(x) = 0, x < 0$ only for the log convexity case is explained in Remark B.9.a.

It follows directly from the definitions that f is log concave if and only if $f(x - y)$ is totally positive of order 2 in x and y, i.e., f is a *Pólya frequency function of order 2* (PF_2). Similarly, f is log convex if and only if $f(x + y)$ is totally positive in x and $y \geq 0$. These facts can be quite useful because the theory of total positivity is well developed; this theory is outlined in Chapter 21.

Examples of log-concave and log-convex densities are familiar; indeed, it is well known and readily verified that

(i) normal densities are log concave,
(ii) exponential densities are both log concave and log convex,
(iii) gamma densities are log concave if $\nu \geq 1$ and log convex if $0 < \nu \leq 1$,
(iv) Weibull densities are log concave if $\alpha \geq 1$ and log convex if $0 < \alpha \leq 1$.

a. Properties of Distributions with Log-Concave and Log-Convex Densities

B.2. Proposition. Log-concave densities are unimodal, that is, they are nondecreasing up to some point and nonincreasing beyond that point.

Proof. To show that a density is unimodal, it is sufficient to show that for any positive constant c, $f(x) - c$ changes sign at most twice, and in the order $-, +, -$ if there are two sign changes (see Notation and Terminology). But this holds if and only if $\log f(x) - d$ has for any constant d, at most two sign changes, in the order $-, +, -$ if there are two sign changes. This sign change pattern holds because $\log f$ is concave. □

B.3. Proposition. If f and g are log-concave densities, then their convolution is log concave.

For an indication of proof, see 21.B.14 or 21.A.14.a.

Partly because of Propositions B.2 and B.3, a log-concave density is sometimes said to be *strongly unimodal*. In general, the convolution of two unimodal densities need not be unimodal; a counterexample is given by Feller (1971, p. 168) who also indicates that another counterexample is due to Kai Lai Chung. The fact that the convolution of two symmetric unimodal densities is unimodal is attributed to Aurel

Wintner (Feller, 1971, p. 167). Ibragimov (1956) shows that the convolution of a log-concave density and a unimodal density is unimodal; of course, this result is stronger than Proposition B.2.

It is not surprising that log-convex densities have properties quite different from those of log-concave densities. A most important property is their closure under mixtures.

B.4. Proposition. Let $\{f_\theta, \theta \in \mathcal{A}\}$ be a family of log-convex densities concentrated on the interval I and suppose that $\mathcal{A}$ is an open convex set. If $f_\theta(x)$ is a measurable function of θ for each fixed x in I, and if G is a probability distribution function on $\mathcal{A}$, then the mixture $f(x) = \int f_\theta(x) \, dG(\theta)$ is a log-convex density.

For a proof, see Marshall and Olkin (1979, p. 452).

b. Completely Monotone Densities

Completely monotone densities form an important subclass of log-convex densities; their properties are reviewed here. According to Proposition 20.D.4, a function ϕ defined on $(0, \infty)$ is said to be *completely monotone* if it possesses derivatives $\phi^{(n)}$ of all orders n, and

$$(-1)^n \phi^{(n)}(x) \geq 0, \quad x > 0.$$

It follows from well-known results about Laplace transforms (see Definition 20.D.5) that a survival function is completely monotone if and only if for some distribution function H, it has the form

$$\bar{F}(x) = \int_0^\infty e^{-\lambda x} \, dH(\lambda), \quad x \geq 0. \qquad (1)$$

Thus, completely monotone survival functions are exactly those that are mixtures of exponential survival functions. From this, or from the definition of complete monotonicity, it follows that a survival function is completely monotone if and only if it has a completely monotone density f. Clearly,

$$f(x) = \int_0^\infty \lambda e^{-\lambda x} \, dH(\lambda), \quad x > 0. \qquad (2)$$

B.5. Example. When H itself is an exponential distribution, say, with parameter $1/\theta$, then (1) becomes

$$\bar{F}(x) = \int_0^\infty e^{-\lambda x}\, \theta^{-1} e^{-\lambda/\theta}\, d\lambda = 1/(1+\theta x), \quad x > 0.$$

This is a special kind of Pareto survival function, encountered in Chapter 11. A more general mixture result is given in Proposition 11.D.3.

B.6. Proposition (Steutel, 1967, 1969). If $\bar{F}$ is completely monotone (or equivalently, if the corresponding density is completely monotone), then it is infinitely divisible.

This result is stated also as Proposition 20.D.10.

B.7. Proposition. If $F(0) = 0$ and $\bar{F}$ is completely monotone, then f is log convex on $[0, \infty)$.

This result is an application of Proposition B.4 because exponential distributions have densities log linear on $[0, \infty)$.

c. Comments About Log Concavity and Log Convexity

The following proposition has a meaning that is relatively easy to understand, but its proof requires some results concerning total positivity; these results have been summarized in Section 21.B. Note that the cases of log convexity and log concavity are not entirely parallel.

B.8. Proposition. If f is log concave, then both $\bar{F}$ and F are log concave. If f is log convex on $[0, \infty)$, then $\bar{F}$ is log convex on $[0, \infty)$.

Proof. First suppose that $\log f$ is concave. As noted in Definition 21.B.8, this property is equivalent to the statement that $f(x-y)$ is totally positive of order 2. Let K be the indicator function given by Example 21.B.6, i.e., $K(y, z) = 1$ if $y \leq z$, and $K(y, z) = 0$ if $y > z$. Compute that

$$\int f(x-y) K(y,z)\, dy = \int_{-\infty}^z f(x-y)\, dy = \int_{x-z}^\infty f(u)\, du = \bar{F}(x-z). \tag{3}$$

Because the function K is totally positive, it follows from Theorem 19.B.11 that $\bar{F}(x-z)$ is totally positive of order 2 in x and z; that is, $\bar{F}$ is log concave. □

To prove that F is log concave, use the same proof with K defined in Example 21.B.6.a, i.e., $K(y,z) = 0$ if $y < z$, and $K(y,z) = 1$ if $y \geq z$. The result concerning log convexity has a similar proof but makes use of the fact that $K(y,z)$ given by Example 21.B.6.a is totally positive in nonnegative y and z, together with the fact that $\log f$ is convex on $[0, \infty)$ if and only if $f(x+y)$ is totally positive of order 2 in $x, y \geq 0$.

B.8.a. Proposition. If f is log concave, then the hazard rate $r = f/\bar{F}$ of F is increasing, and the reverse hazard rate $s = f/F$ of F is decreasing. If f is log convex on $[0, \infty)$, then the hazard rate r is decreasing on $[0, \infty)$.

Proof. If f is log concave, then it follows from Propositions B.8 and 21.B.15 that both $\log \bar{F}$ and $\log F$ have decreasing derivatives. This is equivalent to the condition that r is increasing and s is decreasing. The proof in case f is log convex on $[0, \infty)$ is similar. □

The following example shows that the converse of Proposition B.8.a is false; it may be that a distribution has an increasing hazard rate and a density that is not log concave.

B.8.b. Example. Suppose that $r(x) = x + (1+x)^{-1}, x \geq 0$. It can be verified by differentiation that r is increasing, and thus $R(x) = \int_0^x r(t)\,dt = \frac{1}{2}x^2 + \log(1+x)$ is convex (see Proposition 21.A.3(iv)). By 1.B(1), $\log \bar{F}(x) = -R(x)$, and consequently $\bar{F}$ is log concave.

To show that f is neither log concave nor log convex, use the formula $\log f(x) = \log r(x) + \log \bar{F}(x)$ to check that the second derivative of $\log f(x)$ is positive at 0 and eventually becomes negative.

A somewhat more involved but similar calculation also shows that $\log F(x)$ is neither convex nor concave.

B.9.a. Remark. The treatments of log concavity and log convexity differ in that log concavity can be required on the whole line; if $\log f$ is concave on $[0, \infty)$, and $f(x) = 0$ for $x < 0$, then by the convention that $\log 0$ is taken to be $-\infty$, and $-\infty + x = -\infty$, for all real x, it follows that $\log f$ is concave on $(-\infty, \infty)$. The same cannot be said for log convexity.

According to Proposition B.8.a, if $\log f$ is convex on $[0, \infty)$, then F has a hazard rate decreasing on $[0, \infty)$, and consequently, the density is decreasing on $[0, \infty)$ (see Proposition C.13 for more details). Because a density must integrate to 1 over the real line, it cannot be decreasing on the whole real line, and thus log convexity of a density on the whole real line is not possible. In what follows, log convexity should be interpreted to mean log convexity on $[0, \infty)$ unless some other interval is specified.

B.9.b. Remark. In the application of Theorem 21.B.11 to (3), where $\log f$ is concave, the variables x, y, and z are allowed to take on all values in $(-\infty, \infty)$ in accordance with the above comments. Theorem 21.B.11 is consistent with the conventions stated in Remark B.9.a and remains valid. Because log convexity of f on $[0, \infty)$ cannot be similarly extended to log convexity on $(-\infty, \infty)$, it is necessary to restrict the range of arguments in the functions K when applying Theorem 21.B.11 to the log convexity. Because of this, it does not follow from log convexity of f on $[0, \infty)$ that F is log convex. In fact, quite the opposite is true.

B.10. Proposition. If $f(x) = 0, x < 0$, and $\log f$ is convex on $[0, \infty)$, then $\log F$ is concave on $(-\infty, \infty)$.

Proof. Under the stated hypotheses concerning f, it follows from Proposition B.8.a that F has a decreasing hazard rate, and consequently, f is decreasing (Proposition C.12). This means that F is concave on $[0, \infty)$, and thus by Proposition 21.A.6, F is log concave on $[0, \infty)$. The extension to log concavity on $(-\infty, \infty)$ is explained in Remark B.9.a. □

The conclusions of Proposition B.8 that $\log \bar{F}$ is convex or concave are conditions discussed in some detail in Section C. The interest in these properties stems from the fact that when a density exists, these properties are equivalent to the hazard rate being decreasing or increasing, as indicated in Proposition B.8.a. On the other hand, the log convexity or log concavity of the distribution function is equivalent to the reverse hazard rate being increasing or decreasing; these conditions are of relatively little interest. For a summary of various consequences of these properties, see Sengupta and Nanda (1999).

B.11. Proposition. If f is log concave [convex], then the residual life density $f_t(x) = f(x+t)/\bar{F}(t)$ is log concave [convex] in x, for all $t > 0$.

C. Monotone Hazard Rates

It should not be surprising that the concept of a "monotone hazard rate" stems from the condition that the hazard rate $r = f/\bar{F}$

is monotone, but this idea requires the existence of a density. The concept can be defined in various ways, the best of which do not require the existence of a density; such definitions usefully stretch the meaning of the concept so the name should not be taken too literally. The following definition is a useful working definition, but it is not the best for explaining the meaning of the concept. That meaning is more clearly revealed in the equivalent conditions that follow.

C.1. Definition. Let F be a distribution such that $F(x) = 0, x < 0$. Then F is said to have an *increasing [decreasing] hazard rate* if for all $x \geq 0$ and all t such that $F(t) < 1$,

$$\frac{[F(t+x) - F(t)]}{1 - F(t)} = \frac{[\bar{F}(t) - \bar{F}(t+x)]}{\bar{F}(t)} \tag{1}$$

is increasing in t and $F(0+) = 0$ [decreasing in t and $F(0-) = 0$].

This condition says that the probability of surviving for a time interval of length x is decreasing [increasing] in the initial age.

The concept of an increasing hazard rate can be extended usefully to distributions such as the normal distribution that have positive mass on the interval $(-\infty, 0)$, but a distribution cannot have a decreasing hazard rate in the entire interval $(-\infty, \infty)$; the support of such distributions must have a finite left-hand endpoint, which in accordance with standard usage has been taken to be 0.

Often, the abbreviation "F is IHR" is used in place of the more complete "F has an increasing hazard rate." Similarly, the abbreviation "F is DHR" is used to indicate that F has a decreasing hazard rate. Particularly in the engineering literature where the term "failure rate" is commonly encountered in place of "hazard rate," the abbreviations "IFR" and "DFR" are used in place of "IHR" and "DHR."

Two key results of this section are that convolutions of IHR distributions are IHR and mixtures of DHR distributions are DHR.

a. Equivalent Conditions

The following proposition gives a condition for monotone hazard rate that is usually more convenient to verify than the condition of the definition.

C.1.a. Proposition. A distribution F has an increasing [decreasing] hazard rate if and only if for all t such that $\bar{F}(t) > 0$,

$$\bar{F}_t(x) = \frac{\bar{F}(t+x)}{\bar{F}(t)} \text{ is decreasing [increasing] in } t \text{ for all } x \geq 0, \quad (2)$$

i.e., the residual life distributions are stochastically decreasing [increasing] in t.

Proof. For all $x \geq 0$ and all t such that $F(t) < 1$,

$$\frac{\bar{F}(t) - \bar{F}(t+x)}{\bar{F}(t)} = 1 - \frac{\bar{F}(t+x)}{\bar{F}(t)}$$

is increasing in t if and only if $\bar{F}(t+x)/\bar{F}(t)$ is decreasing. □

C.1.b. Proposition. If F is a distribution such that $F(x) = 0$ for $x < 0$, then F has an *increasing [decreasing] hazard rate* if and only if the hazard function $R = -\log \bar{F}$ is convex where finite [concave on $[0, \infty)$].

Proof. For all t such that $F(t) < 1$, (2) holds if and only if $\log \bar{F}(t+x) - \log \bar{F}(t)$ is decreasing in t such that $\bar{F}(t) > 0$. According to (iii) of Proposition 21.A.3, this is equivalent to the concavity of $\log \bar{F}$. But this means that $R = -\log \bar{F}$ is convex. The proof for the DHR case is similar. □

As already noted, the condition that R is convex is equivalent to the condition that $\bar{F}$ is log concave. Log concavity of $\bar{F}$ is convenient and often used as a proxy for the statement that F has an increasing hazard rate, and it is a condition that can be imposed without first assuming that a density exists.

C.1.c. Proposition. Suppose that $F(0-) = 0$ and F has a density. Then F has an increasing [decreasing] hazard rate if and if only there is a version f of the density such that the corresponding hazard rate $r = f/\bar{F}$ is increasing [decreasing] on $[0, \infty)$.

Proof. This result follows from Proposition C.1.b by differentiating $R = -\log \bar{F}$ and using (iv) of Proposition 21.A.3, which states that a differentiable function is convex if and only if its derivative is increasing. □

Of course, Proposition C.1.c explains the terminology. But the definition used here extends the concept to cases that may not have a density. It is important to recall that in this book, the terms "increasing" and "decreasing" are not used in the strict sense, they are used here to mean, respectively, "nondecreasing" and "nonincreasing."

Ordinarily, the version of the density that can be used in Proposition C.1.c is the derivative of the distribution function.

As noted in Section 1.B, the hazard rate is often interpreted intuitively through the equation

$$\Delta r(x) \approx P\{x < X \leq x + \Delta \mid X > x\}.$$

The concept of an increasing hazard rate is attractive in an engineering context where it has often been likened to a mathematical representation of "wearout," although some problems with this idea are encountered in Chapter 5. In an engineering context, the idea of a decreasing hazard rate has been associated with "work hardening" but it can arise in a variety of other ways. In a medical context, the idea of a decreasing hazard may be intuitively appealing for such waiting times as hospitalization times, or times to illness relapse.

Proposition B.8.a shows that if F has a density f that is log concave, then F has an increasing hazard rate; Example B.8.b shows that the converse is false.

C.1.d. Proposition. The distribution function F has an increasing hazard rate if and only if the determinant

$$\begin{vmatrix} \bar{F}(t_1 - s_1) & \bar{F}(t_1 - s_2) \\ \bar{F}(t_2 - s_1) & \bar{F}(t_2 - s_2) \end{vmatrix} \geq 0,$$

where $s_1 \leq s_2, t_1 \leq t_2$, that is, $\bar{F}$ is a Pólya frequency function of order 2 (see Definition 21.B.7).

The distribution function F has an decreasing hazard rate if and only if the support of F is $[0, \infty)$ and the determinant

$$\begin{vmatrix} \bar{F}(t_1 + s_1) & \bar{F}(t_1 + s_2) \\ \bar{F}(t_2 + s_1) & \bar{F}(t_2 + s_2) \end{vmatrix} \geq 0$$

whenever $s_1 \leq s_2, t_1 \leq t_2$ and $s_1 + t_1 \geq 0$, i.e., $\bar{F}(x+y)$ is totally positive in x and $y, x + y \geq 0$ (see Definition 21.B.1).

Proof. This proposition can be viewed as a restatement of Proposition C.1.e. □

For a given distribution function F, odds ratios are defined in Definition 1.B.11; the following proposition involves the odds ratio of the residual life distribution F_t.

C.1.e. Proposition. A distribution F has an increasing [decreasing] hazard rate if and only if for all $x \geq 0$ and all t such that $\bar{F}(x+t) > 0$, the odds ratio $\mathcal{O}_t^-(x) = [F(x+t) - F(t)]/[\bar{F}(x+t)]$ of the residual life distribution F_t is increasing [decreasing] in t.

Proof. It follows from Proposition C.1.a that F is IHR [DHR] if and only if $\bar{F}_t(x)$ is increasing [decreasing] in t. This is the statement of Proposition C.1.e. □

The following proposition is a direct application of Definition C.1 and the fact that if G is an exponential distribution with parameter λ, then $\bar{G}^{-1}(z) = [-\log z]/\lambda$.

C.1.f. Proposition. The distribution F has an increasing [decreasing] hazard rate if and only if a random variable with distribution F is smaller [larger] than an exponentially distributed random variable in the convex transform order $\leq_c$ of Definition 2.C.7. Here it is clearly sufficient to take the parameter λ of the exponential distribution to be equal to 1.

C.1.g. Proposition. A random variable X has an increasing [decreasing] hazard rate if and only if there exists a continuous increasing concave [convex] function ψ defined on $[0, \infty)$ such that $\psi(Y)$ has the same distribution as X, where Y has an exponential distribution with parameter 1.

It can be seen that the function ψ is the inverse of the hazard function $R = -\log \bar{F}$ of X. This proposition can be viewed as a restatement of Proposition C.1.f; see Proposition 2.C.6 but note that according to 20.A(6), $G^{-1}F = \bar{G}^{-1}\bar{F}$.

C.1.h. Proposition (Barlow and Campo, 1975). Suppose that $F(x) = 0$, for all $x < 0$, and that F has a density. The following are equivalent:

(i) F has an increasing [decreasing] hazard rate.

(ii) The total time on test transform is concave [convex] on $[0, 1]$.

(iii) The total time on test transform distribution is convex [concave] on its support.

Proof. The equivalence of (i) and (ii) follows directly from Proposition C.1.c and Proposition 1.I.5. The equivalence of (ii) and (iii) is a consequence of 21.A.7. □

The following somewhat technical but fundamental result is utilized in C.8.b and in the proof of Proposition 5.D.6.

C.1.i. Proposition (Savits, 1985). The random variable X has an increasing hazard rate if and only if $E[h(t, X)]$ is log concave in t for all functions $h(t, x)$, which are defined and log concave in $(t, x), t \geq 0, x \geq 0$, and are nondecreasing in x, for each fixed t.

Proof. First suppose that X has an increasing hazard rate. By Proposition C.1.g, there exists a continuous increasing concave function ψ defined on $[0, \infty)$ such that $\psi(Y)$ has the same distribution as X, where Y has an exponential distribution with parameter 1. Thus,

$$E[h(t, X)] = E[h(t, \psi(Y))] = \int_0^\infty h(t, \psi(y)) e^{-y} dy.$$

It follows from 21.A.5.a that the composition $h(t, \psi(y))$ is log concave. Hence, by Prékopa's theorem 21.A.14, $E[h(t, X)]$ is log concave. Next, suppose that $E[h(t, X)]$ is log concave for all functions $h(t, x)$, which are defined and log concave in $(t, x), t \geq 0, x \geq 0$, and are nondecreasing in x, for each fixed t. Let $h(t, x) = I_{(t,\infty)}(x)$ be the indicator function that is 1 if $x > t$, and 0 otherwise. It follows that $E[h(t, X)] = \int_t^\infty dF(x) = \bar{F}(t)$ is log concave. □

C.1.j. Proposition. Let X be a positive random variable with survival function $\bar{F}$, and denote the distribution of $1/X$ by G. If $\log \bar{F}$ is concave (F is IHR), then $\log G$ is concave (G has a decreasing reverse hazard rate).

Proof. Because $G(x) = P\{1/X \leq x\} = P\{X \geq 1/x\} = \bar{F}(1/x), x > 0$, $\log G(x) = \log \bar{F}(1/x)$ has the derivative

$$\frac{d}{dx} \log G(x) = \frac{d}{dx} \log \bar{F}(1/x) = \frac{f(1/x)}{\bar{F}(1/x)} \frac{1}{x^2}.$$

If $r(x) = f(x)/\bar{F}(x)$ is increasing, then $r(1/x)$ is decreasing. Consequently, $r(1/x)/x^2$ is decreasing, and hence, $\log G$ has a decreasing derivative. □

b. Increasing Hazard Rates

The intuitive content of an increasing hazard rate stems from the interpretation of $r(t)dt$ as the conditional probability of failure in the interval $[t, t + dt]$ given survival up to time t. Thus, with an increasing hazard rate, the probability of failure in the next instant of time increases as the device or organism ages. In a very real sense this is a mathematical translation of the intuitive concept of "adverse ageing," but it would be unfair to claim that it is the only mathematical translation of this concept.

c. Properties of Distributions with Increasing Hazard Rates

C.2. Proposition. If F has an increasing hazard rate, then F has a density except possibly at the right-hand endpoint of its support, where it may have positive mass.

This result is a consequence of the definition of log concavity. It shows that the use of a definition not requiring the existence of a density has not extended the idea very far. But still, it is a useful extension.

C.3. Proposition. Suppose that F has an increasing hazard rate. Then F has finite moments of all finite positive orders.

This is an immediate consequence of Proposition 20.B.6. There are also various other proofs of this fact; in particular, it follows from the stronger results of Propositions 5.B.6 and 5.C.7.

d. Preservation Theorems

The following theorem states that the class of distributions with increasing hazard rates is closed under convolutions. The elegant proof given here and by Barlow, Marshall and Proschan (1963) is due to Frank Proschan; it depends upon total positivity theory, which is connected to hazard rate behavior by Proposition C.1.d. There is also a critical integration by parts. No real elementary proof is known, but an alternative proof is given in C.9.

The intuitive interpretation of the following theorem is: If each of two devices wears out in the sense that they have an increasing hazard rate, and if one of the devices is used as a spare for the other and put into service at the time the first fails, then this system (a device and its spare) wears out in the sense that it has an increasing hazard rate

110   4. Nonparametric Families: Densities and Hazard Rates

(assuming that the two life lengths are independent). The reader may wish to interpret this theorem in other contexts such as medicine or biology.

C.4. Theorem (Barlow, Marshall and Proschan, 1963). *If F and G have increasing hazard rates, then the convolution $H = F * G$ has an increasing hazard rate.*

Proof. Assume that F has a density f and G has a density g. Recall from Proposition C.1.d that the condition of increasing hazard rate can be written in terms of a determinant. For $t_1 < t_2$, and $u_1 < u_2$, form the determinant

$$D = \begin{vmatrix} \bar{H}(t_1 - u_1) & \bar{H}(t_1 - u_2) \\ \bar{H}(t_2 - u_1) & \bar{H}(t_2 - u_2) \end{vmatrix}$$

$$= \begin{vmatrix} \int \bar{F}(t_1 - s) g(s - u_1) \, ds & \int \bar{F}(t_1 - s) g(s - u_2) \, ds \\ \int \bar{F}(t_2 - s) g(s - u_1) \, ds & \int \bar{F}(t_2 - s) g(s - u_2) \, ds \end{vmatrix}$$

$$= \iint_{s_1 < s_2} \begin{vmatrix} \bar{F}(t_1 - s_1) & \bar{F}(t_1 - s_2) \\ \bar{F}(t_2 - s_1) & \bar{F}(t_2 - s_2) \end{vmatrix} \begin{vmatrix} g(s_1 - u_1) & g(s_1 - u_2) \\ g(s_2 - u_1) & g(s_2 - u_2) \end{vmatrix} ds_2 \, ds_1;$$

the last equality is an application of the basic composition formula 21.B.10. Now integrate the inner integral by parts to obtain

$$D = \iint_{s_1 < s_2} \begin{vmatrix} \bar{F}(t_1 - s_1) & f(t_1 - s_2) \\ \bar{F}(t_2 - s_1) & f(t_2 - s_2) \end{vmatrix} \begin{vmatrix} g(s_1 - u_1) & g(s_1 - u_2) \\ \bar{G}(s_2 - u_1) & \bar{G}(s_2 - u_2) \end{vmatrix} ds_2 \, ds_1.$$

The sign of the first determinant is the same as that of

$$\frac{f(t_2 - s_2)\bar{F}(t_2 - s_2)}{\bar{F}(t_2 - s_2)\bar{F}(t_2 - s_1)} - \frac{f(t_1 - s_2)\bar{F}(t_1 - s_2)}{\bar{F}(t_1 - s_2)\bar{F}(t_1 - s_1)},$$

assuming the denominators are nonzero. But

$$\frac{f(t_2 - s_2)}{\bar{F}(t_2 - s_2)} \geq \frac{f(t_1 - s_2)}{\bar{F}(t_1 - s_2)}$$

by hypothesis and

$$\frac{\bar{F}(t_2 - s_2)}{\bar{F}(t_2 - s_1)} \geq \frac{\bar{F}(t_1 - s_2)}{\bar{F}(t_1 - s_1)}$$

by Proposition C.1.d. Thus, the first determinant is nonnegative. A similar argument applies to show that the second determinant also is nonnegative, so that $D \geq 0$. By Proposition C.1.d, this shows that H is IHR.

If the densities f and/or g do not exist, a limiting argument is required to complete the proof. □

The following result has a clear intuitive meaning: If a device wears adversely as it ages, then it wears adversely even if acquired in a used but unfailed state after a known service time t. This is not an unexpected result.

C.5. Proposition. If F has an increasing hazard rate, then the residual life distribution F_t also has an increasing hazard rate.

Proposition C.5 can be verified directly.

Recall from 1.B(15) that if F has a first moment $\mu = \int_0^\infty \bar{F}(t)\,dt$, then $\bar{F}/\mu$ is a density. The interest of the following proposition lies partly in the fact that the density $f_{(1)} = \bar{F}/\mu$ arises in the study of stationary renewal processes, where it is sometimes called the "equilibrium distribution" or the "stationary renewal distribution" (see Sections 1.B.h and 20.F.b).

C.6. Proposition (Barlow, Marshall and Proschan, 1963). Suppose that F has an increasing hazard rate and first moment μ. Then, the distribution with density $f_{(1)}(x) = \bar{F}(x)/\mu, x \geq 0$, has an increasing hazard rate.

Proof. According to Proposition C.1.b, the distribution F has an increasing hazard rate if and only if $\log \bar{F}(x)$ is concave, i.e., $\log f_{(1)}(x)$ is concave. According to Proposition B.8, this means that $\log \bar{F}_{(1)}(x)$ is concave, that is, $\bar{F}_{(1)}$ is IHR. □

C.6.a. Proposition. Let X be a random variable with distribution F having a finite first moment and let Y have the density $f_{(1)}$ of Proposition C.6. Then, F has an increasing hazard rate if and only if $X \geq_{lr} Y$.

112   4. Nonparametric Families: Densities and Hazard Rates

Proof. With the assumption that F has a density, this result can be directly verified using Proposition 2.A.10. □

For more about the density $f_{(1)}$, see Section 20.B.c.

e. Mixtures of Distributions with Increasing Hazard Rates

Because the mixture of exponential distributions has a decreasing hazard rate (see Propositions B.4 and B.8.a or Theorem C.15), it is clear that the class of distributions with increasing hazard rate is not closed under mixtures. Examples C.7.a and C.7.b show even more; mixtures of distributions with strictly increasing hazard rates can have strictly decreasing hazard rates. The possibility that mixtures of distributions with increasing hazard rates may themselves have decreasing hazard rates has important implications in applications, because nearly always data can be regarded as coming from a mixture. This is the case, for example, when there is an unobserved covariate.

C.7.a. Example. Suppose that $r_1(x) = 1 - e^{-bx}$ and $r_2(x) = a + r_1(x)$. If $a \leq b < a^2/4$, for example, if $a = b > 4$, then the mixture with equal weights of the corresponding distributions F_1 and F_2 has a strictly decreasing hazard rate r even though both r_1 and r_2 are strictly increasing.

To see this, note that the hazard rate r of the mixture (see 3.A(8)) is given by

$$r(x) = \frac{f_1(x)[1 + e^{-ax}] + a e^{-ax} \bar{F}_1(x)}{\bar{F}_1(x)[1 + e^{-ax}]} = r_1(x) + a \frac{e^{-ax}}{1 + e^{-ax}}.$$

From this, it follows that $r'(x) < 0$ for all x if and only if

$$r_1'(x) < a^2 e^{-ax}/(e^{-ax} + 1)^2. \tag{3}$$

Inequality (3) reduces to the condition

$$b(1 + e^{-ax})^2 < a^2 e^{(b-a)x}.$$

The left-hand side of this inequality is strictly decreasing in $x \geq 0$, and because $a \leq b$, the right-hand side of this inequality is increasing in x. The inequality is satisfied at $x = 0$ if $b < a^2/4$. Thus, the inequality is satisfied, for all $x \geq 0$, if $a \leq b < a^2/4$.

In this example, $\lim_{x\to\infty} r_1(x) = 1, \lim_{x\to\infty} r_2(x) = a+1$, and $\lim_{x\to\infty} r(x) = 1$ is the minimum of $\lim_{x\to\infty} r_1(x)$ and $\lim_{x\to\infty} r_2(x)$; this is in accordance with Proposition 3.C.1.

Note that one of the component distributions of the mixture of Example C.7.a has hazard rate that is positive at 0, and the other hazard rate is bounded above. These features are essential for the mixture to have a decreasing hazard rate, as can be seen from Propositions 3.C.1 and 3.D.3 or from 3.A(8). With the assumption that the hazard rate is differentiable, Gurland and Sethuraman (1995) give a necessary and sufficient condition for the mixture of two distributions with increasing hazard rates to have a decreasing hazard rate; however, in Example C.7.a, a direct verification is simpler.

C.7.b. Example. Suppose that

$$\bar{F}(x \mid \lambda, \xi) = \exp\left\{-\xi(e^{\lambda x} - 1)\right\}, \quad x \geq 0, \ \xi, \lambda > 0$$

is the survival function of a Gompertz distribution, the topic of Section 10.A. Treat the parameter ξ as a random variable with the exponential survival function $\bar{G}(\xi) = e^{-\gamma \xi}, \xi \geq 0$, and consider the mixture

$$\bar{H}(x \mid \lambda, \gamma) = \int_0^\infty \bar{F}(x \mid \lambda, \xi) \, dG(\xi)$$
$$= \frac{\gamma}{e^{\lambda x} - \bar{\gamma}}, \quad x \geq 0, \ \lambda > 0, \ \gamma > 0, \ \gamma = 1 - \bar{\gamma}.$$

This survival function is discussed in Section 9.D, where it is termed an *exponential survival function with tilt parameter*. The hazard rate

$$r_H(x \mid \lambda, \gamma) = \frac{\lambda e^{\lambda x}}{e^{\lambda x} - \bar{\gamma}}$$

of such a mixture is strictly increasing when $\gamma > 1$ and it is strictly decreasing when $\gamma < 1$. In either case, $\lim_{x\to\infty} r_H(x \mid \lambda, \gamma) = \lambda$, despite the fact that the hazard rate $r(x \mid \lambda, \xi) = \xi \lambda e^{\lambda x}$ of the Gompertz distribution is strictly increasing and has the limit ∞ at ∞. For a somewhat more general case, see Example 5.M.7.b and Section 10.A.e.

It is of interest to examine the conditional distribution of ξ given survival to time t. This distribution can be obtained from Equation 3.B(4) and is again an exponential distribution, but with parameter $\gamma + e^{\lambda t} - 1$ in place of γ. Thus, the parameter is increasing exponentially, so the conditional distribution of ξ given survival to time t very rapidly puts more and more mass on components of the mixture that have lower hazard rates. This is what causes the hazard rate of the mixture to have a finite limit and to be decreasing when $\gamma < 1$.

Another rather striking example is given by Gurland and Sethuraman (1994); they consider a mixture of two distributions, one an exponential distribution and the other a Gompertz distribution (see Section 10.A), which has the rapidly increasing hazard rate $r(x \mid 1, a, 0) = a\,e^x$ as given in 10.A(3). Gurland and Sethuraman (1994) show that the hazard rate of the mixture can be decreasing, except possibly for a very short interval near the origin, even when the mixture gives little weight to the exponential distribution. They investigate the mixture of a variety of other distributions with an exponential distribution and show that in many cases, the hazard rate is eventually decreasing. See also Proposition 6.C.4.b, where the distribution of the random variable Y is a mixture of two dependent random variables, one with an exponential distribution and the other with a distribution having a strictly increasing hazard rate; but still, Y has an exponential distribution.

Other examples given by Block, Li and Savits (2003b) show that a mixture of distributions, each with increasing hazard rate, can have a hazard rate that is increasing, bathtub shaped, inverted bathtub shaped (see Section J for definitions), or may have a number of points where the direction of monotonicity changes.

The above examples are not to be regarded as typical; the following proposition gives some conditions under which a mixture of IHR distributions is an IHR.

C.8.a. Proposition (Lynch, 1999). Let $\{F(x \mid \theta), \theta \in \Theta\}$ be a family of distribution functions, indexed by a parameter $\theta \in \Theta \subset \mathcal{R}^n$, for some n. If G has a log-concave density g, and $\log \bar{F}(x \mid \theta)$ is concave in the pair (x, θ), then the mixture

$$\bar{F}(x) = \int_\Theta \bar{F}(x \mid \theta)\, dG(\theta) = \int_\Theta \bar{F}(x \mid \theta)\, g(\theta)\, d\theta$$

is log concave; that is, F has an increasing hazard rate.

Proof. This proposition is an immediate application of Prékopa's theorem 21.A.14. □

The requirement of Proposition C.8.a that G has a log-concave density is replaced in the following proposition by the weaker condition that G has an increasing hazard rate.

C.8.b. Proposition (Lynch, 1999). Let $\{F(x \mid \theta), \theta > 0\}$ be a family of distribution functions such that

$$\log \bar{F}(x \mid \theta) \text{ is concave in } (x, \theta), \text{ and increasing in } \theta \text{ for each fixed } x. \tag{4}$$

If G is a distribution with increasing hazard rate and $G(0) = 0$, then the mixture

$$\bar{H}(x) = \int_0^\infty \bar{F}(x \mid \theta) \, dG(\theta) \tag{5}$$

has an increasing hazard rate. Conversely, if H is IHR whenever (4) is satisfied, then G is IHR.

Proof. (Block, Li and Savits, 2003a). Let G be the distribution of Θ; by assumption, G is an IHR. In Proposition C.1.i, take $h(t, \theta) = \bar{F}(t \mid \theta)$ (with Θ in place of X) to conclude that $Eh(t, \Theta) = \bar{H}(t)$ is log concave. □

For an application of Proposition C.8.b, see Proposition 7.D.10.

C.8.c. Example. As an example of a parametric family that satisfies the conditions of Proposition C.8.b, suppose that $\bar{F}(x \mid \theta) = [\bar{F}(x/\theta)]^\theta$, $x, \theta > 0$, where F has the increasing hazard rate r. The required log concavity in (x, θ), that is, the convexity of $-\log \bar{F}(x \mid \theta) = -\theta \log \bar{F}(x/\theta)$, can be verified using Proposition 21.A.3(vi). For this function, the Hessian H is given by

$$H = \begin{vmatrix} -\dfrac{\partial^2 \log F(x \mid \theta)}{\partial x^2} & -\dfrac{\partial^2 \log F(x \mid \theta)}{\partial x \partial \theta} \\ -\dfrac{\partial^2 \log F(x \mid \theta)}{\partial \theta \partial x} & -\dfrac{\partial^2 \log F(x \mid \theta)}{\partial \theta^2} \end{vmatrix} = r'\left(\dfrac{x}{\theta}\right) \begin{vmatrix} \dfrac{1}{\theta} & -\dfrac{x}{\theta^2} \\ -\dfrac{x}{\theta^2} & \dfrac{x^2}{\theta^3} \end{vmatrix}.$$

Because $r' \geq 0$, the diagonal elements of this matrix are nonnegative and the determinant is 0. Consequently, the matrix is positive semidefinite and $-\log \bar{F}(x \mid \theta)$ is convex, that is, $\log \bar{F}(x \mid \theta)$ is concave. Monotonicity in θ follows from the fact that IHR distributions are IHRA (see Section 5.B), and thus satisfy (iii) of Proposition D.3. This example has been given by Block, Li and Savits (2003a) for the case that F is a Gompertz distribution (see Chapter 8). For another example, see Section 7.D.g.

116   4. Nonparametric Families: Densities and Hazard Rates

The following is an application of Proposition C.8.b.

C.8.d. Alternative proof of Theorem C.4. If F is a distribution with increasing hazard rate and $\bar{F}(x \mid \theta) = \bar{F}(x - \theta)$, then (3) is satisfied. Let G be an IHR distribution function such that $G(0-) = 0$. Then, the mixture (4) is the convolution of F with G, that is

$$H(x) = \int_0^\infty F(x \mid \theta)\, dG(\theta) = \int_0^\infty F(x - \theta)\, dG(\theta).$$

It follows from Proposition C.8.b that H is IHR. □

C.9. Remark. Is an IHR distribution unimodal? According to Proposition B.2, log-concave densities are unimodal, and according to Proposition B.8, distributions with log-concave densities are IHR. Might the weaker condition that F is IHR also imply that the density is unimodal? The answer to this question is no. In fact, if the hazard rate is an increasing step function, then the density has a mode at all points where the hazard function jumps. This means that an IHR density can have even an infinite number of modes. However, many of the IHR distributions encountered in later chapters have log-concave densities, and these are unimodal.

f. Decreasing Hazard Rates

An item has a decreasing hazard rate if, as it ages, the chance of failure (death) in the next instant of time decreases. This is the opposite of wearout, and might be called "wearin." Humans might exhibit a decreasing probability of failing at some particular job as they gain experience and practice. But mixtures may be the most important source of distributions with decreasing hazard rates. Example C.7.a exhibits a mixture of two distributions with strictly increasing hazard rates, but still, the mixture has a strictly decreasing hazard rate. More interesting for applications is the following fundamental proposition.

C.10. Proposition. If $\bar{F}$ is a mixture of exponential survival functions, then it has a decreasing hazard rate. More generally, if $\bar{F}$ has a log-convex density, then it has a decreasing hazard rate.

Proof. These results follow directly from Propositions B.4 and B.8.a. □

This proposition has sometimes been viewed as counter-intuitive or paradoxical. To understand why mixtures of exponential distributions have decreasing hazard rates, suppose that an item is selected randomly from a bin that contains items with exponential distributions having various scale parameters. As the chosen item ages without failure, it becomes increasingly likely that the selected item has a particularly small hazard rate (large expectation). This idea, made more precise in Propositions 3.A.2 and 3.B.1, leads to the realization that the probability of failure in the next instant of time is decreasing.

As noted in Chapter 3, Bayesian statisticians refer to the parametric family (in this case the exponential distribution) as the *model distribution*. The mixture does not arise as described in the previous paragraph, but from uncertainty about the parameter. With a *prior distribution* on the parameter, the resulting "mixture" is called a *predictive distribution*. The model may be a constant hazard rate, but the predictive distribution, the mixture, is based on current information concerning survival as expressed by the prior distribution (see Barlow, 1985; Lynn and Singpurwalla, 1997).

Note that mixtures of exponential survival functions can be regarded as Laplace transforms of the mixing (prior) distribution, restricted to nonnegative arguments. As such, the survival functions must be completely monotone; see Proposition 20.D.5. This means, among other conditions, that mixtures of exponential distributions have decreasing convex densities; they are also infinitely divisible (Proposition B.6 or 20.D.10). As an example of how Laplace transforms can be regarded as survival functions, compare the Laplace transform of the gamma distribution, 9.A(5), with the survival function of the Pareto II distribution 11.B(1).

Mixtures of exponential distributions arise in applications and this is an important source of distributions with decreasing hazard rate. See Proschan (1963) for a well-known example involving failures of air conditioning systems.

g. Properties of Distributions with Decreasing Hazard Rates

In this section, it is shown that distributions with decreasing hazard rates have decreasing densities, and may have tails heavy enough to preclude the existence of moments of positive order.

C.11. Proposition. Suppose that $F(0-) = 0$ and that F has a decreasing hazard rate. Then, F has a density except possibly for positive mass at the origin. There is a version f of the density that is decreasing and satisfies $f(x) > 0$, for all $x > 0$.

Proof. A distribution F has a decreasing hazard rate when $R = -\log \bar{F}$ is concave on $[0, \infty)$. Consequently, R has a continuous derivative except for countably many points (Proposition 21.A.4). Where this derivative $f = F'$ exists, it serves as a density; elsewhere, the density can be defined using continuity. This means that $R' = f/\bar{F}$ is decreasing (again by Proposition 21.A.4), and consequently, f is decreasing.

Should there be a point $a < \infty$ such that $\bar{F}(a) = 0$, then $R(a) = \infty$ and concavity of R would be violated. Also, if there were a point $b > 0$ such that $\bar{F}(b) = 1$, then $R(b) = 0$, and again concavity of R would be violated unless $R(x) = 0$, for all $x > 0$, but then F is not a distribution function. □

C.12. Proposition. Suppose that F has a decreasing hazard rate. Then, F need not have finite moments of any positive order.

This fact can be shown using Proposition 20.B.6, and it is in contrast to Proposition C.3. It stems from the fact that DHR survival functions, being log convex, can have tails that decay more slowly than the tails of any exponential distribution.

C.12.a. Example. If $\bar{F}(x) = 1/(1+x), x \geq 0$, then also $r(x) = 1/(1+x), x \geq 0$. This distribution is a special Pareto distribution with decreasing hazard rate, and the first moment fails to exist. The hazard rate of a more general Pareto distribution is given in 11.B(5), namely

$$r(x) = \frac{\lambda\alpha\xi(\lambda x)^{\alpha-1}}{1+(\lambda x)^\alpha}, \quad x \geq 0;$$

this hazard rate is decreasing for $\alpha < 1$. With $\xi = 1$, the rth moment exists finitely only for $-\alpha < r < \alpha$ (see 11.B(6)).

h. Preservation Theorems

C.13. Proposition. If F has a decreasing hazard rate, then the residual life distribution F_t also has a decreasing hazard rate.

This result is a parallel to Proposition C.5 with a similar proof and intuitive meaning.

The following generalization of Proposition C.10, due to Frank Proschan, is fundamental to the understanding of the origins of distributions with decreasing hazard rate.

C.14. Theorem (Barlow, Marshall and Proschan, 1963). The family of distributions with decreasing hazard rate is closed under the formation of mixtures.

Proof. Distributions with decreasing hazard rate have concave hazard functions. In addition, mixtures have concave increasing hazard transforms (Proposition 3.D.2). Thus, the result is an immediate consequence of the fact that an increasing concave function of a concave function is concave (Proposition 21.A.5). □

C.14.a. Remark. If the hazard rates of the mixture components are differentiable, then Theorem C.14 is immediate from Corollary 3.D.4.a.

More complicated proofs of Theorem C.14 not involving the notion of a hazard transform are given by Barlow, Marshall and Proschan (1963) or Marshall and Olkin (1979, p. 452).

C.15. Proposition. Suppose that F has a decreasing hazard rate and first moment μ. Then, the distribution with density $f_{(1)}(x) = \bar{F}(x)/\mu, x \geq 0$, has a decreasing hazard rate.

This fact, a parallel to Proposition C.6, is another special case of Proposition B.8. However, here the assumption of a finite first moment is critical because DHR distributions need not have finite moments (Proposition C.12).

C.15.a. Proposition. Let X be a random variable with distribution F and let Y have the density $f_{(1)}$ of Proposition C.15. Then, F has a decreasing hazard rate if and only if $X \leq_{\mathrm{lr}} Y$.

Proof. With the assumption that F has a density, this result can be directly verified using Proposition 2.A.10. □

C.16. Observation. If F and G have decreasing hazard rates, then the convolution need not have a decreasing hazard rate. To see this, note from Proposition C.4 that the convolution of two exponential distributions is IHR, but such a convolution is not exponential so it cannot also be DHR. More generally, if $F(\cdot) = F(\cdot \mid \lambda, v_1)$ and $G(\cdot) = F(\cdot \mid \lambda, v_2)$ are gamma distributions with densities given by 1.F(6) or 9.A(1) and if $v_1 \leq 1, v_2 \leq 1$ while $v_1 + v_2 > 1$, then F and G have decreasing hazard rates, but their convolution is a gamma distribution with parameters λ and $v_1 + v_2$ and the hazard rate is strictly increasing (see 7.A(11)).

Recall that another characterization of DHR distributions is given by Proposition C.1.f.

C.17. Summary. *Mixtures* of DHR distributions are always DHR, but mixtures of IHR distributions need not be IHR and can even be DHR; *convolutions* of IHR distributions are always IHR, but convolutions of DHR distributions can be IHR.

D. Bathtub Hazard Rates

D.1. Definition. A distribution is said to have a *bathtub hazard rate* if for some $0 \leq a \leq b$, the hazard rate $r(t)$ is decreasing in $t, 0 \leq t \leq a$, is constant in the interval $a \leq t \leq b$, and is increasing in $t, t \geq b$.

From at least two points of view, distributions with a bathtub hazard rate have considerable intuitive appeal. The first and most commonly given idea is based upon the assumption that the device or organism under consideration comes from a mixture of individuals of varying inherent strength. Those individuals with life threatening defects at birth suffer a high rate of early mortality, but as a device ages without failure, the conditional probability that a life-threatening defect is present diminishes and so the hazard rate decreases. There comes a time a when deaths due to birth defects no longer occur and accidents become the only significant cause of death, so the hazard rate becomes constant. But eventually, at time b, the adverse effects of age begin to take their toll and the hazard rate begins to rise. This concept of a mixture is related to a Bayesian concept in which the mixture is not real, but is treated mathematically as such due to uncertainty about the underlying distribution. See Figure D.5 for a graph of a bathtub hazard rate.

A second intuitive basis for bathtub hazard rates applies primarily to biological organisms. When young, such organisms may have immature immune systems, they may have difficulty competing for food, and they may suffer from a number of other disadvantages that diminish as the organism grows and matures. During the period of maturation, the hazard rate decreases. But eventually, the organism fully matures and again, the adverse effects of age take effect and cause the hazard rate to increase. This idea was already apparent in the writings of Price (1771) who wrote that human life, from birth upwards, grows gradually stronger until the age of 10 years, then slowly loses strength until the age of 50, then more rapidly loses strength until, at 70 or 75, it is brought back to all the weakness of the first

D. Bathtub Hazard Rates

month. Bathtub hazard rates for human life lengths are explicitly discussed by Wittstein (1883) who based his ideas on studies of mortality tables.

The just described origin of bathtub hazard rates for biological organisms has its counterpart for mechanical systems. A new system may suffer from "bugs", that is, from errors of design or of construction. Moreover, the operators of the system may be initially inexperienced. As the system ages, the potential for bugs or human error diminishes, causing the hazard rate to decrease. But after a while, the effects of aging cause the hazard rate to rise. As noted in the New York Times Magazine, July 18, 2006, p. 56, David Lochbaum of the Union of Concerned Scientists has pointed out that the bathtub curve applies to the safety of nuclear power plants; they are most dangerous when first brought on line, or at the end of their life cycle. More recent examples where bathtub hazard rates were found useful for fitting data are given by Rajarshi and Rajarshi (1988). For a survey of such hazard rates, see Lai, Xie and Murthy (2001).

Bathtub hazard rates motivate a process called "burn-in" for manufactured items. The idea is to place the device in a simulated service (or more stressful) environment to discover defects before the device is introduced into actual service. For a review of this common practice, see Block and Savits (1997) and the references contained therein. From a Bayesian point of view, burn-in is related to belief; see Lynn and Singpurwalla (1997).

It is notable that none of the standard parametric families contains life distributions with bathtub hazard rates (apart from the case $a = b = 0$ of distributions with increasing hazard rate and the case $a = b = \infty$ of distributions with decreasing hazard rates). However, especially with $a = b$, various constructions of such distributions are possible. Examples are given in this section and in Section 15.G. It is not surprising to find that the most interesting parametric families that allow for bathtub hazard rates have at least three parameters.

It has been pointed out by Klutke, Keissler and Wortman (2003) that an assumption of a bathtub hazard rate must be made with caution. View the hazard rate as the product of the density and the reciprocal of the survival function, that is,

$$r(x) = f(x)[\bar{F}(x)]^{-1}.$$

Note that $[\bar{F}(x)]^{-1}$ is an increasing function to conclude that

(i) if the density is increasing in an interval, then the hazard rate is increasing in that interval,

and consequently

(ii) if the hazard rate is decreasing in an interval, then the density is decreasing in that interval (see also Proposition C.10).

From (ii), it follows that if F has a bathtub-shaped hazard rate as described above, then the density of F is decreasing on the interval $[0, b]$. This is not always a reasonable assumption.

a. Bathtub-Shaped Hazard Rates from Mixtures

Consider the first motivation supporting bathtub hazard rates described above and let F_1 and F_2 be the distributions associated with failures due to defects and those due to eventual wearout, respectively. The survival function of the mixture has the form $[\pi \bar{F}_1(x) + \bar{\pi} \bar{F}_2(x)]$. It is often assumed that there is a time a after which deaths due to birth defects do not occur and there is a time b before which deaths due to wearout do not occur. If $a < b$, then the supports of F_1 and F_2 do not overlap. When densities exist, the density of the mixture is 0 between a and b. Such a density is unrealistic in practice because there is always the possibility of failure due to an accident, no matter what the age of the device is. The waiting time for an accident is usually assumed to be exponentially distributed because the exponential distribution has a constant hazard rate. So the actual time of failure is the minimum of the waiting time for a death due to accident and those due to other causes. Because the survival function of a minimum is the product of individual survival functions, it is natural to consider the model

$$\bar{F}(x) = e^{-\lambda x}[\pi \bar{F}_1(x) + \bar{\pi} \bar{F}_2(x)], \qquad (1)$$

where F_1 has support $[0, a]$ and F_2 has support $[b, \infty), 0 < a \leq b$. Of course, π represents the proportion of the population with birth defects.

It is easy to check that if $\bar{F}$ is given by (1), then its hazard rate is the sum of the hazard rates of the exponential part and the mixture $\pi \bar{F}_1(x) + \bar{\pi} \bar{F}_2(x)$:

$$\begin{aligned} r(x) &= \lambda + \frac{\pi f_1(x)}{\pi \bar{F}_1(x) + \bar{\pi}}, & 0 \leq x < a, \\ &= \lambda, & a \leq x \leq b, \\ &= \lambda + r_2(x), & b < x. \end{aligned} \qquad (2)$$

Because the distribution F_1 has support $[0, a]$, it cannot have a decreasing hazard rate (Proposition C.12). But under certain conditions, r itself can be decreasing on the interval $[0, a)$.

Although the model (1) is often used to model a bathtub-shaped hazard rate and explain its presence, additional conditions are required if (2) is indeed to have a bathtub shape.

D.2.a. Proposition. Suppose that r is given by (2). A necessary (but *not* sufficient) condition for $r(x)$ to be decreasing in x on $[0, a)$ is that the density $f_1(x)$ is decreasing in x on $[0, a)$.

Proof. To avoid the complications of a limiting argument, assume that the density f_1 is differentiable. Then r is differentiable on $(0, a)$, and this derivative is negative if and only if

$$(\pi \bar{F}_1(x) + \bar{\pi}) f_1'(x) + [f_1(x)]^2 < 0. \tag{3}$$

This can happen only if $f_1'(x) < 0$. □

D.2.b. Proposition. Suppose that r is given by (2). If $r(x)$ is decreasing in x on $[0, a)$ and if π is replaced by a smaller value, then $r(x)$ is again decreasing in x on $[0, a)$. Thus, there exists a $\pi^* \geq 0$ such that $r(x)$ is decreasing in x on $[0, a)$ only when $\pi \leq \pi^*$.

Proof. To obtain this result, make use of Proposition D.2.a; where r is decreasing, $f_1' \leq 0$, and with this, it can be checked that the expression on the left-hand side of (3) is increasing in π. □

D.2.c. Remark. Proposition D.2.a is somewhat disturbing, because it says that the most common intuitive idea behind the bathtub hazard rate does not necessarily lead to a bathtub hazard rate; the requirement that the density f_1 is decreasing is a severe restriction that may not be realistic. It may also be somewhat disturbing that a bathtub shape may be lost if proportions in the mix of items with and without birth defects are changed.

There is a possible further difficulty with this model; the hazard rate (2) is discontinuous at a unless $\lim_{x \uparrow a} f_1(x) = 0$, and this is another substantial restriction for a density on $[0, a]$.

D.3. Example. Suppose that the model (1) holds, where

$$\bar{F}_1(x) = \left(1 - \frac{x}{a}\right)^\xi, \quad 0 \leq x \leq a, \xi > 0.$$

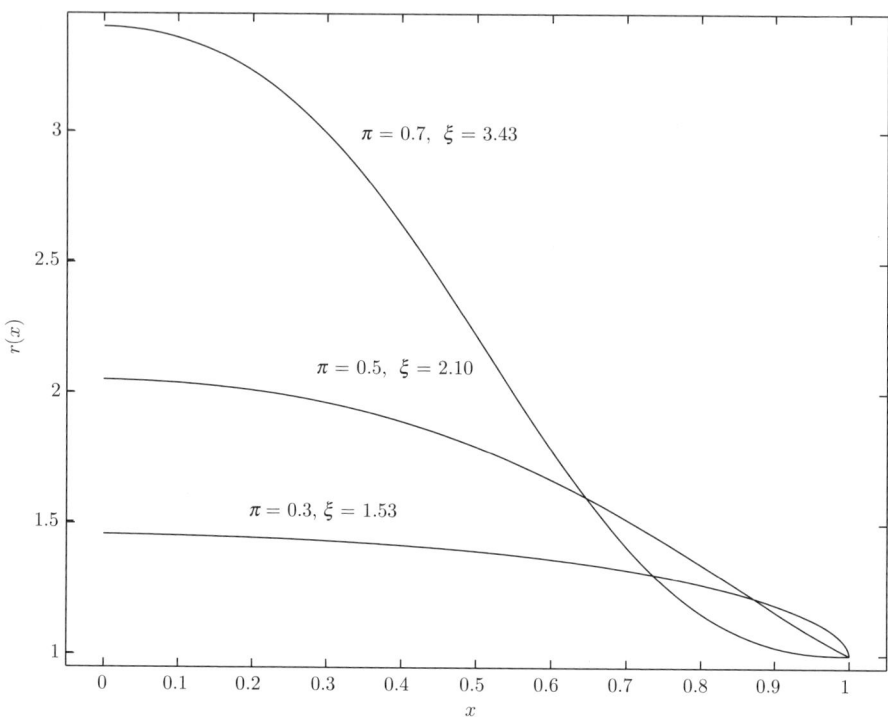

Fig. D.1. Hazard rates of Example D.3 ($\lambda = 1$, $a = 1$)

It follows from (2) that, for $x \leq a$,

$$r(x) = \frac{\pi \xi (a-x)^{\xi-1}}{\pi(a-x)^\xi + \bar{\pi} a^\xi} + \lambda.$$

It can be verified that this hazard rate is decreasing in $x \leq a$ if and only if $\pi \leq (\xi - 1)/\xi = \pi^*$. Moreover, the hazard rate (2) is continuous at a if $\xi > 1$ and $\pi < 1$. See Figure D.1.

Because F_2 has support bounded away from 0, the model (1) is applicable only for devices that, in the absence of defects, have a guaranteed period during which accidents are the only causes of failure. Such conditions are often unrealistic and are not necessary for a mixture to have a bathtub-shaped hazard rate.

Model (1) can be modified to take the simple form

$$\bar{F}(x) = \pi \bar{F}_1(x) + \bar{\pi} \bar{F}_2(x), \tag{4a}$$

where, as before, $\bar{F}_1$ is the survival function of a defective component and $\bar{F}_2$ is the survival function of a component without defects. Here, the exponential component of (1) has been absorbed in $\bar{F}_1$ and/or $\bar{F}_2$, so it is not explicitly present. In (4a), there is no requirement that $\bar{F}_1(a) = 0$ or $\bar{F}_2(b) = 1$, as is the case with (1). But in accordance with the idea of a defective device, it is natural to assume that $\bar{F}_1(x) \leq \bar{F}_2(x)$, for all x. Clearly, $\bar{F}$ has the hazard rate

$$r = \frac{\pi f_1 + \bar{\pi} f_2}{\pi \bar{F}_1 + \bar{\pi} \bar{F}_2}. \tag{4b}$$

A slightly different view of this model can be described in terms of random variables. Let X_0 be the waiting time for failure due to some flaw in a device (when present), and let X_2 be the waiting time for failure of a properly constructed device. Assume that X_0 and X_2 are independent and let $X_1 = \min(X_0, X_2)$. If π is the probability that a flaw is present, then the lifetime of the device is given by

$$X = X_1, \quad \text{with probability } \pi$$
$$ = X_2, \quad \text{with probability } \bar{\pi},$$

and (4a) holds. As above, denote the survival function, density, and hazard rate of X_i, respectively, by $\bar{F}_i = \bar{F}_i(x)$, $f_i = f_i(x)$, and $r_i = r_i(x)$, $i = 0, 1, 2$. With the observation that the survival function of X_1 is given by $\bar{F}_1 = \bar{F}_0 \bar{F}_2$, it is straightforward to show that survival function of X is given by

$$\bar{F} = \pi \bar{F}_0 \bar{F}_2 + \bar{\pi} \bar{F}_2 = \bar{F}_2(\pi \bar{F}_0 + \bar{\pi}) \tag{5a}$$

and the hazard rate r of X is given by

$$r = \frac{f_2(\pi \bar{F}_0 + \bar{\pi}) + \pi \bar{F}_2 f_0}{\bar{F}_2(\pi \bar{F}_0 + \bar{\pi})} = r_2 + \frac{\pi f_0}{\bar{\pi} + \pi \bar{F}_0} = r_2 + \frac{\pi r_0 \bar{F}_0}{\bar{\pi} + \pi \bar{F}_0}. \tag{5b}$$

Because the third expression for r in (5b) is derived using the relationships $f_i(x) = r_i(x) \bar{F}_i(x)$, the quantity $r_0(x) \bar{F}_0(x)$ should be interpreted as 0 when $\bar{F}_0(x) = 0$ even though $r_0(x)$ may be infinite.

The hazard rate (5b) can take various shapes.

D.4. Example. Suppose that $\bar{F}_0(x) = \exp\{-x^\beta\}$ and $\bar{F}_2(x) = \exp\{-(\lambda x)^\alpha\}$, $x \geq 0$, that is, F_0 and F_2 are Weibull distributions. If $\alpha \geq 2$ and $\beta \leq 1$, then the hazard rate (5b) is bathtub shaped

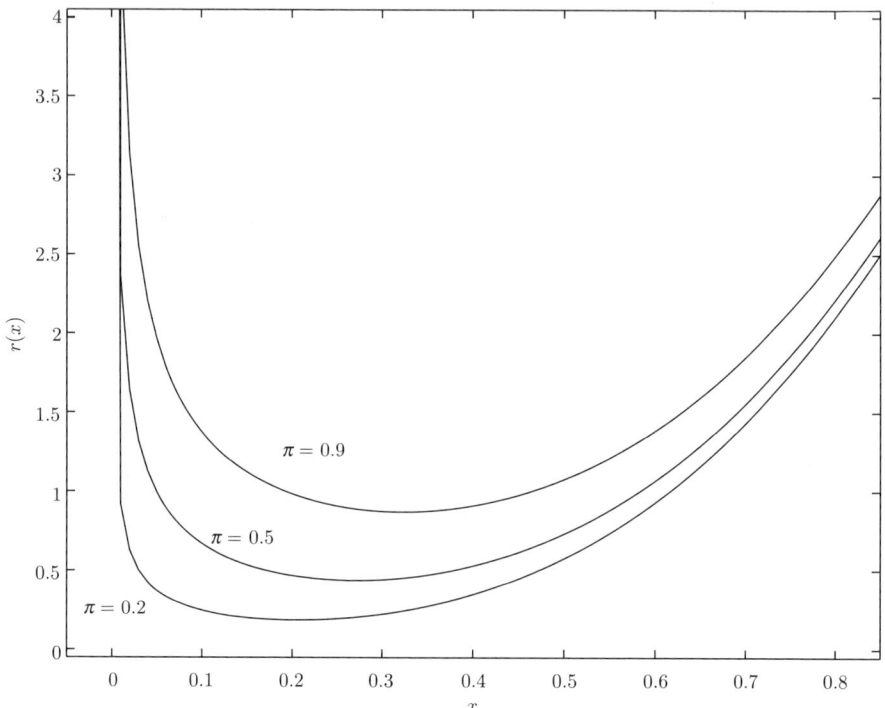

Fig. D.2. Hazard rates of Example D.4 ($\lambda = 1$, $\alpha = 4$, $\beta = 0.5$)

(and convex). Here, F_0 has a decreasing hazard rate and F_2 has an increasing hazard rate; both hazard rates are convex. By differentiating (5b) twice and keeping in mind that $\beta \leq 1$, it can be shown that r is convex. Also, it is necessary to check that r is not monotone because $\lim_{x \to 0} r'(x) = -\infty$ and $\lim_{x \to \infty} r'(x) = \infty$.

Example D.4 includes conditions on α and β in order for the mixture to have a bathtub shape. The shape of the hazard rate when these conditions are violated has been investigated by Jiang and Murthy (1998). See Figure D.2.

D.5. Example. Suppose that $\bar{F}_0(x) = (1-x)^4, 0 \leq x \leq 1, \bar{F}_2(x) = \exp\{-x^2\}, x \geq 0$, and $\pi = 1/2$ in (5a). Then, the hazard rate (5b) becomes

$$r(x) = 2x + \frac{4(1-x)^3}{1+(1-x)^4}, \quad 0 \leq x < 1,$$
$$= 2x, \quad x \geq 1.$$

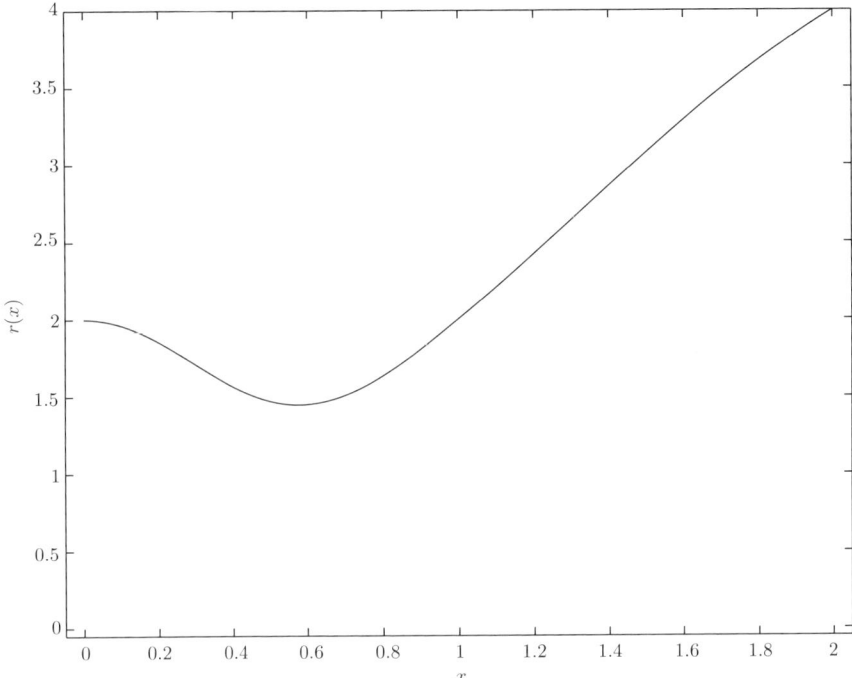

Fig. D.3. Hazard rates of Example D.5

Elementary but tedious calculations can be used to show that this hazard rate is bathtub shaped, although in this case, the hazard rate is not convex. See Figure D.3.

It might be expected that a mixture of a distribution with decreasing hazard rate and a distribution with an increasing hazard rate would lead to a bathtub hazard rate; Example D.6 shows that this need not hold. Hazard rates of mixtures are not mixtures of hazard rates, and in the upper tail, the distribution with a decreasing hazard rate sometimes dominates the mixture, forcing the hazard rate of the mixture to eventually decrease (see Proposition 3.C.1).

D.6. Example. Consider the mixture (4a) where $\pi = 1/2, \bar{F}_1(x) = 1/(1+x), x \geq 0$, and $\bar{F}_2(x) = \exp\{-x^2\}, x \geq 0$. Here, F_2 has an increasing hazard rate and F_1 has a decreasing hazard rate. The hazard rate

$$r(x) = \frac{2x(1+x)^2 + e^{x^2}}{(1+x)^2 + (1+x)e^{x^2}}, \quad x \geq 0,$$

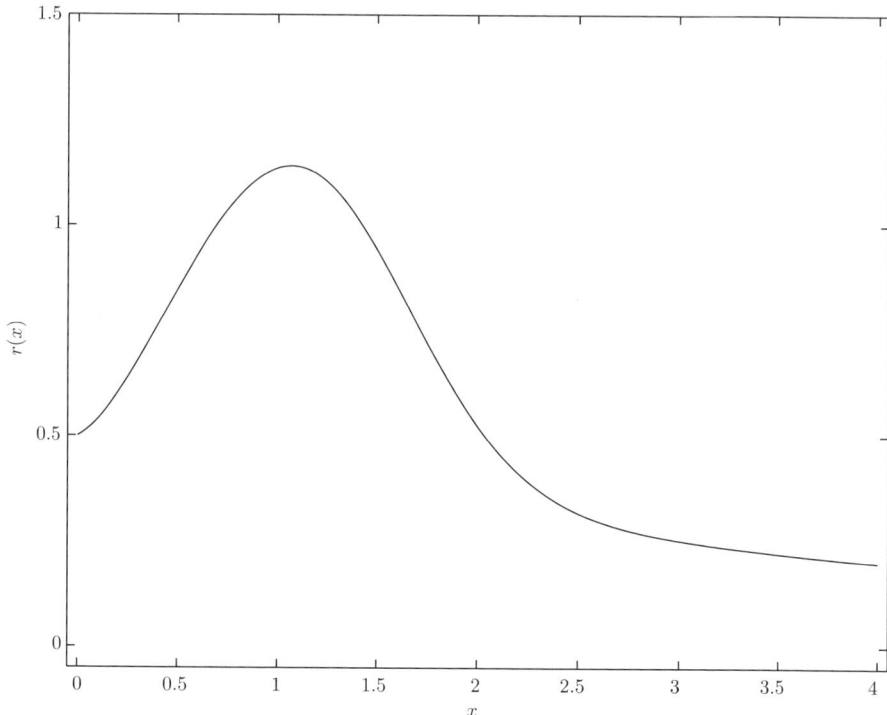

Fig. D.4. Hazard rate of Example D.6

of this mixture is initially increasing and eventually decreasing. See Figure D.4.

The phenomenon illustrated in Example D.6 does not occur when the decreasing hazard rate stays above the increasing hazard rate, as is the case with gamma distributions having the same scale parameter. Mixtures of two gamma distributions one with decreasing hazard rate and one with increasing hazard rate have been discussed by Glaser (1980); they have a bathtub-shaped hazard rate, but the limiting value of this hazard rate is finite. For more discussion of these issues, see Block and Joe (1997).

b. Bathtub-Shaped Hazard Rates from Minima

Another origin for bathtub hazard rates arises from minima; according to Section 3.E, this derivation is mathematically equivalent to the mixture model, but it is conceptually different. Suppose that

$$X = \min(U, V, W), \tag{6}$$

where U, V, and W are independent, U has a decreasing hazard rate r_1, V has an increasing hazard rate r_2, and W has an exponential distribution with parameter λ. Then, the hazard rate of X is $r_1 + r_2 + \lambda$. Because it is the sum of a decreasing and an increasing function, this hazard rate can be bathtub shaped. If the hazard rates r_1 and r_2 are both convex, then $r_1 + r_2 + \lambda$ must be convex, and hence monotone or bathtub shaped (possibly with $a = b$). According to this model, there is no parameter like the mixing proportion of previous models, and even though there is flexibility in choosing r_1 and r_2, the matter is still delicate if the hazard rate is to be bathtub shaped. Of course, in this model, W can be easily dispensed with if convenient; just replace V by $\min(V, W)$, and then r_2 is replaced by $r_2 + \lambda$. Several examples of (6) are given in Section 15.G.

The form (6) calls to mind the concept of competing risks (see Chapter 17), because the device is subject to the risk of death due to a birth defect, to wearout or old age, as well as to an accident.

D.7. Example. Suppose that r is the bathtub hazard rate graphed in Figure D.5 and given by

$$\begin{aligned} r(x) &= \frac{1}{2\sqrt{x}} - \frac{1}{2\sqrt{a}} + \lambda, \quad x \leq a, \\ &= \lambda, \quad a \leq x \leq b, \\ &= \lambda x^2/b^2, \quad x \geq b. \end{aligned} \tag{7}$$

Recall that the survival function corresponding to (7) can be obtained using 1.B(3). That (7) is indeed the hazard rate of a proper distribution follows from the fact that it is nonnegative and that its integral over the interval $(0, t)$ is finite for finite t and infinite for $t = \infty$.

The hazard rate (7) can arise in a variety of ways. In particular, if $\lambda \geq 1/[2\sqrt{a}]$, it is the hazard rate of

$$X = \min(U, V),$$

where U has the hazard rate

$$r_1(x) = 1/[2\sqrt{x}], \quad x > 0,$$

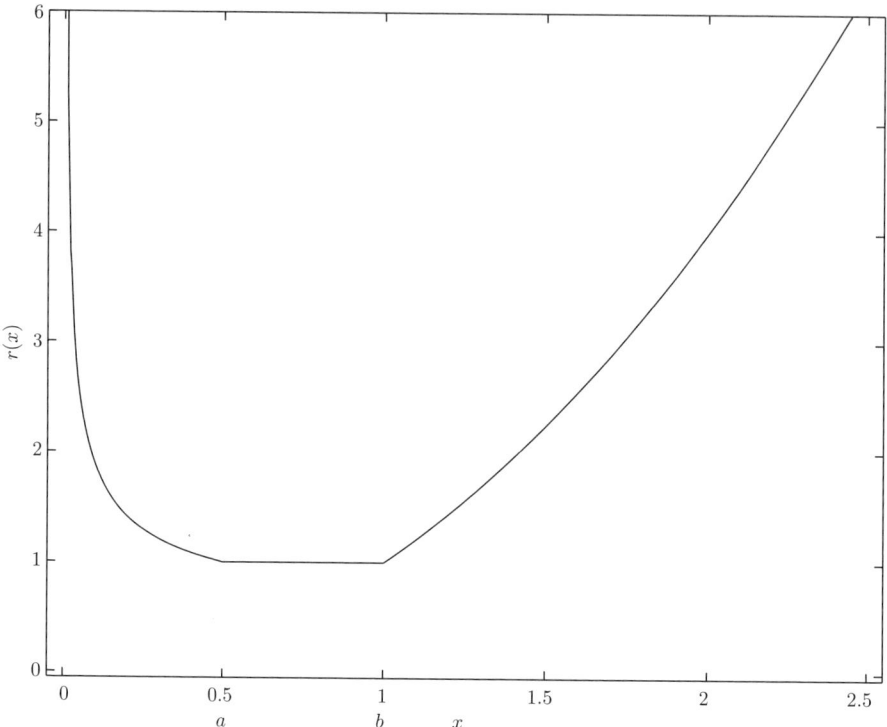

Fig. D.5. Hazard rates of Example D.7 ($\lambda = 1$, $a = 0.5$, $b = 1$)

i.e., U has a Weibull distribution with shape parameter $1/2$ and V has the hazard rate r_2 given by

$$r_2(x) = \lambda - 1/[2\sqrt{a}], \quad x \leq a,$$
$$= \lambda - 1/[2\sqrt{x}], \quad a \leq x \leq b,$$
$$= \lambda x^2/b^2 - 1/[2\sqrt{x}], \quad x \geq b.$$

This example also arises from (1) with

$$\bar{F}_1(x) = \frac{\exp\{-\sqrt{x} + (x/(2\sqrt{a}))\} - \{\exp\{-\sqrt{a}/2\}}{1 - \exp\{-\sqrt{a}/2\}}, \quad 0 \leq x \leq a,$$
$$\bar{F}_2(x) = \exp\{\lambda(x-b) - \lambda(x-b)^3/3b^2\}, \quad x \geq b,$$

and

$$\pi = 1 - \exp\{-\sqrt{a}/2\}.$$

Other somewhat similar examples are given by Rajarshi and Rajarshi (1988). The hazard rates 10.C(4) and 10.C(15) are, in terms of random variables, of the form (6) and may have bathtub hazard rates.

According to Propositions C.5 and 5.D.2, the mean residual life is decreasing whenever the hazard rate is increasing. There are related result for bathtub hazard rates.

D.8. Proposition (Mi, 1995). Suppose that F has a bathtub hazard rate that first achieves a minimum at the point a. Then, the mean residual life has an inverted bathtub shape with a maximum at a point a^*. Moreover, $a^* \leq a$.

c. Delayed Bathtub Hazard Rates

The condition that a density must be decreasing wherever the hazard rate is decreasing is a severe limitation that limits the applicability of bathtub hazard rates. However, many more examples can be encompassed if the concept of a bathtub hazard rate is modified to allow an initial period in which the hazard rate may be increasing before assuming a true bathtub shape. Such hazard rates might be termed "delayed bathtub hazard rates." In model (1), with the assumption that the hazard rate has a delayed bathtub shape, the density f_1 need not be initially decreasing, and may be 0 at the origin. Delayed bathtub hazard rates have not received much attention in the literature, although examples are known (see Jiang and Murthy, 1998).

d. Inverted Bathtub Hazard Rates

In contrast with the bathtub hazard rate, there are a number of well-known parametric families of life distributions with inverted bathtub (unimodal) hazard rates.

D.9. Definition. A distribution is said to have an *inverted bathtub hazard rate* if for some $0 \leq a \leq b$, the hazard rate $r(t)$ is increasing in $t, 0 \leq t \leq a$, is constant in the interval $a \leq t \leq b$, and is decreasing in $t, t \geq b$. Alternatively, such hazard rates are said to be *unimodal*.

In the past, inverted bathtub hazard rates have not attracted much interest, at least in reliability theory, perhaps because the bathtub hazard rates have been a focus of attention. But there are good reasons for giving consideration to inverted bathtub hazard rates.

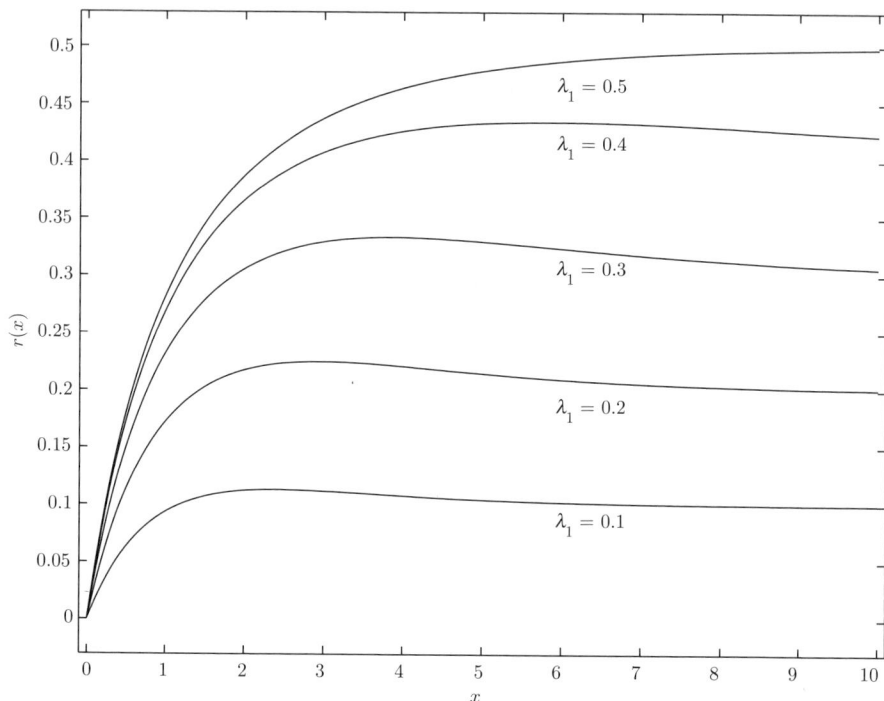

Fig. D.6. Hazard rates for the maximum of two exponentially distributed random variables ($\lambda_2 = 1$)

D.10. Example. Let X_1 and X_2 be independent exponentially distributed life lengths having respective parameters λ_1 and λ_2. Then, $\max(X_1, X_2)$ has survival function

$$\bar{F}(x) = e^{-\lambda_1 x} + e^{-\lambda_2 x} - e^{-(\lambda_1+\lambda_2)x},$$

and hazard rate r given by

$$r(x) = [\lambda_1 \, e^{\lambda_2 x} + \lambda_2 \, e^{\lambda_1 x} - 1]/[e^{\lambda_2 x} + e^{\lambda_1 x} - 1]. \tag{8}$$

This hazard rate is graphed in Figure D.6.

In Example D.10, the hazard rate is inverted bathtub shaped, as is apparent from Figure D.6. This example is examined again in Section 5.A, where an intuitive explanation for the shape of the hazard rate is offered.

This example does not explain why distributions with inverted bathtub hazard rates arise naturally in such examples as the Pareto (IV) distribution of Section 11.B, the generalized F distribution of Section 11.C, the lognormal distribution of Section 12.B, or the log logistic distribution of Section 12.C. These examples stand in contrast to the case of a bathtub hazard rate which is possessed by few, if any, examples known to arise naturally.

The explanation for at least some of these inverted bathtub hazard rates lies in their representations as mixtures. Mixtures sometimes have eventually decreasing hazard rates, as is apparent from Examples C.7.a, C.7.b, and D.6. This fact and numerous other examples are discussed in detail by Gurland and Sethuraman (1995). The tendency for mixtures to produce decreasing hazard rates sometimes becomes stronger as one moves into the right-hand tail of the distribution. Where a mixture is involved, an inverted bathtub hazard rate should not come as a surprise. See Block and Joe (1997) for a careful study of these issues.

Inverted bathtub hazard rates have been encountered as first passage time distributions by Aalen and Gjessing (2001). A well-known example of this phenomenon is the inverse Gaussian distribution, discussed in Chapter 13.

E. Determination of Hazard Rate Shape

Sometimes a distribution is defined in terms of its density but the distribution function and survival function cannot be given in closed form. In addition to the normal distribution, examples include the gamma and lognormal distributions introduced in Chapter 1. When neither the survival function nor the hazard rate can be given in closed form, a direct study of hazard rate behavior may not be particularly easy. The purpose of this section is to provide some methods for the determination of hazard rate behavior useful in later chapters where parametric families of distributions are studied.

It can be seen directly from the definition of the hazard rate that

(i) if the density f is increasing at $x = x_0$, then the hazard rate r is increasing at $x = x_0$;
(ii) if the hazard rate r is decreasing at $x = x_0$, then the density f is decreasing at $x = x_0$.

These observations can often be useful.

The essential fact to be exploited in the following is that the behavior of the hazard rate is related to the behavior of the derivative of the logarithm of the density, namely,

$$\rho(x) = -\frac{d \log f(x)}{dx} = -\frac{f'(x)}{f(x)}, \quad x > x_0. \tag{1}$$

Of course, this requires the density to be differentiable, an implicit assumption in what follows.

E.1. Lemma. Let f be a density strictly positive and differentiable on (x_0, ∞) such that $\lim_{x \to \infty} f(x) = 0$. Denote the hazard rate of f by r and suppose that $\rho(x) - c$ has k sign changes in $x > x_0$. Then,

(i) $r(x) - c$ has at most k sign changes in $x > x_0$ and

(ii) if $r(x) - c$ has k sign changes in $x > x_0$, then they are in the same order as those of $\rho(x) - c$.

Proof. Let $H(x, t) = 1$ if $x \geq t$, and $H(x, t) = 0$ if $x < t$. This function is defined in Example 19.B(5) and noted there to be totally positive of order ∞. Let $g(x) = 0$, $x < x_0$, and $g(x) = f(x)$, $x \geq x_0$. Observe that the sign change pattern of $\rho(x) - c$ in $x > x_0$ is the same as that of $-f'(x) - cf(x)$ in $x > x_0$. By the variation diminishing property Theorem 21.B.13 of totally positive functions,

$$\int_{-\infty}^{\infty} [-g'(x) - cg(x)] H(x, t) \, dx = f(t+) - c\bar{F}(t), \quad t \geq x_0,$$

$$= f(x_0+) - c\bar{F}(x_0), \quad t < x_0$$

has at most k sign changes in $t > x_0$ (here, $f(z+) = \lim_{x \downarrow z} f(x)$). If $f(x) - c\bar{F}(x)$ has k sign changes in $x \geq x_0$, then the sign changes occur in the same order as those of $-g'(x) - cg(x)$, that is, of $-f'(x) - cf(x)$ in $x \geq x_0$. But the sign change pattern of $f(x) - c\bar{F}(x)$ is the same as the sign change pattern of $r(x) - c$. □

E.2. Theorem. Let f be a density function satisfying the conditions of Lemma E.1 with $x_0 = 0$.

(a) If ρ is increasing, then r is increasing.

(b) If ρ is decreasing, then r is decreasing.

(c) If there exists x_1 for which ρ is decreasing in $x \leq x_1$ and increasing in $x \geq x_1$, then there exists x_2, $0 \leq x_2 < x_1$, such that r is decreasing in $x \leq x_2$ and increasing in $x \geq x_2$.

(d) If there exists x_1 for which ρ is increasing in $x \leq x_1$ and decreasing in $x \geq x_1$, then there exists $x_2, 0 \leq x_2 \leq x_1$, such that r is increasing in $x \leq x_2$ and decreasing in $x \geq x_2$.

Proof. If ρ is monotone, it follows directly from Lemma E.1 that r is monotone in the same direction. If there exists x_1 for which ρ is decreasing in $x \leq x_1$ and increasing in $x \geq x_1$, then for every positive constant c, $\rho(x) - c$ has at most two sign changes, in the order $+, -, +$ if there are two changes. By Lemma E.1, this means that $r(x) - c$ has at most two sign changes, in the order $+, -, +$ if there are two changes. Consequently, there exists x_2 such that r is decreasing in $x \leq x_2$ and increasing in $x \geq x_2$. Because ρ is increasing in $x \geq x_1$, it follows similarly from Lemma E.1 that r is increasing in $x \geq x_1$. This means that $0 \leq x_2 \leq x_1$.

The proof of (d) is similar. □

The proof of Theorem E.2 uses mathematical theory beyond that of ordinary calculus because it depends upon Lemma E.1. A result very similar to that of Theorem E.2 has been obtained by Glaser (1980) using basic calculus, but his proof is somewhat more computationally involved than the one given here.

E.3. Proposition (Glaser, 1980).

(i) Suppose that ρ satisfies the conditions of (c) of Theorem E.2. If $\lim_{x \to 0} f(x) = \infty$, then $x_2 > 0$; if, $\lim_{x \to 0} f(x) = 0$, then $x_2 = 0$.

(ii) Suppose that ρ satisfies the conditions of (d) of Theorem E.2. If $\lim_{x \to 0} f(x) = \infty$, then $x_2 = \infty$; if, $\lim_{x \to \infty} f(x) = 0$, then $x_2 < \infty$.

Proof. These results are obtained from the fact that $f(0) = r(0)$. If (i) and $\lim_{x \to 0} f(x) = \infty$, then r must initially decrease so $x_1 > 0$; on the other hand, if $\lim_{x \to 0} f(x) = 0$, then r must initially increase, so $x_1 = 0$. The situation in case (ii) is similar. □

E.4. Example. Consider the lognormal distribution with density given in the form 1.F(12b). To show that this distribution has an inverted bathtub hazard rate, first compute $\rho(x) = [1 + \alpha^2 \log(\lambda x)]/x$. By setting the derivative of ρ equal to 0, it can be determined that ρ is increasing in $x \leq [\exp\{1 - \alpha^{-2}\}/\lambda] = x_0$ and decreasing in $x \geq x_0$. It follows that $r(x)$ is increasing in $x \leq x_1$ and decreasing in $x \geq x_1$, where $x_1 \leq x_0$. Here $x_1 > 0$ because $f(0) = 0$.

To fully appreciate the utility of Theorem E.2, it is helpful to try to analytically determine the hazard rate behavior of the lognormal distribution directly without its aid.

Glaser's Proposition E.3 has been extended by Gupta and Warren (2001) to more fully cover the case that a hazard rate has multiple changes in direction. Their clever proof depends only on basic ideas of calculus.

E.5. Proposition (Gupta and Warren, 2001). *Let f be a twice differentiable density concentrated on the interval $(0, \infty)$ such that $f(x) > 0$, for all $x > 0$. Retain the notation introduced above and suppose that the equation $\rho'(x) = 0$ has n solutions, say $x_1, \ldots, x_n$, where $0 = x_0 < x_1 < \cdots < x_n$. Then, the equation $r'(x) = 0$ has at most one solution in the closed interval $[x_{k-1}, x_k], k = 1, 2, \ldots, n$.*

Proof. Let $g(t) = 1/r(t)$ and let

$$s(t) = g'(t)f(t) = \bar{F}(t)[\rho(t) - r(t)]; \qquad (2)$$

consequently,

$$s'(t) = \rho'(t)\bar{F}(t).$$

Thus, s' and ρ' have a common sign and common zeros. Because ρ is monotonic on $[x_{k-1}, x_k]$, so s is also monotonic on that interval; consequently, it has at most one zero in that interval. From (2) it follows that $s(t) = 0$ if and only if $g'(t) = 0$; because g' and r' have common zeros, it follows that $s(t) = 0$ if and only if $r'(t) = 0$. □

5

Nonparametric Families: Origins in Reliability Theory

> If there is a 50-50 chance that something can go wrong, then 9 times out of 10 it will.
>
> Paul Harvey News, Fall, 1979

As mentioned in the introduction of Chapter 4, nonparametric families of distributions have mostly been studied in the context of reliability theory. The theory that has been developed for these families has thus involved the notion of components and systems, which might be mechanical, electrical, or hydraulic. But the same ideas often can be applied to biological systems also.

This chapter begins with a discussion of a class of systems called "coherent systems." The study of such systems, together with additional ideas that originated in reliability theory, helps to explain the origins and importance of the nonparametric families discussed in this chapter. Also in this chapter are additional results for the basic families introduced in Chapter 4.

Readers should feel free to move on to later chapters before reading this chapter. However, occasional references to this chapter will be encountered.

A. Coherent Systems

Coherent systems were introduced in a classic paper by Birnbaum, Esary and Saunders (1961). This paper, written under the auspices of the Boeing Scientific Research Laboratories, was the beginning of a long series of papers giving results about coherent systems. Only a brief introduction to the theory of coherent systems is given here.

Coherent systems are a fundamental concept in reliability theory even though they may appear to be of limited usefulness. They are based upon the idea that components and systems have but two states, functioning and failed; in practice, it is more often the case that components and systems have a multitude of possible states. But ordinarily, some criterion is imposed to classify the various system states as "functioning" or "failed." Fortunately, it is very often possible to classify the components as "functioning" or "failed" in such a way that the state of the system is determined by the state of the components. Then the notion of a coherent system is useful.

a. Structure Functions

The principal idea is the premise that if the system is "coherent," then the repair of a failed component would not cause the system to fail. Consequently, the function ϕ that indicates the state of the system in terms of the component states would be increasing (nondecreasing). One would not expect the system to work when all components have failed and would not even be interested in a system that did not function when all components function. These ideas are the basis of the following definition.

A.1. Definition. A binary function ϕ of n binary variables is called a *coherent system (of order n)*, or a *coherent structure function* if

(i) $\phi(0,\ldots,0) = 0$,
(ii) $\phi(1,\ldots,1) = 1$,
(iii) ϕ is increasing in each of its arguments.

The intent here is to assume that the components of the system have been labeled by the numbers $1, 2, \ldots, n$, and for the ith argument x_i to indicate whether the ith component of the system is functioning ($x_i = 1$) or failed ($x_i = 0$). The value of ϕ indicates in the same way whether the system is functioning or failed.

A.2. Notation. For any vector $\boldsymbol{x} = (x_1,\ldots,x_n)$, let $(0_i, \boldsymbol{x})$ be the vector $\boldsymbol{x}$ altered by placement of 0 in the ith place; let $(1_i, \boldsymbol{x})$ be the vector $\boldsymbol{x}$ altered by placement of 1 in the ith place. With this notation, it is easy to see that

$$\phi(\boldsymbol{x}) = x_i \phi(1_i, \boldsymbol{x}) + (1 - x_i)\phi(0_i, \boldsymbol{x}). \tag{1}$$

A. Coherent Systems

Fig. A.1. Diagram of a series system

This representation is frequently used in proving results about coherent structure functions using induction of the order n.

It is often required of a coherent system that each component is *relevant*; that is, for each i, there exists $\boldsymbol{x}$ such that $\phi(0_i, \boldsymbol{x}) = 0, \phi(1_i, \boldsymbol{x}) = 1$. That requirement is not imposed in Definition A.1, and indeed it is sometimes useful not to impose even the properties (i) and (ii) of that definition. The reason for this is that the systems $\phi(1_i, \boldsymbol{x})$ and $\phi(0_i, \boldsymbol{x})$ of $n-1$ components need not be coherent when these properties are required. Then proofs by induction on the order of the system using (1) become awkward. The issue is one of semantics rather than substance.

The two simplest coherent systems are the *series system* with

$$\phi(x_1, \ldots, x_n) = \min(x_1, \ldots, x_n) = \Pi_{i=1}^n x_i$$

and the *parallel system* with

$$\phi(x_1, \ldots, x_n) = \max(x_1, \ldots, x_n) = 1 - \Pi_{i=1}^n (1 - x_i).$$

In a human, two eyes or two lungs might be considered to form parallel systems, but two legs would be considered to form a series system because both are needed for walking. It is not hard to show that any coherent system is bounded below by the series system (which works only if all components are working), and is bounded above by the parallel system (which works as long as at least one component works). See Figures A.1 and A.2. For reliability, parallel systems are used because they provide *redundancy* to increase system life length.

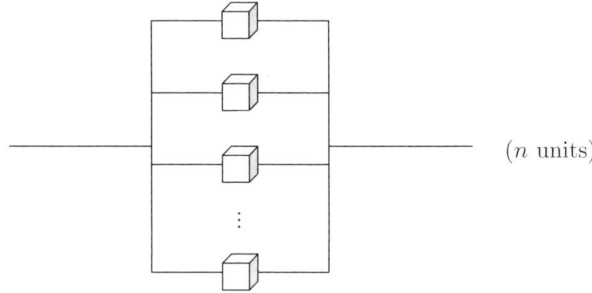

Fig. A.2. Diagram of a parallel system

A more general kind of system is the *k-out-of-n system*, which functions if and only if at least k of the n components function. A series system is an n-out-of-n system and a parallel system is a one-out-of-n-system. A three-engine airplane that can fly on any two of it engines would constitute a 2-out-of-3 system; such a system has the structure function

$$\phi(x_1, x_2, x_3) = 1 - (1 - x_1x_2)(1 - x_1x_3)(1 - x_2x_3)$$
$$= x_1x_2 + x_1x_3 + x_2x_3 - 2x_1x_2x_3. \qquad (2)$$

Note that the structure function is symmetric (invariant under permutations of its arguments) and $\phi(0,0,0) = \phi(1,0,0) = 0, \phi(1,1,0) = \phi(1,1,1) = 1$.

If ϕ is a coherent structure function, then $\phi_D(x_1, \ldots, x_n) = 1 - \phi(1 - x_1, \ldots, 1 - x_n)$ defines another coherent structure function called the *dual* of ϕ. This terminology is appropriate because ϕ is the dual of ϕ_D, that is, $(\phi_D)_D = \phi$. It is intuitively clear, and can be shown that series and parallel systems are dual.

b. Path and Cut Sets

As indicated above, it is assumed that the components of a coherent system have been labeled by the numbers $1, 2, \ldots, n$; consequently, any subset of components can be represented by a subset of the set $\{1, 2, \ldots, n\}$. A subset P is called a *path set* of the coherent structure ϕ if $\phi(\boldsymbol{x}) = 1$ whenever $x_i = 1$, for all $i \in P$. Similarly, a subset C is called a *cut set* of the coherent structure ϕ if $\phi(\boldsymbol{x}) = 0$ whenever $x_i = 0$, for all $i \in C$. The path set P is called a *minimal path set* if no proper subset of P is a path set; a cut set C is called a *minimal cut set* if no proper subset of C is a cut set.

A coherent system can be represented by placing the components of each minimal path in series and then placing these series systems in parallel. Alternatively, the system can be represented by placing the components of each minimal cut in parallel and placing these parallel systems in series. These representations are given by

$$\phi(\boldsymbol{x}) = \max_P \min_{i \in P_j} x_i = \min_C \max_{i \in C_j} x_i, \qquad (3)$$

where P is the set of all path sets and C is the set of all cut sets. In these forms, the structure function (2) of a 2-out-of-3 system

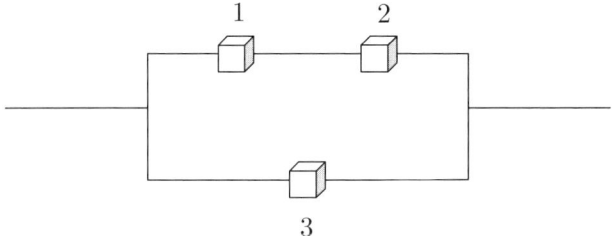

Fig. A.3.a. The diagram of Example A.3 in terms of path sets

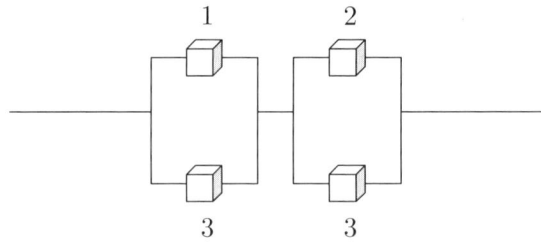

Fig. A.3.b. The diagram of Example A.3 in terms of cut sets

becomes

$$\phi(x) = \max\left[\min(x_1, x_2), \min(x_1, x_3), \min(x_2, x_3)\right]$$
$$= \min\left[\max(x_1, x_2), \max(x_1, x_3), \max(x_2, x_3)\right]. \quad (4)$$

The similarity of these forms is due to the unusual fact that for the 2-out-of-3 system, the minimal path and minimal cut sets coincide.

A.3. Example. Suppose that the minimal path sets are $\{1,2\}$ and $\{3\}$; the minimal cut sets are $\{1,3\}$ and $\{2,3\}$. Equation (3) becomes

$$\phi(x) = \max\{\min[x_1, x_2], x_3\} = \min\{\max[x_1, x_3], \max[x_2, x_3]\}.$$

The first form of this structure function is illustrated in Figure A.3.a. The second form is illustrated in Figure A.3.b, where the component 3 appears twice because it is a member of two minimal cut sets.

c. Reliability Functions

In applications, the state of components of a coherent system are random, and are represented by random variables X_i with Bernoulli distributions. Let $P_i = P\{X_i = 1\} = 1 - P\{X_i = 0\} = EX_i$. When the arguments of a coherent structure function are random, its expected value is a function of the p_i, and this function is important.

A.4. Definition. The *reliability function* h of a coherent system ϕ is defined as

$$h(P_1,\ldots,P_n) = E\phi(X_1,\ldots,X_n), \quad 0 \le p_i \le 1, \quad i = 1, 2, \ldots, n,$$

where $X_1,\ldots,X_n$ are assumed to be independent.

Reliability functions play an important role in the theory of coherent systems. The reliability function of a series system is given by $h(p_1,\ldots,p_n) = \Pi p_i$, the reliability function of a parallel system is $h(p_1,\ldots,p_n) = 1 - \Pi(1 - p_i)$, and the reliability function of a 2-out-of 3 system is $h(p_1, p_2, p_3) = p_1 p_2 + p_1 p_3 + p_2 p_3 - 2 p_1 p_2 p_3$.

When all component life lengths are independent and identically distributed, the reliability function can be written as a function of the common value of the p_i. In this case, the reliability function of a k-out-of-n system is

$$h(p) = \sum_{i=k}^{n} \binom{n}{i} p^i (1-p)^{n-i} = \int_0^p \frac{t^{k-1}(1-t)^{n-k}}{B(k, n-k+1)}\, dt, \quad 0 \le p \le 1, \quad (5)$$

where B is the beta function and the integral comes from 23.B(8).

As an aside, readers familiar with the vector ordering of majorization may note that the reliability function of a series system is Schur-concave and the reliability function of a parallel system is Schur-convex. (See, for example, Marshall and Olkin (1979) for definitions and discussions.) This means that with $\sum p_i$ fixed, a series system has a maximum reliability when all of the p_i are equal but under this condition a parallel system has a minimum reliability. Even without the mathematics to prove these results, they may be intuitively reasonable.

Suppose that $X_i = 0$ or 1, as the random variable $T_i \le t$ or $T_i > t$, i.e., T_i is below or above the threshold t; suppose also that the T_i are independent with common distribution F. Then $\bar{H}_k(t) = h(\bar{F}(t))$ is the survival function of the $(n-k+1)$th order statistic of a random sample of size n from F. Thus, k-out-of-n systems have interest outside the realm of reliability theory.

The following lemma is restated with a different notation as Proposition A.9. In its restated form, it provides the basis for the proof of Theorem B.8.

A.5. Lemma (Esary, Marshall and Proschan, 1970). If h is the reliability function of a coherent system, then

$$h(p_1^\theta, \ldots, p_n^\theta) \geq [h(p_1, \ldots, p_n)]^\theta, \quad 0 \leq \theta \leq 1. \tag{6}$$

To prove this lemma some simplifying notation is useful. For compactness, write

$$\mathbf{p}^\theta = (p_1^\theta, \ldots, p_n^\theta) \quad \text{and in case} \quad \theta = 1, \quad \mathbf{p} = (p_1, \ldots, p_n).$$

Proof of Lemma A.5. (Barlow and Proschan, 1975). To prove (6) by induction on n it is convenient to prove the result for the class of reliability functions of coherent systems augmented by the functions identically 0 or 1.

The lemma is trivially true for $n = 1$, because then $h(p)$ is either identically 0, identically 1, or identically p. Now, assume that the result is true for the augmented class of order $n - 1$. This means that

$$h(1_n, \mathbf{p}^\theta) \geq [h(1_n, \mathbf{p})]^\theta \quad \text{and} \quad h(0_n, \mathbf{p}^\theta) \geq [h(0_n, \mathbf{p})]^\theta. \tag{7}$$

But from (1), it follows that

$$h(\mathbf{p}^\theta) = p_n^\theta h(1_n, \mathbf{p}^\theta) + (1 - p_n^\theta) h(0_n, \mathbf{p}^\theta). \tag{8}$$

Together, (7) and (8) yield

$$h(\mathbf{p}^\theta) \geq p_n^\theta [h(1_n, \mathbf{p})]^\theta + (1 - p_n^\theta)[h(0_n, \mathbf{p})]^\theta. \tag{9}$$

It remains to show that the right-hand side of (9) is greater than or equal to

$$[h(p_1, \ldots, p_n)]^\theta = [p_n h(1_n, \mathbf{p}) + (1 - p_n) h(0_n, \mathbf{p})]^\theta, \quad 0 \leq \theta \leq 1.$$

With the notation $h(1_n, \mathbf{p}) = x, h(0_n, \mathbf{p}) = y$, this means that it is necessary to show that for $0 \leq p_n \leq 1$ and $0 \leq \theta \leq 1$,

$$p_n^\theta x^\theta + (1 - p_n^\theta) y^\theta - (p_n x + (1 - p_n) y)^\theta \geq 0, \quad x \geq y \geq 0. \tag{10}$$

Fix y and regard the left-hand side of (10) as a function $g(x)$ of x. Note that $g(y) = 0$, and that because $\theta \leq 1$, $g'(x) \geq 0$ when $x \geq y$. □

d. Consideration of Time: Coherent Life Functions

In the discussion of coherent systems, the concept of time was introduced for identically distributed components to show the connection between k-out-of-n systems and order statistics. More generally, it is convenient to introduce the notion of "performance processes" for both components and systems.

Let $X(t) = 0$ or 1 according to whether the device is failed or functioning at time t. If there exists a time T such that $X(t) = 1$, for $t < T$, and $X(t) = 0$, for $t > T$, then the *life length* T of the device is well defined, and the device does not operate on an intermittent basis. The definition of the performance process $X(T)$ at the time T of failure is arbitrary, but when the process is chosen to be right continuous, then

$$P\{X(t) = 1\} = \bar{F}(t),$$

where $\bar{F}$ is the survival function of T.

If $X_i(t)$, $i = 1, \ldots, n$, are the performance processes of the components of a coherent system of order n, then $X(t) = \phi(X_1(t), \ldots, X_n(t))$ is the performance process of the system. If component i has the life T_i, $i = 1, \ldots, n$ (the performance processes X_i are decreasing), then because ϕ is monotone, $X(t)$ is decreasing. This means that the system has a life, say T. When the life lengths are independent, the survival function $\bar{F}$ of T is given in terms of the survival functions $\bar{F}_i$ of the component life lengths T_i by the formula

$$\bar{F}(t) = h(\bar{F}_1(t), \ldots, \bar{F}_n(t)). \tag{11}$$

When each component has a well-defined life, there exists a function τ such that $T = \tau(T_1, \ldots, T_n)$. This result is noted by Esary and Marshall (1964), who also note that converses are false; it is possible that a coherent system has a life even though none of the components has a life. For example, a 2-out-of-3 system has a life when components operate intermittently for a period of time, just so long as components remain in a failed state once the system first fails. Moreover, a system and its components can have lives even though the system is not coherent. For example, the system $\phi(x_1, x_2, x_3) = \max[x_1(1 - x_2), x_3]$ is not coherent because $\phi(1,0,0) = 1 > \phi(1,1,0) = 0$, but it has a life if the third component always lives longer than the second.

A.6. Definition. The *life function* τ of a coherent system gives the life length of the system as a function of the component lives. More precisely, τ is a nonnegative function defined on $[0,\infty)^n$ with the property that $t = \tau(t_1,\ldots,t_n)$ is the life length of the coherent system of order n when $t_1,\ldots,t_n$ are the life lengths of the components.

It is apparent that a coherent life function τ is an extension of the underlying structure function ϕ, i.e.,

$$\tau(x_1,\ldots,x_n) = \phi(x_1,\ldots,x_n), \quad \text{each } x_i = 0 \text{ or } 1.$$

Of course, reliability functions also are extensions of the structure function, but of a different kind.

Coherent life functions have been characterized and studied by Esary and Marshall (1970). They show that a coherent life function τ has the following properties:

(i) τ is increasing,

(ii) τ is homogeneous, i.e., $\tau(ct_1,\ldots,ct_n) = c\tau(t_1,\ldots,t_n)$ for all $c \geq 0$ and all $(t_1,\ldots,t_n)$ in $[0,\infty)^n$,

(iii) $\tau(t_1 + \delta,\ldots,t_n + \delta) = \tau(t_1,\ldots,t_n) + \delta$ for all $\delta \geq 0$,

(iv) $\tau(\delta,\ldots,\delta) = \delta$,

(v) τ is continuous,

(vi) If $\phi(x) = \max_P \min_{i \in P_j} x_i = \min_C \max_{i \in C_j} x_i$ is the minimal path and minimal cut representations (3) of the coherent structure, then

$$\tau(t_1,\ldots,t_n) = \max_P \min_{i \in P_j} t_i = \min_C \max_{i \in C_j} t_i. \qquad (12)$$

It is a well-known engineering principle that parallel redundancy at the component level is better than parallel redundancy at the system level. That is, if all components of a system are available in duplicate, it is better to put these component pairs in parallel than it is to build two identical systems and place the systems in parallel. As noted by Boland and El-Neweihi (1995), this principle regarding parallel redundancy does not necessarily apply to spare (standby) redundancy; Example A.7 provides an illustration of the fact that keeping a spare for each component to be used for replacement upon failure may not be as good as building a spare system with the spare components for replacement upon system failure.

With parallel redundancy, the device and spare are placed in parallel, and the engineering principle says that

$$\tau(\max(t_1, u_1), \ldots, \max(t_n, u_n)) \geq \max(\tau(t_1, \ldots, t_n), \tau(u_1, \ldots, u_n))$$
for all $t_i, u_i \geq 0$. (13)

This fact can be verified by noticing that because τ is increasing,

$$\tau(\max(t_1, u_1), \ldots, \max(t_n, u_n)) \geq \tau(t_1, \ldots, t_n)$$

and similarly,

$$\tau(\max(t_1, u_1), \ldots, \max(t_n, u_n)) \geq \tau(u_1, \ldots, u_n).$$

Boland and El-Neweihi (1995) note that because of (13),

$$\tau(\max(T_1, U_1), \ldots, \max(T_n, U_n)) \geq \max(\tau(T_1, \ldots, T_n), \tau(U_1, \ldots, U_n))$$

whatever the (random) life lengths T_i, U_i are, and consequently, the pointwise order can be replaced by stochastic order. But they show that the hazard rate order need not hold.

As noted above, spare (standby) redundancy at the component level may not be as good as spare redundancy at the system level. Intuitively, the reason for this is that, except for series systems, it may not be necessary for a component to be immediately replaced upon failure in order to insure that the system continues to function. But when the component is replaced, the replacement is immediately subject to failure, and if the replacement component fails before it is needed to maintain the system in a functioning state, then the replacement component has served no useful purpose.

A.7. Example. Consider a system consisting of two components in parallel. Denote the component life lengths by X_1 and X_2. Suppose that spares for each component are available, with respective life lengths Y_1 and Y_2. With standby redundancy at the component level, the system life length is given by

$$U = \max(X_1 + Y_1, X_2 + Y_2).$$

With the standby redundancy at the system level, the system life length is

$$V = \max(X_1, X_2) + \max(Y_1, Y_2).$$

Clearly, $V \geq X_1 + Y_1$ and $V \geq X_2 + Y_2$; thus $V \geq U$. It can be shown that the strict inequality $V > U$ holds unless the vectors (X_1, Y_1) and (X_2, Y_2) are similarly ordered, that is, unless

$$(X_1 - Y_1)(X_2 - Y_2) \geq 0$$

with probability one.

For additional discussions regarding comparisons of coherent systems, see Kochar, Mukerjee and Samaniego (1999) and the references contained there.

e. Hazard Transforms of Coherent Systems

Another useful tool in working with coherent systems is the hazard transform, which gives the hazard rate of the system in terms of the hazard transform of the components, still under the assumption that the component life lengths are independent.

A.8. Definition. The function η which gives the system hazard function in terms of the component hazard functions is called the *hazard transform of the coherent system*. More precisely, η is defined by

$$\eta(\rho_1, \ldots, \rho_n) = -\log h(e^{-\rho_1}, \ldots, e^{-\rho_n}), \quad 0 \leq \rho_i \leq \infty.$$

With the notation $\boldsymbol{\rho} = (\rho_1, \ldots, \rho_n)$, Lemma A.5 can be restated compactly in terms of the hazard transform.

A.9. Proposition (Esary, Marshall and Proschan, 1970). The hazard transform η of a coherent system is starshaped on its extended domain; i.e.,

$$\eta(a\boldsymbol{\rho}) \leq a\eta(\boldsymbol{\rho}) \quad \text{for } 0 \leq a \leq 1 \quad \text{and} \quad 0 \leq \rho_i \leq \infty, \quad i = 1, 2, \ldots, n. \tag{14}$$

The following proposition states that the hazard transform of a coherent system is superadditive (see Proposition 21.A.10). According to Proposition 21.A.11, in one dimension a starshaped function must be superadditive. However, this result does not extend to higher dimensions, so the following proposition does not follow from Proposition A.9. Although it may be somewhat obscure in meaning, it is an important tool in proving Proposition C.10.

A.10. Proposition (Esary, Marshall and Proschan, 1970). The hazard transform η of a coherent system is superadditive; i.e.,

$$\eta(\boldsymbol{u}) + \eta(\boldsymbol{v}) \geq \eta(\boldsymbol{u} + \boldsymbol{v}), \tag{15}$$

for all $\boldsymbol{u} = (u_1, \ldots, u_n), \boldsymbol{v} = (v_1, \ldots, v_n)$ such that $u_i, v_i \geq 0$, $i = 1, 2, \ldots, n$.

Proof. Let ϕ be the structure function of a coherent system, h be its reliability function, and η be its hazard transform. Make use of the notation $\mathbf{x} = (x_1, x_2, \ldots, x_n), \mathbf{y} = (y_1, y_2, \ldots, y_n)$, and $\mathbf{x} \times \mathbf{y} = (x_1 y_1, x_2 y_2, \ldots, x_n y_n)$. (For matrices, $\mathbf{x} \times \mathbf{y}$ is called the Schur or Hadamard product, and is sometimes denoted $\mathbf{x} \circ \mathbf{y}$.) Because ϕ is increasing and takes only the values 0 and 1, it follows that

$$\phi(\mathbf{x} \times \mathbf{y}) \leq \phi(\mathbf{x})\phi(\mathbf{y}). \tag{16}$$

Next, note that

$$h(\mathbf{p} \times \mathbf{q}) \leq h(\mathbf{p})h(\mathbf{q}), \tag{17}$$

where $\mathbf{p} = (p_1, p_2, \ldots, p_n)$ and $\mathbf{q} = (q_1, q_2, \ldots, q_n)$. Inequality (17) follows from (16) and the fact that

$$h(\mathbf{p})h(\mathbf{q}) - h(\mathbf{p} \times \mathbf{q}) = \sum_{\mathbf{x},\mathbf{y}} [\phi(\mathbf{x})\phi(\mathbf{y}) - \phi(\mathbf{x} \times \mathbf{y})] P\{X = \mathbf{x}\} P\{Y = \mathbf{y}\},$$

where the vectors $X = (X_1, \ldots, X_n)$, $Y = (Y_1, \ldots, Y_n)$ have independent binary random components such that $P\{X_k = x_i\} = p_i$ and $P\{Y_k = y_i\} = q_i$, $i, k = 1, \ldots, n$. But (17) is just a restatement of (15), as can be seen from Definition A.8. □

f. Wearout

The notion of an increasing hazard rate was initially introduced primarily because of its intuitive appeal as a mathematical representation of "wearout." More about this can be found in Section B.

A.11. Example. Let X_1 and X_2 be independent exponentially distributed life lengths having respective parameters λ_1 and λ_2. The parallel system with components X_1 and X_2 has survival function

$$\bar{F}(x) = e^{-\lambda_1 x} + e^{-\lambda_2 x} - e^{-(\lambda_1+\lambda_2)x}$$

and hazard rate r given by

$$r(x) = [\lambda_1 e^{\lambda_2 x} + \lambda_2 e^{\lambda_1 x} - 1]/[e^{\lambda_2 x} + e^{\lambda_1 x} - 1]. \tag{18}$$

This is a repeat of Example 4.D.10, because the life of a parallel system is the maximum of the component life lengths. It can be verified by differentiation that if $\lambda_1 = \lambda_2 = \lambda$ (the components are identically distributed), then the hazard rate (18) is increasing in x. But if $\lambda_1 \neq \lambda_2$, then the hazard rate (18) is not increasing in x, for example, when $\lambda_1 > \lambda_2 > 0$ and x is large (see Figure 4.D.6).

More generally, suppose that the two components in parallel have life lengths X and Y, and distributions F_X and F_Y. Then, the system life $Z = \max(X, Y)$ has the survival function $\bar{F}_Z(x) = 1 - F_X(x)F_Y(x)$ and hazard rate

$$r_Z(x) = r_X(x)\frac{\bar{F}_X(x)F_Y(x)}{1 - F_X(x)F_Y(x)}$$
$$+ r_Y(x)\frac{F_X(x)\bar{F}_Y(x)}{1 - F_X(x)F_Y(x)} + 0\frac{\bar{F}_X(x)\bar{F}_Y(x)}{1 - F_X(x)F_Y(x)}. \tag{19}$$

Here, the coefficients of $r_X(x), r_Y(x)$, and 0 add to 1. They represent, respectively, the conditional probabilities, given that the system is alive at time x, that the system consists only of the first component, only of the second component, or of both components in parallel. The behavior of the hazard rates in Figure 4.D.6 stems from the fact that these probabilities change with x. It can be verified that as x tends to ∞, the coefficient of 0 tends to 0, the coefficient of $r_X(x)$ tends to $\lim_{x \to \infty} \bar{F}_X(x)/[\bar{F}_X(x) + \bar{F}_Y(x)]$, and the coefficient of $r_Y(x)$ tends to $\lim_{x \to \infty} \bar{F}_Y(x)/[\bar{F}_X(x) + \bar{F}_Y(x)]$.

In case X and Y are identically distributed, say with distribution F and hazard rate r, then (19) becomes

$$r_Z(x) = 2r(x)\frac{F(x)}{1 + F(x)}; \tag{20}$$

if r is increasing, then r_Z is increasing. A more general statement can be made.

A.12. Proposition. If all components of a k-out-of-n system are identically distributed, independent, and have an increasing hazard rate, then the system has an increasing hazard rate.

Proof. Denote the survival function of the system by $\bar{H}$. Monotonicity of the hazard rate is demonstrated here by showing that the derivative of $\log \bar{H}(t)$ is decreasing. To this end, it is convenient to first show that the reliability function h of a k-out-of-n system has the property that $ph'(p)/h(p)$ is decreasing in p. Use the second form of equation A(5) to write

$$\frac{ph'(p)}{h(p)} = \left[\int_0^1 u^{k-1} \left(\frac{1-pu}{1-p} \right)^{n-k} du \right]^{-1},$$

and differentiate the integrand on the right with respect to p. Next, let F be the common life distribution of the components of a k-out-of-n system and H be the distribution of the system life length. Then,

$$\frac{d \log \bar{H}(t)}{dt} = \frac{d \log h(\bar{F}(t))}{dt} = -\left[\frac{\bar{F}(t) h'(\bar{F}(t))}{h(\bar{F}(t))} \right] \frac{f(t)}{\bar{F}(t)}. \tag{21}$$

The first factor on the right of (21) is increasing because both $\bar{F}(t)$ and $ph'(p)/h(p)$ are decreasing. The second factor on the right of (21) is also increasing because $\bar{F}$ is log concave. Thus, (21) is decreasing in t because it is the negative of a product of two nonnegative increasing functions. □

Proposition A.12 has been given by Barlow and Proschan (1975) and further discussed by Samaniego (1985).

A.12.a. Remark. Example A.11 shows that the class of distributions with increasing hazard rate is not closed under the formation of coherent systems. That is, a coherent system can have components with independent life lengths that are all IHR, but that does not mean that the system life length is IHR. Proposition A.12 has been given by Barlow and Proschan (1975). Special conditions under which a system with IHR components is IHR are given by Samaniego (1985).

g. More About Ordering Coherent Systems

Because reliability functions are increasing, it follows that if $F_i \leq_{st} G_i$, $i = 1, 2, \ldots, n$, then

$$h(\bar{F}_1, \ldots, \bar{F}_n) \leq_{st} h(\bar{G}_1, \ldots, \bar{G}_n);$$

that is, stochastic order is preserved under the formation of coherent systems. The same cannot be said about the hazard rate order. However, hazard rate order is preserved under the formation of series systems because the hazard rate of a series system is the sum of the hazard rates of its components. Even two parallel systems with corresponding components ordered by the hazard rate order need not be hazard rate ordered. On the other hand, k-out-of-n systems are hazard rate ordered if all components are identically distributed. See Boland, El-Neweihi and Proschan (1994) and Nanda and Shaked (2001) for discussions of these issues.

B. Monotone Hazard Rate Averages

A class of distributions is said to be closed under the formation of coherent systems if, when all components of the system have life lengths belonging to the class, so does the system life. As illustrated by Example A.11, the class of IHR distributions is not closed under the formation of coherent systems. What is the smallest class of distributions that contains the IHR distributions and is closed under the formation of coherent systems? To answer this question, Birnbaum, Esary, and Marshall (1966) introduced the class of distributions with increasing hazard rate average. This class is closed under the formation of coherent systems; it is the smallest such class that contains the exponential distributions and is closed under weak limits.

B.1. Definition. A distribution F satisfying $F(0) = 0$ is said to have an *increasing hazard rate average* (IHRA) if the hazard function $R = -\log \bar{F}$ of F (Definition 1.B.3) satisfies

$$R(t)/t \text{ is increasing in } t > 0. \tag{1a}$$

Similarly, F is said to have a *decreasing hazard rate average* (DHRA) if

$$R(t)/t \text{ is decreasing in } t > 0. \tag{1b}$$

152   5. Nonparametric Families: Origins in Reliability Theory

The terminology of Definition B.1 comes from the fact that if a hazard rate r exists, then

$$\frac{R(t)}{t} = \frac{1}{t}\int_0^t r(z)\,dz$$

is the average of r over the interval $[0, t]$.

Condition (1a) is the condition that R is starshaped. Starshaped functions are defined in Definition 19.A.8, and the geometric meaning of (1a) is discussed in Chapter 21.

B.2. Example. If

$$\bar{F}(x) = e^{-rx}, \quad 0 \le x < 1,$$
$$= e^{-sx}, \quad x \ge 1,$$

then F is IHRA whenever $r \le s$. When $r < s$, this piecewise exponential distribution has a discrete part, with mass at 1, and F is not IHR.

Example A.11 exhibits another IHRA distribution that is not IHR, an example where the hazard rate is inverted bathtub shaped (unimodal). Examples of parametric families of IHRA distributions that are not IHR are mostly contrived. This may be partially due to the fact that the IHRA condition is often difficult to check analytically, especially in examples where the survival function does not take a simple form.

a. Characterizations of IHRA Distributions

B.3. Proposition. The following are equivalent:

(i) F is IHRA.

(ii) $\bar{F}$ can cross the survival function of any exponential distribution at most once, and only from above.

(iii) $[\bar{F}(x)]^{1/x}$ is decreasing in $x \ge 0$, or equivalently,

$$\bar{F}(ax) \ge [\bar{F}(x)]^a \quad \text{for all } a \text{ in } [0,1] \text{ and } x \ge 0, \tag{2a}$$

that is,

$$R(ax) \le aR(x) \quad \text{for all } a \text{ in } [0,1] \text{ and } x \ge 0, \tag{2b}$$

where R is the hazard function of F.

(iv) F is less than G in the star ordering $\leq_*$ of Definition 2.C.10, where G is an exponential distribution (that is, $\bar{G}^{-1}\bar{F}$ is starshaped, where $\bar{G}(x) = e^{-x}, x \geq 0$).

Proof. Condition (iii) is clearly a restatement of the definition. Condition (ii) is most easily seen by restating the result in terms of the hazard function: The hazard function of an IHRA distribution can cross a ray emanating from the origin at most once and only from below. Verification of (iv) is similar to the verification of Proposition 4.C.1.f. □

B.3.a. Proposition. If F has a density, the following are equivalent:

(i) F is IHRA,
(ii) $r(x) \geq R(x)/x$, $x > 0$.
(iii) $\bar{F}(x) \geq e^{-xr(x)}$, $x \geq 0$.

Proof. Condition (ii) comes from (1a) and is the condition that $R(x)/x$ has a nonnegative derivative. Condition (iii) comes from (ii) by exponentiation, using the fact that $R(x) = -\log \bar{F}(x)$. □

The following fundamental lemma gives yet another characterization of IHRA distributions. It has a rather technical proof requiring some theory of the Lebesgue integral; it is used in the proof of Proposition B.9. However, the meaning of Proposition B.9 is easy to understand without the following lemma and its proof.

B.4. Lemma (Block and Savits, 1976). The distribution F of the random variable X is IHRA if and only if for all nonnegative increasing functions ϕ and $a \in (0, 1)$,

$$E\phi(X) \leq (E\phi(X/a))^{1/a}, \qquad (3)$$

that is,

$$\int \phi(z)\, dF(z) \leq \left\{ \int [\phi(z/a)]^a\, dF(z) \right\}^{1/a}.$$

Proof. First, suppose that (3) holds and take ϕ to be the indicator function $I_{(x,\infty)}$ of the set (x, ∞), that is, $\phi(z) = 1$ if $z > x$, and $\phi(z) = 0$ otherwise. Then (3) reduces to (2a), so (3) implies that F is IHRA. Next, suppose that F is IHRA. It follows directly from (2a) that for $0 < a < 1$,

$$\int I_{(x,\infty)}(z)\, dF(z) \leq \left\{ \int [I_{(x,\infty)}(z/a)]^a\, dF(z) \right\}^{1/a}. \qquad (4)$$

Now, let $\phi(z) = \sum_{i=1}^{n} c_i I_{(x_i,\infty)}(z)$, where each $c_i \geq 0$, and get from (4) that

$$\int \phi(z)\, dF(z) \leq \sum_{i=1}^{n} \left\{ \int [c_i I_{(x_i,\infty)}(z/a)]^a\, dF(z) \right\}^{1/a}.$$

But according to Minkowski's inequality for $\alpha \leq 1$ (see, e.g., Marshall and Olkin, 1979, 16.D.1.g, p. 460),

$$\sum_{i=1}^{n} \left\{ \int [c_i I_{(x_i,\infty)}(z/a)]^a\, dF(z) \right\}^{1/a} \leq \left\{ \int \left[\sum_{i=1}^{n} [c_i I_{(x_i,\infty)}(z/a)]^a \right] dF(z) \right\}^{1/\alpha},$$

and this proves the lemma for finite positive combinations of indicator functions. The proof is completed using Lebesgue's monotone convergence theorem (see 24.B.2). □

Another characterization of distributions with monotone hazard rate average has been given by Rojo (1995); this characterization involves slowly varying functions and is beyond the scope of this book.

b. Basic Properties

If F has an IHRA distribution, then F has an absolutely continuous part because R, and hence F, must be strictly increasing if (1a) is to hold. But F can also have a discrete part. In fact, IHRA distributions can have any countable number of discontinuities.

B.5. Example. If $0 \leq x_1 < x_2 < \cdots$, $\lambda_i > 0, i = 1, 2, \ldots$, and $\bar{F}(x) = \exp\{-(\lambda_1 + \lambda_2 + \cdots + \lambda_i)x\}$, for $x_{i-1} \leq x < x_i, i = 1, 2, \ldots$, then each x_i is a point of discontinuity of F, but F is IHRA.

B.6. Proposition. If F is an IHRA distribution, then F has finite moments of all positive finite orders.

Proof. If F is degenerate, the result clearly holds. If F is not degenerate, then there exists t_0 such that $0 < F(t_0) < 1$. Let $\lambda = R(t_0)/t_0$, so that $\bar{F}(t_0) = \exp\{-\lambda t_0\}$. Then from (ii) of Proposition B.3, it follows that $\bar{F}(t) \leq \exp\{-\lambda t\}$ for all $t \geq t_0$. From this, from 1.C(4), and the fact that exponential distributions have finite moments of all orders, it follows that

$$EX^r = r \int_0^\infty \bar{F}(x) x^{r-1}\, dx = r \int_0^{t_0} \bar{F}(x) x^{r-1}\, dx + r \int_{t_0}^\infty \bar{F}(x) x^{r-1}\, dx$$

$$\leq r \int_0^{t_0} \bar{F}(x) x^{r-1}\, dx + r \int_{t_0}^\infty e^{-\lambda x} x^{r-1}\, dx < \infty. \qquad \square$$

B.6.a. Proposition. If F is an IHRA distribution with hazard rate r, then

$$\liminf_{x \to \infty} r(x) > 0.$$

Proof. This result follows directly from (ii) of Proposition B.3.a and the fact that $R(x)/x$ is increasing. $\square$

In case F has a density (and consequently a hazard rate), Proposition B.6 follows from Proposition B.6.a and Proposition 20.B.6. Note that the proof of Proposition 20.B.6 is not unlike the proof of Proposition B.6.

The following proposition can be useful in verifying the IHRA property because total time on test data plots tend to look like total time on test transforms.

B.7. Proposition (Barlow and Campo, 1975). If F is an IHRA [DHRA] distribution, then the normalized total time on test transform K_F^{-1} of 1.I(10) has the property that $[K_F^{-1}(p)]/p$ is decreasing [increasing] in $p \in [0, 1]$. Moreover, $K_F^{-1}(p) \geq p$.

c. Preservation Properties

As already mentioned in Remark A.12.a, a mathematical characterization or definition of "wear out" is somewhat elusive. One might require of such a definition the property that if the components of a coherent system "wear out" in the adopted sense, then the system would "wear out" in the same sense. Imposition of this property conflicts with the commonly assumed idea that an increasing hazard represents "wear out." To face this disconcerting fact, first adopt the view that exponential distributions represent the case of no "wear"; this is generally accepted and hard to escape because of 1.F(4), which says that for an exponential F, $\bar{F}(x+t)/\bar{F}(t) = \bar{F}(x)$. This question then presents itself: What distributions can possibly arise as life distributions of coherent systems with exponentially distributed components? According to the following theorem, the answer to this question is essentially the distributions with increasing hazard rate average.

B.8. Theorem (Birnbaum, Esary and Marshall, 1966). *The class of IHRA distributions is closed under the formation of coherent systems. Moreover, it is the smallest such class containing the exponential distributions that is closed both under formation of coherent systems and limits in distribution.*

The proof of this result given by Birnbaum, Esary and Marshall (1966) is not particularly simple. Esary, Marshall and Proschan (1970) gave a new proof of the closure of the IHRA class which utilizes Proposition A.10, but their proof of Proposition A.10 was not simple. Fortunately, that proof was simplified by Ross (1979). A generalization of Theorem B.8 was obtained by Marshall (1994).

Proof of Closure. Let R be the hazard function of a coherent system with hazard transform η, and suppose that the components have hazard functions $R_i, i = 1, 2, \ldots, n$. To prove closure, it is sufficient to show that R satisfies (2b). Write $\mathbf{R}(x) = (R_1(x), \ldots, R_n(x))$. Because η is increasing and the R_i are starshaped,

$$\mathbf{R}(ax) = \eta(\mathbf{R}(ax)) \leq \eta(a\mathbf{R}(x)), \quad 0 \leq a \leq 1. \tag{5}$$

By Proposition A.8, η is starshaped, so again, for $0 \leq a \leq 1$,

$$\eta(a\mathbf{R}(x)) \leq a\eta(\mathbf{R}(x)) = a\mathbf{R}(x). \tag{6}$$

By combining (5) and (6), (2b) is obtained. □

To complete the proof of Theorem B.8, it remains to show that the class of IHRA distributions is the smallest class of life distributions containing the exponential distributions, which is also closed under limits in distribution. The idea of the proof, given in detail by Birnbaum, Esary and Marshall (1966), is to show first that degenerate distributions can be obtained as limits in distribution of coherent systems with exponentially distributed components. The next step is to show that IHRA distributions can be approximated by distributions having the form of Example B.5 (see Figure B.1). Such distributions arise as life distributions of coherent systems

B. Monotone Hazard Rate Averages

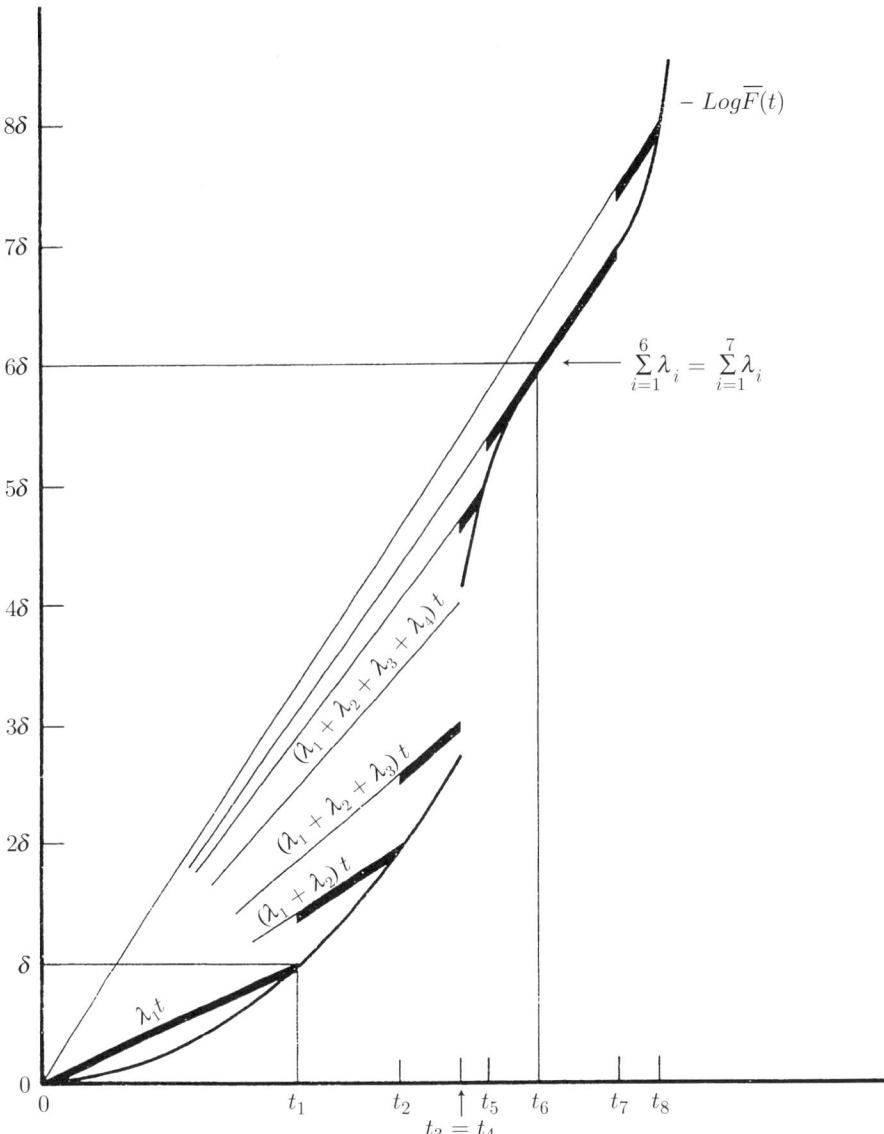

Fig. B.1. Piecewise exponential approximation to an IHRA survival probability

with components having exponential or degenerate distributions (see Figure B.2).

Is the class of IHRA distributions closed under convolutions? This question remained unsettled at the time the classic book of Barlow

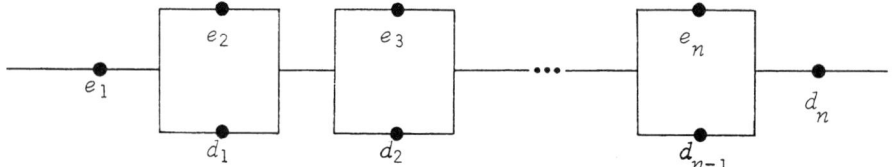

Fig. B.2. Coherent system with piecewise exponential distribution

and Proschan (1975) was first published, and it is posed as an open problem on page 101 of that book. A short time later, separate and quite different proofs were found independently by Block and Savits (1976). One of the proofs is given below.

B.9. Proposition (Block and Savits, 1976). *The class of IHRA distributions is closed under convolutions.*

Proof. Let F and G be IHRA distributions and $H = F * G$ be their convolution. For $0 < a < 1$ and any nonnegative increasing function ϕ, it follows from Lemma B.4 that

$$\int \phi(z)\, dH(z) = \iint \phi(x+y)\, dF(x)\, dG(y)$$
$$\leq \int \left\{ \int \left[\phi\left(\frac{x+y}{a}\right)\right]^a dF(x) \right\}^{1/a} dG(y).$$

But the inner integral on the right-hand side is an increasing function of y, and so another application of Lemma B.4 shows that

$$\int \left\{ \int \left[\phi\left(\frac{x+y}{a}\right)\right]^a dF(x) \right\}^{1/a} dG(y)$$
$$\leq \left\{ \iint \left[\phi\left(\frac{x+y}{a}\right)\right]^a dF(x)\, dG(y) \right\}^{1/a} = \left\{ \int [\phi(z/a)]^a\, dH(z) \right\}^{1/a}.$$

A combination of these inequalities yields

$$\int \phi(z)\, dH(z) \leq \left\{ \int [\phi(z/a)]^a\, dH(z) \right\}^{1/\alpha},$$

which, again by Lemma B.4, shows that H is IHRA. □

B.10. Proposition. *Suppose that the residual life distribution F_t of F has an IHRA survival function for all $t \geq 0$. Then F is IHR.*

This proposition may be a disappointment, because if the IHRA property represents "wearout," then intuition would expect that the residual life distribution of an IHRA distribution to be IHRA, but this is not the case. This apparent conflict appears to have no resolution.

B.11. Observation. The class of IHRA distributions is not closed under mixtures. This is a consequence of the fact that mixtures of exponential distributions are DHR.

B.12. Proposition (Block, Li and Savits, 2003a). Let $\{F(x \mid \theta), \theta \geq 0\}$ be a family of distribution functions such that $\bar{F}(x \mid \theta)$ is increasing in θ for each $t \geq 0$. Suppose further that

$$\bar{F}(ax \mid a\theta) \geq [\bar{F}(x \mid \theta)]^a \quad \text{for all } a \in (0,1),\ x \geq 0,\ \theta \geq 0. \tag{7}$$

If G is an IHRA distribution, then the mixture

$$\bar{H}(x) = \int_0^\infty \bar{F}(x \mid \theta)\, dG(\theta)$$

is IHRA. Conversely, if H is IHRA whenever (7) is satisfied, then G is IHRA.

Proof. First make use of (7) and then apply Lemma B.4 with θ in place of z and with $\phi(\theta) = \bar{F}(x \mid \theta)$. In this way, it follows that

$$\bar{H}(ax) = \int_0^\infty \bar{F}(ax \mid \theta)\, dG(\theta) \geq \int_0^\infty [\bar{F}(x \mid \theta/a)]^a\, dG(\theta)$$
$$\geq \left[\int_0^\infty \bar{F}(x \mid \theta)\, dG(\theta)\right]^a = [\bar{H}(x)]^a.$$

According to Lemma B.4, this means that F is IHRA.

To prove the converse, in (7) take F to be the distribution degenerate at θ, that is, $F(x) = 0, x < \theta$, and $F(x) = 1, x \geq \theta$. □

B.12.a. Example. If F is an IHRA distribution, then by (2a), $\bar{F}(ax) \geq [\bar{F}(x)]^a$, for all a in $[0,1]$ and $x \geq 0$. If $\bar{F}(x \mid \theta) = \bar{F}(x - \theta)$, then it is straightforward to verify that for this parametric family, (7) is satisfied.

B.13. Alternative proof of B.12. If $\bar{F}(x \mid \theta) = \bar{F}(x - \theta)$ for some IHRA distribution F, then according to Example B.12.a, (7) is satisfied. But in this case, the distribution H of Proposition B.12 is the convolution of F and G, and is IHRA by Proposition B.9. □

d. Decreasing Hazard Rate Average

Because the importance of DHRA distributions has never become clear, only a limited discussion of these distributions is offered here. The following is an analog of Proposition B.3.

B.14. Proposition. The following are equivalent:

(i) F is DHRA.

(ii) $\bar{F}$ can cross the survival function of any exponential distribution at most once, and only from below.

(iii) $[\bar{F}(x)]^{1/x}$ is increasing in $x \geq 0$, or equivalently,

$$\bar{F}(ax) \leq [\bar{F}(x)]^a \quad \text{for all } a \text{ in } [0,1] \text{ and } x \geq 0, \tag{8a}$$

that is,

$$R(ax) \geq aR(x) \quad \text{for all } a \text{ in } [0,1] \text{ and } x \geq 0, \tag{8b}$$

where R is the hazard function of F.

(iv) F is greater than G in the star ordering $\leq_*$ of Definition 2.C.7, where G is an exponential distribution (that is, $\bar{F}^{-1}G$ is starshaped, where $\bar{G}(x) = e^{-x}, x \geq 0$).

Because the exponential distribution is both IHRA and DHRA, and the class of IHRA distributions is closed under both convolutions and formation of coherent systems, the class of DHRA distributions cannot be closed under these operations. However, the class is closed under the formation of mixtures.

The proof of the following proposition makes use of the concept of a hazard transform of a mixture, which is defined and discussed in Section 3.D.

B.15. Proposition. The class of DHRA distributions is closed under the formation of mixtures.

Proof. According to Proposition 3.D.2, the hazard transform h of a mixture is concave; thus,

$$\eta(\langle aR(x \mid \theta), \theta \in \Theta \rangle) \geq \eta(\langle R(x \mid \theta), \theta \in \Theta \rangle), \quad 0 \leq a \leq 1. \tag{9}$$

By hypothesis,

$$R(ax \mid \theta) \geq aR(x \mid \theta), \quad 0 \leq a \leq 1. \tag{10}$$

Because η is an increasing function, it follows from (9) and (10) that

$$R(ax) = \eta(\langle R(ax \mid \theta), \ \theta \in \Theta \rangle) \geq \eta(\langle aR(x \mid \theta), \ \theta \in \Theta \rangle)$$
$$\geq a\eta(\langle R(x \mid \theta), \ \theta \in \Theta \rangle) = aR(x), \quad 0 \leq a \leq 1.$$

Thus, F is DHRA. □

This proof is given by Barlow and Proschan (1975). Note that it is analogous to the proof of Theorem B.8, but uses the concavity of the hazard transform of a mixture in place of the starshapedness of the hazard transform of a coherent system. Another proof of Proposition B.15 has been given by Badia, Berrade, Campos and Navascués (2001).

C. New Better (Worse) Than Used Distributions

As might be expected from their name, new better than used distributions were introduced in the context of replacement policy theory. To maintain complex systems where in-service failure has serious consequences, components are sometimes replaced by new ones according to some predetermined schedule. Such replacement policies are not necessarily helpful, but depend upon properties of the underlying life distribution; these issues are discussed in Section I.

C.1. Definition. A distribution F such that $F(0) = 0$ is said to be *new better than used* (NBU) if

$$\bar{F}(x+t) \leq \bar{F}(x)\bar{F}(t) \quad \text{for all } x, t \geq 0, \tag{1a}$$

that is,

$$R(x+t) \leq R(x) + R(t) \quad \text{for all } x, t \geq 0. \tag{1b}$$

If the inequalities (1a) and (1b) are reversed, then F is said to be *new worse than used* (NWU).

In case $\bar{F}(t) > 0$, (1a) can be rewritten as

$$\bar{F}(x) \geq \frac{\bar{F}(x+t)}{\bar{F}(t)} = \bar{F}_t(x) \quad \text{for all } x, t \geq 0,$$

that is, a new item has a stochastically greater life than does an unfailed used one (in the stochastic ordering of Definition 2.A.1). This explains the names given to the NBU and NWU properties.

a. Equivalent Conditions

C.2. Proposition (Kirmani and Gupta, 2001). Let F be a distribution such that $F(0) = 0$. Then F is NBU if and only if, in the notation of Proposition 4.C.1.e,

$$\emptyset_t^-(x) \geq \emptyset(x) \quad \text{for all } x, t \geq 0 \,;$$

the odds ratio of the residual life distribution dominates the odds ratio of the underlying distribution. The distribution is NWU if and only if this odds ratio inequality is reversed.

C.3. Proposition. The distribution F is NBU (NWU) if and only if a random variable with distribution F is smaller (larger) than an exponentially distributed random variable in the superadditive order $\leq_{\text{su}}$ of Definition 2.C.15. Here it is clearly enough to take $\lambda = 1$.

C.4. Proposition (Block, Li and Savits, 2003a). The distribution F is NBU if and only if

$$\int g(\alpha x) h(\bar{\alpha} x) \, dF(x) \leq \int g(x) \, dF(x) \int h(x) \, dF(x) \qquad (2)$$

for all nonnegative increasing functions g and h and all $\alpha \in (0, 1)$. As usual, $\bar{\alpha} = 1 - \alpha$.

Proof. First, note that for $\alpha \in (0, 1)$,

$$a + b \leq \max\left(\frac{a}{\alpha}, \frac{b}{\bar{\alpha}}\right); \qquad (3)$$

to verify (3), let $c = a/\alpha, d = b/\bar{\alpha}$, and write (3) in the familiar form $\alpha c + \bar{\alpha} d \leq \max(c, d)$.

Assume that F is NBU. Because of (3) and because F is NBU, it follows that

$$\bar{F}(\max(a/\alpha, b/\bar{\alpha})) \leq \bar{F}(a+b) \leq \bar{F}(a)\bar{F}(b);$$

this is (2) when g and h are the indicator functions $g = I_{(a,\infty)}$ and $h = I_{(b,\infty)}$. It can also be shown, for example, by using the Lebesgue dominated convergence Theorem 24.B.3, that (2) holds when $g = I_{[a,\infty)}$ and/or $h = I_{[b,\infty)}$. Further, it can be shown that (2) holds when g and h are finite positive linear combinations of such indicator functions; by multiplying out the resulting expressions and making appropriate term by term comparisons inequality (2) is obtained. Consequently it follows from the Lebesgue dominated or Lebesgue monotone convergence theorem that (2) holds for all nonnegative increasing functions g and h. □

For another characterization of the class of NBU distributions that involves slowly varying functions, see Rojo (1995).

b. Basic Properties

C.5. Proposition. If F is NBU, then

$$\bar{F}(kt) \leq [\bar{F}(t)]^k \quad \text{for all } k = 1, 2, \ldots, \quad \text{and } t \geq 0, \tag{4a}$$

or equivalently,

$$\bar{F}(s/k) \geq [\bar{F}(s)]^{1/k} \quad \text{for all } k = 1, 2, \ldots, \quad \text{and } s \geq 0. \tag{4b}$$

This proposition is verified by taking $x = t$ in (1a), then iterating, or by using induction. Note the comparison of (4b) with the corresponding inequality B(2a), which holds for IHRA distributions, namely, $\bar{F}(\alpha t) \geq [\bar{F}(t)]^{\alpha}$, for all α in $[0, 1]$ and $t \geq 0$.

If F is NBU, then F need not have an absolutely continuous part; the following example shows that F can be a discrete distribution.

C.6. Example. Suppose that

$$\bar{F}(x) = p^i, \quad i \leq x < i+1, \ i = 0, 1, \ldots, \text{ where } 0 < p < 1.$$

To see that this distribution is NBU, note first that if $i \leq x < i+1$ and $j \leq t < j+1$, then $i+j \leq x+t < i+j+2$. Thus, $\bar{F}(x+t) = p^{i+j}$ or $\bar{F}(x+t) = p^{i+j+1}$, according as $i+j \leq x+t < i+j+1$ or $i+j+1 \leq x+t < i+j+2$. In either case, $\bar{F}(x+t) \leq \bar{F}(x)\bar{F}(t)$ with equality if and only if $i+j \leq x+t < i+j+1$.

C.7. Proposition. If F is NBU, then it has finite moments of all positive orders.

Proof. Choose $t < \infty$ such that $\bar{F}(t) < 1$. Because $\bar{F}$ is monotone and (4a) holds, it follows that if x and k satisfy $kt \leq x < (k+1)t$, then

$$\bar{F}(x) \leq \bar{F}(kt) \leq [\bar{F}(t)]^k \leq [\bar{F}(t)]^{(x/t)-1}. \tag{5}$$

With the use of successive values of k, (5) yields

$$\bar{F}(x) \leq [\bar{F}(t)]^{(x/t)-1}, \quad x \geq t. \tag{6}$$

From 1.C.(4) and (6), it follows that with $\lambda = -\log \bar{F}(t) > 0$,

$$EX^r = r \int_0^\infty \bar{F}(x) x^{r-1} \, dx \leq r \int_0^t x^{r-1} \, dx + r \int_t^\infty x^{r-1} [\bar{F}(t)]^{(x/t)-1} \, dx. \tag{7}$$

With $\lambda = -\log \bar{F}(t) > 0$, the right-hand side of (7) is equal to

$$t^r + rt^r e^\lambda \int_1^\infty u^{r-1} e^{-\lambda u} \, du < t^r + rt^r e^\lambda \int_0^\infty u^{r-1} e^{-\lambda u} \, du$$
$$= t^r [1 + r e^\lambda \Gamma(r)] < \infty. \quad \square$$

Summary. According to Proposition C.7, NBU distributions have finite moments of all orders. This is stronger than Proposition B.6, which states that IHRA distributions have finite moments of all positive orders. In turn, Proposition B.6 is stronger than Proposition 4.B.3 which gives the same conclusion for the still smaller class of IHR distributions.

C.8. Proposition. If F is NBU and has a density, then the hazard rate satisfies the inequality

$$r(x) \geq r(0). \tag{8}$$

The inequality (8) is reversed if F is NWU.

Proof. From (1b), it follows that

$$\frac{R(x+t) - R(x)}{t} \leq \frac{R(t) - R(0)}{t}, \quad x, t > 0.$$

The result follows upon by letting $t \to 0$. $\square$

Proposition C.8 is a rather weak result that is to be contrasted with (ii) of Proposition B.3.a.

Recall from Definition 1.B.16 that for a distribution function F, the odds ratios $Ø^+(x) = \bar{F}(x)/F(x)$ and $Ø^-(x) = F(x)/\bar{F}(x)$ are defined for values of x for which the denominators are positive.

C.9. Proposition (Kirmani and Gupta, 2001). If F is NBU, that is, $\bar{F}(x+t) \leq \bar{F}(x)\bar{F}(t)$, $x, t \geq 0$, then

$$Ø^+(x+t) \leq Ø^+(x)Ø^+(t), \quad x, t \geq 0.$$

Proof. First, note that for numbers a and b between 0 and 1, $a + b \geq 2\sqrt{ab} \geq 2ab$. Consequently, if F is NBU, it follows that $F(x+t) \geq F(x) + F(t) - F(x)F(t) \geq F(x)F(t)$, $x, t \geq 0$, which together with the NBU property yields the result. □

c. Preservation Properties

C.10. Proposition (Esary, Marshall and Proschan, 1970). If each component of a coherent system has an NBU distribution, then the system life distribution is NBU.

Proof. The fact that components have NBU distributions can be translated to the statement that the component hazard functions R_i are superadditive, i.e., they satisfy (1b). The hazard transform of a coherent system is increasing, and by Proposition A.10 it is superadditive. It follows that the system hazard function satisfies

$$R(s+t) = \eta(R_1(s+t), \ldots, R_n(s+t))$$
$$\leq \eta(R_1(s), \ldots, R_n(s)) + \eta(R_1(t), \ldots, R_n(t)) = R(s) + R(t),$$

which is the statement that the system has an NBU distribution. □

C.11. Proposition (Marshall and Proschan, 1972a). If F and G are NBU distributions, then their convolution H is NBU.

Proof. It is required to show that $\bar{H}(x+y) \leq \bar{H}(x)\bar{H}(y)$ for all $x, y \geq 0$, where

$$H(t) = \int_0^t F(t-z)\,dG(z).$$

Note that

$$\bar{H}(t) = 1 - H(t) = \int_0^t dG(z) + \int_t^\infty dG(z) - \int_0^t F(t-z)\,dG(z)$$
$$= \int_0^t \bar{F}(t-z)\,dG(z) + \bar{G}(t).$$

Suppose first that F and G have densities f and g. Write $\bar{H}(x+y) = I_1 + I_2$, where

$$I_1 = \int_0^x \bar{F}(x+y-z)g(z)\,dz \quad \text{and} \quad I_2 = \int_0^\infty \bar{F}(y-z)g(x+z)\,dz.$$

Because F is NBU,

$$I_1 \leq \bar{F}(y)\int_0^x \bar{F}(x-z)g(z)\,dz = \bar{F}(y)[\bar{H}(x) - \bar{G}(x)].$$

With an integration by parts and the fact that G is NBU, it follows that

$$I_2 = \bar{F}(y)\bar{G}(x) + \int_0^\infty \bar{G}(x+z)f(y-z)\,dz \leq \bar{F}(y)\bar{G}(x)$$
$$+ \bar{G}(x)\int_0^\infty \bar{G}(z)f(y-z)\,dz = \bar{F}(y)\bar{G}(x) + \bar{G}(x)[\bar{H}(y) - \bar{F}(y)].$$

Thus,

$$\bar{H}(x+y) \leq \bar{F}(y)\bar{H}(x) + \bar{G}(x)\bar{H}(y) - \bar{G}(x)\bar{F}(y)$$
$$= \bar{H}(x)\bar{H}(y) - [\bar{H}(x) - \bar{G}(x)][\bar{H}(y) - \bar{F}(y)] \leq \bar{H}(x)\bar{H}(y).$$

The case that densities do not exist poses only notational difficulties: Write "$d_z G(x+z)$" in place of "$g(x+z)\,dz$." □

An alternative proof of Proposition C.11 is given in Example C.16, using Proposition C.15.

C.12. Proposition. The residual life distribution F_t of F is NBU for all t if and only if F is IHR. Thus, the residual life distributions of NBU distributions need not be NBU.

This result can be verified in a straightforward way so the proof is omitted. Note that it is a stronger result than Proposition B.10, which concludes that F is IHR from the assumption that F_t is IHRA for all t.

C.13. Proposition. If F is the mixture of NWU distributions, no two of which cross, then F is NWU.

Proof. This result is a consequence of Proposition 3.D.5, the fact that hazard transforms of mixtures are increasing functions, and the fact that NBU distributions are characterized by having hazard functions R satisfying (1b), that is, $R(x+t) \leq R(x) + R(t)$ for all $x, t \geq 0$. □

An example is given by Barlow and Proschan (1975, p. 187) to show that the no-crossing provision of Proposition C.8 cannot be dispensed with.

A consequence of Proposition C.13 is that a mixture of NWU distributions with proportional hazard functions is NWU (for a discussion of proportional hazards, see Section 7.E). Also, if NWU distributions differ only by a scale parameter, it follows that their mixture is NWU.

The following proposition considers the sum of a random number of random variables each with an NBU distribution.

C.14. Proposition. If $X_1, X_2, \ldots$ is a sequence of independent nonnegative random variables each with NBU distributions, and N is a random variable independent of the X_i taking values on the positive integers, then $S_N = X_1 + \cdots + X_N$ has an NBU distribution.

Proof. The first of the following inequalities follows from Proposition C.11. The second inequality follows from the fact that the covariance of two increasing functions of the same random variable is nonnegative (Proposition 18.H.1). Thus,

$$P\{S_N > x + t\}$$
$$= \sum_{n=1}^{\infty} P\{S_n > x + t\} P\{N = n\}$$
$$\geq \sum_{n=1}^{\infty} P\{S_n > x\} P\{S_n > t\} P\{N = n\}$$
$$\geq \sum_{n=1}^{\infty} P\{S_n > x\} P\{N = n\} \sum_{n=1}^{\infty} P\{S_n > t\} P\{N = n\}$$
$$= P\{S_N > x\} P\{S_N > t\}.$$
□

Sums of the form considered in Proposition C.14 arise most often when N has a geometric distribution, and they are of interest when the X_i are not necessarily NBU. See Gertsbakh (1984) and Brown (1990) for further discussion.

Although a mixture of NBU distributions is in general not NBU, there are conditions under which a mixture of NBU distributions is NBU.

C.15. Proposition (Block, Li and Savits, 2003a). Suppose that $\bar{F}(x \mid \theta), \theta \geq 0$ is a parametric family of survival functions such that

$\bar{F}(x \mid \theta)$ is increasing in θ for each fixed x. Suppose further that

$$\bar{F}(x \mid \theta) \leq \bar{F}(\alpha x \mid \alpha\theta)\bar{F}(\bar{\alpha}x \mid \bar{\alpha}\theta)$$
$$\text{for all } \alpha \in (0,1) \text{ and all } x, \theta \geq 0. \tag{9}$$

If G is NBU, then the mixture $\bar{H}(x) = \int_0^\infty \bar{F}(x \mid \theta) \, dG(\theta)$ is NBU.

Proof. First from (9) and then from Proposition C.4, it follows that

$$\bar{H}(x) = \int_0^\infty \bar{F}(x \mid \theta) \, dG(\theta) \leq \int_0^\infty \bar{F}(\alpha x \mid \alpha\theta)\bar{F}(\bar{\alpha}x \mid \bar{\alpha}\theta) \, dG(\theta)$$
$$\leq \int_0^\infty \bar{F}(\alpha x \mid \theta) \, dG(\theta) \int_0^\infty \bar{F}(\bar{\alpha}x \mid \theta) \, dG(\theta)$$
$$= \bar{H}(\alpha x) \bar{H}(\bar{\alpha}x).$$

□

There is a converse to the above proposition: If $\bar{H}$ is NBU whenever $\bar{F}(\cdot \mid \theta)$ satisfies the conditions of the proposition, then G is NBU.

C.16. Example. If $\bar{F}(x \mid \theta) = \bar{F}(x - \theta)$ for some distribution F that is NBU, then (9) is satisfied, so H is NBU. In this case, H is the convolution of F and G, and Proposition C.15 yields Proposition C.11.

C.17. Example. Suppose that $\bar{F}(x \mid \theta) = [\bar{F}(x)]^\theta$ for some underlying survival function $\bar{F}$. Then for all $\alpha \in (0, 1)$,

$$\bar{F}(x \mid \theta) = [\bar{F}(x)]^{\alpha\theta}[\bar{F}(x)]^{\bar{\alpha}\theta} \leq [\bar{F}(\alpha x)]^{\alpha\theta}[\bar{F}(\bar{\alpha}x)]^{\bar{\alpha}\theta}$$
$$= \bar{F}(\alpha x \mid \alpha\theta)\bar{F}(\bar{\alpha}x \mid \bar{\alpha}\theta),$$

so (9) is satisfied.

C.18. Remark. Proposition C.15 is an example of the following generic problem. Let

$$\psi(x) = \int s(x \mid \theta) \, t(\theta) \, d\theta.$$

For a given class $\mathcal{C}$, what conditions on $s(\cdot \mid \theta)$ are sufficient (or necessary and sufficient) to insure that $t \in \mathcal{C}$ implies $\psi \in \mathcal{C}$? In addition to Proposition C.15, Theorem 21.B.13 is a result of this kind.

D. Decreasing Mean Residual Life Distributions

According to Proposition 4.C.1.a, the notion of an IHR distribution F can be defined in terms of the residual life distributions F_t by the condition that these distributions be stochastically decreasing in t. A weaker condition is that these residual life distributions have decreasing expectations.

For reasons not entirely clear, this class of distributions has not been studied as extensively as those introduced in Sections B and C. Distributions with increasing mean residual life (IMRL) have received even less attention, but see Brown (1981, 1983), and Chen, Hollander and Langberg (1983).

D.1. Definition. A distribution F with finite mean is said to have a *decreasing mean residual life* (F is DMRL) if the mean

$$m(t) = \int_0^\infty \bar{F}_t(x)\,dx \text{ is decreasing in } t \geq 0, \tag{1}$$

where $\bar{F}_t(x) = \bar{F}(x+t)/\bar{F}(t)$ is the residual life distribution at t. Similarly, F has an increasing mean residual life (F is IMRL) if $m(t)$ is increasing in $t \geq 0$.

D.2. Proposition. If F has an increasing hazard rate, then F has a decreasing mean residual life.

Proof. Denote the expectation (mean) of F by μ. According to 20.B(15),

$$m(t) = 1/r_{(1)}(t), \tag{2}$$

where $r_{(1)}$ is the hazard rate of the equilibrium distribution, i.e., the distribution with density $f_{(1)}(t) = \bar{F}(t)/\mu, t > 0$. It follows that F has a decreasing mean residual life if and only if the corresponding equilibrium distribution has an increasing hazard rate; this is a consequence of the assumption that F is IHR (see Propositions 4.C.6 and 4.C.6.a). □

A statement similar to Proposition D.2 holds for the IMRL case; if the expectation required to define $f_{(1)}$ is finite and F has a decreasing hazard rate, then F is IMRL.

Proposition D.2 has been given by Rolski (1975) and by Gupta (1979).

D.2.a. Proposition. Let X have distribution F with finite first moment, and let Y have the density $f_{(1)}$. Then, F is DMRL if and only if $X \geq_{\mathrm{hr}} Y$.

Proof. Assume that F has a density; then the result can be obtained by differentiating $m(t) = \int_t^\infty \bar{F}(x)/\bar{F}(t)\,dx$ with respect to t. $\square$

Proposition D.2.a and a number of related results have been given by Bassan, Rinott and Vardi (2002). It is to be compared with Propositions 4.C.6.a and 4.C.16.a, where the likelihood ratio ordering occurs rather than the hazard rate ordering.

D.2.b. Proposition. The distribution F is DMRL (IMRL) if and only if the hazard rate $r_{(1)}$ of the equilibrium distribution is increasing (decreasing).

D.2.c. Proposition. Let X be a random variable with the distribution F that has a finite first moment. Then, F is DMRL (IMRL) if and only if $E[\phi(X-t) \mid X > t]$ is decreasing (increasing) in $t \geq 0$ for all convex increasing functions ϕ.

Proof. With an integration by parts, it can be verified that

$$E[\phi(X-t) \mid X > t] = \frac{1}{\bar{F}(t)} \int_t^\infty \phi(z-t)\,dF(z)$$

$$= \phi(0+) + \frac{1}{\bar{F}(t)} \int_t^\infty \bar{F}(z)\phi'(z-t)\,dz. \quad (3)$$

Let Y be a random variable with the equilibrium distribution $F_{(1)}$. Compute

$$E(X - t \mid X > t) = \int_t^\infty \frac{\bar{F}(x+t)}{\bar{F}(t)}\,dx, \quad (4)$$

and

$$E[\phi'(Y-t) \mid Y > t] = \frac{\int_t^\infty \phi'(z-t)\bar{F}(z)\,dz}{\int_t^\infty \bar{F}(z)\,dz}$$

$$= \frac{\int_t^\infty \phi'(z-t)\,dF_{(1)}(z)}{\bar{F}_{(1)}t} = \frac{\int_t^\infty \phi'(z)f_{(1)}(z+t)\,dz}{\bar{F}_{(1)}t}. \quad (5)$$

From these quantities, it can be determined that

$$E[\phi(X-t)\,|\,X>t] = \phi(0+) + E(X-t\,|\,X>t)E[\phi'(Y-t)\,|\,Y>t]. \quad (6)$$

Suppose that F is DMRL. Then by hypothesis, $E(X-t\,|\,X>t)$ is decreasing in $t \geq 0$, and furthermore, by Proposition D.2.b, $\bar{F}_{(1)}$ is IHR. This means that the distributions

$$\frac{\bar{F}_{(1)}(z+t)}{\bar{F}_{(1)}(t)}, \quad z \geq 0$$

that appear in (5) are stochastically decreasing in t (Proposition 4.C.1.a). Because ϕ is convex, ϕ' is increasing. By Proposition 2.A.2, this means that $E[\phi'(Y-t)\,|\,Y>t]$ is decreasing in $t \geq 0$, and consequently, $E[\phi(X-t)\,|\,X>t]$ is decreasing in $t \geq 0$.

Next, suppose that $E[\phi(X-t)\,|\,X>t]$ is decreasing in $t \geq 0$ for all convex increasing functions ϕ. Because the function $\phi(x) = x$ is convex and increasing, it follows that $E(X-t\,|\,X>t)$ is increasing in $t \geq 0$, that is, F is DMRL.

The proof for the IMRL case is similar. □

Proposition D.2.c and the given proof is essentially due to Brown (1981), although he treats only the IMRL case and omits the trivial converse.

D.3. Proposition. If $\int_t^\infty \bar{F}(x)\,dx$ is log concave, then F is DMRL.

Proof. The assumed log concavity is equivalent to the assumption that the equilibrium distribution $F_{(1)}$ is IHR. Consequently, the result follows from (2). □

D.4. Proposition (Klefsjö, 1982a). If F is a distribution function with mean μ and normalized total time on test transform K_F^{-1}, then F is DMRL [IMRL] if and only if

$$\frac{\mu[1 - K_F^{-1}(p)]}{1-p} \text{ is decreasing [increasing] in } p, \quad 0 < p < 1. \quad (7)$$

Proof. Recall from 1.I(10) and 1.I(4) that $K_F^{-1}(p) = H_F^{-1}(p)/\mu$ where $H_F^{-1}(p) = \int_0^{F^{-1}(p)} \bar{F}(x)\,dx$. From 1.B(11) and 1.B(10b) it follows that the mean residual life at t is

$$\frac{\int_t^\infty \bar{F}(x)\,dx}{\bar{F}(t)} = \frac{\mu - \int_0^t \bar{F}(x)\,dx}{\bar{F}(t)}. \tag{8}$$

In (8), let $t = F(z)$ to obtain that the mean residual life (8) is decreasing [increasing] if and only if (7) holds. □

For another characterization of the class of DMRL distributions that involves slowly varying functions, see Rojo (1995).

D.5. Proposition (Haines and Singpurwalla, 1974). *If the mixture of IMRL distributions has a finite mean, then it is IMRL. Thus, with the qualification of finite means, the class of distributions with IMRL is closed under mixtures.*

Proof. Suppose that

$$\bar{H}(x) = \int_\Theta \bar{F}(x \mid \theta)\,dG(\theta),$$

where for each θ, $F(\cdot \mid \theta)$ is IMRL. The definition of IMRL requires that the integrals

$$\mu(\theta) = \int_0^\infty \bar{F}(x \mid \theta)\,dx$$

exist finitely for all θ. Note that H has the mean

$$\mu_H = \int_0^\infty \bar{H}(x)\,dx = \int_\Theta \mu(\theta)\,dG(\theta).$$

Now, let

$$h_{(1)}(x) = \frac{\bar{H}(x)}{\mu_H} = \int_\Theta \frac{\bar{F}(x \mid \theta)}{\mu(\theta)} \frac{\mu(\theta)}{\mu_H}\,dG(\theta) = \int_\Theta f_{(1)}(x \mid \theta)\,dG^*(\theta),$$

where $G^*(\theta) = \mu(\theta) G(\theta)/\mu_H$. Because $F(\cdot \mid \theta)$ is IMRL, it follows that $f_{(1)}(\cdot \mid \theta)$ is the density of a DHR distribution. From Theorem 4.C.15, it follows that $h_{(1)}$ is the density of a DHR distribution; according to (1b), this means that H is a DMRL distribution. □

In general, the class of DMRL distributions is not closed under mixtures. However, closure can be obtained in special cases.

D.6. Proposition (Block, Li and Savits, 2003a). If (i) $\bar{F}(x\,|\,\theta), \theta \geq 0$, is increasing in θ for each $x \geq 0$, (ii) $\int_t^\infty \bar{F}(x\,|\,\theta)\,dx$ is log concave in (t,θ), and (iii) G is IHR, then the mixture

$$\bar{H}(x) = \int_0^\infty \bar{F}(x\,|\,\theta)\,dG(\theta)$$

is DMRL.

Proof. Make use of Proposition D.3 and compute that

$$\int_t^\infty \bar{H}(x)\,dx = \int_t^\infty \int_0^\infty \bar{F}(x\,|\,\theta)\,dG(\theta)\,dx = \int_0^\infty \int_t^\infty \bar{F}(x\,|\,\theta)\,dx\,dG(\theta).$$

To complete the proof, apply Proposition 4.C.1.i. □

E. New Better (Worse) Than Used in Expectation Distributions

The comparison $\bar{F}(x) \geq \bar{F}(x+t)/\bar{F}(t) = \bar{F}_t(x)$ for all $x, t \geq 0$, required of NBU distributions is a stochastic ordering between a distribution and a residual life distribution. When stochastic order is too strong, the weaker condition that the expectations of these distributions are ordered may be of interest.

E.1. Definition. A life distribution F is said to be *new better than used in expectation* (NBUE) if it has a finite mean μ that is at least as large as the mean residual life length at time t, for all $t \geq 0$, i.e.,

$$\mu \geq \int_0^\infty \bar{F}_t(x)\,dx = \int_0^\infty \bar{F}(x+t)\,dx/\bar{F}(t)$$

for $t \geq 0$ such that $\bar{F}(t) > 0$. (1a)

The distribution F is *new worse than used in expectation* (NWUE) if the inequality of (1a) is reversed.

The NBUE condition is the condition that a used device of any age has a mean residual life smaller than the mean life of a new device of the same kind.

Inequality (1a) can be rewritten in the form

$$\bar{F}(t) \geq \int_t^\infty [\bar{F}(x)/\mu]\,dx, \quad t \geq 0; \tag{1b}$$

inequality (1b) states that the distribution $F_{(1)}$ with density $f_{(1)}(x) = \bar{F}(x)/\mu, x > 0$, is stochastically less than F, that is,

$$\bar{F}(t) \geq \bar{F}_{(1)}(t), \quad t \geq 0. \tag{1c}$$

Recall that the distribution $F_{(1)}$ was encountered in Propositions 4.C.6 and D.3 and is called the "equilibrium distribution" in Sections 1.B.j and 20.B.c. The inequality (1b) can again be rewritten in the form

$$r_{(1)}(0) \leq r_{(1)}(t), \quad t \geq 0, \tag{1d}$$

where

$$r_{(1)}(t) = \bar{F}(t) / \int_t^\infty \bar{F}(x)\,dx$$

is the hazard rate of the equilibrium distribution $F_{(1)}$.

By assumption, an NBUE distribution has finite first moment. The fact that these distributions have finite moments of all positive orders follows from Proposition 6.A.6.

E.2. Proposition (Bergman, 1979, Klefsjö (1982a)). A distribution F such that $F(x) = 0, x < 0$, is NBUE if and only if the normalized total time on test transform K_F^{-1} satisfies the inequality

$$K_F^{-1}(p) \geq p, \quad 0 \leq p \leq 1.$$

The inequality is reversed if F is NWUE.

Proof. Denote the mean of F by μ. The condition $K_F^{-1}(p) \geq p, 0 \leq p \leq 1$, is the condition that $\int_0^z \bar{F}(x)\,dx \geq F(z)\mu_F$. Subtraction of both sides of this inequality from the identity $\int_0^\infty \bar{F}(x)\,dx = [F(z) + \bar{F}(z)]\mu$ yields the inequality (1a). The proof for the NWUE case is similar. □

a. Preservation Properties

The following example shows that an NBUE distribution need not be NBU; it also shows that the class of NBUE distributions is *not* closed under the formation of coherent systems.

E.3. Example. Suppose that F places mass $1/2$ at 1 and mass $1/2$ at 3. Then, F has mean $\mu = 2$ and

$$\int_0^\infty \bar{F}_t(x)\, dx \leq 2, \quad t \geq 0.$$

Thus, F is NBUE. Now consider a series system consisting of two components each having the distribution of this example. The mean life of this system when new is $\int_0^3 \bar{F}^2(x)\, dx = 3/2$, whereas the mean life of a system of age one is two. Thus, the system life is not NBUE. Because of Proposition C.10, this means that F cannot be NBU. Second, it shows that the class of NBUE distributions is not closed under the formation of coherent systems.

E.4. Proposition (Marshall and Proschan, 1972a). If F and G are NBUE distributions, then their convolution $H = F*G$ is NBUE.

Proof. Because

$$\bar{H}(t+x) = \int_{u=0}^t \bar{F}(t+x-u)\, dG(u) + \int_{u=t}^\infty \bar{F}(t+x-u)\, dG(u),$$

it follows that

$$\int_0^\infty \bar{H}(t+x)\, dx = \int_{x=0}^\infty \left[\int_{u=0}^t \bar{F}(t+x-u)\, dG(u)\right] dx$$

$$+ \int_{x=0}^\infty \left[\int_{u=t}^\infty \bar{F}(t+x-u)\, dG(u)\right] dx.$$

Denote the means of F and G, respectively, by μ_F and μ_G. Because F is NBUE,

$$\int_0^\infty \bar{F}(t+x-u)\, dx \leq \mu_F \bar{F}(t-u), \quad u \leq t.$$

Thus, with a change in the order of integration,

$$\int_{x=0}^\infty \left[\int_{u=0}^t \bar{F}(t+x-u)\, dG(u)\right] dx$$

$$\leq \mu_F \int_{u=0}^t \bar{F}(t-u)\, dG(u) = \mu_F[\bar{H}(t) - \bar{G}(t)].$$

For $u > t$,

$$\int_{x=0}^{\infty} \left[\int_{u=t}^{\infty} \bar{F}(t+x-u)\, dG(u) \right] dx = \int_{u=t}^{\infty} (u-t+\mu_F)\, dG(u).$$

In the right-hand side, let $w = u - t$. Then,

$$\int_{u=t}^{\infty} (u-t+\mu_F)\, dG(u)$$
$$= \mu_F \bar{G}(t) + \int_{u=0}^{\infty} \bar{G}(w+t)\, dw \leq \mu_F \bar{G}(t) + \mu_G \bar{G}(t).$$

By combining these results, it follows that

$$\int_0^{\infty} \bar{H}(t+x)\, dx \leq \mu_F[\bar{H}(t) - \bar{G}(t)] + (\mu_F + \mu_G)\bar{G}(t)$$
$$= \mu_F \bar{H}(t) + \mu_G \bar{G}(t) \leq (\mu_F + \mu_G)\bar{H}(t). \qquad \square$$

E.5. Proposition (Marshall and Proschan, 1972a). If F is the mixture of NWUE distributions, no two of which cross, then F is NWUE.

Proof. Suppose that

$$F(t) = \int F(t \mid \theta)\, dG(\theta)$$

is a mixture of NWUE distributions $F(\cdot \mid \theta)$. Denote the mean of F by μ and the mean of $F(\cdot \mid \theta)$ by μ_θ. Because no two of the $F(\cdot \mid \theta)$ cross, it follows that for fixed t, $\bar{F}(t \mid \theta)$ and μ_θ are similarly ordered in θ. Consequently, by Chebyshev's inequality for similarly ordered functions (Proposition 18.G.1),

$$\mu \bar{F}(t) = \int \mu_\theta\, dG(\theta) \int \bar{F}(t \mid \theta)\, dG(\theta) \leq \int \mu_\theta \bar{F}(t \mid \theta)\, dG(\theta).$$

By using first the NWUE property and then Fubini's theorem (22.B.1), it follows that

$$\int \mu_\theta \bar{F}(t \mid \theta)\, dG(\theta) \leq \iint \bar{F}(t+x \mid \theta)\, dx\, dG(\theta)$$
$$= \iint \bar{F}(t+x \mid \theta)\, dG(\theta)\, dx = \int \bar{F}(t+x)\, dx.$$

Thus, F is NWUE. $\square$

b. Some Additional Inequalities

Denote the mean of F by μ, and let $f_{(1)} = \bar{F}/\mu$, the density that appears in (1b). Similarly, denote by $\bar{F}_{(1)}$ and $r_{(1)}$ the corresponding survival function and hazard rate.

E.6. Proposition (Marshall and Proschan, 1972a). If F is NBUE and has mean μ, then

$$\int_t^\infty \frac{\bar{F}(x)}{\mu} \, dx = \bar{F}_{(1)}(t) \leq e^{-t/\mu}. \tag{2}$$

Proof. From (1b), it follows that $\bar{F}_{(1)}(x) \leq \mu \bar{F}(x)/\mu = \mu f_{(1)}(x)$, that is, $r_{(1)}(x) \geq 1/\mu$. Integrate both sides of this inequality on x from 0 to t to conclude that $-R_{(1)}(t) \geq -t/\mu$; this, when exponentiated, completes the proof because $\bar{F}_{(1)}(t) = e^{-R(t)}$. □

Inequality (2) has also been obtained by Brown and Ge (1984), who show that the inequality is reversed when F is NWUE.

Inequality (2) has been used as a definition of another class of distributions called *harmonic new better than used in expectation* (HNBUE). If (2) is reversed, then F is said to be *harmonic new worse than used in expectation*. As indicated by Proposition E.6, this condition is weaker than NBUE. These distributions do not play a big role in this book. For a survey of these classes, see Klefsjö (1982b) or Johnson, Kotz and Balakrishnan (1995, p. 664). For a shock model derivation, see Klefsjö (1981).

E.7. Proposition (Shaked and Shanthikumar, 1994). If F is NBUE, then in the convex order $\leq_{cx}$ of Definition 2.B.1, F is less than the exponential distribution with the same mean as F.

Proof. This result follows from Propositions E.6 and 2.B.2. □

F. Additional Nonparametric Families of Distributions

A number of nonparametric families of distributions have been introduced in the literature that have not been discussed in previous sections; see Johnson, Kotz and Balakrishnan (1995, p. 664) for definitions and references. A few additional classes are discussed in this section.

a. Concave Distribution Functions (Decreasing Densities)

Suppose that F is a distribution function such that $F(0-) = 0$. Suppose further that F has a density except possibly for positive mass at the origin and that the density is decreasing on $(0, \infty)$. Such distribution functions are concave on $[0, \infty)$ and are referred to here simply as concave distributions. According to Proposition 4.C.12, the class of all such distributions includes the DHR distributions, and the exponential distributions are a prime example. It also includes the distribution of random variables $|X|$, where X has a density with unique mode at 0. But the class includes many other distributions as well.

F.1. Preservation properties.

(i) The class of concave distributions is closed under the formation of mixtures.

(ii) The class of concave distributions is not closed under formation of coherent systems or under convolution.

Lack of closure under the formation of coherent systems can be seen by reference to Example A.11. Lack of closure under convolutions can be seen from the fact that the convolution of an exponential distribution with itself is a gamma distribution with shape parameter 2; this distribution has a unimodal density but not a decreasing density.

(iii) The residual life distribution of a concave distribution is concave.

b. Decreasing Reverse Hazard Rate

For distributions F that have a density f, the ratio

$$s(x) = f(x)/F(x)$$

is called a *reverse hazard rate* for F (See Definition 1.B.10). The concept of a monotone reverse hazard rate can be defined in various ways, analogous to the various equivalent ways of defining a monotone hazard rate given in Section 4.C. Here only one of the possible forms is given.

F.2. Definition. A distribution F is said to have a decreasing reverse hazard rate (DRHR) if the corresponding reverse hazard function $S(x) = \log F(x)$ is concave; similarly, F is said to have an increasing reverse hazard rate (IRHR) if $S(x)$ is convex where finite.

F.3. Observation. The reverse hazard rate of X is the hazard rate of $-X$. By using this observation, some results regarding hazard rates can be converted to results about reverse hazard rates.

F.4. Proposition. Suppose that F has a density. Then, F has a decreasing [increasing] reverse hazard rate if and only if there is a version f of the density such that $s = f/F$ is decreasing [increasing].

F.5. Proposition (Block, Savits and Singh, 1998). If F has an increasing reverse hazard rate, then there exists a real number $a < \infty$ such that $f(x) > 0, x < a, f(x) = 0, x > a$.

This proposition follows from Proposition 4.C.12 and Observation F.3.

F.6. Remark. According to Proposition F.5, nonnegative random variables cannot have distributions with increasing reverse hazard rate; consequently, these distributions are not further discussed in this book.

F.7. Proposition. Suppose that the distribution function F is concave. Then, $\log F$ is concave. If F has a density, then F is DRHR.

Proof. This proposition follows immediately from the definition and Proposition 21.A.5. □

F.8. Proposition. If F has a log-concave density, then F is DRHR.

This proposition is a special case of Proposition 4.B.8.a.

The DRHR property is not discussed in more detail in this book, though some interesting distributions possess the property by virtue of one or another of the propositions mentioned above.

c. Closure Properties of DRHR Distributions

F.9. Proposition (Barlow, Marshall and Proschan, 1963). If F and G are DRHR, then the convolution $H = F * G$ is DRHR.

This result can be obtained from Theorem 4.C.4 and the observation that the reverse hazard rate of the random variable X is the hazard rate of $-X$. For another proof, see Shaked and Shanthikumar (1994, Corollary 1.B.33).

F.10. Example. To see that the class of DRHR distributions is not closed under mixtures, consider the mixture $F(x) = [F_1(x) + F_2(x)]/2$, where, for $i = 1, 2$,

$$F_i(x) = 0, \quad x < 0,$$
$$= e^{\lambda_i x}/2, \quad 0 \le x < (\log 2)/\lambda_i,$$
$$= 1, \quad x \ge (\log 2)/\lambda_i.$$

When $\lambda_1 > \lambda_2 > 0$, this mixture has a reverse hazard rate that is increasing on the interval $0 < x < \log 2/\lambda_1$.

F.11. Proposition. If F is a DRHR distribution, then the residual life distributions F_t of F are DRHR.

Proof. Note that the residual life distribution of F is given by $F_t(x) = [F(t+x) - F(t)]/\bar{F}(t)$; assume that F has a density f such that f/F is decreasing. Then, $\log F_t(x)$ is convex (has a decreasing derivative) if

$$\frac{f(t+x)}{F(t+x)} \frac{1}{1 - [F(t)/F(t+x)]}$$

is decreasing. But both of these factors are decreasing. □

Because the product of log-concave functions is log concave, the class of DRHR distributions is closed under the formation of parallel systems. Whether or not the class is closed under the formation of coherent systems is not known.

F.12. Remark. According to Proposition F.8, if F has a log-concave density, then F is DRHR. It is natural to ask if the weaker condition that F is IHR implies that F is DRHR. That this is not the case is demonstrated by the following example.

F.13. Example. Suppose that

$$\bar{F}(x) = e^{-x}, \quad 0 \le x < 1,$$
$$= 0, \quad x \ge 1.$$

Although F puts positive mass at 1, it is IHR, but not DRHR.

G. Summary of Relationships and Closure Properties

A summary of abbreviations, orderings, relationships, and closures are provided below.

G. Summary of Relationships and Closure Properties

a. Summary of Abbreviations

Abbreviation	Full name	Definition
IHR	Increasing hazard rate	4.C.1.a
DHR	Decreasing hazard rate	4.C.1.a
IHRA	Increasing hazard rate average	B.1
DHRA	Decreasing hazard rate average	B.1
NBU	New better than used	C.1
NWU	New worse than used	C.1
DMRL	Decreasing mean residual life	D.1
IMRL	Increasing mean residual life	F.1
NBUE	New better than used in expectation	E.1
NWUE	New worse than used in expectation	E.1
HNBUE	Harmonic new better than used in expectation	E.7
HNWUE	Harmonic new worse than used in expectation	E.7
DRHR	Decreasing reverse hazard rate	F.2
IRHR	Increasing reverse hazard rate	F.2

b. Summary of Relationships

$$
\begin{array}{ccccccccc}
 & & IHRA & \Longrightarrow & & & NBU & & \\
 & & \Uparrow & & & & \Downarrow & & \\
\text{log concave density} & \Longrightarrow & IHR & \Longrightarrow & DMRL & \Longrightarrow & NBUE & \Longrightarrow & HNBUE \\
\Downarrow & & & & & & & & \\
DRHR & \Longleftarrow & \text{concave distribution} & & & & & & \\
 & & \Uparrow & & & & & & \\
\text{log convex density} & \Longrightarrow & DHR & \Longrightarrow & IMRL & \Longrightarrow & NWUE & \Longrightarrow & HNWUE \\
 & & \Downarrow & & & & \Uparrow & & \\
 & & DHRA & \Longrightarrow & & & NWU & & \\
\end{array}
$$

c. Order Conditions

Let Y be a random variable with the exponential survival function $\bar{G}(x) = \exp\{-x\}, x \geq 0$. Let X be a random variable with survival function $\bar{F}$. Equivalences involving F and G are as follows:

$$F \text{ is IHR} \Leftrightarrow G^{-1}F \text{ is convex, i.e., } X \leq_c Y$$
$$F \text{ is IHRA} \Leftrightarrow G^{-1}F \text{ is starshaped, i.e., } X \leq_* Y$$
$$F \text{ is NBU} \Leftrightarrow G^{-1}F \text{ is superadditive, i.e., } X \leq_{su} Y$$

$$F \text{ is DHR} \Leftrightarrow G^{-1}F \text{ is concave, i.e., } X \geq_c Y$$
$$F \text{ is DHRA} \Leftrightarrow -G^{-1}F \text{ is starshaped, i.e., } X \geq_* Y$$
$$F \text{ is NWU} \Leftrightarrow G^{-1}F \text{ is subadditive, i.e., } X \geq_{su} Y$$

d. Summary of Closure Properties

Class of Distribution	Formation of Coherent systems	Convolution	General Mixtures	Mixtures of noncrossing distributions	Residual life distribution
log-concave density	not closed	closed	not closed	not closed	closed
IHR	not closed	closed	not closed	not closed	closed
IHRA	closed	closed	not closed	not closed	not closed
NBU	closed	closed	not closed	not closed	not closed
DMRL	not closed	not closed	not closed	not closed	closed
NBUE	not closed	closed	not closed	not closed	not closed
log-convex density	not closed	not closed	closed	closed	closed
DHR	not closed	not closed	closed	closed	closed
DHRA	not closed	not closed	closed	closed	not closed
NWU	not closed	not closed	not closed	closed	not closed
IMRL	not closed	not closed	closed	closed	closed
NWUE	not closed	not closed	closed	closed	not closed
Concave distribution	not closed	not closed	closed	closed	closed
DRHR	?	closed	not closed	not closed	closed

H. Shock Models

With the exception of the exponential distribution, which is characterized by 1.F(4), residual life distributions of items change over time, whether they be of devices or systems, biological or not. This change in the residual life distribution reflects a change either within the item or in the environment where it is living. A number of authors have considered the state of an item to be characterized by a real number, and have proposed models representing the item state as it changes over time by a stochastic process. The item is considered to have "failed" or "died" when the process first crosses a specified threshold. A prime example of this kind of process is Brownian motion with drift, where the waiting time to cross a threshold has an inverse Gaussian distribution (see Chapter 13). Various other models are reviewed by Aalen and Gjessing (2001), all of which allow for only very simple state spaces. A model allowing the state of an item to be represented in very general terms has been proposed by Marshall (1994), who also considered stochastic processes taking values in such state spaces. The item is considered to have failed when the process first enters a specified set of "failed" states.

A related notion has been introduced by Kotz and Singpurwalla (1999). They note that if X is the life length of a device with hazard function R, then $Z = R(X)$ is a random variable such that

$$P\{X > t\} = P\{R(t) < Z\}, \quad t \geq 0.$$

One can think of R as a stochastic process with but one sample path, and Z as a random threshold; Kotz and Singpurwalla (1999) call R the *hazard potential*.

Shock models such as those discussed in this section, particularly the cumulative damage threshold models discussed later, are threshold models for the process of item deterioration, represented as a real function of time. Models of various kinds and complexity have been proposed and studied primarily because of their intuitive appeal.

Suppose that the process $\{N(t), t \geq 0\}$ counts the number of shocks to a device in the interval $[0, t]$. Let $\bar{P}_k$ be the probability that the device survives k shocks, $k = 0, 1, \ldots$. Then, the survival function of the device has the form

$$\bar{H}(x) = \sum_{k=0}^{\infty} \bar{P}_k\, P\{N(x) = k\}, \quad x \geq 0.$$

The case that N is a Poisson process has been studied by Esary, Marshall and Proschan (1973) and the cases that N is a nonhomogeneous Poisson process and a birth process have been studied by A-Hameed and Proschan (1973, 1975). Various versions of this model have been studied by a number of other authors.

In this section, the case that N is a Poisson process is briefly described. For more complete details and additional references, see Esary, Marshall and Proschan (1973), the source of all the results of this section. The Poisson process is introduced in Section 20.F.a.

Because $\bar{P}_k$ is the probability of surviving k shocks, it is assumed that $1 \geq \bar{P}_0 \geq \bar{P}_1 \geq \ldots$, so that

$$p_k = \bar{P}_{k-1} - \bar{P}_k \geq 0, \quad k = 1, 2, \ldots.$$

For the Poisson process,

$$\bar{H}(x) = \sum_{k=0}^{\infty} \bar{P}_k\, e^{-\lambda x}(\lambda x)^k/k!, \quad x \geq 0; \qquad (1)$$

if $P_0 = 0$, then H has density

$$h(x) = \lambda \sum_{k=1}^{\infty} p_k\, e^{-\lambda x}(\lambda x)^{k-1}/(k-1)!, \quad x \geq 0. \qquad (2)$$

The hazard rate r can be obtained, but does not take a particularly nice form; because the chance of failure in an interval $(t, t+\Delta)$ cannot be greater than the chance of a shock in that interval, $r \leq \lambda$.

H.1. Example. If $\bar{P}_k = 1, k = 0, 1, \ldots, n, \bar{P}_k = 0, k > n$, then (1) is the survival function of a gamma distribution. See 9.A(2) and Section 9.A.b(i).

In spite of the apparently limited scope of (1), an important fact is that *any* survival function on $[0, \infty)$ can be approximated by survival functions of the form (1). More precisely, if $\bar{H}$ is a survival function of a nonnegative random variable, then

$$\bar{H}(x) = \lim_{\lambda \to \infty} \sum_{k=0}^{\infty} \bar{H}(k/\lambda)\, e^{-\lambda x}(\lambda x)^k/k!$$

at continuity points x of H (see Feller, 1968, p. 219). Because of this fact, interesting properties of H can be expected only with restrictions on the $\bar{P}_k$.

H.2. Theorem. Suppose (1) and $\bar{P}_0 = 1$. Then, with the notation $p_k = \bar{P}_{k-1} - \bar{P}_k \geq 0$, $k = 1, 2, \ldots$,

H has a log-concave density if p_{k+1}/p_k
is decreasing in $k = 1, 2, \ldots$; (3)
H is IHR if $\bar{P}_k/\bar{P}_{k-1}$ is decreasing in $k = 1, 2, \ldots$; (4)
H is IHRA if $\bar{P}_k^{1/k}$ is decreasing in $k = 1, 2, \ldots$; (5)
H is NBU if $\bar{P}_j \bar{P}_k \geq \bar{P}_{j+k}$, $j, k = 0, 1, \ldots$. (6)

Analogous conditions can also be given for a number of other classes of distributions.

The results (3) to (5) can be proved using the variation diminishing property stated in Theorem 19.B.13 and the total positivity of x^k stated in 19.B.5.a. To illustrate this technique, consider condition (5). Because $\bar{P}_k^{1/k}$ is decreasing in k, $\bar{P}_k - \zeta^k$ ($0 \leq \zeta \leq 1$) has, as a function of k, at most one sign change, from + to − if one occurs. Hence,

$$\bar{H}(x) - e^{-(1-\zeta)\lambda x} = \sum_{k=0}^{\infty} (\bar{P}_k - \zeta^k) e^{-\lambda x} (\lambda x)^k / k!$$

has at most one sign change, from + to − if one occurs. By (ii) of Proposition B.3, this means that H is IHRA.

a. Cumulative Damage Threshold Models

Suppose that the ith shock to an item causes a random damage X_i. Damages accumulate additively, and the kth shock is survived by the item if $X_1 + \cdots + X_k \leq z$, where z is the capacity or threshold of the item. If the X_i are independent and identically distributed, then

$$\bar{P}_k = F^{k*}(z), \quad k = 1, 2, \ldots,$$

where F^{k*} is the k fold convolution of F with itself.

H.3. Lemma. If F is a distribution function satisfying $F(0-) = 0$, then

$$[F^{k*}(z)]^{1/k} \text{ is decreasing in } k = 1, 2, \ldots.$$

H.4. Lemma. Let $X_1, X_2, \ldots$ be nonnegative random variables with a joint distribution satisfying

$$P\{X_k \leq u \mid X_1, \ldots, X_{k-1}\} \text{ depends on } X_1, \ldots, X_{k-1}$$
$$\text{only via their sum } S_{k-1}, \tag{7}$$
$$P\{X_k \leq u \mid S_{k-1} = s\} \text{ is decreasing in } s \geq 0, \tag{8}$$
$$P\{X_k \leq u \mid S_{k-1} = s\} \geq P\{X_{k+1} \leq u \mid S_k = s\}, \quad s \geq 0, k = 1, 2, \ldots,$$
$$\text{where } S_0 = 0. \tag{9}$$

Then,

$$[P\{X_1 + \cdots + X_k \leq z\}]^{1/k} \text{ is decreasing in } k = 1, 2, \ldots.$$

The proofs of these lemmas are obtained by induction, and are omitted (see Esary, Marshall, and Proschan 1973 for details).

H.5. Corollary. If $\bar{P}_k = P\{X_1 + \cdots + X_k \leq z\}, k = 1, 2, \ldots$, and if the X_i are identically distributed and independent, or the conditions of Lemma H.3 are satisfied, then the survival function (1) is IHRA.

b. Random Threshold

Suppose that in the cumulative damage model with independent identically distributed damages, the threshold z is random, say with distribution G such that $G(0) = 0$. For convenience, assume that F and G have no common points of discontinuity. Then, one can write

$$\bar{P}_k = \int_0^\infty F^{k*}(z) \, dG(z) = EG(X_1 + \cdots + X_k), \quad k = 0, 1, \ldots. \tag{10}$$

H.6. Proposition. The survival function (1) with the $\bar{P}_k$ given by (10) is exponential for all F if and only if G is exponential. If G is IHR, then the distribution H defined by (1) is IHRA; conversely, if $\bar{P}_k^{1/k}$ is decreasing in $k = 1, 2, \ldots$, then G is IHRA.

The proof of Proposition H.6 is rather lengthy, and is omitted, but see Esary, Marshall, and Proschan (1973).

Some multivariate extensions of the results of this section are given by Esary and Marshall (1974).

I. Replacement Policies: Renewal Theory

The notions of NBU and NWU have intuitive appeal, but it was the results of this section that motivated the introduction of these distributions by Marshall and Proschan (1972a).

a. Planned Replacements

When it is important to sustain the functioning of a component or system, it is a common practice to employ a maintenance policy that calls for replacement upon failure. Light bulbs in domestic use provide a simple and familiar example. If in-service failures are particularly serious, or where unscheduled maintenance (due to a failure) is costly, planned replacements are often scheduled to reduce the number of replacements due to failure. Thus, it is sometimes advantageous to schedule the replacement of an unfailed component or system when it reaches a certain age. On the other hand, planned replacements can be scheduled at regular intervals of time, regardless of their age.

A policy that calls for replacement of an item upon failure or at age T, whichever comes first, is called an *age replacement policy*. A policy that calls for replacement upon failure and at specified times $T, 2T, 3T, \ldots$, is called a *block replacement policy*. Block policies require less record keeping than age policies, and they allow repairs to follow a fixed schedule, but they are intuitively less efficient than an age policy.

For comparison of maintenance policies, the following quantities are of interest:

$N(t)$ = number of failures (renewals) in $[0,t]$ with no planned replacements. $N(t)$ is an ordinary renewal process, as briefly discussed in Section 20.F.b.

$N_A(t,T)$ = number of in-service failures in $[0,t]$ under an age replacement policy that calls for replacement at failure or age T.

$N_B(t,T)$ = number of in-service failures in $[0,t]$ under a block replacement policy with replacement interval T.

Note that the second two quantities do not count planned replacements, but only in-service failures that do not coincide with a planned replacement. Contrary to a common practice in renewal theory, e.g., in the book of Feller (1971), these quantities do not count the origin as a renewal point.

It is often assumed that planned replacements reduce the number of in-service failures; this is not always true, and it involves an assumption about the life distribution of the device under consideration. Likewise, it is often assumed that an age replacement policy calling for replacement at a younger age is better, but this also is true only for a restricted class of life distributions. Specific results here contribute to the understanding of the various classes considered, and they identify conditions under which age and block replacement policies reduce the frequency of in-service failures. It was in this context that Marshall and Proschan (1972a) introduced the NBU and NWU families of distributions. The following propositions characterize these classes of distributions.

I.1. Proposition (Marshall and Proschan, 1972a). The comparison $N(t) \geq_{st} N_A(t,T)$ holds for all t and $T > 0$ if and only if F is NBU. The inequality is reversed if and only if F is NWU.

I.2. Proposition (Marshall and Proschan, 1972a).
The comparison $N(t) \geq_{st} N_B(t,T)$ holds for all t and $T > 0$ if and only if F is NBU. The inequality is reversed if and only if F is NWU.

A characterization of IHR [DHR] distributions in the context of maintenance policies shows the validity of the intuition behind the definition.

I.3. Proposition (Marshall and Proschan, 1972a). $N_A(t,T)$ is stochastically increasing in $T > 0$ for each fixed t if an only if F is IHR. The stochastic monotonicity is reversed if and only if F is DHR.

The proofs of these propositions are somewhat involved and are given, respectively, in Propositions I.5, I.8, and I.9. It is convenient to begin with more notation. Let Y_i denote the length of time between the $(i-1)$th and the ith failure when no planned replacements are made, i.e.,

$$Y_i = \inf\{t : N(t) \geq i\} - \inf\{t : N(t) \geq i-1\}.$$

Similarly, define $Y_{i,A}$ and $Y_{i,B}$, $i = 1, 2, \ldots$ for the respective processes $N_A(t,T)$ and $N_B(t,T)$.

I.4. Proposition (Marshall and Proschan, 1972a). The comparison $Y_i \leq_{st} Y_{i,A}$ holds for all $T > 0, i = 1, 2, \ldots$ if and only if F is NBU. The stochastic order is reversed if and only if F is NWU.

Proof. First note that Y_i and $Y_{i,A}$ have distributions independent of i and

$$P\{Y_1 > t\} = \bar{F}(t),$$
$$P\{Y_{1,A} > t\} = [\bar{F}(T)]^j \bar{F}(t - jT), \quad jT \leq t < (j+1)T, \quad j = 0, 1, \ldots.$$

If F is NBU, then by repeated application of the definition, it follows that

$$\bar{F}(t) \leq [\bar{F}(T)]^j \bar{F}(t - jT), \quad jT \leq t < (j+1)T, \quad j = 0, 1, \ldots,$$

and consequently, $Y_i \leq_{\text{st}} Y_{i,A}$.

If $Y_i \leq_{\text{st}} Y_{i,A}$, that is, $P\{Y_i > t\} \leq P\{Y_{i,A} > t\}$ for all t, T, take $T = \max(x, y)$, $t - T = \min(x, y)$ to conclude that $\bar{F}(x+y) \leq \bar{F}(x)\bar{F}(y)$. The proof for the NWU case is essentially the same, but with the inequalities reversed. □

I.5. Proof of Proposition I.1. If F is NBU, then since $Y_1, Y_2, Y_3, \ldots$ are independent and $Y_{1,A}, Y_{2,A}, Y_{3,A}, \ldots$ are independent, it follows from Proposition I.4 that

$$P\{N(t) \geq n\} = P\{Y_1 + \cdots + Y_n \leq t\} \geq P\{Y_{1,A} + \cdots + Y_{n,A} \leq t\}$$
$$= P\{N_A(t, T) \geq n\}.$$

If $N(t) \geq_{\text{st}} N_A(t, T)$, for all t and $T > 0$, then

$$P\{Y_1 > t\} = P\{N(t) = 0\} \leq P\{N_A(t, T) = 0\} = P\{Y_{1,A} > t\},$$

and hence F is NBU by Proposition I.4. Again, the proof for the NWU case is analogous. □

The process $N_B(t, T)$ has more dependencies than the process $N_A(t, T)$, and consequently, results for block replacement are not as easily obtained as those for age replacement. The following lemma is required.

I.6. Lemma. Let planned replacements occur at fixed time points $0 < t_1 < t_2 < \cdots$ under Policy 1 and at these and the additional time t_0 under Policy 2. Let $N_i(t)$ be the number of failures in $[0, t]$ under Policy $i, i = 1, 2$. Then, $N_1(t) \geq_{\text{st}} N_2(t)$, for all $t > 0$, if and only if F is NBU. The stochastic order is reversed if and only if F is NWU.

Proof. Suppose first that F is NBU. For $t < t_0$, $N_1(t)$ and $N_2(t)$ have the same distribution. Next, assume that $t_0 \leq t \leq t_k$, where t_k is the smallest $t_j > t_0$. Take $t_k = \infty$ if $t_0 > t_j$, for all $j > 0$. Let Z be the age of the unit in operation at time t_0 (if there is no replacement at the time), and let Z be the age of the unit being replaced at time t_0 (if there is such a replacement). Call this the age of the unit at time t_0-. The distribution of Z does not depend upon the policy. Let τ_i denote the interval between t_0 and the time of first failure subsequent to t_0- under Policy $i, i = 1, 2$. Because F is NBU,

$$P\{\tau_1 > t \mid Z\} \leq P\{\tau_2 > t \mid Z\}, \quad \text{for all } t \geq 0.$$

Let U_i be the number of failures in $[t_0, t]$ under policy $i, i = 1, 2$. If $X_1, X_2, \ldots$ are independent and have distribution F, then

$$P\{U_1 \geq n \mid Z\} = P\{t_0 + \tau_1 + X_1 + \cdots + X_{n-1} \leq t \mid Z\}$$
$$\geq P\{t_0 + \tau_2 + X_1 + \cdots + X_{n-1} \leq t \mid Z\} = P\{U_2 \geq n \mid Z\},$$

$n = 1, 2, \ldots$. Thus, $N_1 \geq_{\text{st}} N_2$.

Finally, assume that $t > t_k$. Let $N_i(t_k, t)$ denote the number of failures in the interval $(t_k, t]$ under policy $i, i = 1, 2$. Then, $N_i(t_k) + N_i(t_k, t) = N_i(t)$, with $N_i(t_k)$ and $N_i(t_k, t)$ independent, $i = 1, 2$. Because $N_1(t_k) \geq_{\text{st}} N_2(t_k)$ and $N_1(t_k, t) =_{\text{st}} N_2(t_k, t)$, that is, both $\leq_{\text{st}}$ and $\geq_{\text{st}}$ hold, it follows that if F is NBU, then $N_1(t) \geq_{\text{st}} N_2(t)$.

Next, suppose that $N_1(t) \geq_{\text{st}} N_2(t)$ for all t and all $t_0, t_1, \ldots$. For $0 < t_0 < t_1$,

$$\bar{F}(t_1) = P\{N_1(t_1) = 0\} \leq P\{N_2(t_1) = 0\} = \bar{F}(t_0)\bar{F}(t_1 - t_0).$$

Thus, F is NBU. $\square$

I.7. Proposition. *The comparison $Y_i \leq_{\text{st}} Y_{i,B}(T)$ holds, for all $T > 0, i = 1, 2, \ldots$, if and only if F is NBU. The stochastic order is reversed if and only if F is NWU.*

Proof. Let $S_{i-1,B} = Y_{1,B}(T) + \cdots + Y_{i-1,B}(T)$, and let k be the smallest integer for which $kT \geq S_{i-1,B}$. Then $P\{Y_{i,B}(T) > t \mid S_{i-1,B}\} = P\{Y_1^* > t\}$, where Y_1^* is the first failure when planned replacements are made at $kT - S_{i-1,B} = \delta, \delta + T, \delta + 2T, \ldots$.

If F is NBU, apply Lemma I.6 to compare successive pairs of a sequence of replacement policies in which the ith policy calls for planned replacement at time points $0, \delta, \delta + T, \delta + 2T, \ldots, \delta + iT$. This comparison yields

$$\bar{F}(t) = P\{Y_i > t\} = P\{Y_1 > t\} \leq P\{Y_1^* > t\} \quad \text{for all } t \geq 0,$$

and by unconditioning, it follows that $Y_i \leq_{st} Y_{i,B}(T)$. The converse follows from Proposition I.4, since $Y_{1,B}(T)$ and $Y_{1,A}(T)$ have the same distribution.

The proof for the NWU case is analogous, but with inequalities reversed. □

I.8. Proof of Proposition I.2. If F is NBU, then use Lemma I.6 to compare successive pairs of a sequence of replacement policies in which the jth policy calls for planned replacement at time points $0, T, \ldots, (j-1)T$. The converse follows as in the converse of Proposition I.1 with $i = 1$. □

I.9. Proof of Proposition I.3. For fixed $T > 0$, $\{N_A(t,T), t \geq 0\}$ is a renewal process with underlying distribution

$$S_T(x) = 1 - [\bar{F}(T)]^n \bar{F}(x - nT), \quad nT \leq x < (n+1)T, \ n = 0, 1, \ldots.$$

Under the assumption that F has a density, it follows by differentiating $S_T(x)$ with respect to T that $S_T(x)$ is increasing in $T > 0$ for fixed $x \geq 0$ if and only if F is IHR; if F does not have a density, then a limiting argument is required. It follows that the nth convolution $S_T^{n*}(x) = P\{N_A(t,T) \geq n\}$ is increasing in $T > 0$ for fixed $t \geq 0$ if and only if F is IHR. □

b. Summary of Stochastic Comparisons

F is IHR if and only if $N_A(t,T)$ is stochastically increasing in $T > 0$ for each fixed t.

The following are equivalent:

(i) F is NBU.
(ii) $N(t) \geq_{st} N_A(t,T)$ holds for all t and $T > 0$.
(iii) $N(t) \geq_{st} N_B(t,T)$ holds for all t and $T > 0$.

(iv) $Y_i \leq_{st} Y_{i,A}$ holds for all $T > 0, i = 1, 2, \ldots$.
(v) $Y_i \leq_{st} Y_{i,B}(T)$ holds for all $T > 0, i = 1, 2, \ldots$.

Here $Y_i = \inf\{t : N(t) \geq i\} - \inf\{t : N(t) \geq i - 1\}$ and $Y_{i,A}$ and $Y_{i,B}, i = 1, 2, \ldots$, are defined similarly, but for the respective processes $N_A(t, T)$ and $N_B(t, T)$.

J. Some Additional Families

Wang, Hossain and Zimmer (2003) have proposed some families of life distributions with some similarities to the families of IHR, IHRA, and NBU distributions, as well as the DHR, DHRA, and NWU families. These families are based upon the logarithm of the odds ratio $\emptyset^-(x) = F(x)/\bar{F}(x)$, introduced in 1.B(17b). In particular, they consider properties of the function

$$Q(x) = \log \emptyset^-(e^x),$$

From 1.B(18), it follows that $\bar{F}(x) = 1/[1 + \emptyset^-(e^x)]$, so that F is determined by the corresponding log-odds ratio:

$$\bar{F}(x) = \frac{1}{1 + \emptyset^-(e^x)}.$$

Wang, Hossain and Zimmer (2003) successively consider the conditions that Q is convex, starshaped, and superadditive. These conditions parallel the same conditions imposed upon the hazard function R; convexity, starshapedness, and superadditivity of R are, respectively, the conditions that F is IHR, IHRA, and NBU.

Assuming differentiability, the hazard function R is determined by its derivative r; when integrating r to obtain R, the constant of integration is determined by the condition that $R(0) = 0$. However, such is not the case with the function Q. It is determined only up to a constant by its derivative. This means that for families defined in terms of Q, the derivative of Q cannot play a role parallel to the role that the hazard rate plays for families defined in terms of R, as noted in the following examples.

J.1. Example. Let $\bar{F}(x) = 1/[1+\lambda x], x \geq 0, \lambda > 0$; this is the Pareto distribution of 11.A(1). Direct computations show that in this case, $Q(x) = x + \log \lambda$. This linear function has the derivative 1 and is independent of λ.

J.2. Example. The survival function $\bar{F}(x \mid \gamma) = \gamma \bar{F}(x)/(1 - \bar{\gamma}\bar{F}(x))$, $\gamma, x > 0$ is discussed in Section 7.F. Here, the derivative $Q'(x) = r(x)/F(x)$, where r is the hazard rate of F. This derivative is independent of the parameter γ.

6
Nonparametric Families: Inequalities for Moments and Survival Functions

> I collected statistics, I worked out the golden mean, and never understood that extremes join hands, that the man who goes to bed late meets the man who gets up very early, and he who chooses to take his seat on the golden mean, risks falling between two stools.
>
> André Gide, Prometheus Misbound

A. Results Concerning Moments

With the notation of 1.C(3), let $\lambda_r = \mu_r/\Gamma(r+1), r \geq 0$, denote the normalized moments of the distribution F. An interesting fact is that, when they exist, these normalized moments satisfy inequalities that mimic the properties of the density or survival function for many of the classes of distributions described in Chapters 4 and 5. Even when moments do not all exist, the inequalities hold with natural interpretations.

The proofs of Propositions A.1, A.2, and A.3 are given by Marshall and Olkin (1979, p. 494); these proofs, which are somewhat technical and depend upon total positivity theory, are not reproduced here.

A.1. Proposition (Karlin, Proschan and Barlow, 1961). If $F(0) = 0$ and if $f(x)$ is log concave [convex] in $x \geq 0$, then λ_r is log concave [convex] in $r \geq 0$.

A.2. Proposition (Barlow, Marshall and Proschan, 1963). If $F(0) = 0$ and F is IHR (DHR), then λ_r is log concave [convex] in $r \geq 1$.

A.3. Proposition. If f is a completely monotone density, then λ_r is log convex in $r > -1$.

6. Nonparametric Families

A.4. Proposition. If F is an IHRA (DHRA) distribution, then $\lambda_r^{1/r}$ is decreasing (increasing) in $r \geq 0$.

Proof. According to (iv) of Proposition 5.B.3, F is IHRA if and only if $\bar{G}^{-1}\bar{F}$ is starshaped, where G is an exponential distribution. Recall from Definition 2.C.10 that this means $X \leq_* Y$, where X has the distribution F and Y has the distribution G. According to Proposition 2.C.12, this means that $aX \leq_* Y$, for all $a > 0$; consequently, the parameter of the exponential distribution is not relevant and can be of the form $\bar{G}(x) = \exp\{-x/\lambda_r^{1/r}\}$. Now, make use of 1.C(4), that is, of the formula

$$\mu_r = r \int_0^\infty x^{r-1} \bar{F}(x)\,dx$$

and apply Proposition 2.C.13 with $\psi(x) = x^{s-r}$, where $r \leq s$. This yields

$$\lambda_s = \int_0^\infty \frac{sx^{s-1}\bar{F}(x)}{\Gamma(s+1)}\,dx \leq \int_0^\infty \frac{sx^{s-1}\exp\{-x/\lambda_r^{1/r}\}}{\Gamma(s+1)}\,dx = \lambda_r^{s/r}.$$

The proof for the DHRA case is similar. □

According to Proposition 5.C.7, NBU distributions have finite moments of all positive orders. As would be expected, these moments satisfy special inequalities.

A.5. Proposition. If F is NBU, then

$$\lambda_{r+s} \leq \lambda_r \lambda_s, \quad r,s, \geq 0. \tag{1}$$

If F is NWU and the moments are finite, then the inequality (1) is reversed.

Proof. If F is NBU, then $\bar{F}(x+y) \leq \bar{F}(x)\bar{F}(y)$ so that

$$\frac{x^{r-1}}{\Gamma(r)}\frac{y^{s-1}}{\Gamma(s)}\bar{F}(x+y) \leq \frac{x^{r-1}}{\Gamma(r)}\bar{F}(x)\frac{y^{s-1}}{\Gamma(s)}\bar{F}(y), \quad x,y \geq 0.$$

It follows that

$$\int_0^\infty \int_0^\infty \frac{x^{r-1}}{\Gamma(r)}\frac{y^{s-1}}{\Gamma(s)}\bar{F}(x+y)\,dx\,dy$$

$$\leq \int_0^\infty \frac{x^{r-1}}{\Gamma(r)}\bar{F}(x)\,dx \int_0^\infty \frac{y^{s-1}}{\Gamma(s)}\bar{F}(y)\,dy = \lambda_r \lambda_s. \tag{2a}$$

In the left-hand member of this inequality, first let $x+y=z$, so that $y=z-x$, and then make use of 23.B.1 and 23.B.5 to obtain

$$\int_0^\infty \bar{F}(z) \int_0^z \frac{x^{r-1}}{\Gamma(r)} \frac{(z-x)^{s-1}}{\Gamma(s)} \, dx \, dz = \int_0^\infty \bar{F}(z) \frac{z^{r+s-1}}{\Gamma(r+s)} \, dz = \lambda_{r+s}. \tag{2b}$$

If F is NWU, the proof given above can be modified by reversing the inequalities to show that inequality (1) is reversed. □

A.6. Proposition. If F is NBUE, then

$$\lambda_{r+1} \leq \lambda_r \lambda_1, \quad r > 0. \tag{3}$$

If F is NWUE and the moments exist, then the inequality (3) is reversed.

Proof. If F is NBUE, then according to Definition 5.E.1,

$$\int_t^\infty \bar{F}(x) \, dx = \int_0^\infty \bar{F}(x+t) \, dx \leq \mu_1 \bar{F}(t).$$

Consequently,

$$\int_0^\infty \frac{t^{r-1}}{\Gamma(r)} \int_t^\infty \bar{F}(x) \, dx \, dt \leq \frac{\mu_1}{\Gamma(r)} \int_0^\infty t^{r-1} \bar{F}(t) \, dt = \lambda_1 \lambda_r.$$

By interchanging the order of integration, the left side of the inequality becomes

$$\frac{1}{\Gamma(r)} \int_0^\infty \bar{F}(x) \int_0^x t^{r-1} \, dt \, dx = \frac{\mu_{r+1}}{r(r+1)\Gamma(r)} = \lambda_{r+1}.$$

If F is NWUE, then the proof must be modified by reversing all of the inequalities. □

Note that with $r = 1$, (3) yields the inequality $\mu_2 \leq 2\mu_1^2$; this is to be compared with the inequality $\mu_2 \geq \mu_1^2$, which holds for all distributions. The first of these inequalities can be restated in terms of the *coefficient of variation* $\mathrm{CV}(F) = \sigma/\mu_1$.

A.7. Corollary. Denote by μ and σ^2, respectively, the mean and variance of the distribution F. If F is NBUE, then

$$CV(F) = \frac{\sigma}{\mu} \leq 1.$$

If F is NWUE, then

$$CV(F) = \frac{\sigma}{\mu} \geq 1.$$

The coefficient of variation is sometimes used as a measure of normalized dispersion of a distribution, normalized so as to be scale invariant. It is also used to compare normalized dispersion with that of the exponential distribution, for which the coefficient of variation is 1. By examining the proof of Proposition A.6, it is possible to conclude that among the family of NBUE and NWUE distributions, the exponential distribution is unique in having a coefficient of variation equal to 1.

B. Bounds for Survival Functions

As usual, let

$$\bar{F}(x-) = \lim_{\varepsilon \to 0} \bar{F}(x - \varepsilon)$$

so that if the random variable X has the distribution F, then

$$\bar{F}(x-) = P\{X \geq x\}.$$

If the distribution function F satisfies $F(0-) = 0$ and has rth moment $\mu_r < \infty$, where $r > 0$, then according to Markov's inequality,

$$\begin{aligned} 0 \leq \bar{F}(t-) &\leq \mu_r/t^r, \quad t \geq \mu_r, \\ &\leq 1, \quad t \leq \mu_r. \end{aligned} \quad (1)$$

These inequalities are known to be sharp; indeed, for each positive r and t and for each of the inequalities of (1), there exists a distribution satisfying the conditions of (**1**) and attaining equality. Markov's inequality, although best possible without stronger hypotheses, is useful in proving the weak law of large numbers, but is not very useful for many other purposes.

A number of improvements of Markov's bound have been obtained with stronger hypotheses. Indeed, there is a better bound due to Gauss (1823), which predates the work of Markov (1898) and depends upon the assumption that $\bar{F}$ is convex on $[0, \infty)$. This section offers a brief introduction to some improvements of Markov's inequality for some of the classes of distributions discussed in earlier sections of this chapter.

A number of additional inequalities have been given by Barlow and Marshall (1964), some of which depend upon additional information such as bounds on the hazard rate, or on the value of a Laplace transform at one point. Bounds on interval probabilities, densities, and hazard rates are given by Barlow and Marshall (1967) with various hypotheses.

The methods used in what follows ensure that the inequalities are "sharp," meaning that they cannot be improved without stronger hypotheses. Indeed, there is a distribution satisfying the hypotheses that achieves the bound—that is, the inequality becomes an equality. In most cases, the idea is to first identify a family of "extremal" distributions and then find the most extreme of that family satisfying the stated hypotheses. For example, to obtain an upper bound on $\bar{F}(t)$, the method calls for the identification of an extremal distribution G for which $\bar{F}(t) \leq \bar{G}(t)$ whenever F satisfies the specified conditions. The existence of the distribution G ensures that the inequality is sharp.

Summary of upper bounds for survival functions

Assumptions	Proposition
$\bar{F}$ convex, μ_r given	B.1
f unimodal, μ given	B.1.a
F IHRA, μ_r given	B.2
F DHR, μ_r given	B.3
Log F concave, μ given	B.4
F NWUE, μ given	B.5

Summary of lower bounds for survival functions

Assumptions	Proposition
F IHR, μ_r given	B.6
F IHRA, μ_r given	B.7
F NBUE, μ given	B.8

a. Improvements of Markov's Inequality Under Various Conditions

The following proposition was partially obtained by Camp (1922) and Meidell (1922) and was obtained by Fréchet (1950). The case $r = 2$ is essentially equivalent to the result of Gauss (1823) who used the method of proof offered below.

B.1. Proposition. If $F(0-) = 0$, $\bar{F}$ is convex on $[0, \infty)$, and F has rth moment $\mu_r, r > 0$, then

$$\bar{F}(t) \leq 1 - \frac{t}{(r+1)^{1/r} \mu_r^{1/r}}, \quad t \leq t_0, \qquad (2)$$

$$\leq \frac{\mu_r}{t^r} \left[\frac{r}{r+1} \right]^r, \quad t \geq t_0,$$

where $t_0 = r\mu_r^{1/r}/(r+1)^{1-1/r}$. The inequality is sharp.

Proof. Because $\bar{F}$ is convex on $[0, \infty)$, $\bar{F}$ has a supporting line at t, and consequently, for some $a \in (0, 1]$ and $b > 0$, $\bar{F}(x) \geq \bar{G}(x), x \geq 0$, and $\bar{F}(t) = \bar{G}(t)$, where

$$\bar{G}(x) = a - bx, \quad 0 \leq x \leq a/b, \quad \bar{G}(x) = 0, x \geq a/b.$$

See Figure B.1. Because $\bar{F}(t) = \bar{G}(t)$, it must be that $b = [a - \bar{F}(t)]/t$. Because $\bar{F}(x) \geq \bar{G}(x), x \geq 0$, it follows from 1.C(4) that

$$\mu_r \geq r \int_0^\infty x^{r-1} \bar{G}(x)\, dx = \frac{a^{r+1}}{(r+1)b^r} = \frac{a^{r+1} t^r}{(r+1)[a - \bar{F}(t)]^r}.$$

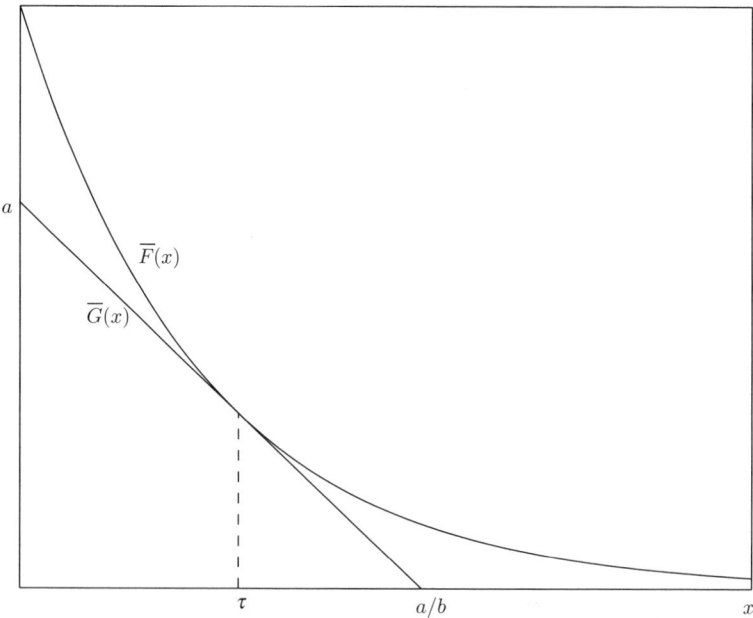

Fig. B.1. The survival functions $\bar{F}$ and $\bar{G}$

This means that for some $a \in (0, 1]$,
$$\bar{F}(t) \leq a - [a^{(r+1)/r}t]/(r+1)^{1/r}\mu_r^{1/r}.$$

Maximization of the bound subject to $a \in (0, 1]$ yields (2). Here, the maximum occurs at $a = r\mu_r^{1/r}/t(r+1)^{(r-1)/r}$ if $t \geq t_0$, and at 1 if $t \leq t_0$. Equality is attained in (2) when $t \leq t_0$ by the distribution
$$F(x) = x/[(r+1)\mu_r]^{1/r}, \quad 0 \leq x \leq [(r+1)\mu_r]^{1/r};$$
when $t \geq t_0$, equality is attained with
$$\bar{F}(x) = \frac{r^r \mu_r}{(r+1)^{r-1}t^r}\left[1 - \frac{rx}{(r+1)t}\right], \quad 0 \leq x \leq \frac{(r+1)t}{r}. \quad \square$$

The above inequality depends upon the knowledge that $\bar{F}$ is convex; when a density exists, this is equivalent to the assumption that the density is decreasing, i.e., the density is unimodal with mode at 0. The following inequality due to Colin Mallows (private communication) assumes that the density is unimodal, but the location of the mode is unknown.

B.1.a. Proposition. If f is a unimodal density with $\mu_1 = 1$ and if $F(0-) = 0$, then
$$\bar{F}(t) \leq 1, \quad 0 \leq t \leq 1,$$
$$\leq (2/t) - 1, \quad 1 \leq t \leq 3/2,$$
$$\leq 1/2t, \quad t \leq 3/2.$$

The proof of this inequality makes use of the methods provided by Mallows (1963), and is omitted.

B.2. Proposition. Suppose that F is IHRA and has rth moment μ_r where $r > 0$. Then,
$$\bar{F}(t-) \leq 1, \quad t \leq \mu_r^{1/r},$$
$$\leq e^{-wt}, \quad t > \mu_r^{1/r}, \quad (3a)$$

where w is determined by the equation
$$\int_0^t rz^{r-1} e^{-wz}\, dz = \mu_r. \quad (3b)$$

(Note that the left-hand side of (3b) is decreasing from t^r at $w = 0$ to 0 at $w = \infty$, and consequently crosses the value μ_r at one point.) The bound is sharp and cannot be improved even with the assumption that F is IHR.

Proof. Let $\bar{G}(z) = e^{-wz}$, $z < t$, $\bar{G}(z) = 0$, $z \geq t$, and choose w so that G has the same rth moment as F. Because e^{-wz} can cross $\bar{F}(z)$ at most once, and only from below, it must be that $\bar{G}(t) \geq \bar{F}(t)$ for otherwise F and G would not cross at all and could not have the same rth moment. Sharpness is obtained because the distribution G attains the bound when it is not trivial. When only the trivial upper bound 1 is given, the distribution degenerate at $\mu_r^{1/r}$ attains the bound. □

This bound has been tabulated for $r = 1$ by Barlow and Marshall (1965). When $r = 1$, (3a) takes the particularly simple form $1 - w\mu = e^{-wt}$. Because the left-hand side of this equation is linear and the right-hand side is convex, the equation has a unique positive solution.

B.3. Proposition. If F is DHR and has rth moment $\mu_r < \infty$, where $r > 0$ and $\lambda_r = \mu_r/\Gamma(r+1)$, then

$$\bar{F}(t-) \leq \exp\{-t/\lambda_r^{1/r}\}, \quad t \leq r\lambda_r^{1/r},$$
$$\leq r^r \lambda_r e^{-r}/t^r, \quad t \geq r\lambda_r^{1/r}.$$

This inequality is sharp.

For a proof of this inequality, see Barlow and Marshall (1964).

B.4. Proposition. Suppose that $F(0-) = 0$, $\log F$ is concave (F is DRHR), and F has first moment μ. Then,

$$\bar{F}(t) \leq 1 - \left(\frac{w}{w-1}\right)^{w-1} e^{-1}, \quad w = \frac{t}{\mu} > 1, \quad (4)$$
$$\leq 1, \quad w \geq 1.$$

The bound is sharp.

Proof. Because $\log F$ is concave, this function has a supporting line at t; thus, for some $a \geq 0$ and $\log c > 0$, $\log F(x) \leq ax - \log c$ with equality at $x = t$. Consequently, $a = \log cF(t)/t > 0$ and

$$F(x) \leq \frac{[cF(t)]^{x/t}}{c}, \quad 0 \leq x \leq \frac{t \log c}{\log cF(t)} = x_0,$$
$$\leq 1, \quad x \geq x_0.$$

It follows from this and from 1.C(4) that

$$\mu \geq \int_0^{x_0} \left[1 - \frac{[cF(t)]^{x/t}}{c}\right] dx = \frac{t}{\log cF(t)} \left[\frac{1}{c} - 1 + \log c\right].$$

This means that for some $c > 1$,

$$\log F(t) \geq \frac{t}{\mu}\left[\frac{1}{c} - 1 + \log c\right] - \log c.$$

Minimization of the lower bound with respect to c leads to the inequality (4). Equality is attained when

$$F(x) = 0, \quad x < 0,$$
$$= \left(\frac{w}{w-1}\right)^{(wx/t)-1} e^{-x/t}, \quad 0 \leq x \leq x_1,$$
$$= 1, \quad x > x_1,$$

where $x_1 = \left[t \log \frac{w}{w-1}\right] / \left[w \left(\log \frac{w}{w-1}\right) - 1\right]$. □

B.5. Proposition (Brown, 2001). If F is NBU and has first moment μ, then

$$\bar{F}(t-) \leq (\mu/t)^2, \quad t \geq \mu,$$
$$\leq 1, \quad t < \mu. \qquad (5)$$

Brown (2001) obtains this bound from general results concerning submultiplicative functions due to Phillips (1954). The bound (5) is not sharp, and for large t, the bound (6) of the following proposition is better. To be more explicit, let $s = t/\mu$. When $2 \log s = s - 1$, the bounds of (5) and (6) are equal. For $s > c \approx 3.51286$, the bound (6) of the following proposition is better.

B.6. Proposition (Brown, 2001). If F is NBUE and has first moment μ, then

$$\bar{F}(t-) \leq \exp\left\{-\left(\frac{t}{\mu} - 1\right)\right\}, \quad t \geq \mu. \qquad (6)$$

This bound is likely not sharp, but no better bound is known; the proof is omitted. This bound is to be compared with 5.E(2) which provides a bound for the equilibrium survival function $\bar{F}_{(1)}(t)$.

B.7. Proposition. If F is NWUE and has first moment μ, then

$$\bar{F}(t-) \leq \mu/(\mu+t), \quad t \geq 0. \tag{7}$$

This inequality is sharp and cannot be improved even with the assumption that F is NWU.

Proof. The condition that F is NWUE can be written in the form

$$\mu \bar{F}(t) \leq \int_t^\infty \bar{F}(z)\,dz, \quad t \geq 0;$$

because

$$\int_t^\infty \bar{F}(z)\,dz = \mu - \int_0^t \bar{F}(z)\,dz \quad \text{and} \quad \int_0^t \bar{F}(z)\,dz \geq t\bar{F}(t),$$

it follows that

$$\mu F(t) \geq t\bar{F}(t),$$

and consequently, $\bar{F}(t) \leq \mu/(\mu+t)$. The proof of (7) is completed by the limiting argument

$$\bar{F}(t-) = \lim_{\varepsilon \to 0} \bar{F}(t-\varepsilon) \leq \lim_{\varepsilon \to 0} \mu/(\mu+t-\varepsilon) = \mu/(\mu+t).$$

Equality in (7) is attained by the distribution G given by

$$\bar{G}(x) = p^{i+1}, \quad ix \leq x < (i+1)x, \quad i = 0, 1, 2, \ldots,$$

where $p = \mu/(\mu+t)$. This distribution is NWU and has first moment μ; consequently, (7) cannot be improved with the stronger hypothesis that F is NWU. □

For a comparison of upper bounds for $\bar{F}(t-)$ under various conditions see Figures B.2 and B.3, both with $\mu = 1$.

b. Reversals of Markov's Inequality

Markov's inequality provides an upper bound in the survival function, but in general, no nontrivial lower bound can be given in terms of one moment. On the other hand, such bounds are possible with additional assumptions.

B. Bounds for Survival Functions 205

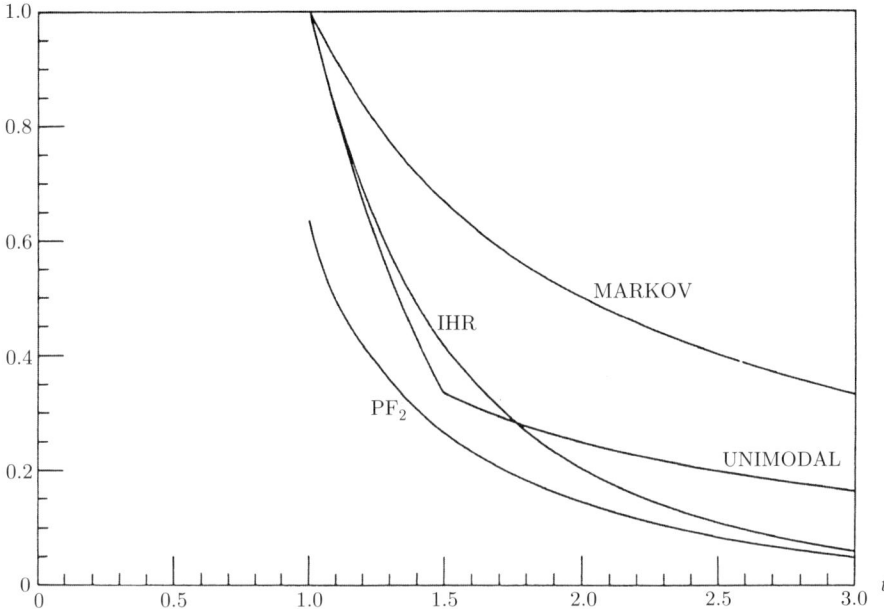

Fig. B.2. Upper bounds for the survival function under various conditions ($\mu = 1$)

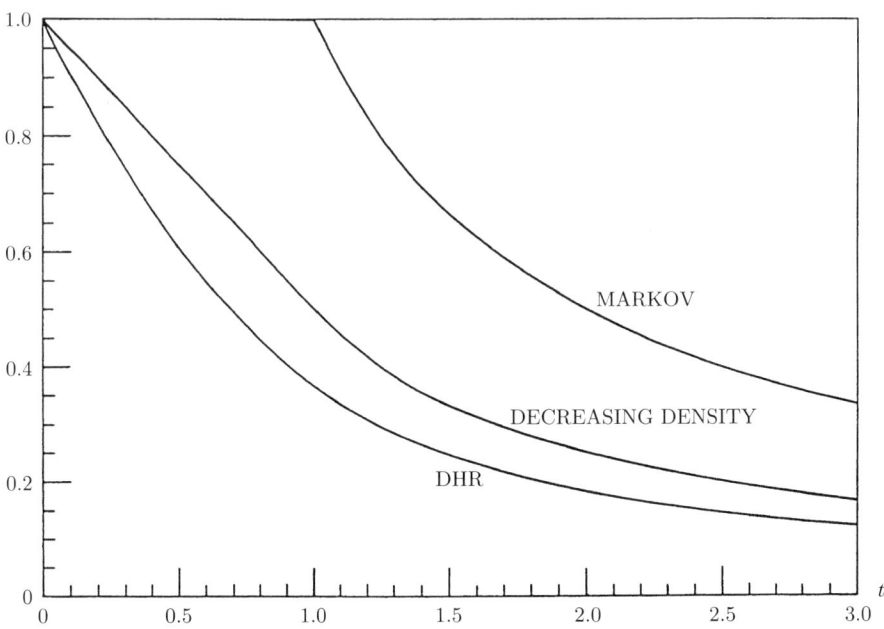

Fig. B.3. Upper bounds for the survival function under various conditions ($\mu = 1$)

B.8. Proposition (Barlow and Marshall, 1964). Suppose that F is IHR and $\lambda_r = \mu_r/\Gamma(r+1)$, where $r \geq 1$. Then,

$$\bar{F}(t) \geq \exp\{-t/\lambda_r^{1/r}\}, \quad t < \mu_r^{1/r}, \tag{8}$$
$$\geq 0, \quad t \geq \mu_r^{1/r}.$$

The bound is sharp.

This result is due to Richard E. Barlow. A more general result is given by Barlow and Marshall (1964), and another generalization is given by Barlow and Proschan (1975).

B.9. Proposition (Barlow and Marshall, 1967). Suppose that F is IHRA and has rth moment μ_r, where $r > 0$. Then,

$$\bar{F}(t) \geq \min\left[e^{-bt}, e^{-ct}\right], \quad t < \mu_r^{1/r}, \tag{9a}$$
$$\geq 0, \quad t \geq \mu_r^{1/r},$$

where $c = 1/\lambda_r^{1/r}$ and $b = b(t)$ is determined as the unique solution of the equation

$$t^r(1 - e^{-bt}) + \int_t^\infty z^r b e^{-bz}\, dz = \mu_r. \tag{9b}$$

For a proof, see Barlow and Marshall (1967, p. 248). For $r = 1$, this bound is tabulated by Barlow and Proschan (1975, p. 117) (see Figure B.4). In case $r = 1$, (9a) takes the relatively simple form

$$b(\mu - t) = e^{-bt}. \tag{9c}$$

B.10. Proposition. If F is NBUE and has first moment μ, then

$$\bar{F}(t) \geq 1 - \frac{t}{\mu}, \quad t \leq \mu,$$
$$\geq 0, \quad t > \mu, \tag{10}$$

This inequality is sharp. Moreover, it cannot be improved even with the stronger hypothesis that F is NBU.

Proof. If F is NBUE, then, for all $t \geq 0$,

$$\mu F(t) \leq \int_0^t \bar{F}(z)\, dz.$$

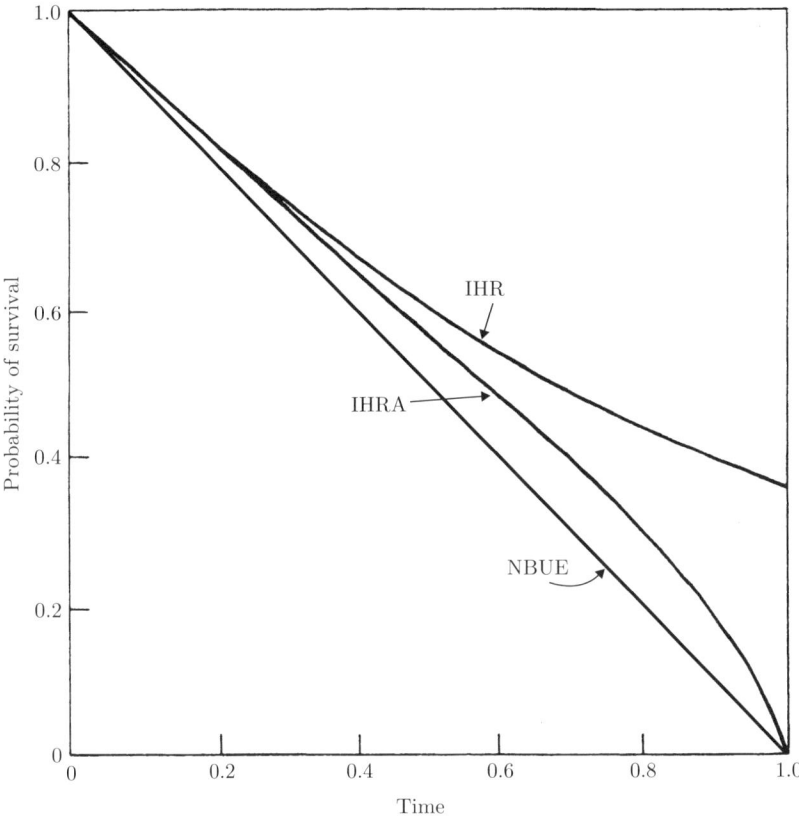

Fig. B.4. Lower bounds for IFR and IFRA survival functions ($\mu = 1$)

But trivially, the integral is less than t. This proves the inequality. To show that it cannot be improved even with the assumption that F is NBU, first consider the case that $x \leq \mu$ and compute that if the hazard function

$$R(z) = -i \log p, \quad ix \leq z < (i+1)x \text{ with } 1 - p = t/\mu,$$

then F has expectation μ and achieves equality in (10). This distribution is that of Example 5.C.6. For the case that $t > \mu$, the distribution degenerate at μ achieves equality. □

With only the first moment known, nontrivial lower bounds on $\bar{F}$ cannot be given under the assumption that F is DHR (or any weaker assumption). To see this note that $\bar{F}(x) = a \exp\{-ax/\mu\}, x > 0$, defines a DHR distribution with mean μ and $\lim_{a \to 0} \bar{F}(x) = 0$, for all $x > 0$.

c. Bounds with Two Moments Given

The above-mentioned bounds all depend upon only one moment, and they can be substantially improved if two moments, say the first and second, are known.

B.11. Proposition (Chebyshev, 1874). Let F be a distribution such that $F(0-) = 0$. If $\mu_1 = 1$ and μ_2 is known, then

$$\bar{F}(t) \geq (1-t)^2/[(\mu_2 - 1) + (1-t)^2], \quad 0 \leq t \leq 1,$$
$$\geq 0, \quad t \geq 1;$$
$$\bar{F}(t) \leq 1, \quad 0 \leq t \leq 1,$$
$$\leq 1/t, \quad 1 < t < \mu_2,$$
$$\leq (\mu_2 - 1)/[\mu_2 - 1 + (t-1)^2], \quad t \geq \mu_2.$$

B.12. Proposition (Royden, 1953). Let F be a distribution concave on $[0, \infty)$ such that $F(0-) = 0$. If $\mu_1 = 1$ and μ_2 is known, then

$$\bar{F}(t) \geq (2-t)^2/(3\mu_2 - 2t), \quad 0 \leq t \leq 2,$$
$$\geq 0, \quad t \geq 2;$$
$$\bar{F}(t) \leq 4(3\mu_2 - 2t)/9\mu_2^2, \quad 3\mu_2/4 \leq t \leq \mu_2,$$
$$\leq (3\mu_2 - 4)/[4(3\alpha^2 - 4\alpha) + 3\mu_2], \quad t \geq \mu_2,$$

where α is the unique root $\geq t/2$ of

$$t = [16\alpha^2(\alpha - 1)]/[4(3\alpha^2 - 4\alpha) + 3\mu_2].$$

For IHR distributions, sharp bounds of this kind have been obtained by Barlow and Marshall (1964), but unfortunately, they are mostly given implicitly as solutions of transcendental equations that have to be solved numerically. This has been done in a number of cases; see Barlow and Marshall (1965). Proofs of these results mostly involve studies of extremal distributions and in spirit are similar to the proof of Proposition B.2 given earlier. They make use of the fact that as functions of r, moments μ_r from two distributions can cross no more times than the survival functions cross (see 21.B.12 for an indication of the proof of this fact). The tightness of the bounds based upon two moments is evidenced by the proximity of upper and lower bounds.

Figures B.5, B.6, B.7, and B.8 illustrate the bounds in special cases, and show the improvement resulting the IHR or DHR assumption by

B. Bounds for Survival Functions 209

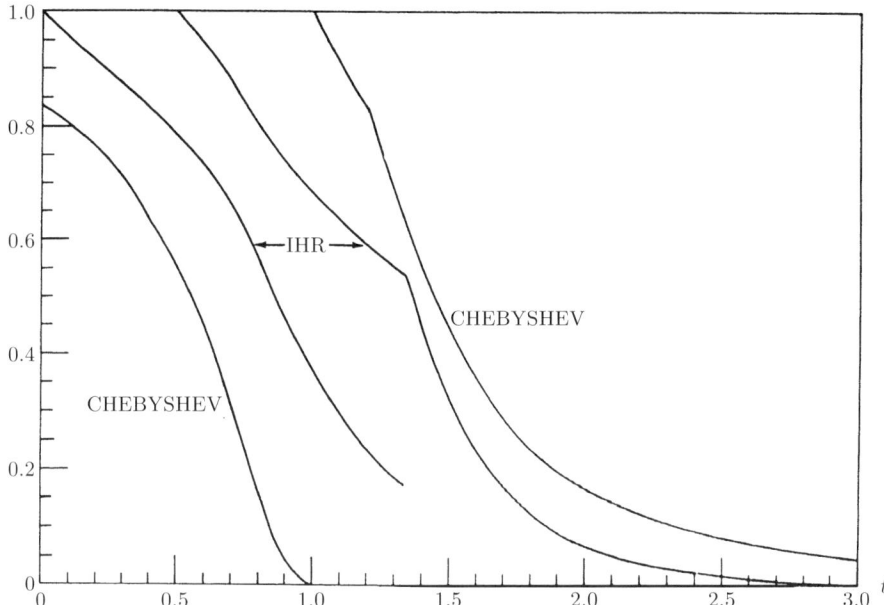

Fig. B.5. Upper and lower bounds for the survival function ($\mu_1 = 1$ and $\mu_2 = 1.2$)

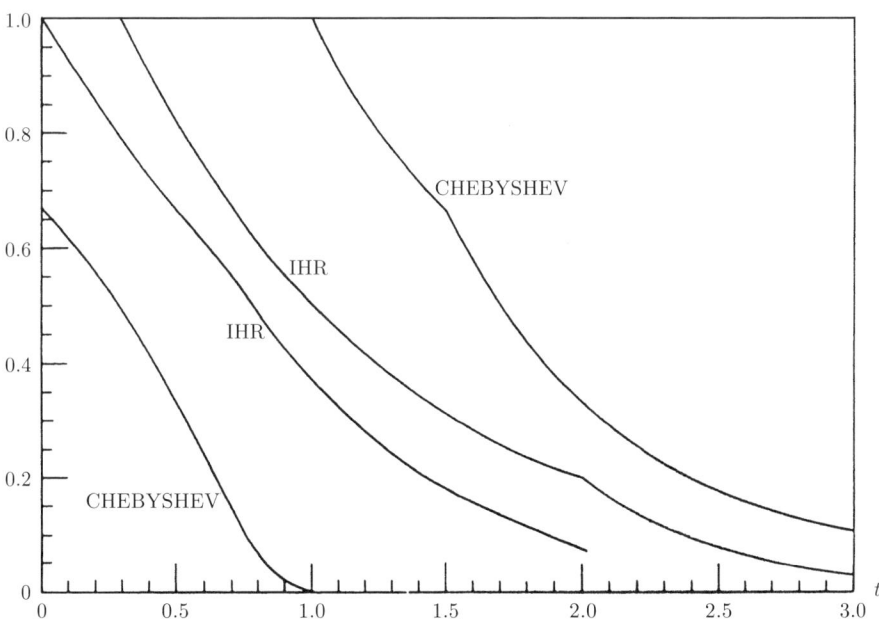

Fig. B.6. Upper and lower bounds for the survival function ($\mu_1 = 1$ and $\mu_2 = 1.5$)

210   6. Nonparametric Families

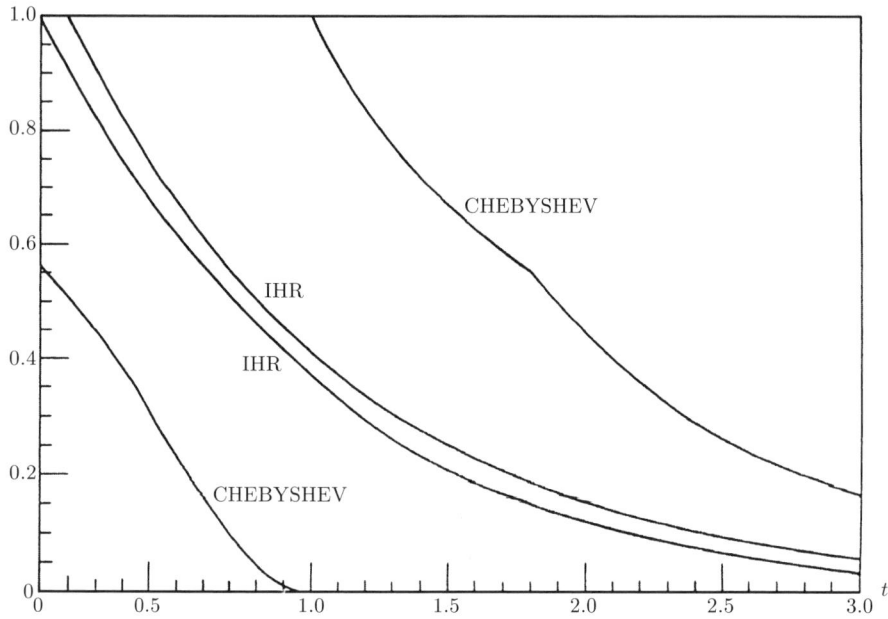

Fig. B.7. Upper and lower bounds for the survival function ($\mu_1 = 1$ and $\mu_2 = 1.8$)

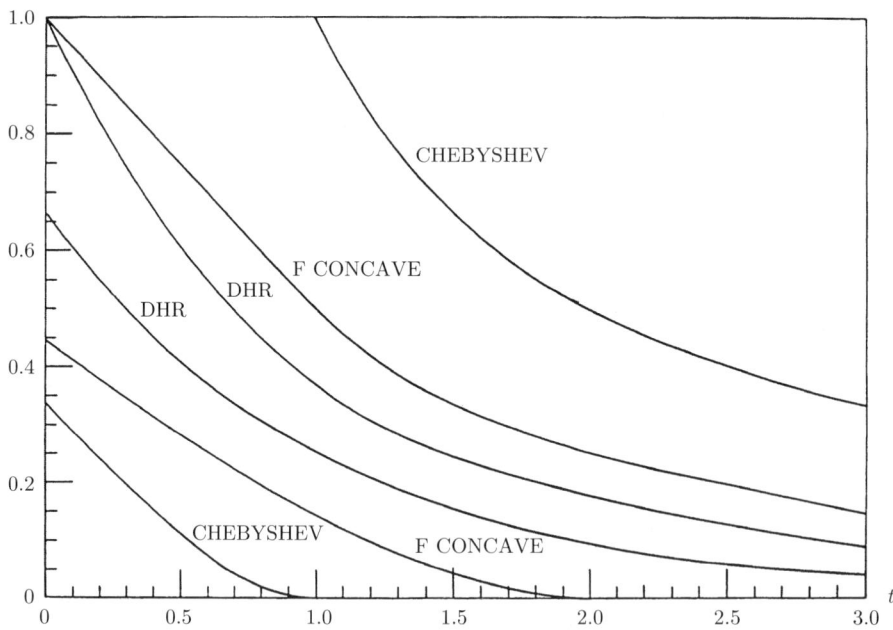

Fig. B.8. Upper and lower bounds for the survival function ($\mu_1 = 1$ and $\mu_2 = 3$)

comparison with more standard Chebyshev bounds. Note the proximity of upper and lower bounds when $\mu_1 = 1$ and μ_2 is close to 2. This is due to the fact that IHR distributions with $\mu_1 = 1$ and $\mu_2 = 2$ must be exponential distributions.

d. Bounds Based on Percentiles

In the previous sections, bounds are based on knowledge of moments. In some applications, there may instead be knowledge of percentiles; for example, the median may be known but not the mean. The following propositions show how to obtain bounds on the survival function in such cases.

B.13. Proposition. Suppose that F is IHRA, and it is known that $\bar{F}(x_0) = p_0$. Let $\bar{G}(x) = p_0^{x/x_0}, x \geq 0$. Then,

$$\bar{F}(x) \geq \bar{G}(x), \quad 0 \leq x \leq x_0, \quad \text{and} \quad \bar{F}(x) \leq \bar{G}(x), \quad x \geq x_0.$$

The bounds are sharp and connot be improved by assuming that F is IHR.

Proof. Because F is IHRA, $[\bar{F}(x)]^{1/x}$ is decreasing in $x > 0$. Because $\bar{F}(x_0) = \bar{G}(x_0)$ and $[\bar{G}(x)]^{1/x}$ is a constant, the result follows. This is just another way of saying that an IHRA survival function can cross an exponential survival function at most once, and only from above (see Proposition 5.B.3(ii)). □

B.14. Proposition. Suppose that F is IHRA, and it is known that $\bar{F}(x_0) = p_0$, $\bar{F}(x_1) = p_1$, where $x_0 < x_1$. Let $\bar{G}_0(x) = p_0^{x/x_0}, x \geq 0$, and $\bar{G}_1(x) = p_1^{x/x_1}, x \geq 0$. Then,

$$\bar{F}(x) \geq \bar{G}_0(x), \quad 0 \leq x \leq x_0,$$
$$\bar{G}_0(x) \geq \bar{F}(x) \geq \bar{G}_1(x), \quad x_0 \leq x \leq x_1,$$
$$\bar{F}(x) \leq \bar{G}_1(x), \quad x \geq x_1.$$

Proof. The proof of these inequalities is similar to the proof of Proposition B.13, in that they all follow from the fact that an IHRA survival function can cross an exponential survival function at most once, and only from above. The fact that the inequalities are sharp can be demonstrated by constructing an IHRA survival function that is made up of segments of the $\bar{G}_i$, equal to the bounds on appropriate intervals. □

B.15. Proposition. Suppose that F is IHR, and it is known that $\bar{F}(x_0) = p_0$, $\bar{F}(x_1) = p_1$, where $x_0 < x_1$. Let $\bar{G}_0$ and $\bar{G}_1$ be as in Proposition B.14, and let

$$\bar{G}_2(x) = 1, \quad 0 \leq x \leq a$$
$$= p_0^{(x_1-x)/(x_1-x_0)} p_1^{(x-x_0)/(x_1-x_0)}, \quad x \geq a,$$

where

$$a = \frac{x_0 \log p_1 - x_1 \log p_0}{\log p_1 - \log p_0}.$$

Then,

$$\bar{F}(x) \geq \bar{G}_0(x), \quad x \leq x_0,$$
$$\geq \bar{G}_2(x), \quad x_0 < x \leq x_1,$$

and

$$\bar{F}(x) \leq \bar{G}_2(x), \quad a \leq x \leq x_0,$$
$$\leq \bar{G}_0(x), \quad x_0 \leq x \leq x_1,$$
$$\leq \bar{G}_1(x), \quad x \geq x_1.$$

These bounds are sharp.

Proof. Because F is IHR, $[\bar{F}(x)]^{1/x}$ is decreasing in $x > 0$; this means that $p_0^{1/x_0} \geq p_1^{1/x_1}$ and consequently, $a > 0$. Because F is IHR, $\log \bar{F}(x)$ is a concave function of x. On the other hand, $\log \bar{G}_i(x)$ is linear in $x \geq 0, i = 0, 1$, and $\log \bar{G}_2(x)$ is linear in $x \geq a$. Consequently the inequalities can be read directly from Figure B.9. The fact that the inequalities are sharp can be demonstrated by constructing IHR survival functions that are piecewise exponential and made up of segments of the $\bar{G}_i$, equal to the bounds on appropriate intervals. □

The above-mentioned inequalities can be easily extended to the case that any finite number of percentiles are given.

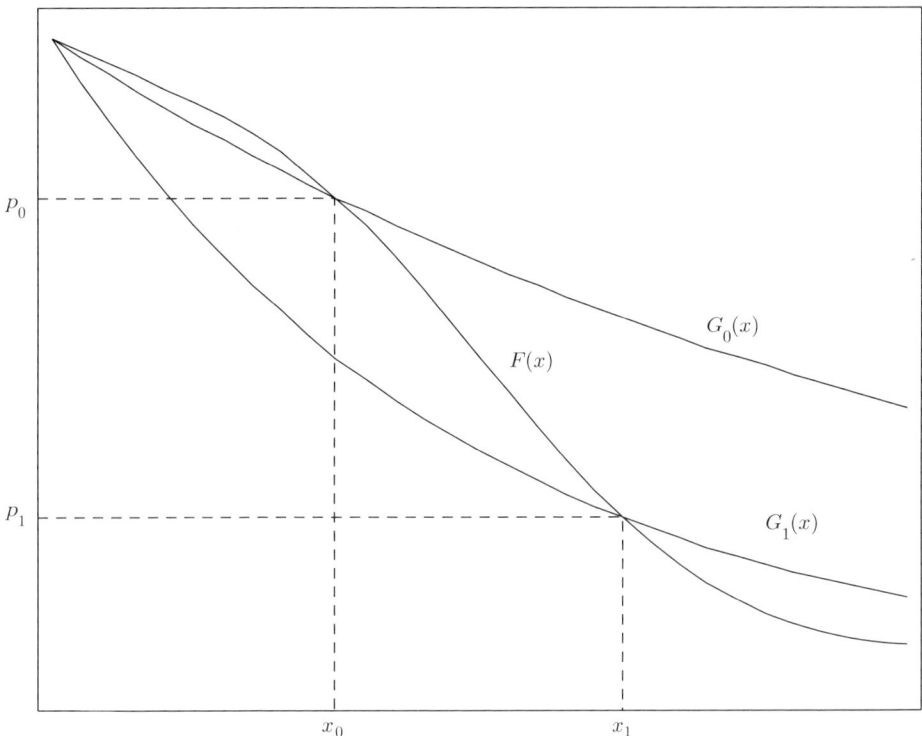

Fig. B.9. The bounds of Proposition B.14

e. Related Bounds

The inequalities considered above provide bounds for distributions or survival functions at a fixed point, say t. The bound is regarded as being "sharp" or best possible if there exists a distribution satisfying the hypotheses that achieves equality. Bounds that apply for all values of t can also be obtained.

For distributions in the various classes discussed in this chapter, the exponential distribution plays a central role; for example, if F is both IHR and DHR or both DMRL and IMRL, then it is an exponential distribution. How far can such distributions deviate from an exponential distribution with the same first moment μ? Recall that the coefficient of variation of an exponential distribution is 1.

To interpret the following proposition, recall that as a consequence of Proposition A.6, F is NBUE implies $\mu_2 \leq 2\mu_1^2$; consequently this moment inequality holds under the stronger condition that F is IHR.

This moment inequality is reversed when F is NWUE, and consequently is reversed if F is IMRL.

B.16.a. Proposition (Brown, 1987). If F has an increasing hazard rate, then
$$\sup_t |\bar{F}(t) - e^{-t/\mu}| \leq 1 - [CV(F)]^2 = 2\left(1 - \frac{\mu_2}{2\mu_1^2}\right),$$
where $CV(F)$ is the coefficient of variation of F.

B.16.b. Proposition (Brown, 1983). If F has an increasing mean residual life, then
$$\sup_t |\bar{F}(t) - e^{-t/\mu}| \leq \frac{[CV(F)]^2 - 1}{[CV(F)]^2 + 1} = 1 - \frac{2\mu_1^2}{\mu_2}.$$

Similar results for NBUE and NWUE distributions have been obtained by Brown and Ge (1984).

Part III

Semiparametric Families

7
Semiparametric Families

A family is a unit composed not only of children, but of men, women, an occasional animal, and the common cold.

Ogden Nash

A. Introduction

Ordinarily, parameters of a distribution are thought of as being real or possibly vector-valued. In this chapter, families of distributions are discussed, which are distinguished by having a parameter that is itself a distribution function. These families also have a real-valued parameter and consequently are said to be *semiparametric*.

A possible procedure making use of a semiparametric model is to first select the parameter that is a distribution function. This distribution function is called the *underlying distribution.* In effect, the choice of an underlying distribution leads to the selection of a parametric model, but with the selection limited to families having the structure of the semiparametric model. Alternatively, the distribution function-valued parameter can be restricted to some nonparametric families such as those discussed in Chapter 4 or 5.

It may be that the underlying distribution F itself already has a parameter or even several parameters; then a semiparametric family may provide a way to add a new parameter θ and thereby to extend the family from which F came. The standard families of gamma distributions and Weibull distributions can be thought of as coming from the exponential distribution by way of semiparametric families that add a second parameter. By the same method, it is possible to find a three-parameter family that includes both the gamma and Weibull families as special cases. So the study of semiparametric families is useful for two purposes: It provides a new understanding of standard families of distributions, and it provides methods of extending families for added flexibility in fitting data.

The various semiparametric families discussed in Sections B through K include many families that can be found in the literature, implicitly or explicitly. Certainly these 10 examples do not exhaust all possibilities (several more examples are noted in Section O), but the intent is to include the most commonly encountered and the most important semiparametric families.

a. Summary/Preview

Table A.1 can be regarded as a summary or preview of what follows in more detail. Here, the name of various kinds of parameters is introduced; the notation is discussed in Section b; it is far from ideal, and the reader is cautioned to note that $\bar{F}(x|\lambda)$, $\bar{F}(x|\alpha)$, $\bar{F}(x|\xi)$, $\bar{F}(x|\gamma)$, and

Table A.1. Summary of some semiparametric families

Location parameter b:	$F(x\,|\,b) = F(x-b), \quad -\infty < b < \infty$
Scale parameter $\lambda > 0$:	$F(x\,|\,\lambda) = F(\lambda x)$
Power parameter $\alpha > 0$:	$F(x\,|\,\alpha) = F(x^\alpha), \quad x \geq 0, \quad F(0-) = 0$
Frailty parameter $\xi > 0$	$\bar{F}(x\,|\,\xi) = [\bar{F}(x)]^\xi$
Resilience parameter $\eta > 0$	$F(x\,|\,\eta) = [F(x)]^\eta$
Tilt parameter $\gamma > 0$	$\bar{F}(x\,|\,\gamma) = \gamma \bar{F}(x)/[F(x) + \gamma \bar{F}(x)]$
Hazard power parameter $\zeta > 0$	$\bar{F}(x\,|\,\zeta) = \exp\{-[R(x)]^\zeta\},$ where $R(x) = -\log \bar{F}(x)$
Moment parameter β	$\bar{F}(x\,|\,\beta) = \dfrac{1}{\mu_\beta} \int_x^\infty z^\beta \, dF(z), x \geq 0,$ where μ_β is the βth moment of F
Laplace transform parameter s	$\bar{F}(x\,|\,s) = \dfrac{1}{\phi(s)} \int_x^\infty e^{-sz} \, dF(z),$ where $\phi(s) = \int e^{-sx} \, dF(x)$
Convolution parameter ν	$\phi(s\,|\,\nu) = [\phi(s)]^\nu,$ where $\phi(s) = \int e^{-sx} \, dF(x)$
(may require F to be infinitely divisible, depending upon the range of ν)	
Age parameter τ	$F(x\,|\,\tau) = \bar{F}(\tau + x)/\bar{F}(\tau)$

so on, mean different things. Consequently, an expression like $\bar F(x\,|\,2)$ has meaning only in terms of the context where it arises. Note that in some cases, the definition is stated most simply in terms of F, in some cases it is most easily stated in terms of $\bar F$, and in other cases there is no difference in simplicity.

b. Comments About Notation

There are two ways to view the semiparametric families discussed in this chapter. As described above, such a family has two parameters, a real-valued parameter, say θ, and a distribution-valued parameter, say F. Then a member of the family might be denoted as $H(\cdot\,|\,\theta, F)$.

When survival functions are more convenient to work with than distribution functions, the survival function can serve as the parameter in place of the distribution function. Then, a member of the family of survival functions and the underlying survival function might be denoted as $\bar H(\cdot\,|\,\theta, \bar F)$. For example, it might be that $\bar H(\cdot\,|\,\theta, \bar F) = [\bar F(\cdot)]^\theta$.

From another point of view, $H(\cdot\,|\,\theta, F)$ is the result of adding a parameter to the underlying distribution F to generate from F a parametric family. As in Table A.1, a member of this parametric family may also be naturally denoted by F_θ or by $F(\cdot\,|\,\theta)$ as an alternative to $\bar H(\cdot\,|\,\theta, \bar F)$. Then, $F(\cdot\,|\,\theta)$ can be distinguished from $F(\cdot)$ by carrying in some way the label θ. These notations carry over to other functions such as survival functions, densities, hazard rates, and Laplace transforms.

In this book, notation of the form $F(\cdot\,|\,\theta) = H(\cdot\,|\,\theta, F)$ is used for many different kinds of parameters θ including location, scale, and all of the other kinds of parameters that appear in Table A.1. The notation $F(\cdot\,|\,\theta)$ alone does not imply anything about the structure of the parametric family. However, as much as possible, a distinctive letter in place of the generic θ is given to each kind of parameter that appears in Table A.1 and is further discussed in this chapter. The notation is intended to indicate the type of family.

c. Criteria for Semiparametric Families

The semiparametric families discussed in this book have two important properties imposed as minimal criteria for judging a family worthy of study.

A.1. Example (Scale parameters). Suppose that $\bar H(x\,|\,\theta, \bar F) = \bar F(\theta x), \theta > 0$, and take the underlying distribution to have the survival function $\bar F(x) = e^{-x}$, $x \geq 0$. Then, $\bar H(x\,|\,\theta, \bar F) = e^{-\theta x}$. Because

$\bar{H}(\cdot\,|\,1,\bar{F}) = \bar{F}(\cdot)$, the underlying distribution is a member of the parametric family.

Criterion 1. The underlying distribution is a member of the parametric family. That is, for some value θ^* of the parameter θ,

$$\bar{H}(\cdot\,|\,\theta^*,\bar{F}) = \bar{F}(\cdot).$$

This means that the semiparametric families include all distributions (or in some cases, only all life distributions) unless the distribution function-valued parameter is restricted.

A.1.a. Continuation of Example A.1. As noted above, if $\bar{H}(x\,|\,\theta,\bar{F}) = \bar{F}(\theta x)$ and $\bar{F}(x) = e^{-x}$, it follows that $\bar{F}(x\,|\,\theta) = e^{-\theta x}$, $x \geq 0, \theta > 0$. If a member $\bar{F}(x) = e^{-\rho x}$ of this family is taken as the underlying survival function and the same semiparametric family $\bar{H}(x\,|\,\theta,\bar{F}) = \bar{F}(\theta x)$ is used, the result is $\bar{H}(x\,|\,\theta,\bar{F}) = e^{-\theta \rho x}$. This family is indexed by $\theta\rho$, but reapplication of the construction has not enlarged the family.

Criterion 2. Suppose that the semiparametric family is used to add a parameter to an underlying distribution F, yielding the family of survival functions of the form $\bar{H}(\cdot\,|\,\theta,\bar{F})$. Now suppose that the same semiparametric family is used with $\bar{H}(\cdot\,|\,\rho,\bar{F})$ in place of $\bar{F}$ as the underlying survival function. This reuse of the same semiparametric family may reparameterize the family, but it should fail to again add a new parameter. That is, the result of reusing the semiparametric family yields survival functions of the form $\bar{H}(\cdot\,|\,\theta,\bar{F})$ but with θ replaced by some function h of ρ and θ. The notation becomes cumbersome, but symbolically, this criterion can be written as

$$\bar{H}(\cdot\,|\,\theta,\bar{H}(\cdot\,|\,\rho,\bar{F})) = \bar{H}(\cdot\,|\,h(\rho,\theta),\bar{F})$$

for some function h. This is a kind of *stability* property discussed further, particularly in Section M and in Chapter 19.

B. Location Parameters

a. Definition and Basic Properties

B.1. Definition. Suppose that $F(\cdot\,|\,b)$ is defined in terms of the distribution function F by the formula

$$F(x\,|\,b) = F(x - b). \tag{1}$$

Then b is called a *location parameter* and $\{F(\cdot\,|\,b)\}$ is a *location parameter family* with underlying distribution F.

Clearly, $F(\cdot\,|\,0) = F(\cdot)$.

B.2. Proposition. A parametric family $\{F(\cdot\,|\,b)\}$ is a location parameter family if and only if

$$F(x\,|\,b) = F(x - b\,|\,0) \quad \text{for all } x \text{ and all } b. \tag{2}$$

B.3. Observation. If F has a density f and hazard rate r, then $F(\cdot\,|\,b)$ has the density $f(\cdot\,|\,b)$ given by

$$f(x\,|\,b) = f(x - b) \tag{3a}$$

and hazard rate $r(\cdot\,|\,b)$ given by

$$r(x\,|\,b) = r(x - b). \tag{3b}$$

B.4. Observation. If the random variable X has the distribution F, and $Y = X + b$, then $F(\cdot\,|\,b)$ given by (1) is the distribution function of Y. Thus, increasing the parameter b by an amount, say Δb, shifts all probability in the underlying distribution to the right by an amount Δb.

For many random variables, location parameters are of considerable importance, and are often taken to be the expected value; the normal distribution is a prime example. But if there is a known smallest possible value for a random variable, that smallest value is the natural point from which to measure the random variable; thus, the smallest value becomes the origin, and it is the origin that serves to locate the distribution. For such nonnegative random variables, the introduction of another location parameter may be inappropriate. This is the reason why none of the parametric families introduced in Section 1.F has a *location* parameter.

In relatively rare applications, an unknown smallest value of a random variable exists; then the introduction of a location parameter may be necessary. Such parameters act to replace the origin as a starting point for the support by a new origin b, and the nonnegativity of the random variable can be lost if $b < 0$. The parameter b is often estimated by the smallest observation; this maximum likelihood estimator is positively biased and not very good for predicting future observations.

There is no doubt that the addition of a location parameter increases flexibility for fitting data. But there is a disturbing aspect of the process: In cases where a rationale is used to derive a distribution, the introduction of a location parameter may not be consistent with the rationale. For example, the lack of memory property

$$\frac{\bar{F}(x+t)}{\bar{F}(t)} = \bar{F}(x), \quad x, t \geq 0 \qquad (4)$$

characterizes the exponential distribution (Chapter 8); it is a fundamental property that forms the basis for understanding the exponential distribution, and is often used to justify an assumption of exponentiality. But this property characterizes the exponential distribution without a location parameter; once a location parameter is introduced, the lack of memory property is lost. So the procedure of adding a location parameter may be useful for fitting data, but not always helpful in understanding underlying structures.

Location parameters are sometimes introduced in the exponential, gamma, Weibull, and lognormal families to yield what are often called "two-parameter" exponential, or "three-parameter" gamma, Weibull, and lognormal distributions; sometimes, the location parameter is required to be positive so as to yield a life distribution. In all of these examples, the case $b = 0$ gives the usual form of the density.

The exponential distribution with location parameter is encountered in Proposition 18.B.15 as the solution to a functional equation. The physical meaning of this functional equation is not readily apparent, but arises as a condition for coincidence of two semiparametric families.

b. Location Parameter Examples

B.5. Two-parameter exponential distribution. The "two-parameter exponential distribution" has density

$$f(x \mid \lambda, b) = \lambda\, e^{-\lambda(x-b)}, \quad x \geq b, \ \lambda > 0. \qquad (5)$$

As noted above, this density possesses the lack of memory property (4) only when $b = 0$.

B.6. Three-parameter gamma distribution. The term "three-parameter gamma distribution" most often is used to mean the distribution with density

$$f(x\,|\,\beta, \lambda, b) = \lambda^\beta (x-b)^{\beta-1} e^{-\lambda(x-b)}/\Gamma(\beta), \quad x \geq b, \ \lambda, \beta > 0. \quad (6)$$

B.7. Three-parameter Weibull density. Here the density and survival functions are given by

$$f(x\,|\,\alpha, \lambda, b) = \alpha\lambda[\lambda(x-b)]^{\alpha-1} \exp\{-[\lambda(x-b)]^\alpha\},$$
$$x \geq b, \ \alpha, \lambda > 0, \quad (7)$$
$$\bar{F}(x, \lambda, b) = \exp\{-[\lambda(x-b)]^\alpha\}, \quad x \geq b. \quad (8)$$

B.8. Three-parameter lognormal distribution. The density of this distribution is readily obtained from 1.F(12a), and is given by

$$f(x\,|\,\mu, \sigma, b) = \frac{1}{\sqrt{2\pi}\,\sigma(x-b)} \exp\{-[\log(x-b) - \mu]^2/2\sigma^2\},$$
$$x > b, \ \sigma > 0. \quad (9)$$

Other forms of this density can also be written directly from 1.F(12b) or 1.F(12c).

c. Inverse Distribution and Total Time on Test Transform for Location Parameter Families

If $F_b(x) = F(x-b), x \geq b$ then

$$F_b^{-1}(p) = F^{-1}(p) + b, \quad 0 \leq p \leq 1. \quad (10)$$

The total time on test transform is given by

$$\psi_{F_b}(p) = \int_0^{F^{-1}(p)+b} \bar{F}(x-b)\,dx = \int_{-b}^{F^{-1}(p)} \bar{F}(x)\,dx = \psi_F(p) + b. \quad (11)$$

From Observation B.4, and the fact that the total time on test transform is a limit of total time on test statistics (Proposition 1.I.4), equation (11) is to be expected.

d. Ordering Location Parameter Families

It is clear from Observation B.4 that if $b_1 \leq b_2$, then $F(\cdot \,|\, b_1) \leq_{st} F(\cdot \,|\, b_2)$; that is, in the usual stochastic ordering, location families are increasing in the location parameter.

Orders stronger than stochastic order require conditions on the underlying distribution F.

B.9. Proposition. Suppose that F has a density f. In the hazard rate ordering, the family $\{F(x \,|\, b)\}$ is increasing (decreasing) in b if and only if the hazard rate of F is decreasing (increasing). The family $\{F(x \,|\, b)\}$ is increasing (decreasing) in the likelihood ratio order if and only if the density is log convex (log concave).

Proof. These results can be obtained using Observation B.3. □

It can be seen from the linearity in (10) that all distributions in a location parameter family are equal in the convex transform order of Definition 2.C.7. This is to be expected for an ordering that reflects the shape of a distribution.

C. Scale Parameters

a. Definition and Basic Properties

For distributions of nonnegative random variables, the most natural and important parameter is a *scale parameter*; such parameters are essential if there is to be freedom in choosing the scale of measurement of the underlying random variable. That is, the flexibility to measure survival times in months, years, or in other time scales requires the presence of a scale parameter. Most of the parametric families of practical importance have a scale parameter.

In this book, the term "scale parameter" is used in the sense of the following definition to avoid confusion with the reciprocal of the parameter, which is also called a "scale parameter".

C.1. Definition. Suppose that $F(\cdot \,|\, \lambda)$ is defined in terms of the distribution function F by the formula

$$F(x \,|\, \lambda) = F(\lambda x), \quad \lambda > 0. \tag{1}$$

Then, λ is called a *scale parameter* and $\{F(\cdot \,|\, \lambda), \lambda > 0\}$ is a *scale parameter family* with underlying distribution F.

Clearly, $F(\cdot \,|\, 1) = F(\cdot)$.

The alternative definition. According to Definition C.1, λ is a scale parameter if the distribution F depends upon λ and x only through their product λx. This definition is common and convenient in disciplines such as reliability theory and survival analysis where the exponential distribution plays a prominent or even dominant role. See, e.g., Barlow and Proschan (1975, p. 73), Kalbfleisch and Prentice (2002, p. 31). But it is also common and correct to call $1/\lambda$ a scale parameter. This is especially natural in parts of statistical theory where the normal distribution plays the central role, in which case the alternative definition is necessary to make the standard deviation σ a scale parameter. When $1/\lambda$ is regarded as a scale parameter, the distribution F depends upon λ and x only through their ratio x/λ. This book is focused on distributions of nonnegative random variables, and Definition C.1 has been adopted because it simplifies many statements and notation. For a further comment, see Definition 20.C.3.

C.2. Proposition. A parametric family $\{F(\cdot\,|\,\lambda), \lambda > 0\}$ is a scale parameter family if and only if

$$F(x\,|\,\lambda) = F(\lambda x\,|\,1) \quad \text{for all } x \text{ and all } \lambda > 0. \tag{2}$$

C.3. Proposition. If F has a density f and hazard rate r, then for $\lambda > 0$, $F(\cdot\,|\,\lambda)$ has the density $f(\cdot\,|\,\lambda)$ given by

$$f(x\,|\,\lambda) = \lambda f(\lambda x) \tag{3a}$$

and hazard rate $r(\cdot\,|\,\lambda)$ given by

$$r(x\,|\,\lambda) = \lambda\, r(\lambda x). \tag{3b}$$

As semiparametric families, families of the form (1) are sometimes termed "accelerated life" models because in the application to life lengths, the scale parameter acts to control the rate at which time passes. Such models are discussed in Chapter 16.

C.4. Observation. If the random variable X has the distribution F, and $Y = X/\lambda$, where $\lambda > 0$, then $F(\cdot\,|\,\lambda)$ given by (1) is the distribution function of Y.

b. Inverse Distribution and Total Time on Test Transform for Scale Parameter Families

Here it is convenient to use the notation $F_\lambda(x)$ in place of $F(x\,|\,\lambda)$, $x \geq 0$. Because $F_\lambda(x) = F(\lambda x)$, it is clear that $F_1(x) = F(x)$. In this case,

$$F_\lambda^{-1}(p) = [F^{-1}(p)]/\lambda, \quad 0 \le p \le 1. \tag{4}$$

From (4), it follows that the total time on test transform for a scale parameter family is

$$\psi_{F_\lambda}(p) = \int_0^{[F^{-1}(p)]/\lambda} \bar{F}(\lambda u)\, du = \int_0^{F^{-1}(p)} \bar{F}(z) \frac{dz}{\lambda} = \frac{\psi_F(p)}{\lambda}. \tag{5}$$

c. Ordering Scale Parameter Families

Scale parameters clearly affect magnitude, and as such, it is to be expected that distributions are ordered in a scale parameter according to one of the magnitude orders of Section 2.A.

C.5. Observation. For any fixed life distribution F, $\bar{F}(x \mid \lambda)$ is decreasing in λ; that is, in the usual stochastic ordering, scale families are decreasing in the scale parameter.

Stronger orderings depend upon properties of F. Propositions C.6 and C.8 are cases where the cautionary note 1.B.2.c applies. There, the nonuniqueness of densities is discussed.

C.6. Proposition. Suppose that F has hazard rate r. In the hazard rate ordering, the family $\{F(x \mid \lambda)\}$ is decreasing in λ if and only if $xr(x)$ is increasing in $x > 0$.

Proof. This result follows directly from (3b). □

Proposition C.6 is essentially a restatement of Example 2.A.6. When derivatives exist, the condition that $xr(x)$ is increasing in $x > 0$ can be written in the form $xr'(x) + r(x) \ge 0$. From this, it is clear that r increasing is a sufficient but not necessary condition for $xr(x)$ to be increasing.

C.6.a. Proposition (Ma, 1999). In the hazard rate order, the family $\{F(x \mid \lambda)\}$ is decreasing in λ if and only if $\log \bar{F}(e^x)$ is concave.

Unlike Proposition C.6, Ma's result does not require the existence of a density. In case a density does exist, the condition that $xr(x)$ is increasing in $x > 0$ is equivalent to the condition that $\log \bar{F}(e^x)$ has a decreasing derivative.

C.7. Example. The survival function $\bar{F}(x) = 1/(1 + \lambda x), x > 0$, is a special form of the Pareto survival function 11.B(1); it has the decreasing hazard rate $r(x) = \lambda/(1 + \lambda x)$, $x > 0$, but $xr(x)$ is increasing in $x > 0$.

C.8. Proposition. Suppose that F has the density f. In the likelihood ratio ordering, the family $\{F(x\,|\,\lambda)\}$ is decreasing in λ if and only if $f(x)/f(x/a)$ is decreasing in $x > 0$ for all $a > 1$.

Proof. This result can be obtained by directly verifying the condition 2.A(11). □

C.8.a. Proposition (Hu, Nanda, Xie and Zhu, 2004). Suppose that F has the density f. In the likelihood ratio ordering, the family $\{F(x\,|\,\lambda)\}$ is decreasing in λ if and only if $\log f(e^x)$ is concave.

Proof. With the assumption that f is differentiable, it can be seen that the derivative of $f(e^x)$ is decreasing in $x \in (-\infty, \infty)$ if and only if f satisfies the condition $f(y)/f(y/a)$ is decreasing in $y > 0$ of Proposition C.8. Alternatively, the result can be obtained by using Example 19.B.7 and Proposition 2.A.11. □

Because the likelihood ratio order implies the hazard rate order, it must be that the condition of Proposition C.8.a that $\log f(e^x)$ is concave implies the condition of Proposition C.6.a that $\log \bar{F}(e^x)$ is concave. A direct proof of this implication can be obtained with a slight modification of the proof of Proposition 21.B.8. First note that $\log f(e^x)$ is concave if and only if $\log e^x f(e^x)$ is concave. According to Proposition 21.B.8, this is equivalent to the total positivity of order 2 in x and y of the function $\log e^{y-x} f(e^{y-x})$. Let K be the indicator function defined in 21.B(4), and compute that

$$\int_{-\infty}^{\infty} e^{y-x} f(e^{y-x}) K(x, z)\, dy = \int_{-\infty}^{z} e^{y-x} f(e^{y-x})\, dy = \bar{F}(e^{x-z}).$$

But again by 21.B.8, this is equivalent to the log concavity of $\bar{F}(e^x)$.

d. Order Preservation with Introduction of a Scale Parameter

Suppose that $F \leq^* G$, where $\leq^*$ is any one of the following orders: stochastic order, hazard rate order, likelihood ratio order, convex order, dispersive order, Lorenz order, convex transform order, star order, or superadditive order (defined and discussed in Chapter 2). Then, in the same order, the distributions $F(\cdot\,|\,\lambda)$ and $G(\cdot\,|\,\lambda)$ obtained from F and G by the introduction of a scale parameter λ satisfy $F(\cdot\,|\,\lambda) \leq^* G(\cdot\,|\,\lambda)$, for all $\lambda > 0$. These facts can be regarded both as a property of scale parameters and as essential properties of the various orders.

The proofs of these results consist of direct verifications of the definitions, and are omitted. For the last three orders listed (convex transform, star, and superadditive), formula (4) is required.

D. Power Parameters

Power parameter families are easily defined, but some motivation and understanding of their function can be gleaned from their appearance with logarithmic transformations.

a. Parameter Relationships Under Logarithmic Transformations

Consider for a moment the case of a distribution such as the normal distribution, which has support $(-\infty, \infty)$. For such distributions, location and scale are quite natural parameters and are nearly always introduced. Thus, starting with a distribution function F having support $(-\infty, \infty)$, the relationship $F(x \mid \mu, \alpha) = F(\alpha(x - \mu))$, $-\infty < x < \infty$, defines a semiparametric family with a real location parameter μ and positive scale parameter α. Now, let X be a random variable with the distribution $F(\cdot \mid \mu, \alpha)$, and let $Y = e^X$ so that Y has support $[0, \infty)$. With the notation $\mu = -\log \lambda$, it follows that

$$P\{Y \leq y\} = F(\alpha(\log y - \mu)) = F(\log (\lambda y)^\alpha).$$

Note that under the log transform, the location parameter μ is replaced by a scale parameter λ, and the scale parameter α has become what is here called a "power parameter." Because of this change, it is not surprising to find that scale and power parameters are often encountered in families of distributions for nonnegative random variables.

Logarithmic transformations are discussed further in Chapter 12.

b. Definitions and Basic Properties

D.1. Definition. Let F be a distribution function such that $F(0-) = 0$. Suppose that $F(\cdot \mid \alpha)$ is defined in terms of F by the formula

$$F(x \mid \alpha) = F(x^\alpha), \quad \alpha > 0. \tag{1}$$

Then, α is called a *power parameter* and $\{F(\cdot\,|\,\alpha), \alpha > 0\}$ is a *power parameter family* with underlying distribution F.

Clearly, for a power parameter family, $F(\cdot\,|\,1) = F(\cdot)$.

The family of Weibull distributions is the prime example of a power parameter family. For this example, the underlying distribution is an exponential distribution.

D.2. Proposition. A parametric family $\{F(\cdot\,|\,\alpha), \alpha > 0\}$ is a power parameter family if and only if

$$F(x\,|\,\alpha) = F(x^\alpha\,|\,1) \quad \text{for all } x \text{ and all } \alpha > 0. \tag{2}$$

D.3. Proposition. If F has a density f and hazard rate r, then for $\alpha > 0$, $F(\cdot\,|\,\alpha)$ has the density

$$f(x\,|\,\alpha) = \alpha x^{\alpha-1} f(x^\alpha) \tag{3}$$

and hazard rate

$$r(x\,|\,\alpha) = \alpha x^{\alpha-1} r(x^\alpha). \tag{4}$$

D.4. Observation. If the nonnegative random variable X has the distribution F, and $Y = X^{1/\alpha}$, where $\alpha > 0$, then $F(\cdot\,|\,\alpha)$ given by (1) is the distribution function of Y.

It is possible to introduce power parameters without the condition that X is nonnegative; this can be done by letting $Y = X^{1/\alpha}$, when $X \geq 0$, and letting $Y = -\,|\,X\,|^{1/\alpha}$, when $X < 0$. However, this possibility is not pursued in this book, which is devoted primarily to distributions of nonnegative random variables.

c. Inverse Distribution and Total Time on Test Transform for Power Parameter Families

If $F(x\,|\,\alpha) = F(x^\alpha), x \geq 0$, then

$$F_\alpha^{-1}(p) = [F^{-1}(p)]^{1/\alpha}, \quad 0 \leq p \leq 1. \tag{5}$$

From (5) and Definition 1.I.2, it follows that

$$\psi_{F_\alpha}(p) = \int_0^{[F^{-1}(p)]^{1/\alpha}} \bar{F}(u^\alpha)\, du = \int_0^{F^{-1}(p)} \bar{F}(z) \frac{1}{\alpha} z^{(1/\alpha)-1}\, dz. \tag{6}$$

d. Ordering Power Parameter Families

D.5. Proposition. Suppose that $F(x \mid \alpha) = F(x^\alpha)$. If X has distribution $F(\cdot \mid \alpha)$ and Y has distribution $F(\cdot \mid \beta) = F(x^\beta)$, where $\alpha < \beta$, then X is greater than Y in the convex transform order.

Proof. The proof is a straightforward verification of the condition of the definition. With the aid of (5) and the notation $F(\cdot \mid \alpha) = F_\alpha(\cdot)$, compute that $\bar{F}_\alpha^{-1} \bar{F}_\beta(x) = x^{\beta/\alpha}$. But this is convex because $\alpha < \beta$. □

Proposition D.5 shows that a power parameter is indeed appropriately called a "shape parameter," as defined in Chapter 3.

The random variables X and Y of Proposition D.5 are not stochastically ordered because $X^{\beta/\alpha}$ has the same distribution as Y and $x^{\beta/\alpha} > x$ when $x > 1$, $x^{\beta/\alpha} < x$ when $x < 1$. Consequently, the tail probabilities of X and Y are not ordered.

e. Order Preservation with Introduction of a Power Parameter

Suppose that in some ordering, F is less than G. For what orderings is it true that F_α is less than G_α for all values or some values of the power parameter α?

D.6. Proposition.

(i) Stochastic order: If $F \leq_{\text{st}} G$, then $F_\alpha \leq_{\text{st}} G_\alpha$.

(ii) Hazard rate order: If $F \leq_{\text{hr}} G$ and $\alpha \geq 1$, then $F_\alpha \leq_{\text{hr}} G_\alpha$.

(iii) Convex transform order: If $F \leq_c G$ and $0 < \alpha \leq 1$, then $F_\alpha \leq_c G_\alpha$.

(iv) Star order: If $F \leq_* G$, then $F_\alpha \leq_* G_\alpha$.

Proof. The implication (i) is an immediate consequence of the definition of F_α, and (ii) follows directly from (4). The proofs of (iii) and (iv) make use of (5); these results follow from Propositions 21.A.15 and 21.A.16. □

f. Properties Preserved with The Introduction of a Power Parameter

Again, suppose that X has distribution F, and $Y = X^{1/\alpha}$, where $\alpha > 0$. What properties of F are inherited by the distribution $F(\cdot \mid \alpha)$ of Y? At least in part, the answer to this question depends upon α. For the abbreviations used below, see Section 4.B.

D.7. Proposition. If X is IHR and $\alpha \geq 1$, then Y is IHR; if X is DHR and $0 < \alpha \leq 1$, then Y is DHR.

Proof. Suppose first that X is IHR, i.e., $-\log \bar{F}$ is convex (and increasing) and $\alpha \geq 1$. Because an increasing convex function of a convex function is convex (see Proposition 21.A.5), $-\log \bar{F}(x^\alpha)$ is a convex function of x. But this is the condition that the survival function of Y is log concave, i.e., Y is IHR. The proof of the DHR case is similar, with concavity replacing convexity. □

In the context of the Weibull distribution, the above-mentioned proposition is well known. For the Weibull distribution, X has an exponential distribution, which is both IHR and DHR; so the hazard rate of the Weibull distribution is increasing for $\alpha \geq 1$ and decreasing for $0 < \alpha \leq 1$.

Note. The fact that the introduction of a power parameter does not preserve hazard rate monotonicity is important for its utility in generating new hazard rate shapes.

D.8. Proposition. If X is IHRA and $\alpha \geq 1$, then $Y = X^{1/\alpha}$ is IHRA; if X is DHRA and $0 < \alpha \leq 1$, then Y is DHRA.

Proof. This result follows immediately from the fact that

$$[\log F(x \mid \alpha)]/x = x^{\alpha-1}[\log F(x^\alpha)]/x^\alpha.$$ □

D.9. Proposition. If X is NBU and $\alpha \geq 1$, then $Y = X^{1/\alpha}$ is NBU; if X is NWU and $\alpha \leq 1$, then Y is NWU.

Proof. According to a very special case of Minkowski's inequality,

$$(x+t)^\alpha \geq x^\alpha + t^\alpha \quad \text{for } x, t \geq 0 \text{ and } \alpha \geq 1; \tag{7}$$

this result can also be obtained from the fact that x^α is a convex function of x, for $\alpha \geq 1$, and according to Proposition 19.A.11, this implies that x^α is superadditive. More directly, (7) can be obtained from the fact that $(\sum a_i^r)^{1/r}$ is decreasing in $r > 0$; see Hardy, Littlewood and Pólya (1952, p. 28).

First because $\bar{F}$ is decreasing and then because of (7), it follows that if X is NBU and $\alpha \geq 1$, then $\bar{F}((x+t)^\alpha) \leq \bar{F}(x^\alpha + t^\alpha) \leq \bar{F}(x^\alpha)\bar{F}(t^\alpha)$. This proves the result for the NBU case and the proof for the NWU case is similar, but uses the fact that the inequality (7) is reversed for $\alpha \leq 1$. □

g. Scale Mixtures of Distributions with a Power Parameter

D.10. Proposition. If F and G have increasing hazard rates and $\alpha \geq 1$, the mixture

$$\bar{H}(x) = \int_0^\infty \bar{F}(x^\alpha/\theta^{\alpha-1}) \, dG(\theta)$$

has an increasing hazard rate.

Proof. This result is an application of Proposition 4.C.8.b. To apply this proposition, it is necessary to show that the function $\log \bar{F}(x^\alpha/\theta^{\alpha-1})$ is concave. According to Example 21.A.3.a, $x^\alpha/\theta^{\alpha-1}$ is convex in the pair x, θ when both variables are positive and $\alpha \geq 1$. An application of Proposition 21.A.6 yields the desired conclusion, because $-\log \bar{F}$ is convex by virtue of F having an increasing hazard rate. □

Block, Li and Savits (2003a) offer as examples three special cases of Proposition D.10. They note that the conclusion applies when F is

(i) a Weibull distribution, in which case $\bar{F}(x) = \exp(-x^\alpha/\theta^{\alpha-1})$;

(ii) a Gompertz distribution with power parameter (Section 10.C.d), in which case

$$\bar{F}(x) = \exp\{-\xi[\exp(x^\alpha/\theta^{\alpha-1}) - 1]\};$$

(iii) a generalized gamma distribution (Section 9.E), in which case the density is

$$f(x) = \frac{\alpha x^{\alpha\beta-1}}{\Gamma(\beta)\theta^{(\alpha-1)\beta}} \exp\left(-\frac{x^\alpha}{\theta^{\alpha-1}}\right).$$

E. Frailty and Resilience Parameters: Proportional Hazards and Reverse Hazards

a. Background and Definitions

Just as linear transformations of random variables lead to the introduction of location and scale parameters, linear transformations of a hazard rate can also introduce parameters.

With $\xi > 0$ and $b > 0$, it is possible to replace a hazard rate r by $\xi r + b$ and thereby introduce the parameters ξ and b. For the family of exponential distributions, this procedure yields nothing new, since in this case the hazard rate r is constant. But the procedure may be useful in some contexts; with $\xi = 1$, it gives rise to the Makeham generalization of the Gompertz distribution discussed in Section 10.B. Nevertheless, we do not consider b to be a fundamental parameter for the following reason. If X is a random variable with hazard rate r, if Y has an exponential distribution with hazard rate b, and if X and Y are independent, then $Z = \min(X, Y)$ is a random variable with hazard rate $r + b$. Generalizations of this kind, where hazard rates have natural expressions as sums of other hazard rates, are discussed in Section 15.G.

With $b = 0$, the hazard rate $\xi r + b$ becomes simply ξr. If r is the hazard rate of the distribution F, then for all $\xi > 0$, ξr is also a hazard rate, and the corresponding survival function is $\bar{F}(\cdot \mid \xi) = [\bar{F}(\cdot)]^\xi$.

E.1. Definition. Let F be a distribution function with hazard function $R = -\log \bar{F}$. Suppose that $F(\cdot \mid \xi)$ is defined in terms of F by the formula

$$\bar{F}(x \mid \xi) = [\bar{F}(x)]^\xi = e^{-\xi R(x)}, \quad \xi > 0. \tag{1}$$

Then, ξ is called a *frailty parameter* and $\{F(\cdot \mid \xi), \xi > 0\}$ is a *frailty parameter family*, or alternatively, a *proportional hazards family*, with underlying distribution F.

Clearly, for a proportional hazards family, $F(\cdot \mid 1) = F(\cdot)$.

E.2. Proposition. A parametric family $\{F(\cdot \mid \xi), \xi > 0\}$ is a frailty parameter family if and only if

$$\bar{F}(x \mid \xi) = [\bar{F}(x \mid 1)]^\xi \quad \text{for all } x \text{ and all } \xi > 0. \tag{2}$$

As noted above, for a proportional hazards family, the hazard rates are also proportional. These survival functions are powers of the survival function F with hazard rate r. Models in which ξ is regarded as a random variable are much used in survival analysis and are often called "Cox models" or "frailty models" in the literature of medical survival analysis, e.g., Vaupel, Manton, and Stallard (1979), Hougard (1984), or Oakes (1989). This is the reason for calling ξ a "frailty parameter." See Section 16.B.b, and in particular, 16.B.5.

When ξ is an integer, (1) has an interpretation familiar in reliability theory; it is the survival function of a series system of ξ independent components each with survival function $\bar{F}$ (see Figure 5.A.1). Of course, series systems like this become more vulnerable to failure as the number of components increases, so the term "frailty" is appropriate from this point of view.

E.3. Proposition. If F has a density f and hazard rate r, then for $\xi > 0, F(\cdot \mid \xi)$ has the density $f(\cdot \mid \xi)$ given by

$$f(x \mid \xi) = \xi f(x)[\bar{F}(x)]^{\xi-1}, \tag{3a}$$

and hazard rate $r(\cdot \mid \xi)$ given by

$$r(x \mid \xi) = \xi r(x). \tag{3b}$$

For exponential and Weibull distributions, introducing powers of the survival function does not introduce a new parameter because these families are already proportional hazards families. But for a number of other families a new parameter is introduced.

There is a parallel semiparametric family obtained by raising not the survival function but the distribution function to a power to obtain the distribution function $F(\cdot \mid \eta) = [F(\cdot)]^\eta$.

E.4. Definition. Suppose that $F(\cdot \mid \eta)$ is defined in terms of the distribution function F by the formula

$$F(x \mid \eta) = [F(x)]^\eta, \quad \eta > 0. \tag{4}$$

Then, η is called a *resilience parameter* and $\{F(\cdot \mid \eta), \eta > 0\}$ is a *resilience parameter family* or alternatively, a *proportional reverse hazards family*, with underlying distribution F.

For a resilience parameter family, $F(\cdot \mid 1) = F(\cdot)$.

E.5. Proposition. A parametric family $\{F(\cdot \mid \eta), \eta > 0\}$ is a resilience parameter family if and only if

$$F(x \mid \eta) = [F(x \mid 1)]^\eta, \quad \text{for all } x \text{ and all } \eta > 0. \tag{5}$$

In the literature, resilience parameters have received less attention than have frailty parameters just as reverse hazard functions have received less attention than hazard functions.

When η is an integer, the distribution function (4) is the distribution function of a parallel system with η independent components all having the survival function $\bar{F}$ (see Figure 5.A.2). Because of their redundancy, parallel systems become less prone to failure as the number of components increases, and hence the term "resilience."

E.6. Proposition. If the underlying distribution F has a density f and hazard rate r, then for $\eta > 0$, $F(\cdot \,|\, \eta)$ has the density $f(\cdot \,|\, \eta)$ given by

$$f(x \,|\, \eta) = \eta [F(x)]^{\eta - 1} f(x), \tag{6a}$$

hazard rate $r(\cdot \,|\, \eta)$ given by

$$r(x \,|\, \eta) = \frac{\eta [F(x)]^{\eta - 1} f(x)}{1 - [F(x)]^{\eta}}, \tag{6b}$$

and reverse hazard rate $s(\cdot \,|\, \eta)$ given by

$$s(x \,|\, \eta) = \eta s(x). \tag{6c}$$

Of course, (6c) is the reason for calling $\{F(\cdot \,|\, \xi), \xi > 0\}$ a proportional reverse hazards family.

Because of Proposition 4.E.2, it is also useful to record that when f is differentiable,

$$\rho(x \,|\, \eta) = -\frac{f'(x \,|\, \eta)}{f(x \,|\, \eta)} = -(\eta - 1)\frac{f(x)}{F(x)} - \frac{f'(x)}{f(x)}. \tag{7}$$

Resilience and frailty parameter families have the stability property Criterion 2 of Section A; once a resilience or frailty parameter has been introduced, the reintroduction of the same kind of parameter does not expand the family.

For those with ideas from reliability theory in mind, seeing both series and parallel systems arise will suggest consideration of the so-called "k-out-of-n" systems (see Section 5.A). Parametric families related to such systems can be generated, but they do not have the stability property unless $k = 1$ or $k = n$.

b. Duality of Frailty and Resilience

Note the parallelism between (3b) and (6c). To see the duality of frailty and resilience in another way, note that the survival function defined in (1) can be written in the form

$$[\bar{F}(\cdot)]^\xi = H(\bar{F}(\cdot) \mid \xi), \text{ where } H(z \mid \xi) = z^\xi, \quad 0 \le z \le 1. \tag{8}$$

Note that H is a distribution function with support $[0, 1]$. In Chapter 14, the dual of such a distribution function H is defined as

$$H_\mathrm{D}(z) = 1 - H(1 - z), \quad 0 \le z \le 1. \tag{9}$$

This distribution is called the "dual" of H because the dual of H_D is H; that is, $(H_\mathrm{D})_\mathrm{D} = H$. If H is replaced by H_D in (8), the equation

$$H_\mathrm{D}(F(\cdot) \mid \xi) = 1 - [1 - F(\cdot)]^\xi = 1 - [\bar{F}(\cdot)]^\xi \tag{10}$$

is obtained. This replacement has caused the parameter to become a resilience parameter; to conform to notation previously introduced, ξ needs to be relabeled as η.

Frailty and resilience are also dual in the following sense. Suppose that X is a random variable with distribution in the frailty family (1), so that for some $\xi > 0$,

$$P\{X > x\} = [\bar{F}(x)]^\xi, \quad x > 0.$$

Then, $Y = 1/X$ has the distribution function

$$P\{Y \le y\} = [\bar{F}(1/y)]^\xi, \quad y > 0.$$

Thus, the parameter that had been a frailty parameter is transformed to a resilience parameter. Of course, if X has a distribution with resilience parameter, then Y has a distribution with frailty parameter.

Furthermore, frailty and resilience are also dual in still another sense. If ξ and η are integers, then $[\bar{F}(\cdot)]^\xi$ is the survival function of a series system with independent components all having the distribution F, and $[F(\cdot)]^\eta$ is the distribution of a parallel system with independent components all having the distribution F. As noted in Section 5.A.a, series and parallel systems are, as coherent structures, dual.

c. Product Families

Families of the form $F(\cdot \,|\, \xi) = [\bar{F}(\cdot)]^\xi$ or $F(\cdot \,|\, \eta) = [F(\cdot)]^\eta$ belong to slightly more general classes of families.

E.7. Definition. Let $\mathcal{F} = \{F(\cdot \,|\, \theta) : \theta \in A\}$ be an indexed family of distributions with index set A satisfying

$$\alpha \in A, \ \beta \in A \ \Rightarrow \ \alpha + \beta \in A. \tag{11}$$

$\mathcal{F}$ is said to be a *survival product family* if

$$\bar{F}(\cdot \,|\, \alpha)\bar{F}(\cdot \,|\, \beta) = \bar{F}(\cdot \,|\, \alpha + \beta), \quad \alpha, \beta \in A; \tag{12}$$

$\mathcal{F}$ is said to be a *distribution product family* if

$$F(\cdot \,|\, \alpha)F(\cdot \,|\, \beta) = F(\cdot \,|\, \alpha + \beta), \quad \alpha, \beta \in A. \tag{13}$$

In the following, only survival product families are discussed, but parallel results can be obtained for distribution product families.

E.8. Lemma. *A survival product family of distributions indexed by $A = (0, \infty)$ must be a proportional hazards family, i.e., the index ξ is a frailty parameter and (2) holds.*

Proof. Fix x and let $\psi(\xi) = \bar{F}(x \,|\, \xi)$. From (12), it follows that

$$\psi(\alpha + \beta) = \psi(\alpha)\psi(\beta), \quad \alpha, \beta > 0.$$

Because ψ is bounded, it follows from Proposition 22.A.2 that this functional equation has, for some real number γ, the solution $\psi(\alpha) = e^{-\gamma\alpha}$, i.e., $\psi(\alpha) = [\psi(1)]^\alpha$. But this is the family $\bar{F}(x \,|\, \xi) = [\bar{F}(x \,|\, 1)]^\xi$. □

E.9. Proposition. *Let $\bar{F}(\cdot \,|\, \theta), \theta \in A \subset (0, \infty)$ be a parametric family of distribution functions and suppose that if X and Y are random variables with distributions in the family, then $\min(X, Y)$ has a distribution in the family. Then, the parameter set A can be extended to become $(0, \infty)$ and the family is a proportional hazards family, i.e., θ is a frailty parameter.*

Proof. This proposition is essentially a restatement of Lemma E.8. That $\min(X, Y)$ must have a distribution in the family means that the family is a survival product family. □

A similar result has been given by Castillo and Ruiz-Cobo (1992, p. 132), but they treat the analogous case of closure under maximums.

d. Frailty and Resilience in Consort

A survival function $\bar{F}^\xi$ with an integer-valued frailty parameter ξ is the survival function of the smallest observation in a random sample of size ξ; similarly, the distribution F^η is the distribution of the largest observation in a random sample of size η. In general, the kth order statistic can be considered in place of the largest or smallest observation. In a sample of size n from a distribution with density f, the kth order statistic has the density

$$f_{k,n}(x) = \frac{n!}{(k-1)!(n-k)!}[F(x)]^{k-1}f(x)[\bar{F}(x)]^{n-k}$$
$$= \frac{[F(x)]^{k-1}[\bar{F}(x)]^{n-k}}{B(k, n-k+1)}f(x), \qquad (14)$$

where B is the beta function defined in Section 23.B. For a discussion of order statistics, see, for example, Balakrishnan and Cohen (1991); David and Nagaraja (2003).

The density can be extended to noninteger n and k using a technique that Ahuja and Nash (1967) use in interesting special cases. Let g be the beta density of 14.C(1) with parameters η and ξ, i.e.,

$$g(u \mid \eta, \xi) = \frac{u^{\eta-1}(1-u)^{\xi-1}}{B(\eta, \xi)}, \quad 0 \le u \le 1, \ \eta, \ \xi > 0. \qquad (15)$$

With the change of variables $u = F(x)$, (15) becomes

$$f(x \mid \eta, \xi) = \frac{[F(x)]^{\eta-1}[\bar{F}(x)]^{\xi-1}}{B(\eta, \xi)}f(x), \quad \eta, \ \xi > 0. \qquad (16)$$

Note that (16) is the same as (14) when $\eta = k$, $\xi = n - k + 1$.

When $\eta = 1$, (16) becomes the density f with frailty parameter ξ, and when $\xi = 1$, it is the density f with a resilience parameter η. So in a sense, the ideas of both frailty and resilience are embodied in the density (16). However, except in the special cases $\eta = 1$ or $\xi = 1$, the parameters are not actually frailty or resilience parameters.

e. Inverse Distributions and Total Time on Test Transforms

With the notation $\bar{F}_\xi(x) = [\bar{F}(x)]^\xi$, $x \geq 0$,

$$\bar{F}_\xi^{-1}(p) = \bar{F}^{-1}(p^{1/\xi}),$$
$$F_\xi^{-1}(1-p) = F^{-1}(1-p^{1/\xi}), \quad 0 \leq p \leq 1. \tag{17}$$

Similarly, if $F_\eta(x) = [F(x)]^\eta$, $x \geq 0$, then

$$F_\eta^{-1}(p) = F^{-1}(p^{1/\eta}), \quad 0 \leq p \leq 1. \tag{18}$$

For these families, there is no significant simplification of the total time on test transform.

f. Ordering Proportional Hazards Families

Because members of a proportional hazards family have the form $\bar{F}(x \,|\, \xi) = [\bar{F}(x)]^\xi$, they are clearly ordered (decreasing in their parameters) according to the usual stochastic order; this means that frailty parameters relate to magnitude. Similarly, powers of the distribution function are decreasing in the power, so in the usual stochastic order, the families are increasing in the resilience parameter. But stronger statements can be made which, unlike the case of a scale parameter (Proposition C.8), require no assumptions on F.

E.10. Proposition. If $\mathcal{F} = \{F(\cdot \,|\, \xi) : \xi > 0\}$ is a proportional hazards family, then $F(\cdot \,|\, \xi)$ is, in the likelihood ratio order, decreasing in ξ. If $\mathcal{F} = \{F(\cdot \,|\, \eta) : \eta > 0\}$ is a resilience parameter family, then $F(\cdot \,|\, \eta)$ is, in the likelihood ratio order, increasing in η.

These results are direct consequences of the definitions, particularly if densities are assumed to exist.

E.10.a. Example. For the exponential survival function $\bar{F}(x) = \exp\{-\lambda x\}$, $x \geq 0$, the parameter λ can be regarded as either a scale parameter or a frailty parameter. When viewed as a frailty parameter, it follows from Proposition E.10 that in the likelihood ratio order, the family is decreasing in λ. This fact is recorded in Section 8.C.d.

g. Order Preservation with Introduction of a Frailty or Resilience Parameter

Recall that survival functions of frailty families have the form $\bar{F}(x \,|\, \xi) = [\bar{F}(x)]^\xi$ for some distribution function F, and distribution functions of resilience families have the form $F(x \mid \eta) = [F(x)]^\eta$ for some distribution function F.

E.11. Proposition.
(i) Stochastic order: If $F \leq_{st} G$, then $F(\cdot \mid \xi) \leq_{st} G(\cdot \mid \xi)$.
(ii) Hazard rate order: If $F \leq_{hr} G$, then $F(\cdot \mid \xi) \leq_{hr} G(\cdot \mid \xi)$.
(iii) Convex transform order: If $F \leq_c G$, then $F(\cdot \mid \xi) \leq_c G(\cdot \mid \xi)$.
(iv) Star order: If $F \leq_* G$, then $F(\cdot \mid \xi) \leq_* G(\cdot \mid \xi)$.

The same statements hold with the frailty parameter replaced by a resilience parameter.

Proof. Both (i) and (ii) are readily verified from the definitions. Both (iii) and (iv) follow from the fact that the quantity $G^{-1}F$ does not change when the same frailty or resilience parameter is introduced to both F and G. □

h. Gini Index for Frailty and Resilience Parameter Families

The Gini index is defined in 1.I(12) as

$$Gini(F) = \mu^{-1} \int_0^\infty F(x) \bar{F}(x) \, dx.$$

To compute this for a frailty parameter family, rewrite the Gini index in the form

$$Gini(F) = \mu^{-1} \left\{ \int_0^\infty \bar{F}(x) \, dx - \int_0^\infty [\bar{F}(x)]^2 \, dx \right\}, \qquad (19)$$

and use 1.C(4) to obtain

$$Gini(F(\cdot \mid \xi)) = 1 - \frac{\mu(2\xi)}{\mu(\xi)}, \qquad (20)$$

where $\mu(\xi)$ is the first moment of the distribution $F(\cdot \mid \xi)$ with frailty parameter ξ.

To compute the Gini index for a resilience parameter family, write the Gini index in the form

$$Gini(F) = \mu^{-1} \left\{ \int_0^\infty [1 - (F(x))^2] \, dx - \int_0^\infty [1 - F(x)] \, dx \right\},$$

but here the integrals are not moments.

i. Mixing Frailty and Resilience Parameter Families

Suppose first that $\bar{F}(x\,|\,\xi) = [\bar{F}(x)]^\xi$; let $R = -\log \bar{F}$ be the hazard function of F and consider the mixture

$$\bar{H}(x) = \int_0^\infty \bar{F}(x\,|\,\xi)\,dG(\xi) = \int_0^\infty [\bar{F}(x)]^\xi\,dG(\xi) = \int_0^\infty e^{-\xi R(x)}\,dG(\xi). \tag{21}$$

Because the Laplace transform of G is given by $\phi(s) = \int_0^\infty e^{-s\xi}\,dG(\xi)$, it follows that

$$\bar{H}(x) = \phi(R(x)). \tag{22}$$

This formula is sometimes convenient. With G a gamma distribution, (22) was obtained by Dubey (1968) who offered several examples including Proposition 11.B.1.

When the distribution G has a parameter, say θ, then (22) can be written in the form

$$\bar{H}(x\,|\,\theta) = \phi_\theta(R(x)); \tag{23}$$

with choices of specific Laplace transforms ϕ_θ and hazard functions R, this formula can lead to a variety of semiparametric families. Example 4.C.7.b is a special case of (23), where G is an exponential distribution. In this case,

$$\bar{H}(x\,|\,\theta) = \frac{\theta}{R(x) + \theta}, \tag{24}$$

where θ is the scale parameter of the exponential distribution. Another example is given in Section F.c; (23) is examined in more detail in Section M.b. For more about mixtures of the form (21), see Sections M.b and M.d.

Next, suppose that $F(x\,|\,\eta) = [F(x)]^\eta$, so that η is a resilience parameter. Let $S = \log F$ be the reverse hazard function of F. Consider the mixture

$$H(x) = \int_0^\infty e^{\eta S(x)}\,dG(\eta) = \int_0^\infty [F(x)]^\eta\,dG(\eta). \tag{25}$$

In this case, (22) is replaced by

$$H(x) = \text{mgf}(S(x)) = \phi(-(S(x)), \tag{26}$$

where mgf is the moment generating function of G and ϕ is its Laplace transform.

It can be seen from (26) that if G has a convolution parameter (Section J), then that parameter becomes a resilience parameter for H.

j. A Majorization Result

Reorder the components of the n-dimensional vectors x and y to obtain $x_{(1)} \leq \cdots \leq x_{(n)}, y_{(1)} \leq \cdots \leq y_{(n)}$. Write $x \prec^w y$ to mean that

$$\sum_{i=1}^{k} x_{(i)} \geq \sum_{i=1}^{k} y_{(i)}, \quad k = 1, \ldots, n.$$

This relationship is called *weak super majorization* by Marshall and Olkin (1979, p. 10).

E.12. Proposition. Let $\{F(\cdot \mid \xi), \xi > 0\}$ be a family of distributions with frailty parameter ξ. Let $U_1, \ldots, U_n$ be independent random variables such that U_i has the distribution $F(\cdot \mid \xi_i), i = 1, \ldots, n$. Let $V_1, \ldots, V_n$ be independent random variables such that V_i has the distribution $F(\cdot \mid \theta_i), i = 1, \ldots, n$. If $(\xi_1, \ldots, \xi_n) \prec^w (\theta_1, \ldots, \theta_n)$, then for all increasing functions h defined on $[0, \infty)^n$,

$$h(U_1, \ldots, U_n) \leq_{\text{st}} h(V_1, \ldots, V_n).$$

For a proof of this result, see Marshall and Olkin (1979, p. 368).

F. Tilt Parameters: Proportional Odds Ratios, Extreme Stable Families

All of the semiparametric families described in previous sections have been broadly used, and well-known examples are readily found. In this section, yet another kind of semiparametric family is introduced, a kind hinted at by Clayton (1974) and discussed by Bennett (1983), Marshall and Olkin (1997), and Kirmani and Gupta (2001). This semiparametric

F. Tilt Parameters: Proportional Odds Ratios, Extreme Stable Families

family is not so well known. In contrast to Section E, where proportional hazards are considered, the family discussed in this section is characterized by having proportional odds ratios.

a. Definitions, Basic Properties, and First Derivation: Proportional Odds

F.1. Definition. Suppose that $F(\cdot \mid \gamma)$ is defined in terms of the underlying distribution F by the formula

$$\frac{F(x \mid \gamma)}{\bar{F}(x \mid \gamma)} = \frac{1}{\gamma}\frac{F(x)}{\bar{F}(x)}, \quad -\infty < x < \infty, \ \gamma > 0, \tag{1}$$

that is,

$$\bar{F}(x \mid \gamma) = \frac{\gamma \bar{F}(x)}{F(x) + \gamma \bar{F}(x)} = \frac{\gamma \bar{F}(x)}{1 - \bar{\gamma}\bar{F}(x)}, \quad -\infty < x < \infty, \ \gamma > 0. \tag{2}$$

Then, γ is called a *tilt parameter* and the family $\{F(\cdot \mid \gamma), \gamma > 0\}$ is said to be a *proportional odds family*, a *tilt parameter family*, or alternatively an *extreme stable family*.

Clearly, $F(\cdot \mid 1) = F(\cdot)$.

F.2. Proposition. If F has a density f and hazard rate r, then for $\gamma > 0$, the distribution $F(\cdot \mid \gamma)$ given by (2) has the density $f(\cdot \mid \gamma)$ given by

$$f(x \mid \gamma) = \frac{\gamma f(x)}{[1 - \bar{\gamma}\bar{F}(x)]^2} \tag{3a}$$

and hazard rate $r(\cdot \mid \gamma)$ given by

$$r(x \mid \gamma) = \frac{1}{[1 - \bar{\gamma}\ \bar{F}(x)]} r(x), \quad -\infty < x < \infty, \ \gamma > 0. \tag{3b}$$

Proposition F.2. can be verified by differentiating (2) with respect to x.
It follows from (3b) that

$$\lim_{x \to -\infty} r(x \mid \gamma) = \lim_{x \to -\infty} r(x)/\gamma$$

and

$$\lim_{x \to \infty} r(x \mid \gamma) = \lim_{x \to \infty} r(x).$$

Note that when $F(0) = 0$, the hazard rate $r(0 \mid \gamma)$ at the origin behaves quite differently than it does for the families of gamma and Weibull distributions, introduced in Sections 1.F.b and 1.F.c; for both these families, either the distribution is an exponential distribution, or $r(0) = 0$, or $r(0) = \infty$ (see 9.A(12) and 9.B(3)), so that $r(0)$ is discontinuous in the shape parameter.

It follows from (3b) that

$$r(x)/\gamma \leq r(x \mid \gamma) \leq r(x), \quad -\infty < x < \infty, \ \gamma \geq 1, \quad (4a)$$
$$r(x) \leq r(x \mid \gamma) \leq r(x)/\gamma, \quad -\infty < x < \infty, \ 0 < \gamma \leq 1. \quad (4b)$$

These inequalities indicate that the hazard rate for the proportional odds family is shifted below ($\gamma \geq 1$) or above ($0 < \gamma \leq 1$) the hazard rate of the underlying distribution by a limited amount; they form the basis for calling γ a "tilt parameter."

Using (4a) and (4b) with 1.B(3) yields

$$\bar{F}(x) \leq \bar{F}(x \mid \gamma) \leq [\bar{F}(x)]^{1/\gamma}, \quad -\infty < x < \infty, \ \gamma \geq 1, \quad (5a)$$
$$[\bar{F}(x)]^{1/\gamma} \leq \bar{F}(x \mid \gamma) \leq \bar{F}(x), \quad -\infty < x < \infty, \ 0 < \gamma \leq 1. \quad (5b)$$

Proportional odds ratios were studied by McCullagh (1980) primarily with ordinal data in mind. The idea was developed by Bennett (1983) for survival data, and further studied by Kirmani and Gupta (2001).

As noted by Bennett (1983), comparison with proportional hazards discussed in Section D is appropriate. Note from (3b) that

$$\frac{r(x \mid \gamma)}{r(x)} = \frac{1}{[1 - \bar{\gamma}\bar{F}(x)]} \quad (6)$$

is increasing in x, for $\gamma \geq 1$, and decreasing in x, for $0 < \gamma \leq 1$; the limit of this ratio as x goes to ∞ is 1. By contrast, with proportional hazards, the ratio of hazard rates is constant. Proportional odds ratios may be appropriate when the effects of differing treatments, say, diminish with time.

The following more complicated derivation offers a different insight, with the geometric distribution playing a key role.

b. Second Derivation: Geometric-Extreme Stability

Extreme value distributions are briefly discussed in Section 20.G. The rationale for considering these distributions is that a life length may be regarded as the extreme of a large number of independent identically distributed random variables. Usually this is an approximation to the actual situation; more often a random variable is the extreme of a finite number of independent identically distributed random variables, and the finite number may be random. Here, it is assumed that the random number has a geometric distribution.

Suppose that $X_1, X_2, \ldots,$ is a sequence of independent random variables with a common distribution F and N is a random variable independent of the X_i's with the geometric (p) distribution $P\{N = n\} = (1-p)^{n-1}p, n = 1, 2, \ldots,$ of 18.E(4). Let

$$U = \min(X_1, \ldots, X_N), \qquad V = \max(X_1, \ldots, X_N). \tag{7}$$

F.3. Definition. If $F \in \mathcal{F}$ implies that the distribution of U is in $\mathcal{F}$, then $\mathcal{F}$ is said to be *geometric-minimum stable*. Similarly, if $F \in \mathcal{F}$ implies that the distribution of V is in $\mathcal{F}$, then $\mathcal{F}$ is said to be *geometric-maximum stable*. If $\mathcal{F}$ is both geometric-minimum and geometric-maximum stable, then $\mathcal{F}$ is said to be *geometric-extreme stable*.

F.3.a. Remark. The term "max-geometric stable" has been used by Rachev and Resnick (1991) to describe a related but more restricted concept. They apply the term not to families of distributions but to individual distributions; in their sense, a distribution is "max-geometric stable" if the location-scale parameter family generated by the distribution is geometric-max stable in our sense. The two ideas essentially coincide for families $\mathcal{F}$ that are parameterized by location and scale. Most of the families considered here are not of that form, a notable exception being the logistic distribution.

F.3.b. Remark. Here, U and V are, respectively, the minimum and maximum of a geometric number of independent identically distributed random variables. The case where the sum takes the place of the minimum or maximum is also of considerable interest, though it is not treated here. See, for example, Feller (1971, Section XI.6) and Brown (1990) for further results and references.

F.4. Example. The family of logistic distributions, with survival functions of the form

$$\bar{F}(x) = \frac{1}{1+\theta\, e^{\lambda x}}, \quad -\infty < x < \infty, \ \theta, \lambda > 0,$$

is an example of a geometric-extreme stable family; indeed, distributions in this family are geometric-extreme stable even in the sense of Rachev and Resnick (1991). The fact that this family is a geometric-minimum stable family was utilized by Arnold (1989) to construct a stationary process with logistic marginals.

For the random variable U of (7),

$$\bar{F}(x\,|\,p) = P\{U > x\} = \sum_{n=1}^{\infty} [\bar{F}(x)]^n (1-p)^{n-1} p \qquad (8)$$

$$= \frac{p\bar{F}(x)}{1-(1-p)\bar{F}(x)}, \quad -\infty < x < \infty.$$

From this derivation, it is apparent that the survival function $\bar{F}(\cdot\,|\,p)$ is a mixture of survival functions $[F(\cdot)]^n$; moreover,

$$\bar{F}(\cdot\,|\,p) = \phi(R(x)\,|\,p),$$

where $\phi(s\,|\,p) = p\, e^{-s}[1 - e^{-s}(1-p)]^{-1}$ is the Laplace transform 20.E(6a), of the geometric distribution of 20.E(4) and $R = -\log \bar{F}$. Thus, the distribution is an example of E(23).

F.5. Proposition. The parametric family of distributions of the form (8) is geometric-minimum stable.

Proof. If $Y_1, Y_2, \ldots$ is a sequence of independent random variables with a common survival function $\bar{F}(\cdot\,|\,p)$ given by (8) and if M is a random variable independent of the Y_i's with a geometric (q) distribution, then $U = \min(Y_1, \ldots, Y_M)$ has a distribution of the form (8) but with p replaced by pq. To see this, write

$$Y_i = \min(X_{i1}, \ldots, X_{iN_i}), \quad i = 1, 2, \ldots,$$

where X_{ij} and N_i are all independent, all the X_{ij} have the distribution F and N_i have a geometric (p) distribution. Then,

$$\min(Y_1, \ldots, Y_M)$$
$$= \min(X_{11}, \ldots, X_{1N_1}, X_{21}, \ldots, X_{2N_2}, \ldots, X_{M1}, \ldots, X_{MN_M}). \qquad (9)$$

F. Tilt Parameters: Proportional Odds Ratios, Extreme Stable Families

By re-indexing the X_{ij}, this can be rewritten as

$$U = \min(Y_1, \ldots, Y_M) = \min(X_1, X_2, \ldots, X_{N_1 + \cdots + N_M}).$$

It is well known that $N_1 + \cdots + N_M$ has a geometric (pq) distribution (see Proposition 18.E.1), so it is immediate that U has the survival function (8) with pq in place of p. □

Arguments similar to those used above show that the distribution function H of the random variable $V = \max(X_1, \ldots, X_N)$ of (7) is given by

$$H(x) = \sum_{n=1}^{\infty} [F(x)]^n (1-p)^{n-1} p = \frac{pF(x)}{1 - (1-p)F(x)}, \quad -\infty < x < \infty. \tag{9a}$$

This distribution can be thought of as a mixture of the distributions $[F(\cdot)]^n$, and H has the form

$$H(x) = \phi(-S(x)),$$

where $S = \log F$ is the reverse hazard function of F and ϕ is the Laplace transform 20.E(6a) of the geometric distribution. This distribution is an example of E(26).

From (9a), it follows that

$$\bar{H}(x) = \frac{\bar{F}(x)}{p + (1-p)\bar{F}(x)}, \quad -\infty < x < \infty. \tag{9b}$$

F.6. Proposition. The parametric family given by (9a) or (9b) is geometric-maximum stable.

This result has a proof similar to that of Proposition F.5.

The families given by (8) and (9b) nicely combine to form a single parametric family $\mathcal{G} = \mathcal{G}(F) = \{F(\cdot \mid \gamma), \gamma > 0\}$, where

$$\bar{F}(x \mid \gamma) = \frac{\gamma \bar{F}(x)}{1 - \bar{\gamma}\bar{F}(x)} = \frac{\gamma \bar{F}(x)}{F(x) + \gamma \bar{F}(x)} = 1 - \frac{F(x)}{F(x) + \gamma \bar{F}(x)},$$
$$-\infty < x < \infty, \quad 0 < \gamma < \infty, \tag{10}$$

and $\bar{\gamma} = 1 - \gamma$; in (8), $0 < \gamma = p \leq 1$, and in (9b), $\gamma = 1/p \geq 1$. As previously noted, $\bar{F}(x \mid 1) = \bar{F}(x)$, so $F \in \mathcal{G}$. Note that (10) is the same as (2).

F.7. Proposition. The parametric family $\mathcal{G}$ of distributions of the form (2) is geometric-extreme stable. That is, the family $\mathcal{G}$ satisfies Criterion 2 of Section A.

Proof. To verify the proposition, it is sufficient to verify closure of $\mathcal{G}$ under a kind of composition, as follows. Suppose that

$$\bar{H}(x) = \frac{\xi \bar{F}(x \mid \gamma)}{[1 - (1 - \xi)\bar{F}(x \mid \gamma)]},$$

where $F(x \mid \gamma)$ is given by (2). Then,

$$\bar{H}(x) = \frac{\xi \gamma \bar{F}(x)}{[1 - (1 - \xi\gamma)\bar{F}(x)]}.$$

This shows that $H \in \mathcal{G}$, and consequently, $\mathcal{G}$ has both geometric maximum and geometric-minimum stability. $\square$

The proof of Proposition F.7 also shows that if F is replaced by any other distribution in $\mathcal{G}$, then that distribution will also generate $\mathcal{G}$.

c. Introduction of Tilt Parameters by Way of Mixtures

As noted in Section E.i, the mixture

$$\bar{H}(x \mid \gamma) = \int_0^\infty [\bar{F}(x)]^\xi \, dG(\xi \mid \gamma) = \int_0^\infty e^{-\xi R(x)} \, dG(\xi \mid \gamma)$$

can be written in the form

$$\bar{H}(x \mid \gamma) = \phi_\gamma(R(x)), \tag{11a}$$

where ϕ_γ is the Laplace transform of $G(\cdot \mid \gamma)$. Let K be the distribution function of a nonnegative random variable and suppose that $R(x) = \emptyset^-(x) = K(x)/\bar{K}(x)$; this odds ratio has the properties required of a hazard function. If $G(\cdot \mid \gamma)$ is an exponential distribution with parameter γ, the Laplace transform ϕ_γ is given in 8.A(11), and in

F. Tilt Parameters: Proportional Odds Ratios, Extreme Stable Families

this case, (11a) takes the form

$$\bar{H}(x\mid\gamma) = \int_0^\infty [\bar{F}(x)]^\xi\, \gamma\, e^{-\gamma\xi}\, d\xi = \phi_\gamma\left(\frac{K(x)}{\bar{K}(x)}\right) = \frac{\gamma \bar{K}(x)}{1 - \bar{\gamma}\bar{K}(x)}, \quad x \geq 0. \tag{11b}$$

This is the survival function given in (2) and (10), with K in place of F. This change of notation was made here to conform to the notation used elsewhere in the context of mixtures.

The mixture representation of (11b) is not unique; a further discussion of this issue can be found in Section M. In particular, see Examples M.4.c, M.5.a, and M.5.b. If the exponential density in (11b) is replaced by the density 1.F(6) of a gamma distribution with parameters γ and ν, then the integral can similarly be obtained from the Laplace transform 9.A(5) of the gamma distribution and is given by

$$\int_0^\infty [\bar{F}(x)]^\xi \frac{\gamma^\nu \xi^{\nu-1}\, e^{-\gamma \xi}}{\Gamma(\nu)}\, d\xi = \left[\phi\left(\frac{K(x)}{\bar{K}(x)}\right)\right]^\nu = \left(\frac{\gamma \bar{K}(x)}{1 - \bar{\gamma}\bar{K}(x)}\right)^\nu, \quad x \geq 0. \tag{11c}$$

Here, the parameter ν of the gamma distribution is called a convolution parameter (see Section J). Note that because ξ is a frailty parameter of $F(\cdot\mid\xi) = [\bar{F}(\cdot)]^\xi$, the convolution parameter ν of the gamma density has become a frailty parameter, but in (11c) γ is no longer a tilt parameter; a different survival function would be obtained if first a frailty and then a tilt parameter is introduced in K. It is not unusual to find that when two different kinds of parameters are introduced successively, the order in which they are introduced is important. See Section 19.B for more discussion of this issue.

d. The Inverse Distribution and the Total Time on Test Transform

It is easy to verify with the notation

$$\bar{F}_\gamma = \bar{F}(\cdot\mid\gamma) = \gamma\bar{F}(\cdot)/[1 - \bar{\gamma}\bar{F}(\cdot)],$$

that

$$F_\gamma^{-1}(p) = F^{-1}\left(\frac{\gamma p}{1 - \bar{\gamma}p}\right) \quad \text{and} \quad \bar{F}_\gamma^{-1}(p) = \bar{F}^{-1}\left(\frac{p}{\gamma + \bar{\gamma}p}\right). \tag{12}$$

From these results, it follows that F_γ has the total time on test transform

$$\psi_{F_\gamma}(p) = \int_0^{F^{-1}(\gamma/(1-\bar\gamma p))} \frac{\gamma \bar F(x)}{1 - \bar\gamma \bar F(x)}\, dx = \int_0^p (1-y)\, dF^{-1}\left(\frac{\gamma y}{1 - \bar\gamma y}\right)$$
$$= \int_0^{\gamma p/(1-\bar\gamma p)} \frac{\gamma(1-z)}{\gamma + \bar\gamma z}\, dF^{-1}(z). \tag{13}$$

See 1.I for a discussion of uses of the total time on test transform.

e. Ordering Tilt Parameter Families

Consider the extreme stable family with survival functions of the form $\bar F(x\,|\,\gamma) = \gamma \bar F(x)/[1 - \bar\gamma \bar F(x)]$ for some fixed underlying distribution function F. It can easily be checked that $F(\cdot\,|\,\gamma)$ is stochastically decreasing in γ, but a stronger statement can be made.

F.8. Proposition. In the likelihood ratio order, the distribution $F(\cdot\,|\,\gamma)$ is increasing in γ. Consequently, $F(\cdot\,|\,\gamma)$ is increasing in γ in the hazard rate order and stochastic order.

Proof. This can be verified directly using (3a) and the characterization 3.A(11) of the likelihood ratio order in terms of densities. □

f. Order Preservation with the Introduction of a Tilt Parameter

F.9. Lemma. If $\bar G_i(x\,|\,\gamma) = [\gamma \bar F_i(x)]/[1 - \bar\gamma \bar F_i(x)]$, $i = 1, 2$, then

$$G_2^{-1} G_1(x) = F_2^{-1} F_1(x) \quad \text{for all } x.$$

Proof. This result can be obtained using the assumed form of G_2 together with G_1^{-1}, which is obtainable from (12). Because the result is independent of γ, the Lemma follows. □

F.10. Proposition (Kirmani and Gupta, 2001). Suppose that X_i has the distribution F_i and Y_i has the distribution $F_i(\cdot\,|\,\gamma), i = 1, 2$. Then, $X_1 \leq X_2$ implies $Y_1 \leq Y_2$ when $\leq$ is any of the following orders: stochastic order $\leq_{\text{st}}$ hazard rate order $\leq_{\text{hr}}$, convex transform order $\leq_{\text{c}}$, star order $\leq_*$, and superadditive order $\leq_{\text{su}}$.

Proof. For stochastic and hazard rate orders, the results are obtainable directly from (2) and (3b). For the other orders, the results follow from Lemma F.9. □

g. Preservation of Distribution Properties with the Introduction of a Tilt Parameter

Because of notational difficulties, it is convenient in this section to use the notation $G = F(\cdot \mid \gamma)$.

F.11. Proposition (Kirmani and Gupta, 2001). If F is IHR and $\gamma \geq 1$, then G is IHR. The same preservation result holds for IHRA and NBU. If F is DHR and $\gamma \leq 1$, then G is DHR. The same preservation result holds for DHRA and NWU.

Proof. Suppose first that F is IHR and $\gamma \geq 1$. If Y has an exponential distribution, then by Proposition 4.C.1.f, it follows that $X \leq_c Y$. By Proposition F.10, $X_\gamma \leq_c Y_\gamma$ where X_γ and Y_γ, respectively, have the distributions of X and Y with the tilt parameter γ introduced. But $Y_\gamma \leq_c Y$, and consequently, Y_γ, or its distribution G, is IHR. The cases of IHRA and NBU have similar proofs, but use 5.B.3(iv) and 5.C.3.

Next, assume that F is DHR and $\gamma \leq 1$. Then the arguments for IHR and IHRA follow through with little alteration, except that they use $Y_\gamma \geq_c Y$ or $Y_\gamma \geq_* Y$, which holds because $\gamma \leq 1$. It follows from the definition of NWU that F is NWU if and only if

$$\bar{G}^{-1}\bar{F}(x+t) \leq \bar{G}^{-1}\bar{F}(x) + \bar{G}^{-1}\bar{F}(t),$$

where G is an exponential distribution with parameter $\lambda = 1$, in which case $G^{-1}(x) = -\log x$. Apply the operator $\bar{F}^{-1}\bar{G}$ to both sides of this inequality, then let $\bar{G}^{-1}\bar{F}(x) = u$ and $\bar{G}^{-1}\bar{F}(t) = v$ to conclude that

$$\bar{F}^{-1}\bar{G}(u+v) \geq \bar{G}^{-1}\bar{G}(u) + \bar{G}^{-1}\bar{G}(v).$$

The preservation property for NWU follows as before, using Lemma F.9. □

The clever idea of using the orders of Proposition F.10 in the above-mentioned proof is due to Kirmani and Gupta (2001).

h. More Properties of Tilt Parameter Families

A number of facts concerning geometric-extreme stable families are evident and may be worth noting; the same properties hold for geometric-minimum and geometric-maximum stable families.

(i) If $\mathcal{F}_1$ and $\mathcal{F}_2$ are geometric-extreme stable families, then $\mathcal{F}_1 \cup \mathcal{F}_2$ and $\mathcal{F}_1 \cap \mathcal{F}_2$ are geometric-extreme stable families. (The empty set is vacuously such a family.)

(ii) Every distribution F determines a geometric-extreme stable family $\mathcal{F}(F)$. If $G \in \mathcal{F}(F)$, then $\mathcal{F}(G) = \mathcal{F}(F)$. Thus, the minimal geometric-extreme stable families form a partition of the set of all distributions into a set of equivalence classes.

(iii) If F and G differ only by a scale (location) parameter, then $\mathcal{F}(G)$ can be obtained from $\mathcal{F}(F)$ by a common scale (location) change.

(iv) Suppose that $F \in \mathcal{F}$ implies that $\bar{F}(0) > 0$, and define F_+ by

$$\bar{F}_+(x) = \bar{F}(x)/\bar{F}(0), \quad x \geq 0.$$

If F is geometric-extreme stable, then $\{F_+ : F \in \mathcal{F}\}$ is geometric-extreme stable.

(v) Let $\mathcal{F}$ be a family of distribution functions and

$$\mathcal{F}_{\theta,\delta} = \{G : G(x) = [F(x - \delta)]^\theta \text{ for some } F \text{ in } \mathcal{F}\}.$$

If $\mathcal{F}$ is geometric-extreme stable, then $\mathcal{F}_{\theta,\delta}$ is geometric-extreme stable, for all $\theta > 0$ and all real δ.

i. Why the Geometric Distribution?

The geometric-extreme stability property of $\mathcal{G} = \mathcal{G}(F)$ is rather remarkable, and it depends upon the fact that a geometric sum of independent identically distributed geometric random variables has a geometric distribution (Proposition 18.E.1). This partially explains why random-minimum stability cannot be expected if the geometric distribution is replaced by some other distribution on $\{1, 2, \ldots\}$. Thus, if the above development is repeated, e.g., with the assumption that $N - 1$ has a Poisson distribution, then $\mathcal{G}$ would be replaced by a family that would not be Poisson-extreme stable.

The following proposition provides a characterization of the geometric distribution.

F.12. Proposition. Suppose that $\{H(\cdot | \theta) : 0 < \theta \leq 1\}$ is a parametric family of distributions supported by the positive integers and parameterized in such a way that H has expected value $\mu(\theta) = 1/\theta$. Denote the corresponding probability mass function by $p_n(\theta), n = 1, 2, \ldots$.

F. Tilt Parameters: Proportional Odds Ratios, Extreme Stable Families   253

Suppose further that the p_n are differentiable and that either

$$\lim_{\theta \to 0} p'_n(\theta) = a > 0, \quad n = 1, 2, \ldots, \tag{14}$$

or

$$\lim_{\theta \to 1} p'_1(\theta) = 1, \quad \lim_{\theta \to 1} p'_2(\theta) = -1, \quad \lim_{\theta \to 1} p'_n(\theta) = 0, \ n = 3, 4, \ldots. \tag{15}$$

If F is a distribution function and

$$\bar{F}(x \mid \theta) = \sum_{n=1}^{\infty} [\bar{F}(x)]^n p_n(\theta) \tag{16}$$

has the stability property

$$\sum_{n=1}^{\infty} [\bar{F}(x \mid \theta)]^n \ p_n(\alpha) = \sum_{n=1}^{\infty} [\bar{F}(x)]^n(x) \ p_n(\xi) \quad \text{for some } \xi = \xi(\theta, \alpha), \tag{17}$$

then $p_n(\theta) = (1 - \theta)^{n-1} \theta, n = 1, 2, \ldots$, i.e., H is a geometric distribution.

F.13. Remark. If H has expected value $\sum_{n=1}^{\infty} np_n(\theta) = \mu(\theta) < \infty$, and μ has an inverse, then a reparameterization is possible to achieve $\mu(\theta) = 1/\theta$. Because of Proposition F.12, it follows that either (14) and (15) both hold or both fail; it is not possible to have one condition without the other.

Proof of Proposition F.12. Let $\bar{F}(x) = z$ and rewrite (17) as

$$\sum_{n=1}^{\infty} \left[\sum_{m=1}^{\infty} z^m p_m(\theta) \right]^n p_n(\alpha) = \sum_{n=1}^{\infty} z^n p_n(\xi), \quad 0 \le z \le 1, \tag{18}$$

where ξ is a function of α and θ. Let

$$\phi(z \mid \alpha) = \sum_{n=1}^{\infty} z^n p_n(\alpha) = E_\alpha z^N, \tag{19}$$

where N has mass function $p_n(\alpha), n = 1, 2, \ldots$. Then (18) yields the functional equation

$$\phi(\phi(z \mid \theta) \mid \alpha) = \phi(z \mid \xi), \quad 0 \le z \le 1, \tag{20}$$

where ϕ satisfies the conditions $\phi(1;\alpha) = 1$, $\phi(0;\alpha) = 0$, and $\phi(z;\alpha)$ is increasing in z.

To determine ξ as a function of α and θ, differentiate in (19) with respect to z to obtain

$$\phi'(z \,|\, \alpha) = \sum_{n=1}^{\infty} n z^{n-1} p_n(\alpha),$$

from which it follows that

$$\phi'(1 \,|\, \alpha) = \sum_{n=1}^{\infty} n p_n(\alpha) = \mu(\alpha). \tag{21}$$

Now differentiate with respect to z in (20) to obtain

$$\phi'(\phi(z \,|\, \theta) \,|\, \alpha) \phi'(z \,|\, \theta) = \phi'(z \,|\, \xi). \tag{22}$$

With $z = 1$ and the fact that $\phi(1 \,|\, \theta) = 1$, (22) becomes $\mu(\alpha)\mu(\theta) = \mu(\xi)$; by assumption, $\mu(\theta) = 1/\theta$, and hence

$$\alpha\theta = \xi. \tag{23}$$

Consequently, (20) becomes

$$\phi(\phi(z \,|\, \theta) \,|\, \alpha) = \phi(z \,|\, \alpha\theta), \quad 0 \le z \le 1. \tag{24}$$

Write $\phi_2(z \,|\, \alpha) = \partial\phi(z \,|\, \alpha)/\partial\alpha$, that is,

$$\phi_2(z \,|\, \alpha) = \sum_{n=1}^{\infty} z^n p'_n(\alpha). \tag{25}$$

From (24), it follows that

$$\phi_2(\phi(z \,|\, \theta) \,|\, \alpha) = \theta \phi_2(z \,|\, \alpha\theta). \tag{26}$$

At this point, the proof must follow two paths assuming that (14) or (15) holds. Suppose first that (14) holds.

From (25), (14), and the fact that $\sum_{n=1}^{\infty} z^n = z/(1-z)$, it follows that $\phi_2(z \,|\, 0) = az/(1-z)$, and then (26) yields

$$\frac{\phi(z \,|\, \theta)}{1 - \phi(z \,|\, \theta)} = \frac{\theta z}{1 - z}. \tag{27}$$

F. Tilt Parameters: Proportional Odds Ratios, Extreme Stable Families

Equation (27) can be solved for ϕ to yield

$$\phi(z\,|\,\theta) = \frac{\theta z}{1 - z(1-\theta)}, \tag{28}$$

the generating function of the geometric distribution.

Now, instead of (14), assume that (15) holds. Then with (15), (26) yields

$$\phi_2(z\,|\,1) = z - z^2, \tag{29}$$

which together with (26) gives

$$\phi(z\,|\,\theta) - \phi^2(z\,|\,\theta) = \theta\phi_2\,(z\,|\,\theta). \tag{30}$$

To solve this differential equation (in θ for z fixed), let $h(\theta) = \phi(z\,|\,\theta)$ and rewrite the equation in the form $\dfrac{d}{d\theta}\dfrac{\theta}{h(\theta)} = 1$. This can be solved to show that $\phi(z\,|\,\theta) = \theta/(c+\theta)$ for some constant c here depending upon z. Consequently, $\phi_2\,(z\,|\,1) = c/(c+1)^2$, which upon substitution in (29), yields the quadratic equation

$$c/(c+1)^2 = z - z^2.$$

Only one root $c(z) = (1-z)/z$ satisfies $\phi(1\,|\,\theta) = 1$, and that root again gives (28). □

F.14. Remark. The general solution to the functional equation (20) is given by Proposition 22.C.2. However, a solution here is required that is a probability generating function. To sort out the probability generating functions from the set of general solutions given by Proposition 22.C.2 does not appear to be an easy problem.

j. More General Models

Dabrowska and Doksum (1988) consider a two-parameter proportional odds model in which

$$\frac{F(x\,|\,\gamma)}{\overline{F}(x\,|\,\gamma)} = \frac{1}{\gamma}\frac{F(x)}{\overline{F}(x)}, \quad -\infty < x < \infty, \ \gamma > 0, \tag{31}$$

is replaced for some $c > 0$, by

$$\frac{1 - [\bar{F}(x\,|\,\gamma)]^c}{[\bar{F}(x\,|\,\gamma)]^c} = \frac{1 - [\bar{F}(x)]^c}{\gamma[\bar{F}(x)]^c}. \tag{32}$$

This yields the semiparametric family

$$\bar{F}(x\,|\,\gamma, c) = \left(\frac{\gamma[\bar{F}(x)]^c}{1 - \bar{\gamma}[\bar{F}(x)]^c}\right)^{1/c}, \quad -\infty < x < \infty. \tag{33}$$

With $c = 1$, this reduces to (2).

A further generalization is obtained by replacing c on the left-hand side of (32) by d. Then, (33) is replaced by

$$\bar{F}(x\,|\,\gamma, c, d) = \left(\frac{\gamma[\bar{F}(x)]^c}{1 - \bar{\gamma}[\bar{F}(x)]^c}\right)^{1/d}. \tag{34}$$

It should be noted, however, that the survival function (34) satisfies Criterion 2 of Section A.c only if $d = 1$.

Sankaran and Jayakumar (2006) propose tilt parameter families with added resilience and frailty parameters, and they extend the idea of proportional odds ratios to the bivariate case.

G. Hazard Power Parameters

According to the Definition 1.B.5, a distribution F and its corresponding hazard function R are related via the formula

$$\bar{F}(x) = \exp\{-R(x)\}$$

for all x. Thus, a function R is the hazard function for some proper distribution function if and only if R is an increasing function and

$$\lim_{x \to -\infty} R(x) = 0 \quad \text{and} \quad \lim_{x \to \infty} R(x) = \infty.$$

It follows that if R is a hazard function, then R^ζ is a hazard function for all $\zeta > 0$. Thus,

$$\bar{F}(x\,|\,\zeta) = \exp\{-[R(x)]^\zeta\} \tag{1}$$

defines a survival function for all $\zeta > 0$, and $\{\bar{F}(\cdot \mid \zeta) : \zeta > 0\}$ is a semi-parametric family. The parameter ζ is called a *hazard power parameter* and the family is called a *hazard power parameter family*.

It is possible to define a *reverse hazard power parameter* by the equation

$$F(x \mid \zeta) = \exp\{-[S(x)]^\zeta\},$$

where $S(x) = -\log F$ is the reverse hazard function defined in 1.B(4). This parameter arises in an occasional example, but it is not investigated in any detail in this book.

With $\bar{F}(x \mid \zeta)$ given by (1), $\bar{F}(\cdot \mid 1) = \bar{F}(\cdot)$, so that the underlying distribution is a member of the family $\{\bar{F}(\cdot \mid \zeta) : \zeta > 0\}$. If a hazard power parameter α is introduced using the underlying survival function $\bar{F}(\cdot \mid \zeta)$, the resulting survival function is $\overline{(F)}(\cdot \mid \zeta + \alpha)$; thus the family has the stability property of Criterion 2 (Section A).

Hazard power parameters were considered by Bradley, Bradley and Naftel (1984).

G.1. Example. Suppose that $\bar{F}$ is the survival function of an exponential distribution with scale parameter λ. Then,

$$\bar{F}(x \mid \zeta) = \exp\{-(\lambda x)^\zeta\}$$

is the survival function of a Weibull distribution with power parameter ζ and scale parameter λ. In this example, ζ can be regarded not only as a hazard power parameter but also as a power parameter (see Proposition 18.B.12).

a. Properties of Hazard Power Parameter Families

If the underlying distribution F has hazard rate r, then by differentiating (1), it can be determined that $F(\cdot \mid \zeta)$ has hazard rate

$$r(x \mid \zeta) = \zeta[R(x)]^{\zeta-1} r(x). \tag{2}$$

It follows that if r is increasing and $\zeta \geq 1$, then $r(\cdot \mid \zeta)$ is increasing; if r is decreasing and $0 < \zeta \leq 1$, then $r(\cdot \mid \zeta)$ is decreasing.

If the underlying distribution F has an increasing hazard rate average (IHRA) and $\zeta \geq 1$, then from (2) it follows that $F(\cdot \mid \zeta)$ is IHRA; if F has a decreasing hazard rate average (DHRA) and $0 < \zeta \leq 1$, then $F(\cdot \mid \zeta)$ is DHRA.

258   7. Semiparametric Families

Furthermore, if F is new better than used (NBU) and $\zeta \geq 1$, then $F(\cdot\,|\,\zeta)$ is NBU; if F is new worse than used (NWU) and $0 < \zeta \leq 1$, then $F(\cdot\,|\,\zeta)$ is NWU. These results follow from Proposition 19.A.11 and the fact that the function $\phi(x) = x^\zeta$ is convex, for $\zeta \geq 1$ (hence superadditive), and concave, for $\zeta \leq 1$ (hence subadditive).

G.1.a. Example G.1 continued. The exponential distribution is both IHR and DHR. It follows from the above comments that the Weibull distribution is IHR when $\zeta \geq 1$ and DHR when $\zeta \leq 1$. This is a familiar fact.

b. The Inverse Distribution and the Total Time on Test Transform

If $\bar{F}_\zeta(x) = \exp\{-[R(x)]^\zeta\}$, $x \geq 0$, then

$$\bar{F}_\zeta^{-1}(p) = R^{-1}([-\log p]^{1/\zeta}), \quad 0 < p \leq 1. \tag{3}$$

Here, the total time on test transform can be written as

$$\psi_{F_\zeta}(p) = \int_0^{-\log(1-p)} \frac{\exp\{-u^\zeta\}}{r(R^{-1}(u))}\, du. \tag{4}$$

c. Ordering Hazard Power Parameter Families

No general results are known about ordering hazard power parameter families. Some results can be obtained for specific underlying distributions; for example, the hazard power parameter of the Weibull distribution is also a power parameter, so results of Section D apply. Specifically when the underlying distribution is Weibull, $F(\cdot\,|\,\xi)$ is, in the convex transform order, increasing in ξ.

H. Moment Parameters

Suppose that F is a distribution function with support in $[0, \infty)$, and finite positive βth moment μ_β for all β in the set B. Then, the function $\bar{F}(\cdot\,|\,\beta)$ defined by

$$\bar{F}(x\,|\,\beta) = \frac{1}{\mu_\beta} \int_x^\infty z^\beta\, dF(z), \quad x \geq 0, \tag{1}$$

is a survival function with a parameter β, called a *moment parameter*. The semiparametric family $\{F(\cdot\,|\,\beta) : \beta \in B\}$ of distribution functions is called a *moment parameter family*. If F has a density f, then $F(\cdot\,|\,\beta)$ has the density

$$f(x\,|\,\beta) = \frac{x^\beta f(x)}{\mu_\beta}, \quad x \geq 0.$$

Cearly, the support of $\{f(\cdot\,|\,\beta) : \beta \in B\}$ is the same as the support of f.

It is always the case that $0 \in B$, and $F(\cdot\,|\,0) = F(\cdot)$ so F is a member of the family $\{F(\cdot\,|\,\beta) : \beta \in B\}$. Stability of the family is easily checked. In fact, if $F(\cdot\,|\,\beta_0)$ is used as an underlying distribution and a new moment parameter β_1 is introduced, then the resulting distribution is $F(\cdot\,|\,\beta_0 + \beta_1)$.

H.1. Remark. It is possible to introduce moment parameters in distributions not supported on $[0, \infty)$ if absolute moments are used. Because this book is primarily focused on distributions of nonnegative random variables, this extension is not considered. But note that a moment parameter cannot be introduced with an underlying distribution degenerate at 0 because all moments of this distribution are 0.

H.2. Example. Suppose that f is the density of an exponential distribution with scale parameter λ. Here, $B = (-1, \infty)$, so that $\beta > -1$ and $f(\cdot\,|\,\beta)$ is the density of a gamma distribution with shape parameter $\nu = \beta + 1$.

a. Inverse Distributions: Total Time on Test Transforms

The inverse of a general distribution with moment parameter cannot be given in closed form. Consequently, the total time on test transform cannot be given in closed form, and must be computed numerically and individually for each underlying distribution.

b. Ordering Moment Parameter Families

H.3. Proposition. In the likelihood ratio order, moment parameter families are increasing in the parameter.

Proof. Because the function e^{-sx} is totally positive in s and x (see 19.B.5), the likelihood ratio ordering follows from Proposition 2.A.10. □

c. The Case $\beta = 1$

The density

$$f(x\,|\,1) = \frac{xf(x)}{\mu}, \quad x \geq 0,$$

is of particular interest because it is the density of the length of the interval covering a fixed point in a stationary renewal process. According to Proposition H.3, such an interval is longer in the likelihood ratio order than an interval with the interarrival time density. This is not surprising because long intervals are more likely to cover fixed points. The density $f(\cdot\,|\,1)$ has been studied in some detail by Brown (2001). See Section 20.B.c.

I. Laplace Transform Parameters

Denote by I the interval for which the Laplace transform $\phi(s) = \int e^{-sx}\,dF(x)$ of the distribution function F is finite. Clearly, $[0,\infty) \subset I$. The function $F(\cdot\,|\,s)$, defined for all s in the interval I by

$$\bar{F}(x\,|\,s) = \frac{1}{\phi(s)}\int_x^\infty e^{-sz}\,dF(z), \tag{1}$$

is a survival function with a parameter s called a *Laplace transform parameter*. The semiparametric family $\{F(\cdot\,|\,s) : s \in I\}$ is called a *Laplace transform parameter family*.

Because $F(\cdot) = F(\cdot\,|\,0)$, the underlying distribution is a member of the parametric family, and it is easily checked that if a new Laplace transform parameter t is introduced in $F(\cdot\,|\,s)$, then the result is $F(\cdot\,|\,s+t)$; this demonstrates stability.

The process of introducing a Laplace transform parameter parallels that of the moment parameter, but with $x^\beta = e^{\beta \log x}$ replaced by e^{-sx}.

If F has a density f, then $F(\cdot\,|\,s)$ has the density

$$f(x\,|\,s) = \frac{e^{-sx}f(x)}{\phi(s)}.$$

This is perhaps the most natural setting in which to introduce a Laplace transform parameter.

a. Inverse Distributions and Total Time on Test Transforms

As with moment parameters, nothing general is known about inverse distributions or total time on test transforms for Laplace transform parameters.

b. Ordering Laplace Transform Parameter Families

I.1. Proposition. Laplace transform parameter families are, in the likelihood ratio order, decreasing in the parameter.

Proof. Because the function e^{-sx} is totally positive in $-s$ and x (see Example 21.B.5), the likelihood ratio ordering follows from Proposition 2.A.9. □

J. Convolution Parameters

For any distribution function F, one can consider the family of ν-fold convolutions of F, where ν is any positive integer. These convolutions have Laplace transforms that are positive integer powers of the Laplace transform of F. In case F is infinitely divisible, any positive power, integer or not, of the Laplace transform of F is again a Laplace transform, so a family indexed by $(0, \infty)$ can be defined. When applied to the exponential distribution, known to be infinitely divisible, this procedure yields the family of gamma distributions. Other examples are given in Chapter 13.

Convolution families have long been of interest, particularly in the study of infinitely divisible distributions. Several results for these families have been given by Marshall and Olkin (1990) for multivariate as well as univariate distributions.

Recall that a distribution F with Laplace transform ϕ is *infinitely divisible* if ϕ^ν is a Laplace transform, for all $\nu > 0$. Denote the corresponding distribution by $F^{\nu*}$. For a discussion of infinitely divisible distributions, see Section 20.D.a. For more general discussions, see, e.g., Feller (1971) or Gut (2005).

J.1. Definition. Let $\mathcal{F} = \{F(\cdot \mid \nu) : \nu \in A\}$ be an indexed family of distributions with index set A satisfying

$$\alpha \in A, \beta \in A \;\Rightarrow\; \alpha + \beta \in A. \tag{1}$$

$\mathcal{F}$ is said to be a *convolution family* if the parameters add under the convolution operation, i.e., if

$$F(\cdot\,|\,\alpha) * F(\cdot\,|\,\beta) = F(\cdot\,|\,\alpha+\beta), \quad \alpha, \beta \in A. \tag{2}$$

In this case, the parameter is called a *convolution parameter*.

Several well-known families of distributions form convolution families; for example, the binomial, negative binomial, Poisson, gamma, normal, and the inverse Gaussian families are all convolution families.

a. Infinite Divisibility in Convolution Families

J.2. Proposition. If $\{F(\cdot\,|\,\nu) : \nu > 0\}$ is a convolution family, then $F(\cdot\,|\,1)$ is infinitely divisible and

$$F(\cdot\,|\,\nu) = F^{\nu*}(\cdot\,|\,1).$$

Proof. Let $\phi(\cdot\,|\,\nu)$ be the Laplace transform of $F(\cdot\,|\,\nu), \nu > 0$. From (2) it follows that

$$\phi(\cdot\,|\,\theta)\,\phi(\cdot\,|\,\nu) = \phi(\cdot\,|\,\theta+\nu), \quad \theta, \nu > 0. \tag{3}$$

This functional equation has the solution $\phi(\cdot\,|\,\nu) = [\phi(\cdot\,|\,1)]^{\nu}$ (Proposition 22.A.2). □

J.3. Example. The gamma distribution, with density $f(\cdot\,|\,\lambda,\nu)$ given by 9.A(1) is a convolution family. The Laplace transform of the gamma distribution is given by

$$E\,e^{-sX} = [\lambda/(\lambda+s)]^{\nu}, \quad s > -\lambda. \tag{4}$$

See 9.A(5). Clearly this Laplace transform satisfies (3).

For a discussion of mixtures involving convolution families, see Section M.c.

b. Ordering Convolution Families

J.4. Proposition. If X and Y have distributions in a convolution family with respective parameters α and β where $\alpha \leq \beta$, then X is less than Y in the stochastic order.

Proof. The random variable Y has the same distribution as $X + Z$, where Z is independent of X and has the distribution in the convolution family with parameter $\beta - \alpha$. Consequently, the result is immediate. □

J.5. Example. If X has a gamma distribution with parameters λ and ν_1, and Y has a gamma distribution with parameters λ and ν_2, where $\nu_1 < \nu_2$, then Y has the same distribution as $X + Z$, where Z is independent of X and has a gamma distribution with parameters λ and $\nu_2 - \nu_1$. This observation can be obtained with the aid of (4).

As noted in Section 20.B.b, the binomial distribution has Laplace transform given by

$$\phi(s \mid p, n) = [(1-p) + p e^{-s}]^n.$$

It follows that $n = 1, 2, \ldots$ is a convolution parameter. Because a random variable with a binomial distribution represents the number of successes in n independent trials, it is clear that the distribution is, in the stochastic order, increasing in n; the number of successes can only increase as the number of trials increases.

Proposition J.4 shows that the parameter of a convolution family orders the family in magnitude. But its role in ordering the family according to shape is much more important.

J.6. Proposition. Suppose that X and Y have distributions in a convolution family with respective parameters α and β. If $\alpha \leq \beta$, then $X \geq_{\text{Lorenz}} Y$, i.e., $X/EX \geq_{\text{cx}} Y/EY$.

Proof. The proof of this proposition makes use of a result of Marshall and Olkin (1979, 11.B.2.b, p. 288) which states that if ϕ is a real continuous convex function of a real variable and $X_1, X_2, \ldots$ is a sequence of independent identically distributed random variables, then

$$E\phi\left(\sum_{i=1}^{n} X_i/n\right) \text{ is nonincreasing in } n = 1, 2, \ldots.$$

To make use of this monotonicity, assume first that there exists $\gamma > 0$ such that α and β are both multiples of γ. Then, the Lorenz ordering follows directly. Otherwise a limiting argument is required. □

K. Age Parameters: Residual Life Families

If F is a distribution function such that $\bar{F}(t) > 0$, then the residual life distribution at time t is defined as

$$\bar{F}_t(x) = \bar{F}(t+x)/\bar{F}(t), \quad x \geq 0,$$
$$= 1, \quad x < 0.$$

These residual life distributions form a parametric family with the new parameter $\tau = t$, called an *age parameter*. Clearly $\bar{F}_0 = \bar{F}$.

In this book, age parameters are introduced only when the underlying distribution F satisfies $F(x) = 0, x < 0$, i.e., F is the distribution of a nonnegative random variable. Only in this case is F itself a member of the parametric family (obtained with $t = 0$).

Age parameters are of interest primarily for distributions such that $\bar{F}(x) > 0$ for all $x \geq 0$, although this restriction is not essential. Age parameters are of particular interest when the survival function $\bar{F}$ has a nice form, as illustrated by application to the Weibull distribution in Section 9.G. Note that if F has hazard rate r, then the hazard rate r_τ of F_τ is given by

$$r_\tau(x) = r(\tau + x). \tag{1}$$

Properties of the semiparametric family $\{F_\tau, \tau \geq 0\}$ are essentially all dependent upon properties of F; the nature of the family's construction by itself does not impart properties to the family. However, see Proposition K.1.

The introduction of an age parameter can be thought of in a somewhat different manner. Suppose that X has the distribution F, and let $Y = X - \tau$. With probability 1, $X \geq 0$, and consequently $Y \geq -\tau$. To obtain a life distribution from the distribution of Y, truncate it at 0 and renormalize. In this way, the residual life distribution is obtained.

a. The Inverse Distribution and the Total Time on Test Transform

If $\bar{F}_\tau(x) = \bar{F}(x+\tau)/\bar{F}(\tau), x \geq 0$, then the inverse is

$$\bar{F}_\tau^{-1}(p) = \bar{F}^{-1}(p\bar{F}(\tau)) - \tau, \quad 0 \leq p \leq 1. \tag{2}$$

It follows with the aid of 1.I(2) that the total time on test transform is

$$\psi_{F_\tau}(p) = \int_0^{F^{-1}(1-\bar{p}\bar{F}(\tau))-\tau} \frac{\bar{F}(u+\tau)}{\bar{F}(\tau)} du,$$

and consequently

$$\psi_{F_\tau}(p) = \frac{\psi_F(1-\bar{p}\bar{F}(\tau)) - \psi_F(F(\tau))}{\bar{F}(\tau)}. \qquad (3)$$

b. Ordering Residual Life Families

The following proposition gives conditions for residual life families to be decreasing in the stochastic order and hazard rate order.

K.1. Proposition. (i) If F has an increasing hazard rate (IHR), then in the hazard rate order, the distributions F_τ are stochastically decreasing in τ. Conversely, if the distributions $\bar{F}(\tau+x)/\bar{F}(\tau)$ are stochastically decreasing in τ, then F is IHR. (ii) In the likelihood ratio order, the distributions F_τ are decreasing in τ if and only if F has a density that is log concave.

The proof of these results use (1) and the fact that F is IHR if and only if $\bar{F}_\tau$ is decreasing in t for all $x > 0$.

L. Successive Additions of Parameters

Successive applications of methods to add parameters may or may not commute; i.e., the order in which the parameters are added may or may not matter. To illustrate this, consider first scale and power parameters.

Suppose that a scale parameter is first introduced in the underlying distribution F, ahead of the introduction of a power parameter. Then after the first step, the distribution $F(\cdot \mid \lambda)$ given by $F(x \mid \lambda) = F(\lambda x)$ is obtained. The introduction next of the power parameter yields

$$F(x \mid \lambda, \alpha) = F(x^\alpha \mid \lambda) = F(\lambda x^\alpha) \quad \text{for all } x \text{ and all } \alpha > 0. \qquad (1a)$$

This can be written schematically as

$$\bar{F}(x) \xrightarrow{\text{scale}} \bar{F}(\lambda x) \xrightarrow{\text{power}} \bar{F}(\lambda x^\alpha).$$

However, if the power parameter is introduced first, then the distribution is given by

$$F(x \mid \lambda, \alpha) = F((\lambda x)^\alpha) \quad \text{for all } x \text{ and all } \alpha > 0. \tag{1b}$$

Schematically, this can be written as

$$\bar{F}(x) \xrightarrow[\text{power}]{} \bar{F}(x^\alpha) \xrightarrow[\text{scale}]{} \bar{F}(\lambda^\alpha x^\alpha).$$

Because the end results differ only in their parameterization, the operations of adding scale and power parameters can be said to *commute*; regardless of the order in which the parameters are introduced, the same family is obtained. In what follows, the parameterization (1b) is used to insure that λ remains a scale parameter.

A well-known example is the Weibull distribution, which is obtained by adding a power parameter to an exponential distribution so that

$$\bar{F}(x \mid \lambda, \alpha) = \exp\{-(\lambda x)^\alpha\}, \quad x \geq 0.$$

Other examples appear in Chapters 11, 12, and 13.

Now consider frailty and resilience parameters. Here, with the same symbolic notation,

$$\bar{F}(x) \xrightarrow[\text{frailty}]{} [\bar{F}(x)]^\xi \xrightarrow[\text{resilience}]{} 1 - \{1 - [\bar{F}(x)]^\xi\}^\eta, \tag{2}$$

whereas

$$\bar{F}(x) \xrightarrow[\text{resilience}]{} 1 - [F(x)]^\eta \xrightarrow[\text{frailty}]{} \{1 - [F(x)]^\eta\}^\xi. \tag{3}$$

Here the resulting families differ. Because these operations do not commute, it means that in the case of (2), the end result is a family that does not have a frailty parameter, so that the family can be expanded by the reintroduction of such a parameter. The result will be a family without a resilience parameter, so the process does not come to a forced end. This is why it is of interest to identify the methods of introducing parameters that commute and those that do not.

Parameter additions of location, scale, and power all result from making parametric transformations of the underlying random variable. These parameters commute with parameters that are obtained from transformations of a survival function such as frailty, resilience, and tilt parameters. Further general statements can be made, but they are not very informative. Consider the parameters discussed in earlier sections, i.e., scale, power, frailty, resilience, tilt, hazard power, moment, Laplace transform, convolution, and age parameters. Scale parameters commute with all of the others in this list; power parameters commute with all but age and convolution. Frailty parameters commute with hazard power and age parameters. Laplace transform parameters commute with moment and convolution parameters. No other pairs commute.

M. Mixing Semiparametric Families

Mixtures of the form

$$H(x) = \int F(x \mid \psi) \, dG(\psi)$$

are discussed in Chapter 3 and are encountered in Sections 7.D.g and 7.E.i. A more involved case arises in Section 7.F.c, where G itself has a parameter. Mixtures of this kind are the subject of this section.

Start with a semiparametric family of distributions $\{F(\cdot \mid \psi), \psi \in \Psi\}$ and treat the parameter ψ as a random variable with a distribution taken from a semiparametric family $\{G(\cdot \mid \theta), \theta \in \Theta\}$. The resulting mixture

$$H(x \mid \theta) = \int F(x \mid \psi) \, dG(\psi \mid \theta) \tag{1}$$

can then be regarded as a member of the semiparametric family $\{H(\cdot \mid \theta), \theta \in \Theta\}$.

a. Mixtures of Scale Parameter Families

M.1. Proposition. Consider two scale parameter families $F(x \mid \psi) = F(\psi x \mid 1)$, for $x, \psi > 0$, and $G(\psi \mid \lambda) = G(\lambda \psi \mid 1)$, for $\psi, \lambda > 0$. The mixture, given by

$$H(x \mid \lambda) = \int F(x \mid \psi) \, dG(\psi \mid \lambda) \quad \text{or} \quad \bar{H}(x \mid \lambda) = \int \bar{F}(x \mid \psi) \, dG(\psi \mid \lambda),$$

is a scale parameter family with scale parameter $1/\lambda$.

Proof. For scale parameter families,

$$H(x \mid \lambda) = \int F(x \mid \psi) \, dG(\psi \mid \lambda) = \int F(\psi x \mid 1) \, dG_\psi(\lambda \psi \mid 1)$$
$$= \int F(\theta x / \lambda \mid 1) \, dG(\theta \mid 1) = H(x/\lambda \mid 1). \qquad \square$$

b. Mixtures of Product and Frailty Parameter Families

The following Lemma shows that convolving mixing distributions leads to a product of mixtures.

M.2. Lemma. If $\bar{H}_i(x) = \int [\bar{F}(x)]^\xi \, dG_i(\xi), i = 1, 2$, then

$$\bar{H}_1(x)\bar{H}_2(x) = \int [\bar{F}(x)]^\xi \, d(G_1 * G_2)(\xi).$$

Proof. Write $[\bar{F}(x)]^\xi = \exp\{-\xi R(x)\}$, where $R(x) = -\log \bar{F}(x)$. Then, the result can be recognized as a reflection of the fact (see Proposition 20.D.6) that the Laplace transform of a convolution is the product of Laplace transforms. □

M.3. Proposition. Let $\{G(\cdot \mid \theta) : \theta \in B\}$ be a convolution family of distributions such that for each $\theta, G(\cdot \mid \theta)$ has support contained in A. Let $\{F(\cdot \mid \xi) : \xi \in A\}$ be a survival product family of distributions such that $F(x \mid \xi)$ is measurable in ξ for each fixed x. If

$$\bar{H}(x \mid \alpha) = \int \bar{F}(x \mid \xi) \, dG(\xi \mid \alpha), \quad \alpha \in B,$$

then

$$\bar{H}(x \mid \alpha)\bar{H}(x \mid \beta) = \bar{H}(x \mid \alpha + \beta), \quad \alpha, \beta \in B.$$

Proof. This result is an immediate consequence of Lemma M.2. □

If $B = (0, \infty)$, then by Lemma E.8, the family $\bar{H}(\cdot \mid \alpha), \alpha \in (0, \infty)$, is a proportional hazards (frailty parameter) family. To see this more clearly, let $\{F(\cdot \mid \xi), \xi > 0\}$ be a frailty parameter family with underlying survival function $\bar{F}$, so that $\bar{F}(\cdot \mid \xi) = [\bar{F}(\cdot)]^\xi$. Let $R(\cdot) = -\log \bar{F}(\cdot)$ be the hazard function of F, and denote the Laplace transform of $G(\cdot \mid \theta)$ by ϕ_θ. As in Sections E.i and F.c, equation (1) can be rewritten in the form

$$\bar{H}(x \mid \theta) = \int [\bar{F}(x)]^\xi \, dG(\xi \mid \theta)$$

$$= \int e^{-\xi R(x)} \, dG(\xi \mid \theta) = \phi_\theta(R(x)), \quad \theta \in \Theta. \tag{2}$$

The hazard rate $r_H(\cdot \mid \theta)$ of $H(\cdot \mid \theta)$ is given by

$$r_H(x \mid \theta) = \frac{\phi'_\theta(R(x))}{\phi_\theta(R(x))} r(x),$$

where $r(x) = R'(x)$ is the hazard rate of the distribution F. Because ϕ_θ is a Laplace transform, it is log concave (Corollary 18.D.5.a), that is, $\phi'_\theta(z)/\phi_\theta(z)$ is decreasing in z. Thus, $r_H(x\,|\,\theta)$ is decreasing in x whenever $r(x)$ is decreasing in x.

With specific choices of the functions ϕ_θ and R, various parametric families emerge from (2). However, if the Criteria 1 and 2 of Section A.c are to be satisfied, much of the apparent freedom of choice is an illusion. Requirements imposed by Criterion 2 are examined in detail and more generally in Section 19.C.a; here, two very special but important cases are considered.

M.4.a. Proposition. Suppose that $\bar{H}(\cdot\,|\,\theta)$ is given by (2) and that, for some distribution function K of a nonnegative random variable, $R(\cdot) = K(\cdot)/\bar{K}(\cdot)$. Then,

$$\bar{H}(x\,|\,\theta) = \phi_\theta(K(x)/\bar{K}(x)), \quad \theta \in \Theta. \tag{2a}$$

Regard K as the underlying distribution of the family $\{\bar{H}(\cdot\,|\,\theta), \theta \in \Theta\}$ and suppose further that ϕ_θ has the form $\phi_\theta(s) = \phi(s/\theta)$, as is the case when θ is a scale parameter for $G(\cdot\,|\,\theta)$. Then the semiparametric family (2a) satisfies criterion 2 of Section A.c if and only if ϕ is the Laplace transform of an exponential distribution. In this case, Criterion 1 of A.c is also satisfied.

Proof. If Criterion 2 is satisfied, then it must be that when the survival function $\bar{K}$ in (2a) is replaced by one of the form $\bar{H}(\cdot\,|\,\tilde{\theta})$, the result retains the form of $\bar{H}(\cdot\,|\,\theta)$, but with a parameter λ that is a function of θ and $\tilde{\theta}$. This means that for some $\lambda = \lambda(\theta, \tilde{\theta})$,

$$\frac{\phi\left(\frac{1}{\theta}\left[1 - \phi\left(\frac{z}{\tilde{\theta}}\right)\right]\right)}{\phi\left(\frac{z}{\tilde{\theta}}\right)} = \phi\left(\frac{z}{\lambda}\right). \tag{3a}$$

where $z = K(x)/\bar{K}(x)$. Note that the arguments of ϕ on both sides of (3a) must be equal. That is,

$$\frac{1}{\theta}\left[1 - \phi\left(\frac{z}{\tilde{\theta}}\right)\right] \bigg/ \phi\left(\frac{z}{\tilde{\theta}}\right) = \frac{z}{\lambda};$$

it follows that

$$\phi(s) = 1 \bigg/ \left(1 + \frac{\theta\tilde{\theta}}{\lambda}s\right).$$

This is the Laplace transform of an exponential distribution with parameter $\lambda/\theta\tilde\theta$, and the underlying distribution is retrieved with $\lambda = \theta\tilde\theta$ so that Criterion 1 is satisfied. □

M.4.b. Proposition. Suppose that $\bar H(\cdot\,|\,\theta)$ is given by (2) and that, for some distribution function F of a nonnegative random variable, $R(\cdot) = -\log \bar F(\cdot)$. Then,

$$\bar H(x\,|\,\theta) = \phi_\theta(-\log \bar F), \quad \theta \in \Theta. \tag{2b}$$

Regard F as the underlying distribution and suppose that ϕ_θ has the form $\phi_\theta(s) = \phi(s/\theta)$. Then the semiparametric family (2b) satisfies Criterion 2 of Section A.c if and only if ϕ is the Laplace transform of a distribution degenerate at some point, say 1. In this case, Criterion 1 is satisfied, and the underlying distribution is retrieved with $\theta = 1$.

Proof. If Criterion 2 is satisfied, then it must be that if the survival function $\bar F$ in (2b) is replaced by one of the form $\bar H(\cdot\,|\,\tilde\theta)$, the result retains the form of $\bar H(\cdot\,|\,\theta)$, but with a parameter λ that is a function of θ and $\tilde\theta$. This means that for some $\lambda = \lambda(\theta,\tilde\theta)$,

$$\phi(-\theta \log [\phi(-\tilde\theta \log \bar F)]) = \phi(-\lambda \log \bar F). \tag{3b}$$

With the notation $R = -\log \bar F$, it follows from (3b) that

$$-\theta \log \phi(\tilde\theta R) = \lambda R;$$

consequently, $\phi(s) = \exp\{-[\lambda(\theta,\tilde\theta)/\theta\tilde\theta]s\}$, and the underlying distribution is retrieved when $\lambda(\theta,\tilde\theta) = \theta\tilde\theta$. □

M.4.c. Example. If $\bar G(x\,|\,\theta) = e^{-\theta x}$, $x,\theta > 0$, is an exponential distribution, then

$$\phi_\theta(s) = \frac{\theta}{\theta + s}, \quad s > -\theta,$$

and (2) yields

$$\bar H(x\,|\,\theta) = \frac{\theta}{\theta + R(x)}, \quad x,\theta > 0. \tag{4}$$

If $R(x) = K(x)/\bar{K}(x)$ for some distribution K concentrated on $[0, \infty)$, then (4) becomes

$$\bar{H}(x \mid \theta) = \frac{\theta \bar{K}(x)}{\theta \bar{K}(x) + K(x)} = \frac{\theta \bar{K}(x)}{1 - \tilde{\theta} \bar{K}(x)}, \quad x, \theta > 0. \tag{5a}$$

This is the tilt parameter family of Section F.

If $R(x) = -\log \bar{F}(x)$ for some distribution F concentrated on $[0, \infty)$, (4) becomes

$$\bar{H}(x \mid \theta) = \frac{\theta}{\theta - \log \bar{F}(x)}, \quad x, \theta > 0. \tag{5b}$$

This defines a semiparametric family of survival functions, but neither Criterion 1 nor Criterion 2 of Section A.c is satisfied.

Mixture representations of the form (2) are not unique. Indeed, the distribution G can be replaced by any other distribution G_1 with support $(0, \infty)$, so long as the hazard function R is replaced by an appropriately chosen hazard function R_1. To see this, solve the equation

$$\bar{H}(x) = \phi(R(x)) = \phi_1(R_1(x))$$

to find that

$$\phi_1^{-1} \phi(R(x)) = R_1(x). \tag{6}$$

With this definition, R_1 is a hazard rate and

$$\bar{H}(x) = \int_0^\infty e^{-\xi R(x)} \, dG(\xi) = \int_0^\infty e^{-\xi R_1(x)} \, dG_1(\xi). \tag{7}$$

M.5.a. Example. First, suppose that in (2), G is a gamma distribution with scale parameter λ and shape parameter ν. According to 9.A(6), this distribution has the Laplace transform

$$\phi_{\lambda, \nu}(s) = \left[\frac{\lambda}{\lambda + s}\right]^\nu, \quad s > -\lambda.$$

As in Proposition M.4.a., take $R(\cdot) = K(\cdot)/\bar{K}(\cdot)$ in (2) to obtain the survival function

$$\bar{H}(x \mid \lambda, \nu) = \phi_{\lambda,\nu}\left(\frac{K(x)}{\bar{K}(x)}\right) = \left[\frac{\lambda \bar{K}(x)}{\lambda \bar{K}(x) + K(x)}\right]^{\nu}$$

$$= \left[\frac{\lambda \bar{K}(x)}{1 - \bar{\lambda}\bar{K}(x)}\right]^{\nu}, \quad x, \lambda, \nu > 0. \tag{8}$$

This is (11c) of Section F.

M.5.b. Example. Replace the gamma distribution used to derive (8) by an inverse Gaussian distribution (Chapter 13). According to 13.A(4), this distribution has the Laplace transform

$$\phi_{\text{IG}}(s) = \exp\left\{\nu\left[1 - \left(1 + \frac{2s}{\lambda}\right)^{1/2}\right]\right\}.$$

By setting

$$\phi_{\text{IG}}(R_1(x)) = \phi_{\lambda,\nu}\left(\frac{K(x)}{\bar{K}(x)}\right),$$

it follows that

$$R_1(x) = \frac{\lambda}{2}\left\{\left[1 - \log\frac{\lambda \bar{K}(x)}{1 - \bar{\lambda}\bar{K}(x)}\right]^2 - 1\right\},$$

and consequently

$$\phi_{\text{IG}}(R_1(x)) = \left[\frac{\lambda \bar{K}(x)}{1 - \bar{\lambda}\bar{K}(x)}\right]^{\nu}. \tag{8a}$$

This does not yield a very convenient mixture representation to replace (8), but it illustrates the lack of uniqueness of the mixture representation. However, in this case it is crucial that the gamma density with its convolution parameter be replaced by another distribution with a convolution parameter. For the inverse Gaussian distribution, this requirement is fulfilled because ν appears as a power in its Laplace transform.

Examples M.5.a and M.5.b illustrate the following proposition.

M.6. Proposition. If $\Theta = (0, \infty)$, then (2) takes the form

$$\bar{H}(x\,|\,\theta) = \phi_\theta(R(x)), \quad x, \theta > 0. \tag{9}$$

In this case, the parameter θ is a frailty parameter for $H(\cdot\,|\,\theta)$ if and only if it is a convolution parameter for $G(\cdot\,|\,\theta)$.

Proof. Suppose first that θ is a convolution parameter for $G(\cdot\,|\,\theta)$. Then, (9) can be written in the form

$$\bar{H}(x\,|\,\theta) = [\phi(R(x))]^\theta, \tag{10}$$

where ϕ is the Laplace transform of $G(\cdot\,|\,1)$. Consequently, θ is a frailty parameter for $H(\cdot\,|\,\theta)$.

Next, suppose that θ is a frailty parameter for $H(\cdot\,|\,\theta)$, and let $\phi(\cdot\,|\,\theta)$ denote the Laplace transform of $G(\cdot\,|\,\theta)$. It follows that

$$\bar{H}(x\,|\,\theta) = [\bar{H}(x\,|\,1)]^\theta = [\phi(R(x)\,|\,1)]^\theta,$$

that is, $\phi(z\,|\,\theta) = [\phi(z\,|\,1)]^\theta$; this means that θ is a convolution parameter for $G(\cdot\,|\,\theta)$. □

M.6.a. Example. In Example M.5.a, consider the case that $R(x\,|\,\xi, \alpha) = \xi x^\alpha$ is the hazard function of a Weibull distribution and ϕ is the Laplace transform of an exponential distribution. Then,

$$\bar{H}(x\,|\,\alpha, \theta) = [\lambda/(\lambda + x^\alpha)]^\theta, \quad \alpha, \theta > 0.$$

Because gamma distributions form a convolution family, and Weibull distributions form a survival product (frailty parameter) family, the outcome here is another frailty parameter family. This family is a form of what is sometimes called *Burr's distribution* (see Johnson, Kotz and Balakrishnan, 1994, p. 54). It is also a form of what Arnold (1983, p. 1) calls a *Pareto Type IV* distribution (see Section 11.B).

Another special case of Example M.5.a is treated in Section 10.A.h.

The following proposition is a consequence of Proposition M.6, but is important enough to warrant repeating.

M.6.b. Proposition. Let $\{G(\cdot \mid \nu), \nu > 0\}$ be a convolution family and suppose that (2) holds (with ν in place of θ). Then,

$$\bar{H}(x \mid \nu) = [\bar{H}(x)]^\nu, \quad \nu, x > 0. \tag{11}$$

Proof. Because $\{G(\cdot \mid \nu), \nu > 0\}$ is a convolution parameter family $\phi_\nu(\cdot) = [\phi(\cdot)]^\nu$ and consequently (10) yields

$$\bar{H}(x \mid \nu) = [\phi_1(\phi_1^{-1}\bar{H}(x))]^\nu = [\bar{H}(x)]^\nu. \qquad \square$$

Equation (11), derived from (2), is independent of the convolution family $\{G(\cdot \mid \nu), \nu > 0\}$; in this respect, convolution families are an exceptional case. In general, the mixture (2) does indeed depend upon the mixing distribution $G(\cdot \mid \theta)$. Consequently, it is sometimes convenient to rewrite (2) in such a manner as to relate the hazard function $R(\cdot)$ to $\bar{H}(\cdot \mid \theta_0)$ for some specific value θ_0 of θ. To this end, fix $\theta_0 \in \Theta$ and regard $G(\cdot \mid \theta_0)$ as the underlying distribution for the semiparametric family $\{G(\cdot \mid \theta), \theta \in \Theta\}$. Write $\bar{H}(\cdot)$ in place of $\bar{H}(\cdot \mid \theta_0)$ and write ϕ in place of ϕ_{θ_0}. With this notation, it follows from (2) that

$$R(x) = \phi^{-1}(\bar{H}(x)).$$

Thus, (2) can be rewritten as

$$\bar{H}(x \mid \theta) = \phi_\theta \phi^{-1} \bar{H}(x), \quad x > 0, \ \theta \in \Theta. \tag{12}$$

In this form, $\{\bar{H}(\cdot \mid \theta), \theta \in \Theta\}$ is a semiparametric family of survival functions with underlying survival function $\bar{H}(\cdot) = \bar{H}(\cdot \mid \theta_0)$. In the following examples, θ_0 is conveniently taken to be 0 or 1.

The form (2) is used above as a step toward the derivation of (12), which clearly exhibits a semiparametric family. But (12) is of interest also because it is suggestive of different examples.

M.7. Proposition. Let $\{G(\cdot \mid \lambda), \lambda > 0\}$ be a scale parameter family with underlying distribution $G(\cdot) = G(\cdot \mid 1)$ such that $G(x) = 0, x < 0$. Denote the Laplace transform of G by ϕ. Then (12) takes the form

$$\bar{H}(x \mid \lambda) = \phi\left(\frac{\phi^{-1}\bar{H}(x)}{\lambda}\right), \quad x, \lambda > 0. \tag{13}$$

Proof. Because $G(\cdot\,|\,\lambda)$ has the Laplace transform $\phi(s\,|\,\lambda) = \phi(s/\lambda)$, (13) follows directly from (12). □

Note that for each underlying distribution G (Laplace transform ϕ), (13) defines a semiparametric family. This is a contrast to Proposition M.6.b where the family (11) is independent of the convolution parameter family $\{G(\cdot\,|\,\nu), \nu > 0\}$.

Remark. It can be verified that any semiparametric family obtained from (13) by choice of ϕ will satisfy Criteria 1 and 2 of Section A.c.

M.7.a. Example. If G is degenerate at 1, that is, $G(x) = 0, x < 1, G(x) = 1, x \geq 1$, then $\phi(s) = e^{-s}$, and (13) yields $\bar{H}(x\,|\,\lambda) = [\bar{H}(x)]^{1/\lambda}$, λ, $x > 0$. In this case, $1/\lambda$ is a frailty parameter. Although λ is introduced in Proposition M.7 as a scale parameter, in this example, $1/\lambda$ also acts as a convolution parameter, which accords with Proposition M.6.

M.7.b. Example. If $\bar{G}(x) = e^{-x}, x \geq 0$, is an exponential survival function, then

$$\phi(s) = \frac{1}{1+s}, \quad s > -1, \text{ and } \phi^{-1}(u) = \frac{1}{u} - 1. \tag{14}$$

In this case, (13) yields

$$\bar{H}(x\,|\,\lambda) = \frac{\lambda \bar{H}(x)}{1 - \bar{\lambda}\bar{H}(x)}, \quad \lambda, x > 0;$$

this is the tilt parameter family of Section F and is also encountered in Example M.4.c.

M.7.c. Example. If G is an inverse Gaussian distribution (Chapter 13) with unit parameters, then

$$\phi(s) = \exp\{1 - (1 + 2s)^{1/2}\}, \quad s > -\frac{1}{2}, \text{ and}$$
$$\phi^{-1}(u) = \frac{1}{2}[(1 - \log u)^2 - 1], \quad u \leq 1. \tag{15}$$

Here, (13) becomes

$$\bar{H}(x\,|\,\lambda) = \exp\left\{1 - \left[1 + \frac{(1 - \log \bar{H}(x))^2 - 1}{\lambda}\right]^{1/2}\right\}, \quad x, \lambda > 0. \quad (16)$$

This semiparametric family is neither familiar nor attractive. When $\bar{H}(x) = e^{-x}, x \geq 0$, (16) becomes

$$\bar{H}(x\,|\,\lambda) = \exp\left\{1 - \left[1 + \frac{2x + x^2}{\lambda}\right]^{1/2}\right\}, \quad x, \lambda > 0.$$

It can be verified that this distribution is DHR.

M.8. Proposition. Let G be a distribution with density g concentrated on $[0, \infty)$, and let I be the interval on which the Laplace transform ϕ of G is finite. Denote by $\{G(\cdot\,|\,s), s \in I\}$ the Laplace transform family with underlying distribution G. In (12), take $\theta_0 = 0$ and write $\bar{H}(x) = \bar{H}(x\,|\,0)$ to obtain

$$\bar{H}(x\,|\,s) = \frac{\phi(s + \phi^{-1}\bar{H}(x\,|\,0))}{\phi(s)}, \quad s \in I. \quad (17)$$

Proof. Here, $\phi_{\theta_0} = \phi$ and $G(\cdot\,|\,s)$ has the density

$$g(x\,|\,s) = \frac{e^{-sx}}{\phi(s)} g(x), \quad s \in I.$$

Thus, $G(\cdot\,|\,s)$ has the Laplace transform

$$\phi_s(t) = \int e^{-tx} \frac{e^{-sx}}{\phi(s)} g(x)\,dx = \frac{\phi(s+t)}{\phi(s)},$$

and (12) yields (17). □

Proposition M.8 can also easily be proved directly from (2).

For any distribution G concentrated on $[0, \infty)$, (17) leads to a semiparametric family $\{H(\cdot\,|\,s), s \in I\}$. Such families satisfy Criterion 1 of

Section A.c, with the underlying distribution $H(\cdot\,|\,0)$. However, these families need not satisfy Criterion 2 of Section A.c.

M.8.a. Example. If G is an exponential distribution as in Example M.4.c, then (17) becomes

$$\bar{H}(x\,|\,s) = \frac{(s+1)\bar{H}(x\,|\,0)}{1+s\bar{H}(x\,|\,0)}, \quad s > -1,\ x > 0.$$

This is the tilt parameter family of Example M.4.c, but with λ replaced by $s+1$.

By way of explanation, note that for the exponential distribution, the introduction of a Laplace transform parameter leads to the same parametric family as the introduction of a scale parameter.

M.8.b. Example. Equation (15) gives the Laplace transform of an inverse Gaussian distribution G with unit parameters. In this case, (17) becomes

$$\bar{H}(x\,|\,s) = \exp\{(2s+1)^{1/2} - [2s + (1 - \log \bar{H}(x\,|\,0))^2]^{1/2}\},$$
$$s > -\frac{1}{2},\ x > 0. \tag{18}$$

Clearly, this semiparametric family is quite unattractive.

With $\bar{H}(x\,|\,0)$ an exponential survival function with unit parameter, (18) becomes

$$\bar{H}(x\,|\,s) = \exp\{(2s+1)^{1/2} - [2s + (1+x)^2]^{1/2}\},$$
$$s > -\frac{1}{2},\ x > 0. \tag{19}$$

an unfamiliar extension of the exponential survival function $\bar{H}(x\,|\,0)$. It can be verified that the survival function (19) is DHR, for $s < 0$, and is IHR, for $s > 0$.

c. Mixtures of Resilience Parameter Families

With the underlying distribution F, let $F(\cdot\,|\,\eta) = [F(\cdot)]^\eta, \eta > 0$, define a resilience parameter family, and let $S(\cdot) = \log F(\cdot)$ be the reverse hazard function of F. Then (1) takes the form

$$H(x\,|\,\theta) = \int [F(x)]^\eta\, dG(\eta\,|\,\theta) = \int e^{\eta S(x)}\, dG(\eta\,|\,\theta)$$
$$= \phi_\theta(-S(x)), \quad \theta \in \Theta, \qquad (20)$$

where ϕ_θ is the Laplace transform of $G(\cdot\,|\,\theta)$.

Caution. Here, the function S is not to be confused with s, often used to denote a Laplace transform parameter.

Because of the similarity of (20) and (2), the propositions and examples of Section M.b all have counterparts for this section. Some of these are given here.

M.9. Example. If $\bar{G}(x\,|\,\theta) = e^{-\theta x}$, $x, \theta > 0$, as in Example M.5.a, then (20) becomes

$$H(x\,|\,\theta) = \frac{\theta}{\theta - S(x)}, \quad \theta, x > 0. \qquad (21)$$

If

$$S(x) = \frac{\bar{K}(x)}{K(x)},$$

for some distribution K concentrated on $[0, \infty)$ then (21) yields

$$H(x\,|\,\theta) = \frac{\theta K(x)}{\theta K(x) + \bar{K}(x)} = \frac{\theta K(x)}{1 - \bar{\theta}\bar{K}(x)}, \quad \theta, x > 0, \qquad (22)$$

which is the distribution defined in terms of the survival function in (5a), but with θ replaced by $1/\theta$.

As in Section M.b, suppose that $\{G(\cdot\,|\,\theta), \theta \in \Theta\}$ is a semiparametric family with underlying distribution $G(\cdot\,|\,\theta_0)$. Write $H(\cdot)$ in place of $H(\cdot\,|\,\theta_0)$ and ϕ in place of ϕ_{θ_0}. With $\theta = \theta_0$ in (20), it follows that

$$S(x) = \phi_\theta(\phi^{-1}H(x)). \qquad (23)$$

M.10. Proposition. In a mixture of the form (20), θ is a resilience parameter for $\{H(\cdot\,|\,\theta), \theta > 0\}$ if and only if it is a convolution parameter for $\{G(\cdot\,|\,\theta), \theta > 0\}$.

The proof of this result is analogous to the proof of Proposition M.6.

M.10.a. Proposition. Let $\{G(\cdot\,|\,\nu), \nu > 0\}$ be a convolution parameter family and suppose that (20) holds. With the notation $H(\cdot) = H(\cdot\,|\,1)$, it follows that

$$H(x\,|\,\nu) = [H(x\,|\,1)]^\nu, \quad \nu, x > 0. \tag{24}$$

This is the general form for a resilience parameter family, and the result is independent of $\{G(\cdot\,|\,\nu), \nu > 0\}$.

M.11. Proposition. Let $\{G(\cdot\,|\,\lambda), \lambda > 0\}$ be a scale parameter family with underlying distribution $G(\cdot) = G(\cdot\,|\,1)$, and let ϕ be the Laplace transform of G. Suppose that $G(x) = 0$, for all $x < 0$. If (20) holds and $H(\cdot) = H(\cdot\,|\,1)$, then

$$H(x\,|\,\lambda) = \phi\left(\frac{\phi^{-1} H(x)}{\lambda}\right), \quad x, \lambda > 0. \tag{25}$$

The proof of this proposition is analogous to that of Proposition M.7 and is omitted. For any choice of G (Laplace transform ϕ), (25) yields a semiparametric family that satisfies Criteria 1 and 2 of Section A.c.

M.11.a. Example. As in Example M.7.a, let $\bar{G}(x) = e^{-x}, x > 0$; then $\phi(s) = \dfrac{1}{1+s}$ and $\phi^{-1}(u) = \dfrac{1}{u} - 1$. In this case, (25) yields the family

$$H(x\,|\,\lambda) = \frac{\lambda H(x)}{1 - \bar{\lambda} H(x)}, \quad \lambda > 0,$$

or with $\lambda = 1/\gamma$ and a change in notation,

$$H(x\,|\,\gamma) = \frac{H(x)}{\gamma + \bar{\gamma} H(x)}, \quad \gamma > 0.$$

It can be verified directly that $\bar{H}(\cdot\,|\,\gamma)$ is the tilt parameter family of Example M.8.a. See also F(10).

If G is replaced by a gamma distribution with Laplace transform $\phi(s\,|\,\nu) = (1+s)^{-\nu}, \nu > 0$, then (25) yields the family

$$H(x\,|\,\gamma,\nu) = \left(\frac{H(x)}{\gamma + \bar{\gamma}H(x)}\right)^{\nu}, \quad \gamma,\nu > 0.$$

This is a tilt parameter family with added resilience parameter ν.

d. Mixtures of Convolution Families

Mixtures of convolution parameter families can more easily be studied in terms of Laplace transforms than in terms of distribution or survival functions. If $\{F(\cdot\,|\,\nu), \nu > 0\}$ is a convolution parameter family, then according to Proposition J.2, $F(\cdot\,|\,1)$ is infinitely divisible. Denote the Laplace transform of $F(\cdot\,|\,1)$ by ϕ, so that ϕ^ν is the Laplace transform of $F(\cdot\,|\,\nu)$. Let $\{G(\cdot\,|\,\theta), \theta \in \Theta\}$ be a semiparametric family of distribution with mass concentrated on $(0,\infty)$, and denote the Laplace transform of $G(\cdot\,|\,\theta)$ by ϕ_θ. Then the mixture

$$H(x\,|\,\theta) = \int F(x\,|\,\nu)\,dG(\nu\,|\,\theta) \tag{26}$$

has the Laplace transform

$$\begin{aligned}\phi_H(s\,|\,\theta) &= \int [\phi(s)]^\nu\,dG(\nu\,|\,\theta) = \int e^{-\nu(\log\phi(s))}\,dG(\nu\,|\,\theta)\\ &= \phi_\theta(-\log\phi(s)).\end{aligned} \tag{27}$$

This equation is similar to (9), but with the survival function $\bar{F}$ and $\bar{H}(\cdot\,|\,\theta)$ of (9) replaced by the corresponding Laplace transforms. Consequently, the equations have parallel special cases.

The following lemma, to be compared with Lemma M.2, states that for convolution families, convolutions of mixtures are mixtures of convolutions. This fact is useful in Bayesian statistical analysis, and it is used here to show that a convolution family mixed by a convolution family yields a convolution family.

M.12. Lemma (Keilson and Steutel, 1974, p. 116). Suppose that $\{F(\cdot \mid \theta) : \theta \in A\}$ is a convolution parameter family and

$$H_i(x) = \int F(x \mid \theta)\, dG_i(\theta), \quad i = 1, 2.$$

Then

$$(H_1 * H_2)(x) = \int F(x \mid \theta)\, d(G_1 * G_2)(\theta).$$

Proof. For convolution parameter families,

$$\begin{aligned}
(H_1 * H_2)(x) &= \int H_1(x-z)\, dH_2(z) = \int_z \int_\theta F_1(x-z)\, dG_1(\theta)\, dH_2(z) \\
&= \int_\theta \int_z F_1(x-z \mid \theta)\, dH_2(z)\, dG_1(\theta) \\
&= \int_\theta \int_z H_2(x-z)\, dF_1(z \mid \theta)\, dG_1(\theta) \\
&= \int_\theta \int_z \int_\eta F_2(x-z \mid \eta)\, dG_2(\eta)\, dF_1(z \mid \theta)\, dG_1(\theta) \\
&= \int_\theta \int_z F(x \mid \theta + \eta)\, dG_2(\eta)\, dG_1(\theta). \quad \square
\end{aligned}$$

M.13. Proposition. If $\{F(\cdot \mid \theta) : \theta \in A\}$ and $\{G(\cdot \mid \alpha) : \alpha \in B\}$ are convolution families, and if

$$H(x \mid \alpha) = \int F(x \mid \theta)\, dG(\theta \mid \alpha), \quad \alpha \in B, \tag{28}$$

then $\{H(\cdot \mid \alpha) : \alpha \in B\}$ is a convolution family.

Proof. This is immediate from Lemma M.12. $\square$

An early important result is that a convolution family mixed by an infinitely divisible distribution is infinitely divisible.

M.14. Proposition (Feller, 1971, p. 538; Kent, 1981). If the family $\{F(\cdot \mid \theta) : \theta \in A\}$ is a convolution family and G is infinitely divisible, then the mixture H given by (28) is infinitely divisible. Moreover,

$$H^{\alpha *}(x) = \int F(x \mid \theta)\, dG^{\alpha *}(\theta), \quad \alpha \in B. \tag{29}$$

Proof. Suppose that for some positive integer m, $\alpha = 1/m$ so that $(G^{\alpha*})^{m*} = G$. If $H^{\alpha*}$ is defined by (29), then by Lemma M.12, $(H^{\alpha*})^{m*} = H$. This proves that H is infinitely divisible and (29) is satisfied for $\alpha = 1/m$, $m = 1, 2, \ldots$. Again from Lemma M.12, it follows immediately that (29) holds for rational α, and the proof is completed by a limiting argument. □

Multivariate versions of the above results have been given by Marshall and Olkin (1990).

M.15. Proposition. If $\{G(\cdot \,|\, \lambda), \lambda > 0\}$ is a scale parameter family and ϕ_λ is the Laplace transform of $G(\cdot \,|\, \lambda)$, then $\phi_\lambda(s) = \phi_1(s/\lambda)$, and (27) takes the form

$$\phi_H(s\,|\,\lambda) = \phi_1\left(\frac{-\log\phi(s)}{\lambda}\right) = \phi_1\left(\frac{\phi_1^{-1}\phi_H(s\,|\,1)}{\lambda}\right),$$

where H is given by (26).

M.15.a. Example. If $\phi_1(s) = 1/(1+s)$, it follows that $\phi_H(s\,|\,\lambda) = \lambda/(\lambda - \log\phi(s))$. If also, $\phi(s) = 1/(1+s)$, then

$$\phi_H(s\,|\,\lambda) = \lambda/(\lambda + \log(1+s)).$$

M.16. Proposition. Let G be a distribution with density g and Laplace transform ϕ_G that is finite on the interval I, and let $\{G(\cdot \,|\, \theta), \theta \in I\}$ be the Laplace transform family with underlying distribution G. Thus, $G(\cdot \,|\, \theta)$ has the density

$$g(\cdot\,|\,\theta) = \frac{e^{-\theta x}}{\phi_G(\theta)} g(x), \quad x > 0, \ \theta \in I.$$

If (27) holds, it follows that

$$\phi_H(s\,|\,\theta) = \int e^{-\nu(-\log\phi(s))} \frac{e^{-\theta\nu}}{\phi_G(\theta)} g(\nu)\,d\nu = \frac{\phi_G(\theta - \log\phi(s))}{\phi_G(\theta)}$$
$$= \frac{\phi_G(\theta + \phi_G^{-1}\phi_H(s\,|\,\theta))}{\phi_G(\theta)}.$$

For each distribution G, this defines a semiparametric family with underlying distribution having the Laplace transform $\phi_H(\cdot\,|\,\theta)$.

N. Summary of Order Properties

In the following summary, only the strongest order known to hold is noted. For example, if a likelihood ratio order is noted, then the hazard rate and stochastic orders also hold.

Scale
$F(x\,|\,\lambda) = F(\lambda x), \quad \lambda > 0$ stochastic order $\leq_{\text{st}}$ decreasing in λ

Power
$F(x\,|\,\alpha) = F(x^\alpha), \quad \alpha > 0$ convex transform order $\leq_{\text{cx}}$ decreasing in α

Frailty
$\bar{F}(x\,|\,\xi) = [\bar{F}(x)]^\xi, \quad \xi > 0$ likelihood ratio order $\leq_{\text{lr}}$ decreasing in ξ

Resilience
$F(x\,|\,\eta) = [F(x)]^\eta, \quad \eta > 0$ likelihood ratio order $\leq_{\text{lr}}$ increasing in η

Tilt
$\bar{F}(x\,|\,\gamma) = \dfrac{\gamma \bar{F}(x)}{1 - \bar{\gamma}\bar{F}(x)}, \quad \gamma > 0$ likelihood ratio order $\leq_{\text{lr}}$ decreasing in γ

Hazard power
$\bar{F}(x\,|\,\zeta) = \exp\{-[R(x)]^\zeta\}, \quad \zeta > 0$ no general results known about orders

Moment
$\bar{F}(x\,|\,\beta) = \frac{1}{\mu_\beta}\int_x^\infty z^\beta\,dF(z)$ likelihood ratio order $\leq_{\text{lr}}$ increasing in β

Laplace transform
$\bar{F}(x\,|\,s) = \frac{1}{\phi(s)}\int_x^\infty e^{-sz}\,dF(z)$ likelihood ratio order $\leq_{\text{lr}}$ decreasing in s

Convolution
$F(x\,|\,\nu) = F^{\nu*}(x), \quad \nu > 0$ stochastic order $\leq_{\text{st}}$ increasing in ν

Age
$\bar{F}_t(x) = \bar{F}(t+x)/\bar{F}(t), \quad t \geq 0$ if F is IHR, hazard rate order $\leq_{\text{hr}}$ decreasing in t

O. Additional Semiparametric Families

The various semiparametric families discussed in earlier sections of this chapter certainly do not exhaust the interesting possibilities. To emphasize this point, some additional families are briefly mentioned here.

O.1. Example. As discussed in Section F·j, Dabrowska and Doksum (1988) study the family with hazard function

$$R(x\,|\,\theta) = \frac{1 - [\bar{F}(x)]^\theta}{\theta[\bar{F}(x)]^\theta}, \quad \theta > 0$$
$$= -\log \bar{F}(x), \quad \theta = 0. \tag{1}$$

This example is motivated when θ is a positive integer by the fact that $\bar{F}^\theta$ is the survival function of a series system of θ components all with survival function $\bar{F}$. Then, $\theta R(x\,|\,\theta)$ is the odds ratio of system failure by time x. This example satisfies Criterion 1 stated at the beginning of this chapter, but fails to satisfy Criterion 2. That is, if $-\log \bar{F}$ has the form of (1) but with some survival function $\bar{G}$ in place of $\bar{F}$, then the resulting hazard function would not be of the form as (1).

O.2. Example. Equilibrium distributions are defined and briefly discussed in Section 20.B.c. These distributions lead to the parametric family with density $f_{(s)}$ given by

$$f_{(s)}(x) = \int_{-\infty}^{\infty} [\gamma^{(s)}(t)/\lambda_s] f(x - t)\, dt, \quad s > 0,$$
$$= f(x), \quad s = 0. \tag{2}$$

Here $\lambda_s = \mu_s/\Gamma(s+1)$ is the normalized sth moment of f and

$$\gamma^{(s)}(t) = (-t)^{s-1}/\Gamma(s), \quad t \le 0, \tag{3a}$$
$$= 0, \quad t > 0.$$

In particular,

$$f_{(1)}(x) = \bar{F}(x)/\mu. \tag{3b}$$

This parametric family clearly satisfies Criterion 1 stated at the beginning of this chapter. That Criterion 2 is satisfied follows from 20.B(13).

Of course the range of possible values for s depends upon the existence of moments.

O.3. Example. According to 1.C(4), if F is the distribution of a non-negative random variable, then the rth moment μ_r of F is given by

$$\mu_r = r \int_0^\infty \bar{F}(x) x^{r-1}\, dx.$$

Consequently,

$$f(x\,|\,r) = \frac{r x^{r-1} \bar{F}(x)}{\mu_r}, \quad x \geq 0, \tag{4}$$

is a density for all r such that μ_r is finite. This density can be obtained by introducing a moment parameter in the equilibrium distribution $f_{(1)}$ given by (2a). It fails to satisfy Criterion 1, but it may be of interest in some circumstances.

O.4. Example. Suppose that the underlying distribution has hazard function R, and consider the hazard function

$$R(x\,|\,\theta) = R(x)\, e^{\theta x}, \quad \theta \geq 0. \tag{5}$$

This semiparametric family satisfies Criterion 1, that the underlying distribution is a member of the parametric family, because $R(x\,|\,0) = R(x)$. Moreover, Criterion 2 is satisfied; reuse of the same parametric family does not add a new parameter. More explicitly, if $R(x\,|\,\rho) = R(x)\, e^{\rho x}$ is used in place of R in (5), then the result is the hazard function $R(x\,|\,\rho+\theta)$, and this is a reparameterization of the family (5).

This example has been proposed by Murthy, Xie and Jiang (2004, pp. 24, 134) for the case that $R(x) = (\lambda x)^\alpha$ is the hazard function of a Weibull distribution.

P. Distributions not Admitting Parameters

In Section A.a, the notion of a semiparametric family is introduced, where there are two parameters; one parameter is real and the other parameter is a survival function. A survival function in this semiparametric family is denoted there with the notation by $\bar{H}(\cdot\,|\,\theta, \bar{F})$.

For each of the semiparametric families discussed above there is some underlying distribution F such that

$$\bar{H}(\cdot \mid \theta, \bar{F}) = \bar{F}, \quad \text{for all } \theta. \tag{1}$$

For example, if the parameter θ is a frailty parameter, that is, if

$$\bar{H}(\cdot \mid \theta, \bar{F}) = [\bar{F}(\cdot)]^\theta, \quad \theta > 0,$$

and if F is a degenerate distribution (taking on only the values 0 and 1), then (1) is satisfied. The introduction of a frailty parameter does not alter a degenerate distribution. In cases of this kind, the semiparametric family cannot be used to introduce a parameter into the distribution F. In most cases, the distributions F with this very stringent property are degenerate, as in the frailty parameter example. In each of the following, it is assumed that F is a proper distribution, i.e., it puts mass 1 on $(-\infty, \infty)$.

P.1. Scale parameters. $F(\lambda x) = F(x)$ for all $\lambda > 0$ if and only if F is degenerate at 0.

P.2. Power parameters. $\bar{F}(x^\alpha) = \bar{F}(x)$ for all, $\alpha > 0$ if and only if F is a Bernoulli distribution, i.e., F puts mass only at 0 and/or 1.

P.3. Frailty parameters. $[\bar{F}(x)]^\eta = \bar{F}(x)$ for all $\eta, x > 0$ if and only if F is degenerate at some point.

P.4. Resilience parameters. $[F(x)]^\xi = F(x)$ for all $\xi, x > 0$ if and only if F is degenerate at some point.

P.5. Tilt parameters. $\dfrac{\gamma \bar{F}(x)}{1 - \bar{\gamma} \bar{F}(x)} = \bar{F}(x)$ for all $\gamma, x > 0$ if and only if F is degenerate at some point.

P.6. Hazard power parameters. $\exp\{-[R(x)]^\zeta\} = \bar{F}(x)$ for all $\zeta, x > 0$ if and only if $\bar{F}(x)$ takes only the values $0, e^{-1}$, and 1. This means that either F is degenerate at some point, or F has mass at only two points; this mass must be e^{-1} at the larger of the two points. This results from the fact that $R(x)$ can take on any of the values $\infty, 1$, or 0.

P.7. Moment parameters. $\int_x^\infty (z^\beta/\mu_\beta)\, dF(z) = \bar{F}(x)$ for all $\beta, x > 0$ if and only if F is degenerate at some point.

P.8. Laplace transform parameters. $\int_x^\infty (e^{-sz}/\phi(s))\,dF(z) = \bar{F}(x)$ for all $s, x > 0$ if and only if F is degenerate at some point.

P.9. Convolution parameters. $[\phi(s)]^\nu = \phi(s)$ for all $s, \nu > 0$ if and only if F is degenerate at 0.

P.10. Age parameters. $\bar{F}(x+t)/\bar{F}(t) = \bar{F}(x)$ for all $x, t > 0$ if and only if F is an exponential distribution.

Part IV

Parametric Families

8
The Exponential Distribution

> "That's not the regular rule: you invented it just now." "It's the oldest rule in the book" said the King. "Then it ought to be Number One." said Alice.
>
> Lewis Carroll, *Alice in Wonderland*

The most important one parameter family of life distributions is the family of exponential distributions. This importance is partly due to the fact that several of the most commonly used families of life distributions are two- or three-parameter extensions of the exponential distributions. But the exponential distributions, with their constant hazard rates, form a baseline for evaluating other families. Because they have only one parameter, they are quite simple to describe and are exceptionally amenable to statistical analyses. A path-breaking paper of Epstein and Sobel (1953) brought new attention to the uses of the exponential distribution.

Because of their remarkable properties, exponential distributions arise naturally in theoretical settings. They have many characterizations of both theoretical and practical importance. It is not surprising, then, that exponential distributions have been overused in applications; but that does not diminish their importance. Some of the reasons why exponential distributions play a central role within the class of all life distributions are discussed in this chapter. See also Balakrishnan and Basu (1995), Johnson, Kotz and Balakrishnan (1994, Chapter 19), Mann, Schafer and Singpurwalla (1974), and Nelson (2004). For characterizations of the exponential distribution, see Azlarov and Volodin (1986) or Galambos and Kotz (1978).

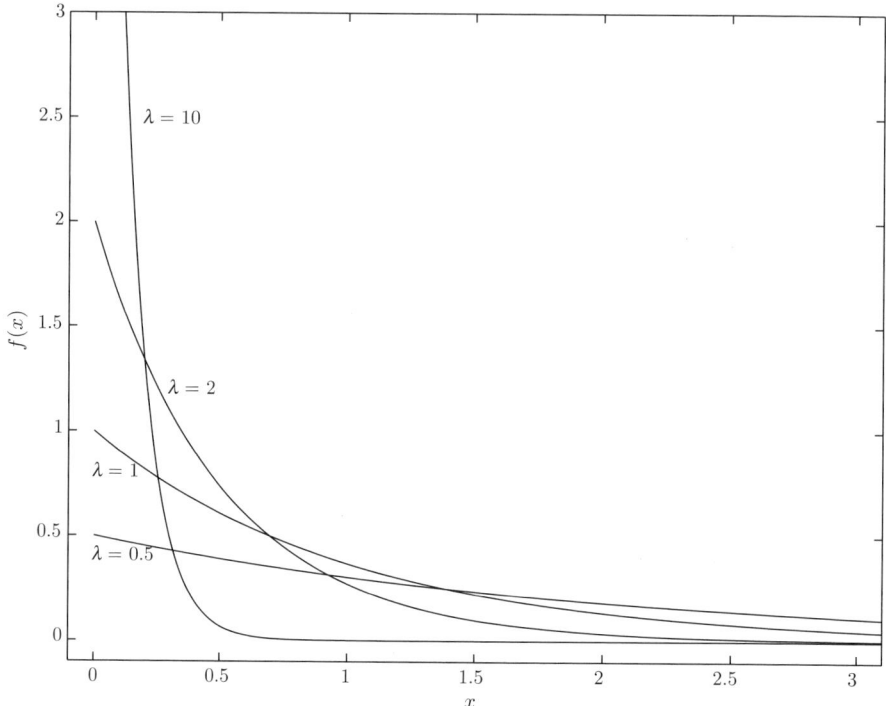

Fig. A.1. Densities of the exponential distribution

A. Defining Functions

The survival function $\bar{F}$, density f, and hazard rate r of the exponential distribution (also given in Section 1.F.a) are, respectively,

$$\bar{F}(x) = e^{-\lambda x}, \quad x \geq 0, \tag{1}$$
$$f(x) = \lambda e^{-\lambda x}, \quad x \geq 0, \tag{2}$$
$$r(x) = \lambda, \quad x \geq 0, \tag{3}$$

where the parameter $\lambda > 0$ acts both as a scale parameter and a frailty parameter. See Figure A.1.

Because so many life distributions are related to the exponential distribution, the expression (1) partially explains why survival functions often take a more convenient form than distribution functions. Clearly, (1) can be rewritten in the form

$$F(x) = 1 - e^{-\lambda x}, \quad x \geq 0. \tag{4}$$

This relatively awkward expression becomes even more awkward in some of the two-parameter extensions discussed in Chapter 9.

For exponential distributions, it is easy to verify that the residual life distribution at t is independent of t. In fact, this characterizes the exponential distributions, as is shown in Section B. As a consequence, the mean residual life of exponential distributions is independent of the age t, another characterizing property.

Singpurwalla (2003) makes extensive use of the fact that for any life distribution G with the hazard function R, $\bar{G}(x) = \exp\{-R(x)\} = P\{X > R(x)\}$, where X has an exponential distribution with parameter 1. From this point of view, it is to be expected that the exponential distribution will play a central role.

Because the survival function has a simple form, it is straightforward to write down the odds ratio

$$\emptyset(x) = \frac{\bar{F}(x)}{F(x)} = \frac{1}{e^{\lambda x} - 1}.$$

To find the total time on test transform, it is first necessary to verify that

$$F^{-1}(p) = [-\log(1-p)]/\lambda, \quad 0 \le p \le 1. \tag{5}$$

From (5) it follows that the total time on test transform of Definition 1.I.2 is given by the simple form

$$\psi(p) = p/\lambda, \quad 0 \le p \le 1. \tag{6}$$

It is verified in Section 1.F.a that the mean of the exponential distribution (1) is $1/\lambda$, so that the normalized total time on test transform $\tilde{\psi} = \psi/\mu$ is given by $\tilde{\psi}(p) = p$, $0 \le p \le 1$.

Direct computations show that the Lorenz curve (Definition 1.I.7) for exponential distributions is given by

$$L(p) = p + (1-p)\log(1-p), \quad 0 \le p \le 1,$$

and that the Gini index is $1/2$.

See Figures A.2 and A.3 for graphs of the inverse and Lorenz curves for the exponential distribution.

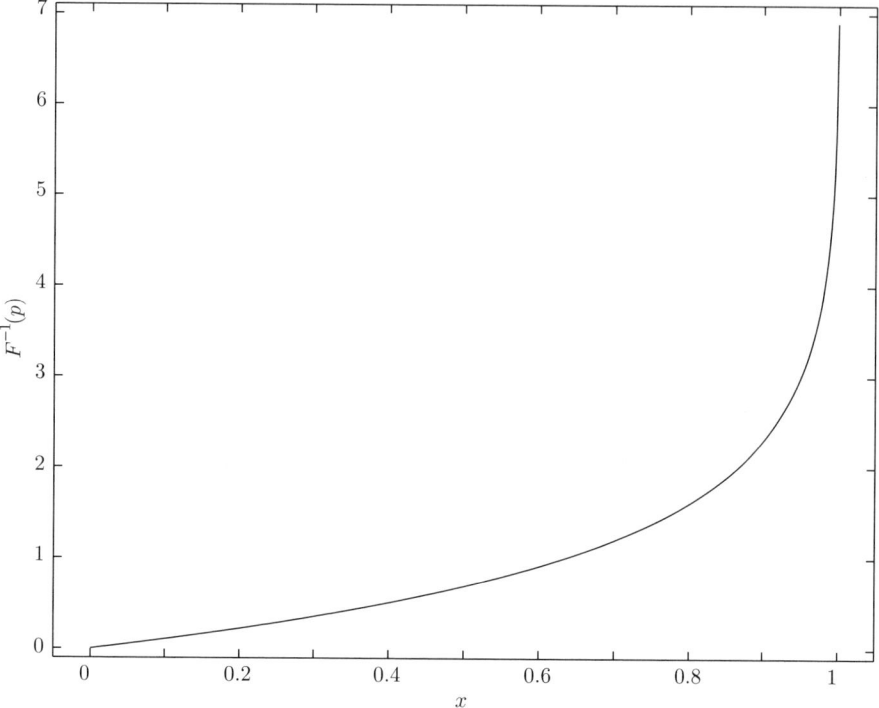

Fig. A.2. Inverse of the exponential distribution function

a. Moments

If X has an exponential distribution with parameter λ, then for $r > -1$,

$$\mu_r = EX^r = \int_0^\infty x^r \lambda e^{-\lambda x}\, dx = \Gamma(r+1)/\lambda^r, \tag{7}$$

where Γ is the usual gamma function discussed in Section 23.A. Thus, exponential distributions have finite moments of all orders greater than -1 and they take a simple form. The *normalized moments*

$$\lambda_r = \mu_r/\Gamma(r+1) = 1/\lambda^r \tag{8}$$

have an even simpler form. Recall that the normalized moments play a special role in Section 6.A.

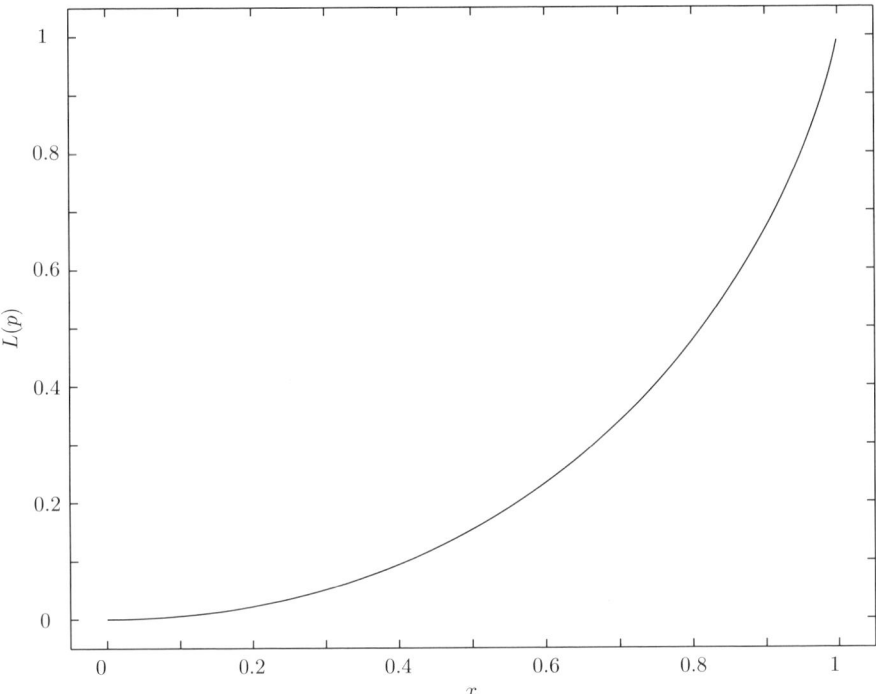

Fig. A.3. The Lorenz curve for the exponential distribution

From (7), it follows that X has the variance

$$\sigma^2 = \text{Var}(X) = 1/\lambda^2 = \mu_1^2. \tag{9}$$

This means that X has coefficient of variation (see 1.C(7))

$$\text{CV}(X) = \mu/\sigma = 1. \tag{10}$$

Thus, for the exponential distribution, the standard deviation and the mean are equal; this is in stark contrast to the normal distribution, where these quantities are quite unrelated. This might be expected from the fact that the normal distribution has more than one parameter.

A direct computation shows that the Laplace transform ϕ of X is given by

$$\phi(s) = E\,e^{-sX} = \frac{\lambda}{\lambda + s}, \quad s > -\lambda. \tag{11}$$

B. Characterizations of the Exponential Distribution

Some of the many characterizations of the exponential distribution are important for the purposes of this book, and are described in this section. For additional characterizations, see, e.g., Arnold and Huang (1975), Azlarov and Volodin (1986) or Galambos and Kotz (1978).

Perhaps the best known characterization of the exponential distribution is the so-called "lack of memory" property.

B.1. Proposition (Lack of memory property). A distribution F is exponential if and only if

$$\bar{F}(x+t) = \bar{F}(x)\bar{F}(t) \quad \text{for all } x, t \geq 0. \tag{1}$$

Equation (1) is known as the "lack of memory" property of the exponential distribution primarily because when $\bar{F}(t) > 0$, it can be rewritten in the form

$$\bar{F}_t(x) = \frac{\bar{F}(x+t)}{\bar{F}(t)} = \bar{F}(x) \quad \text{for all } x, t \geq 0. \tag{2}$$

The obvious interpretation of (2) is that the conditional probability of survival for an additional time x given survival up to time t is independent of the age t, and is the same as the unconditional probability of survival for a time x. This interpretation is most helpful in understanding the exponential distribution.

Equation (1) is a classical functional equation to which Cauchy's name is associated. This equation, discussed more fully in Proposition 22.A.2, is known to have a number of solutions, but the only bounded measurable solutions are exponential survival functions. A simple proof of Proposition B.1, which depends on the assumption of a finite expectation, is given in B.4. Here is another simple proof.

B.1.a. Proof of Proposition B.1. Rewrite (1) in the form

$$R(x+t) - R(t) = R(x) \quad \text{for all } x, t \geq 0, \tag{1a}$$

and set $x = 0$ to see that $R(0) = 0$. Thus, (1a) can be written as

$$\frac{R(x+t) - R(t)}{x} = \frac{R(x) - R(0)}{x}, \quad t \geq 0, \quad x > 0. \tag{1b}$$

Because R is monotone, it is differentiable almost everywhere; if it is differentiable at t, then by (1b), it is differentiable at 0, and from (1b), it follows that R is differentiable everywhere. Let $x \to 0$ in (1b) to conclude that $r(t) = r(0)$, that is, r is a constant. The proof is completed by Proposition B.2. □

B.1.b. Comment. A common interpretation of Proposition B.1 is that an item with an exponential distribution does not "wear out." Indeed, the term "lack of memory" may come from this idea. But an exponential distribution can arise as a mixture of distributions that are not exponential. For example, if

$$\bar{F}(x \mid \theta) = 1, \quad x < \theta,$$
$$= 0, \quad x \geq \theta,$$

then

$$\int_0^\infty \bar{F}(x \mid \theta) \, g(\theta) \, d\theta = \bar{G}(x),$$

and if G is an exponential distribution, then this mixture is an exponential distribution, and exponential distributions can arise in other ways as mixtures. If an item is selected from the mixture of the above example, the item has a degenerate distribution, and consequently, it degrades with time. See Evans (2000) for a thoughtful discussion of this issue.

B.2. Proposition. A distribution has a constant hazard rate if and only if it is an exponential distribution.

Proof. That the exponential distribution given by A(1) has the constant hazard rate λ follows easily from A(1) and A(2). From 1.B(3), that is, from the formula

$$\bar{F}(x) = \exp\left\{-\int_0^x r(z) \, dz\right\},$$

it follows that any distribution with constant hazard rate must be an exponential distribution. □

The intuitive content of Proposition B.2 is as follows: The conditional probability of failure (death) in the interval $(t, t + \Delta t)$ given survival up to time t is independent of the age t. Sometimes this property, or the property (2), is described by saying that "there is no premium for waiting," in accordance with the following example.

B.3. Example. Suppose that you go out on a pier to fish and find someone fishing with the same equipment and bait that you plan to use. You drop your line into the water and begin to fish. Measured from the time you drop your hook into the water, does your waiting time to catch a fish have the same distribution as that of the fisher who has already been fishing for a while? Or does he receive some premium or penalty for waiting? If the waiting time distributions are the same however long the fisher has been fishing, then because of Propositions B.1 or B.2, the waiting time distribution must be an exponential distribution.

If there is a premium for waiting, then the distribution is NBU; if there is a penalty for waiting, then the distribution is NWU.

B.4. Proposition. A distribution F has a mean residual life independent of age if and only if it is an exponential distribution.

Proof. If F is an exponential distribution, then (1) holds. But (1) states that the residual life distribution at t is independent of t, and consequently the mean residual life is independent of t. To prove the converse, suppose that the mean residual life is independent of age. This means that F must have a finite expectation μ, and moreover,

$$\int_t^\infty \bar{F}(x)\, dx = \mu \bar{F}(t) \quad \text{for all } t \geq 0. \tag{3}$$

Now, the left side of (3) is differentiable with respect to t, and consequently the right side must also be differentiable. Perform this differentiation to obtain

$$\frac{f(t)}{\bar{F}(t)} = \frac{1}{\mu}.$$

Thus, the hazard rate is a constant, and by Proposition B.2, the distribution must be exponential. □

An alternative proof of Proposition B.4 can be obtained from 1.B(13), which relates the hazard rate to the mean residual life m via the equation

$$r(t) = \frac{m'(t) + 1}{m(t)}.$$

If m is constant, then $m'(t) = 0$, and consequently r is constant and F is an exponential distribution.

Here is another less well-known origin of the family of exponential distributions; it says that if a residual life distribution has a limit as the age t goes to ∞, then that limit must be an exponential distribution.

B.5. Proposition. If

$$\lim_{t\to\infty} \bar{F}_t(x) = \lim_{t\to\infty} \frac{\bar{F}(x+t)}{\bar{F}(t)} = \psi(x)$$

exists for all $x \geq 0$, then for some $\lambda, 0 \leq \lambda \leq \infty, \psi(x) = \exp\{-\lambda x\}$.

Proof. For $x, y \geq 0$,

$$\begin{aligned}
\psi(x+y) &= \lim_{t\to\infty} \bar{F}_t(x+y) = \lim_{t\to\infty} \frac{\bar{F}(x+y+t)}{\bar{F}(t)} \qquad (4)\\
&= \lim_{t\to\infty} \frac{\bar{F}(x+y+t)}{\bar{F}(y+t)} \frac{\bar{F}(y+t)}{\bar{F}(t)}\\
&= \lim_{t\to\infty} \frac{\bar{F}(x+y+t)}{\bar{F}(y+t)} \lim_{t\to\infty} \frac{\bar{F}(y+t)}{\bar{F}(t)}\\
&= \psi(x)\psi(y).
\end{aligned}$$

Equation (4) is the Cauchy functional equation of Proposition 22.A.2. Because $\psi(0) = 1$ and ψ is decreasing, it follows from Proposition 22.A.2 that ψ is the survival function of a possibly improper exponential distribution. The possibility that λ is 0 or ∞ stems from the fact that no assumption was made here that ψ is a survival function. □

The following is a variant of Proposition B.5.

B.6. Proposition. If the hazard rate r of F has a finite positive limit $\lim_{t\to\infty} r(t) = \lambda$, then F_t converges in distribution to an exponential distribution with parameter λ as $t \to \infty$.

Proof. From 1.B(3), it follows that

$$-\log \bar{F}_t(x) = -\log \bar{F}(x+t) + \log \bar{F}(t) = \int_t^{t+x} r(z)\,dz \to \lambda x. \qquad \square$$

A stronger result than Proposition B.5 can be proved using the following lemma concerning regular variation which is given by Feller (1971, p. 275). This lemma is closely related to the functional equation of Pexider given in Proposition 22.B.1.

B.7. Lemma. If U is a positive monotone function defined on $(0, \infty)$ such that

$$\lim_{v \to \infty} U(uv)/U(v) = \psi(u) \leq \infty$$

on a dense set A of points, then $\psi(u) = u^\rho$, where $-\infty \leq \rho \leq \infty$.

B.8. Proposition. If $\lim_{t \to \infty} \bar{F}_t(x) = \lim_{t \to \infty} \bar{F}(t+x)/\bar{F}(t)$ exists on a dense set, then on $(0, \infty)$, the limit is identically 0, identically 1, or is the survival function of an exponential distribution.

Proof. With $e^t = u, e^x = v$, this result follows directly from Lemma B.7. □

Because their hazard rates are constant, exponential distributions belong to all of the various nonparametric classes discussed in Chapters 4 and 5. This fact leads to several additional characterizations.

B.9. Proposition. A distribution F is an exponential distribution if and only if any one of the following holds:

(i) F has support $[0, \infty)$ and has a density that is both log concave and log convex on its support.

(ii) F is in the intersection of the classes of IHR and DHR distributions.

(iii) F is in the intersection of the classes of IHRA and DHRA distributions.

(iv) F is in the intersection of the classes of NBU and NWU distributions.

(v) F is in the intersection of the classes of NBUE and NWUE distributions.

There are two parts to the characterizations (i) to (v): That only exponential distributions lie in the intersection of corresponding classes in each case follows from the inclusion relations of 5.G.b, if it is shown to hold for the NBUE and NWUE classes. That all exponential distributions lie in these intersections follows from the fact that they lie in the intersection of the classes with log-concave and log-convex densities. This latter fact is immediate because the density is log linear.

B.10. Proof of B.9. that only exponential distributions are both NBUE and NWUE. A distribution that is NBUE or NWUE must have a finite expectation μ, and to be both NBUE and NWUE, the distribution must satisfy (3). According to Proposition B.4, this means that F is an exponential distribution. □

Note: Condition (v) of Proposition B.9 (F is both NBUE and NWUE) would clearly imply condition (iv) of Proposition B.9 (F is both NBU and NWU) if it were not for the fact that the concepts of NBUE and NWUE are defined with the assumption that F has a finite expectation. In fact, NBU distributions must have finite expectations (Proposition 5.C.7), and consequently, condition (v) implies condition (iv). Of course, Proposition B.9 shows that the conditions are equivalent.

Note that (3) can be obtained from (1) by integrating (1) on x from 0 to ∞. Thus, the proof given in Proposition B.4 is an alternative to the functional equation approach to proving Proposition B.1; but it depends upon the knowledge that NBU distributions must have finite expectations.

Equation (1), which says that $\bar{F}(x+t) = \bar{F}(x)\bar{F}(t)$ for all $x, t \geq 0$, can be expanded by iteration as follows:

$$\bar{F}(x + t_1 + t_2 + \cdots + t_k) = \bar{F}(x + t_1 + t_2 + \cdots + t_{k-1})\bar{F}(t_k)$$
$$= \bar{F}(x + t_1 + t_2 + \cdots + t_{k-2})\bar{F}(t_{k-1})\bar{F}(t_k)$$
$$= \ldots = \bar{F}(x)\Pi_{i-1}^{k}\bar{F}(t_i), \quad x, t_i \geq 0, \tag{5}$$

$i = 1, \ldots, k$. An inductive argument shows that if (1), then (5) holds for $k = 1, \ldots$. In particular, if $t_i = x$, for all i, it follows that

$$[\bar{F}(x)]^k = \bar{F}(kx), \quad k = 1, 2, \ldots, \quad x \geq 0. \tag{6}$$

This leads to another characterization of the exponential distribution, related to the lack of memory property (1) but with a somewhat different interpretation. This characterization is based upon the assumption that min $\{X_1, \ldots, X_k\}$ has the same distribution as $X/k, k = 1, 2, \ldots$, where $X_1, X_2, \ldots, X_k$ are independent and distributed as X.

B.11. Proposition (Desu, 1971). A distribution F is exponential if and only if (6) holds.

Proposition B.11 says that in Proposition B.1, equation (1) can be replaced by the apparently weaker equation (6). The solution to the functional equation (6) is given in Proposition 22.A.5, and this yields a proof of Proposition B.11.

a. Additional Characterizations

In addition to Proposition B.11, there are a several characterizations of the exponential distribution that involve order statistics. Here is an important example.

B.12. Proposition (Ferguson, 1965). Let X_1 and X_2 be independent random variables with common absolutely continuous distributions, and let $X_{(1)} \leq X_{(2)}$ be obtained by ordering the X_i. Then, $X_{(1)}$ and $X_{(2)} - X_{(1)}$ are independent if and only if X_1 and X_2 have a common exponential distribution with a location parameter.

The above result has led to a variety of characterizations involving independence properties of order statistics. For other results on characterizations, see also Basu (1965), Crawford (1966), and Galambos and Kotz (1978).

C. Some Basic Properties of Exponential Distributions

a. Closure Under Minima

C.1. Proposition. If X_1 and X_2 have exponential distributions with respective parameters λ_1 and λ_2, and if X_1 and X_2 are independent, then $Z = \min\{X_1, X_2\}$ has an exponential distribution with parameter $\lambda_1 + \lambda_2$.

This fundamental observation is easy to verify because the survival function of Z is the product of the survival functions of X_1 and X_2; it is used by Marshall and Olkin (1967) in the construction of a multivariate exponential distribution.

Proposition C.1 is sometimes viewed as a kind of parallel to the well-known fact that the sum of independent normally distributed random variables has a normal distribution, but here, the sum is replaced by the minimum.

Note that Proposition C.1 holds as a consequence of the fact that the parameter λ is a frailty parameter; see Definition 7.E.1 and note that a survival product family is a parametric family closed under the formation of minima with the parametric structure that leads to Lemma 5.E.8.

b. Infinite Divisibility

C.2. Proposition. The exponential distribution is infinitely divisible.

Proof. To prove this result, Proposition 20.D.8, a proposition stated without proof, has been used here. According to that proposition, a function ϕ is the Laplace transform of an infinitely divisible distribution if and only if $\phi = e^{-\psi}$, where $\psi(0) = 0$ and ψ has a completely monotone derivative (Definition 18.D.5). The Laplace transform ϕ of the exponential distribution, given by A(11), is

$$\phi(s) = E e^{-sX} = \frac{\lambda}{\lambda + s}, \quad s > -\lambda,$$

and in this case,

$$\psi(s) = \log(\lambda + s) - \log \lambda.$$

Because $\psi'(s) = 1/(\lambda + s) = \phi(s)/\lambda$, it is sufficient to show that ϕ is completely monotone. It can easily be verified using induction that the nth derivative $\phi^{(n)}$ of ϕ is

$$\phi^{(n)}(s) = \frac{(-1)^n n! \lambda}{(\lambda + s)^{n+1}}.$$

Because $(-1)^n \phi^{(n)}(s) \geq 0, s > 0$, it follows that the exponential distribution is infinitely divisible. □

c. The Poisson Process

A brief introduction to the Poisson process is given in Section 20.F.a. Note that the waiting time X to the first event in a Poisson process with rate λ can easily be obtained from the relationship $P\{X > t\} = P\{N(t) = 0\} = e^{-\lambda t}$, i.e., X has an exponential distribution with parameter λ. From the derivation of the process as a renewal process, it can be seen that the waiting times between jumps in a Poisson process are all independent and all have an exponential distribution with parameter λ. The fact that the waiting times between events in a Poisson process have exponential distributions is a prime origin of the exponential distribution in applications, and the main reason for introducing the process here. Events may be accidents or episodes that occur at a constant rate λ, and the waiting time to the first such event has an exponential distribution with parameter λ, as do the waiting times between successive events. Thus, in mechanical or electrical devices, the time to failure (age at "death" or life length) has an exponential distribution when deaths are the result of accidents occurring as a Poisson process.

d. Ordering Exponential Distributions

It is easy to verify that if X_1 and X_2 have exponential distributions with respective parameters λ_1 and λ_2, and if $\lambda_1 > \lambda_2$, then $X_1 \leq_{\mathrm{lr}} X_2$; consequently, $X_1 \leq_{\mathrm{hr}} X_2$ and $X_1 \leq_{\mathrm{st}} X_2$. However, the convex order is not applicable because it compares only distributions with equal expectations; the Lorenz order is not of interest because exponentially distributed random variables divided by their expectations are all exponentially distributed with expectation 1. In the convex transform and star orders, exponential distributions with different parameters are equivalent, i.e., they are ordered in both directions.

e. Connections Between Exponential and Geometric Distributions

C.3. Exponential distribution as a weak limit. In a Bernoulli process where trials are made at times $1, 2, \ldots$, the waiting times between successes have a geometric distribution (see Sections 20.E and 20.F). Because rescaled Bernoulli processes can have a Poisson process as a limit, it is not surprising that a sequence of rescaled geometric distributions can have an exponential distribution as a limit.

For $n \geq \lambda$, let Y_n have the geometric distribution 20.E(4) with parameter $p = \lambda/n$. Denote the integer part of z (greatest integer $\leq z$) by $[z]$. Then,

$$\bar{F}_n(x) = P\left\{\frac{Y_n}{n} > x\right\} = \left(1 - \frac{\lambda}{n}\right)^{[nx]}, \quad x \geq 0.$$

Clearly,

$$\lim_{n \to \infty} \bar{F}_n(x) = e^{-\lambda x}, \quad x \geq 0.$$

So the exponential distribution has been obtained as a limit of rescaled geometric distributions.

The derivation of the exponential distribution as a limit of geometric distributions has considerable intuitive appeal, at least in some contexts. Indeed, it is the basis of a derivation used to introduce the distribution by Gertsbakh and Kordonsky (1969) and Gertsbakh (1989).

In C.3, the exponential distribution is obtained from a geometric distribution, but it is also possible to go in the other direction to obtain the geometric distribution from an exponential distribution.

C.4. Proposition. Denote the smallest integer not less than x by $\langle x \rangle$. If X has an exponential distribution with parameter λ, then $\langle X \rangle$ has a geometric distribution with parameter $1 - p = e^{-\lambda}$.

There is a converse to C.4 due to Bosch (1977); if $\langle \alpha X \rangle$ has a geometric distribution for all $\alpha > 0$, then X has an exponential distribution.

f. Random Sums

As already noted, Proposition C.1 has been used by Marshall and Olkin (1967) to construct multivariate exponential distributions. Several multivariate exponential distributions have been constructed with the aid of other univariate results. The following proposition is of particular interest for its use in the construction of multivariate exponential distributions. Such constructions are outside the scope of this book, but see, for example, Arnold (1975).

C.5.a. Proposition. Let $X_1, X_2 \ldots$ be a sequence of independent random variables having a common exponential distribution with parameter λ. If N is independent of the X_i and has the geometric distribution 20.E(4) with parameter p and support $\{1, 2, \ldots\}$, then

$$Z = X_1 + X_2 + \cdots + X_N \qquad (1)$$

has an exponential distribution with parameter λp.

Proof. The Laplace transform of each X_i is $\phi(t) = \lambda/(\lambda + t), t > -\lambda$. Consequently, the Laplace transform of Z is

$$\sum_{n=1}^{\infty} \left[\frac{\lambda}{\lambda + t} \right]^n p(1-p)^{n-1} = \frac{\lambda p}{\lambda p + t}. \qquad \square$$

Proposition C.5.a shows that even though the sum of a *fixed* number of independent exponentially distributed random variables does not have an exponential distribution, the sum of a *random* number of such variables can be exponentially distributed. For a related result, see Proposition 5.C.14.

The next proposition is closely related to Proposition C.5.a; it too has been used to construct multivariate exponential distributions. See Lawrance and Lewis (1977, 1980, 1983); Raftery (1984); O'Cinneide and Raftery (1989); Hutchinson and Lai (1990).

C.5.b. Proposition. Suppose that X has an exponential distribution with parameter λ, I has a Bernoulli distribution with parameter $1-p$, and Z has an exponential distribution with parameter λp. If X, Y, and I are independent, then

$$Y = X + IZ \tag{2}$$

has an exponential distribution with parameter λp.

Proof. Consider the representation of Z in Proposition C.5.a; there $Z = X_1$ with probability p, and is $X_1 + Z^*$ with probability $(1-p)$, where Z^* is independent of X_1 and has the same distribution as Z. So this proposition follows directly from Proposition C.5.a. □

Alternative Proof. The distribution of Y can be regarded as a mixture of two distributions: If $I = 1$, then $Y = X + Z$; if $I = 0$, then $Y = X$. By convolving the densities of X and Z, it can be determined that the density g of $X + Z$ is given by

$$g(x) = \lambda p[e^{-\lambda p x} - e^{-\lambda x}]/[1-p], \quad x \geq 0,$$

whereas X has the density $f(x) = \lambda \exp\{-\lambda x\}$, $x \geq 0$. It is straightforward to verify that

$$pf(x) + (1-p)g(x) = \lambda p\, e^{-\lambda p x}. \tag{3}$$

□

Note that the component f of the mixture (3) has a constant hazard rate, and the component g has a strictly increasing hazard rate (as can be verified directly, but also see Theorem 4.C.4), yet the mixture has a constant hazard rate. Because of (1), the density of (2) can also be represented as a mixture of a countably infinite number of gamma densities, all convolutions of some integer order of the exponential density f, and all having increasing hazard rates.

C.6. Proposition (Esary and Marshall 1973). Let $Y, Z_1, Z_2, \ldots$ be a sequence of independent random variables, where the Z_i are identically distributed and nonnegative but not degenerate at 0. If Y has an exponential distribution, then

$$N = \min\{k : Z_1 + \cdots + Z_k \geq Y\}$$

has a geometric distribution on the positive integers 1, 2, ...

Proof. Compute

$$P\{N > j\} = P\{Z_1 + \cdots + Z_j < Y\}$$
$$= EP\{z_1 + \cdots + z_j < Y \mid Z_i = z_i, \ i = 1, \ldots, j\}$$
$$= E\left\{\exp\left[-\sum_{i=1}^{j} z_i\right] \mid Z_i = z_i, \ i = 1, \ldots, j\right\} = [Ee^{-Z}]^j. \quad \square$$

The Z_i can be viewed as damages due to shocks that accumulate additively over time. If a device has an exponentially distributed threshold of damage that can be withstood without failure, then N is the number of shocks required to cause failure. Shock models are discussed in Section 5.H.

Proposition C.6 has been used to develop bivariate geometric distributions which, in conjunction with Propositions C.3 and C.4, have been used to develop models for bivariate exponential distributions. See Esary and Marshall (1974); Marshall and Olkin (1985).

9

Parametric Extensions of the Exponential Distribution

> According to the classical theory the ultimate strength of a material is determined by the internal stresses in a point ..... Experimental measurements give many results which may hardly be brought to agree with this theory ..... by taking the elementary laws of probability as a starting-point, a theory may be developed, whose formulae may be readily brought to agree with the measuring results inconsistent with the classical theory.
>
> W. Weibull (1939a)

The exponential distribution has a single parameter that serves both as a scale and as a frailty parameter. Moreover, if an age parameter or a Laplace transform parameter is introduced, the distribution remains an exponential distribution and only the parameter is changed. This means that of the various parameters discussed in Chapter 7, only power, convolution, moment, tilt, and resilience can be used to generate two parameter extensions of the exponential distribution. It is shown below that the introduction of moment and convolution parameters both lead to the gamma family, and consequently only four of these extensions are distinct. These four extensions with two parameters are discussed in this chapter along with their further extensions to three-parameter families.

The following diagram illustrates the build-up of families by successive introductions of parameters starting with a basic exponential distribution.

$\bar{F}(x) = e^{-x}$, $\quad\xrightarrow[\text{scale}]{}\quad$ {exp} $\quad\xrightarrow[\text{convolution or moment}]{\text{power}}\quad$ {Weibull}

{gamma} $\quad\xrightarrow[\text{power}]{\text{moment}}\quad$ {generalized gamma}

$x \geq 0$

$\xrightarrow[\text{resilience}]{}\quad \bar{F}(x) = 1 - (1 - e^{-\lambda x})^\eta$

$\xrightarrow[\text{tilt}]{}\quad$ {exponentials with tilt} $\quad\xrightarrow[]{\text{power}}\quad$ {Weibull with tilt}

A. The Gamma Distribution

The *gamma distribution* has the density

$$f(x \mid \lambda, \nu) = \frac{\lambda^\nu x^{\nu-1} e^{-\lambda x}}{\Gamma(\nu)}, \quad x \geq 0, \tag{1}$$

but the survival function can be given in closed form only when ν is an integer. In that case,

$$\bar{F}(x \mid \lambda, \nu\}) = \sum_{k=0}^{\nu-1} e^{-\lambda x}(\lambda x)^k/k!. \tag{2}$$

Clearly, the hazard rate of the gamma distribution does not take a convenient form even when ν is an integer. The parameter ν is sometimes called a "shape parameter," but it is also known as the "index" of the distribution. As can be seen from (1) or (2), λ is a scale parameter. See Figure A.1 for graphs of the density.

a. Moments

A.1. Proposition. If F is a gamma distribution with density (1), then

$$EX^r = \frac{\Gamma(r+\nu)}{\Gamma(\nu)\lambda^r}, \quad r > -\nu. \tag{3}$$

This result can be obtained directly from the definition of the gamma function as an integral, or from the fact that the density (1) integrates to one.

A. The Gamma Distribution

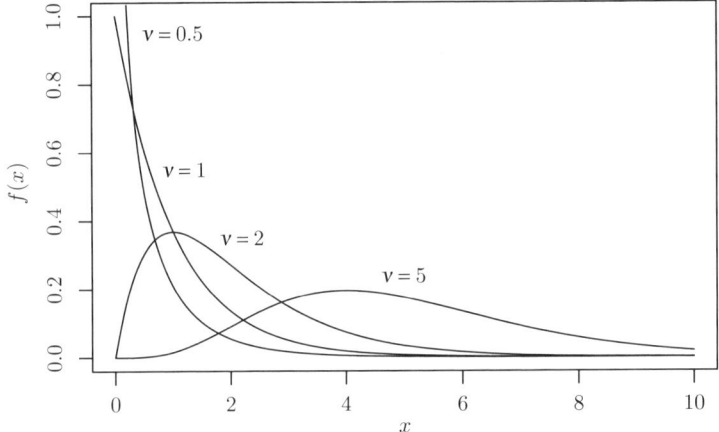

Fig. A.1. Densities of the gamma distribution ($\lambda = 1$)

From (3), it follows that

$$EX = \nu/\lambda \text{ and Var } X = \nu/\lambda^2, \tag{4}$$

so that the coefficient of variation is $CV(X) = \sqrt{\text{Var } X}/EX = 1/\sqrt{\nu}$.

In a similar manner, it can be verified that F has the Laplace transform

$$E e^{-sX} = [\lambda/(\lambda + s)]^\nu, \quad s > -\lambda. \tag{5}$$

b. Derivations

There are several ways to derive the gamma distribution, and each is instructive in its own way. All of those described here are based in some way upon the exponential distribution.

(i) From a Poisson process. A standard derivation of the gamma distribution is via a Poisson process. Let X be the waiting time for the νth jump in the Poisson process $\{N(t), t \geq 0\}$ with parameter λ (see 20.E for a brief review). The survival function of X can be obtained directly:

$$P\{X > x\} = P\{N(x) < \nu\} = \sum_{k=0}^{\nu-1} e^{-\lambda x} (\lambda x)^k / k!;$$

that is, X has the survival function (2). Here, $\lambda > 0$ and ν is a positive integer.

(ii) By introduction of a moment parameter. The simplest derivation of the gamma distribution via the exponential distribution seems to be by the introduction of a moment parameter β in the exponential density 8.A(2). This is straightforward because the moments of the exponential distribution, given by 8.C(1) have the simple form $\mu_r = \Gamma(r+1)/\lambda^r$. When this is done, the resulting density

$$f(x \mid \lambda, \beta) = \frac{\lambda^{\beta+1} x^{\beta} e^{-\lambda x}}{\Gamma(\beta+1)}, \quad x \geq 0, \ \lambda > 0, \ \beta > -1, \tag{1a}$$

is obtained. By setting $\beta = \nu - 1$, the density (1) is obtained. Thus, $\nu - 1$ in (1) is a moment parameter.

(iii) By introduction of a convolution parameter. The Laplace transform ϕ of the exponential distribution is given in 8.A(11) as

$$\phi(s) = \frac{\lambda}{\lambda + s}, \quad s > -\lambda.$$

Because the exponential distribution is infinitely divisible (Proposition 8.C.2), it follows that

$$\phi(s \mid \nu) = [\phi(s)]^{\nu} = \left(\frac{\lambda}{\lambda + s}\right)^{\nu}, \quad s > -\lambda \tag{6}$$

is a Laplace transform, for all $\nu > 0$. In this construction, ν is a convolution parameter. Inversion formulas (beyond the scope of this book) can be used to obtain from (6) the corresponding density; alternatively, the inversion can be obtained from a table of Laplace transforms. Here, these approaches are unnecessary. Because Laplace transforms are unique (Proposition 20.D.2), it is already clear from (5) that inversion would yield (1).

(iv) As a mixture. For $\nu < 1$, the density (1) arises as a mixture of exponential distributions. The following representation is due to Gleser (1989):

$$f(x \mid \lambda, \nu) = \int \theta e^{-\theta x} g(\theta \mid \lambda, \nu) \, d\theta, \quad 0 < \nu < 1, \tag{7}$$

where

$$g(\theta \mid \lambda, \nu) = \frac{(\theta - \lambda)^{-\nu} \lambda^{\nu}}{\theta \Gamma(1-\nu)\Gamma(\nu)}, \quad \theta \geq \lambda,$$
$$= 0, \text{ otherwise.} \tag{8}$$

If g given by (8) is the density of the random variable Θ, then the density of $\Theta - \lambda$ is a special form of the F distribution; see 11.C(3).

The gamma density (1) has no representation as a mixture of exponential distributions when $\nu > 1$ because such mixtures have a decreasing hazard rate (Proposition 4.C.10), and it is shown below that the hazard rate of (1) is increasing when $\nu > 1$.

When ν is a positive integer, the gamma distribution is sometimes called the *Erlang distribution*. This distribution was derived by Agner Krarup Erlang (1878–1929) while working for the Copenhagen Telephone Company (see Erlang (1920); see also Brockmeyer, Halstrøm and Jensen (1960) for a history of Erlang's work).

From the above derivation of the survival function of the Erlang distribution and the fact that the Poisson process is a renewal process, it follows that the Erlang distribution is the convolution of the exponential distribution with itself ν times. From the Laplace transform (5), it follows that $\nu > 0$ is a convolution parameter, ν is also known as the "index" of the distribution.

When ν is a half integer and $\lambda = 1/2$, the gamma distribution is known as a chi-square distribution; this distribution is discussed in Section A.i.

The gamma density (1) was obtained by Pearson (1895) and is known as a Type III Pearson curve. Pearson derived the density from a differential equation; see Johnson, Kotz and Balakrishnan (1994, Chapter 12). If X has the gamma density (1), then it can be shown that $1/X$ has the density g given by

$$g(x) = \frac{\lambda^{\nu}}{\Gamma(\nu) x^{\nu-1}} e^{-\lambda/x}, \quad x > 0. \tag{9}$$

This density was also obtained by Pearson (1895) and is known as a Type V Pearson curve. For further discussion of the Pearson curves, see also Elderton and Johnson (1969) or Kendall and Stuart (1963).

c. Density Properties

A.2. Proposition. The density (1) of the gamma distribution is

(i) completely monotone, log convex, and decreasing, for $0 < \nu \leq 1$,

(ii) log concave and unimodal, for $\nu \geq 1$, with mode at the point

$$x = (\nu - 1)/\lambda.$$

Proof. For $0 < \nu \leq 1$, the fact that the density is decreasing can be verified directly by showing that $\log f(\cdot \mid \lambda, \nu)$ has a negative derivative; convexity follows by showing that the second derivative of $\log f(\cdot \mid \lambda, \nu)$ is positive. Complete monotonicity of the density (1) can be verified directly using the conditions of Definition 20.D.4; or for $\nu < 1$, it can be obtained from Proposition 18.D.5 and the mixture representation (7). It follows from Proposition 4.B.7 and the complete monotonicity of the density that the density is log convex, and by Propositions 4.B.2, 4.B.8.a, and 4.C.11, this means that the density is decreasing.

For $\nu \geq 1$, the log concavity can be checked directly using the fact that the function $\log x$ is concave. By Proposition 4.B.2, this means that the density is unimodal. By setting the derivative of the density or its logarithm equal to 0, the location of the mode is easily determined. □

d. Distribution and Survival Function Properties

In general, neither the distribution function nor the survival function of the gamma distribution has a simple expression; they can be given only in terms of the incomplete gamma function when ν is not an integer. However, properties of these functions can still be determined.

A.3. Proposition. The survival function $\bar{F}$ corresponding to the gamma density (1) is log concave, for $\nu \geq 1$, and log convex on $[0, \infty)$, for $\nu \leq 1$. The gamma distribution function F is log concave, for all ν.

Proof. The properties of the survival function follow from Proposition A.2 and Proposition 4.B.8. The log concavity of the distribution function, noted by Bondesson (1992), follows from Proposition 4B.10. □

See Figure A.2 for graphs of the survival function.

e. The Hazard Rate

Because the survival function of the gamma distribution can be given only in terms of the incomplete gamma function when ν is not an

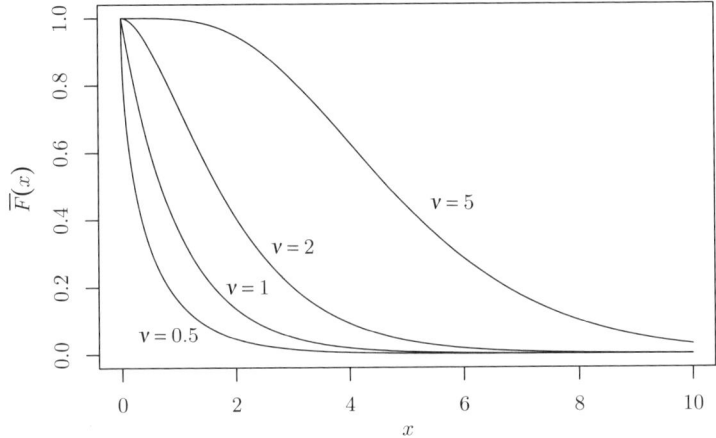

Fig. A.2. Survival functions of the gamma distribution ($\lambda = 1$)

integer, neither the hazard rate nor the reverse hazard rate can be expressed in closed form. Even so, a form of the hazard rate can be given that is useful for the identification of its properties. Following the approach of Barlow and Proschan (1975, p. 74), write the survival function as an integral of the density to obtain

$$\frac{1}{r(x)} = \int_x^\infty \left(\frac{z}{x}\right)^{\nu-1} e^{-\lambda(z-x)}\, dz = \int_0^\infty \left[1 + \frac{u}{x}\right]^{\nu-1} e^{-\lambda u}\, du, \quad (10)$$

where the second integral is obtained from the first by the change of variable $u = z - x$. Now it is clear that

$$\begin{aligned} & r(x) \text{ is increasing in } x \geq 0 \text{ when } \nu > 1, \\ & r(x) \text{ is the constant } \lambda \text{ when } \nu = 1, \text{ and} \\ & r(x) \text{ is decreasing in } x \geq 0 \text{ when } 0 < \nu < 1. \end{aligned} \quad (11)$$

The hazard rate results (11) also follow directly from the log concavity and log-convexity of the survival function given in Proposition A.3.

From (10), it follows directly that

$$\begin{aligned} \lim_{x \to 0} r(x) &= 0, \quad \text{for } \nu > 1, \\ &= \lambda, \quad \text{for } \nu = 1, \\ &= \infty, \quad \text{for } 0 < \nu < 1. \end{aligned} \quad (12)$$

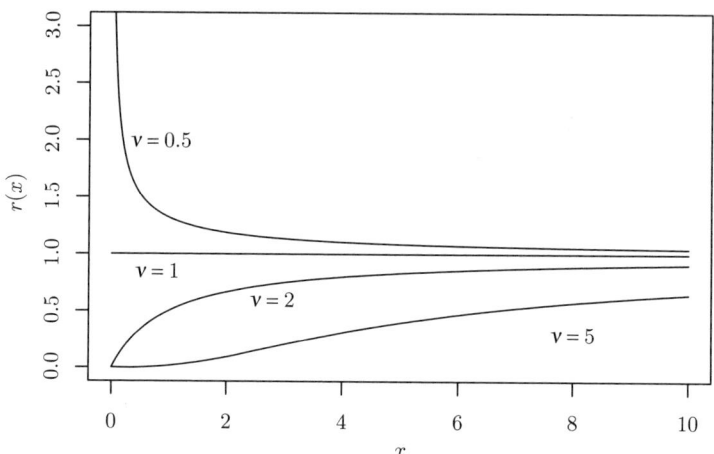

Fig. A.3. Hazard rates of the gamma distribution ($\lambda = 1$)

From the second form of $r(x)$ in (10), it follows directly that

$$\lim_{x \to \infty} r(x) = \lambda \quad \text{for all } \nu > 0. \tag{13}$$

Note that for fixed λ, the hazard rate at any fixed x is decreasing in ν; this is a consequence of the fact that ν is a convolution parameter (see Figure A.3).

Because the gamma distribution function F is log concave, the derivative of $\log F$ is decreasing; this means that the reverse hazard rate is decreasing for all values of ν. This fact stands in contrast to the hazard rate properties that depend upon whether $\nu \geq 1$ or $\nu \leq 1$.

f. Residual Life Distribution

A.4. Proposition. If F is a gamma distribution with density (1), then for all ν, the residual life survival function $\bar{F}_\tau(x) = \bar{F}(x + \tau)/\bar{F}(\tau)$ satisfies

$$\lim_{\tau \to \infty} \bar{F}_\tau(x) = e^{-\lambda x}.$$

Proof. According to Proposition 8.B.b, a residual life distribution converges in distribution to an exponential distribution whenever the hazard rate has a finite positive limit. According to (13), this condition holds. Alternatively, by writing the residual life survival function as a ratio of integrals of the density, the proposition can be verified directly using l'Hospital's rule. □

For a more general result, see also Proposition 8.B.8.

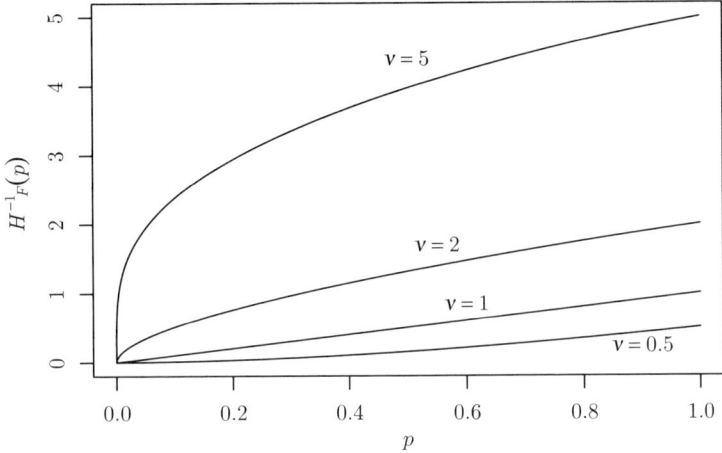

Fig. A.4. Total time on test transform for the gamma distribution ($\lambda = 1$)

g. Convolutions and Infinite Divisibility

It has already been noted that ν is a convolution parameter, a fact that follows directly from (5). Consequently, gamma distributions are infinitely divisible (see Definition 20.D.7 and Remark 18.D.9.a). For $\nu < 1$, infinite divisibility also follows from the representation (7) and Proposition 20.D.10.

If X and Y are independent, have gamma distributions with common scale parameter λ, and have respective shape parameters ν and γ, then $X + Y$ has a gamma distribution with scale parameter λ and shape parameter $\nu + \gamma$.

h. Total Time on Test Transform

The total time on test transform for the gamma distribution can be obtained numerically and is given in Figure A.4.

i. The Chi-Square Distribution

The chi-square distribution has a long and illustrious history during the 50-year period 1835–1885 primarily because of its importance in statistics. This history has been surveyed by Lancaster (1966; 1969, pp. 1–16); see also Johnson, Kotz and Balakrishnan (1994, p. 415).

A.5. Definition. A *chi-square distribution with n degrees of freedom* is a gamma distribution with scale parameter $1/2$ and shape parameter $n/2$. More explicitly, the chi-square distribution with n degrees of freedom has density

$$f(x) = \frac{1}{2\Gamma(n/2)} \left(\frac{x}{2}\right)^{(n/2)-1} e^{-x/2}. \tag{14}$$

Because the shape parameter ν of the gamma distribution is a convolution parameter, it follows that if X and Y are independent, have chi-square distributions with common scale parameter λ, and respective degrees of freedom n and m, then $X + Y$ has a chi-square distribution with scale parameter λ and $n + m$ degrees of freedom. This fact, central in statistics, can also be obtained from the following proposition.

A.6. Proposition. If $X_1, \ldots, X_n$ are independent random variables having a common normal distribution with 0 expectation and unit variance, then

$$X_1^2 + \cdots + X_n^2$$

has a chi-square distribution with n degrees of freedom.

With $n = 1$ in Proposition A.6, it follows that the square of a random variable with a standard normal distribution has a chi-square distribution with one degree of freedom. Directly from the definition, it follows that with $n = 2$, the chi-square distribution is an exponential distribution.

Ratios of random variables with chi-square or gamma distributions are discussed in Section 11.D.f.

j. A Characterization of the Gamma Distribution

A.7. Proposition (Lukacs, 1955). Suppose that X_1 and X_2 are independent, positive nondegenerate random variables. Then $X_1 + X_2$ and X_1/X_2 are independent if and only if X_1 and X_2 have gamma distributions with the same scale parameter.

For a relatively simple proof of this fact, see Findeisen (1978) or Marsaglia (1989). Marsaglia (1974, 1989) shows that without the assumption that X_1 and X_2 are positive, the independence implies that either X_1 and X_2 or $-X_1$ and $-X_2$ have gamma distributions with the same scale parameter.

k. Ordering Gamma Distributions

A.8. Proposition. In the likelihood ratio order, the gamma distribution is increasing in the shape parameter ν and decreasing in the scale parameter λ.

These facts can be verified using 2.A(11). As a consequence, the same monotonicity holds for the hazard rate ordering and the usual stochastic ordering. The hazard rate ordering can be checked directly from (10).

A.9. Proposition (van Zwet, 1964). Suppose that X and Y have gamma distributions with respective shape parameters θ and γ. If $\theta \leq \gamma$, then in the convex transform order, X is greater than Y, or symbolically, $X \geq_c Y$.

The proof of this result is omitted; it is complicated by the fact that the survival functions of X and Y do not have simple forms.

As noted in Section 2.C.j (without proof), the convex transform order is stronger than the Lorenz order. It follows that under the conditions of Proposition A.9,

$$X \geq_{\text{Lorenz}} Y,$$

that is,

$$X/EX \geq_{\text{cx}} Y/EY.$$

The fact that $X \geq_{\text{Lorenz}} Y$ is a result of Taillie (1981).

1. Limits of Gamma Distributions

It is interesting to determine the limit of the survival function as the shape parameter ν tends to ∞.

A.10. Proposition.

$$\lim_{\nu \to \infty} \bar{F}\left(\frac{\nu x}{\lambda} \mid \lambda, \nu\right) = 1, \ x < 1,$$
$$= 0, \ x > 1.$$

Of course, this states that there is convergence in distribution to the distribution degenerate at 1. To prove Proposition A.10, use (5) to show convergence of the Laplace transform of X/EX to e^{-s}, the Laplace transform of the distribution degenerate at 1. Alternatively, this result can be proved using the weak law of large numbers 18.C.6.b and the fact that ν is a convolution parameter.

A.11. Proposition. If $Y = (X - (\nu/\lambda))/(\sqrt{\nu}/\lambda)$, then the limiting distribution of Y as $\nu \to \infty$ is a normal distribution with mean 0 and variance 1.

With (4) and the fact that ν is a convolution parameter, it can be seen that this proposition is a special case of the central limit theorem 20.C.8. More direct proofs can be obtained by the use of Laplace transforms, or by taking the limit of the density of Y.

m. Convolutions of Scaled Exponential Distributions

Suppose that $X_1, X_2, \ldots, X_n$ are independent random variables all with the same exponential survival function $\bar{G}(x) = e^{-\lambda x}$, $x \geq 0$. As previously noted and easily verified from (5), the random variable

$$X = \sum_{i=1}^{n} X_i$$

has a gamma distribution with shape parameter n. Sums of independent exponentially distributed random variables with different parameters arise in the study of birth and death processes. It has been shown by Palm (1946), Good (1955), McGill and Gibbon (1965), and by Likeš (1967) that if $a_i > 0$, $i = 1, 2, \ldots, n$, and $a_i \neq a_j$, $i \neq j$, then the survival function $\bar{F}$ of the random variable

$$X = \sum_{i=1}^{n} a_i X_i$$

is given by

$$\bar{F}(x) = \sum_{i=1}^{n} A_i \exp\left\{\frac{-\lambda x}{a_i}\right\}, \quad x \geq 0 \quad (15)$$

where

$$A_i = \frac{a_i^{n-1}}{\prod_{\substack{k=1 \\ k \neq i}}^{n}(a_i - a_k)}.$$

Note that $X = \sum_{i=1}^{n} Y_i$, where Y_i has an exponential distribution with parameter $\lambda_i = \lambda/a_i$. With this notation, (15) can be rewritten as

$$\bar{F}(x) = \sum_{i=1}^{n} \frac{\lambda_k^{n-1}}{\prod_{k \neq i}(\lambda_k - \lambda_i)} e^{-\lambda_i x} = \sum_{i=1}^{n} \prod_{k \neq i} \frac{\lambda_k}{\lambda_k - \lambda_i} e^{-\lambda_i x}. \quad (15a)$$

The removal of the condition that the λ_i be distinct requires a more complex analysis, and is beyond the scope of this book.

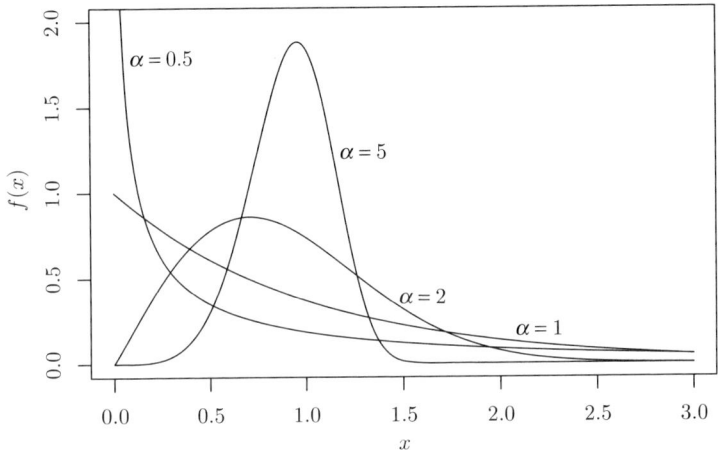

Fig. B.1. Densities of the Weibull distribution ($\lambda = 1$)

B. The Weibull Distribution

With the introduction of a power parameter α, the exponential survival function becomes

$$\bar{F}(x \mid \lambda, \alpha) = \exp\{-(\lambda x)^\alpha\}, \quad \lambda, \alpha, x \geq 0; \tag{1}$$

this is the survival function of the *Weibull distribution.*

From (1), it follows directly that the density f and hazard rate r of the Weibull distribution are given by

$$f(x \mid \lambda, \alpha) = \alpha\lambda(\lambda x)^{\alpha-1} \exp\{-(\lambda x)^\alpha\}, \quad x \geq 0, \tag{2}$$

$$r(x \mid \lambda, \alpha) = \alpha\lambda(\lambda x)^{\alpha-1}, \quad x \geq 0. \tag{3}$$

See Figures B.1, B.2, and B.3 for graphs of the density, survival function and hazard rate. For this distribution, the power parameter α is usually referred to as the "shape parameter." Unfortunately, no parameterization of the Weibull family has been standardized; some authors write λ in place of λ^α. Various parameterizations of the Weibull distribution are discussed in a review paper by Hallinan (1993).

An early mention of the Weibull distribution was made by Thiele (1872) quoting Oppermann from the *Insurance Record* of 11 Feb, 1870.

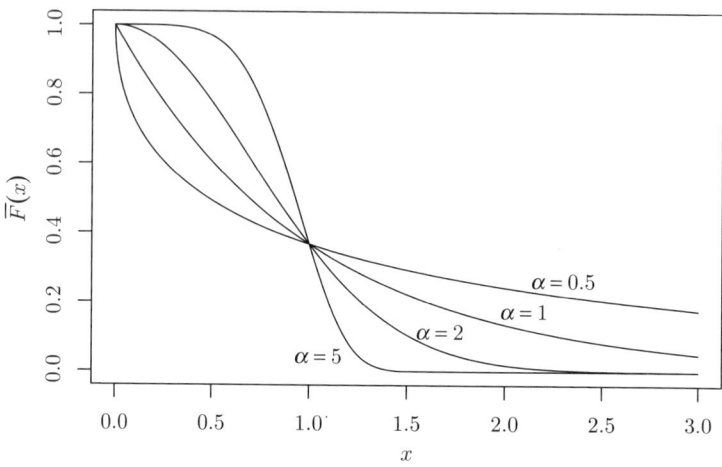

Fig. B.2. Survival functions of the Weibull distribution ($\lambda = 1$)

Oppermann proposed using

$$r(x) = r(x \mid \lambda_1, 1/2) + r(x \mid \lambda_2, 1) + r(x \mid \lambda_3, 3/2)$$

as the hazard rate of young people, up to age 20. This sum of three Weibull hazard rates produces a survival function that is the product of three survival functions of the form (1). In a similar context, Makeham (1890) derived the Weibull distribution with power parameter 2.

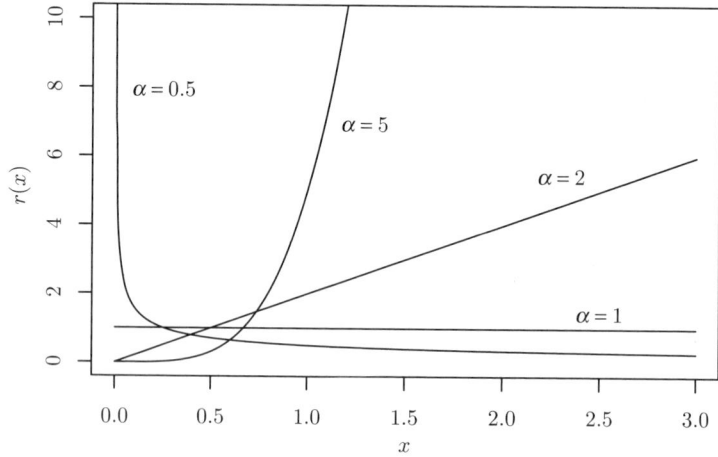

Fig. B.3. Hazard rates for the Weibull distribution ($\lambda = 1$)

B. The Weibull Distribution

The Weibull distribution was encountered by Fisher and Tippett (1928) as an extreme value distribution, a derivation discussed below and in Section 20.G. In a study of the particle size distribution in powdered coal, it was again derived by Rosin and Rammler (1933) using entirely different methods. The Weibull distribution was proposed by Weibull (1939a,b) in the context of strength of materials and ruptures in solids. But later, in more accessible papers (Weibull, 1951, 1952), he promoted the distribution for use in analyzing a much greater variety of data. In those papers, he starts by saying, in effect, that a hazard function can be any nonnegative, increasing function, 0 at the left-hand endpoint of the support of the distribution. Then, he proposes the hazard function $R(x) = (\lambda x)^\alpha$ as the simplest such function. Further, he gives a number of data sets, and for each is able to obtain a good fit with this distribution. However, these data sets are all rather small, and it is only with large data sets that poor fits can be detected. For a more detailed history, see Hallinan (1993) or Johnson, Kotz and Balakrishnan (1994, p. 628).

The Weibull distribution has been used extensively in medical studies, and it has gained considerable popularity as a model in several other fields, including forestry and engineering; a number of papers employing the Weibull distribution in a wide variety of applications is listed by Murthy, Xie and Jiang (2004, pp. 13, 174, 175, 189). Because of its rather pervasive popularity, articles warning against indiscriminate use of the Weibull distribution have appeared (see, e.g., Gorski, 1968; Evans, 1990). For a detailed and comprehensive study of the Weibull and related distributions see Murthy, Xie and Jiang (2004).

a. Extreme Value Distributions

The Weibull distribution was derived by Fisher and Tippett (1928) and again by Gnedenko (1943) as an extreme value distribution; this is perhaps the most important theoretical justification for the distribution, and it provides some intuitive guidance to its applicability. Extreme value distributions are reviewed in Section 20.G; for a more thorough survey, see Kotz and Nadarajah (2000). Extreme value distributions (for minima) are the possible nondegenerate limiting distributions for sequences

$$a_n \min(X_1, \ldots, X_n) + b_n,$$

where $X_1, X_2, \ldots$ are independent identically distributed random variables, and $\{a_n\}, \{b_n\}$ are sequences of real numbers. Of the three such limiting distributions, the Weibull distribution is the only one that has, or can be made by a change in location and scale, to have support $[0, \infty)$. This explains a number of applications of the Weibull distribution; one example of Weibull (1951) has to do with the strength of a chain. The distribution is widely used in hydrology with a clear motivation. The reasons for its wide use in forestry is less obvious, and some usage there may be due simply to the facts that the distribution is amenable to statistical analyses and has been found to provide a good fit to data.

b. Density and Hazard Rate Behavior

From (2), it follows directly that

$\log f$ is convex for $0 < \alpha < 1$, linear for $\alpha = 1$, and concave for $\alpha > 1$.
(4)

By Proposition 4.B.6, (4) implies that

$r(x)$ is increasing in $x \geq 0$ when $\alpha > 1$,
$r(x)$ is the constant λ when $\alpha = 1$, and
$r(x)$ is decreasing in $x \geq 0$ when $0 < \alpha < 1$.

Of course, these results are also easy to see from (3).

For $\alpha \geq 1$, it follows from Proposition 4.B.2 and (4) that f is unimodal; by differentiating the density (or more simply, by differentiating the logarithm of the density), it can be determined that the mode m is given by

$$m = \frac{1}{\lambda}\left(1 - \frac{1}{\alpha}\right)^{1/\alpha}.$$

For $\alpha < 1$, the hazard rate is decreasing, and hence the density is decreasing; that is, there is a unique mode at 0.

It follows directly from (3) that

$$\lim_{x \to \infty} r(x) = 0 \text{ for } \alpha < 1, \text{ and } \lim_{x \to \infty} r(x) = \infty \text{ for } \alpha > 1.$$

Clearly, the hazard rate at 0 is ∞ for $0 < \alpha < 1$; it is equal to λ for

$\alpha = 1$, and it is equal to 0 for $\alpha > 1$. Thus, for the Weibull distribution there is a curious lack of flexibility in choosing the hazard rate at 0.

Note also from (3) that the hazard rate is concave when $1 \le \alpha \le 2$ and is otherwise convex. Use is made of this fact in Example 15.G.3 where a distribution of Hjorth (1980) is discussed.

c. Moments

B.1. Observations. If Y has an exponential distribution with parameter 1, then $X = Y^{1/\alpha}/\lambda$ has a Weibull distribution with shape parameter α and scale parameter λ.

This observation is straightforward to verify. Because of Observation B.1, the moments of the Weibull distribution can be easily retrieved from the moments of the exponential distribution given in 8.A(7). Thus, if X has a Weibull distribution,

$$EX^r = \frac{\Gamma\left(\frac{r}{\alpha} + 1\right)}{\lambda^r}, \quad r > -\alpha, \tag{5}$$

and in particular, the first moment is

$$EX = \frac{\Gamma\left(\frac{1}{\alpha} + 1\right)}{\lambda}.$$

The variance of a Weibull distribution does not take a particularly convenient form even though it can be explicitly expressed in terms of gamma functions using (5) and the definition 1.C(6):

$$Var(X) = \frac{\Gamma\left(\frac{2}{\alpha} + 1\right) - \Gamma^2\left(\frac{1}{\alpha} + 1\right)}{\lambda^2}.$$

With the scale parameter $\lambda = 1$ and large values of the power parameter α, McEwen and Parresol (1991) have shown that

$$EX \approx 1 - \frac{\gamma}{\alpha} + \frac{\gamma^2 + \pi/6}{2\alpha^2}, \quad Var(X) \approx \frac{\pi^2}{6\alpha^2},$$

where $\gamma \approx .577256649$ is Euler's constant. Approximations are also known for small values of α. In particular, if α is small, then

$$EX \approx \sqrt{2\pi}\, e^{-1/\alpha} \alpha^{-(1/\alpha)-(1/2)};$$

see Abramowitz and Stegun (1964, p. 257).

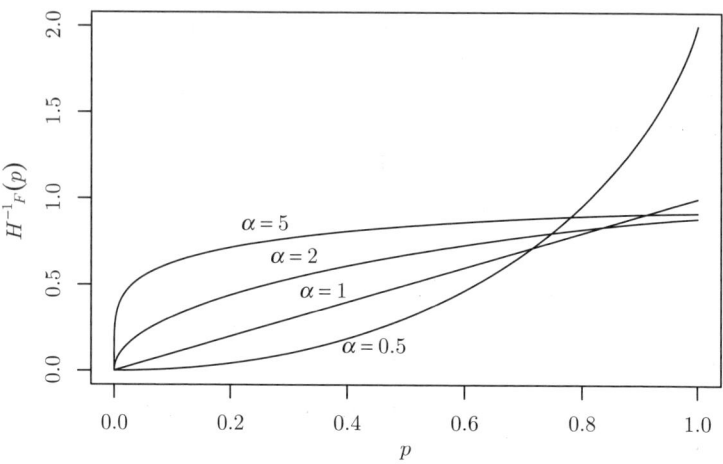

Fig. B.4. Total time on test transforms for Weibull distributions ($\lambda = 1$)

d. Total Time on Test Transform

For the Weibull survival function (1), the total time on test transform cannot be expressed in closed form, but it can be expressed in terms of the incomplete gamma function. It can be verified directly that for this survival function,

$$F^{-1}(p) = \frac{[-\log(1-p)]^{1/\alpha}}{\lambda}, \quad 0 \leq p < 1. \tag{6}$$

See Figure B.4 for a graph of the total time on test transform. From (6), the total time on test transform ψ, defined in Definition 1.I.2, can be calculated as:

$$\psi(p) = \int_0^{F^{-1}(p)} \bar{F}(x)\,dx = \int_0^{F^{-1}(p)} \exp\{-(\lambda x)^\alpha\}\,dx$$
$$= \int_0^{F^{-1}(p)} z^{(1/\alpha)-1} e^{-z}\,dz. \tag{7a}$$

Now, make the change of variables $(\lambda x)^\alpha = z$ and for notational convenience, write $F^{-1}(p) = b$. Then, ψ takes the form of the incomplete gamma function (see 23.A.c):

$$\psi(p) = \frac{1}{\lambda\alpha}\Gamma_{F^{-1}(p)}\left(\frac{1}{\alpha}\right) = \frac{1}{\lambda\alpha}\int_0^{(\lambda b)^\alpha} e^{-z} z^{(1/\alpha)-1}\,dz. \tag{7b}$$

From (5) and (7b), it follows that the normalized total time on test transform $\tilde{\psi}$ of 1.I(9) is given by

$$\tilde{\psi}(p) = [\Gamma_{F^{-1}(p)}(1/\alpha)]/\alpha\Gamma(1/\alpha). \qquad (8)$$

Of course, this function cannot be given in closed form but can be obtained numerically.

e. The Gini Index and Coefficient of Variation

First, let $\lambda^\alpha = \xi$ and rewrite the survival function (1) in the form

$$\bar{F}(x) = \exp\{-\xi x^\alpha\}, \quad x \geq 0.$$

Here, ξ is a frailty parameter, and it follows from 5.E(20) that the Gini index for the Weibull distribution can be written in terms of the moments. Using this fact and (5), the Gini index as a function of α can be determined to be

$$Gini(\alpha) = 1 - \frac{1}{2^{1/\alpha}}.$$

The Gini index can also be obtained by a straightforward integration using 1.I(12)

For the Weibull distribution, the coefficient of variation (the ratio of standard deviation to mean) can be determined directly from (5) to be

$$CV(F) = \left[2\alpha\frac{\gamma(2/\alpha)}{\Gamma^2(1/\alpha)} - 1\right]^{1/2} = \left[\frac{2\alpha}{B(1/\alpha, 1/\alpha)} - 1\right]^{1/2},$$

where $B(a,b)$ is the beta function (see 23.B).

f. Ordering Weibull Distributions

B.2. Proposition. In the likelihood ratio order, the Weibull distribution is decreasing in the scale parameter λ for each fixed power parameter α.

Unlike the gamma distribution, the Weibull distribution is not likelihood ratio ordered in the shape parameter; in fact, it is not even stochastically ordered in the shape parameter (see the comment following Proposition 7.D.5.

B.3. Proposition (Chandra and Singpurwalla, 1981). In the convex transform order $\leq_c$ of Definition 2.C.7, the Weibull distribution is decreasing in the power parameter α.

Proof. This result is a special case of Proposition 7.D.5. Or, to directly verify the conditions of the Definition 2.C.7, obtain from (1) that

$$\bar{F}^{-1}(z) = [-\log z]^{1/\alpha}/\lambda, \quad 0 < z \leq 1.$$

The result follows from the fact that the function x^r is convex for $r \geq 1$. □

From the summary 2.C.j, it is seen that other orderings follow from Proposition B.3. Propositions B.2 and B.3 together show that λ is a magnitude parameter and α is a power parameter that could be called a skewness parameter. Thus, the parameters complement each other in their actions on the distribution.

g. Infinite Divisibility

The Weibull distribution is a special case of the extended gamma–Weibull distribution introduced in Section E.a. As such, it is known to be infinitely divisible, as noted in Section E.b.

h. The Rayleigh Distribution

Suppose that X and Y are normally distributed random variables with mean 0 and variance σ^2 (so that X^2 and Y^2 have chi-square distributions with 1 degree of freedom), and let

$$Z = [X^2 + Y^2]^{1/2}.$$

This random variable is of special interest in electrical engineering; other areas of application are outlined by Johnson, Kotz and Balakrishnan (1994, p. 456). From Proposition A.6, it follows that Z^2 has an exponential distribution with scale parameter $1/(2\sigma^2)$. Consequently, Z has a Weibull distribution with scale parameter $1/\sqrt{2\sigma^2}$ and shape parameter 2. This distribution was derived by Rayleigh (1880) as a distribution of amplitudes, and has come to be known as the *Rayleigh distribution*. It has survival function $\bar{F}$, density f, and linear hazard rate r given by

$$\bar{F}(x) = \exp\left\{-\frac{x^2}{2\sigma^2}\right\}, \quad x \geq 0, \tag{9}$$

$$f(x) = \frac{x}{\sigma^2} \exp\left\{-\frac{x^2}{2\sigma^2}\right\}, \quad x \geq 0, \tag{10}$$

$$r(x) = \frac{x}{\sigma^2}, \quad x \geq 0. \tag{11}$$

Rayleigh (1919) discussed the more general distribution than the one that now bears his name. In particular, he obtained the density

$$f(x) = \frac{2x^{n-1}}{(2\sigma^2)^{n/2}\Gamma(n/2)} \exp\left\{-\frac{x^2}{2\sigma^2}\right\}. \tag{12}$$

For a random variable X with this density, Rayleigh (1919) found that

$$EX^r = \frac{2^{r/2}\sigma^r \Gamma\left(\frac{n+r}{2}\right)}{\Gamma\left(\frac{n}{2}\right)}.$$

The density is unimodal with mode at $\sqrt{n-1}\,\sigma$.

Special cases of (12) include

$n = 1$: this is a "folded" Gaussian or "half normal" (a normal distribution truncated at 0);
$n = 2$: this is the Rayleigh distribution with density (10);
$n = 3$: this case is known as the *Maxwell-Boltzman distribution*.

Both of the cases $n = 2$ and $n = 3$ are of particular interest in physics.

It can be shown with the aid of Eq. 23.A(2) that the parameter n of the density (11) need not be an integer. With the notation $\lambda = 1/\sqrt{2}\sigma$, the resulting density is

$$f(x \mid \lambda, \theta) = \frac{2}{\Gamma(\theta/2)} \lambda(\lambda x)^{\theta-1} e^{-(\lambda x)^2}.$$

i. Characterizations of the Weibull Distribution

If X has a Weibull distribution with power parameter α, then X^α has an exponential distribution. This fact can be used to translate any characterization of the exponential distribution to a characterization of the Weibull distribution with fixed power parameter. For example, Wang (1976) has shown that a random variable X satisfies

$$P\{X > \sqrt[\alpha]{x^\alpha + y^\alpha} \mid X > y\} = P\{X > x\}, \quad x, y > 0, \tag{13}$$

if and only if X has a Weibull distribution with power parameter α. To prove this, rewrite (13) in the form

$$P\{X^\alpha > x^\alpha + y^\alpha \mid X^\alpha > y^\alpha\} = P\{X^\alpha > x^\alpha\}, \quad x, y > 0;$$

this form of the "lack of memory" property leads to the conclusion (Proposition 8.B.1) that X^α has an exponential distribution.

For another characterization of the Weibull distribution, see Arnold and Isaacson (1976).

j. A Mixture Representation

Walker and Stephens (1999) have observed that if

$$g(x \mid u) = \frac{\alpha x^{\alpha - 1}}{u}, \quad 0 < x < u^{1/\alpha},$$

is the density of a uniform distribution on $[0, u^{1/\alpha}]$ with a power parameter (see 14.B.1), and if

$$h(u) = c^2 u\, e^{-cu}, \quad u > 0$$

is a gamma density with shape parameter 2, then the mixture

$$f(x) = \int_{x^\alpha}^{\infty} g(x \mid u) h(u)\, du$$

is a Weibull density; with $c = \lambda^\alpha$; this density is given by (2).

k. Mixtures of Weibull Distributions

Mixtures of Weibull distributions have been used in a variety of applications, a large number of which are listed by Murthy, Xie and Jiang (2004, p. 174). Hazard rates of even the mixture of two Weibull distributions can take a variety of shapes; these hazard rates can have up to four relative maxima and minima. For a catalog of these hazard rates, see Jiang and Murthy (1998).

If X has a Weibull distribution with $\alpha = 2$ and Y has an exponential distribution, then min (X, Y) has an increasing linear hazard rate. Mixtures of distributions with such hazard rates can have a variety of hazard rate shapes; for a detailed study of such mixtures, see Block, Savits and Wondmagegnehu (2003).

l. Inverse and Generalized Weibull Distributions

If X has a Weibull distribution, then the distribution of $1/X$ is sometimes called an *inverse Weibull distribution*. A direct calculation shows that if X has the survival function (1), then $1/X$ has the distribution function

$$W_2^*(x) = \exp\{-(\theta x)^{-\alpha}\}, \quad \theta, \alpha > 0, \ x > 0.$$

This extreme value distribution for maxima, given here in the notation of Section 20.G, was found already by Fréchet (1927).

The density (2) of the Weibull distribution and the density of W_2^* combine nicely to form what might be called the density of an *extended Weibull family*. For this purpose, it is convenient to reparameterize W_2^* by replacing θ by λ (not $1/\lambda$), and replace α by $-\alpha$, so that $\alpha < 0$. Then, W_2^* takes the form

$$F(x \mid \lambda, \alpha) = \exp\{-(\lambda x)^\alpha\}, \quad a < 0, \ \lambda > 0, \ x > 0. \qquad (14)$$

This distribution has the density, say f, which combines with (2) to become

$$f(x \mid \lambda, \alpha) = |\alpha|\lambda(\lambda x)^{\alpha-1} \exp\{-(\lambda x)^\alpha\}, \quad \lambda > 0, \ \alpha \neq 0, \ x > 0. \qquad (15)$$

For $\alpha < 0$, this density is unimodal, with mode at the point

$$x = \frac{1}{\lambda}\left(\frac{\alpha-1}{\alpha}\right)^{1/\alpha};$$

this can easily be checked by setting the derivative of the logarithm of the density equal to 0.

When $\alpha < 0$ the hazard rate

$$r(x \mid \lambda, \alpha) = \frac{|\alpha|\lambda(\lambda x)^{\alpha-1}\exp\{-(\lambda x)^\alpha\}}{1 - \exp\{-(\lambda x)^\alpha\}}, \quad x > 0, \tag{16}$$

satisfies

$$\lim_{x \to \infty} r(x \mid \lambda, \alpha) = \lim_{x \to 0} r(x \mid \lambda, \alpha) = 0.$$

It can be seen from Theorem 4.F.2 that when $\alpha < 0$, the hazard rate r of (16) has a unique mode; that is, r has an inverted bathtub shape. This distribution has been studied by Jiang, Murthy and Ji (2001, p. 115), who determine that the mode of the hazard rate is given by the solution of the equation

$$z(t) = \left(1 - \frac{1}{\alpha}\right)\left(1 - e^{-z(t)}\right) \tag{17}$$

where $z(t) = (\lambda t)^\alpha$. Because the left side of (17) is linear in $z(t)$ and the right side is increasing and convex in $z(t)$, there is but one positive solution.

m. The Weibull Residual Life Distribution

The Weibull residual life distribution has survival function

$$\bar{F}_t(x) = \exp\{-[\lambda(x+t)]^\alpha + (\lambda t)^\alpha\}, \quad x, \alpha, \lambda > 0, \tag{18}$$

and hazard rate

$$r_t(x) = \alpha\lambda(x+t)^{\alpha-1}, \quad x, \alpha, \lambda > 0. \tag{19}$$

Note that this hazard rate satisfies (4), but that unlike the hazard rate of the Weibull distribution, $r_t(0) = \alpha\lambda t^{\alpha-1}$ can take on values other than 0 and ∞.

Note that for $\alpha > 1$, $\bar{F}_t(x) < \bar{F}(x)$, that is,

$$\bar{F}(x+t) < \bar{F}(x)\bar{F}(t), \quad x, t > 0,$$

so that F is NBU. This inequality is reversed if $\alpha < 1$, and then F is NWU. These results can be obtained easily from the fact that for positive numbers $a_1, \ldots, a_n$, $(\sum_{i=1}^n a_i^r)^{1/r}$ is decreasing in $r > 0$. See Hardy, Littlewood and Pólya (1934, 1952, p. 28).

C. Exponential Distributions with a Resilience Parameter

The two parameter family obtained from the exponential distribution by introducing a resilience parameter has not received much attention in the literature for reasons that are not entirely clear. Although the distribution appears to be without a name, it may be appropriate to call it the *Verhulst distribution*, because it was discussed by Verhulst (1838, 1845). A discussion of this and some related distributions has been given by Ahuja and Nash (1967). The hazard rates and the survival functions have explicit expressions, so that these distributions can provide useful models for cases where data are missing.

a. Distribution Function, Density, and Hazard Rate

The exponential distribution with resilience parameter has the distribution function

$$F(x) = (1 - e^{-\lambda x})^\eta, \quad \eta, \lambda, x > 0; \tag{1}$$

the density is given by

$$f(x) = \lambda \eta \, e^{-\lambda x}(1 - e^{-\lambda x})^{\eta - 1}, \quad x > 0, \tag{2}$$

and consequently, the hazard rate is

$$r(x) = \frac{\lambda \eta e^{-\lambda x}(1 - e^{-\lambda x})^{\eta-1}}{1 - (1 - e^{-\lambda x})^\eta}, \quad x > 0. \tag{3a}$$

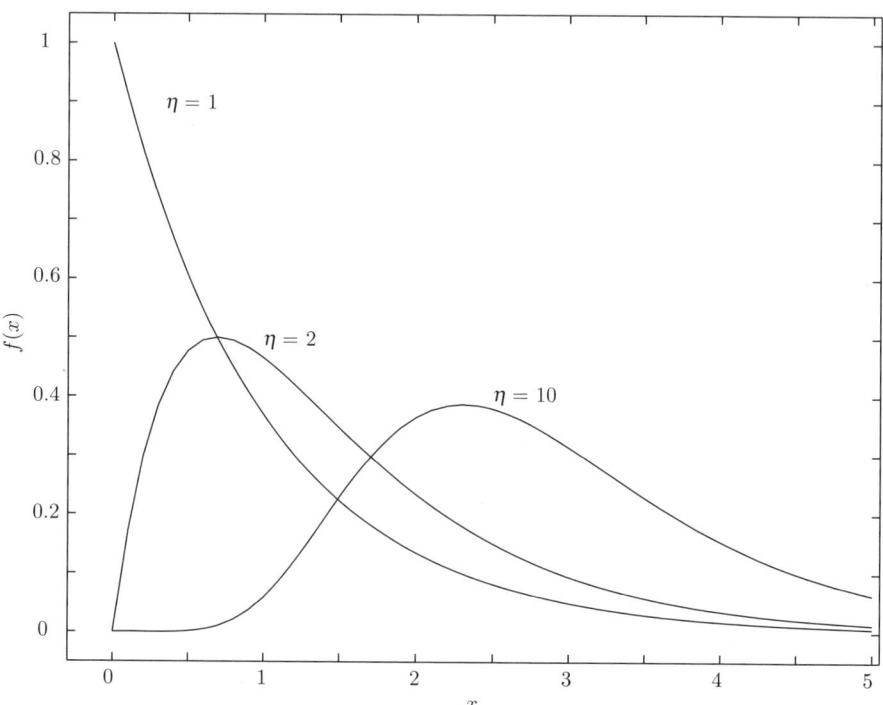

Fig. C.1. Densities of the exponential distribution with resilience paremeter ($\lambda = 1$)

The hazard rate is somewhat more complicated than that of the Weibull distribution primarily because here it is the distribution function that has a simple form, not the survival function. For the same reason, the reverse hazard rate s does take a simple form, namely,

$$s(x) = \lambda\eta/(e^{\lambda x} - 1), \quad x > 0. \tag{3b}$$

See Figures C.1, C.2, C.3, and C.4 for graphs of the density, distribution function, and hazard rate of the exponential distribution with a resilience parameter.

Note that when η is an integer (1) is the distribution of the maximum in a sample of size η from an exponential distribution with parameter λ.

C.1. Example. Suppose that $X_1, X_2, \ldots, X_n$ are independent and all have the exponential distribution 8.A(4). Let

$$W = \max\{X_1, X_2, \ldots, X_n\} - \min\{X_1, X_2, \ldots, X_n\}.$$

C. Exponential Distributions with a Resilience Parameter

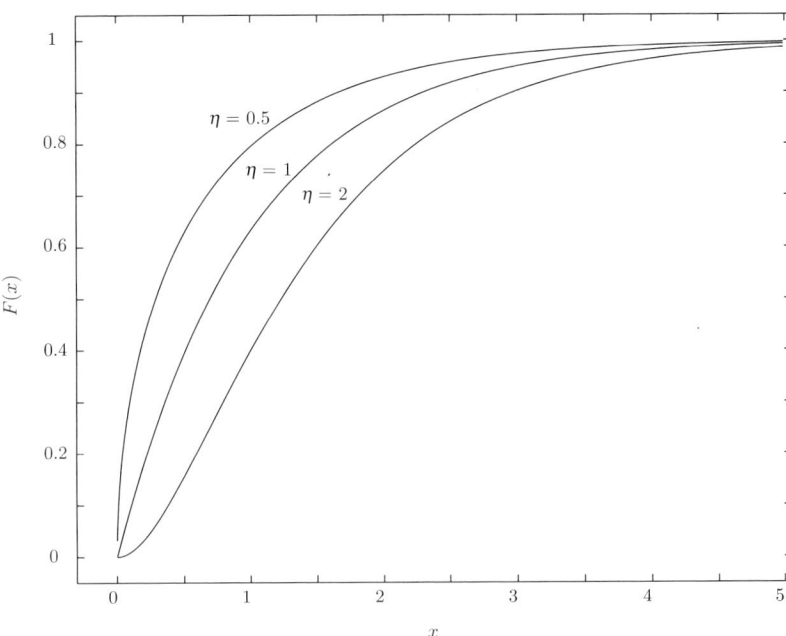

Fig. C.2. Distribution functions of the exponential distribution with resilience parameter

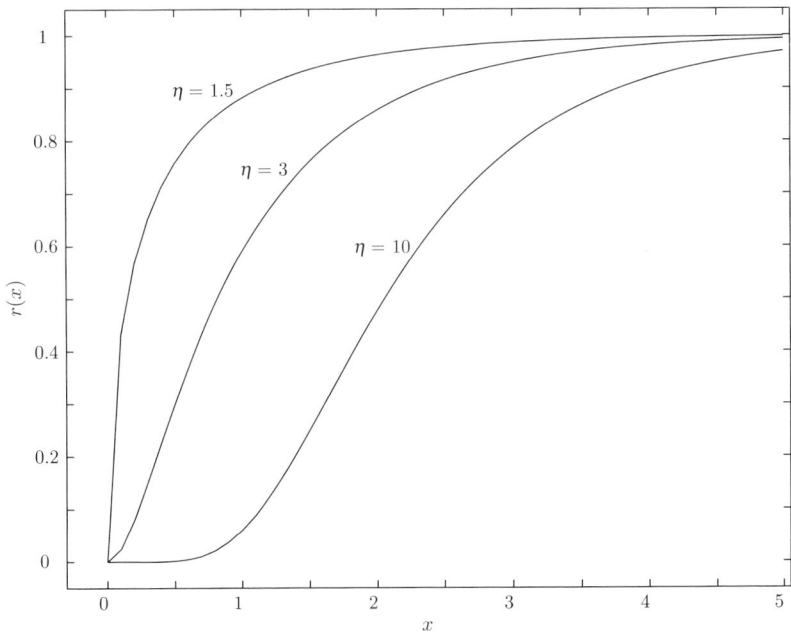

Fig. C.3. Hazard rates of exponential distributions with resilience parameter ($\lambda = 1, \eta \geq 1$)

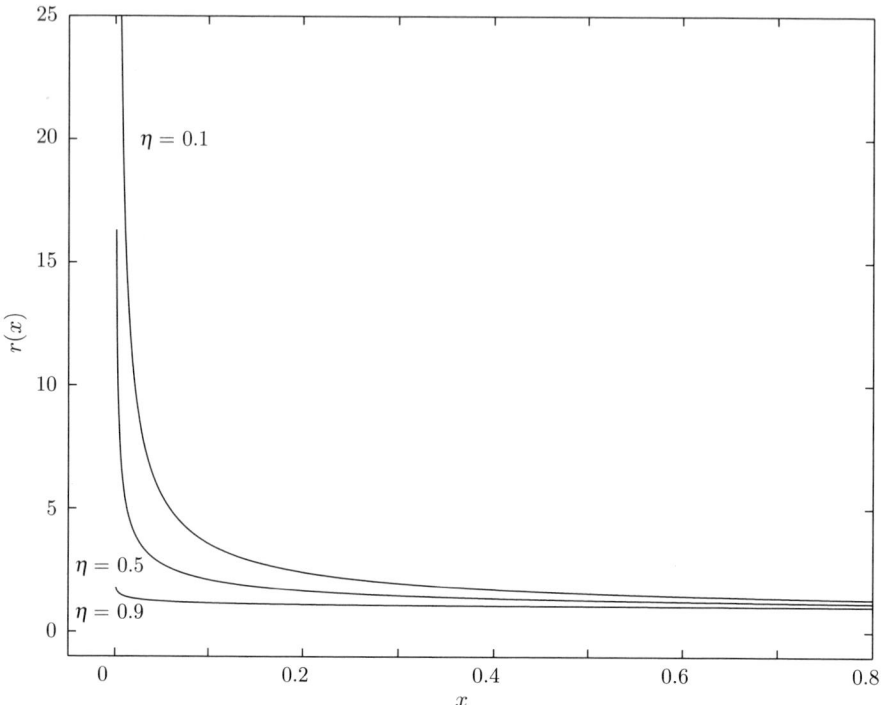

Fig. C.4. Hazard rates of exponential distributions with resilience parameter ($\lambda = 1$, $\eta \leq 1$)

Then W, called the *range* of $X_1, X_2, \ldots, X_n$, has the density (2) and distribution function (1) with $\eta = n - 1$.

C.2. Proposition. The density (2) is log concave for $\eta \geq 1$ and log convex for $\eta \leq 1$.

Proof. From (2) it follows that

$$\frac{d^2}{dx^2} \log f(x) = -\frac{(\eta - 1)\lambda\, e^{\lambda x}}{(e^{\lambda x} - 1)^2}, \qquad (4)$$

which is negative, for $\eta > 1$, and positive, for $\eta < 1$. □

Among the consequences of Proposition C.2 is the fact that the density f is decreasing when $\eta \leq 1$ and is unimodal when $\eta > 1$. If $\eta > 1$, the mode $x_m = [\log \eta]/\lambda$ can be found by setting the expressions in (4) equal to 0.

C. Exponential Distributions with a Resilience Parameter

It is easy to see that when $\eta > 1, \lim_{x \to 0} f(x) = 0$, whereas when $\eta < 1, \lim_{x \to 0} f(x) = \infty$. Of course, $f(0) = \lambda$ when $\eta = 1$.

In case $\eta > 1$, it can be found using (4) that

$$\lim_{x \to 0} f'(x) = \lim_{x \to 0} \lambda \left[\frac{(\eta - 1)\, e^{-\lambda x}}{1 - e^{-\lambda x}} - 1 \right] f(x) \qquad (5)$$

$$= \lim_{x \to 0} \lambda^2 \eta (\eta - 1)(1 - e^{-\lambda x})^{\eta - 2}.$$

This limit is infinite for $1 < \eta < 2$; it is equal to $2\lambda^2$ for $\eta = 2$; and equal to 0 for $\eta > 2$.

According to Proposition 4.B.8.a, another consequence of Proposition C.2 is that the hazard rate r is increasing for $\eta \geq 1$, and decreasing for $\eta \leq 1$. This fact can also be obtained by differentiating the hazard rate. In this way, it can be shown after algebraic simplification that the hazard rate is increasing if

$$(1 - e^{-\lambda x})^\eta \geq 1 - \eta\, e^{-\lambda x}, \qquad (6)$$

and decreasing if (6) is reversed. But (6) can be written in the form

$$(1 - p)^\eta \geq 1 - \eta p, \quad 0 \leq p \leq 1; \qquad (7)$$

this is a well-known inequality which requires $\eta \geq 1$, and which is reversed if $\eta \leq 1$. Inequality (7) can be easily verified by noting that $g(p) = (1 - p)^\eta - 1 + \eta p$ has the properties that $g(0) = 0$ and g is increasing if $\eta > 1$, decreasing if $0 < \eta < 1$. See Beckenbach and Bellman (1961, p. 12, Section 14) for a discussion of this inequality and its use in proving the arithmetic–geometric mean inequality.

Other interesting properties of the hazard rate (3a) are worth noting.

C.3. Proposition.

(i) For a $\eta > 0, \lim_{x \to \infty} r(x) = \lambda$.

(ii) For $0 < \eta < 1, \lim_{x \to 0} r(x) = \infty$.

(iii) For $\eta > 1, \lim_{x \to 0} r(x) = 0$, and $\lim_{x \to 0} r'(x) = \lim_{x \to 0} f'(x)$, so that the limit of the hazard rate derivative at 0 is given by (5).

Proof. That the limiting value of r at ∞ is λ can be verified by a direct computation of the limit using l'Hospital's rule. The value of the hazard rate at 0 is straightforward to compute. To see that $\lim_{x \to 0} r'(x) = \lim_{x \to 0} f'(x)$ when $f(0) = 0$, note that this means

$r(0) = 0$ and check that

$$r'(x) = \frac{f'(x)}{\bar{F}(x)} + r^2(x).$$

One consequence of Proposition C.3 is that the limiting residual life distribution $\lim_{t \to \infty} \bar{F}(x+t)/\bar{F}(t) = e^{-\lambda x}$ is an exponential distribution; this follows from Proposition 1.B.13.

D. Exponential Distributions with a Tilt Parameter

When $\bar{F}(x) = \exp\{-\lambda x\}$, the two parameter family

$$\bar{F}(x \mid \lambda, \gamma) = \frac{\gamma}{e^{\lambda x} - \bar{\gamma}}, \quad x \geq 0, \ \lambda > 0, \ \gamma > 0, \ \bar{\gamma} = 1 - \gamma, \tag{1}$$

is obtained from the semiparametric family 7.F(2) with tilt parameter γ. The case $\gamma = 1$ is the exponential distribution.

The survival function (1) has been investigated by Marshall and Olkin (1997), and by Adamidis and Loukas (1998) for the case that $0 < \gamma \leq 1$. It has also been encountered in Example 4.C.7.b as the mixture

$$\int_0^\infty \bar{F}^*(x \mid \lambda, \xi) \, dG(\xi) = \bar{F}(x \mid \lambda, \gamma),$$

where

$$\bar{F}^*(x \mid \lambda, \xi) = \exp\{-\xi(e^{\lambda x} - 1)\}, \quad x \geq 0, \ \xi, \lambda > 0,$$

is a Gompertz survival function (see Chapter 10), and

$$\bar{G}(\xi) = e^{-\gamma \xi}, \quad \xi \geq 0$$

is an exponential survival function.

In case $0 < \gamma \leq 1$, there is also another mixture representation. When $\gamma = 1$, (1) is encountered as the Laplace transform of a geometric distribution in 20.E(6a); when $\gamma \neq 1$, it is the Laplace transform of a geometric distribution altered by the introduction of a scale parameter. As a scaled geometric mixture of exponential distributions, it follows that (1) is infinitely divisible when $0 < \gamma \leq 1$ (see Proposition 20.D.10).

For analyses of failure time data, the parametric family (1) may sometimes be a competitor to the families of two-parameter Weibull and gamma distributions. In this section, some of the properties of the distribution given by (1) are enumerated.

As special cases of 7.F(3a), it follows that $F(\cdot \,|\, \lambda, \gamma)$ has the density $f(\cdot \,|\, \lambda, \gamma)$ given by

$$f(x \,|\, \lambda, \gamma) = \frac{\gamma \lambda \, e^{-\lambda x}}{(1 - \bar{\gamma} \, e^{-\lambda x})^2} = \frac{\gamma \lambda \, e^{\lambda x}}{(e^{\lambda x} - \bar{\gamma})^2}, \quad x > 0, \ \lambda > 0, \ \gamma > 0; \quad (2)$$

either directly or from 7.F(3b) it follows that the hazard rate r is given by

$$r(x \,|\, \lambda, \gamma) = \frac{\lambda}{1 - \bar{\gamma} \, e^{-\lambda x}} = \frac{\lambda \, e^{\lambda x}}{e^{\lambda x} - \bar{\gamma}}, \quad x > 0, \ \lambda > 0, \ \gamma > 0. \quad (3)$$

See Figures D.1, D.2, and D.3 for the case $\gamma \geq 1$, and Figures D.4, D.5, and D.6 for the case $\gamma \leq 1$. Note that $r(x \,|\, \lambda, 1) = \lambda$, and that $r(x \,|\, \lambda, \gamma)$ is decreasing in x for $0 < \gamma \leq 1$, and $r(x \,|\, \lambda, \gamma)$ is increasing in x for $\gamma \geq 1$.

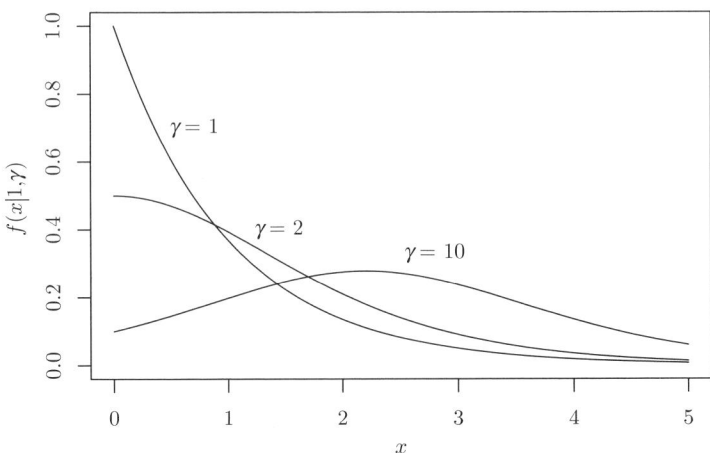

Fig. D.1. Densities of exponential distributions with tilt parameter ($\lambda = 1, \gamma \geq 1$)

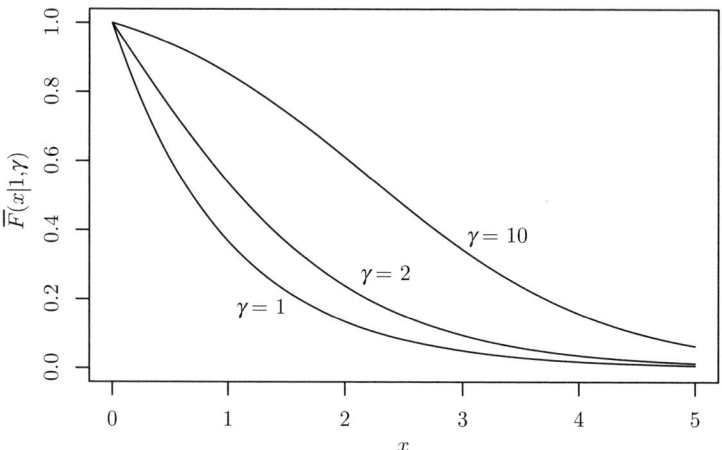

Fig. D.2. Survival functions of exponential distributions with tilt parameter ($\lambda = 1$, $\gamma \geq 1$)

From 7.F(4a), 7.F(4b), 7.F(5a), and 7.F(5b), it follows that

$$\lambda/\gamma \leq r(x \mid \lambda, \gamma) \leq \lambda, \quad -\infty < x < \infty, \ \gamma \geq 1, \quad (4a)$$
$$\lambda \leq r(x \mid \lambda, \gamma) \leq \lambda/\alpha, \quad -\infty < x < \infty, \ 0 \leq \gamma \leq 1, \quad (4b)$$
$$e^{-\lambda x} \leq \bar{F}(x \mid \lambda, \gamma) \leq e^{-\lambda x/\gamma}, \quad -\infty < x < \infty, \ \gamma \geq 1, \quad (5a)$$
$$e^{-\lambda x/\gamma} \leq \bar{F}(x \mid \lambda, \gamma) \leq e^{-\lambda x}, \quad -\infty < x < \infty, \ 0 \leq \gamma \leq 1. \quad (5b)$$

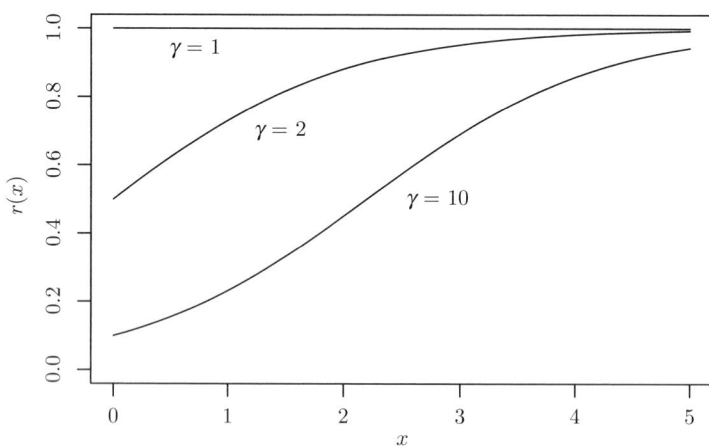

Fig. D.3. Hazard rates of exponential distributions with tilt parameter ($\lambda = 1, \gamma \geq 1$)

D. Exponential Distributions with a Tilt Parameter

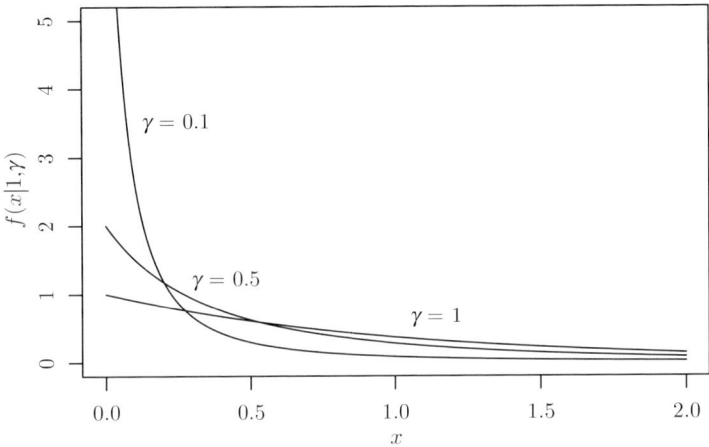

Fig. D.4. Densities of exponential distributions with tilt parameter ($\lambda = 1, \gamma \leq 1$)

D.1. Proposition. The function $\log f(\cdot \mid \lambda, \gamma)$ is convex, for $0 < \gamma \leq 1$, and concave, for $\gamma \geq 1$.

This result can be verified by differentiating $\log f(x \mid \lambda, \gamma)$ with respect to x. Of course, this means that for $\gamma \leq 1$, $f(\cdot \mid \lambda, \gamma)$ is decreasing, and for $\gamma \geq 1$, $f(\cdot \mid \lambda, \gamma)$ is unimodal. By solving $d \, \log f(\cdot \mid \lambda, \gamma)/dx = 0$, it is readily verified that a random variable X with density $f(\cdot \mid \lambda, \gamma)$

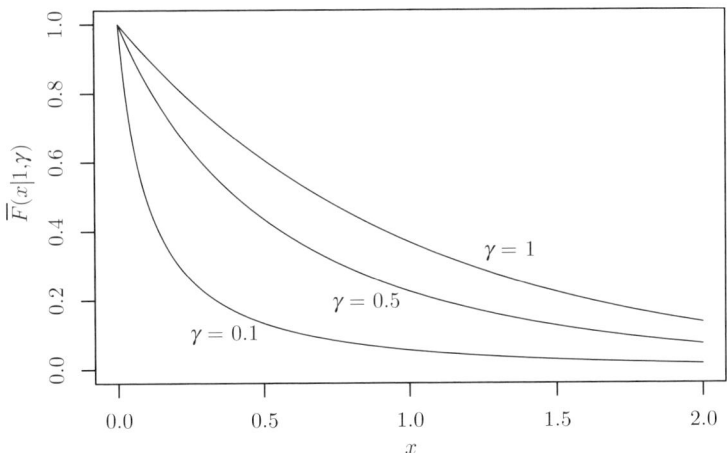

Fig. D.5. Survival functions of exponential distributions with tilt parameter ($\lambda = 1, \gamma \leq 1$)

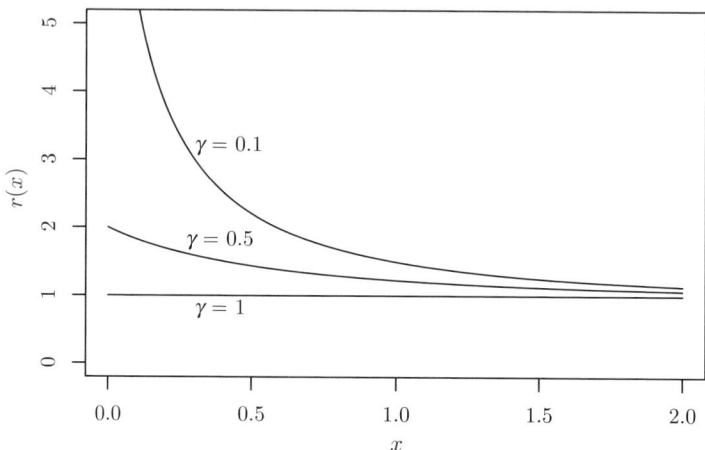

Fig. D.6. Hazard rates of exponential distributions with tilt parameter ($\lambda = 1, \gamma \leq 1$)

has the mode

$$\text{mode}(X) = 0, \quad \gamma \leq 2; \quad \text{mode}(X) = \lambda^{-1} \log(\gamma - 1), \quad \gamma \geq 2. \quad (6)$$

It follows from (5a) and (5b) that $F(\cdot \mid \lambda, \gamma)$ has finite moments of all positive orders. Direct computations show that if X has distribution $F(\cdot \mid \lambda, \gamma)$, then

$$EX = -\frac{\gamma \log \gamma}{\lambda \bar{\gamma}}, \quad \gamma \neq 1, \ \gamma > 0; \quad EX = 1/\lambda, \quad \gamma = 1. \quad (7)$$

Note that this quantity is always positive. More generally, for $r > -1$,

$$EX^r = r \int_0^\infty \bar{F}(x \mid \lambda, \gamma) x^{r-1} \, dx = r\gamma \int_0^\infty \frac{x^{r-1} \, e^{-\lambda x}}{1 - \bar{\gamma} \, e^{-\lambda x}} \, dx \quad (8)$$

$$= \frac{r\gamma}{\lambda^r} \int_0^1 \frac{(-\log y)^{r-1}}{1 - \bar{\gamma} y} \, dy.$$

For $r = 1$, this yields (7).

The Laplace transform of $f(\cdot \mid \lambda, \gamma)$ is given by

$$E \, e^{-sX} = \int_0^1 \left[\frac{\gamma y^{s/\lambda}}{(1 - \bar{\gamma} y)^2} \right] dy. \quad (9)$$

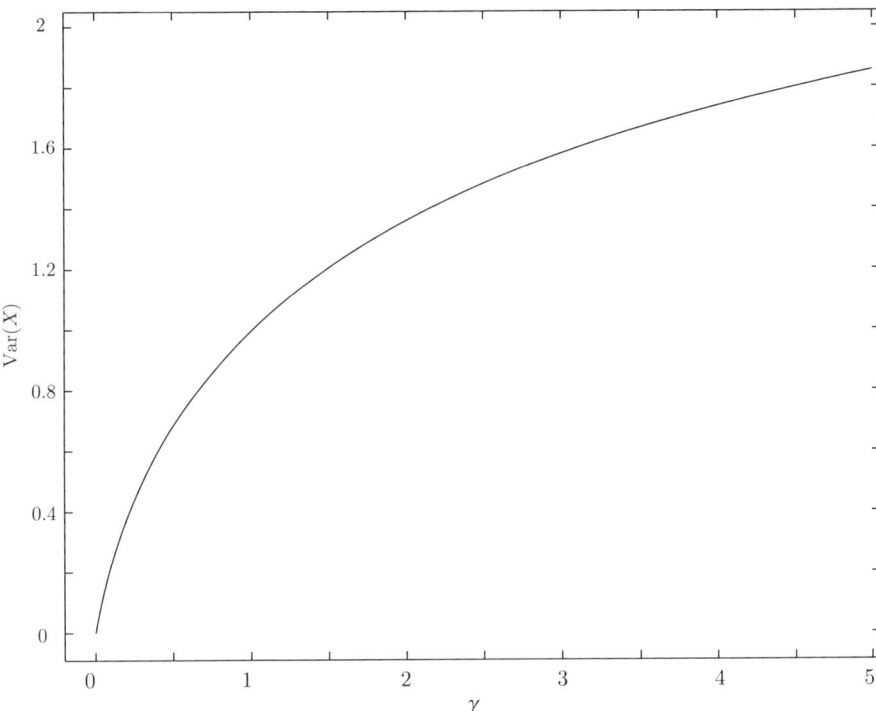

Fig. D.7. Variance of an Exponential distribution with tilt parameter as a function of γ ($\lambda = 1$)

Both (8) and (9) can be expressed as infinite series when $|\bar{\gamma}| \leq 1$. Then, the integrands of (8) and (9) can be expanded in a power series and the result integrated term by term to yield

$$EX^r = \frac{r\gamma}{\lambda^r} \int_0^\infty x^{r-1} e^{-x} \sum_{j=0}^\infty \bar{\gamma}^j \, e^{-jx} \, dx = \frac{r\gamma \Gamma(r)}{\lambda^r} \sum_{j=0}^\infty \frac{\bar{\gamma}^j}{(j+1)^r}, \quad |\bar{\gamma}| \leq 1. \tag{10}$$

$$Ee^{-sX} = \gamma \int_0^1 y^s \sum_{j=0}^\infty (j+1) y^j \bar{\gamma}^j \, dy = \gamma \sum_{j=0}^\infty \bar{\gamma}^j \frac{j+1}{s+j+1}, \quad |\bar{\gamma}| \leq 1. \tag{11}$$

For numerical evaluation, (8) may be preferable to (10). A numerical evaluation of the variance (with the scale parameter $\lambda = 1$) as a function of γ is provided in Figure D.7.

Because of Proposition D.1, total positivity properties yield moment inequalities that are not true in general; see Proposition 6.A.1. In particular, the coefficient of variation σ/μ is less than 1 for $\gamma \geq 1$, and is greater than 1 for $\gamma \leq 1$. The pth percentile x_p of $F(\cdot \mid \lambda, \gamma)$ is obtained by solving the equation $F(x \mid \lambda, \gamma) = p$, and is given by

$$x_p = \frac{1}{\lambda} \log \frac{\bar{p} + \gamma p}{\bar{p}}, \quad \bar{p} = 1 - p. \tag{12}$$

In particular, the median (fiftieth percentile) of X is given by

$$\mathrm{med}(X) = \frac{1}{\lambda} \log(1 + \gamma). \tag{13}$$

It is easy to see that $\mathrm{med}(X), \mathrm{mode}(X)$, and EX are all increasing in γ and decreasing in the scale parameter λ. From the monotonicity of $\log x$ and the fact that $\log x \leq x - 1, x > 0$, it follows that

$$\mathrm{mode}(X) \leq \mathrm{med}(X) \leq \gamma/\lambda \leq EX, \tag{14}$$

but note that

$$\lim_{\gamma \to \infty} \mathrm{mode}(X)/EX = 1.$$

a. Residual Life Distribution of an Exponential Distribution with Tilt Parameter

If $F(\cdot \mid \lambda, \gamma)$ is given by (1), then the residual life distribution at t has the survival function

$$\bar{F}_t(x \mid \lambda, \gamma) = \frac{e^{\lambda t} - \bar{\gamma}}{e^{\lambda(x+t)} - \bar{\gamma}} = \frac{\theta}{e^{\lambda x} - \bar{\theta}},$$

where $\theta = \theta(t) = 1 - \bar{\gamma} \, e^{-\lambda t}$. Thus, the residual life distribution is another exponential distribution with tilt parameter depending upon t. The limit distribution as $t \to \infty$ is an ordinary exponential distribution because the limit of $\theta(t)$ is 1.

It can be verified by direct calculations that $\bar{F}_t(x \mid \lambda, \gamma) > \bar{F}(x \mid \lambda, \gamma)$ when $\gamma < 1$ (F is NWU), and this inequality is reversed when $\gamma > 1$ (F is NBU).

b. A Connection with the Logistic Distribution

The logistic distribution is a distribution with support $(-\infty, \infty)$ and so is beyond the scope of this book, but it is introduced briefly in Section 12.C. It is of some interest to note that the exponential distribution with tilt parameter is the conditional distribution of a random variable Z with a logistic distribution given that $Z > 0$.

The logistic distribution can be derived as the solution of the differential equation

$$\frac{dF(x)}{dx} = cF(x)\bar{F}(x),$$

which is known as the *logistic law of growth*. This differential equation has clear biological meaning, but its meaning in the context of probability is a mystery discussed briefly by Feller (1971, p. 52). However, a number of other derivations of the logistic distribution have been given (see, e.g., Johnson, Kotz and Balakrishnan, 1995).

The logistic distribution may have been found to be useful in part because it has a survival function that can be expressed in closed form. Specifically, the survival function is given by

$$\bar{F}(x) = \frac{1}{1 + c\, e^{\lambda x}}, \quad \lambda, c > 0, \quad -\infty < x < \infty. \tag{15}$$

This survival function truncated at 0 becomes

$$\bar{F}(x) = \frac{1+c}{1 + c\, e^{\lambda x}}, \quad \lambda, c > 0, \quad 0 < x < \infty. \tag{16}$$

Alternatively, (16) can be thought of as the conditional distribution of a random variable X having the distribution (15), given that $X > 0$. With the identification $\gamma = (c+1)/c$, (16) becomes (1). Thus, the exponential distribution with tilt parameter is a truncated logistic distribution.

c. Ordering Distributions with Tilt Parameter

The distributions $F(\cdot \mid \lambda, \gamma)$ of (1) are, for fixed λ, likelihood ratio decreasing in γ as a special case of Proposition 7.F.8. They are clearly stochastically ordered in λ for all γ. Tedious but routine calculus shows that these distributions are also likelihood ratio ordered in λ for fixed $\gamma \leq 2$, but not ordered in λ even in the hazard rate order for fixed $\gamma > 2$.

In the convex transform order, the distributions $F(\cdot \mid \lambda, \gamma)$ are decreasing in γ for all fixed λ. As usual, this order can be established by directly verifying the conditions of the definition. Of course, this means that the distributions are ordered in the Lorenz order and have a coefficient of variation decreasing in γ.

The above-mentioned discussion shows that the scale parameter of (1) is a magnitude parameter, as expected, but only in the relatively weak sense of stochastic order when $\gamma > 2$. The tilt parameter γ is a shape parameter that orders the distributions according to skewness, but it is like the shape parameter of the gamma distribution in that it also orders the distributions according to the likelihood ratio order, a magnitude order.

d. Limits

Suppose that $F(\cdot \mid \lambda, \gamma)$ is the survival function given by (1), and let $\lambda = \gamma \delta$. Then, using L'Hospital's rule it can be determined that

$$\lim_{\gamma \to 0} \bar{F}(x \mid \delta \gamma, \gamma) = \lim_{\gamma \to 0} \frac{\gamma}{e^{\delta \gamma x} - \bar{\gamma}} = \lim_{\gamma \to 0} \frac{1}{\delta x e^{\delta \gamma x} + 1} = \frac{1}{1 + \delta x}$$

is the survival function of a Pareto II distribution 11.B(1) with frailty parameter 1. Thus, this basic Pareto distribution is a weak limit of distributions that are exponential distributions with tilt parameter.

If EX is fixed, say $EX = 1$, then $\lambda = -(\gamma \log \gamma)/\bar{\gamma}$. One can verify that, with this value of λ,

$$\lim_{\gamma \to \infty} \bar{F}(x \mid \lambda, \gamma) = 1, \quad x < 1,$$
$$= 1/2, \quad x = 1,$$
$$= 0, \quad x > 1;$$

this shows that distributions degenerate at the point $\mu = EX > 0$ are weak limits of distributions of the form (1). A further computation shows that

$$\lim_{\gamma \to 0} \bar{F}(x \mid \lambda, \gamma) = 1, \quad x \leq 0,$$
$$= 0, \quad x > 0,$$

so the distribution degenerate at 0 is a weak limit of distributions of the form (1) with unit expectation (and decreasing hazard rate).

e. Comparisons with the Weibull and Gamma Families

For analyses of failure time data, exponential distributions with a tilt parameter may be a competitor to the families of two parameter gamma and Weibull distributions. These families share certain similarities; for example, all have monotone hazard rates: A gamma distribution is IHR if $\nu > 1$ and DHR if $\nu < 1$; a Weibull distribution is IHR if $\alpha > 1$ and DHR if $\alpha < 1$; an exponential distrtibution with tilt is IHR if $\gamma > 1$ and DHR if $\gamma < 1$. In each case, if the parameter is 1, then the distribution is exponential.

A comparison of the gamma, Weibull, and exponential distributions with tilt using simulations is made by Marshall, Meza, and Olkin (2001). Here are some theoretical comparisons.

First, note from (3) that $r(0 \mid \lambda, \gamma) = \lambda \gamma$, so that at the origin this hazard rate varies continuously with the parameters. This is in contrast with the family of Weibull or gamma distributions; for both of those families, either the distribution is an exponential distribution, or $r(0) = 0$, or $r(0) = \infty$, so that $r(0)$ is discontinuous in the shape parameter, and neither family allows real choice for $r(0)$. Of course, for all of these families, $r(0) = f(0)$, so these comments also apply to the density at 0.

For exponential distributions with tilt parameter,

$$\lim_{x \to \infty} r(x \mid \lambda, \gamma) = \lambda$$

is bounded and continuous in the parameters, like the gamma distribution but unlike the Weibull distribution.

f. An Alternative Derivation

Suppose that X has the Pareto II distribution 11.B(1), i.e., X has the survival function

$$\bar{F}_X(x) = [1 + (\lambda x)]^{-\xi}, \quad x \geq 0, \ \lambda, \xi > 0.$$

If $X = e^Y - 1$, then Y has the survival function

$$\bar{F}_Y(y) = [1 + \lambda(e^y - 1)]^{-\xi}, \quad y \geq 0, \ \lambda, \xi > 0. \tag{17a}$$

Now, λ is no longer a scale parameter; let $\lambda = 1/\gamma$, and rewrite (17a) as

$$\bar{F}_Y(y) = \frac{1}{[1 + \gamma^{-1}(e^y - 1)]^\xi} = \left(\frac{\gamma \, e^{-y}}{1 - \bar{\gamma} \, e^{-y}} \right)^\xi, \quad y \geq 0, \ \gamma, \xi > 0. \tag{17b}$$

If a scale parameter is introduced in (17b), the result is the survival function

$$\bar{F}(y\mid \lambda,\gamma,\xi) = \frac{1}{[1+\gamma^{-1}(e^{\lambda y}-1)]^{\xi}} = \left(\frac{\gamma\, e^{-\lambda y}}{1-\bar{\gamma}\, e^{-\lambda y}}\right)^{\xi},$$
$$y \geq 0, \quad \lambda, \gamma, \xi > 0. \tag{17c}$$

For comparison with (17c), rewrite (1) in the form

$$\bar{F}(x\mid \lambda,\gamma) = [1+\gamma^{-1}(e^{\lambda x}-1)]^{-1}, \quad x > 0, \;\lambda > 0, \;\gamma > 0,$$

to see that (17c) is an exponential distribution with both tilt and frailty parameters.

There are no other two-parameter extensions of the exponential distribution obtainable using the parameters discussed in Chapter 7. But of course there are other two-parameter extensions, an example being the density f_Y of 15.E.1. The remainder of this chapter is devoted to three-parameter extensions of the exponential distribution obtainable using the parameters of Chapter 7. Three-parameter families may serve as an umbrella for several two-parameter families, and results for them specialize to the two-parameter subfamilies. From a statistical point of view, estimation with a three-parameter family may lead to the determination that some particular two-parameter subfamily provides an appropriate model.

E. Generalized Gamma (Gamma–Weibull) Distribution

Recall that the Weibull distribution can be obtained from the exponential distribution by a power transformation of the underlying random variable. A similar transformation can be applied to a random variable with a gamma distribution to obtain a three-parameter generalization of both the Weibull and the gamma distribution, sometimes called the "generalized gamma distribution" or the "Stacy distribution." This distribution, with density (2a), was introduced by Amoroso (1925) and studied by Stacy (1962). It is also the subject of a substantial study by Bondesson (1992).

Alternatively, the gamma–Weibull distribution can be obtained by introducing a moment parameter in the Weibull distribution, although this derivation results in a different parameterization, given in (2b).

E. Generalized Gamma (Gamma–Weibull) Distribution

Suppose that the random variable Y has a gamma distribution with unit scale parameter and convolution parameter ν, and let

$$X = Y^{1/\alpha}/\lambda. \tag{1}$$

Then X has density f given by

$$f(x \mid \lambda, \alpha, \nu) = \frac{\lambda \alpha (\lambda x)^{\nu \alpha - 1} \exp\left\{-(\lambda x)^\alpha\right\}}{\Gamma(\nu)}, \quad x > 0, \tag{2a}$$

where $\lambda, \alpha, \nu > 0$. Here, it should be noted that ν started as a convolution parameter, but does not retain that identity after the introduction of a power parameter because convolution parameters and power parameters do not commute (see Section 5.L). Consequently, another parameterization of (2a) is desirable. With $\beta = \alpha(\nu - 1)$, (2a) becomes

$$f(x \mid \lambda, \alpha, \beta) = \frac{\lambda \alpha (\lambda x)^{\alpha + \beta - 1} \exp\left\{-(\lambda x)^\alpha\right\}}{\Gamma((\beta/\alpha) + 1)}, \quad x > 0, \tag{2b}$$

where $\lambda, \alpha > 0$ and $\beta > -\alpha$.

Here, β is a moment parameter. The case that $\beta = 1$ is called a "pseudo-Weibull distribution" by Voda (1989).

Recall that the gamma distribution can be obtained from the exponential distribution by introducing a moment parameter; because moment parameters and power parameters commute, the possibility of parameterizing the generalized gamma distribution with scale, power, and moment parameters is apparent. The progression from the exponential distribution with scale parameter is indicated in the following diagram:

$$
\begin{array}{ccccc}
& \xrightarrow{\text{moment}} & \text{gamma } [\lambda, \beta] & \xrightarrow{\text{power}} & \\
\exp[\lambda] & & & & \text{generalized gamma } [\lambda, \alpha, \beta] \\
& \xrightarrow{\text{power}} & \text{Weibull } [\lambda, \alpha] & \xrightarrow{\text{moment}} & \\
\end{array}
$$

Following Voda (1989), the special case that $\beta = 1$ has been called a "pseudo-Weibull distribution" by Murthy, Xie and Jiang (2004, pp. 23, 122). These authors derive the distribution as though they were

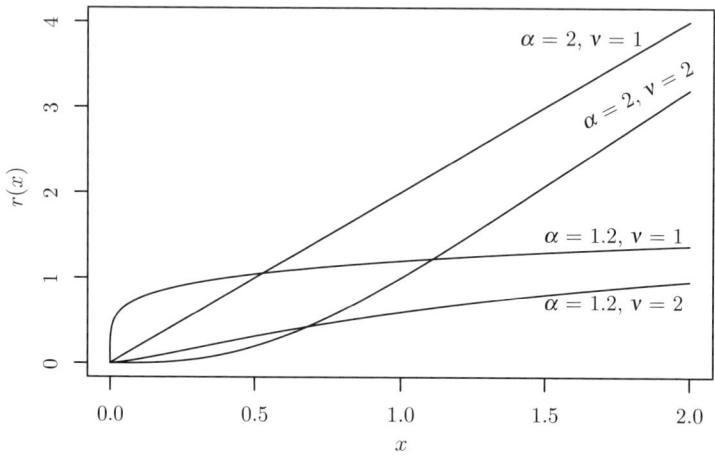

Fig. E.1. Hazard rates of the gamma–Weibull distribution ($\alpha \geq 1$, $\alpha\nu \geq 1$)

introducing a moment parameter β in the Weibull distribution, but they allow only the value $\beta = 1$.

As would be expected, the three-parameter family of densities (2b) offers considerable flexibility for fitting data. Moreover, a variety of hazard rate shapes are possible. Although the hazard rate cannot in general be expressed in closed form, the shape of the hazard rate can be determined using Proposition 4.E.2. Essentially this was done by Glaser (1980) and again by McDonald and Richards (1987), who determined that the hazard rate r of the density (2a) is

(i) increasing if $\alpha \geq 1$ and $\nu\alpha \geq 1$,
(ii) decreasing if $\alpha \leq 1$ and $\nu\alpha \leq 1$,
(iii) bathtub shaped if $\alpha > 1$ and $\nu\alpha < 1$,
(iv) inverted bathtub shaped if $\alpha < 1$ and $\nu\alpha > 1$.

See Figures E.1, E.2, E.3, and E.4. Prior knowledge of the hazard rate shape can be reflected as a limitation on the parameter values of interest. For example, if it is known that the hazard rate is increasing, parameter estimation would be done with the constraints that $\alpha \geq 1$ and $\nu\alpha \geq 1$.

Clearly, f is a gamma density when $\alpha = 1$, and a Weibull density when $\nu = 1$; what is less clear is that the limiting density as $\nu \to \infty$ is a lognormal distribution (see Section 12.B).

E. Generalized Gamma (Gamma–Weibull) Distribution

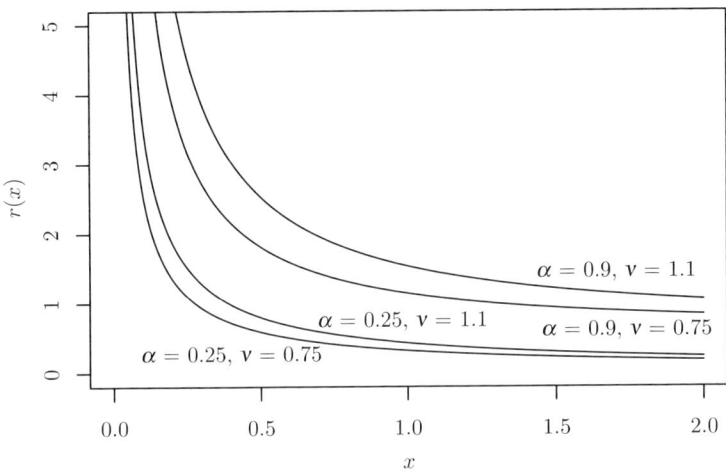

Fig. E.2. Hazard rates of the gamma–Weibull distribution ($\alpha \leq 1$, $\alpha\nu \leq 1$)

By using (1) and A(6), the moments of the generalized gamma distribution can be obtained directly; if X has the density (2a), then

$$EX^r = \frac{\Gamma\left(\dfrac{r}{\alpha} + \nu\right)}{\lambda^r \Gamma(\nu)}, \quad r > -\alpha\nu. \tag{3}$$

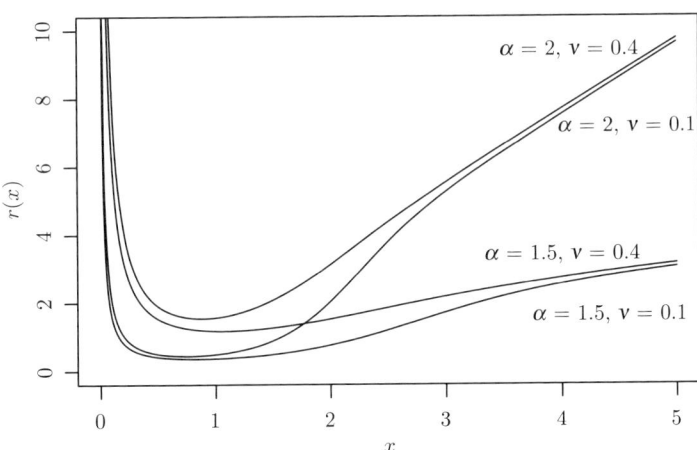

Fig. E.3. Hazard rates of the gamma–Weibull distribution ($\alpha \geq 1$, $\alpha\nu < 1$)

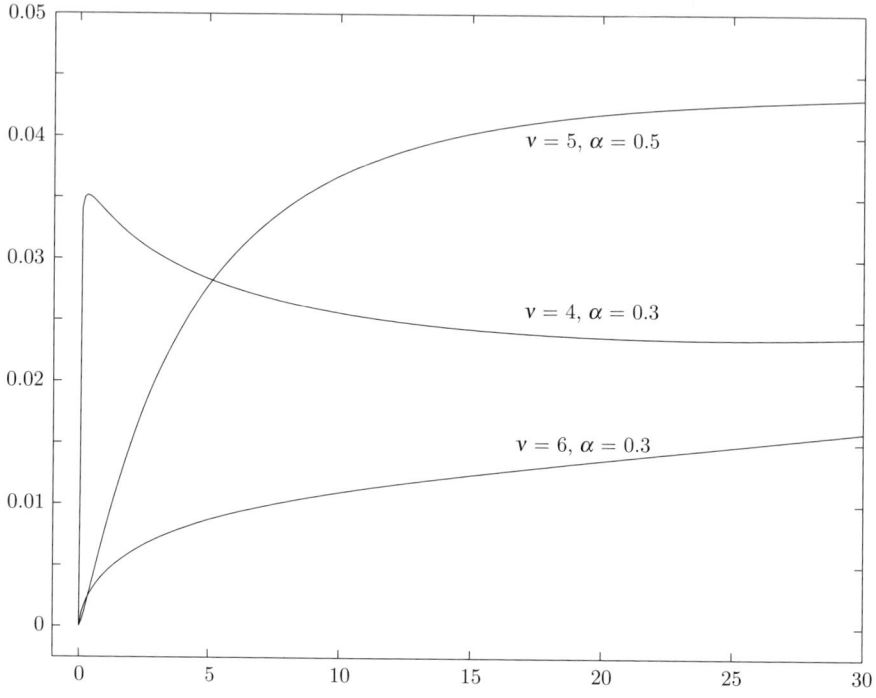

Fig. E.4. Hazard rates of the gamma–Weibull distribution ($\alpha < 1$, $\alpha\nu > 1$)

Thus,
$$EX = \frac{\Gamma\left(\frac{1}{\alpha} + \nu\right)}{\lambda \Gamma(\nu)},$$

and
$$Var(X) = \frac{\Gamma(\frac{2}{\alpha} + \nu)}{\lambda^2 \Gamma(\nu)} - \left[\Gamma\left(\frac{1}{\alpha} + \nu\right) \bigg/ \lambda \Gamma(\nu)\right]^2.$$

a. Extended Gamma–Weibull Distribution

The density B(15) is an extension of the usual Weibull distribution with the shape parameter allowed to be negative. A similar extension of the gamma–Weibull distribution is possible. Such an extension would have a density of the form

$$f(x \mid \lambda, \nu, \alpha) = \frac{\lambda |\alpha| (\lambda x)^{\nu\alpha-1} \exp\{-(\lambda x)^\alpha\}}{\Gamma(\nu)}, \quad x > 0, \quad \lambda, \nu > 0, \quad \alpha \neq 0. \tag{4}$$

b. Infinite Divisibility

The extended gamma–Weibull distribution was conjectured to be infinitely divisible by Bondesson (1978), who proved the result for some special cases. Infinite divisibility for the general case is due to Thorin (1978), who showed that the extended gamma–Weibull distribution is a generalized gamma convolution, discussed in Section I.

F. Weibull Distribution with a Resilience Parameter

A special case of the Weibull distribution (shape parameter 2) with resilience parameter was introduced by Burr (1942) as his Type X distribution. The general Weibull distribution with resilience has been introduced and studied under the name "exponentiated Weibull family" by Mudholkar and Srivastava (1993), Mudholkar, Srivastava and Freimer (1995), and Mudholkar and Hutson (1996). This family has distribution functions of the form

$$F(x \mid \lambda, \alpha, \eta) = [1 - \exp\{-(\lambda x)^\alpha\}]^\eta, \quad \lambda, \alpha, \eta > 0, \quad x \geq 0, \tag{1}$$

and for $\lambda, \alpha, \eta > 0, x \geq 0$, densities given by

$$f(x \mid \lambda, \alpha, \eta) = \lambda \alpha \eta (\lambda x)^{\alpha-1} [1 - \exp\{-(\lambda x)^\alpha\}]^{\eta-1} [\exp\{-(\lambda x)^\alpha\}]. \tag{2}$$

Mudholkar and Hutson (1996) show that this density is decreasing if $\alpha\eta \leq 1$, and is unimodal if $\alpha\eta > 1$. These results can be verified by examining the derivative of $\log f(x \mid \lambda, \alpha, \eta)$.

From (1) and (2) it follows directly that the hazard rate is

$$r(x \mid \lambda, \alpha, \eta) = \frac{\lambda \alpha \eta (\lambda x)^{\alpha-1} [1 - \exp\{-(\lambda x)^\alpha\}]^{\eta-1} [\exp\{-(\lambda x)^\alpha\}]}{1 - [1 - \exp\{-(\lambda x)^\alpha\}]^\eta},$$

$$\lambda, \alpha, \eta > 0, \quad x \geq 0. \tag{3}$$

This hazard rate is not particularly simple primarily because there is no cancellation when the ratio of density to survival function is formed.

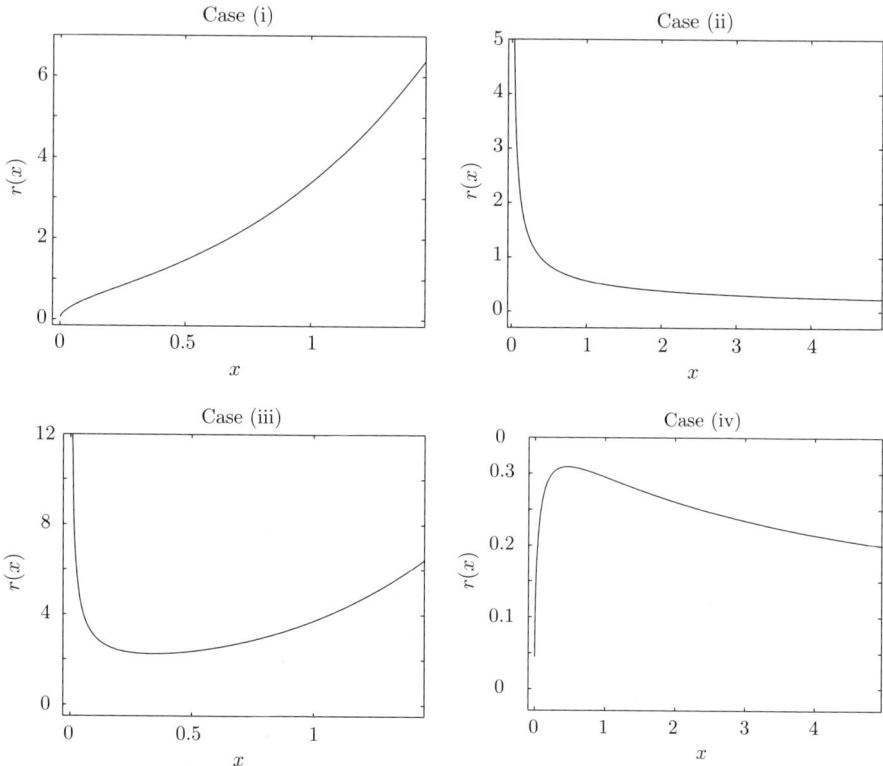

Fig. F.1. Hazard rates of the Weibull distribution with resilience (i) ($\alpha = 3$, $\eta = 0.5$); (ii) ($\alpha = 0.5$, $\eta = 0.5$); (iii) ($\alpha = 3$, $\eta = 0.1$); (iv) ($\alpha = 0.5$, $\eta = 3$)

This is the price paid for the introduction of a resilience parameter, and the reward is that the hazard rate takes on a variety of forms. In particular, Mudholkar and Hutson (1996) show that the hazard rate is

(i) increasing for $\alpha \geq 1$ and $\alpha\eta \geq 1$,
(ii) decreasing for $\alpha \leq 1$ and $\alpha\eta \leq 1$,
(iii) bathtub shaped for $\alpha > 1$ and $\alpha\eta < 1$,
(iv) Inverted bathtub shaped (unimodal) for $\alpha < 1$ and $\alpha\eta > 1$.

See Figure F.1.

When η is an integer, (1) can be regarded as a product of η identical Weibull distributions. The case that the component Weibull distributions are not identical has been studied by Jiang, Murthy and Ji (2001).

G. Residual Life of the Weibull Distribution

If F is a Weibull distribution, the residual life distribution has survival function, density, and hazard rate given by

$$\bar{F}_t(x) = \exp\{-[\lambda(x+t)]^\alpha + [\lambda t]^\alpha\}, \tag{1}$$

$$f_t(x) = \bar{F}_t(x)\lambda[\alpha\lambda(x+t)]^{\alpha-1}, \tag{2}$$

$$r_t(x) = \lambda[\alpha\lambda(x+t)]^{\alpha-1}. \tag{3}$$

This distribution has not found a place in the literature, and its importance is unclear. But here, there is new flexibility in choosing the hazard rate at 0, so in some circumstances, the distribution may be a competitor to other three-parameter extensions of the Weibull distribution. These residual life distributions have monotone hazard rates, and of course, these hazard rates are just those of the Weibull distribution but with the origin moved to t.

If the random variable X has the survival function of (1), then $Y = \log[1 + (X/t)]$ has a Gompertz survival function

$$\bar{G}(y) = \exp\{-\xi(e^{\alpha y} - 1)\}, y \geq 0,$$

where $\xi = (\lambda t)^\alpha$. The Gompertz distribution is discussed in Chapter 10.

G.1. Proposition. In the convex transform order, the residual life distribution (1) is decreasing in t for $\alpha \leq 1$, and increasing in t for $\alpha \geq 1$, the parameters α and λ being fixed.

Proof. Using the fact that

$$\bar{F}_t^{-1}(z) = \{[(\lambda t)^\alpha - \log z]^{1/\alpha}/\lambda\} - t,$$

the convexity required can be verified by differentiation. □

H. Weibull Distribution with a Tilt Parameter

When F is a Weibull distribution function, then 7.F(2) yields the geometric-extreme stable extension of the Weibull distribution with

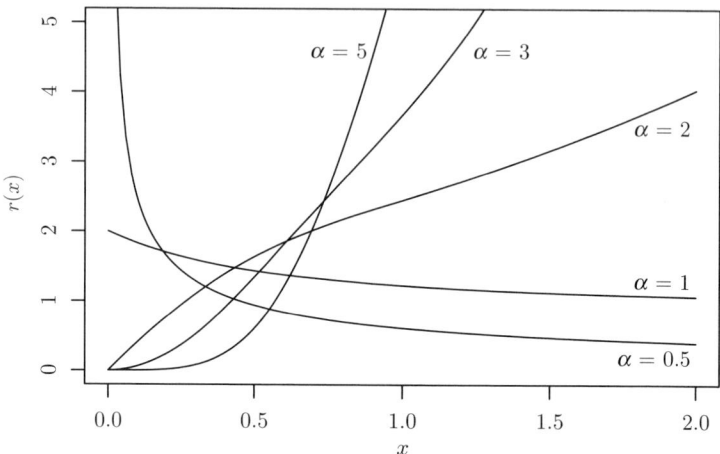

Fig. H.1. Hazard rates of the Weibull distribution with tilt parameter ($\lambda = 1$, $\gamma = 0.5$)

survival function

$$\bar{F}(x \mid \lambda, \gamma, \alpha) = \frac{\gamma \exp\left[-(\lambda x)^\alpha\right]}{1 - \bar{\gamma} \exp\left[-(\lambda x)^\alpha\right]}, \quad x \geq 0, \ \lambda, \alpha, \gamma > 0. \quad (1)$$

The density and hazard rate of the distribution given by (1) can be obtained directly from 7.F(3a) and 7.F(3b). In particular, the hazard rate is

$$r(x \mid \lambda, \alpha, \gamma) = \frac{\lambda \alpha (\lambda x)^{\alpha-1}}{1 - \bar{\gamma} \exp\left[-(\lambda x)^\alpha\right]}, \quad x \geq 0, \ \lambda, \alpha, \gamma > 0. \quad (2)$$

This function is graphed in Figures H.1, H.2, and H.3.

It can be verified using elementary methods that this hazard rate is increasing if $\gamma \geq 1, \alpha \geq 1$, and decreasing if $\gamma \leq 1, \alpha \leq 1$. If $\alpha > 1$ and $\gamma < 1$, then the hazard rate is initially increasing and eventually increasing, but there may be one interval where it is decreasing. Similarly, if $\alpha < 1$ and $\gamma > 1$, then the hazard rate is initially decreasing and eventually decreasing, but there may be one interval where it is increasing.

If Y has an exponential distribution with parameter 1, then $X = Y^{1/\alpha}/\lambda$ has the survival function B(1) of the Weibull distribution. Similarly, if Y has the survival function D(1) with $\lambda = 1$, then $X = Y^{1/\alpha}/\lambda$ has the survival function (1). Consequently, moments of (1)

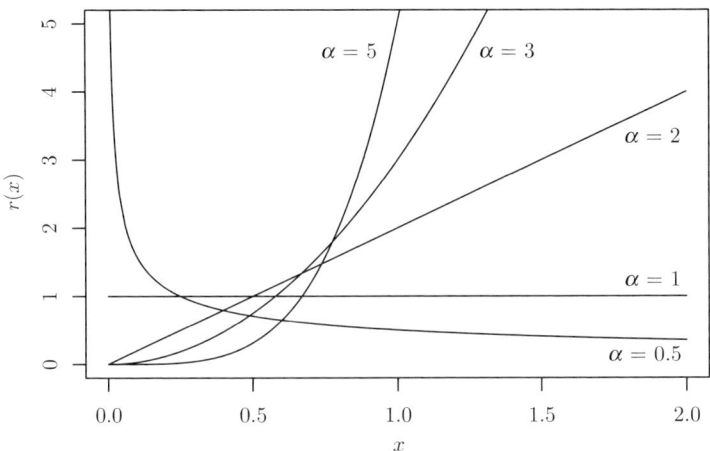

Fig. H.2. Hazard rates of the Weibull distribution with tilt parameter ($\lambda = 1$, $\gamma = 1$)

can be obtained from (noninteger) moments of D(1) given by D(10) using $EX^r = [EY^{r/\alpha}]/\lambda^r$. Thus, if X has the survival function (1), then

$$EX^r = \frac{r\gamma}{\alpha\lambda^r} \sum_{j=0}^{\infty} \frac{\bar\gamma^j}{(j+1)^{r/\alpha}} \Gamma\left(\frac{r}{\alpha}\right), \quad |\bar\gamma| \leq 1. \tag{3}$$

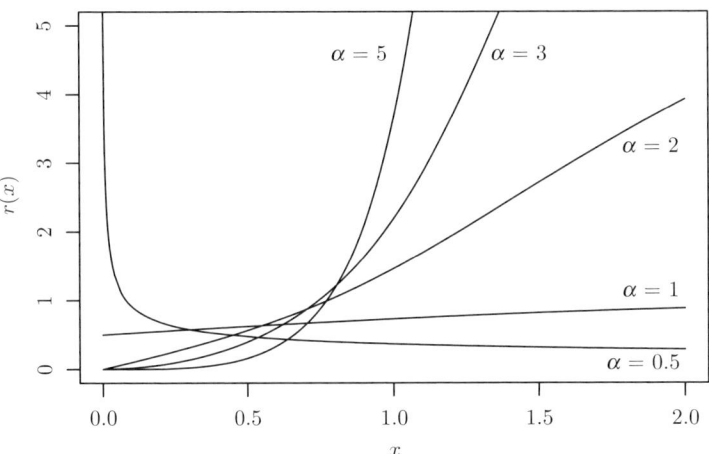

Fig. H.3. Hazard rates of the Weibull distribution with tilt parameter ($\lambda = 1$, $\gamma = 2$)

Table H.1. Expected values of X when X has a Weibull distribution with tilt parameter γ and $\lambda = 1$

α \ γ	0.2	0.5	1.0	1.5	2.0	10	20
0.2	24.68	61.01	120.0	177.4	233.3	1014	1828
0.3	2.046	4.897	9.261	13.28	17.03	61.14	99.29
0.4	0.807	1.844	3.323	4.609	5.764	17.54	26.38
0.5	0.537	1.164	2.000	2.690	3.290	8.779	12.48
0.6	0.446	0.915	1.505	1.971	2.365	5.681	7.739
0.7	0.410	0.801	1.266	1.621	1.913	4.222	5.563
0.8	0.398	0.742	1.133	1.422	1.657	3.408	4.373
0.9	0.397	0.710	1.052	1.299	1.495	2.900	3.641
1	0.402	0.693	1.000	1.216	1.386	2.558	3.153
2	0.518	0.714	0.886	0.994	1.072	1.522	1.710
5	0.725	0.835	0.918	0.966	0.999	1.165	1.225
10	0.842	0.906	0.951	0.977	0.994	1.076	1.105
100	0.982	0.989	0.994	0.997	0.999	1.007	1.010

However, moments cannot be given in closed form; thus, even the first moment of (1) must be obtained numerically. For $\lambda = 1$, Table H.1 shows values of the first moment for various combinations of α and γ.

Note that for fixed α, the expected value is increasing in γ, this is a consequence of Proposition 7.F.8 and the fact that a stochastic ordering implies the ordering of expectations.

By writing

$$EX^s = \int_0^\infty sx^{s-1}\bar{F}(x\,|\,\lambda,\gamma,\alpha)\,dx, \quad s > 0,$$

it can be shown with the change of variables $y = (\lambda x)^\alpha$ in (1) and an integration by parts that

$$\lim_{\alpha \to \infty} EX^s = \lambda^{-s}, \quad s > 0.$$

These are the moments of a random variable degenerate at $1/\lambda$. It can also be verified that the limit of the survival function (1) as $\alpha \to \infty$ is degenerate.

I. Generalized Gamma Convolutions

Thorin (1977a) gives a representation for the Laplace transform of distributions that are limits of convolutions of gamma distributions. This large class of infinitely divisible distributions includes the extended gamma–Weibull distributions and a number of other distributions encountered in later chapters.

The study of generalized gamma convolutions involves complex variable theory and is beyond the scope of this book. What follows is but a brief introduction; for a more complete discussion of generalized gamma convolutions, see Bondesson (1992).

I.1. Definition (Thorin, 1977a). A distribution F is said to be a *generalized gamma convolution* if its Laplace transform ϕ has the form

$$\phi(s) = \exp\{-\int_0^\infty \log\left[1 + \frac{s}{z}\right] dU(z)\}, \tag{1}$$

where U is a nondecreasing function such that $U(0) = 0$,

$$\int_0^1 |\log z| \, dU(z) < \infty \quad \text{and} \quad \int_1^\infty \frac{1}{z} \, dU(z) < \infty.$$

Thorin (1977a) shows that these distributions comprise all limits of convolutions of gamma distributions. The proof is obtained by first assuming that U is a step function, in which case (1) becomes a product of Laplace transforms of the form A(6).

I.1.a. Remark. Thorin (1977a) gives the name "generalized gamma convolution" to a somewhat larger class of distributions that includes translations of those defined above. This extension of the definition is not adopted here. Thorin (1977b) shows that the representation (1) is unique, and that the class of distributions is closed under weak limits.

I.2. Proposition (Thorin, 1977a,b). *Generalized gamma convolutions can be written as mixtures of exponential distributions, and they are infinitely divisible.*

Because of Proposition 20.D.10, the infinite divisibility follows from the mixture representation. Proposition I.2 has been used by Thorin and Bondesson to prove infinite divisibility of a number of distributions, including the extended gamma–Weibull distribution as indicated in Section E.

I.3. Proposition (Bondesson, 1992, p. 53). If the density f is a mixture

$$f(x) = \int \lambda\, e^{-\lambda x} g(\lambda)\, d\lambda, \quad x > 0,$$

of exponential densities where $zg(z)$ is log concave, then f is a generalized gamma convolution.

I.4. Example (Bondesson, 1979). All densities f of the form

$$f(x) = c x^{\beta-1} \prod_{j=1}^{M} \frac{1}{[1 + c_{j1} x^{a_{j1}} + \cdots + c_{jN_j} x^{a_{jN_j}}]^{\gamma_j}}, \quad x > 0,$$

or

$$f(x) = c x^{\beta-1} \exp\{-c_1 x^{a_1} - c_2 x^{a_2} - \cdots - c_N x^{a_N}\}, \quad x > 0,$$

are generalized gamma convolutions.

J. Summary of Distributions and Hazard Rates

Gamma

$$[r(x\mid \lambda,\nu)]^{-1} = \int_x^\infty \left(\frac{z}{x}\right)^{\nu-1} e^{-\lambda(z-x)}\, dz = \int_0^\infty [1 + \frac{u}{x}]^{\nu-1} e^{-\lambda u}\, du,$$

$x, \lambda, \nu > 0$.

Weibull

$$r(x\mid \lambda,\alpha) = \alpha\lambda(\lambda x)^{\alpha-1}, \quad x,\lambda,\alpha > 0.$$

Generalized Weibull

$$r(x\mid \lambda,\alpha) = \frac{|\alpha|\lambda(\lambda x)^{\alpha-1}\, \exp\{-(\lambda x)^{-\alpha}\}}{1 - \exp\{-(\lambda x)^\alpha\}}, \quad x,\lambda > 0,\ \alpha \neq 0.$$

Exponential with resilience

$$r(x\mid \lambda,\eta) = \frac{\lambda\eta\, e^{-\lambda x}(1 - e^{-\lambda x})^{\eta-1}}{1 - (1 - e^{-\lambda x})^\eta}, \quad x,\lambda,\eta > 0.$$

Exponential with tilt

$$r(x\mid\lambda,\gamma) = \frac{\lambda}{1-\bar{\gamma}\ e^{-\lambda x}} = \frac{\lambda e^{\lambda x}}{e^{\lambda x}-\bar{\gamma}}, \quad x,\lambda,\gamma>0.$$

Weibull with resilience

$$r(x\mid\lambda,\alpha,\eta) = \frac{\lambda\alpha\eta(\lambda x)^{\alpha-1}[1-\exp\{-(\lambda x)^{\alpha}\}]^{\eta-1}[\exp\{-(\lambda x)^{\alpha}\}]}{1-[1-\exp\{-(\lambda x)^{\alpha}\}]^{\eta}},$$
$$x,\lambda,\alpha,\eta>0.$$

Weibull with tilt

$$r(x\mid\lambda,\alpha,\gamma) = \frac{\lambda\alpha(\lambda x)^{\alpha-1}}{1-\bar{\gamma}\ \exp[-(\lambda x)^{\alpha}]}, \quad x,\lambda,\alpha,\gamma>0.$$

10

Gompertz and Gompertz–Makeham Distributions

Methods of smoothing mortality tables have long been of interest to actuaries. Early methods were primarily nonparametric, although de Moivre (1724) found a uniform distribution to be useful over short time periods and other parametric methods were occasionally suggested. The first parametric family to gain wide attention was that of Gompertz (1825). This family was extended with the addition of a parameter by Makeham (1860), and subsequently by various other authors. An historical review of considerable interest has been given by Ogborn (1953).

For a time, the work of Gompertz was not well recognized. This was noted by Gray (1858), who wrote as follows:

> "Although long before the public, it is far from being so well known as it deserves to be. This may have arisen from various causes. The work containing it is not very accessible; the form in which the investigation of it is given is rendered forbidding by the employment of the obsolete fluxional (ed. derivative) notation, and a degree of brevity which renders it difficult to be followed; while the whole of the paper containing it is so disfigured by typographical errors as to be in many places almost unintelligible."

The Gompertz distribution was discussed by Edmonds (1832), who claimed credit for its discovery. A rather spirited exchange concerning priority appeared in the *Assurance Magazine and Journal of the Institute of Actuaries*. This exchange involved De Morgan (1861a,b), Edmonds (1861a,b), Gompertz (1861), and Sprague (1861); Edmonds had no supporters, whereas both De Morgan and Sprague strongly defended the priority of Gompertz.

Outside the actuarial community, the Gompertz distribution continues to receive minimal attention; Johnson, Kotz and Balakrishnan (1994, pp. 25, 640) note that the Gompertz distribution is a truncated extreme value distribution.

The applicability of the Gompertz distribution remains a topic of interest. Particularly in the right-hand tail of a distribution, available data becomes thin and consequently any proposed model would be difficult to evaluate. For a recent discussion of this issue as it relates to the Gompertz distribution, see Wang, Müller and Capra (1998).

Gompertz distributions can be viewed as extensions of the exponential distributions because exponential distributions are limits of sequences of Gompertz distributions. Of course, the gamma and Weibull distributions discussed in Chapter 9 are more readily recognized as extensions of the exponential distribution because the exponential distribution is obtained from them by choice of parameter.

A. The Gompertz Distribution

a. The Derivation of Gompertz

Gompertz (1825) investigated the consequences of supposing that "the average exhaustion of a man's power to avoid death be such that at the end of infinitely small intervals of time he lost equal proportions of his remaining power to oppose destruction which he had at the commencement of these intervals." It is apparent that Gompertz regarded the hazard rate $r(t)$ as a representation of vulnerability to death at time t. Consequently, the supposition of Gompertz can be formalized by the differential equation

$$\frac{dr(t)}{dt} = \lambda r(t), \quad t > 0. \tag{1}$$

Gompertz assumed that $\lambda > 0$ and this led him to the conclusion that the "intensity of mortality" (i.e., the hazard rate) increases in geometrical progression, or in other words,

$$r(x \mid \lambda, \xi) = \xi \lambda e^{\lambda x}, \quad \xi, \lambda > 0, \ x \geq 0. \tag{2}$$

The case that $\lambda = 0$ in (1) leads to the exponential distribution, and the case that $\lambda < 0$ and $\xi < 0$ in (2) leads to the negative Gompertz distribution discussed in Section A.c.

A. The Gompertz Distribution

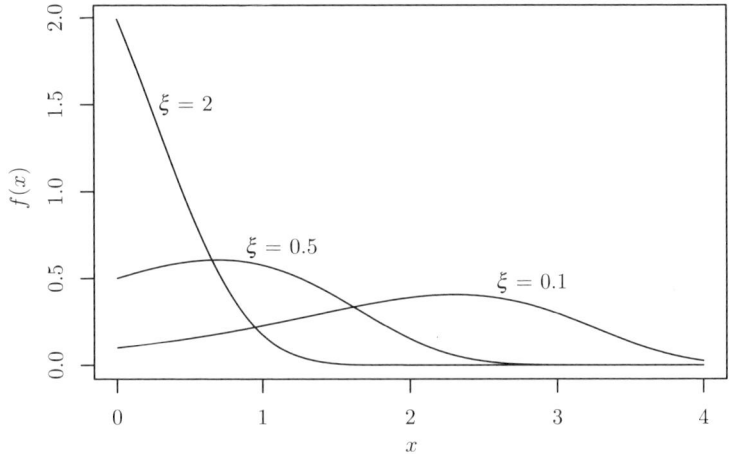

Fig. A.1. Densities of the Gompertz distribution ($\lambda = 1$)

With the aid of the fundamental formula 1.B(3) the hazard rate (2) yields the survival function

$$\bar{F}(x \mid \lambda, \xi) = \exp\left\{-\int_0^x r(t \mid \lambda, \xi)\, dt\right\} = \exp\{-\xi(e^{\lambda x} - 1)\},$$
$$x \geq 0, \quad \xi, \lambda > 0 \tag{3}$$

of the *Gompertz distribution* (Gompertz, 1825), and the corresponding density is

$$f(x \mid \lambda, \xi)] = \lambda \xi \exp\{\lambda x - \xi(e^{\lambda x} - 1)\}, \quad x, \lambda, \xi > 0. \tag{4}$$

See Figures A.1, A.2 and A.3.

Having derived (3) from a physical rationale, Gompertz proceeded to check it against available data. In this, the degree of success might not satisfy a present day statistician because Gompertz found that to obtain a good fit, it was necessary to divide mortality tables into three age groups and to use different values of the parameters in each group. This might be regarded as an admission of model inadequacy, but it could also be regarded as anticipating the development of covariate models by 150 years (see Chapter 16).

The applicability of the Gompertz distribution has continued to be questioned, and was the motivation for Makeham's investigations

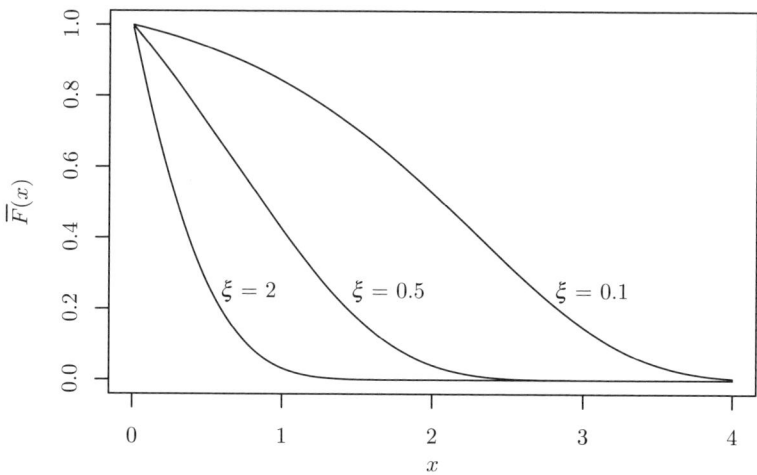

Fig. A.2. Survival functions of the Gompertz distribution ($\lambda = 1$)

described in Section B. For more recent investigations of this issue, see Wang, Müller and Carpa (1998) and the references contained therein.

The hazard rate (2) of the Gompertz distribution is convex and more strongly increasing than any other example considered in this book. Clearly, the Gompertz distribution has a scale parameter λ and a frailty parameter ξ.

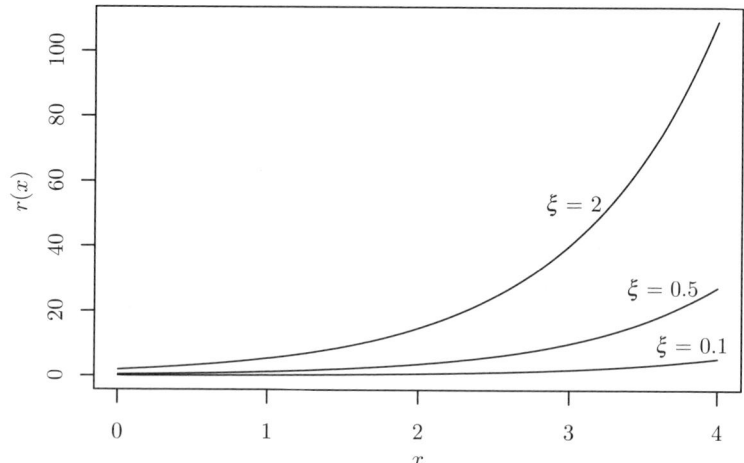

Fig. A.3. Hazard rates of the Gompertz distribution ($\lambda = 1$)

With $\theta = \xi\lambda$, (2) can be replaced by

$$r(x\,|\,\lambda,\theta) = \theta\,e^{\lambda x},$$

and in this form, it can be seen that the exponential distribution is the special case $\lambda = 0$. Alternatively the exponential distribution can be obtained directly from (2) by letting $\lambda \to 0, \xi \to \infty$ while $\theta = \xi\lambda$ is fixed. As noted above, the Gompertz distribution can be regarded, along with the distributions of Chapter 9, as an extension of the exponential distribution.

b. Moments

Because the hazard rate of the Gompertz distribution is increasing, it follows from Proposition 4.C.3 that the positive moments of the Gompertz distribution are finite. Unfortunately, neither these finite moments nor the Laplace transform can be given in closed form. The expected value, variance, and coefficient of variation are graphed, respectively, in Figures A.4, A.5, and A.6 with the scale parameter $\lambda = 1$.

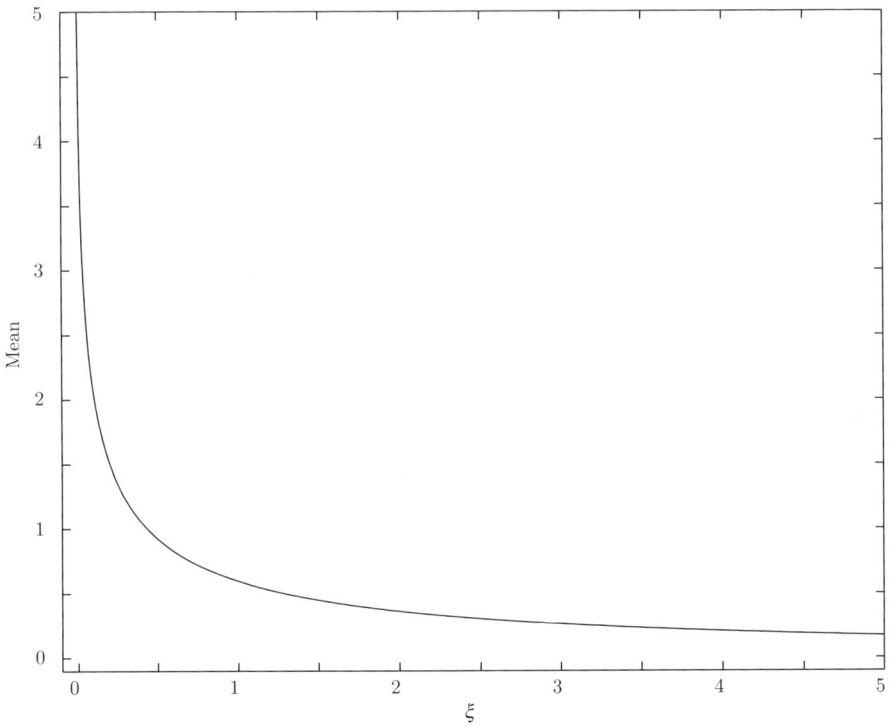

Fig. A.4. Mean of the Gompertz distribution ($\lambda = 1$)

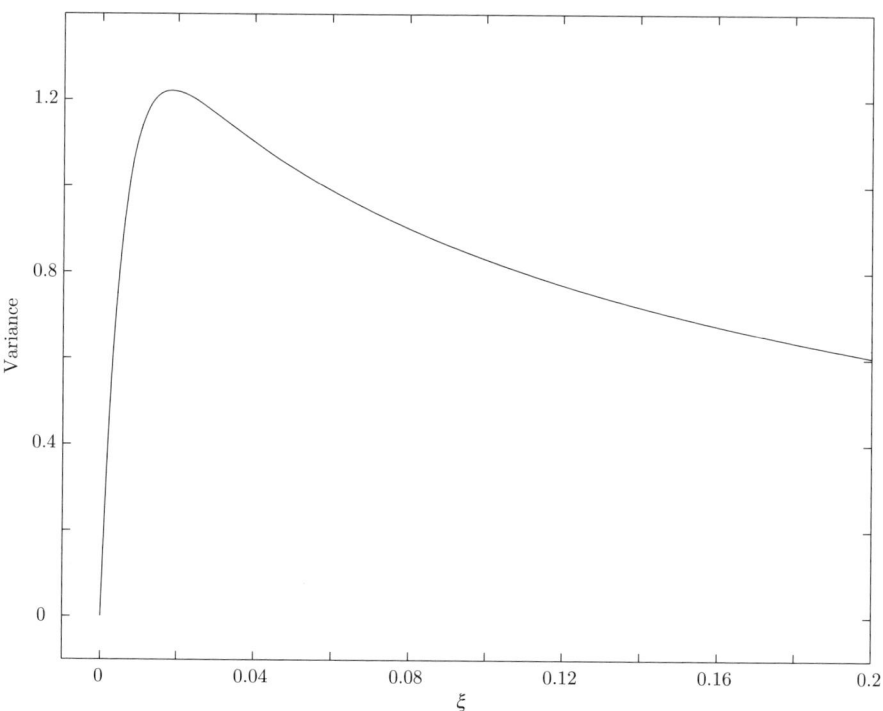

Fig. A.5. Variance of the Gompertz distribution ($\lambda = 1$)

c. The Negative Gompertz Distribution: $\lambda < 0$

If ξ and λ are replaced, respectively, by $-\xi$ and $-\lambda$ in (3), the resulting hazard rate is given by

$$r(x \mid \lambda, \xi) = \xi \lambda e^{-\lambda x}, \quad x, \lambda, \xi > 0; \qquad (5)$$

the distribution with this hazard rate is termed the *negative Gompertz distribution*. The integral over $[0, \infty)$ of this decreasing function is finite, and consequently, the corresponding survival function

$$\begin{aligned}
\bar{F}(x \mid \lambda, \xi) &= \exp\left\{-\int_0^x r(t \mid \lambda, \xi)\, dt\right\} \\
&= \exp\{\xi(e^{-\lambda x} - 1)\}, \quad x, \lambda, \xi > 0,
\end{aligned} \qquad (6)$$

is not proper. If $\bar{F}$ is given by (6), then $\bar{F}(0/\lambda, \xi) = 1$, and $\bar{F}$ is

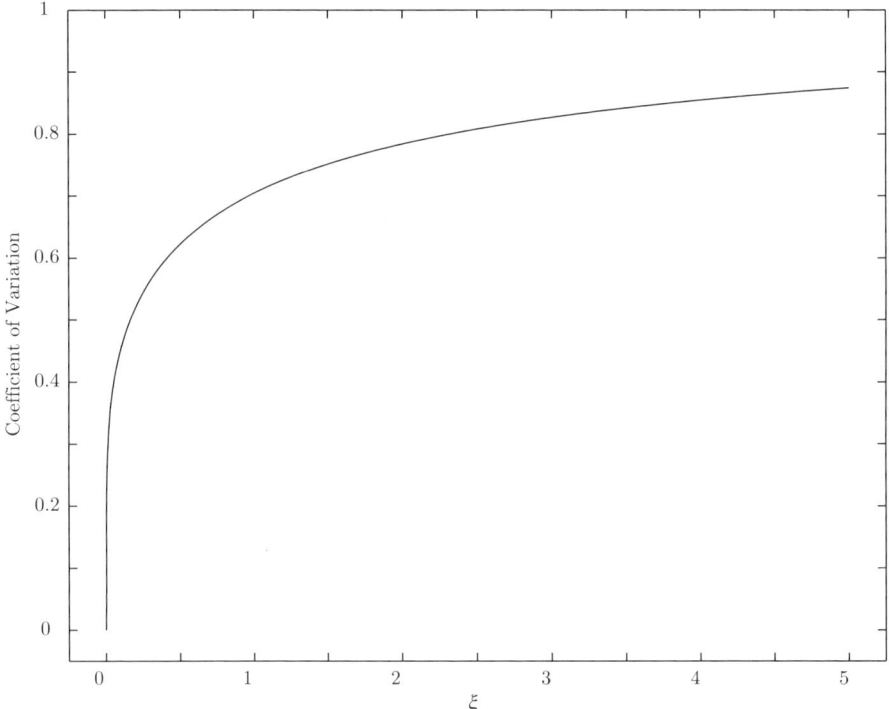

Fig. A.6. Coefficient of variation of the Gompertz distribution ($\lambda = 1$)

decreasing, but

$$\lim_{x \to \infty} \bar{F}(x \mid \lambda, \xi) = e^{-\xi} > 0.$$

This improper distribution has been used by Lomax (1954) to fit certain data on business failures. In this application, the mass that F places at ∞ is the probability that the business does not fail. Additional applications of the negative Gompertz are given below.

Note that the survival function (6) is also a Laplace trasform; when $\lambda = 1$, it is the Laplace transform of a Poisson distribution with parameter ξ (see Section 20.E.c).

d. Functional Equations

In the mid-nineteenth century, a number of years after Gompertz proposed his distribution, some derivations via functional equations were found. In particular, De Morgan (1860) and Woolhouse (1863) gave the derivations presented here.

Suppose that two items have identical distributions F at age zero. Suppose further that the items are unfailed at respective ages x and y, and are then placed in series. Under what conditions does this new series system (of aged components) have the same life distribution as a single component of some age $z = z(x, y)$ depending upon x and y? The requirement of this rather curious question can be written in terms of the functional equation

$$\frac{\bar{F}(x+t)}{\bar{F}(x)} \frac{\bar{F}(y+t)}{\bar{F}(y)} = \frac{\bar{F}(z(x,y)+t)}{\bar{F}(z(x,y))} \quad \text{for all } x, y, t \geq 0. \tag{7}$$

If the common distribution of the items is a Gompertz distribution with scale parameter λ and if $z(x, y) = e^{\lambda x} + e^{\lambda y}$, then it can be verified that (7) is satisfied. With some additional qualitative conditions on F, there are no other solutions to (7). This is the content of the following proposition.

A.1. Proposition (De Morgan, 1860; Aczél, 1999). Suppose that F is an absolutely continuous distribution with a strictly increasing hazard rate such that $F(0) = 0$. If the survival function $\bar{F}$ satisfies (7) where the function $z = z(x, y)$ does not depend upon t, then F is a Gompertz distribution.

Proof. Take logarithms in (7) and differentiate with respect to t to obtain

$$r(x+t) + r(y+t) = r(z(x,y)+t). \tag{8}$$

Now, set $t = 0$ in (8) to obtain

$$z(x, y) = r^{-1}(r(x) + r(y)). \tag{9}$$

With $r(x) = u, r(y) = v$, substitution of z from (9) into (8) yields

$$r(r^{-1}(u) + t) + r(r^{-1}(v) + t) = r(r^{-1}(u+v) + t).$$

This equation has the form

$$\phi_t(u) + \phi_t(v) = \phi_t(u+v), \quad u, v \geq 0, \tag{10}$$

of Cauchy's functional equation 22.A(1) where $\phi_t(u) = r(r^{-1}(u) + t)$. It follows from Proposition 22.A.1 that for some function c of t, $\phi_t(u) = r(r^{-1}(u) + t) = uc(t), u, t \geq 0$. Set $r^{-1}(u) = w$ and rewrite the equation

$r(r^{-1}(u) + t) = uc(t)$ as

$$r(w+t) = r(w)c(t), \quad w, t \geq 0. \tag{11}$$

Equation (11) is the Pexider equation 22.B(2b), and it follows that r has the form (2) for some $\xi, \lambda \geq 0$. □

A.1.a. Remark. If in Proposition A.1 the hypothesis that r is strictly increasing is replaced by the hypothesis that r is strictly monotone (increasing or decreasing), then the negative Gompertz distribution is also a solution of (7).

Comment. Proposition A.1 was stated by De Morgan (1860) without the hypothesis that r is monotone. De Morgan writes that "I do not think it right to occupy space by a very full development of the demonstration: the following will be enough for anyone who has an ordinary acquaintance with functional algebra and the differential calculus." The details of De Morgan's proof are elusive; the proof given above and missing hypothesis were supplied by J. Aczél (1999).

The following proposition suggests that there is a close relationship between the Gompertz and the exponential distributions. This relationship is further illuminated in Proposition A.3.

A.2. Proposition (Woolhouse, 1863). Suppose that F is a proper absolutely continuous distribution such that $F(0) = 0$. If the survival function $\bar{F}$ satisfies the equation

$$\frac{\bar{F}(x+t)}{\bar{F}(x)} = \left[\frac{\bar{F}(x+1)}{\bar{F}(x)}\right]^{\phi(t)} \quad \text{for all } x, t > 0, \tag{12}$$

where the function $\phi = \phi(t)$ is positive and does not depend upon x, then either F is an exponential distribution, a Gompertz distribution, or a negative Gompertz distribution.

Proof. Take logarithms in (12) to obtain

$$R(x+t) - R(x) = \phi(t)[R(x+1) - R(x)]. \tag{13}$$

Next, differentiate (13) with respect to x and set $x = 0$ to obtain

$$r(t) - r(0) = \phi(t)[r(1) - r(0)]. \tag{14}$$

If $r(1) - r(0) \neq 0$, then (14) yields

$$\phi(t) = \frac{r(t) - r(0)}{r(1) - r(0)}.$$

Substitute this in (13) with $x = 0$, and use $R(0) = 0$ to obtain

$$R(t) = \phi(t)R(1) = \frac{r(t) - r(0)}{r(1) - r(0)} R(1). \tag{15}$$

Because the left side of (15) is differentiable with respect to t, so is the right side; this differentiation yields

$$[r'(t)/r(t)] = \lambda \tag{16}$$

for some constant λ; this is the differential equation (1) of Gompertz, and consequently, either F is an exponential distribution, a Gompertz distribution, or a negative Gompertz distribution.

It remains to consider the case that $r(1) - r(0) = 0$. In this case, it follows from (14) that $r(t) - r(0) = 0$, for all $t > 0$, and consequently F is an exponential distribution. □

In equation (13), the function ϕ is given by $\phi(t) = t$ if F is an exponential distribution, and by $\phi(t) = (e^{\lambda t} - 1)/(e^\lambda - 1)$ if F is a Gompertz distribution.

The following functional equation (17) requires that the residual life distribution at age t be the same as the underlying distribution raised to a power depending only on t. A more restrictive requirement is imposed by the lack of memory property of the exponential distribution displayed in 8.B(2), namely that the power depending upon t be identically one. Here, more solutions can be expected.

A.3. Proposition (Kaminsky, 1983). Suppose that F is an absloutely continuous distribution function such that $F(0) = 0$. The survival function $\bar{F}$ satisfies the equation

$$\frac{\bar{F}(x+t)}{\bar{F}(t)} = [\bar{F}(x)]^{\phi(t)}, \quad x, t > 0, \tag{17}$$

for some function ϕ that does not depend on x if and only if either

(i) F is an exponential distribution and $\phi(t) = 1, t > 0$,
(ii) F is a Gompertz distribution and $\phi(t) = e^{\lambda t}$ for some $\lambda, t > 0$, or
(iii) F is a negative Gompertz distribution and $\phi(t) = e^{-\lambda t}$ for some $\lambda, t > 0$.

Proof. By taking logarithms in (17) and by interchanging x and t, it follows that

$$R(x+t) = R(t) + \phi(t)R(x) = R(x) + \phi(x)R(t),$$

and consequently

$$R(x+t) = R(x)[1 - \phi(t)], \quad t, x \geq 0. \tag{18}$$

It follows from Proposition 22.B.2.b that

$$r(x) = r(0) e^{\lambda x} \quad \text{and} \quad \phi(t) = r(t)/r(0) \tag{19}$$

for some λ. If $\lambda = 0$, F is an exponential distribution; if $\lambda > 0$, F is a Gompertz distribution, and if $\lambda < 0$, F is a negative Gompertz distribution. □

Note $\bar{F}_t(x) \leq \bar{F}(x)$ if $\phi(t) > 1$, and $\bar{F}_t(x) \geq \bar{F}(x)$ if $\phi(t) < 1$.

e. A Derivation of the Gompertz Distribution Based on an Odds Ratio

For any distribution function K concentrated on $(0, \infty)$, the odds ratio $Ø^-(x) = K(x)/\bar{K}(x)$ has all the properties of a hazard function. Thus,

$$\bar{F}(x \mid \xi) = e^{-\xi K(x)/\bar{K}(x)}, \quad x \geq 0, \tag{20}$$

is a survival function with frailty parameter ξ. If $\bar{K}(x) = e^{-\lambda x}, x \geq 0$, is an exponential survival function, then (20) yields the Gompertz survival function (3). The fact that $Ø^-(x)$ can be regarded as a hazard function is used in 1.B(21) and in Example 7.M.4.a.

f. A Derivation of the Gompertz Distribution Based on an Exponential Distribution with Tilt and Frailty Parameters

As noted in Section a, if in the Gompertz distribution, $\xi\lambda$ is set equal to θ, then the limit as $\lambda \to 0$ is an exponential distribution with parameter θ. This fact is easily seen through the hazard rate $r(x) = \xi\lambda e^{\lambda x}$. On the other hand, the Gompertz distribution can be obtained from limits based upon the exponential distribution.

Start with an exponential distribution; add first a tilt parameter γ, and then add a frailty parameter θ to obtain from 9.D(1) the survival function

$$\bar{F}(x\mid\lambda,\gamma,\theta)=\left(\frac{\gamma e^{-\lambda x}}{1-\bar{\gamma}e^{-\lambda x}}\right)^{\theta}=\left(\frac{\gamma}{e^{\lambda x}-1+\gamma}\right)^{\theta}=\left(\frac{e^{\lambda x}-1}{\gamma}+1\right)^{-\theta}.$$

Note: This survival function also appears as 9.D(17c), but with a different parameterization. Now, set $\theta=\gamma\xi$, fix ξ, and compute

$$\lim_{\gamma\to\infty}\left(\frac{e^{\lambda x}-1}{\gamma}+1\right)^{-\gamma\xi}=\exp\left\{-\xi(e^{\lambda x}-1)\right\},\quad x\geq 0.$$

This shows that the Gompertz distribution is a limit of exponential distributions with tilt and frailty parameters. Because an exponential distribution with tilt parameter is a truncated logistic distribution (7.D.b), this also exhibits the Gompertz distribution as a limit of truncated logistic distributions with frailty parameter.

g. Ordering Gompertz Distributions

Gompertz distributions are ordered in several ways.

Hazard rate order: It is clear from (2) that in the hazard rate order, the Gompertz distribution is decreasing in λ and ξ.

Likelihood ratio order: A tedious but straightforward calculation, using 2.A(11) with logarithms, shows that the Gompertz distribution is decreasing in λ when $\xi \leq 1$ is fixed.

Convex transform order: The Gompertz survival function has the inverse

$$\bar{F}^{-1}(z)=\frac{1}{\lambda}\log\left(1-\frac{\log z}{\xi}\right),\quad 0<z<1,$$

and the simplicity of this form makes it possible to verify from the definition that the Gompertz family is, in the convex transform order, increasing in ξ, with λ being fixed.

h. A Mixture of Gompertz Distributions

In Section 7.F.c, mixtures of the form $\bar{H}(x\mid\theta)=\int \bar{F}(x\mid\xi)\,dG(\xi\mid\theta)$ are considered, where ξ is a frailty parameter for $\bar{F}(\cdot\mid\xi)$ and G is a gamma distribution with scale and convolution parameters denoted, respectively, by ℓ and θ. The case that $\bar{F}(\cdot\,|\,\xi)$ is a Gompertz distribution is of interest; see, for example, Yashin (2004) and the references therein. This mixture takes the form

$$\bar{H}(x\,|\,\theta) = \int \exp\left\{-\xi(e^{\lambda x} - 1)\right\} dG(\xi\,|\,\ell, \theta) = \left(\frac{\ell}{\ell + e^{\lambda x} - 1}\right)^\theta$$

of an exponential distribution with tilt parameter ℓ and added frailty parameter θ. By differentiating the hazard rate

$$r_H(x) = \frac{\theta \lambda e^{\lambda x}}{\ell + e^{\lambda x} - 1},$$

it can be seen that r_H is increasing when $\ell > 1$, constant when $\ell = 1$, and is strictly decreasing when $\ell < 1$. This provides another example of a parametric family of distributions with strictly increasing hazard rates, the mixture of which has a strictly decreasing hazard rate.

B. The Extensions of Makeham

Gompertz (1825) writes that

> "It is possible that death may be the consequence of two generally coexisting causes; the one, chance, without previous disposition to death or deterioration; the other, a deterioration, or increased inability to withstand destruction."

The distribution of Gompertz described in Section A is derived by consideration alone of death due to deterioration. In addition, Gompertz separately considered deaths due to diseases that affect young and old alike, and he recognized that this kind of cause leads to a constant hazard rate. However, he did not put the two causes of death together. In effect, this is what Makeham (1860) did, although Makeham's motivation was somewhat different.

Makeham (1860) wrote as follows:

> "It seems to be generally admitted, that the *theoretical* law of mortality propounded by Mr. Gompertz, although by no means a perfect representation of the *actual* law, at the same time is so nearly borne out by the facts, as to render it highly probable

that further progress in the investigation will be made in the tract thus opened up; in other words, that practical improvements in the construction of mortality tables may be looked for in some modification of Mr. Gompertz's formula."

Makeham (1860) examined the fit to actuarial data provided by the Gompertz distribution and observed with specific examples that he could improve the fit with the modification now known as the Gompertz–Makeham distribution. The notation of Makeham (1860) is not sufficiently developed to give the hazard rate of the Gompertz–Makeham distribution explicitly, but is given by Makeham (1867, 1890). This hazard rate is

$$r(x) = \zeta + \xi \lambda e^{\lambda x}, \quad \zeta \geq 0, \ \xi, \lambda > 0, \tag{1}$$

or with $\zeta = \xi \lambda \theta$,

$$r(x) = r(x \mid \lambda, \xi, \theta) = \xi \theta \lambda + \xi \lambda e^{\lambda x}, \quad \lambda, \xi > 0, \ x, \theta \geq 0. \tag{2}$$

Thus, the survival function is given by

$$\bar{F}(x \mid \lambda, \xi, \theta) = \exp\{-\xi(e^{\lambda x} - 1) - \xi \theta \lambda x\}, \quad \lambda, \xi > 0, \ x, \theta \geq 0, \tag{3}$$

and the density is

$$f(x \mid \lambda, \xi, \theta) = (\xi \theta \lambda + \xi \lambda e^{\lambda x}) \bar{F}(x \mid \lambda, \xi, \theta), \quad x \geq 0. \tag{4}$$

See Figures B.1, B.2, and B.3. The parameterization of (2), (3), and (4) can be simplified, but it is constructed to insure that λ remains a scale parameter and ξ remains a frailty parameter. However, this parameterization does not immediately yield the exponential distribution as a special case. To obtain the exponential distribution, let $\lambda \to 0, \xi \to \infty$, while $\lambda \xi$ is a positive constant; this is the same process that was used above to obtain the exponential distribution as a limit of Gompertz distributions.

If X and Y are independent random variables, where X has the Gompertz distribution A(3) and Y has an exponential distribution with parameter $\xi \lambda \theta$, then min (X, Y) has the Gompertz–Makeham distribution (3). As noted above, this derivation is implicitly suggested by the writings of Gompertz but not pursued by him, or initially by Makeham (1860). Again in the context of explicit numerical data, Makeham

B. The Extensions of Makeham 377

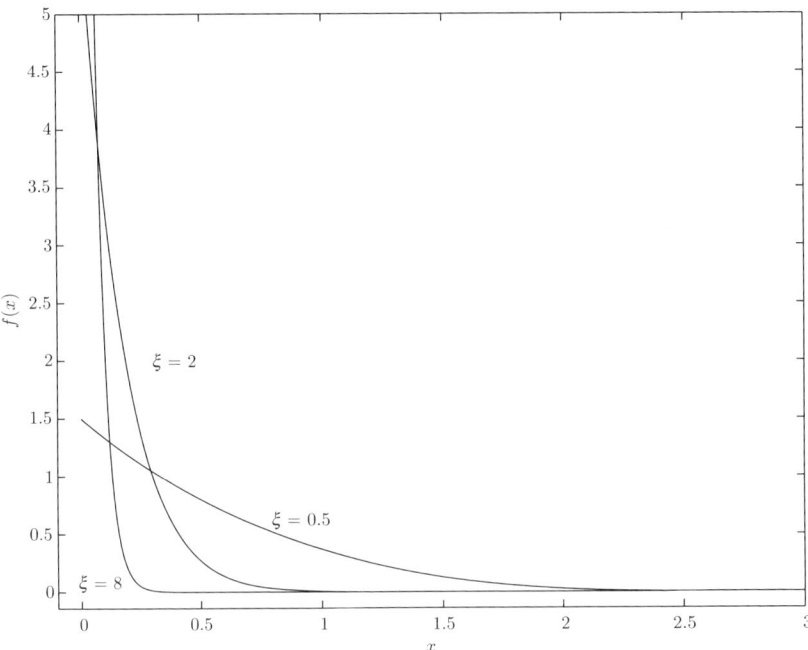

Fig. B.1a. Densities of the Gompertz–Makeham distribution ($\lambda = 1, \theta = 2$)

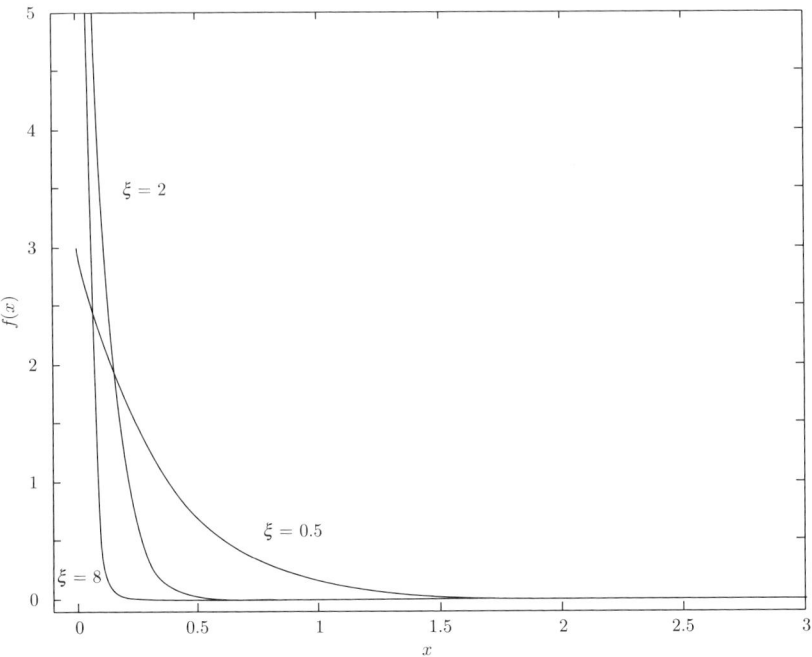

Fig. B.1b. Densities of the Gompertz–Makeham distribution ($\lambda = 1, \theta = 5$)

378   10. Gompertz and Gompertz–Makeham Distributions

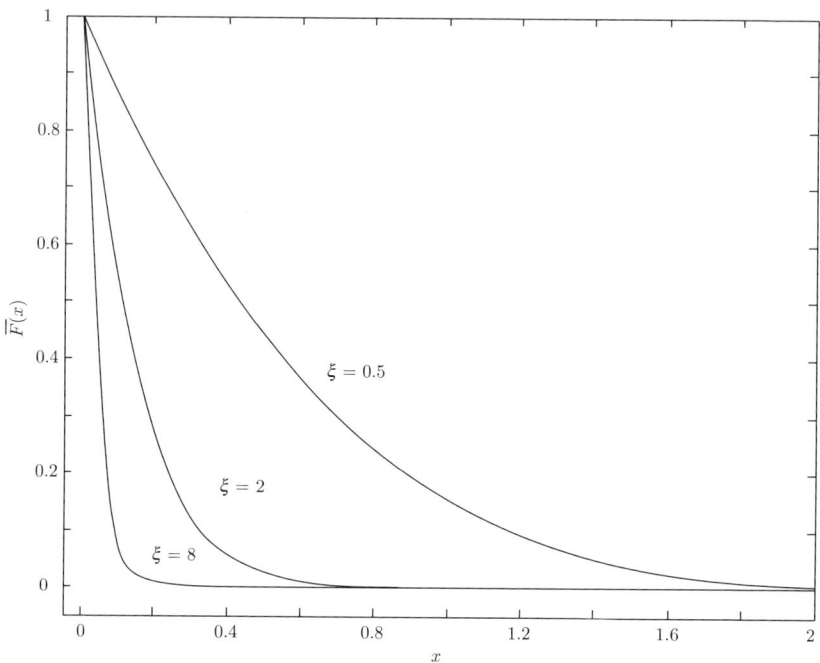

Fig. B.2a. Survival functions of the Gompertz–Makeham distribution ($\lambda = 1, \theta = 2$)

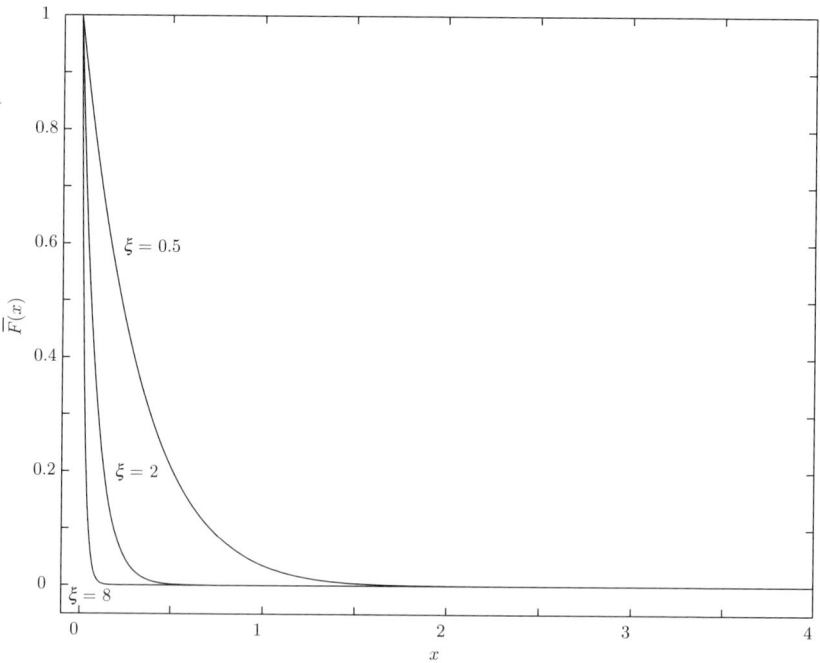

Fig. B.2b. Survival functions of the Gompertz–Makeham distribution ($\lambda = 1, \theta = 5$)

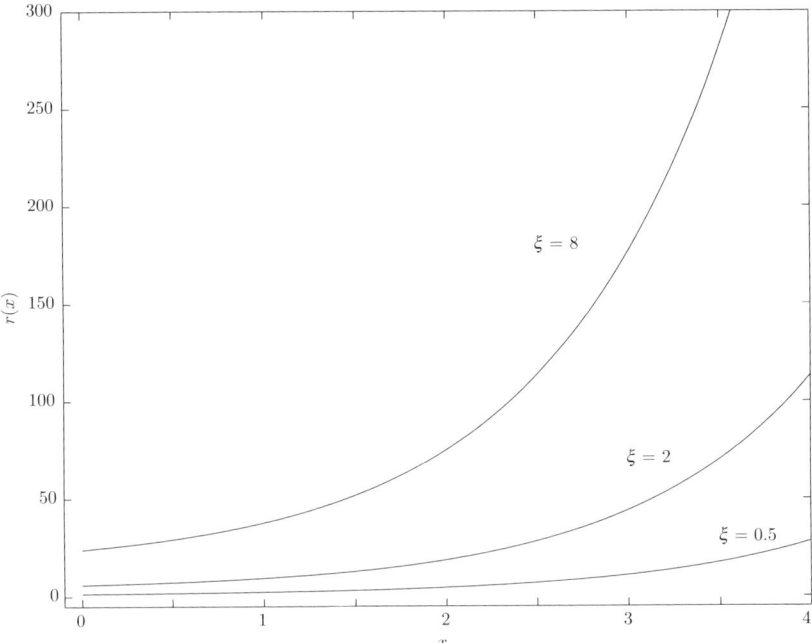

Fig. B.3a. Hazard rates of the Gompertz–Makeham distribution ($\lambda = 1, \theta = 2$)

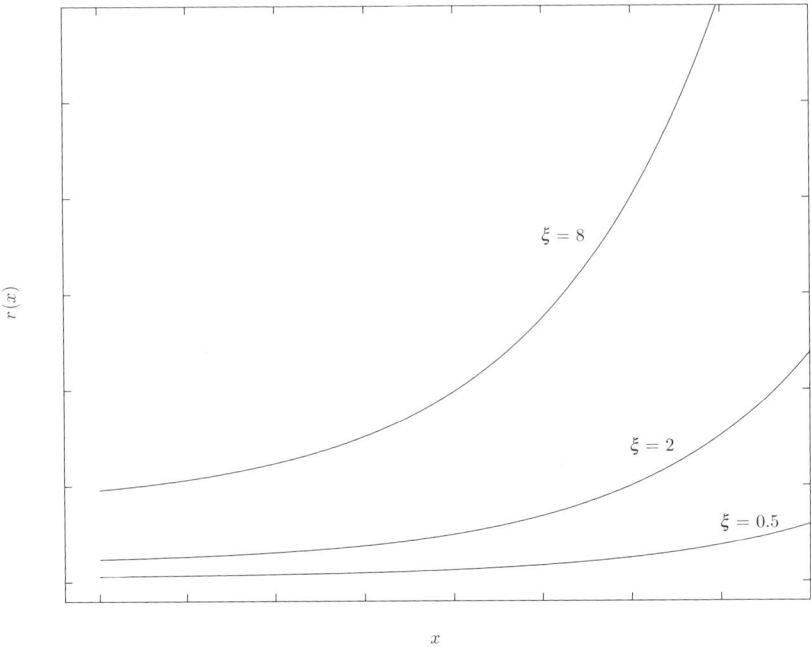

Fig. B.3b. Hazard rates of the Gompertz–Makeham distribution ($\lambda = 1, \theta = 5$)

(1867, 1890) subsequently explored the idea of competing risks, which was earlier suggested by Gompertz (1825).

Although the practice of the period was to work with differences rather than derivatives, Makeham (1890) makes it clear that his approach to finding a suitable modification of "Mr. Gompertz's formula" was essentially to replace A(1) by the differential equation

$$\frac{d^2}{dt^2} r(t) = \lambda \frac{d}{dt} r(t), \quad t, \lambda > 0. \tag{5}$$

With the condition that $\bar{F}(0) = 1$, this equation leads to the hazard rate (1).

B.1. Proposition. The density (4) of the Gompertz–Makeham distribution is log concave. That is, f is a Pólya frequency function of order 2.

Proof. This can be directly verified by checking that the second derivative of $\log f$ is negative. □

a. Moments

Although all of the positive moments of the Gompertz–Makeham distribution are finite, but like the Gompertz distribution, they cannot be given in closed form. The expected value, variance, and coefficient of variation are given for selected values of the parameters in Figures B.4, B.5, and B.6.

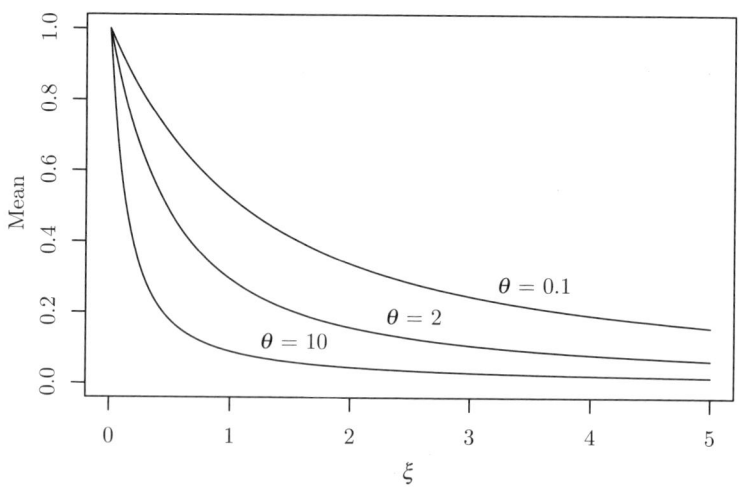

Fig. B.4. Mean of the Gompertz-Makeham distribution ($\lambda = 1$)

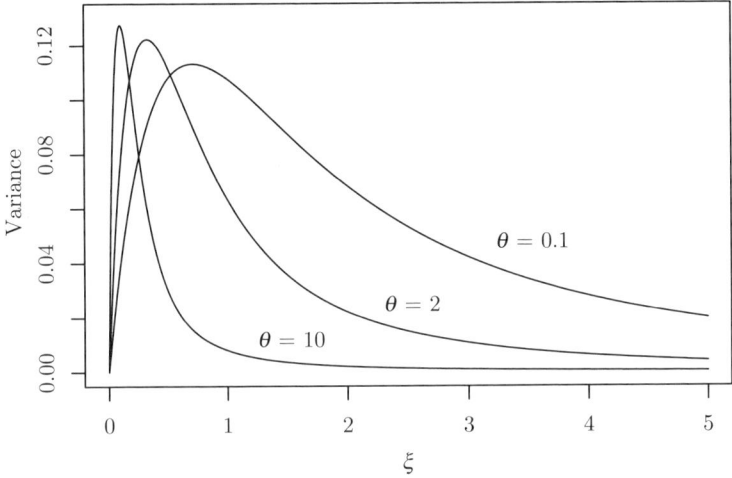

Fig. B.5. Variance of the Gompertz–Makeham distribution ($\lambda = 1$)

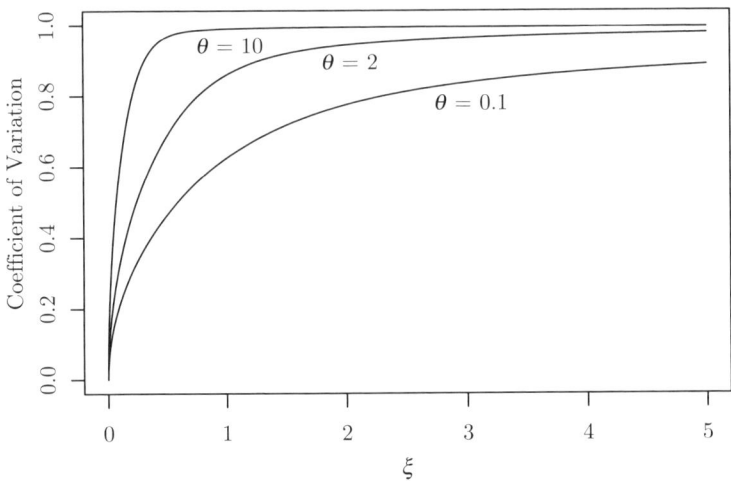

Fig. B.6. Coefficient of variation of the Gompertz–Makeham distribution ($\lambda = 1$)

b. Extended Parameter Range for the Gompertz–Makeham Distribution

The parameter range of the Gompertz–Makeham distribution can be extended beyond $\lambda, \xi > 0, \theta \geq 0$. In fact, (3) is a survival function in all of the following cases:

(i) $\lambda, \xi > 0, \theta \geq 0$; this is the usual Gompertz–Makeham distribution. The case $\theta = 0$ yields the Gompertz distribution.

(ii) $-1 \leq \theta < 0, \lambda, \xi > 0$; this is similar to the Gompertz–Makeham distribution, but it does not arise as the distribution of min (X, Y), where X has a Gompertz distribution and Y has an exponential distribution, X and Y independent.

(iii) $\theta \geq 0, \lambda, \xi < 0$.

(iv) $\theta \leq -1, \lambda < 0, \xi > 0$.

The last two cases here deserve to be spelled out in more detail. *In each case, the sign of the negative parameters has been changed.*

Case (iii): $\theta \geq 0, \lambda, \xi < 0$.

$$\bar{F}(x \mid \lambda, \xi, \theta) = \exp\{\xi(e^{-\lambda x} - 1) - \xi\theta\lambda x\},$$
$$\lambda, \xi > 0, \quad x, \theta \geq 0, \tag{6}$$

$$r(x \mid \lambda, \xi, \theta) = \xi\theta\lambda + \xi\lambda e^{-\lambda x}. \tag{7}$$

The survival function (6) is the survival function of $Z = \min(X, Y)$, where X and Y are independent, X has a negative Gompertz distribution, and Y has an exponential distribution; this proper distribution might be called the *negative Gompertz–Makeham distribution*. Here, r is decreasing to the positive limit $\xi\theta\lambda$. This means that even with this extended parameter space, all moments are finite.

Case (iv): Because the negative parameters of this case can be confusing, their signs have been changed in (8) and (9). For $\lambda > 0, \xi > 0$, $\theta \geq 1$,

$$\bar{F}(x \mid \lambda, \xi, \theta) = \exp\{\xi(1 - e^{-\lambda x}) + \xi\theta\lambda x\}, \quad x \geq 0, \tag{8}$$
$$r(x \mid \lambda, \xi, \theta) = \xi\theta\lambda - \xi\lambda e^{-\lambda x}, \quad x \geq 0. \tag{9}$$

Case (iv) of $\lambda < 0$ and $\theta < 0$ in (2) (or alternatively, the survival function (8)) has not received much attention in the literature, but is included by Lee (1992), though her treatment excludes the frailty parameter.

In the context of kidney transplantation, Bailey and Homer (1977) and Bailey, Homer and Summe (1977) make use of the three parameter hazard rate (7); see Example 16.C.1.

It is likely that case (iv) has appeared in the literature, but the authors have encountered it only as a solution to the functional equation

(11) below. In this case, the hazard rate is increasing, but only to the finite limit $\xi\theta\lambda$.

B.2. Proposition. Suppose that $X_0, X_1, X_2, \ldots$ is a sequence of exponentially distributed random variables, X_0 with parameter δ, and $X_1, X_2, \ldots$ each with parameter λ. Suppose that N is a random variable with a Poisson distribution with parameter ξ. If all of these random variables are independent, then the random variable

$$Y = \text{Min}\,(X_0, X_1, \ldots, X_N)$$

has the Gompertz–Makeham distribution with survival function

$$\bar{F}(x) = \exp\{-\delta x + \xi(e^{-\lambda x} - 1)\}, \quad \lambda, \delta, \xi > 0, \quad x \geq 0.$$

Proof. For $x \geq 0$, $P\{\text{Min}(X_0, X_1, \ldots, X_k) > x\} = e^{-(\delta + \lambda k)x}$ and consequently

$$\bar{F}(x) = \sum_{k=0}^{\infty} e^{-(\delta + \lambda k x)}\, e^{-\xi}\, \frac{\xi^k}{k!} = \exp\{-\delta x + \xi(e^{-\lambda x} - 1)\}. \tag{10}$$

□

With $\lambda = \delta$, equation (10) was given by Seal (1969); in this case, (10) is essentially the Laplace transform of a Poisson distribution, given in Section 20.E.c.

B.3. Corollary. The survival function of the Gompertz–Makeham distribution (6) of case (iii) is completely monotone.

Proof. From (10) with $\delta = 0$, it follows that the survival function (6) of the Gompertz–Makeham distribution with $\lambda < 0$ and $\xi\theta\lambda = 0$ is the Laplace transform of a random variable Z with a Poisson distribution with parameter ξ. The factor $e^{-\xi\theta\lambda x}$ in (6) is the Laplace transform of a random variable degenerate at $\xi\theta\lambda$. Because the product of Laplace transforms is a Laplace transform (see the comments following Proposition 20.D.6), it follows that (6) is a Laplace transform. According to Proposition 20.D.5 this means that $\bar{F}$ is completely monotone. □

According to Proposition 4.B.7, it follows from Corollary C.3 that the density f of the Gompertz–Makeham distribution (6) is log convex. This is in contrast to the log concavity of the density of B(4) given in Proposition B.1. The log convexity of the density of (6) is easily verified directly by showing that the logarithm of f has an increasing derivative.

c. A Functional Equation

Two functional equations for the Gompertz distribution were offered by Govan (1899); one of these is stronger than that of De Morgan (1860), and the other one, given in the following proposition, turns out to have somewhat more general solutions than those found by Govan.

B.4. Proposition. Let F be an absolutely continuous distribution function with strictly monotone hazard rate. If the survival function satisfies the functional equation

$$\frac{\bar{F}(x+t)}{\bar{F}(x)} \frac{\bar{F}(y+t)}{\bar{F}(y)} = \left[\frac{\bar{F}(w(x,y)+t)}{\bar{F}(w(x,y))}\right]^2, \quad x,y,t \geq 0, \quad (11)$$

for some nonnegative function $w = w(x,y)$ not depending on t, then either the hazard rate r has the form (2) with the extended parameter range $\xi, \lambda > 0$, $\theta \geq -1$, that is,

$$r(x) = r(x \mid \lambda, \xi, \theta) = \xi\theta\lambda + \xi\lambda e^{\lambda x}, \quad \lambda, \xi > 0, \; \theta \geq -1, \; x \geq 0, \quad (12a)$$

or it has the form (7), that is,

$$r(x \mid \lambda, \xi, \theta) = \xi\theta\lambda + \xi\lambda e^{-\lambda x}, \quad \lambda, \xi > 0, \; \theta \geq 0, \; x \geq 0, \quad (12b)$$

or it has the form (9), that is,

$$r(x \mid \lambda, \xi, \theta) = \xi\theta\lambda - \xi\lambda e^{-\lambda x}, \quad \lambda, \xi, > 0, \; \theta \geq 1, \; x \geq 0. \quad (12c)$$

Proof. Take logarithms in (11) and differentiate with respect to t to obtain

$$r(x+t) + r(y+t) = 2r(w(x,y)+t). \quad (13)$$

With $t = 0$ in (13), it follows that

$$w(x,y) = r^{-1}\left(\frac{r(x)+r(y)}{2}\right).$$

Now, substitute $u = r(x), v = r(y)$ in (13) to obtain

$$r(r^{-1}(u) + t) + r(r^{-1}(v) + t) = 2r\left(r^{-1}\left(\frac{u+v}{2}\right) + t\right).$$

With $\phi_t(u) = r(r^{-1}(u) + t)$, this equation has the form

$$\frac{\phi_t(u) + \phi_t(v)}{2} = \phi_t\left(\frac{u+v}{2}\right),$$

so that ϕ_t is linear (Proposition 22.A.1.a), and thus for some functions $a(t), b(t)$,

$$\phi_t(u) = ua(t) + b(t);$$

with $x = r^{-1}(u)$, this gives

$$r(x + t) = a(t)r(x) + b(t). \tag{14}$$

This functional equation is discussed in Proposition 22.B.4.a where it is shown that with the notation $g(x) = r(x) - r(0)$,

$$g(x + t) = a(t)g(x) + g(t). \tag{15}$$

The solutions of the functional equation (15) are provided in Proposition 22.B.4; either

(i) $g(x) = \alpha(1 - e^{ax})$, $x \geq 0$,
(ii) $g(x) = e^{ax}$, $x \geq 0$,
(iii) $g(x) = \alpha \neq 0$, or
(iv) $g(x) = 0$.

Cases (ii) and (iii) are not possible because they violate the condition that $g(0) = 0$. Case (iv) can be subsumed by case (i).

Case (i) gives

$$r(x) = r(0) + \alpha(1 - e^{ax}), \quad x \geq 0. \tag{16}$$

This takes the form of (12a), (12b), or (12c); the constraints on the parameters come from the requirement that $r(x) \geq 0$. □

Remarks. (a) The increasing hazard rate (12c) is encountered in Section C.b. (b) The functional equation (11) is adapted from an equation of Govan (1899). Govan's equation is given by

$$\prod_{i=1}^{n} \frac{\bar{F}(x_i + t)}{\bar{F}(x_i)} = \left[\frac{\bar{F}(w + t)}{\bar{F}(w)}\right]^n, \quad t, x_1, \ldots, x_n \geq 0, \quad n = 1, 2, \ldots, \tag{17}$$

where $w = w(x_1, x_2 \ldots, x_n)$ is a function independent of t. It can be verified that the solutions of (11) coincide with the solutions of (17).

d. Residual Life Distribution

Both the Gompertz distribution A(1) and the Gompertz–Makeham distribution (1) have residual life distributions F_t that remain in their respective family, but with the parameter ξ replaced by $\xi e^{\lambda t}$. More explicitly, the residual survival function of the Gompertz–Makeham distribution has hazard function R_t given by

$$R_t(x) = \xi e^{\lambda t}(e^{\lambda x} - 1) + \xi \theta \lambda x.$$

Clearly, $\lim_{t \to \infty} R_t(x) = \infty$, and consequently $\lim_{t \to \infty} \bar{F}_t(x) = 0$ for all $x > 0$. This means that as $t \to \infty$, all of the probability is "piling up" at the origin. However, if the scale is expanded (stretched out) at the right rate as $t \to \infty$, a nondegenerate limit can be obtained. By using l'Hospital's rule, it can be determined that

$$\lim_{t \to \infty} R_t(x e^{-\lambda t}) = \xi \lambda x,$$

and thus, the limiting residual life distribution can be approximated by an exponential distribution (see Proposition 6.B.6). If $\lambda < 0$ and $\xi \theta \lambda > 0$, then $\lim_{t \to \infty} R_t(x) = \xi \theta \lambda x$, so the limiting residual life distribution is exponential.

Because both the Gompertz and the Gompertz–Makeham distributions have increasing hazard rates, they are NBU, that is, $\bar{F}_t(x) \leq \bar{F}(x)$ for all $x, t \geq 0$.

e. Ordering Gompertz–Makeham Distributions

As with the Gompertz distribution, several orderings of the Gompertz–Makeham distribution are known.

Hazard rate order: It is apparent from (2) that the Gompertz–Makeham distribution is decreasing in λ, ξ, and θ when these parameters are nonnegative.

Likelihood ratio order: A tedious but straightforward calculation using 2.A(11) with logarithms shows that the Gompertz distribution is decreasing in λ when $\xi \geq 1$ and $\xi\lambda > 0$.

Convex transform order: The inverse of the Gompertz–Makeham survival function cannot be expressed in closed form (there is a transcendental equation to solve). Consequently, nothing is known about the convex transform ordering beyond that given in Section A.g for the Gompertz distribution.

f. The Second Extension of Makeham

Gompertz derived his distribution with the assumption that the hazard rate is "in geometrical progression"; that is, he made use of the differential equation

$$\frac{d}{dt}r(t) = \lambda r(t)$$

of A(1).

The Gompertz–Makeham distribution can be derived from the assumption that the derivative of the hazard rate is "in geometrical progression." In effect, Makeham (1890) replaced the Gompertz differential equation by

$$\frac{d^2}{dt^2}r(t) = \lambda \frac{d}{dt}r(t), \quad t > 0, \ \lambda > 0,$$

to obtain his first extension (3) of the Gompertz distribution. Makeham (1890) also proposed further modifications based on the assumption that higher order derivatives are "in geometrical progression." Working with differences rather than derivatives, Makeham (1890) found that third differences of empirical hazard functions were much closer to being in geometrical progression than were second differences. This assumption leads to the differential equation

$$\frac{d^3}{dt^3}r(t) = \lambda \frac{d^2}{dt^2}r(t), \quad t > 0, \ \lambda > 0,$$

With the condition that $R(0) = 0$, this equation has the solution

$$r(x) = \xi\lambda e^{\lambda x} + \xi\theta\lambda + 2\xi\alpha\lambda^2 x, \quad \lambda, \xi, \theta, \alpha > 0, \quad x > 0, \qquad (18)$$

so that

$$\bar{F}(x \mid \lambda, \xi, \theta, \alpha) = \exp\{-\xi(e^{\lambda x} - 1) - \xi\theta\lambda x - \xi\alpha(\lambda x)^2\},$$
$$\lambda, \xi, \theta, \alpha > 0, \quad x \geq 0. \qquad (19)$$

This is Makeham's second extension of the Gompertz distribution.

Note that if $X = \min(U, V, W)$ where U has a Gompertz distribution, V has an exponential distribution, and W has a Weibull distribution with shape parameter 2, then X has the survival function (19). This is an early appearance of the Rayleigh distribution (Weibull distribution with shape parameter 2).

g. Extended Parameter Range for the Second Gompertz–Makeham Distribution

Makeham's second extension (19) of the Gompertz distribution can also have an extended parameter range. The allowable range can be determined by finding conditions required for the hazard rate

$$r(x) = \xi\lambda e^{\lambda x} + \xi\theta\lambda + 2\xi\alpha\lambda^2 x = \xi\lambda(e^{\lambda x} + \theta + 2\alpha\lambda x), \quad x > 0, \qquad (20)$$

to be nonnegative.

The following three cases are considered here:

Case (i): $\lambda > 0, \xi > 0, \theta > -1$ and if $\alpha < -1/2$, then also

$$-2\alpha[1 - \log(-2\alpha)] + \theta \geq 0; \qquad (21)$$

Case (ii): $\lambda < 0, \xi > 0, \alpha \geq 0$, and $\theta \leq -1$;

Case (iii): $\lambda < 0, \xi < 0, \alpha \leq 0, \theta \geq -1$, and if $\alpha > -1/2$, then also (21).

To examine these cases, it is convenient to note that the hazard rate (20) has the derivatives

$$r'(x \mid \lambda, \xi, \theta, \alpha) = \xi\lambda^2 e^{\lambda x} + 2\alpha\xi\lambda^2, \qquad (22)$$

and

$$r''(x \mid \lambda, \xi, \theta, \alpha) = \xi\lambda^3 e^{\lambda x}. \qquad (23)$$

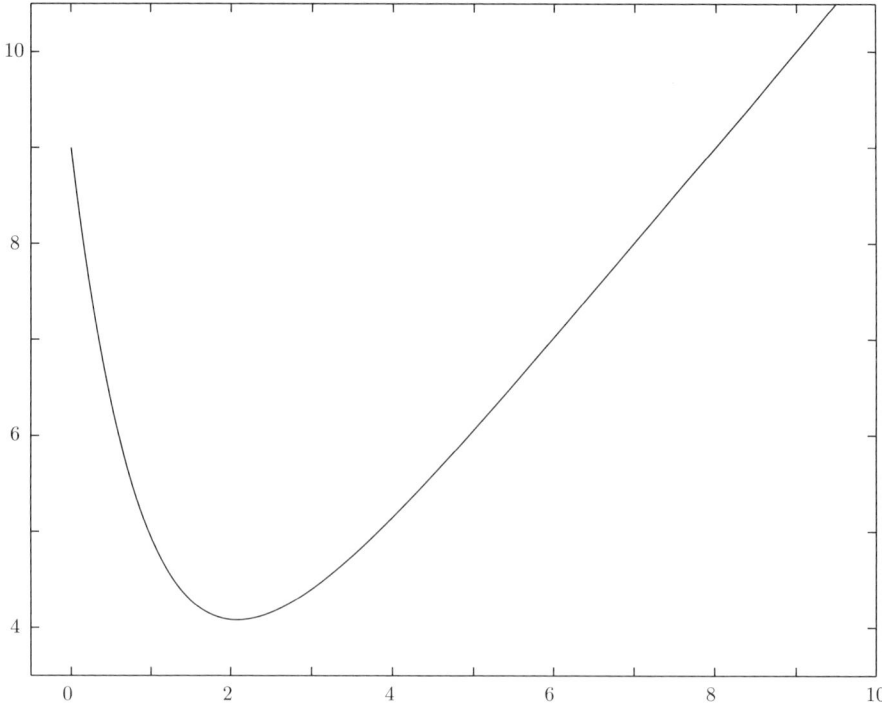

Fig. B.7. Hazard rate $r(x) = 8e^{-x} + 1 + x$

(i) If $\lambda > 0$, then because the exponential term in (20) dominates for large x, it must be that $\xi > 0$ and it follows from (23) that r is convex. For nonnegativity at $x = 0$, it must be that $\theta \geq -1$. If $\alpha \geq -1/2$, the hazard rate is increasing and no additional conditions are required. But if $\alpha < -1/2$, then the hazard rate is bathtub shaped with minimum at $x_0 = [\log(-2\alpha)]/\lambda$. To insure that $r(x_0) \geq 0$, α and θ must satisfy (21)

(ii) If $\lambda < 0, \xi > 0$, then in order that $r(0) \geq 0$ it must be that $\theta \leq -1$; in order that $r(x) > 0$ for large x, it must be that $\alpha \geq 0$. In this case, r is concave and increasing.

(iii) If $\lambda < 0, \xi < 0$, then r is convex. Because $r(0) > 0$, it must be that $\theta \geq -1$, and if $r(x) > 0$ for large x, it must be that $\alpha \leq 0$. When $\alpha \leq -1/2, r$ is increasing but when $\alpha > -1/2, r$ is bathtub shaped and as in case (i), it is necessary that (22) hold.

The case that $\lambda = -1, \xi = -8, \theta = 1/8$, and $\alpha = -1/16$ is graphed in Figure B.7.

C. Further Extensions of the Gompertz Distribution

A number of extensions and variations that have been made to the Gompertz distribution are collected in this section.

a. Modified Negative Gompertz Distribution

The improper survival function A(5) can be modified to yield the proper survival function

$$\bar{G}(x \mid \lambda, \xi) = \frac{\bar{F}(x \mid \lambda, \xi) - e^{-\xi}}{1 - e^{-\xi}} = \frac{\exp\{\xi e^{-\lambda x}\} - 1}{e^{\xi} - 1},$$
$$\xi, \lambda > 0, \quad x \geq 0. \tag{1}$$

This survival function has been proposed by Dahiya and Hossain (1996) in the context of software reliability. The hazard rate corresponding to (1) is

$$r_G(x) = \frac{\xi \lambda e^{-\lambda x} \exp\{\xi e^{-\lambda x}\}}{\exp\{\xi e^{-\lambda x}\} - 1}, \quad \xi, \lambda > 0, \quad x \geq 0, \tag{2}$$

and the density is

$$g(x) = \frac{\xi \lambda e^{-\lambda x} \exp\{\xi e^{-\lambda x}\}}{e^{\xi} - 1}, \quad \xi, \lambda > 0, \quad x \geq 0. \tag{3}$$

Because $e^{-\lambda x}$ is a convex function, it is easy to verify that g is log convex. Consequently, the hazard rate is decreasing. According to Proposition 4.C.13, this means that the residual life distributions of G all have a decreasing hazard rate, a result verified directly by Dahiya and Hossain (1996). Note that $\lim_{x \to \infty} r(x) = \lambda$, so that the limiting residual life distribution is exponential with scale parameter λ.

b. The Negative–Positive Gompertz Distribution

If X has the Gompertz survival function A(3) and Y has the negative Gompertz survival function A(6) then the distribution of $Z = \min(X, Y)$ is called a *negative–positive Gompertz distribution*. This distribution has the hazard rate

$$r(x \mid \lambda_1, \lambda_2, \xi_1, \xi_2) = \lambda_1 \xi_1 \, e^{\lambda_1 x} + \lambda_2 \xi_2 \, e^{-\lambda_2 x},$$
$$\xi_1, \xi_2, \lambda_1, \lambda_2 \geq 0, \quad \lambda_1 + \lambda_2 > 0, \quad x \geq 0, \tag{4}$$

C. Further Extensions of the Gompertz Distribution 391

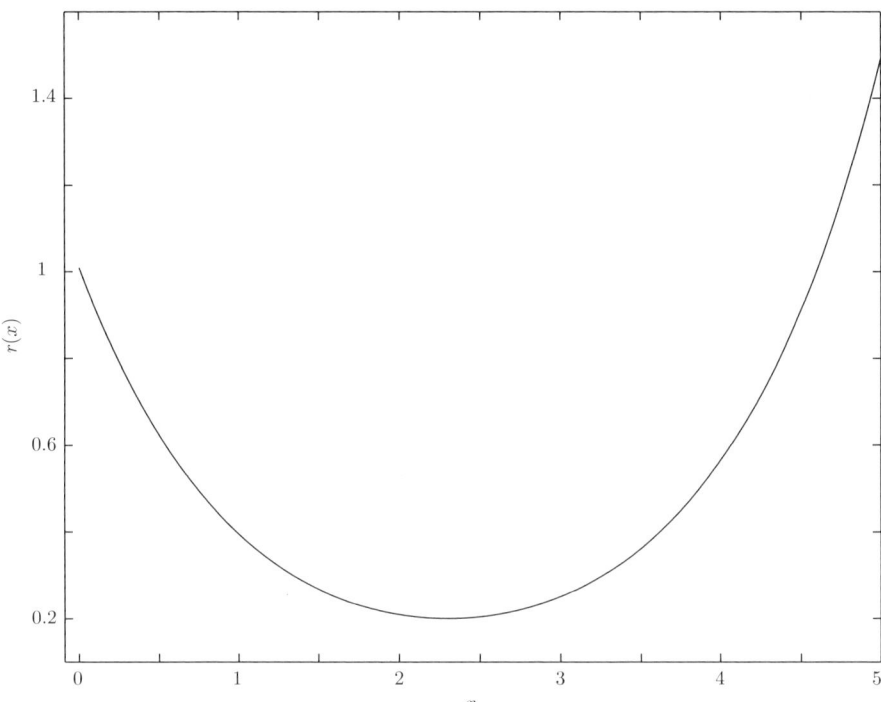

Fig. C.1. Hazard rate of a positive/negative Gompertz ($\lambda_1 = \lambda_2 = 1, \xi_1 = 0.01$, $\xi_2 = 1$)

and survival function

$$\bar{F}(x \mid \lambda_1, \lambda_2, \xi_1, \xi_2) = \exp\{-\xi_1(e^{\lambda_1 x} - 1) + \xi_2(e^{-\lambda_2 x} - 1)\}, \quad (5)$$
$$\xi_1, \xi_2, \lambda_1, \lambda_2 \geq 0, \ \lambda_1 + \lambda_2 > 0, \ x \geq 0.$$

The hazard rate (4) is increasing and convex when $\lambda_1^2 \xi_1 \geq \lambda_2^2 \xi_2$, and otherwise is bathtub shaped with a minimum at

$$x_0 = \frac{1}{\lambda_1 + \lambda_2} \log \frac{\lambda_2^2 \xi_2}{\lambda_1^2 \xi_1}. \quad (6)$$

It follows from (4) that the shape of the hazard rate depends upon the parameters ξ_1 and ξ_2 through their ratio. Figure C.1 shows a bathtub hazard rate.

The Gompertz distribution has been applied to human mortality tables only for deaths at ages greater than 18 or 20 because infant mortality can be high, leading to an initially decreasing hazard rate. The survival function (5) with bathtub hazard rate may sometimes be useful in this regard.

Clearly, a negative–positive Gompertz–Makeham distribution can be defined, that is, in terms of the hazard rates B(12a), and B(12c).

c. The Extension of Perks

Perks (1932) proposed the four-parameter extension of the Gompertz–Makeham distribution that has hazard rate of the form

$$r(x) = \frac{A + B e^{\lambda x}}{K e^{-\lambda x} + 1 + D e^{\lambda x}}, \quad x > 0. \tag{7}$$

The choice $K = D = 0$ yields the Gompertz–Makeham hazard rate.

It appears that Perks intended the parameters of (7) to be nonnegative. In case $4KD < 1$ and $D > 0$, (7) is the hazard rate of the survival function

$$\bar{F}(x) = \left(\frac{\alpha e^{-\lambda x}}{1 - \bar{\alpha} e^{-\lambda x}}\right)^\xi \left(\frac{\beta e^{-\lambda x}}{1 - \bar{\beta} e^{-\lambda x}}\right)^\theta = \left(\frac{\alpha}{e^{\lambda x} - \bar{\alpha}}\right)^\xi \left(\frac{\beta}{e^{\lambda x} - \bar{\beta}}\right)^\theta,$$

$$\alpha, \beta > 0, \ \xi + \theta \geq 0, \ \alpha\theta + \beta\xi \geq 0, \ x \geq 0, \tag{8}$$

where

$$\alpha - 1 = \frac{1 + \sqrt{1 - 4KD}}{2D}, \quad \beta - 1 = \frac{1 - \sqrt{1 - 4KD}}{2D},$$

$$\xi = \frac{B[\sqrt{1 - 4KD} - 2AD + 1]}{2D\sqrt{1 - 4KD}}, \quad \theta = \frac{B[\sqrt{1 - 4KD} + 2AD - 1]}{2D\sqrt{1 - 4KD}}.$$

It is not possible to take $D = 0$ in (8). However, to obtain the Gompertz–Makeham distribution from (8), first take $K = 0$ so that $\alpha - 1 = 1/D, \beta - 1 = 0$ and replace ξ by $\alpha\xi$. Next, take the limit as $D \to 0$, that is, $\alpha \to \infty$.

Note that if $\xi, \theta > 0$, then (8) is a product of two exponential survival functions with tilt and frailty parameters. Of course, products of survival functions arise as survival functions of minima of independent random variables.

d. Gompertz Distribution with Power Parameter

The Gompertz distribution with power parameter has survival function

$$\bar{F}(x\,|\,\lambda,\xi,\alpha) = \exp\left\{-\xi(e^{(\lambda x)^\alpha} - 1)\right\}, \quad x \geq 0, \ \lambda,\xi,\alpha > 0, \qquad (9)$$

hazard rate

$$r(x\,|\,\lambda,\xi,\alpha) = \xi\lambda\alpha(\lambda x)^{\alpha-1}\,e^{(\lambda x)^\alpha}, \quad \xi,\lambda,\alpha > 0, \ x > 0, \qquad (10)$$

and density

$$f(x\,|\,\lambda,\xi,\alpha) = \xi\lambda\alpha(\lambda x)^{\alpha-1}\,e^{(\lambda x)^\alpha}\exp\left\{-\xi(e^{(\lambda x)^\alpha} - 1)\right\}. \qquad (11)$$

With the frailty parameter $\xi = 1$, this distribution is discussed by Dhillon (1981), Leemis (1986), and Kunitz (1989). With $\lambda = 1$, the distribution is proposed by Chen (2000); the general case has been called a "modified Weibull extension" and is studied by Xie, Goh and Tang (2002) as well as by Murthy, Xie and Jiang (2004, pp. 151–154). The various authors note that the hazard rate of this distribution is bathtub shaped when $\alpha < 1$. In this case, the hazard rate has a minimum at the point

$$x_0 = \frac{1}{\lambda}\left(\frac{1-\alpha}{\alpha}\right)^{1/\alpha}.$$

When $\alpha \geq 1$, the hazard rate is increasing. The hazard rate is convex for all $\alpha > 0$, as can be seen by computing its second derivative. However, the stronger result of log convexity fails to hold. See Figure C.2.

As noted in Section A, the Gompertz distribution has a limiting exponential distribution; similarly, the Gompertz distribution with power parameter has a limiting Weibull distribution. This is most easily seen by considering the hazard rate (10). Set $\xi\lambda^\alpha = \theta^{\alpha-1}$, and let $\lambda \to 0$ to obtain the limiting hazard rate $\alpha(\theta x)^{\alpha-1}$, the hazard rate of the Weibull distribution; moreover, this convergence is monotone. It follows from the Lebesgue Monotone Convergence Theorem 24.B.2 that the corresponding hazard functions converge to the hazard function of the Weibull distribution, and hence the survival functions converge. Because of this convergence, the Gompertz distribution with power parameter is called a "Weibull extension" by Murthy, Xie, and Jiang (2004, p 151).

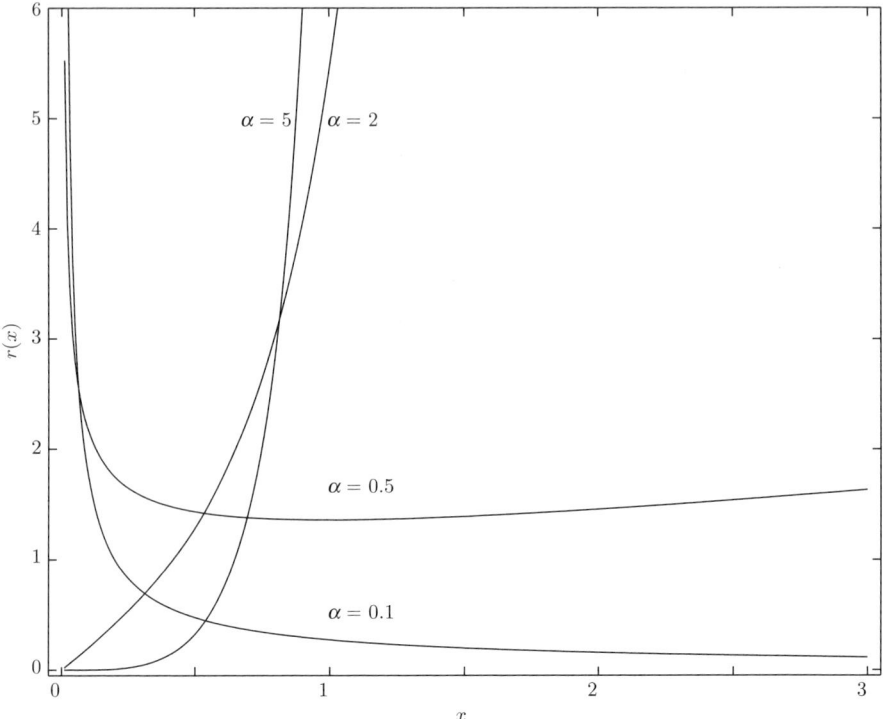

Fig. C.2. Hazard rates of the Gompertz distribution with power parameter ($\lambda = 1$)

e. Gompertz Distribution with Hazard Power Parameter

The Gompertz distribution with hazard power parameter has survival function

$$\bar{F}(x \mid \lambda, \xi, \zeta) = \exp\{-\xi(e^{\lambda x} - 1)^\zeta\}, \quad \lambda, \xi, \zeta > 0, \ x \geq 0, \qquad (12)$$

and hazard rate

$$r(x \mid \lambda, \xi, \zeta) = \lambda \xi \zeta \, e^{\lambda x}(e^{\lambda x} - 1)^{\zeta-1}, \quad \lambda, \xi, \zeta > 0, \ x > 0. \qquad (13)$$

See Figures C.3, C.4, and C.5. By differentiation, it can be verified that this hazard rate is convex. It is increasing when $\zeta \geq 1$, and when $\zeta < 1$ the hazard rate has a minimum at $x = [-\log \zeta]/\lambda$.

The improper distribution obtained by changing the signs of λ and ξ in (12) has been discussed by Bradley, Bradley and Naftel (1984)

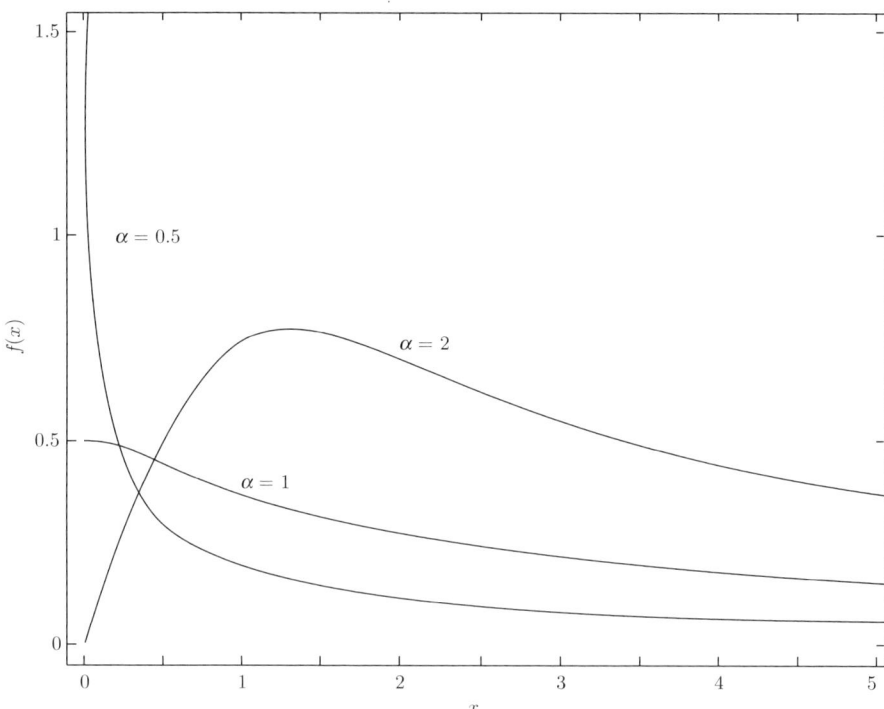

Fig. C.3. Densities of the Gompertz distribution with hazard power ($\lambda = 1$)

with negative λ and ξ as part of a generalization of Makeham's second extension of the Gompertz distribution which is discussed below.

For a derivation of the survival function (12) as a transformed extreme value survival function, see Example 12.G.1.

f. Second Gompertz–Makeham Distribution with Gompertz Power Parameter

The survival function

$$\bar{F}(x \mid \lambda, \xi, \theta, \alpha) = \exp\{-\xi(1 - e^{-\lambda x})^\zeta - \xi\theta\lambda x - \xi\alpha(\lambda x)^2\},$$
$$\lambda, \xi, \theta, \alpha, \zeta > 0, \quad x \geq 0, \tag{14}$$

was introduced by Bradley, Bradley and Naftel (1984). This survival function will be recognized as a product; the first factor is a negative Gompertz survival function with a hazard power parameter ζ. The second factor is the survival function of an exponential distribution,

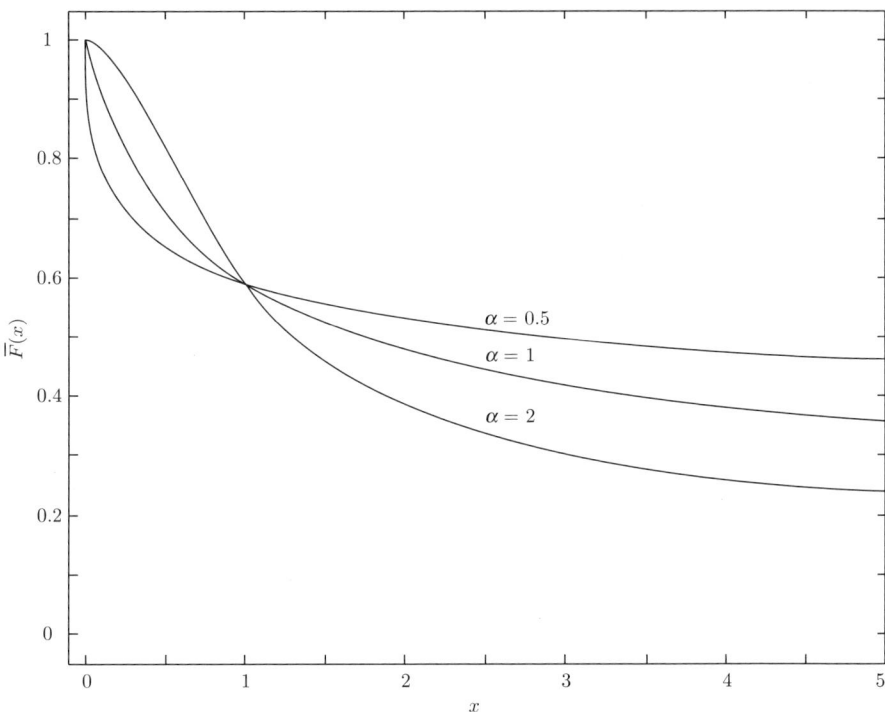

Fig. C.4. Survival functions of the Gompertz distribution with hazard power ($\lambda = 1$)

and the third factor is the survival function of a Rayleigh distribution, i.e., a Weibull distribution with shape parameter 2. As noted in Section A.a, the negative Gompertz survival function is an improper survival function in that its limit as $x \to \infty$ is not 0. Bradley, Bradley and Naftel (1984) call the survival function (14) a "Gompertz–Rayleigh" survival function.

The hazard rate corresponding to (14) is

$$r(x \mid \lambda, \xi, \theta, \alpha) = \xi\zeta\lambda(1 - e^{-\lambda x})^{\zeta-1} e^{-\lambda x} + \xi\theta\lambda + 2\xi\alpha\lambda^2 x,$$
$$\lambda, \xi, \theta, \alpha, \zeta > 0, \quad x > 0. \tag{15}$$

D. Summary of Distributions and Hazard Rates

Gompertz

$$r(x \mid \lambda, \xi) = \xi\lambda e^{\lambda x}, \quad \xi, \lambda > 0, \quad x > 0.$$

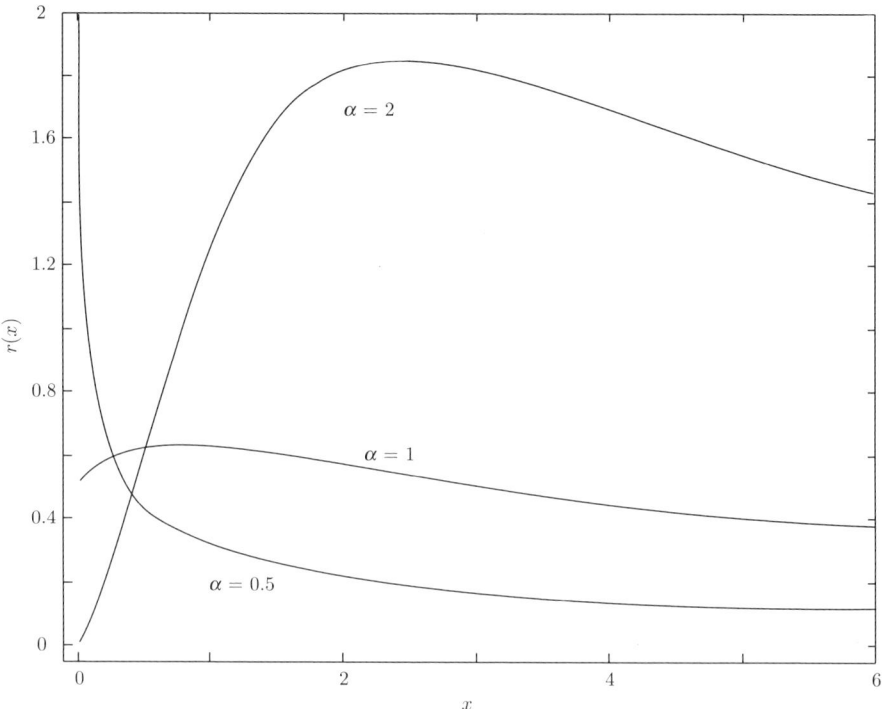

Fig. C.5. Hazard rates of the Gompertz distribution with hazard power ($\lambda = 1$)

Gompertz–Makeham (1)

$$r(x) = r(x \mid \lambda, \xi, \theta) = \xi\theta\lambda + \xi\lambda e^{\lambda x}, \quad \lambda, \xi, \theta > 0, \ x > 0,$$
$$\text{or more generally, } \xi, \lambda > 0, \theta \geq -1.$$

Gompertz–Makeham (2)

$$r(x) = \xi\lambda e^{\lambda x} + \xi\theta\lambda + 2\xi\alpha\lambda^2 x, \quad \lambda, \xi, \theta, \alpha > 0, \ x > 0.$$

Negative Gompertz

$$r(x \mid \lambda, \xi) = \xi\lambda e^{-\lambda x}, \quad \xi, \lambda > 0, \ x > 0.$$

Negative Gompertz–Makeham

$$r(x \mid \lambda, \xi, \theta) = \xi\theta\lambda + \xi\lambda e^{-\lambda x}, \quad \lambda, \xi > 0, \ \theta \geq 0, \ x \geq 0, \text{or}$$
$$r(x \mid \lambda, \xi, \theta) = \xi\theta\lambda - \xi\lambda e^{-\lambda x}, \quad \lambda, \xi > 0, \ \theta \geq 1, \ x \geq 0.$$

Negative–positive Gompertz

$$r(x\,|\,\lambda_1,\lambda_2,\xi_1,\xi_2) = \lambda_1\xi_1\,e^{\lambda_1 x} + \lambda_2\xi_2\,e^{-\lambda_2 x}, \quad \xi_1,\xi_2,\lambda_1,\lambda_2 > 0, \ x > 0.$$

Perks' extension

$$r(x) = \frac{A + B\,e^{\lambda x}}{K\,e^{-\lambda x} + 1 + D\,e^{\lambda x}}, \quad x > 0.$$

Gompertz with power parameter

$$r(x\,|\,\lambda,\xi,\alpha) = \xi\lambda\alpha(\lambda x)^{\alpha-1}\,e^{(\lambda x)^\alpha}, \quad \xi,\lambda,\alpha > 0, \ x > 0.$$

Gompertz with hazard power parameter

$$r(x\,|\,\lambda,\xi,\zeta) = \lambda\xi\zeta\,e^{\lambda x}(e^{\lambda x} - 1)^{\zeta-1}, \quad \lambda,\xi,\zeta > 0, \ x > 0.$$

Gompertz–Makeham with Gompertz power parameter

$$r(x\,|\,\lambda,\xi,\theta,\alpha) = \xi\zeta\lambda(1 - e^{-\lambda x})^{\zeta-1}\,e^{-\lambda x} + \xi\theta\lambda + 2\xi\alpha\lambda^2 x,$$
$$\lambda,\xi,\theta,\alpha,\zeta > 0, \ x > 0.$$

11

The Pareto and F Distributions and Their Parametric Extensions

A. Introduction

Extensions of the exponential distribution are obtained in Chapter 9 through use of the various semiparametric families discussed in Chapter 7. Here, the approach is to again start with the exponential distribution having parameter θ, but now let the parameter be random, with another exponential distribution $G(\cdot \mid 1/\lambda)$ having parameter $1/\lambda$. This way, the survival function

$$\bar{F}(x \mid \lambda) = \int_0^\infty e^{-\theta x} \, dG(\theta \mid 1/\lambda) = \int_0^\infty e^{-\theta x} \frac{1}{\lambda} e^{-\theta/\lambda} \, d\theta = \frac{1}{1 + \lambda x}, \quad x \geq 0, \tag{1}$$

is obtained. The focus of this chapter is to see what emerges from (1) with the addition of other parameters discussed in Chapter 7, and to examine the properties of the distributions.

Because (1) is a mixture of exponential survival functions, it follows from Proposition 4.C.10 that the hazard rate is decreasing. This fact suggests a "heavy right tail," and indeed, because the mixture includes distributions with arbitrarily small hazard rates, the right tail of (1) is "heavier" than that of any exponential distribution. Distributions with heavy tails are of particular interest in economics, and it is in this context that distributions related to (1) were proposed by Vilfredo Pareto (1897). In fact, Pareto's name is attached to several distributions related to (1); these are defined in this chapter using the terminology of Arnold (1983), whose book is devoted to Pareto distributions. Pareto was interested in describing income distributions, and the distributions also arise from economic models. See Cirillo (1979). But the results can be regarded as life distributions. As such, translation parameters are of little interest, and are generally omitted in what follows.

The F distribution and the generalized F distribution are generalizations of particular kinds of Pareto distributions obtained by the introduction of a moment parameter; this origin of the F distribution is the reason for its inclusion in this chapter. But the F distribution is much better known as the distribution of the ratio of two independent chi-square distributed random variables normalized by their degrees of freedom; in this context, it was named in honor of Sir Ronald Aylmer Fisher by Snedecor (1934). The name "F distribution" is well entrenched in the statistical literature, but the distribution is also known as a beta distribution of the second kind, and as a Pearson Type VI distribution. Several origins of the F distribution are discussed in this chapter.

B. Pareto Distributions

a. Basic Definitions

Several variations of the Pareto distribution are discussed in detail by Arnold (1983). In Arnold's terminology, the *Pareto I distribution* has survival function

$$\bar{F}(x) = 1, \quad x \leq 1/\lambda,$$
$$= 1/\lambda x, \quad x \geq 1/\lambda.$$

Note that if X has the distribution F, then $1/X$ has a distribution uniform on $(0, \lambda)$. The Pareto I distribution is encountered with scale parameter $\lambda = 1$ in Proposition 18.B.9 as a special log Weibull distribution ($b = c = 1$ in 18.B(42b)). Because the support of this distribution is bounded away from 0, as it stands it may be of minimal interest as a life distribution. However, if the left-hand endpoint $1/\lambda$ of its support is translated to the origin, the more natural life distribution $\bar{F}(x \mid \lambda) = (1 + \lambda x)^{-1}, x > 0$, of A(1), is obtained. This distribution already has a scale parameter, and other parameters can be introduced.

The *Pareto II distribution* has survival function

$$\bar{F}(x \mid \lambda, \xi) = [1 + (\lambda x)]^{-\xi}, \quad x \geq 0, \ \lambda, \xi > 0, \tag{1}$$

and is obtained by introducing a frailty parameter in the survival function A(1). This distribution has the density

$$f(x\mid \lambda, \xi) = \frac{\lambda \xi}{(1+\lambda x)^{\xi+1}}, \quad x \geq 0, \ \lambda, \xi > 0. \tag{1a}$$

It was used by Lomax (1954) to fit data in business failure, and has sometimes been called the "Lomax" distribution. This distribution has been identified by Pickands (1975) as one of the three distributions that can approximate residual life distributions (see Proposition 20.G.4). The Pareto II distribution is also encountered in Proposition 18.B.8.

By introducing a power parameter in the basic survival function (1), the survival function

$$\bar{F}(x\mid \lambda, \alpha) = [1+(\lambda x)^\alpha]^{-1}, \quad x \geq 0, \ \alpha, \lambda > 0, \tag{2}$$

of the *Pareto III distribution* is obtained. The corresponding density is

$$f(x\mid \lambda, \alpha) = \frac{\alpha \lambda (\lambda x)^{\alpha-1}}{[1+(\lambda x)^\alpha]^2}, \quad x \geq 0, \ \lambda, \alpha > 0. \tag{2a}$$

This distribution is discussed by Fisk (1961) and has at times been called the "Fisk" distribution. It is also discussed by Kalbfleisch and Prentice (2002, p. 37), and it is the generalization of A(1) encountered in Proposition 18.B.3.

The more general three-parameter *Pareto IV distribution*, obtained by introducing both a frailty and a power parameter in A(1), has the survival function

$$\bar{F}(x\mid \lambda, \alpha, \xi) = [1+(\lambda x)^\alpha]^{-\xi}, \quad x \geq 0, \ \lambda, \alpha, \xi > 0, \tag{3}$$

and density

$$f(x\mid \lambda, \alpha, \xi) = \frac{\lambda \alpha \xi (\lambda x)^{\alpha-1}}{[1+(\lambda x)^\alpha]^{\xi+1}}, \quad x \geq 0, \ \lambda, \alpha, \xi > 0. \tag{3a}$$

This distribution is also known as *Burr's distribution*, and is called *Burr's Type XII distribution* by Johnson, and Kotz (1970a, p. 31) or Johnson, Kotz and Balakrishnan (1994, p. 54). The case $\alpha = 1$ in (3a) yields the density (1a) of the Pareto II distribution, and the case $\xi = 1$ in (3a) yields the density (2a) of the Pareto III distribution.

Another extension of the Pareto III distribution, obtained by introducing a resilience parameter, is given by

$$F(x \mid \lambda, \alpha, \eta) = \left(\frac{(\lambda x)^\alpha}{1 + (\lambda x)^\alpha} \right)^\eta. \tag{4}$$

This distribution, sometimes called a Burr Type 3 distribution, has received less attention in the literature than has the Pareto IV distribution. As expected from the duality of resilience and frailty parameters as discussed in Section 7.E.b, the survival function (3) is, after replacing λ by $1/\lambda$, the survival function of $1/X$, where X has the distribution (4).

b. The Hierarchy of Pareto and Related Distributions

The most basic Pareto distribution is obtained from (1) with the parameters set to 1. Then, the introduction of certain parameters discussed in Chapter 7 shows clearly how some of these families are connected. The descendants of the basic Pareto distribution are illustrated in the following diagram.

Descendants of a basic Pareto distribution

$$\overline{F}(x) = \frac{1}{(1+x)}$$

$\downarrow$ scale

$$\overline{F}(x) = \frac{1}{(1 + \lambda x)}$$

$\downarrow$ power $\qquad\qquad$ $\downarrow$ frailty $\qquad\qquad$ $\downarrow$ resilience

$\{Pareto\ III\}$ $\qquad\;$ $\{Pareto\ II\}$ $\qquad$ $\overline{F}(x) = 1 - \left(\dfrac{\lambda x}{1 + \lambda x} \right)^\eta$

$\downarrow$ frailty $\qquad\qquad$ $\downarrow$ moment $\qquad\qquad$ $\downarrow$ moment

$\{Pareto\ IV\}$ $\qquad\qquad\qquad\quad\;\;$ $\{F\}$

$\downarrow$ moment $\qquad\qquad\qquad\qquad\;$ $\downarrow$ power

$\qquad\qquad\qquad\quad\;\;$ $\{Gen\ F\}$

c. The Pareto IV Distribution

The Pareto IV distribution, with survival function (3), includes both the Pareto II and Pareto III distributions as special cases so details for these distributions are not given separately.

By differentiation of the Pareto IV density (3a) or its logarithm, it can be determined that if $\alpha \leq 1$, this density is decreasing and log convex on $[0, \infty)$; if $\alpha > 1$, this density is unimodal with mode at

$$x = [(\alpha - 1)/(1 + \xi\alpha)]^{1/\alpha}/\lambda.$$

The hazard rate

$$r(x) = \lambda\alpha\xi(\lambda x)^{\alpha-1}[1 + (\lambda x)^\alpha]^{-1} \tag{5}$$

is increasing in $x \leq (\alpha - 1)^{1/\alpha}/\lambda$ and decreasing in $x \geq (\alpha - 1)^{1/\alpha}/\lambda$ when $\alpha > 1$, and hence the hazard rate has an inverted bathtub shape. For $\alpha \leq 1$, the hazard rate is decreasing. These results follow from computation of the hazard rate derivative. For $\alpha \leq 1$, the result also follows from the log convexity of the density (Proposition 4.C.11).

Note that $r(0) = 0$, for $\alpha > 1, r(0) = \lambda\alpha\xi$ for $\alpha = 1$, and $r(0) = \infty$ for $\alpha < 1$. For all values of α, $\lim_{x\to\infty} r(x) = 0$.

See Figure B.1.

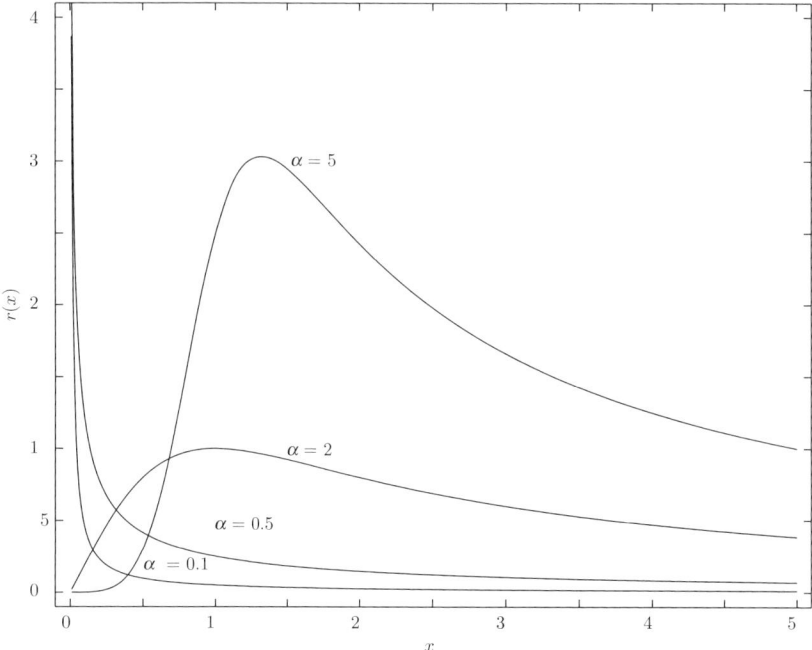

Fig. B.1. Hazard rates of the Pareto IV distribution ($\lambda = \xi = 1$)

d. Moments and Moment Parameters of the Pareto IV Distribution

It follows directly from Proposition 23.B.3 that the rth moment of the Pareto IV distribution (3) with density (4) is given by

$$\mu_r = \frac{\xi}{\lambda^r} B\left(\frac{r}{\alpha} + 1, \xi - \frac{r}{\alpha}\right), \quad -\alpha < r < \alpha\xi, \tag{6}$$

where $B(a, b)$ is the beta function defined and discussed in Section 23.B. Outside the interval $-\alpha < r < \alpha\xi$, the rth moment fails to exist finitely. Thus, the mean (first moment) fails to exist when $\alpha\xi \leq 1$.

e. The Total Time on Test Transform

Direct computation shows that the Pareto IV distribution function has inverse

$$F^{-1}(p) = \frac{1}{\lambda}\left(\frac{1}{(1-p)^{1/\xi}} - 1\right)^{1/\alpha}, \quad 0 \leq p \leq 1. \tag{7}$$

From this, the normalized total time on test transform as defined in 1.I(10) can be determined to be

$$K_F^{-1}(p) = I_{F^{-1}(p)}\left(\frac{1}{\alpha} + 1, \xi - \frac{1}{\alpha}\right), \quad \alpha, \xi > 1,$$

where I_x is the incomplete beta function defined in 23.B(4).

Some numerical evaluations of the total time on test transform are given in Figure B.2.

f. The Gini Index and Coefficient of Variation

Because the Pareto IV distribution has a frailty parameter ξ, the Gini index can be written in terms of moments using 7.E(20) as

$$\text{Gini}(F) = 1 - \frac{\mu(2\xi)}{\mu(\xi)},$$

where $\mu(\xi)$ is the first moment when the frailty parameter is ξ. From

B. Pareto Distributions 405

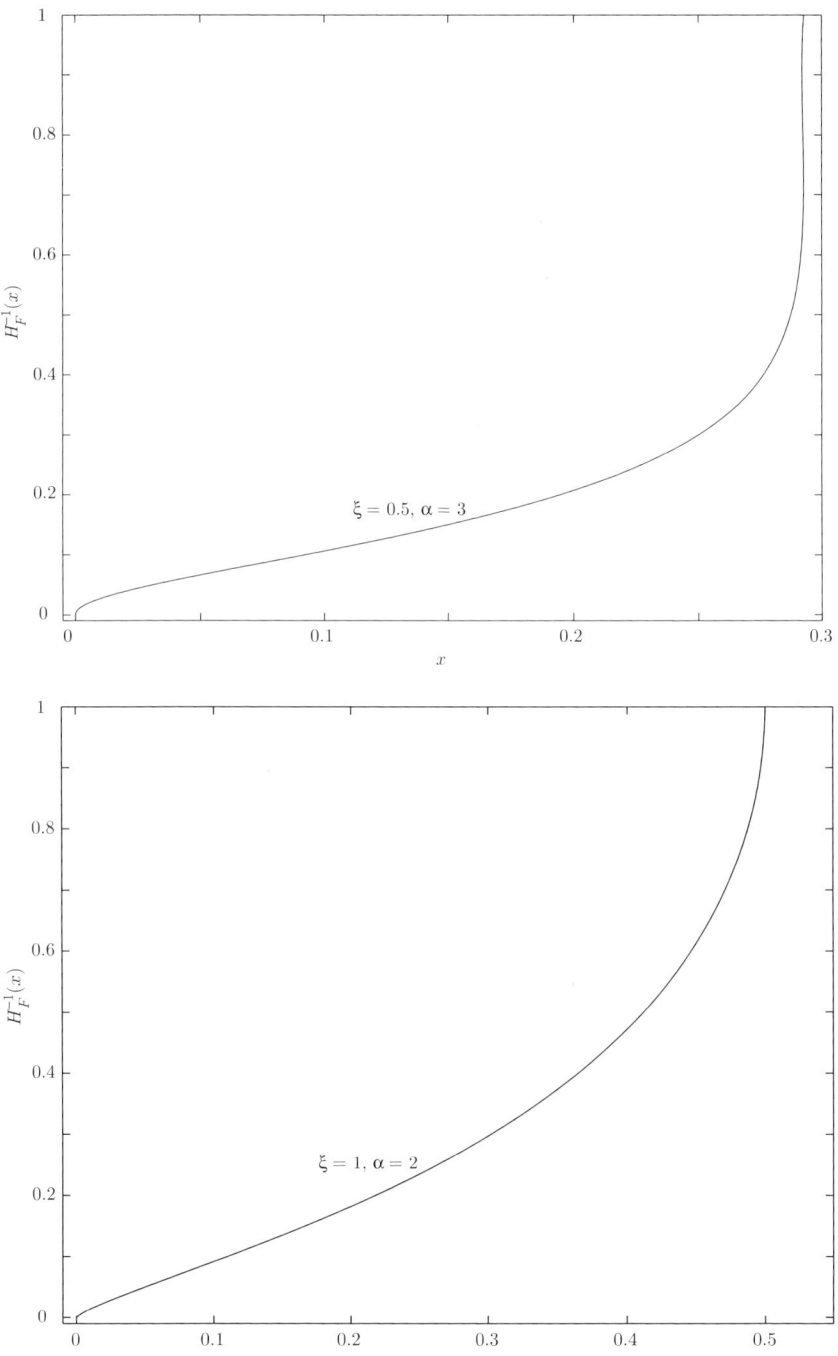

Fig. B.2. Total time on test transform of the Pareto IV distribution

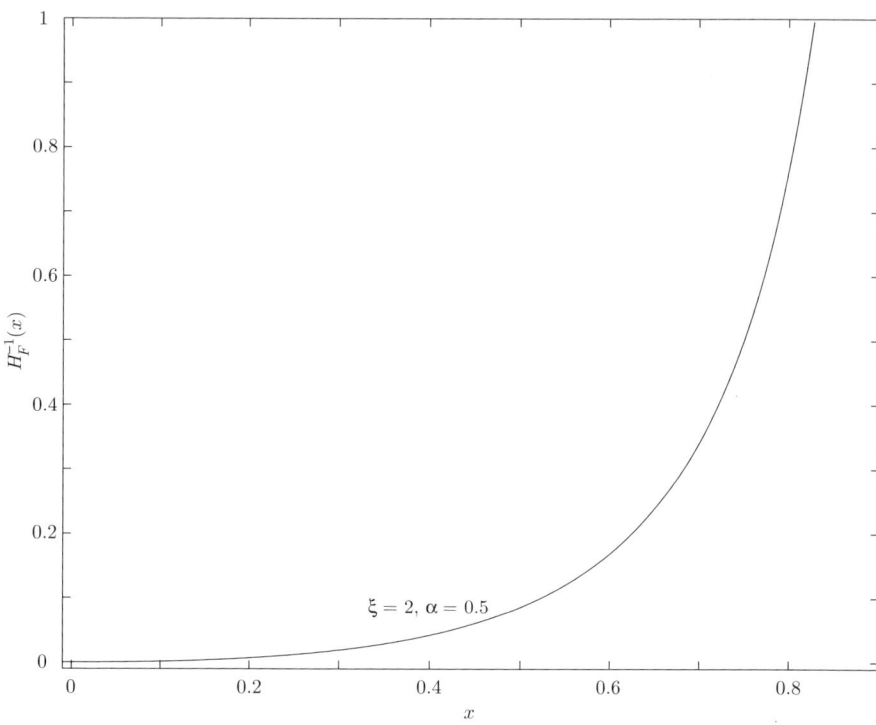

Fig. B.2. (*Continued*)

this and (6), it follows that

$$\text{Gini}(F) = 1 - 2\frac{\Gamma(\xi+1)\,\Gamma(2\xi-\alpha^{-1})}{\Gamma(2\xi+1)\,\Gamma(\xi-\alpha^{-1})} = 1 - \frac{\Gamma(\xi)\,\Gamma(2\xi-\alpha^{-1})}{\Gamma(2\xi)\,\Gamma(\xi-\alpha^{-1})}$$

$$= 1 - \frac{B(\alpha^{-1}, 2\xi-\alpha^{-1})}{B(\alpha^{-1},\ \xi-\alpha^{-1})}, \quad \alpha\xi > 1.$$

It also follows from (6) that the coefficient of variation of F is given by

$$CV(F) = \left[2\alpha\frac{\Gamma(2\alpha^{-1})\,\Gamma(\xi-2\alpha^{-1})\,\Gamma(\xi)}{\Gamma^2(\alpha^{-1})\,\Gamma^2(\xi-\alpha^{-1})} - 1\right]^{1/2}$$

$$= 2\alpha\frac{B(2\alpha^{-1},\xi-2\alpha^{-1})}{B^2(\alpha^{-1},\xi-\alpha^{-1})}, \quad \alpha\xi > 2.$$

g. Residual Life Distribution

The residual life survival function of the Pareto IV distribution is given by

$$\bar{F}_t(x) = \frac{[1 + (\lambda t)^\alpha]^\xi}{\{1 + [\lambda(x+t)]^\alpha\}^\xi}, \quad x, t \geq 0, \ \lambda, \alpha, \xi > 0; \tag{8}$$

clearly, $\lim_{t \to \infty} \bar{F}_t(x) = 1$ for all $x > 0$. This means that in the limit, all of the mass has escaped to $t \to \infty$. However, by shrinking the scale of the axis at an appropriate rate as $t \to \infty$, this mass can be retained, and a proper limiting distribution can be obtained. To do this, replace x by tx in (8) and compute

$$\lim_{t \to \infty} \bar{F}_t(tx) = \lim_{t \to \infty} \frac{\{1 + (\lambda t)^\alpha\}^\xi}{\{1 + [\lambda(tx + t)]^\alpha\}^\xi} = \frac{1}{(1+x)^{\alpha\xi}}, \quad x \geq 0. \tag{9}$$

This limiting distribution is a Pareto II distribution (see Section 20.G.b).

h. Pareto IV Distribution from Mixtures

A basic Pareto distribution was derived as a mixture of exponential distributions in Section A. More general Pareto distributions also arise as mixtures.

B.1. Proposition (Dubey, 1968; Harris and Singpurwalla, 1968; Thyrion, 1964). The Pareto IV distribution is a gamma mixture of Weibull distributions. More precisely,

$$\int_0^\infty \exp\{-z(\lambda x)^\alpha\} \frac{z^{\xi-1} e^{-z}}{\Gamma(\xi)} dz = \frac{1}{[1 + (\lambda x)^\alpha]^\xi}.$$

This fact is a direct consequence of 9.A(5). The result for $\alpha = 1$ is due to Thyrion (1964); the extension to general α was given by Dubey (1968) and by Harris and Singpurwalla (1968) (see also Gurland and Sethuraman (1994)).

Because of the uniqueness of the Laplace transform (Proposition 20.D.2), the gamma distribution here cannot be replaced by any other distribution. This proposition is an application of 7.M.6a.

i. Infinite Divisibility

Proposition B.1 shows that the Pareto II distribution is a mixture of exponential distributions. It follows from Proposition 20.D.10 that the Pareto II distribution is infinitely divisible. This fact and the method of proof was given by Thorin (1977a). As noted by Thorin (1977a), the infinite divisibility of the Pareto II distribution may have been first obtained by Steutel (1969) using Propositions 20.D.5 and 20.D.8, making the explicit representation of Proposition B.1 unnecessary.

j. Pareto Distributions as Limiting Distributions

The Pareto IV distribution can be obtained as a limit, given in the following proposition.

B.2. Proposition (Canfield and Borgman, 1975). Suppose that $Y_1, Y_2, \ldots$ is a sequence of independent identically distributed random variables with limiting extreme value survival function $\bar{W}_1(x) = \exp\{-x^b\}, x \geq 0$, given by 20.G(2) with parameter $\alpha = b$. To state this more precisely, suppose that the common distribution F of the Y_i has the property that, for $x \geq 0$,

$$\lim_{n \to \infty}[\bar{F}(a_n x + b_n)]^n = \exp\{-x^b\}$$

for appropriate norming sequences a_n and b_n.

Let N have a negative binomial distribution 20.E(7) with parameters p and $r = \alpha$, so that $EN = \alpha(1-p)/p = \alpha\theta$, where $\theta = (1-p)/p$. Let

$$X_\theta = \min(Y_1, \ldots, Y_N) \text{ if } N > 0,$$
$$= 0 \text{ if } N = 0.$$

Then as θ tends to 0, the distribution of $(X_\theta - b_\theta)/a_\theta$ converges weakly (i.e., in distribution) to the Pareto IV distribution (3) with $\lambda = 1$ as $\theta \to \infty$ through integer values.

Canfield and Borgman (1975) actually give a more general result than the one quoted here (but take care to correct the typographical errors when reading that paper).

k. Limits of Pareto Distributions

Suppose that F is a Pareto IV distribution with scale parameter λ and frailty parameter $\xi = (\delta/\lambda)^\alpha$. Then, the limit of F as $\lambda \to 0, \xi \to \infty$

(δ fixed) is a Weibull distribution with scale parameter δ. To see this, use the well known fact that $\lim_{x\to\infty}(1+\frac{a}{x})^x = e^a$.

If F is a Pareto III distribution with added resilience parameter η, then in a similar manner, it can be shown that the inverse Weibull distribution, defined in Section 9.B.1, is a limiting distribution of F.

l. The Pareto Distribution with a Hazard Power Parameter

Other extensions of Pareto distributions have appeared in the literature. In particular, Dhillon (1981) introduced a hazard power parameter in the Pareto distribution A(1) to obtain the survival function

$$\bar{F}(x \mid \lambda, \zeta) = \exp\{-[\log(\lambda x + 1)]^\zeta\}, \quad x \geq 0, \ \lambda, \zeta > 0. \qquad (10)$$

This distribution has the hazard rate

$$r(x \mid \lambda, \zeta) = \frac{\lambda \zeta}{\lambda x + 1}[\log(\lambda x + 1)]^{\zeta - 1}, \quad x \geq 0, \ \lambda, \zeta > 0, \qquad (11)$$

which is clearly decreasing for $\zeta \leq 1$. For $\zeta > 1$, it can be verified by setting the derivative of (11) equal to zero that the hazard rate is unimodal with mode at $(e^{\zeta-1})/\lambda$. See Figure B.3 for graphs of the hazard rates.

Of course, various other parameters can be introduced in (10), or alternatively a hazard power parameter can be introduced in any Pareto distribution that extends A(1).

m. Transformations and Pareto Distributions

Exponential transforms. Suppose that the random variables X and Y satisfy $X = e^Y - 1$. Direct calculations show that X has the survival function in the left-hand column of Table B.1 whenever Y has the survival function in the right-hand column of the table.

Table B.1 shows that through the transformation $X = e^Y - 1$ or $Y = \log(X + 1)$, the Pareto distribution with frailty parameter is related to the exponential distribution, the Pareto distribution with scale parameter is related to the exponential distribution with tilt parameter, and the Pareto distribution with hazard power parameter is related to the Weibull distribution.

It is shown in Section 12.C that the Pareto III distribution is a log logistic distribution.

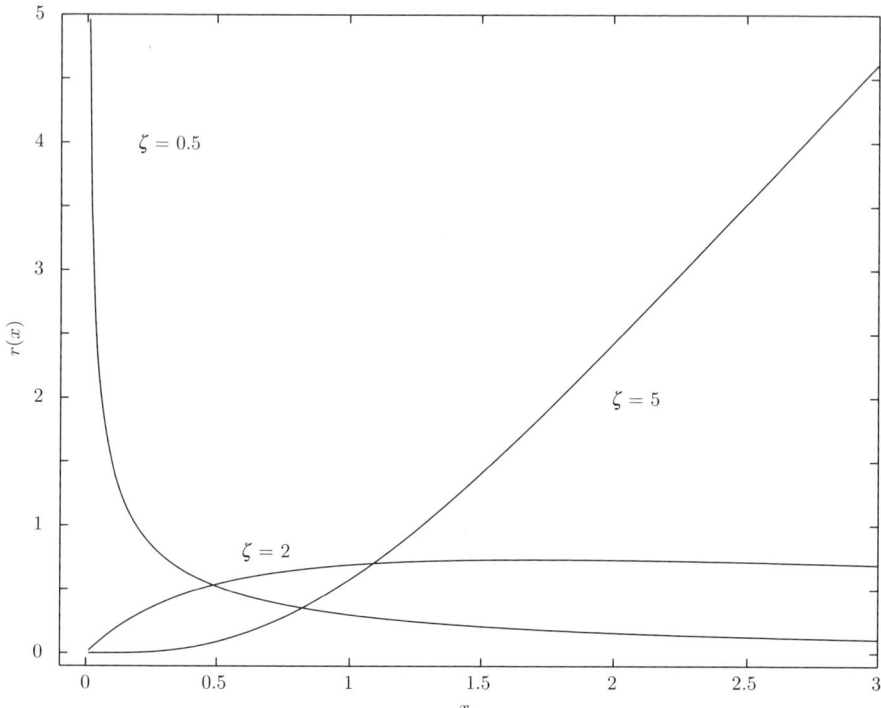

Fig. B.3. Hazard rates of the Pareto distribution with hazard power parameter ($\lambda = 1$)

Table B.1. Correspondence between the Pareto and exponential types

X	Y
$\bar{F}(x) = 1/(1+x)^\lambda$	$\bar{G}(x) = e^{-\lambda x}$
$\bar{F}(x) = 1/(1+\lambda x)$	$\bar{G}(x) = \dfrac{\gamma}{e^x - \gamma}, \quad \gamma = 1/\lambda$
$\bar{F}(x) = \exp\{-[\log(1+x)]^\alpha\}$	$\bar{G}(x) = \exp\{-x^\alpha\}$

n. Pareto Distributions with Tilt Parameter

If a tilt parameter is introduced in the survival function A(2) of a Pareto III distribution, say by way of 7.F(2), then a new Pareto III distribution is obtained. This new distribution has an unchanged power parameter, but the scale parameter is replaced by $\lambda/\gamma^{1/\alpha}$. Such a result is to be expected because of Proposition 18.B.3. Note also that Example 7.F.4 shows that the logistic distribution is geometric-extreme stable, and that the Pareto III distribution is a log logistic distribution (see Section 12.C).

The Pareto II distribution with tilt parameter is introduced by Ghitany, Al-Awadhi and Khalfan (2006); they study this three-parameter distribution in considerable detail. Among their results is the mixture representation that can be obtained from Example 7.M.4.c by taking K to be a Pareto II distribution.

o. Another Connection with the Exponential Distribution

Equation (1), that is,

$$\bar{F}(x \mid \lambda) = \int_0^\infty e^{-\theta x} \frac{1}{\lambda} e^{-\theta/\lambda} \, d\theta = \frac{1}{1 + \lambda x}, \quad x \geq 0, \tag{12}$$

exhibits a basic Pareto distribution as a mixture of exponential distributions. Now, start with the survival function (12), and consider the survival function $\bar{H}(x) = e^{-F(x)/\bar{F}(x)}$ of 1.B(21). A simple calculation shows that in this case, $\bar{H}(x) = e^{-\lambda x}, x \geq 0$, so that the exponential distribution is retrieved from the Pareto distribution.

C. Generalized F Distribution

The generalized F distribution is a four-parameter family, and thereby subsumes several distributions as special cases. As the name indicates, one special case is the F distribution, which is well known and of considerable importance in statistical contexts. Although the generalized F distribution is not nearly as well known or important, it can be studied nearly as easily as the F distribution. To avoid repetition of arguments, results for the generalized F distribution are given below, with the F distribution following as a special case in Section D.

The term "generalized F distribution" as used here refers to one specific generalization of the F distribution. The noncentral F distribution of Section 15.B is a generalization of a very different kind.

From B(6), it follows that a moment parameter β can be introduced in the density B(3a) of the Pareto IV distribution provided $-\alpha < \beta < \alpha\xi$. Replace ξ by ρ in B(3a), and introduce the moment parameter β

to obtain the density

$$f(x) = \frac{\lambda \alpha (\lambda x)^{\alpha+\beta-1}}{B\left(\frac{\beta}{\alpha}+1, \rho - \frac{\beta}{\alpha}\right)[1+(\lambda x)^{\alpha}]^{\rho+1}}, \quad x \geq 0, -\alpha < \beta < \alpha\rho. \quad (1)$$

This density has four parameters, the form of which is not particularly convenient. In (1), make the parameter change

$$\theta = 1 + \frac{\beta}{\alpha}, \quad \xi = \rho - \frac{\beta}{\alpha},$$

so that the conditions $-\alpha < \beta < \alpha\rho$ become simply $\xi, \theta > 0$. Then (1) becomes

$$f(x \mid \lambda, \alpha, \xi, \theta) = \frac{\lambda \alpha (\lambda x)^{\alpha\theta - 1}}{B(\xi, \theta)[1+(\lambda x)^{\alpha}]^{\xi+\theta}}, \quad x \geq 0, \ \lambda, \alpha, \xi, \theta > 0. \quad (2)$$

This is the density of the *generalized F distribution* (see Kalbfleisch and Prentice, 1980, p. 28); the distribution is also called the "Feller–Pareto distribution" (Arnold, 1983). The special case of (2) where $\theta = 1$, or of (1) where $\beta = 0$, is the Pareto IV density B(3a).

In the Pareto IV density B(3a), ξ is a frailty parameter. But in (2), without $\theta = 1$, ξ is no longer a frailty parameter because the introduction of frailty and moment parameters do not commute (see Section 7.L). In (2), α is a power parameter only if $\theta = 1$ because the introduction of a moment parameter and a power parameter do not commute (again, see Section 7.L). Only λ retains its nature as a scale parameter. So the parameterization of (2) is simple and conventional, but not very illuminating.

The case $\alpha = 1$ in (2) is the density

$$f(x \mid \lambda, \xi, \theta) = \frac{\lambda (\lambda x)^{\theta - 1}}{B(\xi, \theta)[1+(\lambda x)]^{\xi+\theta}}, \quad x \geq 0, \ \lambda, \xi, \theta > 0, \quad (3)$$

of the F distribution, which is discussed in Section D.

a. Density Characteristics

C.1. Proposition. When $\alpha\theta < 1$, the density (2) of the generalized F distribution is decreasing. When $\alpha\theta > 1$, the density is unimodal with

mode at
$$x = \frac{1}{\lambda}\left(\frac{\alpha\theta - 1}{\alpha\xi + 1}\right)^{1/\alpha}.$$

When $\alpha \leq 1$ and $\alpha\theta < 1$, the density is log convex. Under no conditions is $\log f$ concave.

These results can be obtained by differentiating f and $\log f$. It follows from the log convexity of f and Proposition 4.B.8.a (or Proposition 4.C.1.c) that the hazard rate is decreasing when $\alpha \leq 1$ and $\alpha\theta \leq 1$.

b. Moments

The moments of the generalized F distribution (2) can be determined from the fact that (2) is a density whenever the parameters are positive. This leads to the conclusion that

$$EX^r = \frac{B\left(\xi - \frac{r}{\alpha}, \theta + \frac{r}{\alpha}\right)}{\lambda^r B(\xi, \theta)}, \quad -\theta\alpha < r < \alpha\xi. \tag{4}$$

c. The Distribution Function

The distribution function of the generalized F distribution cannot be given in closed form, but can be expressed in terms of the incomplete beta function I_x (see 23.B(4)):

$$F(x) = I_{h(x)}(\xi, \theta), \text{ where } h(x) = (\lambda x)^\alpha [1 + (\lambda x)^\alpha]. \tag{5}$$

For the case $\alpha = 1$ of the F distribution, (5) does not significantly simplify.

d. The Hazard Rate

Because the survival functions of the F and generalized F distributions cannot be written in closed form, the same is true of the hazard rates; this makes the hazard rates troublesome to study. However, if the survival function is written as an integral, the hazard rate of the generalized F distribution can be obtained in the form

$$r(x \mid \lambda, \alpha, \xi, \theta) = \left[\frac{(1 + \lambda x)^{\xi+\theta}}{\alpha\lambda(\lambda x)^{\alpha\theta-1}} \int_{(\lambda x)^\alpha}^\infty \frac{z^{\theta-1}}{(1+z)^{\xi+\theta}} dz\right]^{-1}, \quad \xi, \theta > 0, x \geq 0. \tag{6}$$

This in turn leads to the forms

$$r(x \mid \lambda, \alpha, \xi, \theta)$$
$$= \left[\frac{1}{\alpha\lambda} \int_0^\infty \frac{[w + (\lambda x)^\alpha]^{\theta-1}}{(\lambda x)^{\alpha\theta-1}} \left(1 + \frac{w}{1+(\lambda x)^\alpha}\right)^{-(\theta+\xi)} dw \right]^{-1}$$
$$= \left[\frac{1}{\alpha(\lambda x)^{\alpha-1}} \int_0^\infty \left(1 + \frac{\lambda t}{(\lambda x)^\alpha}\right)^{\theta-1} \left(1 + \frac{\lambda t}{1+(\lambda x)^\alpha}\right)^{-(\theta+\xi)} dt \right]^{-1},$$
$$x \geq 0. \tag{7}$$

See Figure C.1.

C.2. Proposition.

$$\lim_{x \to 0} r(x \mid \lambda, \alpha, \xi, \theta) = 0 \text{ if } \alpha\theta > 1,$$
$$= \alpha\lambda/B(\theta, \xi) \text{ if } \alpha\theta = 1,$$
$$= \infty \text{ if } \alpha\theta < 1.$$

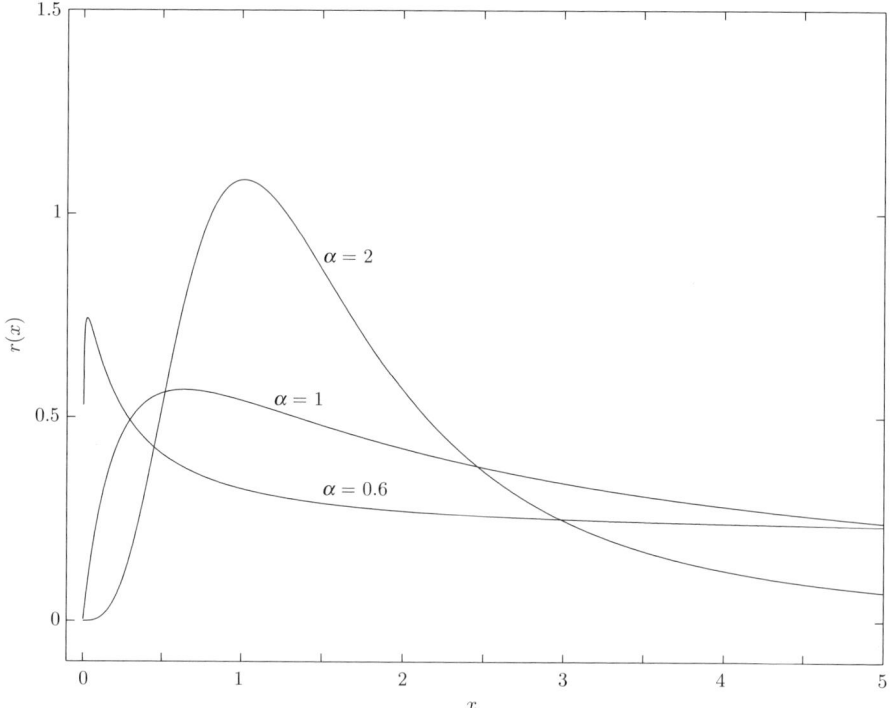

Fig. C.1a. Hazard rates of the generalized F distribution ($\theta = 2, \xi = 1.5$)

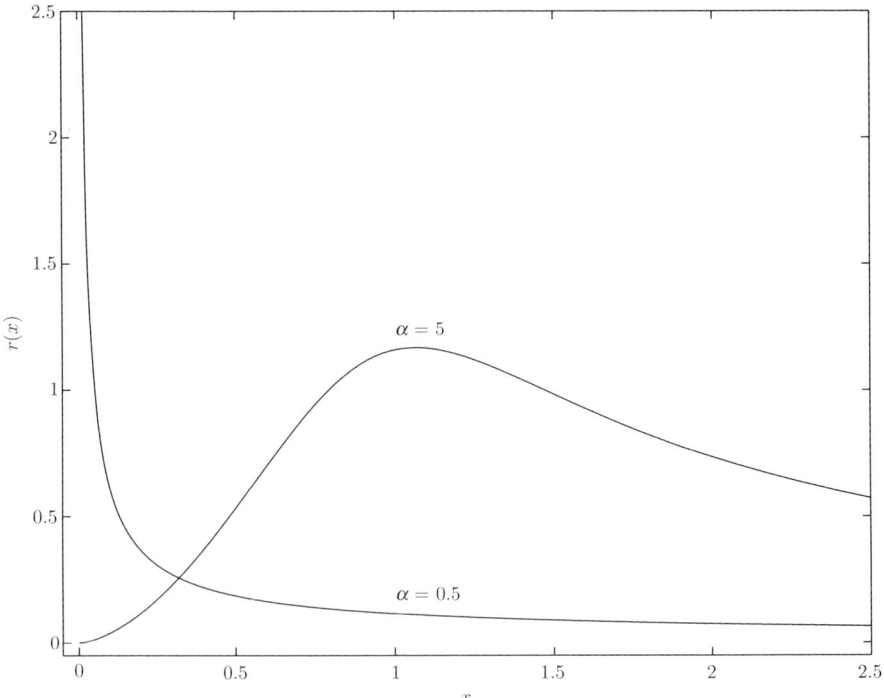

Fig. C.1b. Hazard rates of the generalized F distribution ($\theta = 0.5, \xi = 0.1$)

Proof. These limits can be obtained directly from (6), with the aid of Proposition 23.B.3. □

C.3. Proposition. $\lim_{x \to \infty} r(x \mid \lambda, \alpha, \xi, \theta) = 0.$

Proof. In the second form of (7), let $z = \lambda t/(\lambda x)^\alpha$ to obtain

$$[r(x)]^{-1} = \frac{1}{\alpha(\lambda x)^{\alpha-1}} \int_0^\infty (1+z)^{\theta-1} \left(1 + \frac{z}{1+(\lambda x)^{-\alpha}}\right)^{-\theta-\xi} \frac{(\lambda x)^\alpha}{\lambda}\, dz.$$

Because

$$\lim_{x \to \infty} \int_0^\infty (1+z)^{\theta-1} \left(1 + \frac{z}{1+(\lambda x)^{-\alpha}}\right)^{-\theta-\xi} dz = \int_0^\infty (1+z)^{-\xi-1}\, dz,$$

it follows that

$$\lim_{x \to \infty} [r(x)]^{-1} = \lim_{x \to \infty} \frac{x}{\alpha} \int_0^\infty (1+z)^{-\xi-1}\, dz = \infty. \qquad \square$$

It is noted above that the hazard rate is decreasing when $\alpha \leq 1$ and $\alpha\theta \leq 1$. A more detailed examination of the hazard rate behavior is given in the following proposition.

C.4. Proposition. In (6), reparameterize to replace α, θ, and ξ by a, b, and c, where

$$a = -(1 + \alpha\xi),$$
$$b = \alpha^2(\theta + \xi) + \alpha(\theta - \xi) - 2 = 2(\alpha\theta - 1) + \alpha(\alpha - 1)(\theta + \xi), \text{ and}$$
$$c = \alpha\theta - 1,$$

so that

$$b^2 - 4ac = \alpha^2(\theta + \xi)[4(\alpha\theta - 1) + (\alpha - 1)^2(\theta + \xi)].$$

(i) If $c > 0$, or $c = 0$ and $b > 0$, then the hazard rate (6) of the generalized F distribution is unimodal.

(ii) If $c \leq 0$ and $b^2 - 4ac < 0$, or if $c \leq 0, b^2 - 4ac > 0$, and $b \leq 0$, then the hazard rate is decreasing.

(iii) If $c < 0, b^2 - 4ac > 0$, and $b > 0$, then the hazard rate is initially decreasing and eventually decreasing, but there may be one interval in which the hazard rate is increasing.

Proof. In the notation of Theorem 4.E.2, if $\rho(x) = -d \log f(x)/dx$ and f is given by (2), then

$$\rho(x) = \frac{\alpha\theta - 1}{x} + \frac{(\xi + \theta)\alpha\lambda(\lambda x)^{\alpha-1}}{1 + (\lambda x)^\alpha}. \qquad (8)$$

The sign of the derivative of ρ is the sign of the quadratic form $q(z) = az^2 + bz + c$, where $z = (\lambda x)^\alpha$; because $a < 0, q$ is negative for large z, and consequently, ρ is eventually decreasing; if $q(z) = 0$ has no positive roots, then ρ is decreasing and that means r is decreasing. If $q(z) = 0$ has but one positive solution, then by Theorem 4.E.2(d), r is unimodal. If $q(z) = 0$ has two positive solutions, then the hazard rate r may change from decreasing to increasing to decreasing (it must eventually decrease), but according to Lemma 4.E.1, r can have at most two changes of direction. □

From (7), it follows that the hazard rate of the usual F distribution is given by

$$r(x\,|\,\lambda,\xi,\theta) = \left[\int_0^\infty \left(1+\frac{z}{x}\right)^{\theta-1}\left(1+\frac{\lambda z}{1+\lambda x}\right)^{-(\xi+\theta)} dz\right]^{-1}, \quad x \geq 0. \tag{9}$$

Here, for $\theta > 1$, $\lim_{x\to 0} r(x) = \lim_{x\to\infty} r(x) = 0$, and r is unimodal.

e. The Generalized F Distribution from Mixtures

C.5. Proposition. The generalized F-distribution is a gamma mixture of generalized gamma distributions. That is, with

$$f(x\,|\,z^{1/\alpha},\lambda,\theta,\alpha) = \frac{\alpha z^{1/\alpha}\lambda(z^{1/\alpha}\lambda x)^{\alpha\theta-1}}{\Gamma(\theta)}\exp\{-(z^{1/\alpha}\lambda x)^\alpha\}$$

given by 9.E(2a) and

$$g(x\,|\,1,\xi) = \frac{z^{\xi-1}e^{-z}}{\Gamma(\xi)}$$

given by 9.A(1),

$$\int_0^\infty f(x\,|\,z^{1/\alpha}\lambda,\theta,\alpha)\,g(z\,|\,1,\theta)\,dz = \frac{\lambda\alpha(\lambda x)^{\alpha\theta-1}}{B(\xi,\theta)[1+(\lambda x)^\alpha]^{\xi+\theta}},$$

which is the density (2).

f. Ratios of Generalized Gamma Variates

The generalized F density (2) is also the density of a ratio; in fact, if $X = U/V$ where U and V have generalized gamma distributions (Section 9.E) with the same power parameter α, then X has the density (2). This is the content of the following proposition. See Malik (1967).

C.6. Proposition. Let $X = U/V$, where U and V are independent and have respective generalized gamma densities $f(\cdot\,|\,\lambda_1,\alpha,\nu_1)$ and $f(\cdot\,|\,\lambda_2,\alpha,\nu_2)$ defined by

$$f(x\,|\,\lambda,\alpha,\nu) = \lambda\alpha(\lambda x)^{\nu\alpha-1}\exp\{-(\lambda x)^\alpha\}/\Gamma(\nu), \quad x > 0,\ \lambda,\alpha,\nu > 0,$$

as in 9.E(2a). Then, X has the generalized F distribution (2) with parameters $\lambda = \lambda_1/\lambda_2, \theta = \nu_1,$ and $\xi = \nu_2$.

Proof. By hypotheses, the densities of U and V are generalized gamma densities obtained from usual gamma densities of the form 7.A(1) by inserting the same power parameter. Because $X^{1/\alpha} = (U/V)^{1/\alpha} = U^{1/\alpha}/V^{1/\alpha}$, it is convenient to prove the theorem without the power parameter, then insert the power parameter at the end.

In forming the ratio $X = U/V$, division by the shape parameters of the gamma distributions is superfluous because that would only affect the scale parameter of the ratio, which is already arbitrary.

The distribution H of X is given by

$$H(x) = \int_0^\infty P\{U \leq xz \mid V = z\} f(z \mid \lambda_2, \nu_2) \, dz$$
$$= \int_0^\infty F(xz \mid \lambda_1, \nu_1) f(z \mid \lambda_2, \nu_2) \, dz$$

and density h given by

$$h(x \mid \lambda, \xi, \theta) = \int_0^\theta f(xz \mid \lambda_1, \nu_1) f(z \mid \lambda_2, \nu_2) z \, dz = \frac{\lambda^\theta x^{\theta-1}}{B(\xi, \theta)(1 + \lambda x)^{\xi+\theta}}, \quad (10)$$

where $\lambda = \lambda_1/\lambda_2, \theta = \nu_1,$ and $\xi = \nu_2$. This is the density (3a), and the proof is completed by the insertion of the power parameter α. □

g. Limits of Generalized F Distributions

It is apparent from Proposition C.6 that the generalized gamma distribution is a limit of generalized F distributions. This can be verified by setting $\lambda^\alpha \xi = \delta^\alpha$ in (2) and letting $\xi \to \infty, \lambda \to 0$.

D. The F Distribution

The case $\alpha = 1$ in C(2),

$$f(x \mid \lambda, \xi, \theta) = \frac{\lambda(\lambda x)^{\theta-1}}{B(\xi, \theta)(1 + \lambda x)^{\xi+\theta}}, \quad x \geq 0, \ \lambda, \xi, \theta > 0, \quad (1)$$

is called the *F density* in this book, although statisticians usually reserve that term for the special case $\theta = k_1/2, \xi = k_2/2$, and $\lambda = k_1/k_2$, where k_1 and k_2 are integers called the *degrees of freedom*. Because of its importance, the F density is treated in some detail here. The results are obtained as specializations of the generalized F distribution with $\alpha = 1$.

The F density can be obtained directly by introducing a moment parameter in a Pareto II density. The density C(2) is an F density with power parameter α as well as a Pareto IV density with moment parameter θ (see Section D.f for further details).

a. Density Characteristics

D.1. Proposition. When $\theta \leq 1$, the density (1) of the F distribution is decreasing. When $\theta > 1$, the density is unimodal with mode at

$$x = \frac{1}{\lambda}\left(\frac{\theta - 1}{\xi + 1}\right).$$

When $\theta \leq 1$, the density is log convex. Under no conditions on the parameters λ, ξ, θ is log f concave.

These results can be obtained from Proposition C.1 by taking $\alpha = 1$.

b. Moments

With $\alpha = 1$, the expression C(4) simplifies somewhat. Thus, for the F distribution (1),

$$EX^r = B(\xi - r, \theta + r)/\lambda^r B(\xi, \theta), \quad -\theta < r < \xi. \qquad (2)$$

By taking $r = 1$ and $r = 2$ in (2), by using the fact 21.A.2.a that $z\Gamma(z) = \Gamma(z + 1)$ it follows that

$$EX = \theta/\lambda(\xi - 1), \quad \xi > 1. \qquad (3)$$

$$Var(X) = \frac{\theta(\xi + \theta - 1)}{\lambda^2(\xi - 1)^2(\xi - 2)}, \quad \xi > 2. \qquad (4)$$

c. The Distribution Function

The distribution function of the F distribution is, in terms of the incomplete beta function I_x (see 23.B(4)), given by

$$F(x) = I_{h(x)}(\xi, \theta), \quad \text{where } h(x) = (\lambda x)/[1 + (\lambda x)]. \tag{5}$$

d. The Hazard Rate

If the survival function of the F distribution is written as an integral, the hazard rate takes the form

$$r(x \mid \lambda, \xi, \theta) = \left[\int_0^\infty \left(1 + \frac{t}{x}\right)^{\theta-1} \left(1 + \frac{\lambda t}{1 + \lambda x}\right)^{-(\theta+\xi)} dt \right]^{-1}, \quad x \geq 0. \tag{6}$$

See Figure D.1.

D.2. Proposition. If $\theta > 1$, the hazard rate is unimodal; if $\theta \leq 1$, the hazard rate is decreasing. Moreover,

(i) $\lim_{x \to 0} r(x) = 0$ if $\theta > 1$,
$\qquad = \lambda / B(\theta, \xi)$ if $\theta = 1$,
$\qquad = \infty$ if $\theta < 1$.

(ii) $\lim_{x \to \infty} r(x) = 0$.

e. The F Distribution from Mixtures

D.3. Proposition. The F-distribution is a gamma mixture of gamma distributions. That is, with f and g given by 9.A(1),

$$\int_0^\infty f(x \mid z\lambda, \theta, \alpha) g(z \mid 1, \theta) \, dz = \int_0^\infty \left(\frac{z\lambda(z\lambda x)^{\theta-1} e^{-z\lambda x}}{\Gamma(\theta)} \right) \left(\frac{z^{\xi-1} e^{-z}}{\Gamma(\xi)} \right) dz$$

$$= \frac{\lambda(\lambda x)^{\theta-1}}{B(\xi, \theta)(1 + \lambda x)^{\xi+\theta}},$$

which is the density (1).

f. Ratios of Gamma Variates

Consider two independent random variables U and V, each with a chi-square distribution having respective degrees of freedom m and n. It is well known in statistical theory that the ratio

$$X = \frac{U/m}{V/n}$$

D. The F Distribution

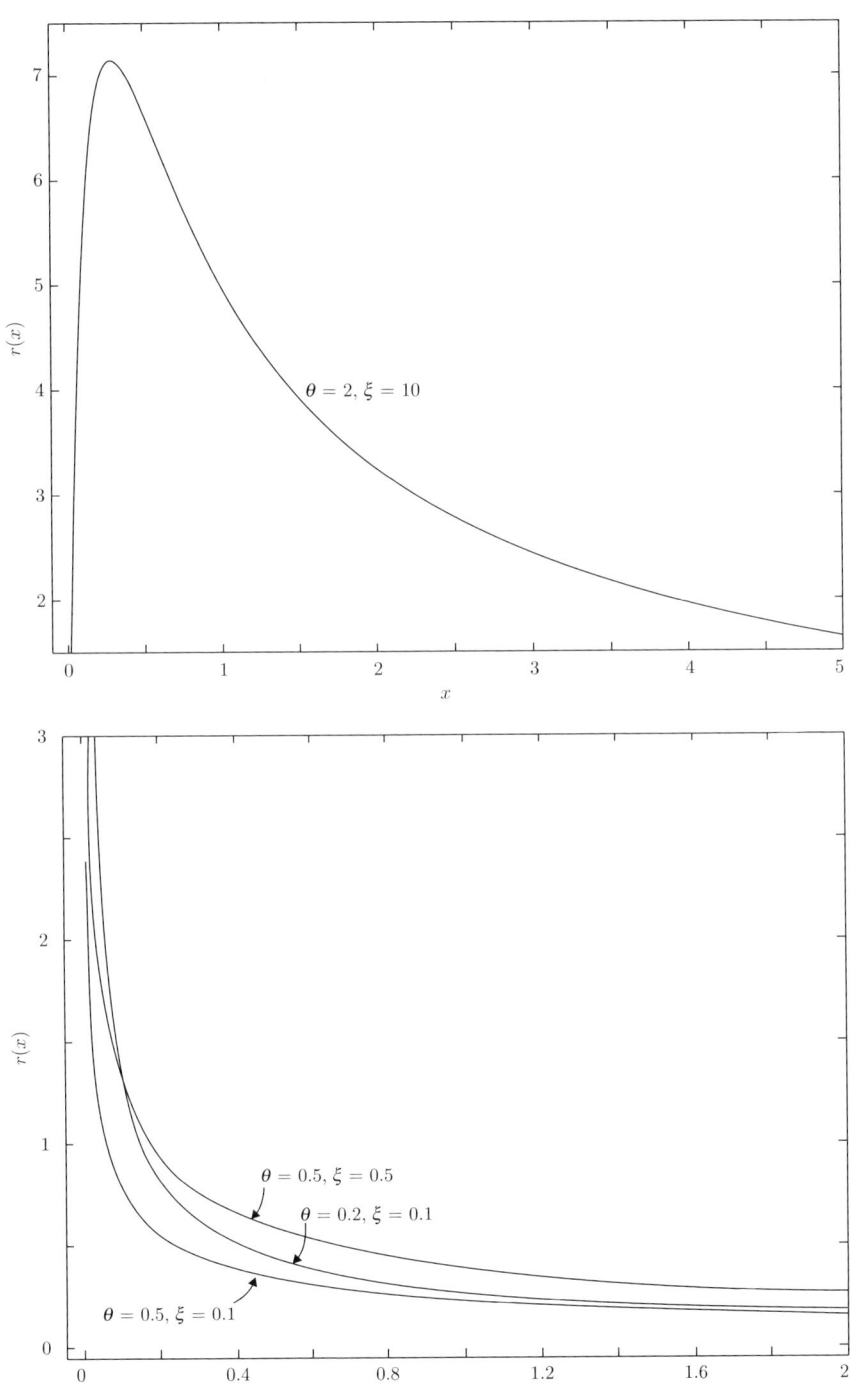

Fig. D.1. Hazard rates of the F distribution ($\lambda = 1$)

has the density f given by

$$f(x) = \frac{(m/n)^{m/2}}{B(m/2, n/2)} \frac{x^{(m/2)-1}}{[1+(m/n)x]^{(m+n)/2}}, \qquad x \geq 0. \tag{7}$$

In this context, the density (7) is called an *F density with m, n degrees of freedom*.

More generally, ratios of independent gamma distributed variates also have an F distribution, as is proved in Proposition C.6.

g. Waiting Times in the Pólya Process

The wear or degradation of a device at time t can sometimes be thought of as a stochastic process $\{X(t), t > 0\}$. If the device "fails" when the degradation exceeds a threshold, then the time of failure is the time that the process first exceeds the threshold. Such *first passage times* have been encountered in the derivation of the exponential and gamma distributions (where the process is a Poisson process); the inverse Gaussian distribution is similarly a first passage time for Brownian motion.

The Pólya process, introduced in Section 20.F.c, results from the sampling scheme known as Pólya's urn scheme. This model, introduced by Eggenberger and Pólya (1923) as a model for contagion, leads to the Pólya process. The Pólya process also arises as a Poisson process with a random parameter that has a gamma distribution.

The probability mass function of $N(t)$, where $N(t), t \geq 0$, is a Pólya process, is derived in Section 20.F.c and is given by

$$g^+(k \mid \xi, \beta, x) = \int_0^\infty \frac{(\lambda x)^k}{k!} e^{-\lambda x} \frac{\beta^\xi \lambda^{\xi-1}}{\Gamma(\xi)} e^{-\beta\lambda} d\lambda.$$

Let X be the waiting time for the θth jump in a Pólya process $\{N(t), t \geq 0\}$ with parameters λ, β, and x. The survival function of X can be written down quite directly:

$$P\{X > x\} = P\{N(x) < \theta\} = \sum_{k=0}^{\theta-1} g^+(k \mid \xi, \beta, x)$$

$$= \int_0^\infty \sum_{k=0}^{\theta-1} \frac{(\lambda x)^k}{k!} e^{-\lambda x} \frac{\beta^\xi \lambda^{\xi-1}}{\Gamma(\xi)} e^{-\beta\lambda} d\lambda$$

$$= \int_0^\infty \bar{G}(x \mid \lambda, \theta) \frac{\beta^\xi \lambda^{\xi-1}}{\Gamma(\xi)} e^{-\beta\lambda} d\lambda,$$

where $\bar{G}(\cdot \mid \lambda, \theta)$ is the survival function of a gamma distribution. But this yields the density (9) of the F distribution with $\lambda = 1/\beta$.

E. Ordering Pareto and F Distributions

E.1. Proposition. The generalized F density of C(2) is, in the likelihood ratio order, decreasing in λ and ξ and increasing in θ, the other parameters being fixed. Thus, the Pareto IV density B(3) is, in the likelihood ratio order, decreasing in λ and ξ.

Proof. By using C(2), the condition of 2.A(11) for likelihood ratio order can be directly verified. For the Pareto IV distribution, the monotonicity in the parameter ξ follows from Proposition 5.E.10 because ξ is a frailty parameter. □

Recall that the likelihood ratio order appearing in Proposition E.1 is a magnitude order. The following Propositions E.2 and E.3 involve the convex transform order, which can be regarded as a skewness order. Implications of the convex transform ordering $\leq_c$ are summarized in Section 2.C.j.

E.2. Proposition. The generalized F density is, in the convex transform order, decreasing in α, the other parameters being fixed.

This Proposition is a special case of Proposition 7.D.2.

E.3. Proposition. Pareto IV distributions are, in the convex transform order, decreasing in ξ when $\alpha \leq 1$ and λ are fixed.

Proof. From D(3) and the relationship $\bar{F}^{-1}(p) = F^{-1}(1-p)$, it follows that

$$\bar{F}^{-1}(p) = \frac{1}{\lambda}\left(\frac{1}{p^{1/\xi}} - 1\right)^{1/\alpha}.$$

First, fix α and λ. Then,

$$\bar{F}^{-1}(\bar{F}(x \mid \lambda, \alpha, \xi^*) \mid \lambda, \alpha, \xi) = \frac{1}{\lambda}\{[1 + (\lambda x)^\alpha]^\rho - 1\}^{1/\alpha}, \qquad (1)$$

where $\rho = \xi^*/\xi$. To show that this is convex when $\rho \geq 1$ and $\alpha \leq 1$, it is sufficient to take $\lambda = 1$; then the result follows from Proposition 21.A.15.

Next compute

$$\bar{F}^{-1}(\bar{F}(x\mid\lambda,\alpha^*,\xi\mid\lambda,\alpha,\xi) = (\lambda x)^{\alpha^*/\alpha}/\lambda; \qquad (2)$$

this is clearly convex when $\alpha^* \geq \alpha$. □

For the case that $\alpha > 1$, the convex transform ordering in ξ is not resolved. The ordering in α is given by Arnold and Groeneveld (1995) for the case that $\xi = 1$, that is, for the Pareto III distribution.

E.4. Proposition. Pareto IV distributions are, in the star order, decreasing in ξ when α and λ are fixed.

Proof. Use (1) to obtain

$$\bar{F}^{-1}(\bar{F}(x\mid\lambda^*,\alpha,\xi)\mid\lambda,\alpha,\xi)/x = \{[1+(\lambda x)^\alpha]^\alpha - 1\}^{1/\alpha}/\lambda x.$$

Differentiate with respect to x to find that, when $\rho \geq 1$, this expression is increasing in x if and only if

$$\rho y(1+y)^{\rho-1} - (1+y)^\rho + 1 \geq 0, \quad y \geq 0. \qquad (3)$$

Here, $y = x^\alpha$. The left-hand side of (3) is 0 at $y = 0$, and can be seen to be increasing in y by noting that its derivative $\rho(\rho-1)y(y+1)^{\rho-2}$ is nonnegative. □

The various orderings are summarized in the following diagram.

Order	Pareto IV Distribution
Likelihood ratio	decreasing in λ, ξ
Convex transform	decreasing in ξ if $\alpha \leq 1$ and λ is fixed
Star	decreasing in ξ if α, λ are fixed

F. Another Generalization of the Pareto Distribution

The density of the Pareto III distribution can be written in the form

$$f(x\mid\lambda,\alpha) = \frac{\alpha}{x[(\lambda x)^\alpha + 2 + (\lambda x)^{-\alpha}]}.$$

F. Another Generalization of the Pareto Distribution

For fitting income distributions, Champernowne (1937, 1952) proposed a generalization with added parameter δ.

$$f(x \mid \lambda, \alpha, \delta) = \frac{c(\delta)\,\alpha}{x[(\lambda x)^\alpha + 2\delta + (\lambda x)^{-\alpha}]}, \tag{1}$$

where $c(\delta)$ is a normalizing constant depending upon δ. This distribution is discussed by Fisk (1961).

For $-1 < \delta < 1, c(\delta) = 1/\theta$ where $\cos\theta = \delta$, and

$$\bar{F}(x \mid \lambda, \alpha, \delta) = \frac{1}{\theta} \tan^{-1}\left(\frac{\sin\theta}{\cos\theta + (\lambda x)^\alpha}\right), \quad x \geq 0. \tag{2}$$

For $\delta = 1$, the survival function is the Pareto III survival function and $c(\delta) = 1$. For $\delta > 1, c(\delta) = (\theta - \theta^{-1})/2\log\theta$ where $\theta = \delta + \sqrt{\delta^2 - 1}$, and

$$\bar{F}(x \mid \lambda, \alpha, \theta) = \frac{1}{2\log\theta} \log\frac{(\lambda x)^\alpha + \theta}{(\lambda x)^\alpha + \theta^{-1}}, \quad \theta > 1, \ x \geq 0. \tag{3}$$

12
Logarithmic Distributions

A. Introduction

The lognormal distribution, introduced in Chapter 1, can be obtained from the normal distribution by means of a transformation. Specifically, if $Y = \log X$ has a normal distribution, then $X = e^Y$ has a lognormal distribution.

Besides the normal distribution, there are several other standard distributions with support $(-\infty, \infty)$ that arise from natural considerations; the same transformation yields a random variable with support $(0, \infty)$ for these other distributions as well. The procedure allows the generation of life distributions from distributions with support $(-\infty, \infty)$. It is perhaps surprising that distributions for nonnegative random variables $X = e^Y$ that arise in a seemingly arbitrary manner are often interesting. Distributions of this kind are here termed *logarithmic distributions*.

Among the various logarithmic distributions, the lognormal occupies a special place. Discussion of the lognormal distribution is deferred to Section B; this section is devoted to a general discussion of logarithmic distributions. As discussed in Section B, the lognormal distribution has a direct derivation of its own that is quite as convincing as the derivation of the normal distribution via the central limit theorem.

Denote the distribution function of X by F and the distribution function of $Y = \log X$ by H. If densities exist, denote them, respectively, by f and h and use subscripts to distinguish the hazard rates. It is straightforward to verify that

$$F(x) = H(\log x), \quad f(x) = x^{-1} h(\log x),$$
$$r_F(x) = x^{-1} r_H(\log x), \quad x > 0. \qquad (1)$$

The following proposition follows immediately from (1).

A.1. Proposition. Suppose that the random variable Y has a density h. Then $X = e^Y$ has the density f given by (1). Moreover,

(i) If $f(x)$ is increasing at $x = x_0$, then $h(x)$ is increasing at $x = e^{x_0}$.
(ii) If $h(x)$ is decreasing at $x = e^{x_0}$, then $f(x)$ is decreasing at $x = x_0$.
(iii) If $r_F(x)$ is increasing at $x = x_0$, then $r_H(x)$ is increasing at $x = e^{x_0}$.
(iv) If $r_H(x)$ is decreasing at $x = e^{x_0}$, then $r_F(x)$ is decreasing at $x = x_0$.

In case Y is a nonnegative random variable, the transformation $X = e^Y$ leads to a random variable that is always greater than or equal to one; such distributions are discussed in Section 15.D. More generally, if the support of Y is bounded below, say by x_0, then the exponential transformation will not lead to a distribution with 0 as the left-hand endpoint of support. However, a simple modification of the transformation removes this difficulty. Specifically if

$$X = \exp\{Y\} - \exp\{x_0\}, \qquad (2)$$

then the support of X has left-hand endpoint 0.

a. Parameter Changes Under a Log Transformation

For distributions on $(-\infty, \infty)$, location and scale are often the appropriate parameters. For the transformed distributions on $(0, \infty)$, these parameters become, respectively, scale and power parameters, as indicated in the following proposition.

A.2. Proposition. Suppose that $\{H(\cdot \mid \mu, \sigma), -\infty < \mu < \infty, \sigma > 0\}$ is the family of distributions of the form $H(x \mid \mu, \sigma) = G((x-\mu)/\sigma)$ for some underlying basic distribution G that has support $(-\infty, \infty)$. If X has the distribution $H(\cdot \mid \mu, \sigma)$, then $Y = e^X$ has the distribution $F(\cdot \mid \lambda, \alpha)$, where $\lambda = e^{-\mu}$, $\alpha = 1/\sigma$, and

$$F(x \mid \lambda, \alpha) = H(\log x \mid -\log \lambda, 1/\alpha) = G(\log(\lambda x)^\alpha). \qquad (3)$$

Thus, λ is a scale parameter and α is a power parameter for the transformed distribution. Note that in the above notation,

$$G((x-\mu)/\sigma) = H((x-\mu)/\sigma \mid 0, 1).$$

A.3. Proposition. If Y has the distribution (3), then for $\theta, \zeta > 0$, $(\theta Y)^\zeta$ has the distribution

$$F\left(x \mid \frac{\lambda}{\theta}, \frac{\alpha}{\zeta}\right) = G\left(\frac{\alpha}{\zeta} \log \frac{\lambda x}{\theta}\right).$$

This fact is straightforward to verify; it depends not on G, but only on the way G is parameterized.

From (3), it follows that if G has a density g, then F has a density f given by

$$f(x \mid \lambda, \alpha) = \frac{\alpha}{x} g(\alpha \log(\lambda x)). \tag{4}$$

Moreover, the hazard rate r_F of F is given in terms of the hazard rate r_G of G by

$$r_F(x \mid \lambda, \alpha) = (\alpha/x)\, r_G(\alpha \log(\lambda x)). \tag{5}$$

To express the moments of $F(\cdot \mid \lambda, \alpha)$ in terms of the moment generating function of G, note that with a change of variables it follows from (3) that whenever the integrals exist,

$$\int_0^\infty x^s \, dF(x \mid \lambda, \alpha) = \lambda^{-s} \int_{-\infty}^\infty \exp\left\{\frac{sy}{\alpha}\right\} dG(y). \tag{6}$$

In view of (6), it is of interest to see what happens if, instead of location and scale parameters, a Laplace transform parameter s is introduced in G. When this is done, the parametric family

$$H(x \mid s) = \frac{1}{\phi(s)} \int_{-\infty}^x e^{-sz} \, dG(z) \tag{7}$$

is obtained, and exists for all s such that the Laplace transform ϕ of G exists.

A.4. Proposition. If X has the distribution $H(\cdot \mid s)$ given by (7), then $Y = e^X$ has the distribution $F(\cdot \mid s)$ given by

$$F(x \mid s) = \frac{1}{\phi(s)} \int_0^x y^{-s} \, dF(y), \tag{8}$$

where $F(x) = G(\log x)$. Thus, $-s$ has become a moment parameter.

b. Negative Logarithmic Distributions

For a nonnegative random variable Y, the random variable $Z = e^{-Y}$ takes values in $[0, 1]$. Distributions of such random variables are termed *negative logarithmic distributions*, examples of which are discussed in 15.D. Of course, the distribution of $Z = e^{-Y}$ can also be regarded as a logarithmic distribution based upon the random variable $-Y$. In this case, it can be easily verified that the random variable Z has the distribution function F given by

$$F(x) = \bar{G}(-\log x), \quad 0 \leq x \leq 1.$$

Suppose that the random variable Y takes values in $[0, \infty)$ and has the distribution function G. It can be verified that in the negative log distribution obtained from (7), it is the Laplace transform parameter s that becomes a moment parameter rather than $-s$ as in Proposition A.4.

c. A Common Distribution for a Random Variable and Its Reciprocal

In some applications, it is natural to impose the condition that X and $1/X$ have the same distribution. Clearly, this condition is preserved when scale and power parameters are introduced; that is, if X and $1/X$ have the same distribution, then $(\lambda X)^\alpha$ and $1/(\lambda X)^\alpha$ have the same distribution, $\lambda, \alpha > 0$.

A.5. Proposition. The distribution of X and $1/X$ are the same if and only if $Y = \log X$ has a distribution symmetric about 0. More generally, X/e^m and e^m/X have the same distribution if and only if $Y = \log X$ has a distribution symmetric about m.

Proof. Consider first the case of symmetry about 0. From the relation $Y = \log X$, it follows that $-Y = \log(1/X)$. But Y and $-Y$ have the same distribution if and only if Y has a distribution symmetric about 0. The general case follows by applying the special case to the random variable X/e^m. □

d. Monotone Hazard Rates Under Logarithmic Transformations

The following proposition is closely related to Proposition A.1 but does not depend upon the existence of a density.

A.6. Proposition. Suppose that $Y = \log X$, and denote the survival functions of X and Y by $\bar{F}_X$ and $\bar{F}_Y$.

(i) If $\bar{F}_X$ is IHR, then $\bar{F}_Y$ is IHR.
(ii) If $\bar{F}_Y$ is DHR, then $\bar{F}_X$ is DHR.

The converses of these statements are false.

Proof. As noted in (1),

$$\log \bar{F}_Y(y) = \log \bar{F}_X(e^y).$$

From Proposition 21.A.5(iii), it follows that log concavity of $\bar{F}_X$ implies the log concavity of $\bar{F}_Y$. This is the first statement of Proposition A.6. The second statement follows similarly from Proposition 21.A.5(iv). □

Because the converses of Proposition A.6 are false, it is to be expected that the lognormal distribution may not be IHR even though the normal distribution has a log concave survival function. Similar statements can be made about other logarithmic distributions.

B. The Lognormal Distribution

Certainly, the most important logarithmic distribution is the lognormal distribution. This distribution has been extensively studied by many authors, the landmark study being that of Aitchison and Brown (1957). A history and genesis of the lognormal distribution is given by Shimizu and Crow (1988). The collection of papers edited by Crow and Shimizu (1988) is also devoted to the subject; see also Johnson, Kotz and Balakrishnan (1994, Chapter 14). All of these books give extensive historical accounts and bibliographies.

a. Appropriate Physical Models

Additive models are addressed by the central limit theorem and lead to the normal distribution. In some circumstances, multiplicative models are more appropriate. Nevertheless, multiplicative models have been overshadowed by additive models just as the geometric mean is often neglected in favor of the arithmetic mean. Concern about this and an interest in multiplicative models seem to have prompted Galton (1879) to have encouraged McAlister (1879) to undertake the first detailed study of the lognormal distribution in a paper entitled "The law of the geometric mean." For a more detailed account, see Aitchison and Brown (1957).

Multiplicative models are versions of what, in an economic context, was called the "law of proportional effects" by Gibrat (1930, 1931). Because of Gibrat's work, the lognormal distribution has sometimes been called "Gibrat's distribution."

Because multiplicative models can be transformed to additive models by use of the logarithm, the theory of multiplicative models can be obtained from that of additive models, and of course, the reverse is also true.

a.1. Growth Models

In both economics and biology, models of growth of the form

$$X_i - X_{i-1} = Z_i X_{i-1}, \text{ or equivalently, } X_i = X_{i-1}(1 + Z_i)$$

arise from the idea that growth is proportional to size. These models lead to the conclusion that

$$X_n = X_0 \prod_{i=1}^{n} (1 + Z_i).$$

When the Z_i are all independent and nonnegative and the individual increments $X_i - X_{i-1}$ are small, the normalized X_n will, for large n, be approximately lognormally distributed. For further details of these models, see Cramér (1946, p. 220), Crow and Shimizu (1988, p. 4), or Johnson, Kotz and Balakrishnan (1994, p. 210).

a.2. Breakage Theory

The following breakage model can be regarded as an analog of the growth model. A particle is subjected to a series of breakages, where at the ith breakage it suffers a loss of a random proportion T_i of its mass. Thus, if X_0 is the initial mass, the mass after n breakages is $X_n = X_0 \prod_{i=1}^{n} T_i$. If the random proportions are independent and identically distributed, then the limiting distribution of X_n is, under mild conditions, a lognormal distribution. For some history of this model and results, see Aitchison and Brown (1957, p. 26).

a.3. Physics of Failure

Arguments have been made concerning the failure mechanism of solid state devices that lead to the lognormal distribution as the distribution of life length. Such arguments are beyond the scope of this book, but see Joyce and Anthony (1988).

There are a large number of papers in which the lognormal distribution has been used to fit data without reference to physical models such as those mentioned here. The success of many of these fits for the most part has not been judged by comparison with possible fits of other distribution families. An exception to this is the application of the lognormal distribution to air pollutant concentration; Bencala and Seinfeld (1976) have compared fits using several other standard distributions, and found that in general the lognormal distribution provided a best fit to their data. Such statistical applications of the lognormal distribution are not discussed here, but can be found in the books by Aitchison and Brown (1957) and Johnson, Kotz and Balakrishnan (1994). The lognormal distribution is often used in settings where there may be extremes in the right-hand tail, for example with data about wealth.

b. Mathematical Derivation

Because the normal distribution arises as a limiting distribution for sums, the lognormal distribution arises as a limiting distribution for products. More explicitly, suppose that $U_1, U_2, \ldots$ is a sequence of independent identically distributed random variables with finite expectation μ and variance σ^2. Then according to the central limit theorem 20.C.8, $[U_1 + U_2 + \cdots + U_n - n\mu]/\sqrt{n}\sigma$ has a limiting standard normal distribution. If $V_i = \exp\{U_i\}, i = 1, 2, \ldots$, it follows that $[\Pi_i^n V_i]^{1/\sqrt{n}}$ has what is termed a limiting "standard lognormal distribution." This was the motivation of Galton (1879) and McAlister (1879).

c. Survival Function

If Y is normally distributed with expectation μ and variance σ^2, then $X = e^Y$ has the survival function

$$\bar{F}(x) = \bar{\Phi}\left(\frac{\log x - \mu}{\sigma}\right) = \bar{\Phi}(\log(\lambda x)^\alpha) = \bar{\Phi}\left((\alpha \log x) - \left(\frac{\beta}{\alpha}\right)\right), \quad (1)$$

where Φ is the standard normal distribution function, $-\infty < \mu, \beta < \infty$, $x, \lambda, \sigma, \alpha > 0$, and

$$\mu = -\log \lambda, \ \sigma = 1/\alpha, \ \text{and} \ \beta = \alpha^2 \mu = \mu/\sigma^2. \quad (2)$$

d. Density

Three useful ways to parameterize the lognormal distribution are noted here. The densities are given subscripts to indicate differing parameterizations, but are all representations of the same density. When no confusion is possible, the subscripts are omitted.

The density f corresponding to the survival function (1) is given by

$$f(x) = \frac{1}{\sigma x}\phi\left(\frac{\log x - \mu}{\sigma}\right) = \frac{\alpha}{x}\phi\left(\log(\lambda x)^{\alpha}\right)$$

$$= \frac{\alpha}{x}\phi\left(\alpha \log x - \frac{\beta}{\alpha}\right), \quad x > 0, \tag{3}$$

where ϕ is the density of the standard normal distribution. More explicitly, for $x > 0$, with parameters related by (2),

$$f_1(x \mid \mu, \sigma) = \frac{1}{\sigma x \sqrt{2\pi}} \exp\{-(\log x - \mu)^2/2\sigma^2\}$$

$$= f_2(x \mid \lambda, \alpha) = \frac{\alpha}{x\sqrt{2\pi}} \exp\{-[\log(\lambda x)^{\alpha}]^2/2\}$$

$$= f_3(x \mid \alpha, \beta) = x^{\beta} \exp\{-\beta^2/2\alpha^2\} \frac{\alpha}{x\sqrt{2\pi}} \exp\{-[\log x^{\alpha}]^2/2\}$$

$$= x^{\beta} \exp\{-\beta^2/2\alpha^2\} f_2(x \mid 1, \alpha). \tag{4}$$

The form of the distribution that appears to be standard in books is f_1 with parameters μ and σ. These parameters are inherited from the normal distribution, but for the lognormal distribution, their meaning is less clear. As noted above, it can be seen from (1) or (4) that in the parameterization of f_2, λ is a scale parameter. With the parameterization of f_3, it follows from (4) that β is a moment parameter, but α is no longer a power parameter because of its connection with β in (2). It is demonstrated in Proposition 18.B.5 that the lognormal distribution is the only distribution that can be parameterized by either a scale parameter or a moment parameter. See Figures B.1 and B.2.

At both 0 and ∞, the density of the lognormal distribution tends to 0 faster than any power of x. This fact may partially explain the usefulness of the distribution in fitting some types of data.

B. The Lognormal Distribution   435

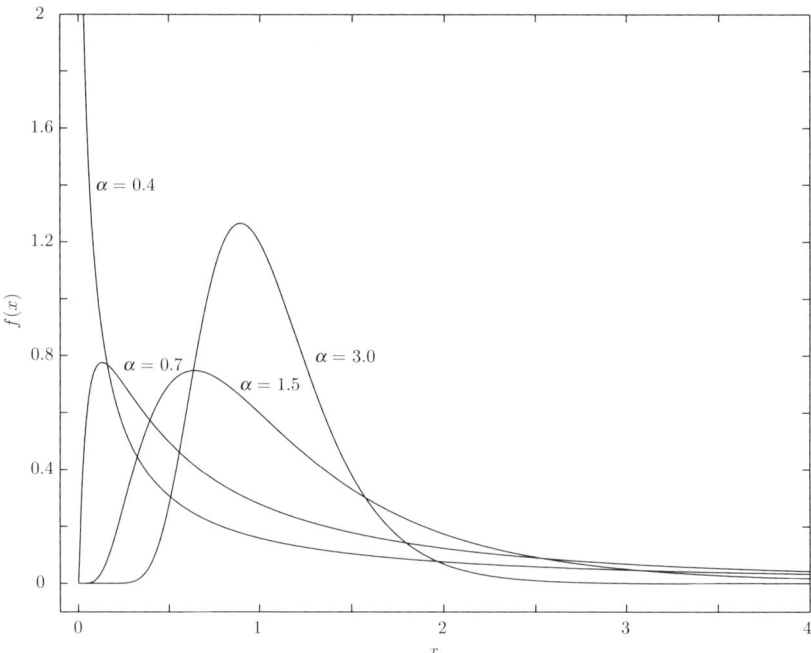

Fig. B.1. Densities of the lognormal distribution ($\lambda = 1$)

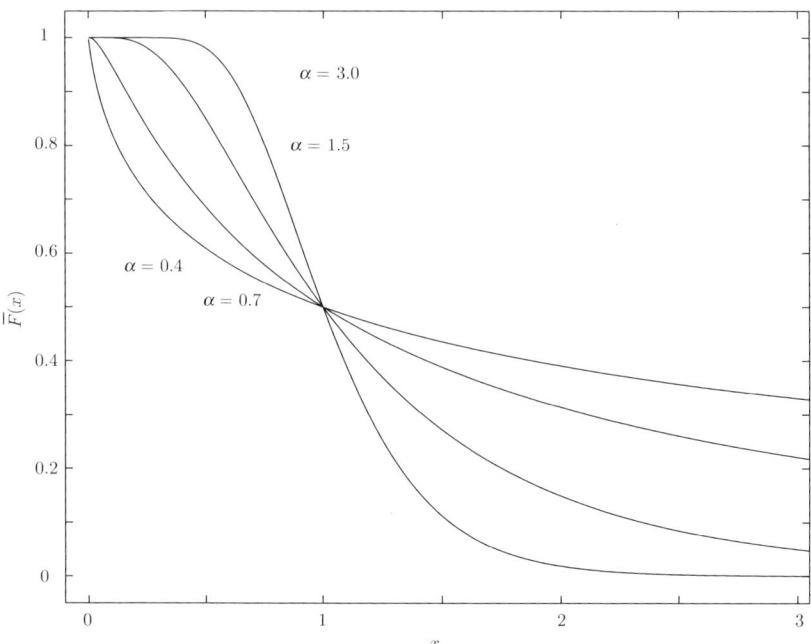

Fig. B.2. Survival functions of the lognormal distribution ($\lambda = 1$)

B.1. Proposition. For any real θ,

(i) $$\lim_{x \to \infty} x^\theta f(x) = \lim_{x \to 0} x^\theta f(x) = 0.$$

Moreover, the kth derivative $f^{(k)}$ of f satisfies

(ii) $$\lim_{x \to 0} x^\theta f^{(k)}(x) = 0.$$

Proof. The proofs of the results concerning f itself can be accomplished by making the change of variables $z = \log x$, and then finding the limit of the logarithm of the transformed $x^\theta f(x)$. Alternatively, the result can be obtained from the form f_3 of the density in (4) once it is known for $\theta = 1$. For the results concerning derivatives, make use of (i) with θ replaced by $\theta + k + 1$, that is,

$$\lim_{x \to \infty} x^{\theta+k+1} f(x) = \lim_{x \to 0} x^{\theta+k+1} f(x) = 0,$$

and apply l'Hospital's rule k times. □

B.2. Proposition. The lognormal density is unimodal, with mode at the point

$$\text{Mode} = \exp\{\mu - \sigma^2\} = \lambda^{-1} \exp\{-1/\alpha^2\} = \exp\{(\beta - 1)/\alpha^2\}. \tag{5}$$

This result can be verified by differentiating the density.

e. Hazard Rate

The hazard rate of the lognormal distribution cannot be expressed in closed form because the survival function of the normal distribution does not have a closed form. This makes the study of the lognormal hazard rate somewhat troublesome, and the first careful study of it appears to be that of Sweet (1990). As has long been noted from numerical work, the hazard rate is zero at both 0 and ∞, and it has a unique mode. See Figure B.3.

The hazard rate r of the lognormal distribution can be written in terms of the hazard rate $r_\mathcal{N} = \phi/\bar{\Phi}$ of the standard normal distribution as follows:

$$r(x) = \frac{r_\mathcal{N}(w)}{\sigma} \exp\{-\sigma w - \mu\}, \tag{6}$$

where $w = ((\log x) - \mu)/\sigma$.

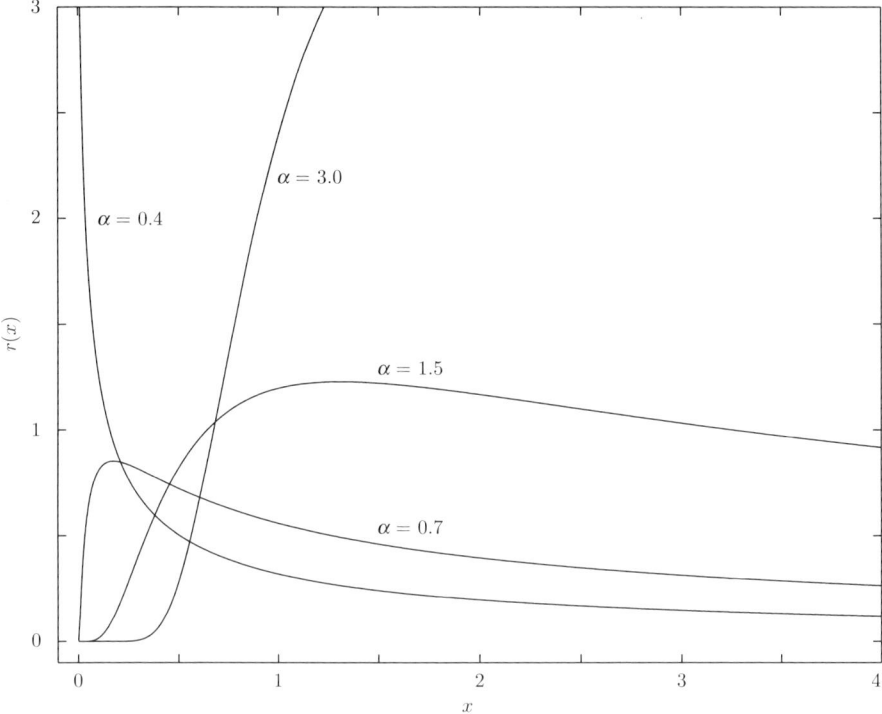

Fig. B.3. Hazard rates of the lognormal distribution ($\lambda = 1$)

B.3. Proposition. For all real θ, $\lim_{x \to 0} x^\theta r(x) = 0$. Moreover, the kth derivative $r^{(k)}$ of the hazard rate r satisfies $\lim_{x \to 0} x^\theta r^{(k)}(x) = 0$. Additionally, $\lim_{x \to \infty} r(x) = 0$.

Proof. The first part of this proposition follows directly from Proposition B.1. The limit at infinity can be computed directly with the aid of l'Hospital's rule. □

A proof of unimodality is somewhat more troublesome, but can be accomplished with the aid of the following lemma.

B.4. Lemma. The hazard rate $r_\mathcal{N}$ of the standard normal distribution is log concave.

Proof. The concavity of $\log r_\mathcal{N}$ can be demonstrated by showing that it has a negative second derivative, i.e., that $r''_\mathcal{N}(x) r_\mathcal{N}(x) \leq [r'_\mathcal{N}(x)]^2$.

By definition, $r_\mathcal{N} = \phi/\bar\Phi$, and so $\log r_\mathcal{N} = \log \phi - \log \bar\Phi$ and

$$\frac{r'_\mathcal{N}(x)}{r_\mathcal{N}(x)} = \frac{d \log r_\mathcal{N}(x)}{dx} = -\frac{x\phi(x)}{\phi(x)} + \frac{\phi(x)}{\bar\Phi(x)} = -x + r_\mathcal{N}(x).$$

This yields the equality $r'_\mathcal{N}(x) = r_\mathcal{N}(x)[r_\mathcal{N}(x) - x]$. Further,

$$\begin{aligned}r''_\mathcal{N}(x) &= 2r_\mathcal{N}(x)r'_\mathcal{N}(x) - xr'_\mathcal{N}(x) - r_\mathcal{N}(x)\\ &= r_\mathcal{N}(x)[2r^2_\mathcal{N}(x) - 3xr_\mathcal{N}(x) + x^2 - 1].\end{aligned}$$

With these derivatives, the condition $r''_\mathcal{N}(x)r_\mathcal{N}(x) \leq [r'_\mathcal{N}(x)]^2$ for log concavity can be rewritten as

$$r_\mathcal{N}(x) < [x + \sqrt{x^2 + 4}]/2. \tag{6a}$$

But this inequality is a result of Birnbaum (1942) which gives a bound for the hazard rate of the normal distribution. □

B.5. Proposition. The hazard rate of the lognormal distribution is unimodal, with mode at $\exp\{\sigma z + \mu\}$, where z is the unique solution of the equation

$$r_\mathcal{N}(z) = z + \sigma. \tag{6b}$$

This solution is less than $\exp\{1 + \mu - \sigma^2\}$.

Proof. The proof follows a sequence of steps. First, with $w = ((\log x) - \mu)/\sigma$, write the hazard rate r of the lognormal distribution in terms of the hazard rate of the normal distribution as in (6b); that is,

$$r(x) = e^{-\sigma w} r_\mathcal{N}(w)/\sigma e^\mu = e^{-\sigma w - \mu} r_\mathcal{N}(w)/\sigma, \quad 0 < x < \infty.$$

From this it follows that

$$r'(x) = \frac{dr(x)}{dx} = \frac{e^{-\sigma w - \mu}}{\sigma}[r'_\mathcal{N}(w) - \sigma r_\mathcal{N}(w)]\frac{dw}{dx}, \quad 0 < x < \infty.$$

Consequently, $r'(x) = 0$ if and only if

$$r'_\mathcal{N}(w)/r_\mathcal{N}(w) = \sigma. \tag{7}$$

By Lemma B.4, the ratio $r'_\mathcal{N}(w)/r_\mathcal{N}(w)$ is decreasing in w, and ranges from ∞ to $-\infty$ as w ranges from $-\infty$ to ∞ (x ranges from 0 to ∞). Thus, (7) must have one and only one solution say $w = z$. From the proof of Lemma B.4, $r'_\mathcal{N}(z) = r_\mathcal{N}(z)[r_\mathcal{N}(z) - z]$, so that (6b) follows from (7).

To show that the hazard rate mode is less than $\exp\{1 + \mu - \sigma^2\}$, it is necessary to show that $\exp\{\sigma z + \mu\} < \exp\{1 + \mu - \sigma^2\}$, that is,

$$z < (1 - \sigma^2)/\sigma.$$

According to (6a) and (6b),

$$\sigma = r_\mathcal{N}(z) - z < (\sqrt{z^2 + 4} - z)/2,$$

which reduces to the required inequality $z < (1 - \sigma^2)/\sigma$. □

An alternative proof of the unimodality of the hazard rate of the lognormal distribution can be obtained using Theorem 4.E.2.

The fact that the hazard rate of the lognormal distribution is eventually decreasing may not be of practical significance. Gottfried (1990) comments on that "the relationship between a statistical model and the real world almost always becomes tenuous in the tails." He points out that the mode of the lognormal hazard rate may be well out in the right-hand tail of the distribution.

f. Moments and Related Characteristics

In the above discussion, μ and σ^2 are, respectively, the mean and variance of the normally distributed random variable $Y = \log X$, and this notation is retained here. To distinguish between the moments of X and Y, the rth moment of the lognormally distributed random variable X is denoted below by μ_r even when $r = 1$.

A number of characteristics of the lognormal distribution can be expressed in several ways using the various related parameters that appear in the three forms of the density. It has already been observed that moments of $f(\cdot \mid 1, \alpha)$ can be obtained directly from the form (3) of the density. It follows that the rth moment $EX^r = \mu_r$ of the lognormal distribution is given by

$$\mu_r = \frac{1}{\lambda^r} \exp\left\{\frac{r^2}{2\alpha^2}\right\} = \exp\left\{\frac{r\beta}{\alpha^2} + \frac{r^2}{2\alpha^2}\right\} = \exp\left\{r\mu + \frac{r^2\sigma^2}{2}\right\}. \quad (8)$$

The moments can also be obtained from the moment generating function of the normal distribution, as indicated by A(6).

In particular, it follows from (8) that the expected value μ_1, variance Var X, and coefficient of variation $CV(X)$ are given by

$$\mu_1 = \frac{1}{\lambda} \exp\left\{\frac{1}{2\alpha^2}\right\} = \exp\left\{\frac{\beta}{\alpha^2} + \frac{1}{2\alpha^2}\right\} = \exp\left\{\mu + \frac{\sigma^2}{2}\right\}, \quad (9)$$

$$\text{Var } X = \exp\{2\mu_1 + \sigma^2\}[e^{\sigma^2} - 1] = e^{1/\alpha^2}(1 - e^{1/\alpha^2})/\lambda^2$$
$$= \exp\{(1 - 2\beta)/\alpha^2\}[e^{1/\alpha^2} - 1], \quad (10)$$

$$CV(X) = [e^{\sigma^2} - 1]^{1/2} = [e^{1/\alpha^2} - 1]^{1/2}. \quad (11)$$

Feller (1971, p. 227) notes that the lognormal distribution is not determined by its moments. He gives an example of another density with exactly the same moments as the lognormal distribution, with credit to C. C. Heyde. This example is the density

$$f_\alpha(x) = \frac{1}{x\sqrt{2\pi}} \exp\left\{-\frac{1}{2}(\log x)^2\right\}[1 + \alpha \sin(2\pi \log x],$$
$$-1 \leq \alpha \leq 1, \quad x \geq 0.$$

As noted in Proposition B.2, the lognormal distribution is unimodal with mode $\exp\{\mu - \sigma^2\}$ given by (5). Because the median of the normal distribution is μ, it follows that the median Med of the lognormal distribution is

$$\text{Med} = e^\mu = 1/\lambda. \quad (12)$$

From (5), (9), and (12), it can be seen that for the lognormal distribution,

$$\text{Mode} < \text{Med} < \text{Mean}.$$

This indicates that the density is "stretched to the right," or in other words, has a long right tail.

g. Infinite Divisibility

The lognormal distribution was shown to be infinitely divisible by Thorin (1977b). Thorin's methods, involving Laplace transforms, are involved and are beyond the scope of this book.

h. Ordering Lognormal Distributions

Because λ is a scale parameter, it follows from Observation 7.C.5 that lognormal distributions are, in the usual stochastic ordering, decreasing in the scale parameter λ. However, these distributions are not decreasing in the hazard rate order because the hazard rates are not decreasing. Of course, this means that the likelihood ratio order also does not order lognormal distributions.

Because the parameter α in (4) is a power parameter, it follows from Proposition 7.D.5 that lognormal distributions are in the convex transform order, decreasing in the parameter α.

C. Log Logistic Distributions

The standard logistic distribution has distribution and survival functions given by

$$H(x) = [1 + \exp\{-(x-a)/b\}]^{-1}, \quad \bar{H}(x) = [1 + \exp\{(x-a)/b\}]^{-1},$$
$$-\infty < x < \infty. \tag{1}$$

This distribution is briefly discussed in Section 9.D.b, and extensive studies of it are reviewed by Johnson, Kotz and Balakrishnan (1995, Chapter 23).

If Y has the distribution (1) and $X = e^Y$, then with $\lambda = e^{-a}, \alpha = 1/b$, it follows in a straightforward way that X has the Pareto III distribution F given by

$$\bar{F}(x) = 1/[1 + (\lambda x)^\alpha], \quad x > 0. \tag{2}$$

See Section 11.B for a discussion of this distribution.

The form of the distribution function and survival function of the logistic distribution make the introduction of frailty and resilience parameters inviting. The case of a resilience parameter was studied by Ahuja and Nash (1967) and Dubey (1968). This extension is called the Type I generalized logistic distribution by Johnson and Kotz (1970b, p. 140). They call the extension with frailty parameter a Type II generalized logistic distribution. The logarithmic Type I and Type II generalized logistic distributions are obtained from (2) by adding, respectively, resilience and frailty parameters. These logarithmic distributions are introduced in Section 11.B.a.

Another generalization of the logistic distribution has the density

$$h(y\,|\,p,q) = \frac{e^{py}}{B(p,q)(1+e^y)^{p+q}}, \quad -\infty < y < \infty, \quad p,q > 0,$$

the logarithmic form of which is the F distribution of Section 11.D.

D. Log Extreme Value Distributions

Extreme value distributions are discussed in Section 20.G. Two of these distributions, G and H, have support $(-\infty, \infty)$ and are defined, respectively, by

$$G(y) = \exp\{-e^{-(y-a)/b}\}, \quad -\infty < y < \infty, \qquad (1)$$

and

$$\bar{H}(y) = \exp\{-e^{-(y-a)/b}\}, \quad -\infty < y < \infty. \qquad (2)$$

The first of these distributions is an extreme value distribution for a maximum, and the second is an extreme value distribution for a minimum. They are closely related; if Y has the distribution (1), then $Z = 2a - Y$ has the survival function (2). If Y has distribution G, then $X = e^Y$ has the log extreme value distribution F_+ of the form

$$F_+(x) = \exp\{-1/(\lambda x)^\alpha\}, \quad x \geq 0, \qquad (3)$$

and on the other hand, $X = e^{-Y}$ has the negative log extreme value distribution $\bar{F}_-$ of the form

$$\bar{F}_-(x) = \exp\{-(\lambda x)^\alpha\}, \quad x \geq 0. \qquad (4)$$

Of course, F_- is a Weibull distribution and F_+ is the distribution of a variable with reciprocal having a Weibull distribution. Because of the relationship between G and H, similar results are obtained by starting with (2) rather than with (1). Only in this case, the log and negative log distributions are interchanged.

As noted by Johnson and Kotz (1970b, p. 3), the distribution (1) has been called a "log Weibull distribution." However, in this book, the pattern set by the long entrenched terminology for the lognormal distribution has been followed.

E. The Log Cauchy Distribution

The log Cauchy distribution is an unusual distribution, perhaps more curious than useful; no moments apart from the zero-th moment are finite. The basic log Cauchy distribution (without parameters) is given by

$$f(x) = \{\pi x[1 + (\log x)^2]\}^{-1}, \quad \bar{F}(x) = \frac{1}{2} - \frac{1}{\pi}\tan^{-1}\log x, \quad x > 0. \quad (1)$$

From (1), it follows that the hazard rate is

$$r(x) = \left\{\pi x[1 + (\log x)^2]\left[\frac{1}{2} - \frac{1}{\pi}\tan^{-1}\log x\right]\right\}^{-1}. \quad (2)$$

See Figure E.1.

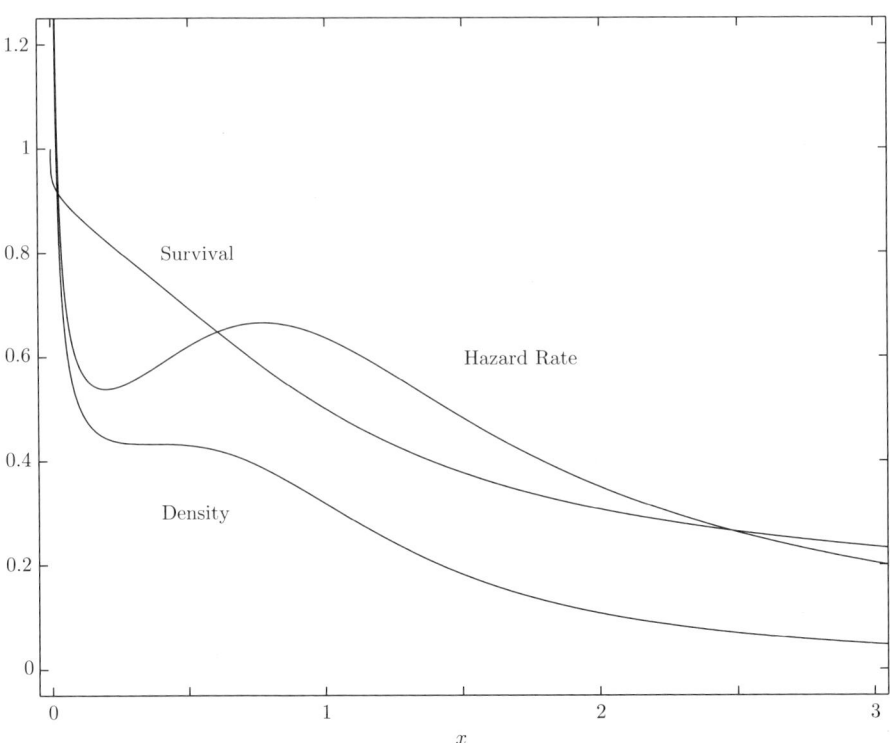

Fig. E.1. Density, survival function and hazard rate of the log Cauchy distribution

E.1. Proposition. The log Cauchy distribution has a decreasing density, and a hazard rate that is initially decreasing, eventually decreasing, but may increase on some interval.

Proof. The density f has a derivative given by

$$f'(x) = -\frac{(1+\log x)^2}{\pi x[1+(\log x)^2]}, \tag{3}$$

from which it is easy to see that f is decreasing. To determine the nature of the hazard rate, it is convenient to compute

$$\rho(x) = -f'(x)/f(x)$$
$$= (1+\log x)^2/x[1+(\log x)^2],$$

and consequently,

$$\rho'(x) = (1+\log x)[1 - 3\log x - (\log x)^2 - (\log x)^3]/x^2[1+(\log x)^2]^2.$$

The equation $\rho'(x) = 0$ has the root $x = 1/e$. It also has one real root from the factor $z^3 + z^2 + 3z - 1 = 0$, where $z = \log x$ (see, e.g., Abramowitz and Stegun (1964, Paragraph 3.8.2, p. 17). The conclusion follows from Theorem 4.E.2 □

E.2. Proposition. If X has a log Cauchy distribution then EX^r exists finitely if and only if $r = 0$.

Proof. Rewrite EX^r in the form

$$EX^r = \int_{-\infty}^{\infty} \frac{e^{rx}}{\pi(1+x^2)} dx;$$

if $r < 0$, then the integrand has the limit ∞ at $-\infty$, and if $r > 0$, the integrand has the limit ∞ at ∞. Thus, the integral cannot be finite. □

The results of the above propositions do not change if a scale parameter is introduced in the distribution (1). On the other hand, the behavior of the density and hazard rate may change with the introduction of a power parameter.

F. The Log Student's t Distribution

The generally unfamiliar log t distribution can be regarded as a generalization of the lognormal distribution. In its simplest form, this distribution has the density

$$f(x\,|\,\nu) = \left[\sqrt{\nu}B\left(\frac{1}{2},\frac{\nu}{2}\right)\left(1+\frac{(\log x)^2}{\nu}\right)^{(\nu+1)/2} x\right]^{-1}, \quad x,\nu > 0. \qquad (1)$$

Because the normal distribution is the limit of Student's t distribution when $\nu \to \infty$, it follows that the lognormal distribution is the limit of the log t distribution. For full generality, this requires the introduction of additional parameters in (1), as can be done by introducing a location and scale parameter in the t distribution.

The hazard rates of the log t distributions do not take particularly nice forms.

G. Alternatives for the Logarithm Function

In the relationship $F(x) = H(\log x)$ of A(1), the logarithm plays a key role in transforming distributions with support $(-\infty, \infty)$ to distributions concentrated on $[0, \infty)$. The essential properties of the logarithm function for this change of support is that the logarithm is a strictly increasing function, and

$$\lim_{x \to 0} \log x = -\infty, \quad \lim_{x \to \infty} \log x = \infty.$$

Any other function, say w, can be substituted for the logarithm providing only that

(i) $w(x)$ is strictly increasing in $x > 0$,

and

(ii) $\lim_{x \to 0} w(x) = -\infty, \quad \lim_{x \to \infty} w(x) = \infty.$

This idea is embedded in the work of Slymen and Lachenbruch (1984), who in addition to the logarithm, consider the three functions

$$w_1(x\,|\,\theta) = \frac{x^\theta - x^{-\theta}}{2\theta}, \quad \theta > 0, \qquad (1)$$

$$w_2(x\,|\,\lambda) = \log(e^{\lambda x} - 1), \quad \lambda > 0, \qquad (2)$$

$$w_3(x\,|\,\lambda,\alpha) = \log\log[1 + (\lambda t)^\alpha], \quad \lambda,\alpha > 0. \qquad (3)$$

Note that these functions satisfy (i) and (ii).

For any function w satisfying (i) and (ii), $cw + d$ satisfies these conditions providing only that $c > 0$. This offers a way to introduce more parameters in the transformed distribution.

Because

$$\lim_{\theta \to 0} w_1(x \mid \theta) = \log x,$$

the results of employing w_1 in place of the logarithm can be interpreted as generalizing the logarithmic distributions through the introduction of the additional parameter θ.

G.1. Example. Transformed extreme value distributions (modified Weibull distributions) (Slymen and Lachenbruch, 1984). As noted in Section D, the logarithmically transformed extreme value distribution H of D(2) is a Weibull distribution.

If the log function is replaced by w_1, the survival function

$$\bar{F}(x) = \bar{H}(w_1(x)) = \exp\left\{-\exp\left[\frac{x^\theta - x^{-\theta}}{\theta b} - \frac{a}{b}\right]\right\}, \quad x, \theta > 0, \quad (4)$$

is obtained. This survival function has been called the *modified Weibull survival function* by Slymen and Lachenbruch (1984). For most values of the parameters, this distribution has a hazard rate that is a delayed bathtub shape. See Figures G.1, G.2, G.3a, and G.3b.

If the log function is replaced by $\zeta w_2 + \log \xi$, the survival function 10.C(12) of a Gompertz distribution with hazard power parameter is obtained.

If the log function is replaced by w_3 and $a = 0, b = 1$ in D(2), the resulting survival function is

$$\bar{F}(x) = \{1 + \log[1 + (\lambda x)^\alpha]\}^{-1}, \quad x > 0 \quad (5)$$

This survival function has a decreasing hazard rate.

G.2. Example (Slymen and Lachenbruch, 1984). As noted in Section C, the log logistic distribution is a Pareto III distribution. If the log function used in Section C to transform the logistic distribution is replaced by w_2, the resulting survival function is

$$\bar{F}(x) = \bar{H}(w_2(x \mid \theta)), \quad x > 0,$$

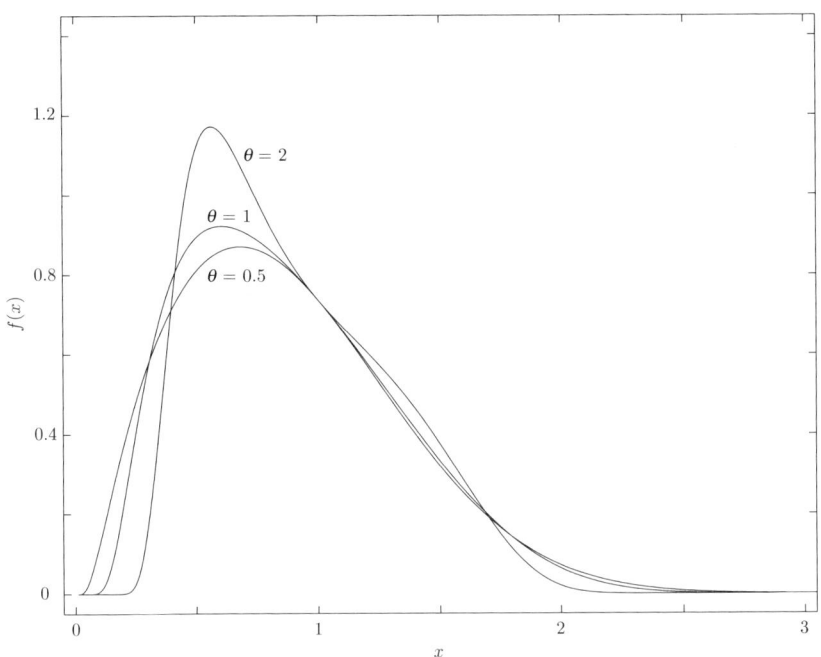

Fig. G.1. Densities of the modified Weibull distribution $(a = 0, b = 1)$

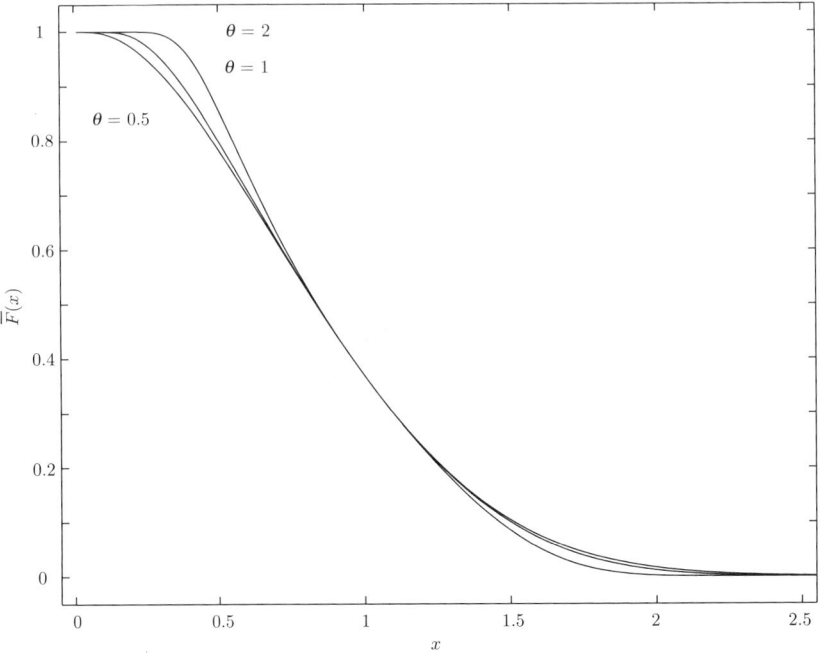

Fig. G.2. Survival functions of the modified Weibull distribution $(a = 0, b = 1)$

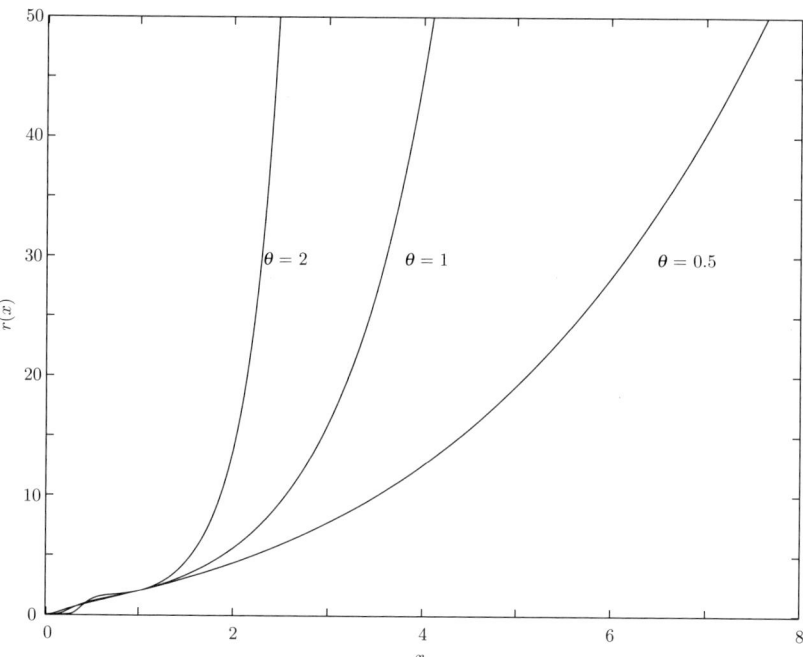

Fig. G.3a. Hazard rates of the modified Weibull distribution ($a = 0, b = 1$)

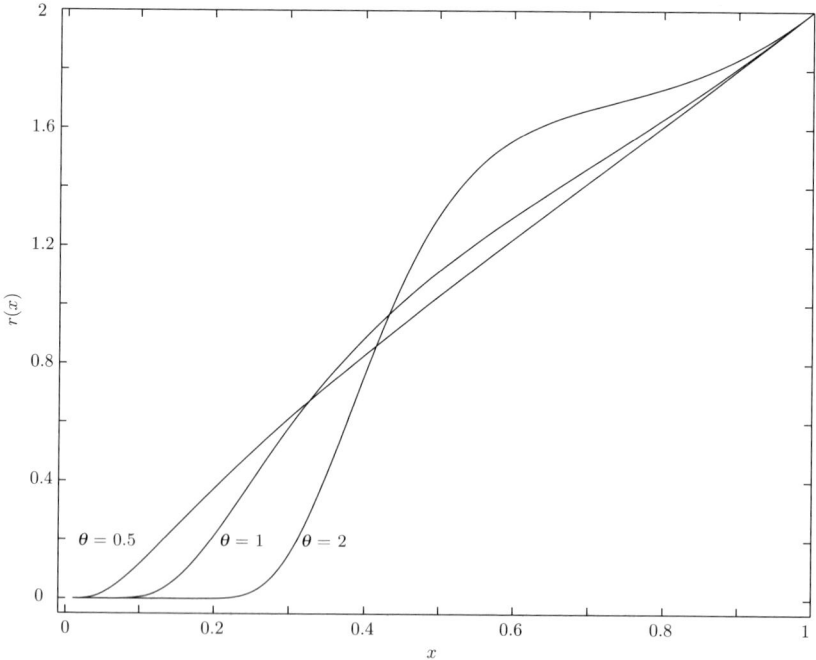

Fig. G.3b. Hazard rates of the modified Weibull distribution ($a = 0, b = 1$) in the neighborhood of the origin

where $\bar{H}$ is given by C(1). This survival function is somewhat awkward. However, when the parameters of $\bar{H}$ are $a = 0$ and $b = 1/\theta$, then

$$\bar{F}(x) = [1 + (e^{\lambda x} - 1)^\theta]^{-1}, \quad \lambda,\ \theta > 0,\ x > 0. \tag{6}$$

Here is another two-parameter extension of the exponential distribution which is obtained with $\theta = 1$.

If the log function used in Section C to transform the logistic distribution is replaced by w_1, the resulting distribution is called the *modified log-logistic distribution* by Slymen and Lachenbruch (1984). The survival function of this distribution is given by

$$\bar{F}(x) = \bar{H}(w_1(x \mid \theta)), \quad x > 0,$$

where $\bar{H}$ is given by C(1). This survival function does not take a particularly attractive form. However, the hazard rate has been plotted by Slymen and Lachenbruch (1984) and found to take a variety of shapes.

If the log function used in Section C to transform the logistic distribution is replaced by w_3, the result is a Pareto III distribution.

Box and Cox (1964) introduced the transformation

$$\begin{aligned} w(x) &= (x^\lambda - 1)/\lambda, \quad x, \lambda > 0, \\ &= \log x, \quad x > 0,\ \lambda = 0. \end{aligned}$$

This can be viewed as a parametric extension of the logarithm.

See also MacGillivray (1992) for other transformations of the normal distribution.

13

The Inverse Gaussian Distribution

The inverse Gaussian distribution was derived by Schrödinger (1915) and Smoluchowski (1915) as the first passage time distribution of Brownian motion with a drift. The distribution subsequently arose in the related contexts of population growth studies by Hadwiger (1940), in an early application to clinical trials by Tweedie (1941), and in the context of the sequential analysis by Wald (1947); all of these are in the context of Brownian motion. The distribution has also been called the "Wald distribution." The inverse Gaussian distribution is the subject of books by Chhikara and Folks (1989) and Seshadri (1993, 1999). Saunders (2007) offers a relatively brief, but inciteful treatment. For a review of the inverse Gaussian distribution, see Folks and Chhikara (1978). A particularly interesting history of the distribution and its close relatives is given by Seshadri (1993). Some of the essential definitions and properties are presented in Section A.

The inverse Gaussian distribution has two parameters that come from the drift parameter and the variance of the Brownian motion. Etienne Halphen proposed a three-parameter distribution for fitting hydrological data. As noted by Seshadri (1993, p. 2), because of anti-Jewish regulations imposed by the Nazis, Halphen's work was published under the name "Dugué" (1941). Halphen's three-parameter distribution, later named the "generalized inverse Gaussian distribution" by Barndorff-Nielsen and Halgreen (1977), can be obtained from the inverse Gaussian distribution through the introduction of a moment parameter. However, Halphen arrived at the generalized inverse Gaussian distribution through a two-parameter special case different from the usual two-parameter inverse Gaussian distribution. The generalized inverse Gaussian distribution has been extensively studied by Jørgensen (1982).

The term "inverse Gaussian" is due to Tweedie (1945, 1956), and is based upon a relationship between the cumulant generating functions of the Gaussian and inverse Gaussian distributions. The Gaussian and inverse Gaussian distributions have several similarities; both families are closed under convolutions and have similar characterizations (see Chhikara and Folks, 1989, Section 4.4; Seshadri, 1993, Chapter 3).

A. The Inverse Gaussian Distribution

a. The Density Function

The inverse Gaussian distribution is, like the gamma distribution, most easily defined in terms of the density because the survival function cannot be expressed in closed form. With the usual parameterization, the density is given by

$$f_a(x) = \frac{\sqrt{\theta}}{\sqrt{2\pi x^3}} \exp\left\{-\frac{\theta(x-m)^2}{2m^2 x}\right\}$$
$$= \frac{\sqrt{\theta}}{\sqrt{2\pi x^3}} e^{\theta/m} \exp\left\{-\frac{\theta}{2}\left(\frac{x}{m^2} + \frac{1}{x}\right)\right\}, \quad x, \theta, m > 0. \quad (1a)$$

The distribution with this density is sometimes labeled the "IG(m, θ) distribution," and is the form used by Tweedie (1945, 1956).

The density (1a) is sometimes given with the alternative parameterization $m = \beta, \theta = \beta/\alpha^2$. Then, (1a) can be written in the form

$$f_b(x) = \frac{\sqrt{\beta}}{\alpha\sqrt{2\pi x^3}} \exp\left\{-\frac{1}{\alpha^2}\right\} \exp\left\{-\frac{1}{2\alpha^2}\left(\frac{x}{\beta} + \frac{\beta}{x}\right)\right\}; \quad (1b)$$

it is with this parameterization and this form of the density that Wald's name is associated.

Yet another parameterization of the inverse Gaussian distribution is of interest. With $\alpha = (\lambda\chi)^{-1/4}$ and $\beta = (\chi/\lambda)^{1/2}$, (1b) becomes

$$f_c(x) = \frac{\lambda}{\sqrt{2\pi(\lambda x)^3}} \exp\left\{\sqrt{\lambda\chi}\right\} \exp\left\{-\frac{1}{2}\left(\lambda x + \frac{\chi}{x}\right)\right\}. \quad (1c)$$

A. The Inverse Gaussian Distribution

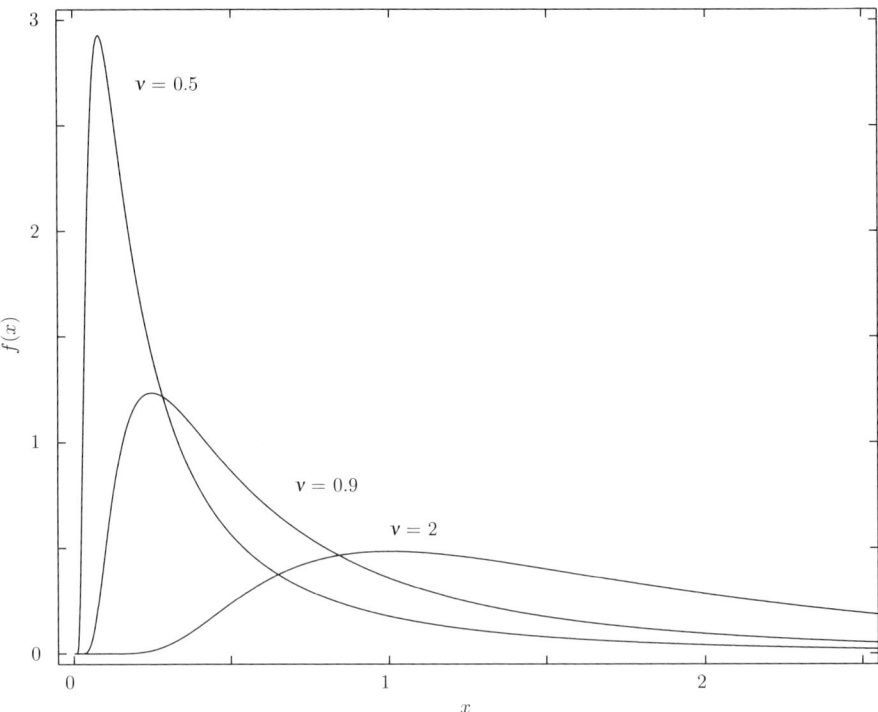

Fig. A.1. Densities of the inverse Gaussian distribution ($\lambda = 1$)

The roles of the parameters θ, m, α, and β are not entirely transparent, so let $m = \beta = \eta/\lambda$ and $\theta = \beta/\alpha^2 = \nu^2/\lambda$; then $\nu = \theta/m = 1/\alpha^2$ and $\lambda = \theta/m^2 = 1/\alpha^2\beta$. With this parameterization, the density (1b) takes the form

$$f(x \mid \lambda, \nu) = \frac{\lambda\nu}{\sqrt{2\pi(\lambda x)^3}} \exp\left\{-\frac{(\lambda x - \nu)^2}{2\lambda x}\right\}, \quad x, \lambda, \nu > 0. \qquad (2)$$

Now it is clear that λ is a scale parameter, and consequently

$$f(x \mid \lambda, \nu) = \lambda\, f(\lambda x \mid 1, \nu). \qquad (3)$$

See Figure A.1.

To understand the role of the parameter ν, it is helpful to compute the Laplace transform of f. The following can be derived with some algebra to complete the square in the exponent of the density;

alternatively, it can be directly verified using (2) that

$$\phi(s) = E\, e^{-sX} = \int_0^\infty e^{-sx} f(x\,|\,\lambda,\nu)\,dx$$

$$= \exp\left\{\nu\left[1-\left(1+\frac{2s}{\lambda}\right)^{1/2}\right]\right\} \int_0^\infty f\left(x\,\Big|\,\lambda+2s, \nu\left(1+\frac{2s}{\lambda}\right)^{1/2}\right) dx$$

$$= \exp\left\{\nu\left[1-\left(1+\frac{2s}{\lambda}\right)^{1/2}\right]\right\}. \tag{4}$$

Note that ν appears as an exponent in this transform; this means that ν is a convolution parameter, and the following proposition is a consequence.

A.1. Proposition. The density f of (2) is infinitely divisible and moreover,

$$\int f(x-t\,|\,\lambda,\nu_1) f(t\,|\,\lambda,\nu_2)\,dt = f(x\,|\,\lambda,\nu_1+\nu_2). \tag{5}$$

By differentiating the logarithm of density (2), it is straightforward to verify that the density is unimodal with

$$\text{mode} = [-3 + (9+4\nu^2)^{1/2}]/2\lambda. \tag{6}$$

With $r > 0$, successive applications of l'Hospital's rule yield

$$\lim_{x\to 0} f(x)/x^r = \lim_{x\to\infty} x^r f(x) = 0. \tag{7}$$

b. The Survival Function

The survival function $\bar{F}$ corresponding to the density (2) can be given in two different forms. First, in terms of the distribution function Φ of the standard normal distribution,

$$\bar{F}(x\,|\,\lambda,\nu) = \bar{\Phi}\left(\frac{\lambda x - \nu}{\sqrt{\lambda x}}\right) - e^{2\nu}\Phi\left(-\frac{\lambda x + \nu}{\sqrt{\lambda x}}\right); \tag{8}$$

second, in terms of the survival function

$$\bar{G}(z) = \int_z^\infty (2\pi t)^{-1/2} \exp\left\{\frac{-t}{2}\right\} dt$$

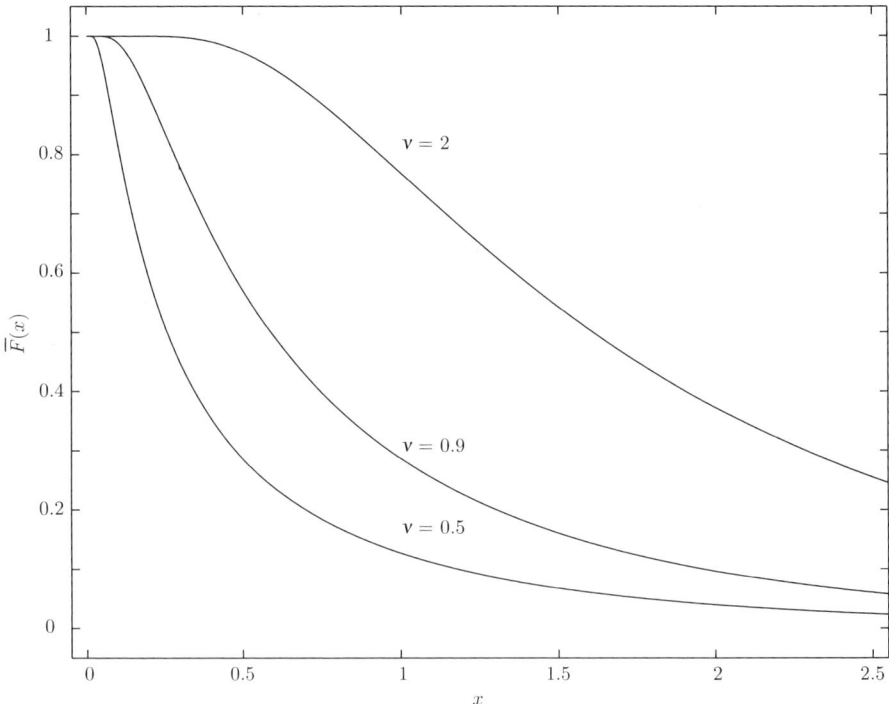

Fig. A.2. Survival functions of the inverse Gaussian distribution ($\lambda = 1$)

of a chi-squared distribution with one degree of freedom (a gamma distribution with scale parameter $1/2$ and shape parameter $1/2$),

$$\begin{aligned}
\bar{F}(x \mid \lambda, \nu) &= 1 - \frac{1}{2}\bar{G}\left(\frac{(\lambda x - \nu)^2}{\lambda x}\right) - \frac{1}{2} e^{2\nu}\bar{G}\left(\frac{(\lambda x + \nu)^2}{\lambda x}\right), \quad 0 \leq x \leq \nu/\lambda, \\
&= \frac{1}{2}\bar{G}\left(\frac{(\lambda x - \nu)^2}{\lambda x}\right) - \frac{1}{2} e^{2\nu}\bar{G}\left(\frac{(\lambda x + \nu)^2}{\lambda x}\right), \quad x \geq \nu/\lambda. \quad (9)
\end{aligned}$$

The form (9) follows from (8), but derivations of (8) and (9) are rather involved and are not given here. The form (8) was found independently by Zigangirov (1962) and Shuster (1968), but see Seshadri (1993, Section 2.9) for further elucidation. Fortunately, a relatively straightforward differentiation of (8) or (9) can be carried out to yield (2), so (8) and (9) can be verified even if they are not easy to derive. See Figure A.2.

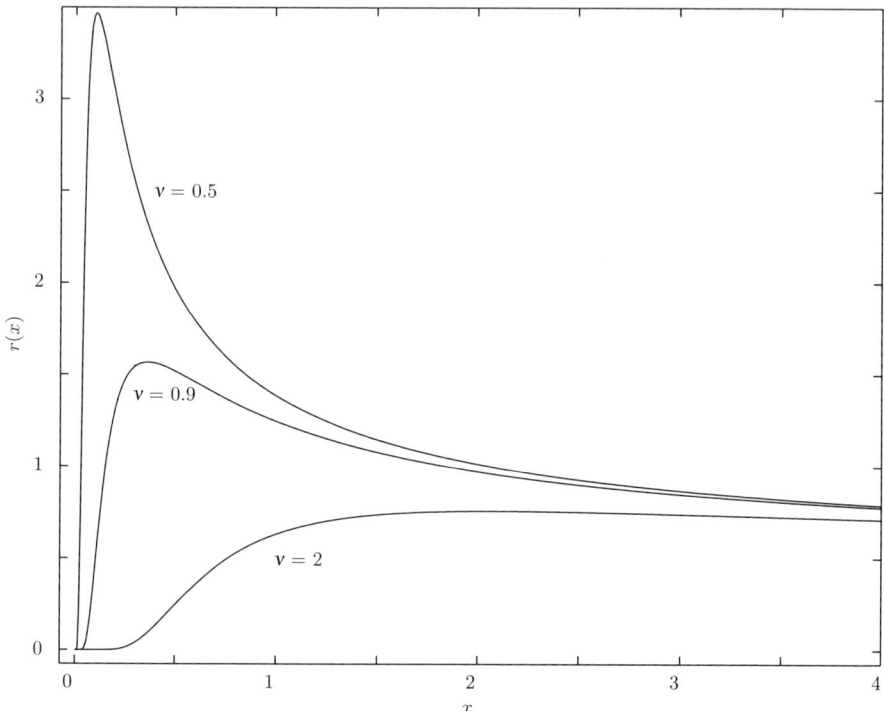

Fig. A.3. Hazard rates of the inverse Gaussian distribution ($\lambda = 1$)

c. The Hazard Rate

The hazard rate of the inverse Gaussian distribution is the ratio of the density (2) to the survival function (8), but the resulting expression is somewhat awkward. Fortunately some properties of the hazard rate can be found without direct examination of any analytical expression for it. See Figure A.3.

The goal of the following is to show that the hazard rate starts off at 0 and increases to a unique maximum, and then decreases to a limiting value $\lambda/2$. Also, some bounds are obtained for the hazard rate mode. These results are all obtained by Chhikara and Folks (1977, 1989) with an ingenious but very involved use of elementary calculus and algebra. Here, a derivation is given that is simpler and more direct, but it makes use of Theorem 4.E.2, and thus it depends on some theory of total positivity.

To apply Theorem 4.E.2, first compute

$$\rho(x) = -\frac{f'(x)}{f(x)} = -\frac{d \log f(x)}{dx} = \frac{3}{2x} + \frac{\lambda}{2} - \frac{\nu^2}{2\lambda x^2}. \tag{10}$$

As indicated by (7), $\lim_{x\to\infty} f(x) = \lim_{x\to 0} f(x) = 0$; thus, $\lim_{x\to 0} r(x) = 0$. By using the definition of r, then l'Hospital's rule, and finally (10), it follows that

$$\lim_{x\to\infty} r(x) = \lim_{x\to\infty} \frac{f(x)}{\overline{F}(x)} = \lim_{x\to\infty} -\frac{f'(x)}{f(x)} = \frac{\lambda}{2}.$$

By computing $d\rho(x)/dx = (2\nu^2 - 3\lambda x)/2\lambda x^3$, it can be seen that $\rho(x)$ is increasing for $x \le 2\nu^2/3\lambda = x_0$ and decreasing for $x \ge x_0$. Thus, by Theorem 4.E.2, there exists $x_1 \le x_0$ such that the hazard rate $r(x)$ is increasing $x \le x_1$ and decreasing in $x \ge x_1$. The actual hazard rate mode is the solution to the equation

$$r(x) = \frac{3}{2x} + \frac{\lambda}{2} - \frac{\nu^2}{2\lambda x^2}, \quad \text{that is,} \quad r(x) = \rho(x).$$

d. Moments

The moment generating function mgf$(s) = \phi(-s)$ of the inverse Gaussian distribution can be obtained directly from (4). By differentiating this moment generating function r times and setting $s = 0$, the integer moments of the inverse Gaussian distribution can be obtained. Specifically,

$$\begin{aligned} EX^r &= \frac{\nu^r}{\lambda^r} \sum_{j=0}^{r-1} \frac{(r-1+j)!}{j!(r-1-j)!} (2\nu)^{-j} \\ &= \frac{\nu^r}{\lambda^r} \sum_{j=0}^{r-1} \frac{\Gamma(r+j)}{\Gamma(r-j)} \frac{(2\nu)^{-j}}{j!}, \quad r = 1, 2, \ldots. \end{aligned} \quad (11)$$

In particular,

$$EX = \frac{\nu}{\lambda} = \mu \quad \text{and} \quad \mathrm{Var}(X) = \frac{\nu}{\lambda^2} = \frac{\mu^3}{\theta}.$$

Thus, it is apparent that the parameter m in (1a) is the first moment μ, and the coefficient of variation is

$$CV(X) = \frac{\sqrt{\mathrm{Var}(X)}}{EX} = \frac{1}{\sqrt{\nu}} = \sqrt{\frac{\mu}{\theta}}.$$

Note that the mode of the density, given by (6), is less than the expected value.

There is a simple relationship between positive and negative moments given by

$$E\left(\frac{\lambda X}{\nu}\right)^{-r} = E\left(\frac{\lambda X}{\nu}\right)^{r+1}, \quad r = 1, 2, \ldots.$$

This relationship can be obtained from the integral $\int_0^\infty (\lambda x/\nu)^{-r} f(x)\,dx$ with the change of variables $y = \nu/\lambda x$ (for details see Chhikara and Folks (1977, 1989) or Seshadri 1993, p. 52).

e. Ordering Inverse Gaussian Distributions

Because the parameter λ in the density (2) is a scale parameter, the corresponding distribution is, in the sense of the usual stochastic order, decreasing in λ; in fact, the stronger likelihood ratio ordering also holds. This result follows from a straightforward but somewhat tedious verification of condition 2.A(11). That condition can here be written as

$$\frac{f(x\mid\lambda_1,\nu)}{f(x\mid\lambda_2,\nu)} \leq \frac{f(y\mid\lambda_1,\nu)}{f(y\mid\lambda_2,\nu)}, \quad x < y,\ \lambda_1 < \lambda_2,$$

which reduces to $(y - x)(\lambda_2 - \lambda_1)(\nu^2 + \lambda_1\lambda_2 xy) \geq 0$.

From Proposition A.1, it follows that ν is a convolution parameter, and consequently (from Propositions 7.J.4) the inverse Gaussian distribution is increasing in ν in the sense of stochastic order. However, condition 2.A(11) can again be used to show the stronger result that the inverse Gaussian distribution is likelihood ratio increasing in ν.

Because ν is a convolution parameter, it follows from Proposition 7.J.6 the inverse Gaussian distribution is increasing in ν in the sense of Lorenz order.

f. Limiting Normal Distribution

Because ν is a convolution parameter, it follows from the central limit theorem 20.C.8 that if X has the inverse Gaussian density (2), then $(X - EX)/\sqrt{Var(X)} = (X - (\nu/\lambda))/\sqrt{\nu}/\lambda$ has a limiting standard normal distribution as $\nu \to \infty$. This result is given by Whitmore and Yalovsky (1978), who note that the convergence is quite slow.

g. A Relationship with the Chi-Square Distribution

It is shown by Shuster (1968) that if X has the inverse Gaussian distribution, then

$$W = (\lambda X - \nu)^2/\lambda X$$

has a chi-square distribution with one degree of freedom. Shuster's proof involves a two-step transformation, first to $Y = \min(X, \nu^2/\lambda^2 X)$, and then to $Z = (\lambda Y - \nu)^2/\lambda Y$. This avoids consideration separately of the cases $\lambda Y \leq \nu$ and $\lambda Y > \nu$.

h. Density of the Reciprocal Inverse Gaussian Variate

As indicated in the next section, the density of $1/X$, where X has an inverse Gaussian distribution is sometimes of interest. This density is a special case of the generalized inverse Gaussian distribution of Section B.

If X has the density (1), then a straightforward calculation shows that the density g of $Y = 1/X$ is given by

$$g(y) = \frac{\sqrt{\theta}}{\sqrt{2\pi y}} \exp\left\{-\frac{\theta(my-1)^2}{2m^2 y}\right\}. \tag{12}$$

Similarly, with the parameterization of (2), Y has the density

$$g(y \mid \lambda, \nu) = \frac{\nu}{\sqrt{2\pi \lambda y}} \exp\left\{-\frac{(\nu y - \lambda)^2}{2\lambda y}\right\}. \tag{13}$$

B. The Generalized Inverse Gaussian Distribution

Halphen, publishing through Dugué (1941), proposed a density of the form

$$f(x) = Cx^{-1}\exp\left\{-ax - \frac{b}{x}\right\}, \quad x > 0, \tag{1}$$

which he called the "harmonic type." Halphen, who had developed this density for use in hydrology, remarked that this distribution can be generalized in many ways, but he stated that the most interesting generalization is of the form

$$f(x) = C_1 x^{\alpha-1} \exp\left\{-ax - \frac{b}{x}\right\}, \qquad (2)$$

where $C_1 = C_1(a, b, \alpha)$ depends upon $a, b,$ and α. Note that (2) not only generalizes (1) but it also generalizes A(1), the density of the inverse Gaussian distribution. The density (2) was named the "generalized inverse Gaussian" distribution by Barndorff-Nielsen and Halgreen (1977).

It is clear that (2) can be obtained from either (1) or A(2) by the introduction of a moment parameter and appropriate change of parameters. To be more specific, introduce a moment parameter β in A(2) to obtain the density

$$f(x) = \frac{C_2(\lambda x)^{\beta}}{\sqrt{(\lambda x)^3}} \exp\left\{-\frac{(\lambda x - \nu)^2}{2\lambda x}\right\}, \qquad x, \lambda, \nu > 0,\ -\infty < \beta < \infty, \qquad (3)$$

where the norming constant $C_2 = C_2(\lambda, \nu, \beta)$ depends upon the three-parameters. This density is not yet in a particularly convenient form, but with $\theta = \beta - (1/2)$, it can be rewritten as

$$f(x) = \frac{C_3}{\lambda}(\lambda x)^{\theta-1} \exp\left\{-\frac{1}{2}\left(\lambda x + \frac{\nu^2}{\lambda x}\right)\right\}, \qquad x, \lambda, \nu > 0,\ -\infty < \theta < \infty, \qquad (4)$$

and $C_3 = C_3(\lambda, \nu, \theta)$. This density has the scale parameter λ, and θ can still be regarded as a moment parameter. [Note: the parameter θ here is not related to the parameter θ of Section A.] With the introduction of a moment parameter, ν is no longer a convolution parameter because the actions of introducing a moment parameter and convolution parameter do not commute (see Section 7.L).

There is a standard parameterization of the density (4) that is obtained from (4) by setting $\nu^2/\lambda = \chi$. With this change and an evaluation of the norming constant for all parameter values where it exists, (4) takes the form

$$f(x \mid \lambda, \chi, \theta) = C_0 x^{\theta-1} \exp\left\{-\frac{1}{2}\left(\lambda x + \frac{\chi}{x}\right)\right\}, \qquad x > 0, \qquad (5)$$

where the normalizing constant $C_0 = C_0(\lambda, \chi, \theta)$. The distribution

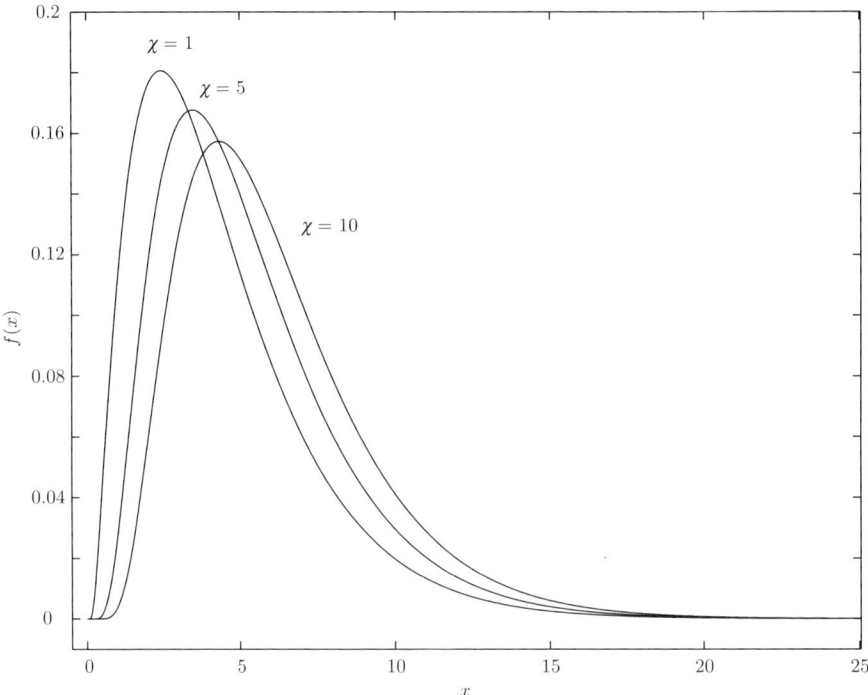

Fig. B.1. Densities of the generalized inverse Gaussian distribution ($\lambda = 1$)

with this density is sometimes denoted in shorthand notation as the "$GIG(\lambda, \chi, \theta)$ distribution." See Figures B.1 and B.2.

a. Evaluation of the Normalizing Constants

First, consider the normalizing constant C_0 of (5). In terms of the modified Bessel function K_θ of the third kind (also called *Macdonald's function*, and even the "modified Bessel function of the second kind" in the statistical literature) C_0 can be written as

$$
\begin{aligned}
C_0 &= \frac{(\lambda/\chi)^{\theta/2}}{2K_\theta(\sqrt{\lambda\chi})} && \text{if } \chi, \lambda > 0, \quad -\infty < \theta < \infty \quad\quad (6) \\
&= \frac{\lambda^\theta}{2^\theta \Gamma(\theta)} && \text{if } \chi = 0, \lambda, \theta > 0, \\
&= \frac{2^\theta}{\chi^\theta \Gamma(-\theta)} && \text{if } \chi > 0, \lambda = 0, \theta < 0.
\end{aligned}
$$

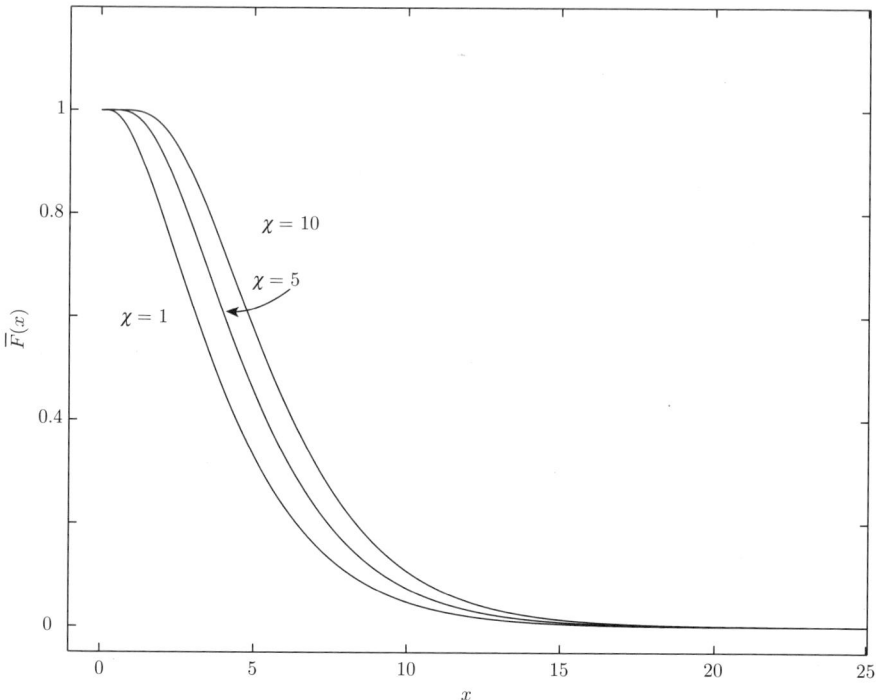

Fig. B.2. Survival functions of the generalized inverse Gaussian distribution ($\lambda = 1$)

The Bessel function K_θ can be defined in several equivalent forms; see 24.B.4.

The constant C_0 for χ or $\lambda = 0$ is obtained from the facts that

$$K_\theta(z) \approx \Gamma(\theta) 2^{\theta-1} z^{-\theta} \quad \text{as} \quad z \downarrow 0, \theta > 0,$$

and $K_\theta(z) = K_{-\theta}(z)$.

Bessel functions have played an important role in classical analysis, where they arise as solutions to certain differential equations. These functions possess a number of very convenient properties; see, e.g., Erdélyi, Magnus, Oberhettinger, and Tricomi (1953, vol. 2); Gradshteyn and Ryzhik (1994, Sections 8.4, 8.5).

It can be verified directly that

$$C_1 = C_0(2a, 2b, \alpha).$$
$$C_2 = \lambda^{\frac{3}{2}-\beta} e^{-\nu} C_0(\lambda, \nu^2/\lambda, \beta - \tfrac{1}{2}), \text{ and}$$
$$C_3 = \lambda^{2-\theta} C_0(\lambda, \nu^2/\lambda, \theta).$$

b. Special Cases of the Generalized Inverse Gaussian Distribution

Four special cases of the density (5) of the $GIG(\lambda, \chi, \theta)$ distribution are worth noting:

(i) If $\chi = 0$, then the generalized inverse Gaussian density (5) reduces to the gamma density. The shape parameter of this gamma distribution was introduced here as a moment parameter but it is shown in Proposition 18.B.20 that the gamma (and only the gamma distribution) has a parameter that is simultaneously both a convolution and a moment parameter. Because of this special case, (5) is a three-parameter extension of the exponential distribution, an extension not discussed in Chapter 9. If both $\chi = 0$ and $\theta = 1$, (5) is the density of an exponential distribution.

(ii) If $\lambda = 0$, then (5) becomes the distribution of the reciprocal of a gamma variate.

(iii) With $\theta = -1/2$, the generalized inverse Gaussian distribution reduces to the nongeneralized case of Section A. This is most clearly seen using the parameterization of A(1c), using the fact that

$$K_{-1/2}(z) = \frac{\sqrt{\pi}}{\sqrt{2z}} e^{-z}. \qquad (7)$$

(iv) The case $\theta = 1/2$ is also of particular interest, and arises in Section C, and is the distribution of the reciprocal of a random variable with an inverse Gaussian distribution. With (7) and the fact that $K_\theta(z) = K_{-\theta}(z)$, it can be shown that

$$f(x \mid \lambda, \chi, 1/2) = \frac{\lambda}{\sqrt{2\pi\lambda x}} e^{\sqrt{\lambda\chi}} \exp\left\{-\frac{1}{2}\left(\lambda x + \frac{\chi}{x}\right)\right\}$$

$$= \frac{\lambda}{\sqrt{2\pi\lambda x}} \exp\left\{-\frac{1}{2}\left(\sqrt{\lambda x} - \sqrt{\chi/x}\right)^2\right\}, \quad x > 0. \qquad (8)$$

c. The Hazard Rate

The hazard rate of the generalized inverse Gaussian distribution has properties that can be determined in the same manner as those of the nongeneralized case, using Theorem 4.E.2. Here, direct computations show that

$$\rho(x) = -\frac{d \log f(x)}{dx} = \frac{\lambda}{2} - \frac{\theta - 1}{x} - \frac{\chi}{2x^2}.$$

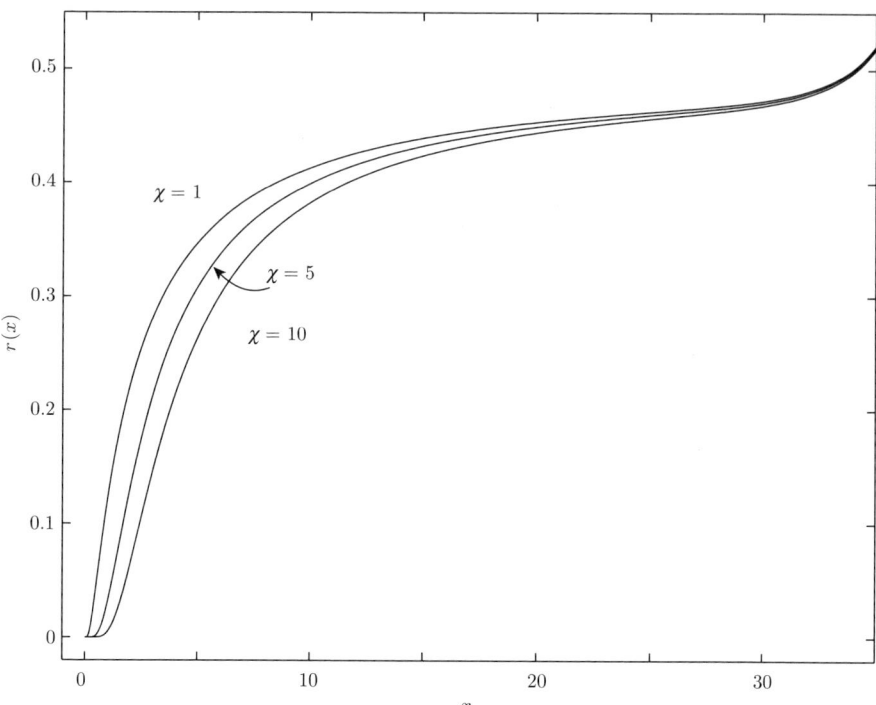

Fig. B.3. Hazard rates of the generalized inverse Gaussian distribution ($\lambda = 1$)

For $\theta \geq 1, \rho(x)$ is increasing, and so the hazard rate is also increasing. For $\theta < 1, \rho(x)$ is increasing in $x \leq x_1 = \chi/(1-\theta)$ and decreasing in $x \geq x_1$. This means that there exists $x_2, 0 \leq x_2 \leq x_1$, such that the hazard rate $r(x)$ is increasing in $x \leq x_2$ and decreasing in $x \geq x_2$. See Figure B.3.

d. Moments and Laplace Transform

As with any density having a moment parameter, the moments are easily obtained from the norming constant, which is a function of the moment parameter. Thus, it follows with the aid of (6) that if X has the density (5), then

$$EX^r = \frac{K_{\theta+r}(\sqrt{\lambda\chi})}{K_\theta(\sqrt{\lambda\chi})} \left(\frac{\chi}{\lambda}\right)^{r/2} = \frac{K_{\theta+r}(\nu)}{K_\theta(\nu)} \left(\frac{\nu}{\lambda}\right)^r, \quad \lambda, \chi > 0. \quad (9)$$

For this distribution, moments of all orders, positive and negative, exist.

The Laplace transform can be determined in a similar manner to be given by

$$E\,e^{-sX} = \left(\frac{\lambda}{\lambda+2s}\right)^{\theta/2} \frac{K_\theta(\sqrt{(\lambda+2s)\chi})}{K_\theta(\sqrt{\lambda\chi})}$$
$$= \left(\frac{\lambda}{\lambda+2s}\right)^{\theta/2} \frac{K_\theta(\nu\sqrt{1+(2s/\lambda)})}{K_\theta(\nu)}. \qquad (10)$$

The moments (9) of the generalized inverse Gaussian distribution are ratios involving Bessel functions and are not easy to study. With a considerable amount of calculations that are beyond the scope of this book, Nguyen, Chen and Gupta (2003) have shown that for any real θ and positive λ, χ, the skewness measure $E(X-\mu)^3/\sigma^3$ of the generalized inverse Gaussian distribution is positive. For comments about skewness, see Section 2.C.e.

e. Ordering Generalized Inverse Gaussian Distributions

The introduction of a moment parameter does not affect the likelihood ratio order; because the inverse Gaussian distribution is decreasing in the scale parameter λ in the sense of the likelihood ratio order, the same is true for the generalized inverse Gaussian distribution. But the introduction of a moment parameter changes the character of the parameter ν; it does not remain a convolution parameter. It is not known how ν orders the generalized Gaussian distribution.

f. Convolutions

Although the generalized inverse Gaussian distribution does not have a convolution parameter, several convolution results were given by Barndorff-Nielsen (1978). See also Jørgensen (1982). These results can be compactly written by denoting the distribution with density (5) by $GIG(\theta, \chi, \lambda)$ and by using the symbol $*$ to denote convolution. The results are as follows:

$$GIG\left(-\tfrac{1}{2}, \chi_1, \lambda\right) * GIG\left(-\tfrac{1}{2}, \chi_2, \lambda\right) = GIG\left(-\tfrac{1}{2}, (\sqrt{\chi_1}+\sqrt{\chi_2})^2, \lambda\right), \qquad (11)$$

$$GIG\left(-\tfrac{1}{2}, \chi_1, \lambda\right) * GIG\left(\tfrac{1}{2}, \chi_2, \lambda\right) = GIG\left(\tfrac{1}{2}, (\sqrt{\chi_1}+\sqrt{\chi_2})^2, \lambda\right), \qquad (12)$$

$$GIG(-\theta, \chi, \lambda) * GIG(\theta, 0, \lambda) = GIG(\theta, \chi\lambda), \qquad (13)$$

$$GIG(\theta_1, 0, \lambda) * GIG(\theta_2, 0, \lambda) = GIG(\theta_1+\theta_2, 0, \lambda). \qquad (14)$$

The last of these results is just a statement about the gamma distribution.

g. Infinite Divisibility

The fact that the inverse Gaussian distribution is infinitely divisible follows from the fact that the distribution has a convolution parameter. But the generalized inverse Gaussian distribution does not have a convolution parameter and infinite divisibility is not so transparent.

Barndorff-Nielsen and Halgreen (1977) have determined that the generalized inverse Gaussian distribution is infinitely divisible by using Proposition 20.D.8. This theorem asserts that a distribution is infinitely divisible if and only if $-\log E\, e^{-sX}$ has a completely monotone derivative. By using the formulas (see, e.g., Erdélyi, Magnus, Oberhettinger, and Tricomi, 1953)

$$K_\theta(z) = K_{-\theta}(z),$$
$$K_{\theta+1}(z) = 2(\theta/z)K_\theta(z) + K_{\theta-1}(z),$$
$$K_{\theta-1}(z) + K_{\theta+1}(z) = -2K_\theta'(z).$$

Barndorff-Nielsen and Halgreen (1977) compute

$$\frac{d}{ds}(-\log E\, e^{-sX}) = \frac{2\theta}{\lambda+2s} + Q_\theta(\chi(\lambda+2s)), \qquad \theta \geq 0,$$
$$= \chi Q_{-\theta}(\chi(\lambda+2s)), \qquad \theta \leq 0,$$

where

$$Q_\nu(z) = \frac{K_{\nu-1}(\sqrt{z})}{\sqrt{z}K_\nu(\sqrt{z})}, \qquad \nu \geq 0,\ z > 0.$$

The desired result follows from the fact (Grosswald, 1976) that Q_ν is completely monotone for all $\nu \geq 0$, $2\theta/(\lambda+2s)$ is completely monotone, and the sum of completely monotone functions is completely monotone.

C. The Birnbaum–Saunders Distribution

a. General Description

The Birnbaum–Saunders distribution has appeared in several contexts, with various derivations. It was given by Fletcher (1911), and according

C. The Birnbaum–Saunders Distribution

to Schrödinger (1915) it was obtained by Konstantinowsky (1914). Subsequently, it was obtained by Freudenthal and Shinozuka (1961), but it was the derivation of Birnbaum and Saunders (1969) that brought the usefulness of the distribution into clear focus. The distribution is included here because it is a mixture of an inverse Gaussian distribution and a generalized inverse Gaussian distribution. For a comparison of the Birnbaum–Saunders distribution and the inverse Gaussian distribution, see Bhattacharya and Fries (1982).

Birnbaum and Saunders (1969) introduced the distribution that has come to bear their names specifically for the purpose of modeling fatigue life of metals subject to periodic stress; consequently, the distribution is sometimes called the *fatigue–life distribution*. Birnbaum and Saunders apply the central limit theorem to the crack-growth process and approximate the number of stress cycles to failure by a continuous random variable. The distribution function of the Birnbaum–Saunders distribution is given by

$$F(x\,|\,\lambda, \alpha) = \Phi(\alpha^{-1} h(\lambda x)), \qquad x > 0, \qquad (1)$$

where $\lambda, \alpha > 0$, and Φ is the standard normal distribution function and

$$h(x) = x^{1/2} - x^{-1/2}. \qquad (2)$$

The form of this distribution function bears considerable resemblance to the lognormal distribution function which can be written as (1) but with $h(x)$ replaced by $\log x$. Both $h(x)$ and $\log x$ are increasing and concave functions that map $(0, \infty)$ onto the interval $(-\infty, \infty)$. However, for the Birnbaum–Saunders distribution, the shape parameter α is not any of the standard parameters discussed in Chapter 7.

By differentiating the distribution function (1), the corresponding density f can be found:

$$\begin{aligned}
f(x\,&|\,\lambda, \alpha) \\
&= \frac{\lambda}{2\alpha\sqrt{2\pi}} \left[\frac{1}{\sqrt{\lambda x}} \left(1 + \frac{1}{\lambda x}\right) \right] \exp\left\{ -\frac{1}{2\alpha^2} \left(\lambda x - 2 + \frac{1}{\lambda x}\right) \right\} \\
&= \left[\frac{\sqrt{\lambda}}{2\alpha\sqrt{2\pi x}} + \frac{1}{2\alpha\sqrt{2\pi \lambda x^3}} \right] \exp\left\{ -\frac{1}{2\alpha^2} \left(\lambda x - 2 + \frac{1}{\lambda x}\right) \right\}. \quad (3)
\end{aligned}$$

See Figures C.1, C.2, and C.3.

13. Inverse Gaussian Distribution

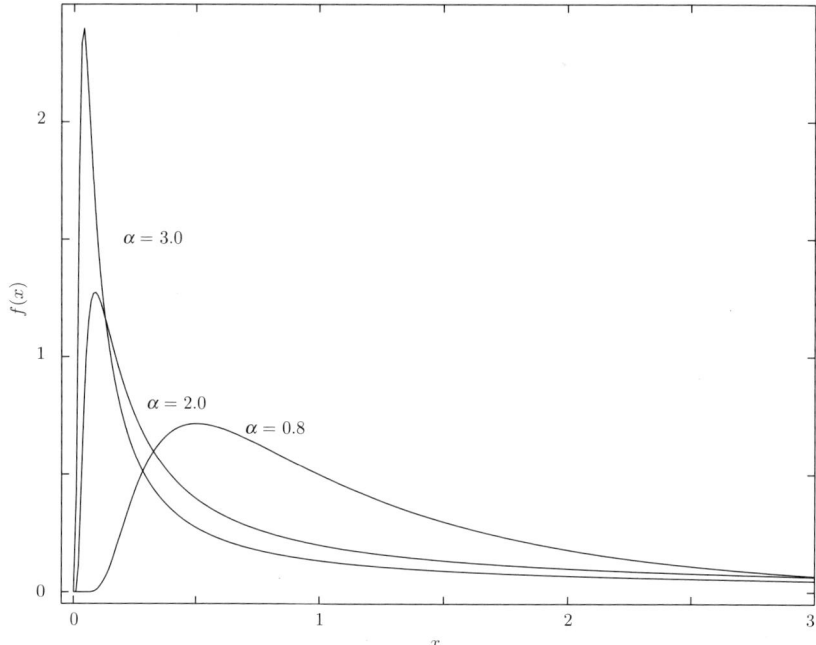

Fig. C.1. Densities of the Birnbaum-Saunders distribution ($\lambda = 1$)

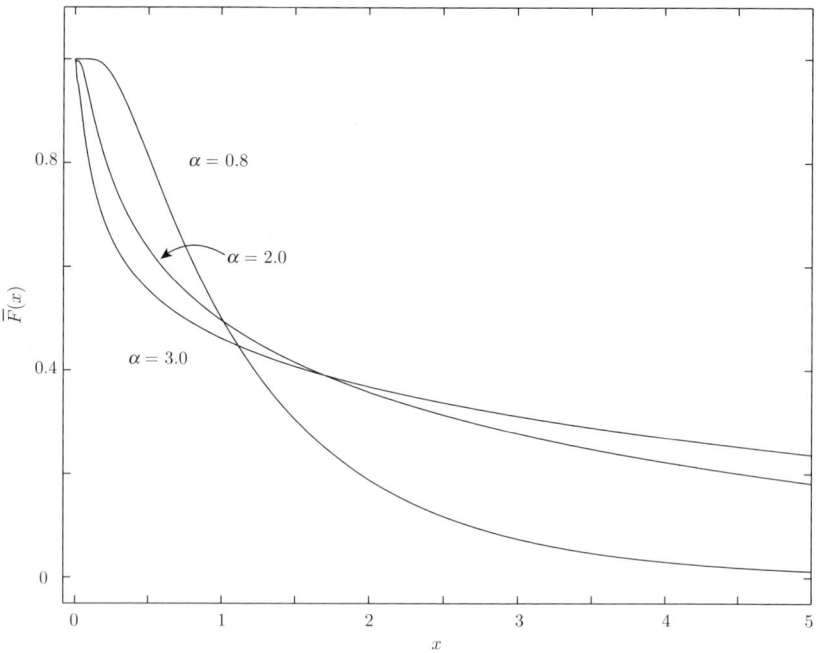

Fig. C.2. Survival functions of the Birnbaum-Saunders distribution ($\lambda = 1$)

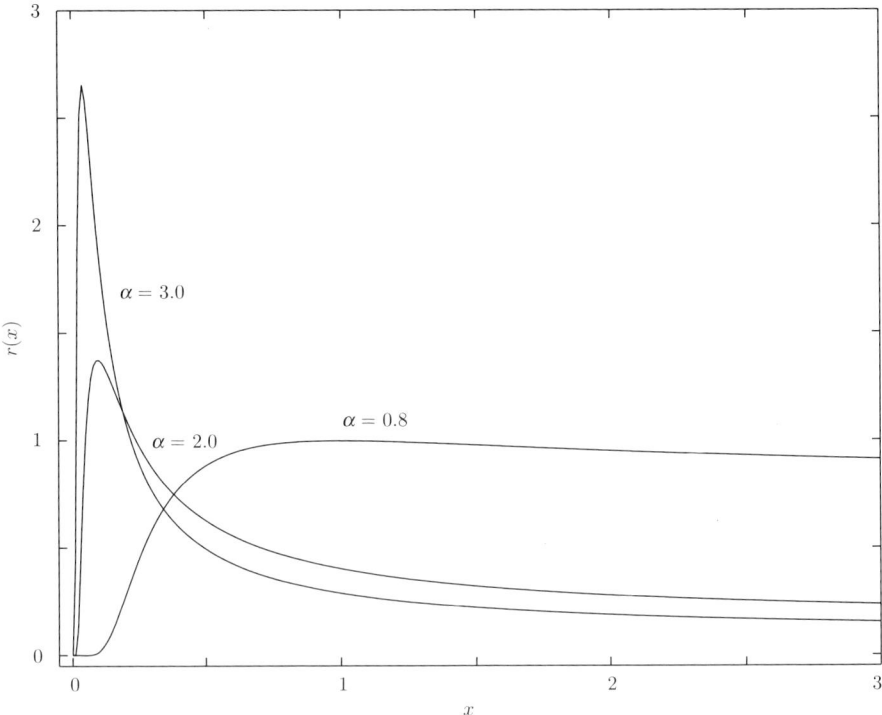

Fig. C.3. Hazard rates of the Birnbaum-Saunders distribution ($\lambda = 1$)

To see that (3) is a mixture of inverse Gaussian and generalized inverse Gaussian distributions, let

$$\lambda = \sqrt{\ell/\chi}, \quad \text{and} \quad \alpha = 1/[\ell\chi]^{1/4};$$

with this new parameterization, the density (3) can be written in the form

$$\frac{1}{2} f\left(x \mid \ell, \chi, -\tfrac{1}{2}\right) + \frac{1}{2} f\left(x \mid \ell, \chi, \tfrac{1}{2}\right), \tag{4}$$

where $f(x \mid \ell, \chi, \theta)$ is given by B(5), that is,

$$f(x \mid \ell, \chi, \theta) = C_2(\ell, \chi, \theta) \, x^{\theta-1} \exp\left\{-\frac{1}{2}\left(\ell x + \frac{\chi}{x}\right)\right\}, \quad x > 0.$$

The mixture representation (4) was given in a slightly different form by Desmond (1986).

The mixture of the form (4), but with general mixing weights in place of 1/2 is called Schrödinger's distribution by Saunders (2007); a still more general distribution is studied by Jørgensen, Seshadri and Whitmore (1991).

The moments and moment generating function of the Birnbaum–Saunders distribution have been found by Rieck (1999) in terms of Bessel functions. The particular cases of the mean and variance take a simple form: If X is a random variable with the density $f(\cdot \mid \lambda, \alpha)$ of (3), then

$$EX = \frac{1}{\lambda}\left(1 + \frac{\alpha^2}{2}\right), \; Var(X) = \left(\frac{\alpha}{\lambda}\right)^2 \left(1 + \frac{5}{6}\alpha^2\right). \tag{5}$$

b. Derivation of the Birnbaum–Saunders Distribution

Recall the shock model of Section 5.H, but suppose that the shocks occur not as events in a Poisson process, but rather at regular intervals, say at times $\Delta, 2\Delta, \ldots$. As with the cumulative damage model of Section 5.H.a, suppose that the ith shock to an item causes a random damage X_i. Suppose further that the X_i are identically distributed, mutually independent, and $EX_i = \mu\Delta$, $Var(X_i) = \Delta\sigma^2 < \infty$. Damages accumulate additively, and the kth shock is survived by the item if $S_k = X_1 + \cdots + X_k \leq z$, where z is the capacity or threshold of the item. With this model, the time T_Δ of failure of the item undergoing shocks has the survival function $P\{T_\Delta > k\Delta\} = P\{S_k \leq z\}$. With the intent of letting $\Delta \to 0$, set $k\Delta = t$ to obtain

$$P\{T_\Delta > t\} = P\{S_k \leq z\} = P\left\{\frac{S_{t/\Delta} - \mu t}{\sqrt{t\sigma}} \leq \frac{z - \mu t}{\sqrt{t\sigma}}\right\}. \tag{6}$$

Set $\alpha = \sigma/\sqrt{\mu z}$ and recall from (2) that $h(x) = x^{1/2} - x^{-1/2}$. Let $\lambda = \mu/z$ and make use of the fact that

$$\frac{1}{\alpha}h\left(\frac{z}{\mu t}\right) = \frac{z - \mu t}{\alpha\sqrt{z\mu t}} = \frac{z - \mu t}{\sigma\sqrt{t}}$$

to rewrite (6) in the form

$$P\{T_\Delta > t\} = P\left\{\frac{S_{t/\Delta} - \mu t}{\sqrt{t\sigma}} \leq \frac{1}{\alpha}h\left(\frac{z}{\mu t}\right)\right\}$$
$$= P\left\{\frac{S_{t/\Delta} - \mu t}{\sqrt{t\sigma}} \leq \frac{1}{\alpha}h\left(\frac{1}{\lambda t}\right)\right\}. \tag{7}$$

C. The Birnbaum–Saunders Distribution

Let $\Delta \to 0$, and apply the central limit theorem 20.C.8 to conclude that

$$P\{T > t\} = \lim_{\Delta \to 0} P\{T_\Delta > t\}$$

$$= \lim_{\Delta \to 0} P\left\{\frac{S_{t/\Delta} - \mu t}{\sqrt{t\sigma}} \leq \frac{1}{\alpha}\xi h\left(\frac{1}{\lambda t}\right)\right\} = \Phi\left(\frac{1}{\alpha}h\left(\frac{1}{\lambda t}\right)\right).$$

Because Φ is symmetric about 0 and because $h(x) = -h(1/x)$, it follows that

$$P\{T \leq t\} = 1 - \Phi\left(\frac{1}{\alpha}h\left(\frac{1}{\lambda t}\right)\right) = \Phi\left(-\frac{1}{\alpha}h\left(\frac{1}{\lambda t}\right)\right) = \Phi\left(\frac{1}{\alpha}\xi h(\lambda t)\right),$$

which is (1). For an additional discussion of fatigue life and the Birnbaum–Saunders distribution, see Desmond (1985).

14
Distributions with Bounded Support

A. Introduction

Distributions that have support contained in a known finite interval can be translated using scale and location parameters so that the support of the distribution is contained in the interval [0, 1], but in no closed subinterval of [0, 1]. More precisely, these are distributions F of nonnegative random variables X that can take values arbitrarily close to 0 and 1, but have the property that $P\{0 \leq X \leq 1\} = 1$. Such distributions are identified by the conditions

$$F(0-) = 0, \quad 0 < F(x) < 1 \quad \text{for} \quad 0 < x < 1, \quad \text{and} \quad F(1) = 1. \quad (1)$$

Sections B, C, and D deal with distributions that satisfy (1). Section E is concerned with the introduction of a scale parameter λ and the resulting distributions which have support $[0, \theta]$, where $\theta = 1/\lambda$.

Parametric families of distributions that satisfy (1) cannot have location or scale parameters because such parameters alter the support of the distribution. But these families can have any of the other parameters discussed in Chapter 7.

The most familiar and basic distributions satisfying (1) include the distribution uniform on [0, 1] and its two-parameter extension, the beta distribution. Occasionally, a distribution with bounded support will be found in the literature that is obtained by truncating some distribution. In addition, the distribution of the sample correlation coefficient for a sample from a bivariate normal distribution is well known; of course this distribution has support $[-1, 1]$. The distribution of the squared multiple correlation coefficient has support [0, 1]. Other distributions with bounded support are not commonly found in the literature. Nevertheless, there are a number of one- or two-parameter families of distributions with support [0, 1], and there are several methods for deriving them.

One method for deriving distributions with support [0, 1] is to obtain the distribution of the random variable $X/(1+X)$ when the distribution of X has support $[0, \infty)$. By this procedure, the beta distribution defined in Section C can be obtained from the F distribution with density 11.C(1). But this derivation does not seem to have many interesting examples.

a. Duality

To every distribution F with support [0, 1], there corresponds a dual distribution F_D.

A.1. Definition. The distribution F_D defined by

$$F_D(x) = \bar{F}(1-x) \qquad (2)$$

is called the *dual* of F.

If X has the distribution F, then $1 - X$ has the distribution F_D. Thus, the term "dual" is justified by the fact that $F_{DD}(x) = F(x)$. A distribution with support [0, 1] is self-dual, i.e., $F_D = F$ if and only if it is symmetric about the point $1/2$. The notion of duality arises whatever method is used to derive a distribution, and consequently, it arises in each of the sections that follow.

b. Moments

Necessary and sufficient conditions for a sequence $\{\mu_n, n = 0, 1, 2, \ldots\}$ to be moments of a distribution with support [0, 1] are given in Proposition 20.B.10. It can be verified directly that if X is a random variable for which $0 \leq X \leq 1$ with probability 1, then $X^2 \leq X$ with probability 1, and consequently,

$$EX^2 \leq EX. \qquad (3)$$

Thus, the variance of X satisfies the inequalities

$$0 \leq \operatorname{Var}(X) = EX^2 - (EX)^2 \leq EX - (EX)^2 = EX(1 - EX) \leq 1/4. \qquad (4)$$

The last inequality holds because the function $h(\mu) = \mu(1 - \mu)$ has the maximum of $1/4$ at $\mu = 1/2$.

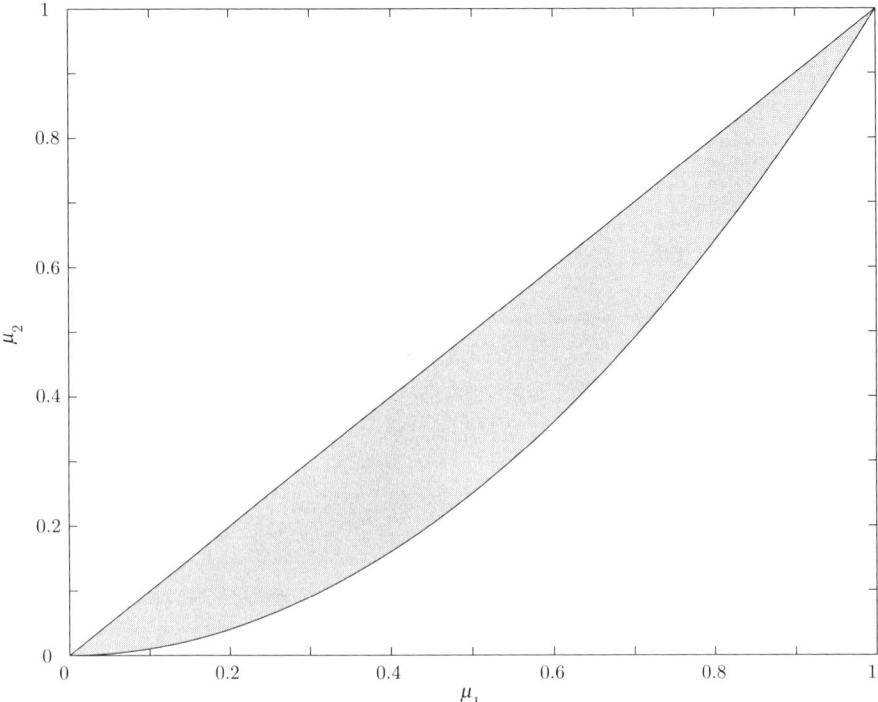

Fig. A.1. Possible μ_1, μ_2 pairs (shaded region)

Equality holds in (3), that is, $\text{Var}(X) = EX - (EX)^2$ if and only if $P\{X = 0 \text{ or } X = 1\} = 1$; $\text{Var}(X) = 1/4$ if and only if $P\{X = 0\} = P\{X = 1\} = 1/2$, and $\text{Var}(X) = 0$ if and only if $P\{X = EX\} = 1$.

Figure A.1 shows the points in the $(\mu_1, \mu_2) = (EX, EX^2)$ plane that can be the first two moments of a random variable X that satisfies $P\{0 \leq X \leq 1\} = 1$. Figure A.2 shows the possible values in the (μ_1, σ^2) plane. The areas shaded in these figures are often referred to as "moment spaces." As noted above, points on the boundaries of these spaces are achieved only by discrete distributions.

B. The Uniform Distribution and One-Parameter Extensions

The uniform distribution

$$F(x) = x, \quad 0 \leq x \leq 1, \tag{1}$$

plays a central role in the class of distributions with support [0, 1]. It

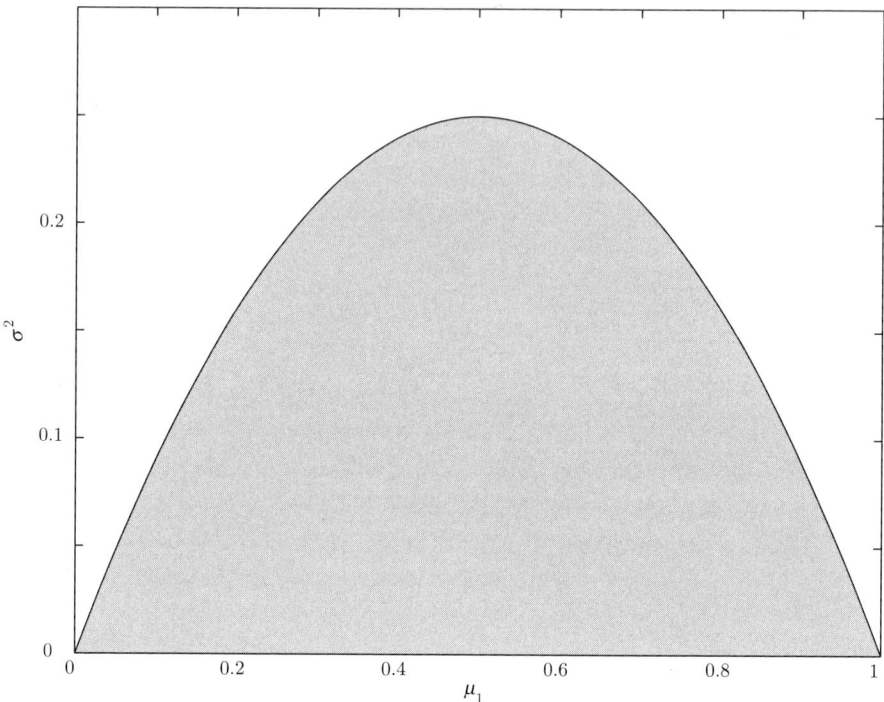

Fig. A.2. Possible μ_1, σ^2 pairs (shaded region)

arises naturally in a variety of contexts; for example, it is the distribution of $F(X)$, where X is a random variable with strictly increasing distribution function F.

With the exception of a convolution parameter, the parameters discussed in Chapter 7 can all be introduced using the uniform distribution as the underlying distribution. For each type of parameter other than location and scale, this procedure yields a one-parameter family of distributions on $[0, 1]$, which contains the uniform distribution.

B.1. Power, resilience, and moment parameters. With a power or resilience parameter α, the distribution (1) becomes

$$F(x \mid \alpha) = x^\alpha, \quad 0 \leq x \leq 1, \quad \alpha > 0. \qquad (2)$$

Because $F(\cdot \mid \alpha)$ has the density $f(x \mid \alpha) = \alpha x^{\alpha-1}, 0 \leq x \leq 1$, it is clear that $\beta = \alpha - 1$ in (2) is a moment parameter.

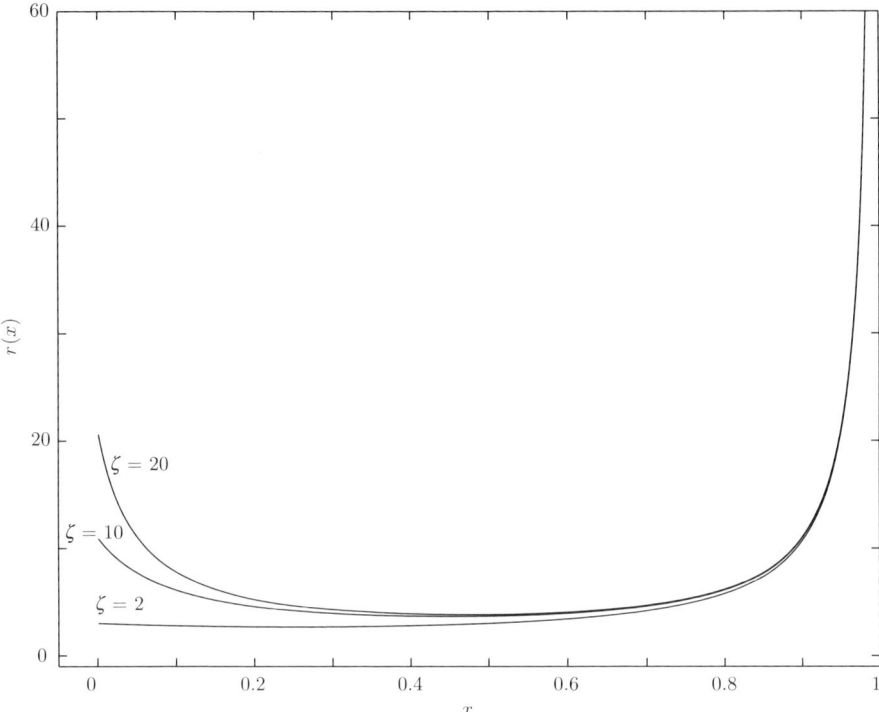

Fig. B.1. Hazard rates of the uniform distribution with introduced frailty parameter

The case where power and resilience parameter families coincide is determined in Proposition 18.B.10; the case where power and moment parameters coincide is determined in Proposition 18.B.13; and the case where resilience and moment parameters coincide is determined in Proposition 18.B.17. Each of these coincidences include the distribution (2).

B.2. Frailty parameter. With the introduction of a frailty parameter, the survival function corresponding to (1) becomes

$$\bar{F}(x \mid \xi) = (1-x)^\xi, \quad 0 \le x \le 1, \ \xi > 0. \tag{3}$$

Note that this is the dual of (2). This distribution has been identified by Pickands (1975) as one of the possible limits of residual life distributions, and it remains a possible limit if a scale parameter is added (see Proposition 20.G.4).

14. Distributions with Bounded Support

B.3. Tilt parameter. If a tilt parameter is introduced in (1), the distribution

$$\bar{F}(x\,|\,\gamma) = \frac{\gamma(1-x)}{1-\bar{\gamma}(1-x)} = \frac{\gamma(1-x)}{\gamma+\bar{\gamma}x}$$
$$= \frac{1-x}{1-\delta x}, \quad 0 \leq x \leq 1,\ \gamma > 0,\ \delta = \bar{\gamma}/\gamma, \qquad (4a)$$

is obtained. This somewhat unfamiliar distribution has the dual $\bar{F}_D(x\,|\,\gamma) = \bar{F}(x\,|\,\gamma^{-1})$ and the density

$$f(x\,|\,\gamma) = \frac{1+\delta}{(1+\delta x)^2}, \quad 0 \leq x \leq 1,\ \delta = \bar{\gamma}/\gamma. \qquad (4b)$$

B.4. Hazard power parameter. With a hazard power parameter, (1) becomes

$$\bar{F}(x\,|\,\zeta) = \exp\{-[-\log(1-x)]^\zeta\}, \quad 0 \leq x \leq 1,\ \zeta > 0. \qquad (5a)$$

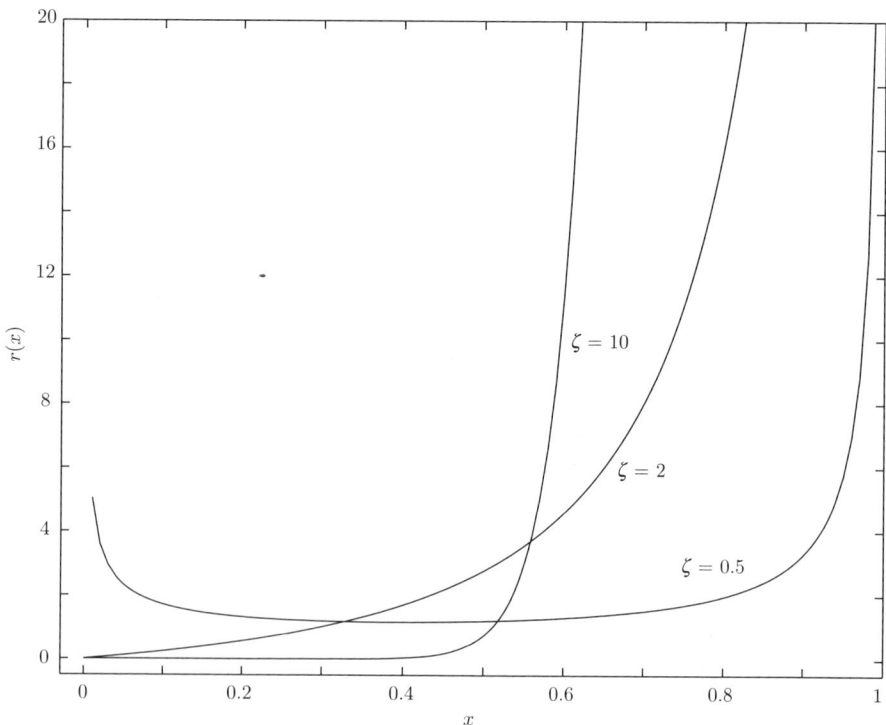

Fig. B.2. Hazard rates of the uniform distribution with hazard power parameter

This distribution has the dual

$$F_D(x\,|\,\zeta) = \exp\{-[-\log x]^\zeta\}, \quad 0 \le x \le 1, \quad \zeta > 0, \tag{5b}$$

which, like (2), is a special case of the two-parameter distribution encountered in Proposition 18.B.10. The distribution (5b) is obtained directly from (1) by the introduction of a reverse hazard power parameter (defined in Section 7.G).

B.5. Laplace transform parameter. The distribution (1) has the Laplace transform

$$\phi(s) = (1 - e^{-s})/s,$$

and with the introduction of a Laplace transform parameter, (1) yields

$$F(x\,|\,s) = \frac{1 - e^{-sx}}{1 - e^{-s}}, \quad 0 \le x \le 1. \tag{6}$$

This truncated exponential distribution has the dual $F_D(x\,|\,s) = F(x\,|\,-s)$.

C. The Beta Distribution

The family of beta distributions is the best known and most widely used two-parameter family with support [0, 1]. It has several derivations, a number of desirable properties, and the density takes on a variety of shapes. It is often used in Bayesian statistical analysis as the prior distribution for the parameter of the binomial distribution.

a. Defining Functions

The two-parameter family of beta distributions is defined in terms of the density

$$f(x\,|\,a, b) = \frac{x^{a-1}(1-x)^{b-1}}{B(a,b)}, \quad 0 \le x \le 1,\ a, b > 0. \tag{1}$$

The corresponding distribution function

$$F(x\,|\,a, b) = B_x(a, b)/B(a, b) \tag{2}$$

480   14. Distributions with Bounded Support

involves the incomplete beta function $B_x(a,b)$, defined in Section 23.B.a. It cannot be written in closed form and consequently, neither it, the hazard rate, nor the total time on test transform have particularly simple forms. See Figures C.1 and C.2.

b. Special Cases

When $a = 1$, (1) is the density of the distribution with survival function B(3). With $b = 1$, the density of the distribution B(2) is obtained. When $a = b = 1$, the beta distribution is a uniform distribution.

The dual of $f(x \,|\, a, b)$ is $f(1-x \,|\, a, b)$. But note that $f(x \,|\, a, b) = f(1-x \,|\, b, a)$, thus when $a = b$, the density is symmetric about $1/2$; consequently, it is self-dual.

When a is an integer, the survival function $\bar{F}$ of f can be given explicitly as a finite sum; see 23.B(9).

c. Shape of the Beta Density

In terms of monotonicity and convexity, the shape of the beta density (1) is determined using standard methods of calculus. The results of this somewhat tedious investigation are summarized here.

The shape of the beta density (1) is identical to the shape of

$$g(x \,|\, a, b) = B(a,b) f(x \,|\, a, b) = x^{a-1}(1-x)^{b-1}, \quad 0 \leq x \leq 1.$$

To determine this shape, it is convenient to start by computing

$$g'(x \,|\, a, b) = x^{a-2}(1-x)^{b-2}[(a-1) - x(a+b-2)],$$
$$g''(x \,|\, a, b) = x^{a-3}(1-x)^{b-3}[(a-1)(a-2)$$
$$- 2(a-1)(a+b-3)x + (a+b-2)(a+b-3)x^2].$$

Thus, $g'(x \,|\, a, b) = 0$ at the point

$$x_0 = \frac{a-1}{a+b-2},$$

whereas $g''(x \,|\, a, b) = 0$ at the points

$$x_- = \frac{(a-1)(a+b-3) - \sqrt{(a-1)(b-1)(a+b-3)}}{(a+b-2)(a+b-3)},$$

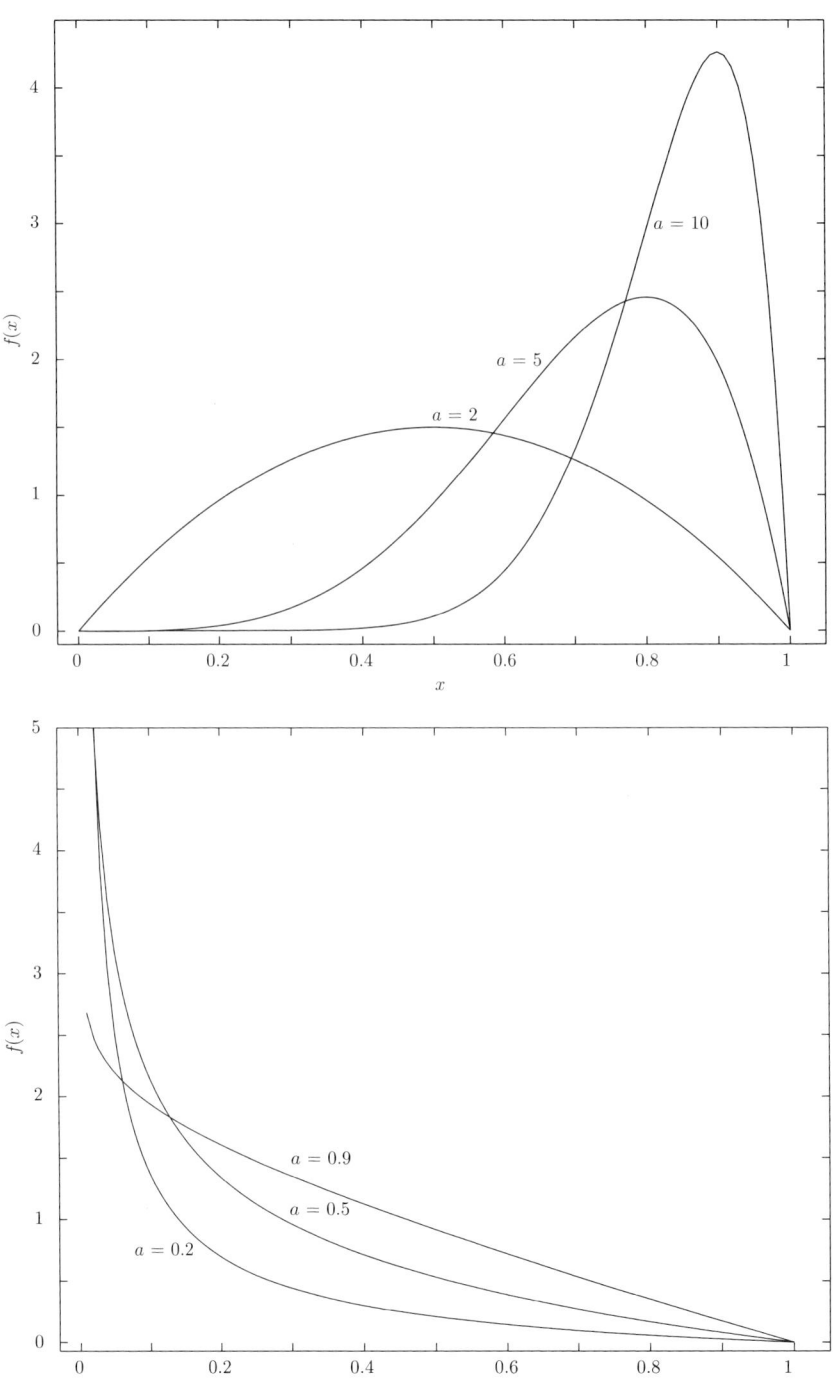

Fig. C.1. Densities of the beta distribution ($b = 2$)

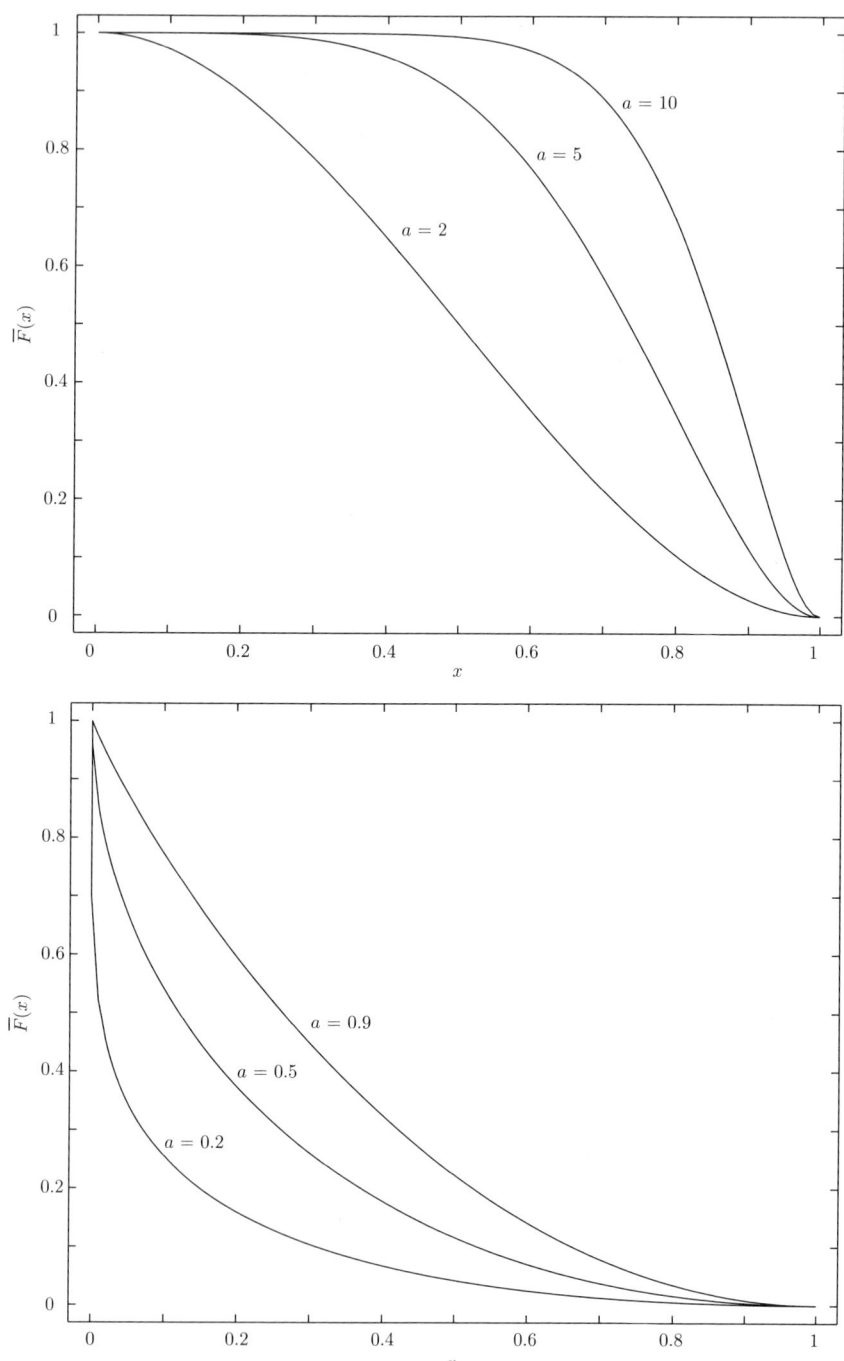

Fig. C.2. Survival of the beta distribution ($b = 2$)

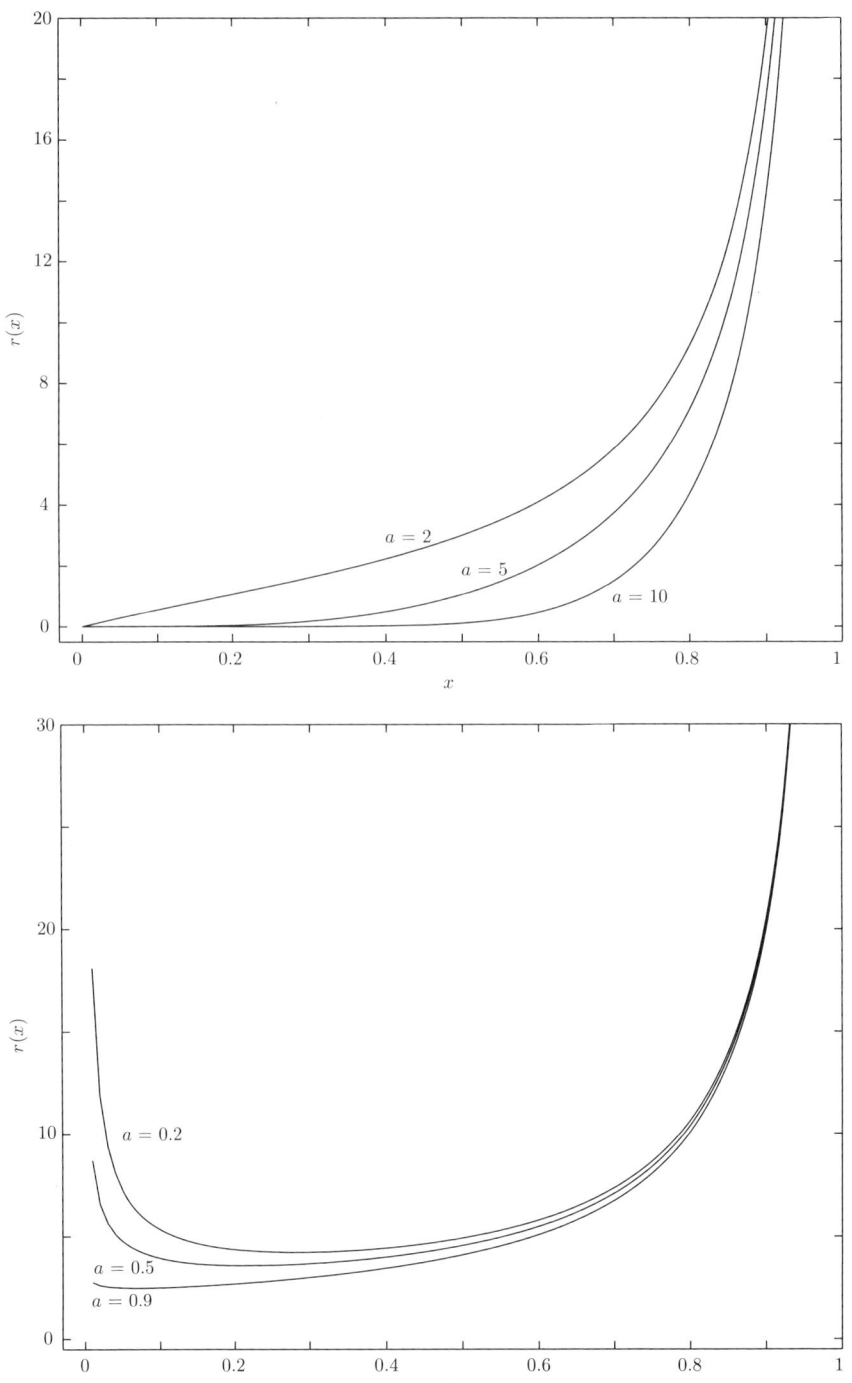

Fig. C.3. Hazard rates of the beta distribution ($b = 2$)

and

$$x_+ = \frac{(a-1)(a+b-3) + \sqrt{(a-1)(b-1)(a+b-3)}}{(a+b-2)(a+b-3)}.$$

Depending upon the values of a and b, x_0, x_-, and x_+ may or may not lie in the interval $(0, 1)$. Additionally, x_- and x_+ may or may not be real.

Because $g'(x \mid a, b) = 0$ for at most one point in $(0, 1)$, it changes sign at most once in the interval $(0, 1)$. Because there are at most two points in $(0, 1)$ where $g''(x \mid a, b) = 0$, there are at most two points in $(0, 1)$ where it changes sign.

Case: $a \leq 1, b \leq 1$. In this case, x_- and x_+ are not real so g'' is of one sign, which can easily be seen to be positive because it is positive for x near 0. Moreover, $x_0 \in (0, 1)$. Thus, g is decreasing in $(0, x_0)$, increasing in $(x_0, 1)$, and convex on $[0, 1]$. In this case, the convexity is also easily obtained by observing that $\log g$ is convex, and hence g is convex (Proposition 21.A.5). In this case, $f(0 \mid a, b) = f(1 \mid a, b) = \infty$.

Case: $a \geq 1, b \geq 1$. In this case, it follows from the above discussion that $f(x \mid a, b)$ is unimodal with mode at x_0. Alternatively, when $a \geq 1, b \geq 1$, it follows from the concavity of the logarithm function that $\log f(x \mid a, b)$ is concave. By Proposition 4.B.2, the log concavity implies that f is unimodal. In this case, it is clear that $f(0 \mid a, b) = f(1 \mid a, b) = 0$.

Case: $a \leq 1, b \geq 1$. Using similar methods, it can be determined that the density (1) is decreasing. If $1 < b < 2$, the density is first concave, then convex; if $b \geq 2$, then the density is convex.

Case: $a \geq 1, b \leq 1$. As in the previous case, it can be determined that the density (1) is increasing. If $1 < a < 2$, then the density is first concave then convex; if $a \geq 2$, the density is convex.

d. Hazard Rate Behavior

For $a \geq 1, b \geq 1$, it follows from the log concavity of $f(x \mid a, b)$ and Proposition 4.B.8.a that the hazard rate $r(x \mid a, b)$ is increasing. If $a \geq 1, b \leq 1$, then again the hazard rate is increasing because the density is increasing. However, if $a < 1$, then $f(0 \mid a, b) = \infty$, and so $r(0 \mid a, b) = \infty$. Thus, when $a < 1$, the hazard rate is initially decreasing, but when $x_0 = (a-1)/(a+b-2) < 1$, $r(x \mid a, b)$ is increasing in the interval $x_0 < x < 1$ because $f(x \mid a, b)$ is increasing in the same interval. By using Theorem 4.E.2, it can be verified that the hazard rate derivative has at most one sign change. It is possible to draw some of these conclusions directly by writing the hazard rate in the form

$$r(x \mid a, b) = \left[\int_x^1 \left(\frac{t}{x}\right)^{a-1} \left(\frac{1-t}{1-x}\right)^{b-1} dt\right]^{-1}$$
$$= \left[\int_0^{1-x} \left(1 + \frac{z}{x}\right)^{a-1} \left(1 - \frac{z}{1-x}\right)^{b-1} dz\right]^{-1}. \quad (3)$$

This representation can be obtained by writing the survival function as an integral; the second form follows with a change of variables.

e. Moments of the Beta Distribution

For notational purposes, let X be a random variable with the density $f(\cdot \mid a, b)$. Because it is known that the density of the beta distribution integrates to 1, it is easy to determine that $f(\cdot \mid a, b)$ has the rth moment

$$EX^r = \mu_r = B(a + r, b)/B(a, b).$$

Thus,

$$\mu_1 = \frac{a}{a+b}, \quad \mu_2 = \frac{a(a+1)}{(a+b)(a+b+1)}, \quad \text{and}$$

$$Var(X) = \frac{ab}{(a+b)^2(a+b+1)}.$$

Because X takes on only values in $[0, 1]$, it follows from A(3) that

$$0 \leq Var(X) = \frac{\mu_1(1-\mu_1)}{a+b+1} \leq \mu_1(1-\mu_1).$$

By fixing $\mu_1 = a/(a+b)$ and letting $a \to 0, b \to 0$ together, the upper bound for the variance is approached; by letting $a \to \infty, b \to \infty$, the lower bound for the variance is approached. Thus, the beta distribution allows for all pairs (μ_1, σ^2) possible under the constraint that $P\{0 \leq X \leq 1\} = 1$ apart from those values on the boundary which are achieved only by discrete distributions.

f. Residual Life Distribution

The residual life distribution of the beta distribution can be written in the form

$$\bar{F}_t(x) = \frac{\bar{F}(x+t)}{\bar{F}(t)} = \frac{\int_{x+t}^1 z^{a-1}(1-z)^{b-1}\,dz}{\int_t^1 z^{a-1}(1-z)^{b-1}\,dz}$$

$$= \frac{1 - I_{x+t}(a,b)}{1 - I_t(a,b)}, \quad 0 \le x \le 1-t,$$

where $I_x(a,b)$ is the incomplete beta function defined in Section 23.B.a. By rescaling the distribution $\bar{F}_t$ to have support $[0,1]$, the survival function becomes

$$\bar{G}_t(x) = \bar{F}_t(x(1-t)) = \frac{\int_{x(1-t)+t}^1 z^{a-1}(1-z)^{b-1}\,dz}{\int_t^1 z^{a-1}(1-z)^{b-1}\,dz}, \quad 0 \le x \le 1. \quad (4)$$

The limit of this survival function as $t \to 0$ is the beta survival function, and the limit as $t \to 1$ is given by

$$\lim_{t \uparrow 1} \bar{F}_t(x(1-t)) = (1-x)^b, \quad 0 \le x \le 1. \quad (5)$$

This limiting distribution is a uniform distribution with frailty parameter (see B(3) and 21.G.b).

g. Derivations of the Family of Beta Distributions

C.1. Uniform distribution with added parameters. The distribution B(3) was obtained from the uniform distribution by introducing a frailty parameter. This distribution has the βth moment

$$\mu_\beta = \int_0^1 x^\beta \xi (1-x)^{\xi-1}\,dx = \xi B(\beta+1, \xi),$$

where B is the beta function. With an added moment parameter, B(3) becomes

$$f(x\mid\xi,\beta) = \frac{x^\beta(1-x)^{\xi-1}}{B(\beta+1,\xi)}, \quad 0 \le x \le 1, \ \xi > 0, \ \beta > -1. \tag{6}$$

Now make the change of parameters $\beta = a - 1, \xi = b$ to rewrite (6) as (1), the density of a standard beta distribution.

C.2. Ratios of random variables. Suppose that U and V are independent random variables having the gamma distributions $f(\cdot\mid\lambda,\nu_1)$ and $f(\cdot\mid\lambda,\nu_2)$ given by 9.A(1). If

$$X = \frac{V}{U+V} = \frac{1}{1+(U/V)}, \tag{7}$$

then X has a beta distribution of (4) with parameters $a = \nu_2$ and $b = \nu_1$. One way to see this is to use the results of Section 11.D.f to conclude that U/V has the F density 11.D(1) with $\lambda = 1$.

Proposition C.2 can be recast as follows: Suppose that X has the F density 11.D(1) with $\lambda = 1$, $\theta = a$, and $\xi = b$, then $1/(1+X)$ has the beta density (1).

C.3. Waiting times in a Pólya process. The Pólya process is a counting process that arises as a limit of an urn process as discussed in Section 20.F.c. The waiting time for the kth jump in the process is derived in Section 20.F.c; the case $s < 0$ leads to the beta distribution with parameters $a = k, b = \theta - k + 1$, and having a scale parameter β. This is not the general case of a beta distribution because in this derivation, a must be an integer.

h. Ordering Beta Distributions

It follows directly from C.2 that the beta distribution is stochastically increasing in a and decreasing in b. However, a stronger result can be obtained using 2.A(11).

C.4. Proposition. In the likelihood ratio order, the beta distribution is increasing in a and decreasing in b.

i. The Generalized Beta Distribution

As noted in Section 11.C.f, random variables that are ratios of generalized gamma distributed random variables have a generalized F

distribution. Suppose that (7) holds, but that U and V have generalized gamma densities so that U/V has the density 11.C(2), of the generalized F distribution. Then, X has the density

$$f(x\mid \xi, \theta, \lambda, \alpha) = \frac{\lambda \alpha x^{\alpha\xi-1}[\lambda(1-x)]^{\alpha\theta-1}}{B(\xi,\theta)[x^\alpha + \lambda^\alpha(1-x)^\alpha]^{\xi+\theta}},$$
$$0 \leq x \leq 1, \quad \alpha, \lambda, \xi, \theta > 0. \tag{8}$$

This is the density of the four-parameter *generalized beta distribution*. With $\alpha = \lambda = 1, \xi = a$, and $\theta = b$, the density reduces to the standard beta density. Other special cases with fewer parameters may also be of interest.

Suppose that X has the density (1) of the standard beta distribution (with $\xi = a, \theta = b$) and let

$$Y = \frac{\lambda X^{1/\alpha}}{(1-X)^{1/\alpha} + \lambda X^{1/\alpha}}.$$

Then Y has the distribution (8) of the generalized beta distribution. This change of variables introduces two new parameters in the standard beta distribution using the method described in Section 19.A.a. Of course, the parameters λ and α are not, respectively, scale and power parameters as they are in the generalized F distribution from which (8) was derived.

A straightforward calculation shows that in the dual of the density (8) the roles of θ and ξ are interchanged

$$f_D(\cdot \mid \xi, \theta, \lambda, \alpha) = f(\cdot \mid \theta, \xi, 1/\lambda, \alpha);$$

thus, the distribution is self-dual when $\lambda = 1$. This can easily be obtained directly from (8).

j. Generalized Gamma Distribution as a Limit of the Generalized Beta Distribution

Suppose that X has the density (6), and let $Y = \theta^{1/\alpha} X$. Then the density of Y is

$$f_Y(x) = \theta^{-1/\alpha} f(\theta^{-1/\alpha} x) \mid \xi, \theta, \lambda, \alpha), \quad 0 \leq x \leq \theta^{1/\alpha}. \tag{9}$$

A direct computation using the approximation 23.A(5) and the fact that $\lim_{z\to\infty}(1+(a/z))^z = e^a$ shows that the limit as $\theta \to \infty$ of the density (9) is given by 9.E(2a) with ξ in place of ν and $1/\lambda$ in place of λ. Thus, the generalized gamma distribution is a limit of the generalized beta distribution. Of course, this means that the ordinary gamma distribution is a limit of the ordinary rescaled beta distribution. This is related to the fact that the Poisson distribution is a limit of binomial distributions (see 20.E.c).

k. Beta Distribution with Power Parameter

If a power parameter is introduced in the density (1), the density

$$f(x \mid a, b, \alpha) = \frac{\alpha x^{a\alpha-1}(1-x^\alpha)^{b-1}}{B(a,b)}, \quad 0 \leq x \leq 1, \; a, b, \alpha > 0, \quad (10)$$

is obtained. This density can also be obtained from (8) by setting $\lambda = 1$ (with $\xi = a$, $\theta = b$). The density (10) with an added scale parameter has been discussed by McDonald and Richards (1987), who call it a "generalized beta type I density" and investigate the hazard rate behavior using the version of Theorem 4.E.2 offered by Glaser (1980).

D. Additional Two-Parameter Extensions of the Uniform Distribution

The uniform distribution can be extended to two-parameter families by the successive application of the methods introduced in Chapter 7.

D.1. Frailty parameter then moment parameter. This is the procedure used in C.1 to derive the beta distribution.

D.2. Moment parameter then frailty parameter. As noted in Section 7.L, the results of introducing these two-parameters depends upon the order in which they are introduced. If first a moment parameter (or equivalently, a power parameter) and then a frailty parameter are introduced in (1), the result is the survival function

$$\bar{F}(x \mid \beta, \xi) = (1 - x^{\beta+1})^\xi, \quad 0 \leq x \leq 1, \; \xi > 0, \; \beta > -1. \quad (1)$$

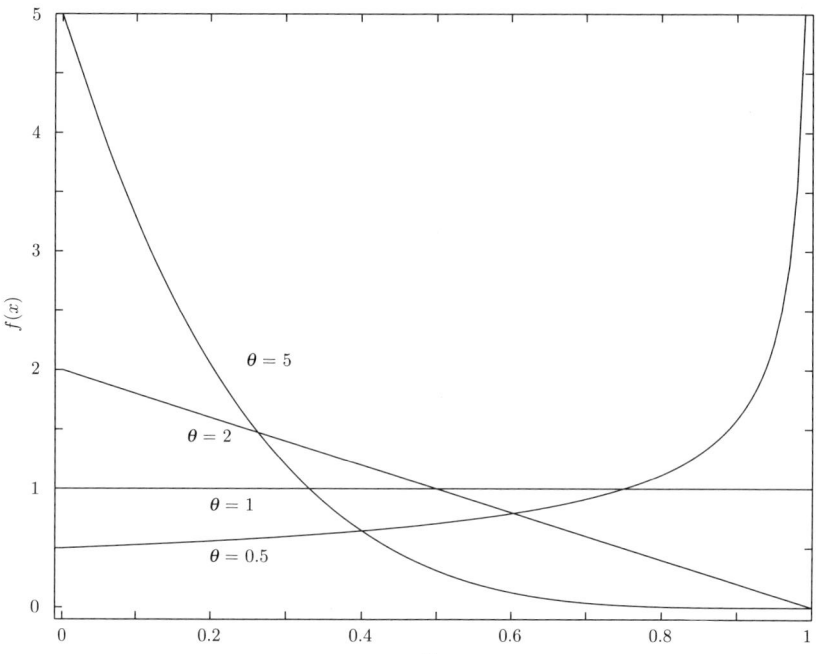

Fig. D.1a. Densities of the generalized beta distribution ($\alpha = 1, \xi = 1$)

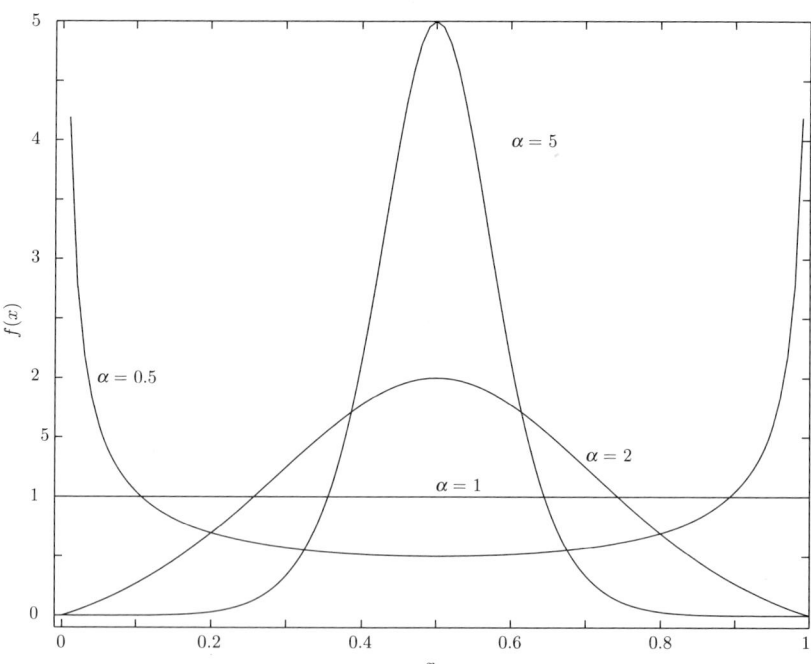

Fig. D.1b. Densities of the generalized beta distribution ($\theta = 1, \xi = 1$)

D. Additional Two-Parameter Extensions of the Uniform Distribution 491

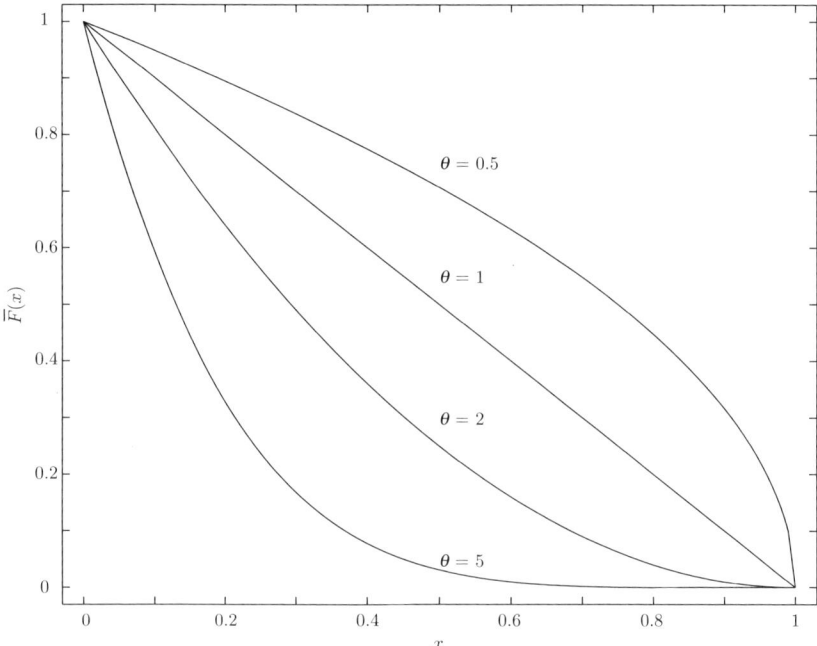

Fig. D.2a. Survival functions of the generalized beta distribution ($\alpha = 1, \xi = 1$)

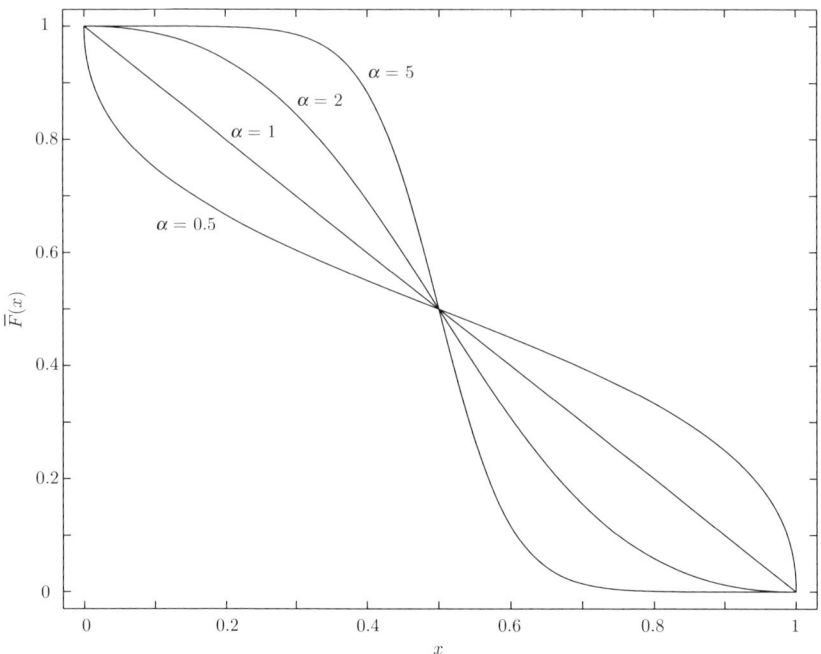

Fig. D.2b. Survival functions of the generalized beta distribution ($\theta = 1, \xi = 1$)

492   14. Distributions with Bounded Support

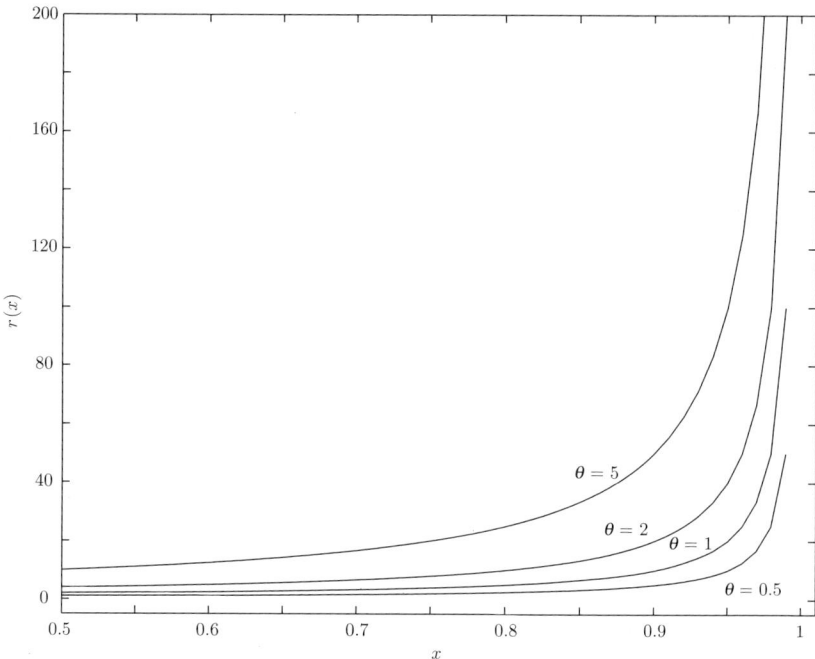

Fig. D.3a. Hazard rates of the generalized beta distribution ($\alpha = 1, \xi = 1$)

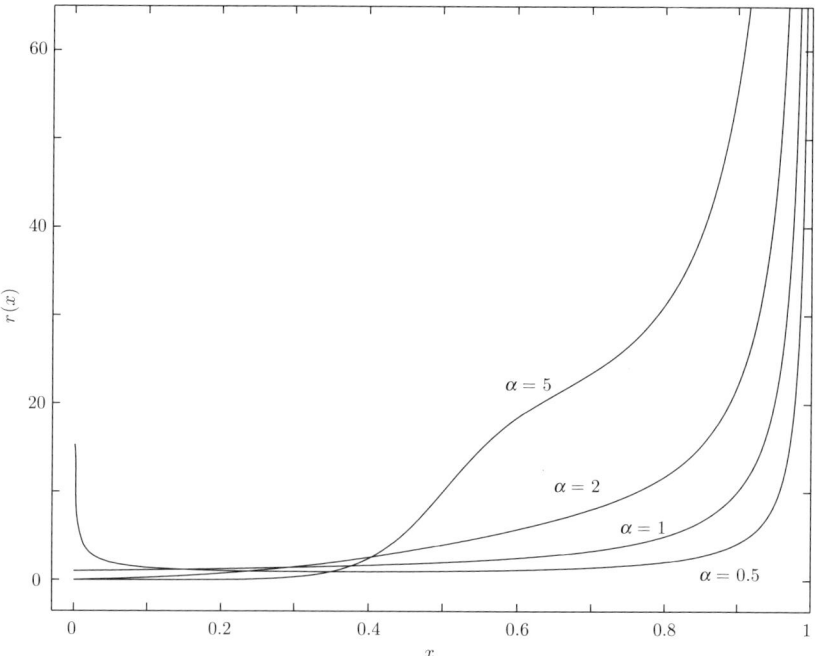

Fig. D.3b. Hazard rates of the generalized beta distribution ($\theta = 1, \xi = 1$)

This distribution does not have a moment parameter because the introduction of the frailty parameter ξ changes the character of β, but $\beta + 1$ is still a power parameter.

If (1) is rewritten in the more familiar form

$$\bar{F}(x \mid \alpha, \xi) = (1 - x^\alpha)^\xi, \quad 0 \leq x \leq 1, \; \xi > 0, \; \alpha > 0, \qquad (2)$$

then the corresponding hazard rate is

$$r(x) = \xi \frac{\alpha x^{\alpha-1}}{1 - x^\alpha}, \quad 0 < x < 1. \qquad (3)$$

It can be determined by differentiating r that when $a \geq 1$, this hazard rate is increasing, but when $\alpha < 1$, the hazard rate is bathtub shaped, with a minimum at the point $x = (1 - \alpha)^{1/\alpha}$.

D.3. Reverse hazard power then power parameter. The introduction of a power parameter in B(5b) yields the distribution

$$F(x \mid \zeta, \alpha) = \exp\{-[-\log x^\alpha]^\zeta\}, \quad 0 \leq x \leq 1, \; \alpha, \zeta > 0. \qquad (4)$$

Because the operations of introducing a reverse hazard power parameter and a power parameter commute (see Section 7.L), the same distribution is obtained if the power parameter is introduced first.

E. Introduction of a Scale Parameter

It is noted in Section 7.L that the scale parameters commute with all the other parameter types discussed in Chapter 7. This means that a scale parameter λ can be introduced in all the distributions appearing above, and the result is the same as would have been obtained if the scale parameter was first introduced.

When $F(x \mid \lambda) = F(\lambda x)$ where F satisfies A(1), the interval $[0, 1]$ is replaced by the interval $[0, 1/\lambda]$. To avoid this awkward typography, the notation

$$\theta = 1/\lambda$$

is used in what follows.

The notion of duality is still viable for a distribution with support $[0, \theta]$, and the following definition can be regarded as a more general version of Definition A.1.

E.1. Definition. Let F be a distribution with support $[0,1]$ and define $F(\cdot\,|\,\lambda)$ by $F(x\,|\,\lambda) = F(\lambda x)$. The distribution $F_D(\cdot\,|\,\lambda)$ defined by

$$F_D(x\,|\,\lambda) = 1 - F(\theta - x\,|\,\lambda) = 1 - F(1 - \lambda x) \tag{1}$$

is called the *dual* of F.

The dual of F is the distribution of $\theta - X$, where X has the distribution F. Note that $F_D(x\,|\,\lambda) = F_D(\lambda x)$, so that the operations of taking the dual and introducing a scale parameter commute.

Because the introduction of a scale parameter in any of the densities given above is straightforward, details are provided here only for the beta distribution.

E.2. Beta distribution with scale. The introduction of a scale parameter in the beta density C(1) yields the density

$$f(x\,|\,a,b,\lambda) = \frac{\lambda(\lambda x)^{a-1}(1-\lambda x)^{b-1}}{B(a,b)}, \quad 0 \le x \le \theta, \ a,b,\lambda > 0. \tag{2}$$

The case $a = 1$ in (2) is B(3) with a scale parameter.

F. Algebraic Structure of the Distributions on [0, 1]

The class $\mathcal{F}$ of distributions F such that $F(0-) = 0, 0 < F(x) < 1$ for $0 < x < 1$, and $F(1) = 1$ has interesting properties unique to the class. In particular, if $F \in \mathcal{F}$, then its inverse $F^{-1} \in \mathcal{F}$. Also, the class is closed under composition; similarly, the composition to two survival functions is in $\mathcal{F}$ whenever the corresponding distributions are in $\mathcal{F}$. This section can be considered a diversion allowed by the above facts; it may have appeal to some readers with a background in algebra.

Two binary operations on $\mathcal{F} \times \mathcal{F} \to \mathcal{F}$ are of interest: The composition $F \bullet G = F(G)$ and the composition $F \bar{\bullet} G = \bar{F}(\bar{G})$. Note that this composition is a distribution function, not a survival function. Under the $\bullet$ operation $\mathcal{F}$ forms a group; the operation is associative. The distribution uniform on $[0,1]$ is the identity, and the usual inverse F^{-1} is the inverse of F.

The situation is somewhat different under the $\bar{\bullet}$ operation because this operation is not associative; it can be verified that

$$F \bar{\bullet} (G \bar{\bullet} H) = \bar{F}(\overline{G \bar{\bullet} H}) = \bar{F}(G(\bar{H}))$$
$$\ne (F \bar{\bullet} G) \bar{\bullet} H = [\bar{F}(\bar{G})] \bar{\bullet} H = [\overline{\bar{F}(\bar{G})}](\bar{H}) = F(\bar{G}(\bar{H})).$$

F. Algebraic Structure of the Distributions on [0, 1]

Under the binary $\bar{\bullet}$ operation the distribution U uniform on $[0, 1]$ is a left identity, i.e., $U \,\bar{\bullet}\, F = F$, but U is not a right identity because $F \,\bar{\bullet}\, U = F_D$. It can be shown that U is the unique left identity. It is easily checked that if $G = 1 - F^{-1}(1 - F)$, then $F \,\bar{\bullet}\, G = F$, but of course, G is not a right identity because it depends upon F; no right identity exists for this binary operation.

Under the $\bar{\bullet}$ operation there is a left inverse of F, but it is not the usual inverse F^{-1}. (Here, it is important to recognize that $\bar{F}^{-1}$, the inverse of $\bar{F}$, is not the same as $1 - F^{-1} = \overline{F^{-1}}$.) To show that the left inverse of F is F_D^{-1}, the notation $\bar{F}_D = 1 - F_D$ is used in place of the more accurate but typographically awkward $\overline{F_D}$.

F.1. Proposition. $(F^{-1})_D = (F_D)^{-1}$ and $(\bar{F}^{-1})_D = (\bar{F}_D)^{-1}$.

Thus, the expression $\bar{F}_D^{-1}$ can be written without ambiguity.

F.2. Proposition. Under the $\bar{\bullet}$ operation, the inverse of F is $\bar{F}_D^{-1}$. That is,

$$\bar{F}(1 - \bar{F}_D^{-1}(x)) = \bar{F}_D^{-1}(1 - F(x)) = U(x) = x,$$

and U is the left identity.

By interchanging F and F_D, Proposition F.2 can be rewritten as follows.

F.3. Proposition. $\overline{F^{-1}} = (\bar{F}_D)^{-1}$.

F.4. Example. If $F(x \,|\, a) = x^a$, then

$$F^{-1}(x \,|\, a) = x^{1/a},$$

$$\bar{F}_D(x \,|\, a) = (1 - x)^a, \text{ and}$$

$$\bar{F}_D^{-1}(x \,|\, a) = 1 - x^{1/a} = \overline{F^{-1}}(x \,|\, a).$$

15

Additional Parametric Families

The purpose of models is not to fit data but to sharpen the questions.
                                                                Samuel Karlin

In this chapter, several distributions are discussed which, although well-known, are not often encountered as life distributions. These distributions may have survival functions that arise as products of those from earlier chapters, or they can be obtained by introducing a new parameter in a better-known distribution.

A. Noncentral Chi-Square Distributions

The central chi-square distribution with k degrees of freedom arises in statistics as the distribution of $\sum_{i=1}^{k} X_i^2$, where $X_1, X_2, \ldots, X_k$ are independent random variables having a normal distribution with mean 0 and variance 1. If the means are not 0, but rather $EX_i = \mu_i, i = 1, \ldots, k$, then $\sum_{i=1}^{k} X_i^2$ has a *noncentral chi-square distribution*. This distribution depends upon the μ_i only through $\delta = (\sum_{i=1}^{k} \mu_i^2)/2$. This quantity is called the *noncentrality parameter* (the term has not been standardized, and sometimes is applied to $\sqrt{\delta}$ or 2δ). The noncentral chi-square distribution is a Poisson mixture of central chi-square distributions, and has the density

$$f(x \mid \delta, k) = \sum_{j=0}^{\infty} \frac{e^{-\delta}\delta^j}{j!} \frac{x^{(k/2)+j-1}e^{-x/2}}{2^{(k/2)+j}\Gamma((k/2)+j)}, \quad x > 0, \ \delta \geq 0, \ k = 1, 2, \ldots \tag{1}$$

498   15. Additional Parametric Families

The density (1) is a special case of the *noncentral gamma distribution*, which has the density, denoted with the usual abuse of notation again by f, and given by

$$f(x\,|\,\lambda,\delta,\nu) = \sum_{j=0}^{\infty} \frac{e^{-\delta}\delta^j}{j!} \frac{\lambda(\lambda x)^{\nu+j-1}e^{-\lambda x}}{\Gamma(\nu+j)}, \qquad x,\nu > 0,\ \delta \geq 0. \qquad (2)$$

This density can be written in the form of a mixture;

$$f(x\,|\,\lambda,\delta,\nu) = \sum_{j=0}^{\infty} g(\delta,j) h(j,x\,|\,\lambda,\nu), \qquad (3)$$

where

$$g(\delta,j) = \frac{e^{-\delta}\delta^j}{j!}, \qquad h(j,x\,|\,\lambda,\nu) = \frac{\lambda(\lambda x)^{\nu+j-1}e^{-\lambda x}}{\Gamma(\nu+j)},$$
$$x,\nu > 0,\ \delta \geq 0,\ j = 0,1,2,\ldots.$$

a. Laplace Transform and Moments

It follows from 9.A(5) and the representation (3) that (2) has the Laplace transform

$$\phi(s\,|\,\lambda,\delta,\nu) = \sum_{j=0}^{\infty} g(\delta,j)\left(\frac{\lambda}{\lambda+s}\right)^{\nu+j} = \left(\frac{\lambda}{\lambda+s}\right)^{\nu}\exp\left\{-\frac{s}{\lambda+s}\delta\right\}. \qquad (4)$$

The moments can also be obtained from (3) and the moments 9.A(3) of the gamma distribution. In particular,

$$EX = (\delta+\nu)/\lambda, \qquad Var(X) = (\nu+2\delta)/\lambda^2.$$

b. Log Concavity and Unimodality

The following propositions concerning log concavity and unimodality of the density (3) have rather technical proofs based on the theory of total positivity. It is possible to prove unimodality directly without using log concavity, but the proof is similar in method and complexity to the proof of log concavity.

A. Noncentral Chi-Square Distributions   499

According to Proposition 21.A.3(iv), a differentiable density f is log concave if $f'(x)/f(x)$ is decreasing in x, that is, if for every constant c,

$$\frac{f'(x)}{f(x)} - c$$

has at most one sign change, $+$ to $-$ if one occurs. Equivalently, f is log concave if $f'(x) - cf(x)$ has at most one sign change, $+$ to $-$ if one occurs.

A.1. Proposition. *The noncentral gamma density (2) is log concave when $\nu \geq 1$.*

Proof. For notational simplicity here, write $f(x)$ in place of $f(x\,|\,\lambda,\delta,\nu)$. Moreover, it is sufficient and notationally convenient to prove the theorem with the scale parameter $\lambda = 1$. The condition $\nu \geq 1$ is required in the following to insure that $\Gamma(\nu-1)$ is positive; for the case that $\nu = 1$, take $1/\Gamma(0) = 0$. Use $\sum_{j=0}^{\infty} b_j = \sum_{j=1}^{\infty} b_{j-1}$ to compute

$$\begin{aligned}
&f'(x) - cf(x) \\
&= \sum_{j=0}^{\infty} \frac{e^{-\delta}\delta^j}{j!} \frac{x^{\nu+j-2}}{\Gamma(\nu+j-1)} e^{-x} - (c+1) \sum_{j=0}^{\infty} \frac{e^{-\delta}\delta^j}{j!} \frac{x^{\nu+j-1}}{\Gamma(\nu+j)} e^{-x} \\
&= \sum_{j=0}^{\infty} \frac{e^{-\delta}\delta^{j-1}}{j!} \frac{x^{\nu+j-2}}{\Gamma(\nu+j-1)} e^{-x} - (c+1) \sum_{j=1}^{\infty} \frac{e^{-\delta}\delta^{j-1}}{(j-1)!} \frac{x^{\nu+j-2}}{\Gamma(\nu+j-1)} e^{-x} \\
&= e^{-\delta} \frac{x^{\nu-2}}{\Gamma(\nu-1)} e^{-x} + \sum_{j=1}^{\infty} \left[\frac{e^{-\delta}\delta^j}{j!} - (c+1)\frac{e^{-\delta}\delta^{j-1}}{(j-1)!}\right] \frac{x^{\nu+j-2}}{\Gamma(\nu+j-1)} e^{-x} \\
&= \sum_{j=0}^{\infty} a_j \frac{e^{-\nu+j-2}e^{-x}}{\Gamma(\nu+j-1)},
\end{aligned}$$

where

$$a_0 = e^{-\delta}, \quad a_j = \frac{e^{-\delta}\delta^{j-1}}{(j-1)!}\left[\frac{\delta}{j} - (c+1)\right], \quad j = 1, 2, \ldots.$$

Because $a_0 > 0$ and a_j is decreasing in j, it follows that the sequence $\{a_j\}_{j=0}^{\infty}$ can have at most one sign change, $+$ to $-$ if one occurs. When $\nu > 1$, it has such a sign change because $a_0 > 0$ and $a_j < 0$ for large j.

Furthermore, the function

$$\frac{x^{\nu+j-2}e^{-x}}{\Gamma(\nu+j-1)}$$

is totally positive of order ∞ in x and j (Propositions 21.B.2 and 21.B.5.a), and hence it follows from Theorem 21.B.13 that $f'(x) - cf(x)$ has at most one sign change, $+$ to $-$ if one occurs. □

A.2. Proposition. The noncentral gamma distribution has an increasing hazard rate and a unimodal density (2), with mode at the unique solution of the equation

$$f(x\,|\,\lambda,\delta,\nu) = f(x\,|\,\lambda,\delta,\nu-1). \tag{5}$$

Proof. The monotonicity of the hazard rate and unimodality of the density follow immediately from Propositions A.1, 4.B.8.a, and 4.B.2. Because the density is unimodal and not constant over any interval, the equation $df(x)/dx = 0$ has a unique solution; this equation can be written as (5). □

c. Ordering Noncentral Gamma Distributions

A.3. Proposition. The noncentral gamma distribution with density (2) is, in the likelihood ratio ordering, increasing in δ.

Proof. Equation (3) exhibits the noncentral gamma distribution as a convolution of g and h; g is totally positive of order ∞ in δ and j, h is totally positive of order ∞ in j and x. It follows from Theorem 21.B.11 that the noncentral gamma density is totally positive of order ∞ in δ and x. The likelihood ratio ordering follows from this via Proposition 2.A.11. □

It follows from Proposition A.3 and the implications 2.A(6f) and 2.A(17) that the noncentral gamma distribution is, in the hazard rate ordering, and also the usual stochastic ordering, increasing in δ.

A.4. Proposition. The noncentral gamma distribution with density (2) is, in stochastic ordering, increasing in ν and decreasing in λ.

Proof. The stochastic ordering in ν follows from the fact that each central gamma distribution is stochastically increasing in ν, and stochastic ordering is preserved under mixtures (Proposition 3.F.1). The ordering in λ follows directly from the fact that λ is a scale parameter. □

B. Noncentral F Distributions

The stochastic ordering in ν was proved by Ghosh (1973) and the stochastic ordering in both ν and δ was proved by Ruben (1974).

B. Noncentral F Distributions

The noncentral F distribution arises in statistics as the distribution of a ratio of independent random variables, the numerator with a noncentral chi-square distribution having k_1 degrees of freedom and noncentrality parameter δ, and the denominator having a central chi-square distribution with k_2 degrees of freedom. From this origin, it can be shown that the noncentral F distribution with k_1 and k_2 degrees of freedom and noncentrality parameter δ has the density

$$f(x \mid \delta, k_1, k_2) = \sum_{j=0}^{\infty} \frac{e^{-\delta} \delta^j}{j!} \frac{x^{(k_1/2)+j-1}}{(1+x)^{[(k_1+k_2)/2]+j} B((k_1/2)+j, k_2/2)},$$
$$x > 0, \quad \delta \geq 0, \quad k_1, k_2 = 1, 2, \ldots. \tag{1}$$

A note of caution: in the statistical literature, the term "F distribution" is used to denote the distribution of the ratio of chi-square distributed random variables *normalized by their degrees of freedom*. Here, as in Section 11.D, that normalization has been omitted.

If the chi-square distributed random variables are replaced by gamma-distributed random variables in forming the ratio that leads to (1), then the resulting density is given by

$$f(x \mid \delta; \nu_1, \nu_2) = \sum_{j=0}^{\infty} \frac{e^{-\delta} \delta^j}{j!} \frac{x^{\nu_1+j-1}}{(1+x)^{\nu_1+\nu_2+j} B(\nu_1+j, \nu_2)},$$
$$x > 0, \quad \delta \geq 0, \quad \nu_1, \nu_2 > 0. \tag{2}$$

a. Moments

The rth moment of the density (2) is finite for $r < \nu_2$ and can be obtained from 11.D(2) and (2); these moments are

$$\mu_r = \sum_{j=0}^{\infty} \frac{e^{-\delta} \delta^j}{j!} \frac{B(\nu_2 - r, \nu_1 + r + j)}{B(\nu_1 + j, \nu_2)}.$$

Alternatively, the moments can be obtained using the origin of (2) as the ratio of two independent random variables with noncentral and gamma distributions. In either way, it can be verified that if X has the density (2), then for $\nu_2 > 2$,

$$EX = \frac{\nu_1 + \delta}{\nu_2 - 1}, \quad Var(X) = \frac{(\nu_1 + \delta)^2 + (\nu_1 + 2\delta)(\nu_2 - 1)}{(\nu_2 - 1)^2(\nu_2 - 2)}.$$

b. Unimodality

The following proof that the noncentral F density is unimodal is very similar to the proof of Proposition A.1 that the noncentral gamma density is log concave. The noncentral F density cannot be expected to be log concave because the central F density does not have that property.

B.1. Proposition. For $\nu_1 > 1$, the noncentral F density (2) is unimodal; if $0 < \nu_2 \leq 1$ and $\delta < 1$, then the density is decreasing.

Proof. Assume $\nu_2 > 1$. For notational simplicity write $f(x)$ in place of $f(x \mid \delta; \nu_1, \nu_2)$. The idea of this proof is to show that the derivative f' of f changes sign at most once, from $+$ to $-$ if a sign change occurs. Use $\sum_{j=0}^{\infty} b_j = \sum_{j=1}^{\infty} b_{j-1}$ compute

$$f'(x) = \sum_{j=0}^{\infty} \frac{e^{-\delta}\delta^j \Gamma(\nu_1 + \nu_2 + j)}{j!\Gamma(\nu_1 + j)\Gamma(\nu_2)} \frac{x^{\nu_1+j-2}}{(1-x)^{\nu_1+\nu_2+j+1}}$$
$$\times [(1+x)(\nu_1 + j - 1) - x(\nu_1 + \nu_2 + j)]$$

$$= \sum_{j=0}^{\infty} \frac{e^{-\delta}\delta^j \Gamma(\nu_1 + \nu_2 + j)}{j!\Gamma(\nu_1 + j - 1)\Gamma(\nu_2)} \frac{x^{\nu_1+j-2}}{(1+x)^{\nu_1+\nu_2+j}}$$
$$- \sum_{j=0}^{\infty} \frac{e^{-\delta}\delta^j \Gamma(\nu_1 + \nu_2 + j + 1)}{j!\Gamma(\nu_1 + j)\Gamma(\nu_2)} \frac{x^{\nu_1+j-1}}{(1+x)^{\nu_1+\nu_2+j+1}}$$

$$= \frac{e^{-\delta}\Gamma(\nu_1 + \nu_2)}{\Gamma(\nu_1 - 1)\Gamma(\nu_2)} \frac{x^{\nu_1-2}}{(1+x)^{\nu_1+\nu_2}}$$
$$+ \sum_{j=1}^{\infty} \left[\frac{e^{-\delta}\delta^j \Gamma(\nu_1 + \nu_2 + j)}{j!\Gamma(\nu_1 + j - 1)\Gamma(\nu_2)} - \frac{e^{-\delta}\delta^{j-1}\Gamma(\nu_1 + \nu_2 + j)}{(j-1)!\Gamma(\nu_1 + j - 1)\Gamma(\nu_2)} \right]$$
$$\times \frac{x^{\nu_1+j-2}}{(1-x)^{\nu_1+\nu_2+j}}$$

$$= \sum_{j=0}^{\infty} a_j \frac{x^{\nu_1+j-2}}{(1-x)^{\nu_1+\nu_2+j}},$$

where

$$a_0 = \frac{e^{-\delta}\Gamma(\nu_1+\nu_2)}{\Gamma(\nu_1-1)\Gamma(\nu_2)} = \frac{e^{-\delta}\Gamma(\nu_1+\nu_2)(\nu_1-1)}{\Gamma(\nu_1)\Gamma(\nu_2)} \geq 0$$

and for $j = 1, 2, \ldots$,

$$\begin{aligned}a_j &= \frac{e^{-\delta}\delta^j\Gamma(\nu_1+\nu_2+j)}{j!\Gamma(\nu_1+j-1)\Gamma(\nu_2)} - \frac{e^{-\delta}\delta^{j-1}\Gamma(\nu_1+\nu_2+j)}{(j-1)!\Gamma(\nu_1+j-1)\Gamma(\nu_2)} \\ &= \frac{e^{-\delta}\delta^{j-1}\Gamma(\nu_1+\nu_2+j)}{(j-1)!\Gamma(\nu_1+j-1)\Gamma(\nu_2)}\left[\frac{\delta}{j}-1\right].\end{aligned}$$

Note that the sequence $\{a_j\}_{j=0}^{\infty}$ has exactly one sign change, from $+$ to $-$. Moreover, by 21.B.3 and 21.B.5.a, the function $[x/(1+x)]^j$ is totally positive in x and j. From the variation diminishing property of totally positive functions (Theorem 19.B.13) it follows that f' has at most one sign change, from $+$ to $-$ if one occurs.

If $0 < \nu_1 \leq 1$ and $\delta < 1$, then $a_j \leq 0$, for all j, and consequently, the noncentral F density is decreasing. □

c. Ordering Noncentral F Distributions

B.2. Proposition. In the likelihood ratio ordering, the noncentral F distribution is increasing in δ. In the stochastic ordering, the noncentral F distribution is increasing in ν_1 and decreasing in both λ and ν_2.

Proof. The proof of the likelihood ratio order is similar to the proof of Proposition A.2, but uses the total positivity of $[x/(1+x)]^j$ in x and j. The stochastic ordering follows from Proposition 11.E.1 and the fact that the stochastic order is preserved under mixtures (Proposition 3.F.1). □

d. Doubly Noncentral F Distribution

The doubly noncentral F distribution is discussed by Johnson, Kotz and Balakrishnan (1995, p. 499). Here only a brief discussion is given.

The noncentral F density (2) is the density of a ratio of independent random variables, the numerator with a noncentral gamma distribution having shape parameter ν_1 and noncentrality parameter δ, and the denominator having a central gamma distribution with ν_2 degrees of freedom. For the doubly noncentral F, both numerator and denominator have noncentral distributions, and the resulting density is given by

$$f(x \mid \delta_1, \delta_2; \nu_1, \nu_2)$$
$$= \sum_{i=0}^{\infty} \sum_{j=0}^{\infty} \frac{e^{-\delta_1} \delta_1^i}{i!} \frac{e^{-\delta_2} \delta_2^j}{j!} \frac{x^{\nu_1+i-1}}{(1+x)^{\nu_1+\nu_2+i+j} B(\nu_1+i, \nu_2+j)},$$
$$\delta_1, \delta_2 \geq 0, \quad \nu_1, \nu_2 > 0, \quad x > 0. \tag{3}$$

This somewhat formidable looking density could be further generalized by using a bivariate Poisson distribution in place of the two independent Poisson distributions present in (3); e.g., $[e^{-\delta_1} \delta_1^i / i!][e^{-\delta_2} \delta_2^j / j!]$ in (4) can be replaced by

$$\sum_{i=0}^{\min(x,y)} \frac{\delta_1^{x-i} \delta_2^{y-i} \delta_{12}^i}{(x-i)!(y-i)!i!} e^{-(\delta_1+\delta_2+\delta_{12})};$$

similarly, other bivariate distributions with Poisson marginals could be used.

C. A Noncentral Beta Distribution and the Noncentral Squared Multiple Correlation Distribution

The noncentral beta distribution and the noncentral squared multiple correlation distribution arise in quite different contexts, but they are treated here together because they are quite similar in form and consequently have similar properties.

a. The Noncentral Beta Distribution

If X has the noncentral F density B(2), then $Y = X/(1+X)$ has the density

$$f(x \mid \delta, \nu_1, \nu_2) = \sum_{j=0}^{\infty} \frac{e^{-\delta} \delta^j}{j!} \frac{x^{\nu_1+j-1}(1-x)^{\nu_2-1}}{B(\nu_1+j, \nu_2)},$$
$$\delta, \nu_1, \nu_2 > 0, \quad 0 < x < 1; \tag{1}$$

this is the density of the *noncentral beta distribution*. The noncentral beta density has the form

$$f(x \mid \delta; \nu_1, \nu_2) = \sum_{j=0}^{\infty} g(j \mid \delta) h(x \mid \nu_1+j, \nu_2), \quad 0 < x < 1,$$

where

$$g(j\,|\,\delta) = e^{-\delta}\frac{\delta^j}{j!}, \quad \delta > 0, \ j = 0, 1, 2, \ldots \qquad (2)$$

is a Poisson mass function and

$$h(x\,|\,\nu_1 + j, \nu_2) = \frac{x^{\nu_1+j-1}(1-x)^{\nu_2-1}}{B(\nu_1 + j, \nu_2)}, \quad 0 < x < 1, \ \nu_1, \nu_2 > 0. \qquad (3)$$

is a beta distribution of the form 14.C(1).

b. The Noncentral Squared Multiple Correlation Distribution

The noncentral squared multiple correlation distribution arises as follows. Let $Y, X_1, X_2, \ldots, X_p$ have a multivariate normal distribution with mean vector 0 and covariance matrix

$$\Sigma = \begin{pmatrix} \sigma_{00} & \sigma_{01} \\ \sigma'_{01} & \Sigma_{11} \end{pmatrix};$$

if $X = (X_1, X_2, \ldots, X_p)$, then $\sigma_{00} = \text{Var } Y$, σ_{01} is the p-dimensional vector Cov (Y, X), and Σ_{11} is the $p \times p$ covariance matrix of X. With the notation $a = (a_1, a_2, \ldots, a_p)$, the multiple correlation coefficient ρ is defined by

$$\rho = \underset{a}{\text{Max}}\,\text{Corr}(Y, aX') = \underset{a}{\text{Max}}\,\frac{a\sigma'_{01}}{\sqrt{\sigma_{00}\,a\Sigma_{11}a'}}.$$

For a sample of size n, and sample covariance S, the sample multiple correlation is

$$R = \underset{a}{\text{Max}}\,\frac{as'_{01}}{\sqrt{s_{00}\,aS_{11}a'}},$$

where S is partitioned as was Σ. Because of invariance under linear transformations, it can be shown that the distribution of R depends only on ρ.

Set $\theta = \rho^2$, $\kappa = p/2$, and $\nu = (n-1)/2$ and assume that $\rho \neq 0$. With the usual overuse of the letter "f" to denote a density, *the noncentral squared multiple correlation distribution* has the density

$$f(x\,|\,\theta, \kappa, \nu) = \sum_{j=0}^{\infty} g(j\,|\,\theta, \nu) h(x\,|\,\kappa + j, \nu - \kappa),$$

$$0 < x < 1, \ 0 \leq \theta \leq 1, \ \nu > \kappa > 0, \qquad (4)$$

where

$$g(j\mid\theta,\nu) = \theta^j(1-\theta)^\nu \frac{\Gamma(\nu+j)}{j!\Gamma(\nu)}, \quad 0 \leq \theta \leq 1, \ \nu > 0, \ j = 0, 1, 2, \ldots, \tag{5}$$

is a negative binomial probability mass function, and

$$h(x\mid \kappa+j, \nu-\kappa) = \frac{x^{\kappa+j-1}(1-x)^{\nu-\kappa-1}}{B(\kappa+j, \nu-\kappa)}, \quad 0 < x < 1, \ \nu > \kappa > 0, \tag{6}$$

is again a beta density of the form 14.C(1). The densities (1) and (4) are both mixtures of beta densities, one mixed using the Poisson distribution and the other mixed using the negative binomial distribution.

c. Unimodality

C.1. Proposition. The noncentral beta density (1) is unimodal when $\nu_1 > 1$.

Proof. The derivative $f'(x\mid \delta, \nu_1, \nu_2)$ of the noncentral beta density is given by

$$f'(x\mid \delta,\nu_1,\nu_2) = \sum_{j=0}^{\infty} \frac{e^{-\delta}\delta^j}{j!B(\nu_1+j,\nu_2)} q(x,j\mid \nu_1,\nu_2), \tag{7}$$

where

$$\begin{aligned}
q(x,j\mid \nu_1,\nu_2) &= (\nu_1+j-1)x^{\nu_1+j-2}(1-x)^{\nu_2-1} \\
&\quad - (\nu_2-1)x^{\nu_1+j-1}(1-x)^{\nu_2-2} \\
&= (\nu_1+j-1)x^{\nu_1+j-2}(1-x)^{\nu_2-2}(1-x) \\
&\quad - (\nu_2-1)x^{\nu_1+j-1}(1-x)^{\nu_2-2} \\
&= (\nu_1+j-1)x^{\nu_1+j-2}(1-x)^{\nu_2-2} \\
&\quad - (\nu_1+\nu_2+j-2)x^{\nu_1+j-1}(1-x)^{\nu_2-2}. \tag{8}
\end{aligned}$$

Thus,

$$\begin{aligned}
&f'(x\mid \delta,\nu_1,\nu_1) \\
&= \sum_{j=0}^{\infty} \frac{e^{-\delta}\delta^j}{j!} \left[\frac{\Gamma(\nu_1+\nu_2+j)}{\Gamma(\nu_1+j)\Gamma(\nu_2)}(\nu_1+j-1)x^{\nu_1+j-2}(1-x)^{\nu_2-2} \right.\\
&\quad \left. - \frac{\Gamma(\nu_1+\nu_2+j)}{\Gamma(\nu_1+j)\Gamma(\nu_2)}(\nu_1+\nu_2+j-2)x^{\nu_1+j-1}(1-x)^{\nu_2-2} \right]
\end{aligned}$$

$$= e^{-\delta} \frac{\Gamma(\nu_1 + \nu_2)}{\Gamma(\nu_1)\Gamma(\nu_2)} (\nu_1 - 1) x^{\nu_1 - 2} (1-x)^{\nu_2 - 2}$$
$$+ \sum_{j=1}^{\infty} \frac{e^{-\delta} \delta^j}{j!} \frac{\Gamma(\nu_1 + \nu_2 + j)}{\Gamma(\nu_1 + j - 1)\Gamma(\nu_2)}$$
$$- \sum_{j=0}^{\infty} \frac{e^{-\delta} \delta^j}{j!} \frac{\Gamma(\nu_1 + \nu_2 + j)(\nu_1 + \nu_2 + j - 2)}{\Gamma(\nu_1 + j)\Gamma(\nu_2)} x^{\nu_1 + j - 1}(1-x)^{\nu_2 - 2}$$

$$= e^{-\delta} \frac{\Gamma(\nu_1 + \nu_2)}{\Gamma(\nu_1)\Gamma(\nu_2)} (\nu_1 - 1) x^{\nu_1 - 2}(1-x)^{\nu_2 - 2}$$
$$+ \sum_{j=1}^{\infty} \frac{e^{-\delta} \delta^j}{j!} \frac{\Gamma(\nu_1 + \nu_2 + j)}{\Gamma(\nu_1 + j - 1)\Gamma(\nu_2)} x^{\nu_1 + j - 2}(1-x)^{\nu_2 - 2}$$
$$- \sum_{j=1}^{\infty} \frac{e^{-\delta} \delta^{j-1}}{(j-1)!} \frac{\Gamma(\nu_1 + \nu_2 + j - 1)(\nu_1 + \nu_2 + j - 3)}{\Gamma(\nu_1 + j - 1)\Gamma(\nu_2)} x^{\nu_1 + j - 2}(1-x)^{\nu_2 - 2}$$

$$= \sum_{j=0}^{\infty} a_j x^{\nu_1 + j - 2}(1-x)^{\nu_2 - 2},$$

where $a_0 = e^{-\delta} \dfrac{\Gamma(\nu_1 + \nu_2)}{\Gamma(\nu_1)\Gamma(\nu_2)} (\nu_1 - 1)$, and for $j = 1, 2, \ldots$,

$$a_j = \frac{e^{-\delta} \delta^{j-1}}{(j-1)!} \frac{\Gamma(\nu_1 + \nu_2 + j - 1)}{\Gamma(\nu_1 + j - 1)\Gamma(\nu_2)} \left[\frac{(\nu_1 + \nu_2 + j)\delta}{j} - (\nu_1 + \nu_2 + j - 3) \right].$$

It can be shown that $a_j < 0$ implies $a_{j+1} < 0$ by verifying that the function

$$\phi(z) = \frac{(\nu_1 + \nu_2 + z)\delta}{z} - (\nu_1 + \nu_2 + z - 3)$$

has the derivative $\phi'(z) = -\dfrac{\delta(\nu_1 + \nu_2)}{z^2} - 1 < 0$. Because $\nu_1 > 1, a_0 > 0$ and thus the sequence $\{a_j\}_{j=0}^{\infty}$ has at most one sign change, from $+$ to $-$ if one occurs. With the aid of Theorem 21.B.13, the proposition follows from the fact that $K(j, x) = x^{\nu_1 + j - 2}(1-x)^{\nu_2 - 2}$ is totally positive in j and x. □

C.2. Proposition. The noncentral squared multiple correlation density (4) is unimodal when $\kappa \geq 1$.

Proof. Compute

$$f'(x \mid \theta, \kappa, \nu) = \sum_{j=0}^{\infty} \theta^j (1-\theta)^\nu \frac{[\Gamma(\nu+j)]^2}{j! \Gamma(\nu) \Gamma(j+\kappa) \Gamma(\nu-\kappa)} q(x, j \mid \kappa, \nu),$$

where apart from parameterization q is the same as q given by (8). As in the proof of Proposition C.1, it can be determined that

$$f'(x \mid \theta, \kappa, \nu) = \sum_{j=0}^{\infty} a_j x^{\kappa+j-1} (1-x)^{\nu+\kappa-2},$$

where

$$a_0 = \frac{(1-\theta) \Gamma(\nu) (\kappa - 1)}{\Gamma(\kappa) \Gamma(\nu - \kappa)},$$

and for $j = 1, 2, \ldots$,

$$a_j = \frac{\theta^j (1-\theta)^\nu [\Gamma(\nu+j-1)]^2}{(j-1)! \Gamma(\kappa) \Gamma(\nu-\kappa)} \left[\frac{\theta(\nu+j-1)^2}{j} - (j+\nu-3) \right].$$

Because $\kappa \geq 1$, $a_0 \geq 0$. As in the proof of Proposition C.1, it can be shown that the function

$$\phi(z) = \frac{\theta(\nu + z - 1)^2}{z} - (z + \nu - 3)$$

has a negative derivative, and consequently $a_j < 0$ implies $a_{j+1} < 0$. Thus, the sequence $\{a_j\}_{j=0}^{\infty}$ changes sign at most once, from $+$ to $-$ if a sign change occurs. Moreover, the function $x^{\kappa+j-1}(1-x)^{\nu+\kappa-2}$ is totally positive in x and j. Consequently, it follows from Theorem 21.C.13 that $f'(x \mid \theta, \kappa, \nu)$ can change sign at most once, from $+$ to $-$ if a sign change occurs. □

d. Doubly Noncentral Beta Distribution

The doubly noncentral beta distribution is discussed by Johnson, Kotz and Balakrishnan (1995, p. 499). In the same way that the noncentral beta density (1) can be derived from the noncentral F density B(1), so a doubly noncentral beta density can be derived from the doubly noncentral F density B(3), this density is

$$f(x \mid \delta_1, \delta_2; \nu_1, \nu_2) = \sum_{i=0}^{\infty} \sum_{j=0}^{\infty} \frac{e^{-\delta_1} \delta_1^i}{i!} \frac{e^{-\delta_2} \delta_2^j}{j!} \frac{x^{\nu_1+i-1}(1-x)^{\nu_2+j-1}}{B(\nu_1+i, \nu_2+j)}, \quad (9)$$

$$\delta_1, \delta_2, \nu_1, \nu_1 > 0, \quad 0 < x < 1.$$

e. Ordering Noncentral Beta Distributions and Noncentral Multiple Squared Correlation Distributions

Because the noncentral beta density (1) is a Poisson mixture of beta densities, it has total positivity properties that can be obtained using Theorem 21.B.11. In particular, the density is totally positive in the pairs δ and x, ν_1 and x, and ν_2 and $-x$. According to Proposition 2.A.11, this means that noncentral beta distributions are likelihood ratio increasing in δ. A random variable Y with a noncentral beta distribution has the representation $Y = X/(1+X)$, where X has the noncentral F density. Because the noncentral F distribution is stochastically increasing in ν_1 and stochastically decreasing ν_2, it follows that the noncentral beta distribution is ordered in the same way. The same conclusion can be obtained from 14.C.2.

It follows from a similar argument that the noncentral multiple squared correlation density is totally positive in θ and x, and so these distributions are likelihood ratio increasing in θ. From 14.C.2, it follows that the beta distribution (6) is stochastically increasing in κ when $\nu - \kappa$ is fixed, and stochastically decreasing in ν.

D. Log Distributions from Nonnegative Random Variables

As mentioned in Section 12.A, for a distribution with support $[0, \infty]$ the logarithmic version has support $[1, \infty)$; that is to say, if Y is a nonnegative random variable, then $X = e^Y$ takes on values in $[1, \infty)$. On the other hand, negative logarithmic distributions that arise from the transformation $X = e^{-Y}$ have support $[0, 1]$. Several such distributions are mentioned here.

a. The Log Gamma Distribution

The log gamma distribution has support $[1, \infty]$ and the density

$$f(x \mid \lambda, \nu) = \lambda^\nu (\log x)^{\nu-1} x^{-(\lambda+1)}/\Gamma(\nu), \quad x > 1, \quad \lambda, \nu > 0. \tag{1}$$

This rather unfamiliar density is encountered in Proposition 18.B.13. See Figure D.1.

By computing the derivative f', it can be determined that when $\nu > 1, f$ is unimodal with mode at $x = \exp\{(\nu-1)/(\lambda+1)\}$. When

510   15. Additional Parametric Families

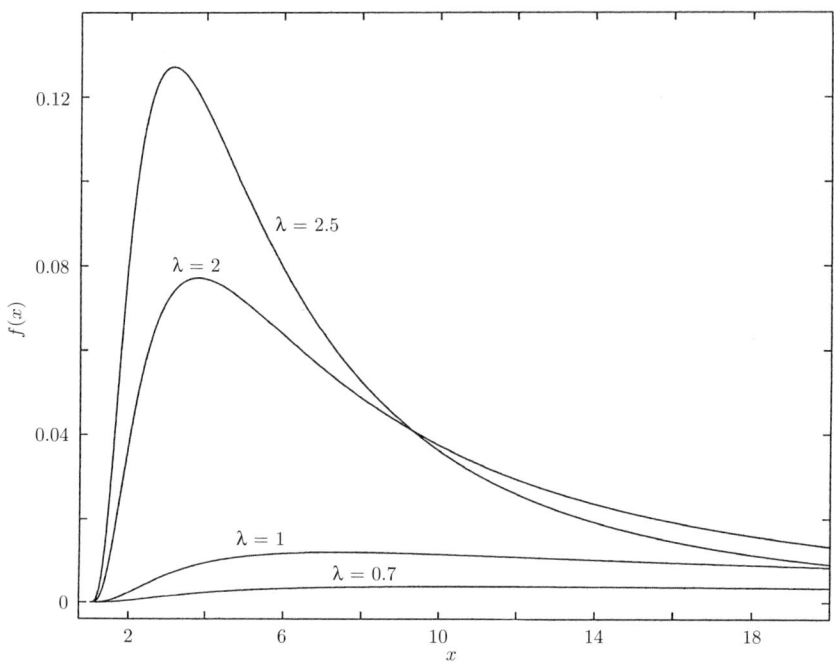

Fig. D.1a-1. Densities of the log gamma distribution ($\nu = 5$)

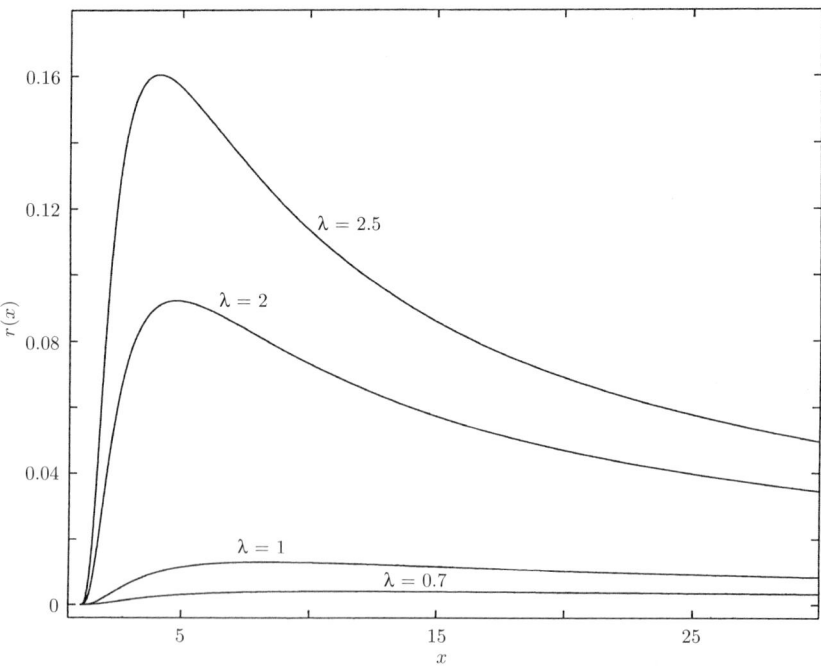

Fig. D.1a-2. Hazard rates of the log gamma distribution ($\nu = 5$)

D. Log Distributions from Nonnegative Random Variables   511

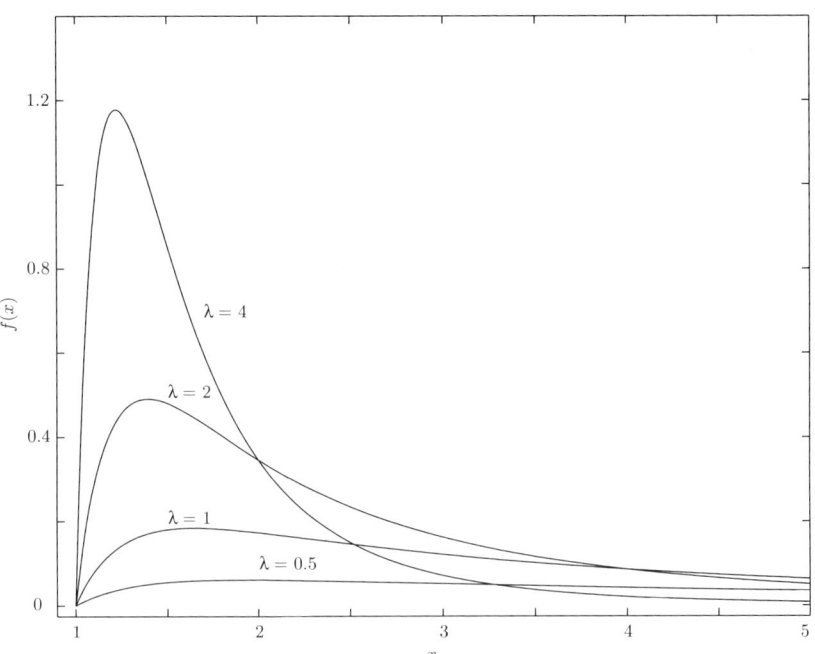

Fig. D.1b-1. Densities of the log gamma distribution ($\nu = 2$)

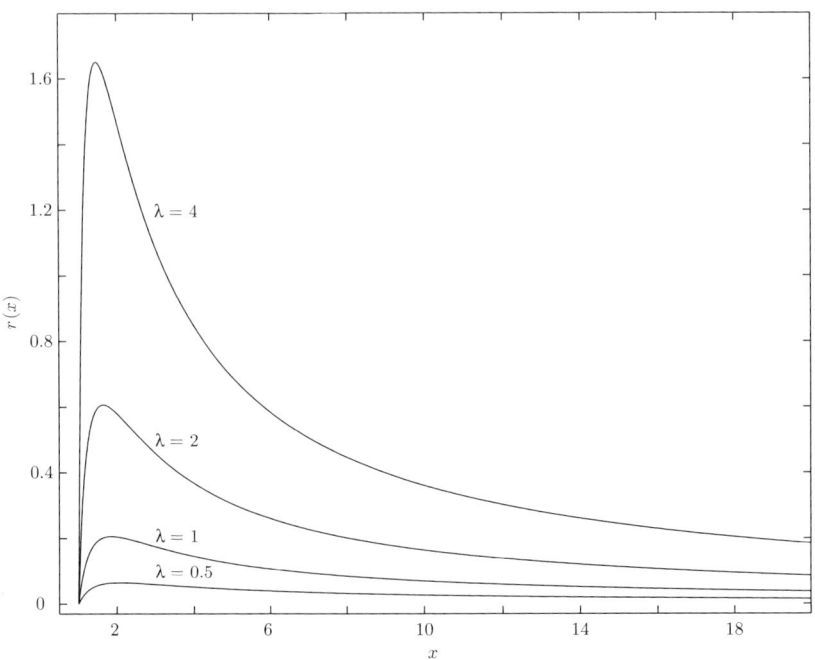

Fig. D.1b-2. Hazard rates of the log gamma distribution ($\nu = 2$)

512   15. Additional Parametric Families

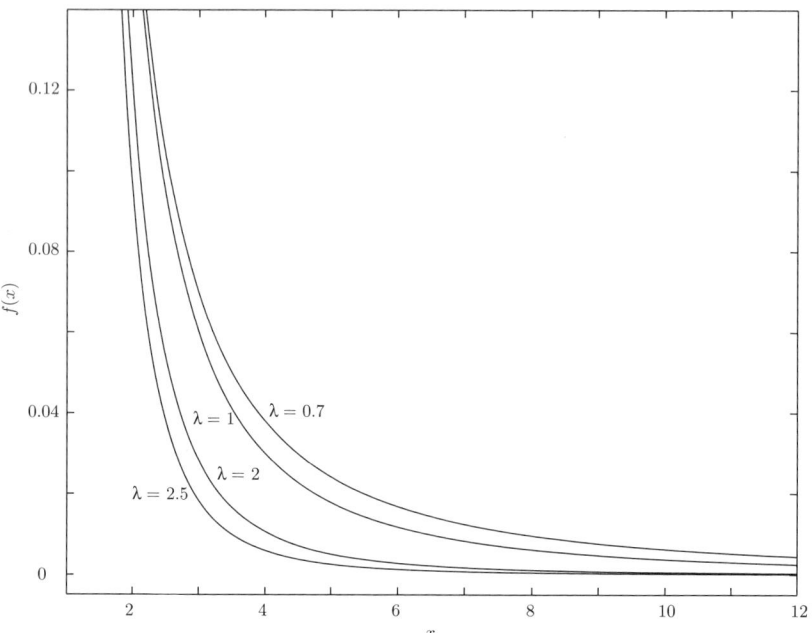

Fig. D.1c-1. Densities of the log gamma distribution ($\nu = 0.5$)

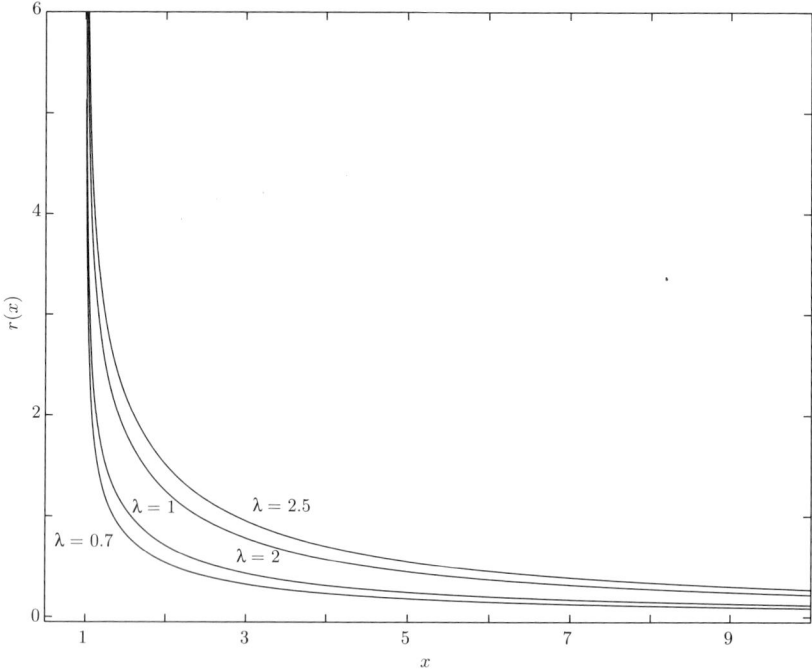

Fig. D.1c-2. Hazard rates of the log gamma distribution ($\nu = 0.5$)

$\nu \le 1$, the density is decreasing. For this distribution the moments are

$$\mu_r = \left(\frac{\lambda}{\lambda - r}\right)^{\nu}, \quad \lambda > r.$$

Neither the survival function nor the hazard rate of the log gamma distribution can be written in closed form.

b. The Hazard Rate of the Log Gamma Distribution

To obtain the shape of the hazard rate of the log gamma distribution, note from 12.A.1(iv) that because the hazard rate of the gamma distribution is decreasing for $\nu < 1$, the same is true for the hazard rate of the log gamma distribution. When $\nu > 1$, the hazard rate shape can be examined using Lemma 4.E.1. To do this, it is necessary first to compute, for the density (1), that

$$\rho(x) = -\frac{d \log f(x)}{dx} = \frac{\lambda + 1}{x} - \frac{\nu - 1}{x \log x}, \quad x > 1.$$

For $\nu > 1$, the equation $\rho'(x) = 0$ is quadratic with one negative root. Because $\rho'(x) > 0$ for x just larger than 1 and $\rho'(x) < 0$ for large x, it follows that $\rho(x) = 0$ has exactly one root > 1 and moreover, $\rho(x)$ is initially increasing to a unique maximum, say at $x = z$, and then is decreasing. It follows from Lemma 4.E.1 that when $\nu > 1$, the log gamma distribution has either a decreasing hazard rate or a unimodal hazard rate with mode at some point in the interval $(0, z]$. When $\nu > 1$, the hazard rate of the gamma distribution has the limiting value $1/\lambda$ as $x \to \infty$. It follows from 12.A(5) that the hazard rate of the log gamma distribution behaves like $1/\lambda$ for large x.

c. Negative Log Gamma Distribution

If Y has a gamma distribution with density 9.A(1), then $X = e^{-Y}$ has the density

$$\begin{aligned} f(x \mid \alpha, \nu) &= \frac{\alpha^{\nu-1} x^{\alpha-1}(-\log x)^{\nu-1}}{\Gamma(\nu)} \\ &= \frac{x^{\alpha-1}(-\log x^{\alpha})^{\nu-1}}{\Gamma(\nu)}, \quad 0 < x < 1, \ \nu, \alpha > 0; \end{aligned} \quad (2)$$

here, the parameter λ has been changed to α because the log transformation has changed the scale parameter to a power parameter. This is the density of the *negative log gamma distribution*, which is encountered in Proposition 18.B.13. When $\alpha, \nu > 1$, this density is unimodal with mode at $z = \exp\{-(\nu-1)/(\alpha-1)\}$. When $\alpha, \nu < 1$, the density is U-shaped with a minimum at z. When $\alpha < 1, \nu > 1$, the density is decreasing, and when $\alpha > 1, \nu < 1$, the density is increasing.

The special case $\nu = 1$ is the density 14.B(2a), so the density (2) can be regarded as a two-parameter extension of the uniform distribution, but it is not any of those encountered in Chapter 14.

d. The Log Weibull and Negative Log Weibull Distributions

The log Weibull distribution has survival function

$$\bar{F}(x) = \exp\{-(\alpha \log x)^\zeta\} = \exp\{-(\log x^\alpha)^\zeta\}, \quad x \geq 1, \ \zeta, \alpha > 0,$$
$$= 1, \quad x < 1; \tag{3}$$

here the parameter λ of the Weibull distribution has been replaced by α because the log transformation has changed the scale parameter into a power parameter. Likewise, the power parameter α of the Weibull distribution has been replaced by ζ because it has become a hazard power parameter.

The hazard rate of the log Weibull distribution is

$$r(x) = \frac{\zeta \alpha^\zeta (\log x)^{\zeta-1}}{x}, \quad x > 1. \tag{4}$$

See Figure D.2. With $\zeta = 1$, this survival function reduces to $\bar{F}(x) = x^{-\alpha}, x \geq 1$, which is the survival function of a Pareto I distribution with unit scale and a power parameter λ (see Section 11.B).

The negative log Weibull distribution has the distribution function

$$F(x) = \exp\{-(-\alpha \log x)^\zeta\}, \quad 0 < x < 1, \ \alpha, \zeta > 0, \tag{5}$$

which is 18.B(44b) apart from parameterization.

D. Log Distributions from Nonnegative Random Variables 515

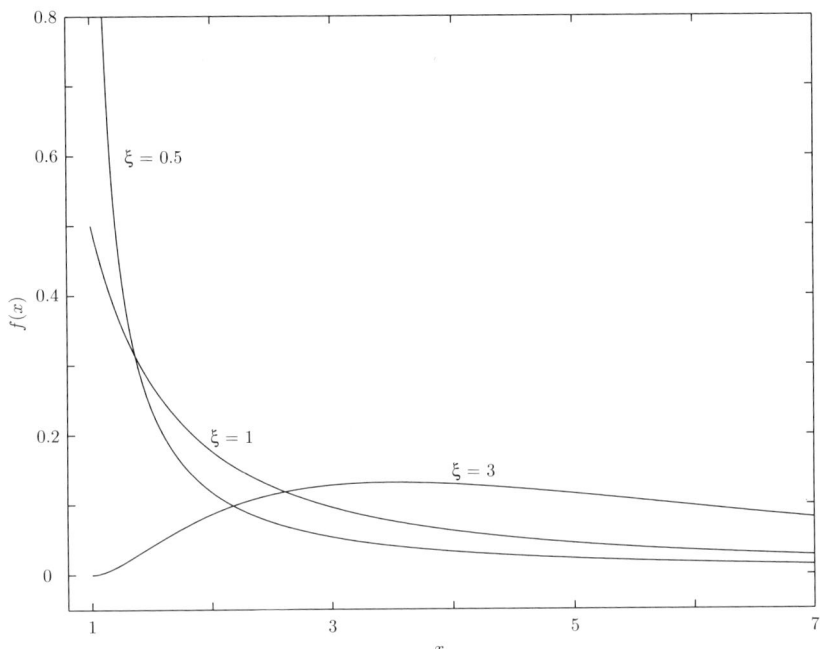

Fig. D.2a-1. Densities of the log Weibull distribution ($\alpha = 0.5$)

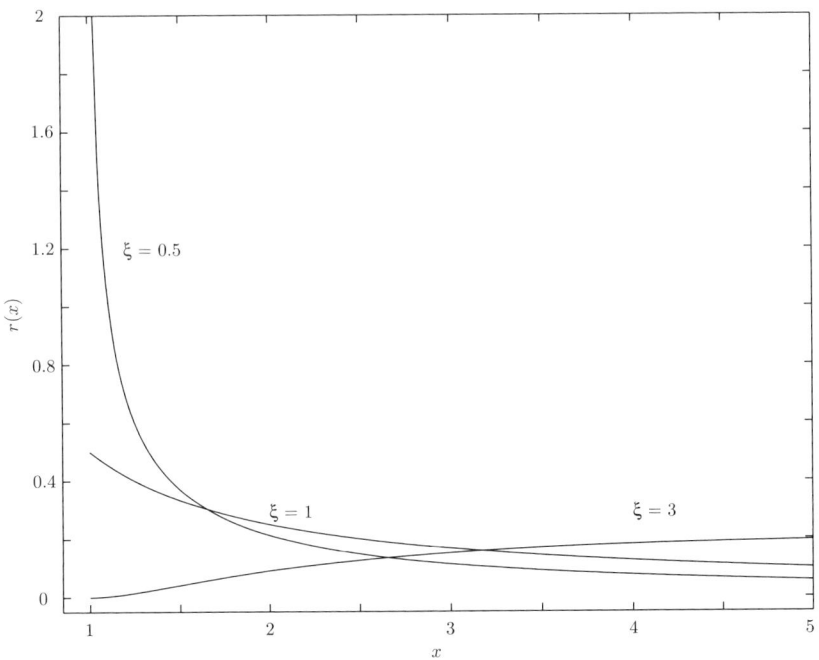

Fig. D.2a-2. Hazard rates of the log Weibull distribution ($\alpha = 0.5$)

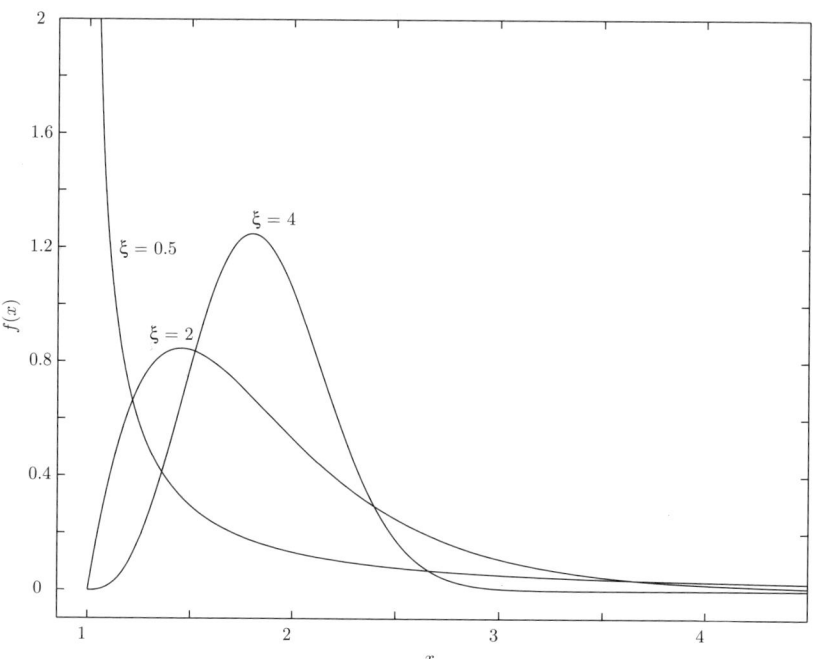

Fig. D.2b-1. Densities of the log Weibull distribution ($\alpha = 1.5$)

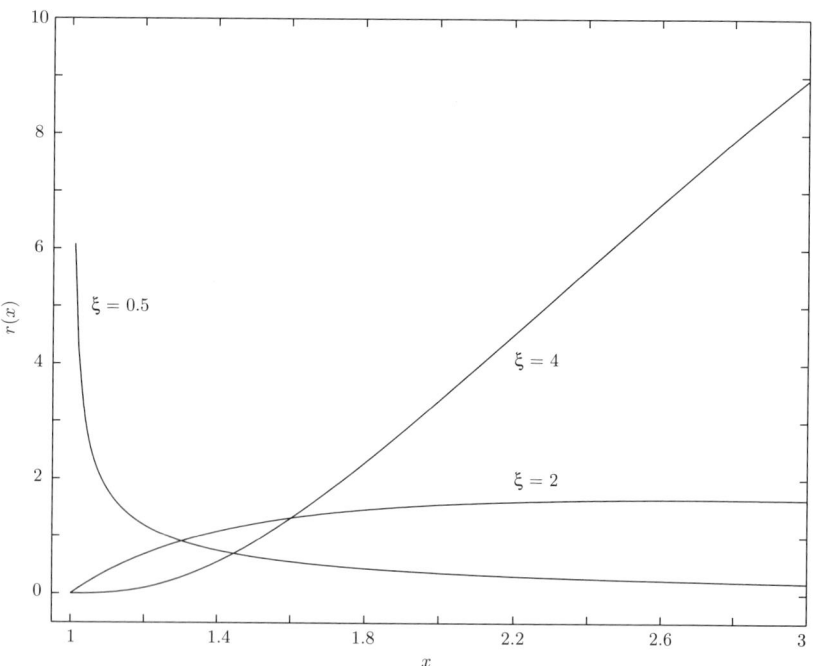

Fig. D.2b-2. Hazard rates of the log Weibull distribution ($\alpha = 1.5$)

D. Log Distributions from Nonnegative Random Variables 517

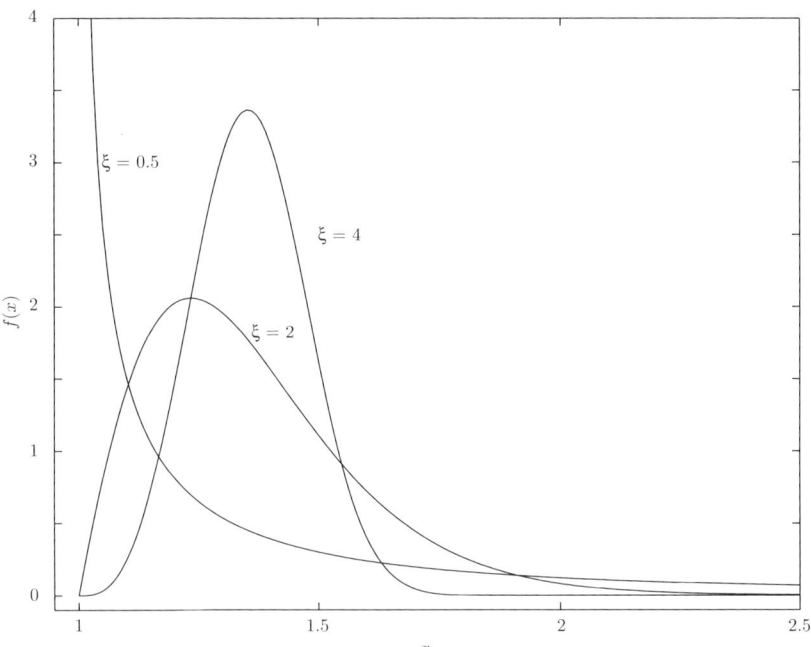

Fig. D.2c-1. Densities rates of the log Weibull distribution ($\alpha = 3$)

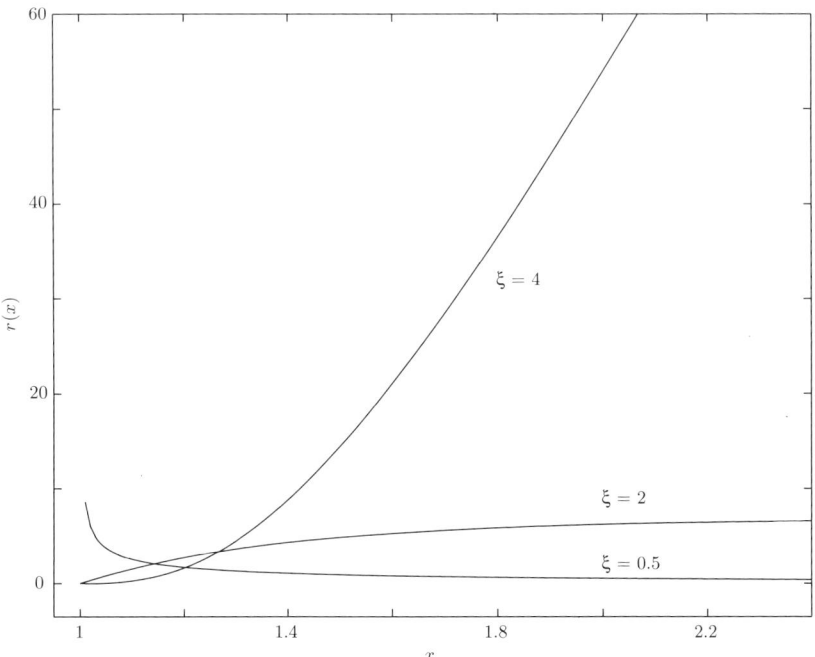

Fig. D.2c-2. Hazard rates of the log Weibull distribution ($\alpha = 3$)

e. The Log Gompertz Distribution

The log Gompertz distribution function is given by

$$\bar{F}(x \mid \alpha, \xi) = \exp\{-\xi(x^\alpha - 1)\}, \quad x > 1, \ \alpha, \xi > 0, \qquad (6)$$

and density

$$f(x \mid \alpha, \xi) = \alpha \xi x^{\alpha-1} \exp\{-\xi(x^\alpha - 1)\}, \quad x > 1. \qquad (7)$$

If X has a log Gompertz distribution and $Y = X - 1$, then Y has the survival function

$$\bar{F}(x \mid \alpha, \xi) = \exp\{-\xi[(x+1)^\alpha - 1]\}, \quad x > 0, \ \alpha, \xi > 0; \qquad (8)$$

note that α is no longer a power parameter.

It can be verified that the survival function (8) has the hazard rate

$$r(x) = \xi\alpha(x+1)^{\alpha-1}, \quad x > 0, \ \alpha, \xi > 0. \qquad (9)$$

Apart from parameterization, this is the hazard rate 9.B(19) of the Weibull residual life distribution F_t with $t = 1$. See Figure D.3.

The negative log Gompertz distribution has survival function

$$\bar{F}(x \mid \alpha, \xi) = \exp\{-\xi(x^{-\alpha} - 1)\}, \quad 0 < x < 1, \ \alpha, \xi > 0. \qquad (10)$$

For negative logarithmic distributions such as (10), a connection between moment and Laplace transform parameters is discussed in Section 12.A.b.

E. Another Extension of the Exponential Distribution

In Sections D.c and D.d, several distributions having support $[0, 1]$ are derived from distributions having support $[0, \infty]$ through use of the transformation of the random variables $X = e^{-Y}$ of the corresponding random variables. The following uses the inverse of this transformation to obtain a distribution with support $[0, \infty]$ from the beta distribution.

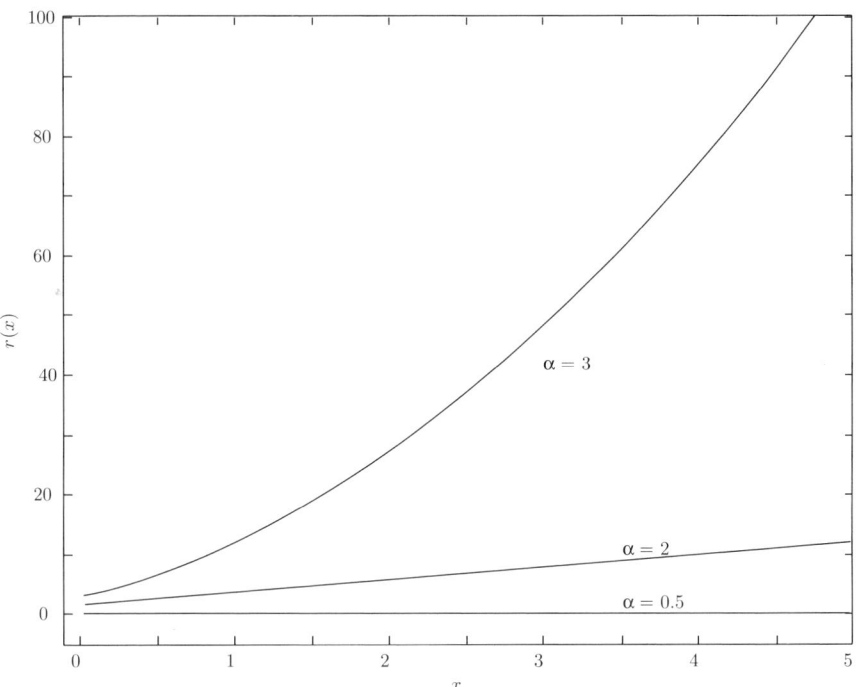

Fig. D.3. Hazard rates of the translated log Gompertz distribution

E.1. Transformed beta distribution. Suppose that X has the beta distribution of Section 12.C, i.e., X has the density

$$f_X(x) = x^{a-1}(1-x)^{b-1}/B(a,b), \quad 0 \le x \le 1.$$

If $Y = -\log X$, then Y has the density

$$f_Y(y) = e^{-ay}(1 - e^{-y})^{b-1}/B(a,b), \quad y > 0.$$

The exponential density is the case $b = 1$. If $b < 1$, then $\log f$ is convex, and hence f has a decreasing hazard rate; if $b > 1$, then $\log f$ is concave, and hence f is unimodal and the distribution has an increasing hazard rate.

F. Weibull–Pareto–Beta Distribution

A distribution that includes the Pareto IV distribution, the Weibull distribution, and the uniform distribution with power and frailty parameters has been proposed by Mudholkar and Kollia (1994), and further discussed by Mudholkar, Srivastava and Kollia (1996) and Mudholkar and Sarkar (1999). These authors call the distribution the "generalized Weibull distribution." This three-parameter distribution offers flexibility in fitting data when the proper choice of a more specialized model is unclear. Such an approach was taken by Prentice (1975) with a four-parameter distribution for random variables taking values on $(-\infty, \infty)$.

Start with a distribution uniform on $[0,1]$. Then add successively a scale, power, and frailty parameter to obtain the survival function

$$\bar{G}(x \mid \lambda, \alpha, \xi) = [1 - (\lambda x)^\alpha]^\xi, \quad 0 \leq x \leq 1/\lambda, \quad \alpha > 0, \; \theta > 0.$$

Mudholkar and Kollia (1994) make the following clever reparameterization: Replace ξ by $1/\theta$ and replace λ by $\lambda \theta^{1/\alpha}$. Then the survival function

$$\bar{F}(x \mid \lambda, \alpha, \theta) = [1 - \theta(\lambda x)^\alpha]^{1/\theta}, \quad 0 < x < 1/(\theta^{1/\alpha} \lambda), \quad \alpha > 0, \; \theta > 0, \tag{1a}$$

is obtained, and the limiting form of this survival function as $\theta \to 0$ is given by

$$\bar{F}(x \mid \lambda, \alpha, \theta) = \exp\{-(\lambda x)^\alpha\}, \quad x > 0, \; \alpha > 0, \; \theta = 0, \tag{1b}$$

which is the familiar survival function of a Weibull distribution.

Finally, observe that if $\theta < 0$, the survival function (1a) remains a survival function but the support becomes $[0, \infty)$; that is,

$$\bar{F}(x \mid \lambda, \alpha, \theta) = [1 - \theta(\lambda x)^\alpha]^{1/\theta}, \quad x > 0, \; \alpha > 0, \; \theta < 0. \tag{1c}$$

Together, (1a) to (1c) constitute what Mudholkar and Kollia (1994) call the "generalized Weibull distribution."

The generalized Weibull distribution has the hazard rate

$$r(x \mid \lambda, \alpha, \theta) = \frac{\alpha \lambda (\lambda x)^{\alpha-1}}{1 - \theta(\lambda x)^\alpha}, \tag{2}$$

this holds for $x > 0$ when $\theta \leq 0$, and it holds for $0 < x < \theta^{-1/\alpha}$ when $\theta > 0$. See Figures D.4a,b,c.

a. Hazard Rate Shape

(i) When $\theta < 0$, the hazard rate (2) becomes, with a straightforward reparameterization, the hazard rate 11.B(5), and so is an inverted-bathtub (unimodal) shape for $\alpha > 1$, and otherwise is decreasing.

(ii) When $\theta = 0$, this hazard rate is the hazard rate 9.B(3) of a Weibull distribution, and as such is increasing or decreasing according to $\alpha \geq 1$ or $\alpha \leq 1$.

(iii) When $\theta > 0$, the distribution is a rescaled uniform distribution with power and frailty parameters as discussed in Paragraph 14.D.2. This distribution has an increasing hazard rate when $\alpha \geq 1$, but a bathtub shape when $\alpha < 1$.

The conclusion is that the generalized Weibull distribution has hazard rates that are increasing when $\alpha \geq 1, \theta \geq 0$, decreasing when $\alpha < 1, \theta > 0$, bathtub shaped when $\alpha < 1, \theta > 0$, and inverted bathtub shape when $\alpha > 1, \theta < 0$. However, all shapes of the hazard rate are not attainable without mixing the cases of bounded and unbounded support. See Figure F.1.

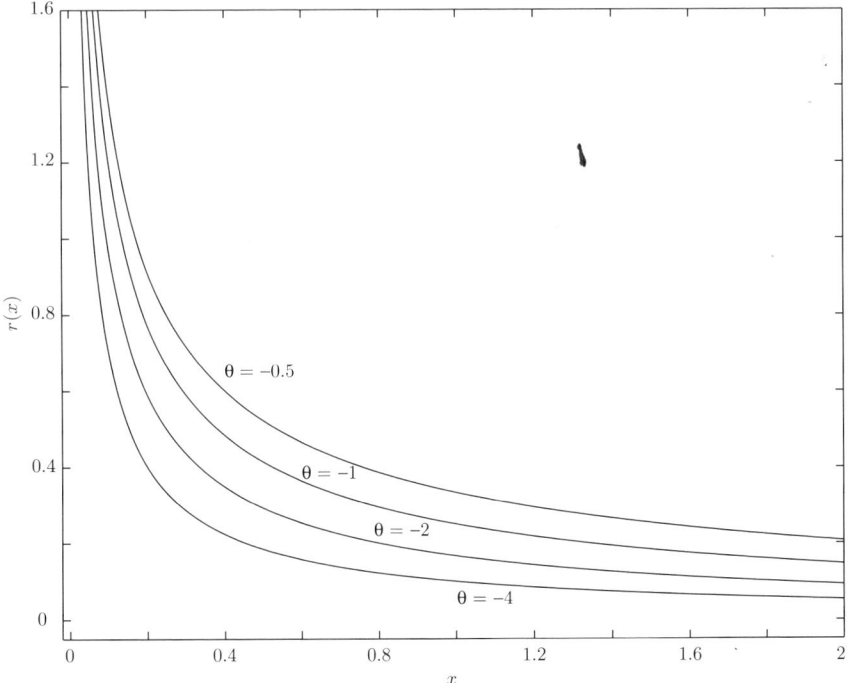

Fig. F.1a. Hazard rates of the Weibull-Pareto-Beta distribution ($\alpha = 0.5, \lambda = 1$)

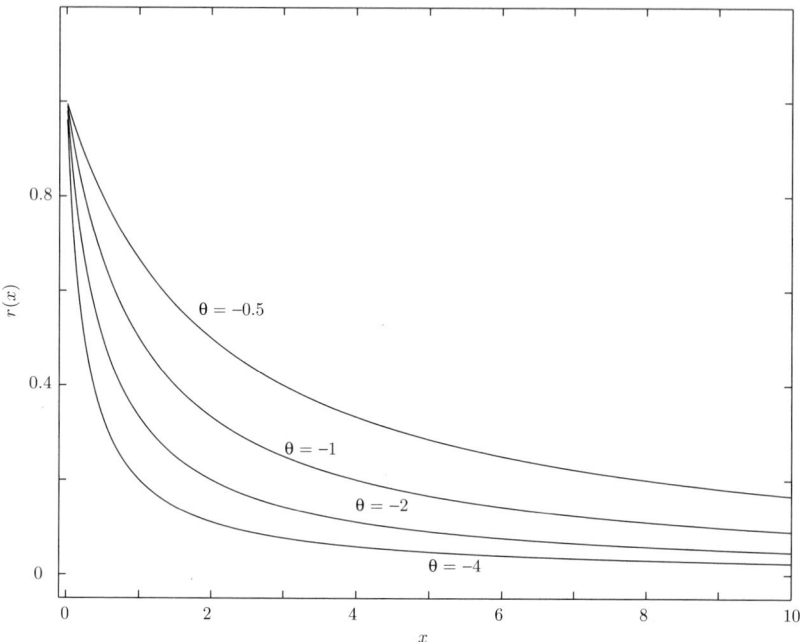

Fig. F.1b. Hazard rates of the Weibull-Pareto-Beta distribution ($\alpha = 1, \lambda = 1$)

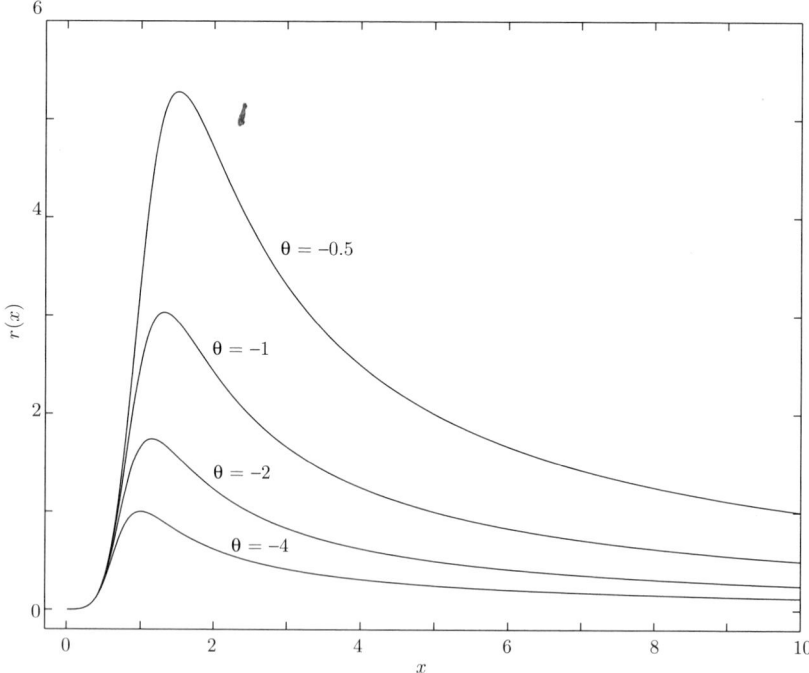

Fig. F.1c. Hazard rates of the Weibull-Pareto-Beta distribution ($\alpha = 5, \lambda = 1$)

b. A Further Extension

The generalized Weibull distribution was derived starting with a uniform distribution; it can be further extended by starting with the beta density 14.C(1). Addition of a scale and power parameters leads to the density

$$f(x \mid a, b, \lambda, \alpha) = \frac{\alpha \lambda (\lambda x)^{\alpha a - 1}[1 - (\lambda x)^\alpha]^{b-1}}{B(a,b)}, \quad 0 \le x \le 1/\lambda. \quad (3)$$

Now, follow the lead of Mudholkar and Kollia (1994) and replace b by $1/\theta$ and λ by $\lambda \theta^{1/\alpha}$ to obtain the density

$$f(x \mid a, \theta, \lambda, \alpha) = \frac{\alpha \lambda \theta^a (\lambda x)^{\alpha a - 1}[1 - \theta(\lambda x)^\alpha]^{(1/\theta)-1}}{B(a, 1/\theta)}, \quad 0 \le x \le 1/(\theta^{1/\alpha}\lambda). \quad (4)$$

Here, the parameters are positive, but the change of scale allows θ to be negative if the normalizing constants are appropriately changed. When this is done and $\delta = -\theta$, the resulting density is

$$f(x \mid a, \delta, \lambda, \alpha) = \frac{\alpha \lambda \delta^a (\lambda x)^{\alpha a - 1}}{B(a, \delta^{-1} - a + 1)[1 + \delta(\lambda x)^\alpha]^{(1/\delta)+1}}, \quad 0 \le x < \infty. \quad (5)$$

Here, it is necessary that $\delta^{-1} - a + 1 > 0$, a condition that reduces to $\delta > 0$ when $a = 1$. In both cases (4) and (5), as θ or δ approaches 0 the limiting distribution is a Weibull distribution with moment parameter. An alternate derivation of these densities is to introduce a moment parameter in the generalized Weibull distribution of Mudholkar and Kollia (1994).

G. Composite Distributions

A number of survival functions have been constructed that are products of two or more other survival functions. These products might be called *composite survival functions*. When U and V are independent, the survival function of $X = \min(U, V)$ is a composite, the product of the survival functions of U and V.

The hazard rate of X is the sum of the hazard rates of U and V. Thus, if U has an increasing hazard rate and V has a decreasing hazard rate, it is not surprising to find that in many special cases, X has a monotone or bathtub-shaped hazard rate, depending upon the parameters. The construction of families of distributions with such a variety of hazard rate behaviors has sometimes been the motivation for introducing composite distributions. Distributions that have been used by various authors to construct composite distributions often include the exponential, Weibull, and/or Pareto distributions. Composite distributions also arise when there are independent competing risks (see Chapter 17).

a. Weibull, Pareto II, and Exponential Components

Suppose that

$$r(x) = r_1(x) + r_2(x) + r_3(x), \qquad (1)$$

where for $x > 0$,

$$r_1(x \mid \lambda_1, \xi) = \frac{\lambda_1 \xi}{1 + \lambda_1 x} \qquad (2)$$

is the hazard rate (11.B(5) with $\alpha = 1$) of a Pareto II distribution,

$$r_2(x \mid \lambda_2, \alpha) = \lambda_2 \alpha (\lambda_2 x)^{\alpha - 1} \qquad (3)$$

is the hazard rate 9.B(3) of a Weibull distribution, and

$$r_3(x \mid \lambda_3) = \lambda_3 \qquad (4)$$

is the constant hazard rate of an exponential distribution. Because of the additive nature of (1), the corresponding survival function is the product

$$\bar{F}(x) = \bar{F}_1(x) \bar{F}_2(x) \bar{F}_3(x), \qquad (5)$$

where $\bar{F}_i$ has the hazard rate $r_i, i = 1, 2, 3$.

The distribution (5) has five parameters, which may be excessive for many purposes, and various special cases have been discussed in the literature. Special cases of (1) include $r_1(x) = 0, r_2(x) = 0$, and $r_3(x) = 0$, so long as the three hazard rates are not all 0.

G. Composite Distributions

G.1. Example (Ljubo, 1965). Suppose that the Weibull component of (1) is omitted, and

$$r(x) = r_1(x \mid \lambda, \theta\xi) + r_3(x \mid \lambda\xi). \tag{6}$$

The corresponding survival function is given by

$$\bar{F}(x) = [e^{-\lambda x}/(1 + \lambda x)^\theta]^\xi, \quad x \geq 0, \ \lambda, \xi, \theta > 0. \tag{7}$$

An interesting feature of this survival function is that the presence of the exponential factor allows an expansion of the range of θ beyond what is allowed in the Pareto distribution; (7) is a survival function whenever $\theta \geq -1$; then, the hazard rate is decreasing for $\theta \geq 0$ and increasing for $-1 \leq \theta \leq 0$. The parameterization here is such that λ is a scale parameter and ξ is a frailty parameter. The survival function (7) has been used by Davis and Feldstein (1979) to model some censored survival data.

G.2. Example. Murthy, Swartz, and Yuen (1973) study the hazard rate

$$r(x) = \frac{\lambda_1}{1 + \lambda_1 x} + \alpha\lambda_2(\lambda_2 x)^{\alpha-1}, \quad x \geq 0, \ \lambda_1, \lambda_2, \alpha > 0$$

which is (1) with $\lambda_3 = 0$. The resulting survival function is the product

$$\bar{F}(x) = \frac{1}{(1 + \lambda_1 x)^\xi} \exp\{-(\lambda_2 x)^\alpha)\}, \quad x \geq 0, \ \lambda_1, \lambda_2, \alpha, \xi > 0 \tag{8}$$

of Pareto and Weibull survival functions. Here, the hazard rate is convex and decreasing for $\alpha \leq 1$, is an inverted bathtub for $1 < \alpha < 2$, is increasing and convex for $\alpha = 2$, and is bathtub shaped for $\alpha > 2$. These results can be obtained by examining the hazard rate together with its derivative.

526   15. Additional Parametric Families

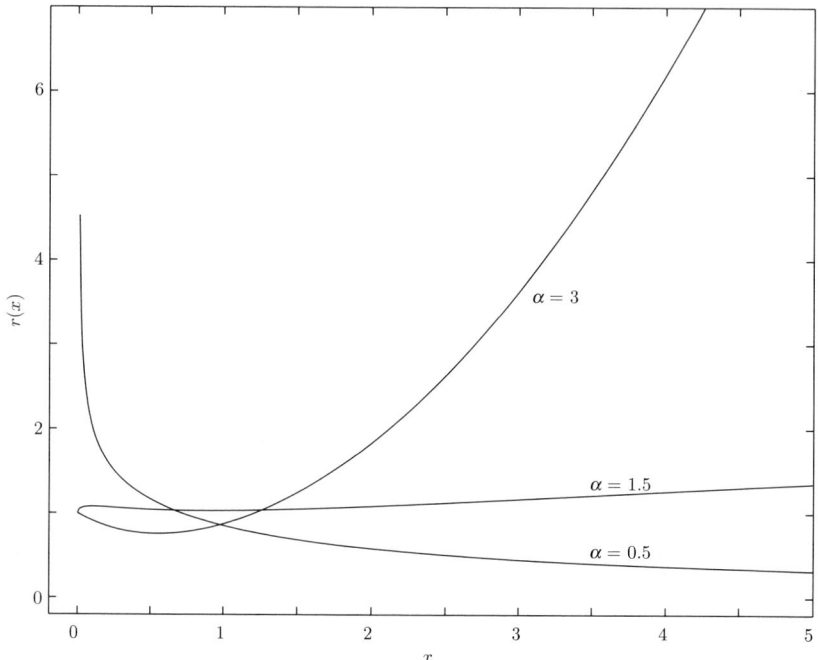

Fig. G.1a. Some hazard rates for example G.2 ($\lambda_1 = 1, \lambda_2 = 0.5$)

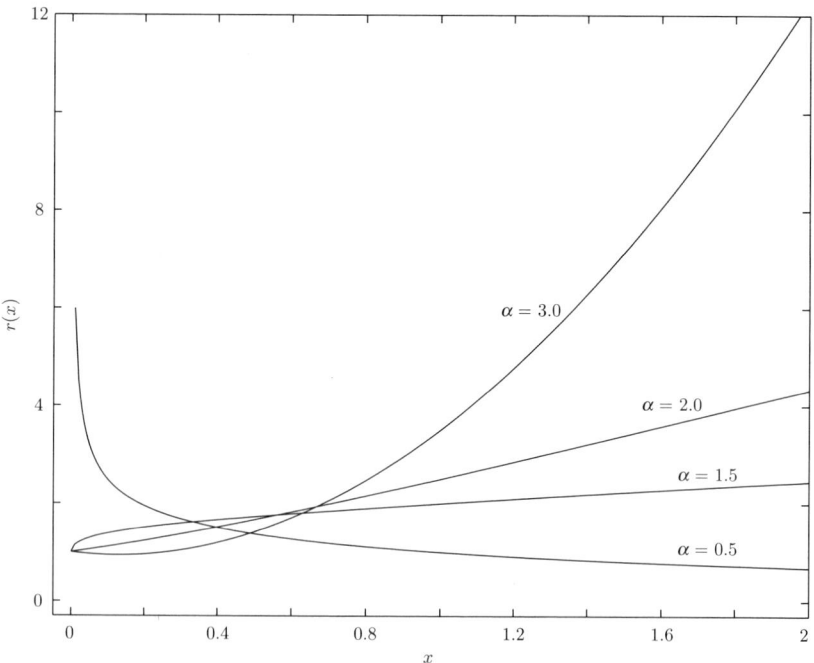

Fig. G.1b. Some hazard rates for example G.2 ($\lambda_1 = 1, \lambda_2 = 1$)

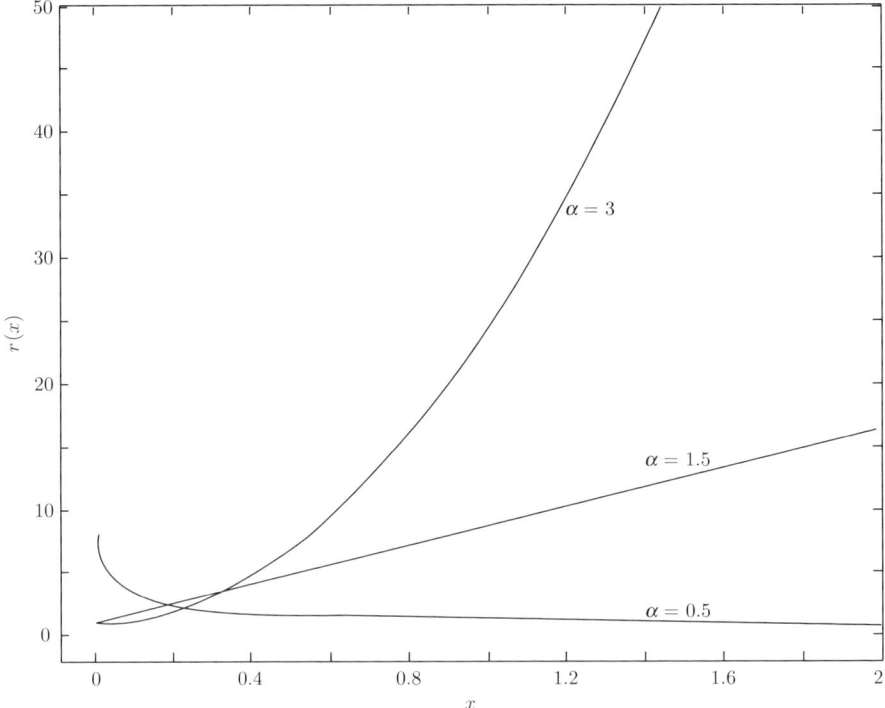

Fig. G.1c. Some hazard rates for example G.2 ($\lambda_1 = 1, \lambda_2 = 2$)

G.3. Example. Gaver and Acar (1979) considered the hazard rate (1) with r_1 given by (2), $r_2 = r_2(x \mid \sqrt{\delta/2}, 2) = \delta x$ given by (3), and r_3 given by (4); Hjorth (1980) studied the same hazard rate with $r_2 = 0$. In either case, the hazard rates have the same qualitative behavior. With the notation $\lambda_1 = \theta$, Hjorth's distribution has the survival function

$$\bar{F}(x) = e^{-\delta x^2/2}/(1+\theta x)^\xi, \quad x, \delta, \theta, \xi \geq 0, \text{ but either } \delta > 0 \text{ or } \theta \xi > 0, \tag{9}$$

and hazard rate

$$r(x) = \delta x + \frac{\xi \theta}{1 + \theta x}. \tag{10}$$

The parameter ξ is a frailty parameter of the Pareto II part, but ceases to be a frailty parameter in the composite. Because each of the components of the hazard rate (10) is convex, the hazard rate is convex. It is increasing if $\delta \geq \xi \theta^2$; otherwise, it is bathtub shaped with a minimum at the point $x = [\theta\sqrt{\xi/\delta} - 1]/\theta$. However, in the bathtub case, there is no interval over which the hazard rate is constant.

Further details concerning this distribution are given by Hjorth (1980).

The following example is another composite which is constructed in a manner similar to that of Example G.1, and has many of the same features.

G.4. Example (Two Weibull distributions). Let

$$\bar{F}(x) = \exp\{-(\lambda x)^\alpha - (\lambda x)^\beta\}, \quad x \geq 0, \ \alpha, \beta, \lambda > 0. \tag{11}$$

This survival function is the product of two Weibull survival functions, and has the hazard rate

$$r(x) = \alpha\lambda(\lambda x)^{\alpha-1} + \beta\lambda(\lambda x)^{\beta-1}. \tag{12}$$

When

$\alpha, \beta \leq 1$, then r is decreasing and convex;

$\alpha < 1 < \beta < 2$, then r is bathtub shaped but not convex;

$\alpha \leq 1$ and $2 \leq \beta$, then r is bathtub shaped and convex;

$1 \leq \alpha, \beta \leq 2$, then r is increasing and concave;

$1 < \alpha < 2 < \beta$, then r is increasing but not convex;

$\alpha, \beta \geq 1$, then r is increasing and convex.

Motivated by a competing risk model, Canfield and Borgman (1975) consider the more general product of K different Weibull survival functions; this more general case is also discussed by Murthy, Xie and Jiang (2004).

The case that $\beta = 1$ in (11) is particularly interesting; this arises when F is the distribution of $\min[X, Y]$, where X has a Weibull distribution and Y has an exponential distribution. The random variable X can be thought of as the waiting time to death due to wearout or other internal causes, and Y can be thought of as the waiting time for death due to an accident. The resulting distribution function modifies the Weibull distribution in the same way that the Gompertz–Makeham distribution modified the Gompertz distribution. This important special case has been studied in detail by Bertholon, Bousquet and Celeux (2004).

G.5. Example (Quadratic hazard rate). Hall and Wang (2006) have investigated the hazard rate (1) with $r_1(x) = \lambda_1$, $r_2(x) = 2\lambda_2 x$, and $r_3(x) = 3\lambda_3 x_2^2$. This yields the survival function

$$\bar{F}(x) = \exp\{-\lambda_1 x - \lambda_2 x^2 - \lambda_3 x^3\}, \quad x \geq 0, \tag{13}$$

which is the product of an exponential survival function and two Weibull survival functions. This survival function is similar to the second Gompertz–Makeham distribution of Section 10.C.e, but the Gompertz component is replaced by the Weibull survival function having shape parameter 3.

For survival data studied by Hall and Wang (2006), (13) provided a better fit than did the Gompertz–Makeham survival function 10.B(3), and the process of finding the best fit was simpler. Consequently, (13) can be regarded as a serious competitor to the Gompertz–Makeham distribution.

b. Gompertz–Makeham Distribution

The Gompertz–Makeham distribution discussed in Chapter 10 is a composite distribution with a Gompertz and an exponential component. Its hazard rate 10.B(1) is the sum of the hazard rate of the Gompertz distribution and a constant.

H. Stable Distributions

Although stable distributions (see Definition 20.D.11) can be regarded as a natural extension of the normal distribution, some of these distributions have support $[0, \infty)$, which makes them quite unlike the normal distribution. The best known example is the case that the stable distribution with index (characteristic exponent) $\alpha = 1/2$. This strictly stable distribution takes the form

$$F(x \mid \lambda) = 2[1 - \Phi((\lambda x)^{-1/2})], \quad x \geq 0, \tag{1}$$

and has the density

$$f(x \mid \lambda) = \frac{1}{\sqrt{2\pi}} \frac{\lambda}{\sqrt{(\lambda x)^3}} \exp\{-1/(2\lambda x)\}, \quad x \geq 0. \tag{2}$$

15. Additional Parametric Families

From the distribution function or density, it is clear that λ is a scale parameter. Note also that directly from the definition it follows that stable distributions remain stable if a scale parameter is introduced.

The density (2) is unimodal, as can be verified by setting the derivative of $\log f$ equal to 0. The resulting equation has but one solution in $(0, \infty)$. This solution, $1/(3\lambda)$, is the location of the mode.

From the derivative of $\log f$, the function ρ of Theorem 4.E.2 is immediately obtained. From this theorem, it follows that the hazard rate of the distribution (1) has an inverted bathtub shape, with mode in the interval $[0, 2/3\lambda]$.

It is known that stable distributions with support $[0, \infty)$ have an exponent α in the interval $(0, 1]$. Moreover, their Laplace transforms, apart from scale parameters, are given by

$$L(s \mid \alpha) = \exp(-s^\alpha). \qquad (3)$$

These distributions have densities

$$f(x \mid \alpha) = -\frac{1}{\pi x} \sum_{k=1}^{\infty} \frac{\Gamma(\alpha k + 1)}{k!} (-x^{-\alpha})^k \sin(\alpha k \pi), \quad x > 0; \qquad (4)$$

see Feller (1971, p. 538).

It has been shown by Gawronski (1984) that these densities are all unimodal.

Hougaard (1986) adds scale and Laplace transform parameters to the density (4). He shows that the resulting family is infinitely divisible and closed under convolution, and finds other properties.

Part V

Models Involving Several Variables

16
Covariate Models

The cause is hidden, but the result is known.
Ovid (Publius Ovidius Naso, 43 B.C.–17A.D.)

A. Introduction

In many populations, factors can be recognized the presence of which increases the population heterogeneity. Especially when the factors are categorical (say, male/female or species of trees), it is common to break the population into subpopulations. But for factors such as blood pressure or age, the consideration of subpopulations may not be practical, and an alternative approach is to make use of covariate models. The use of these models can lead to more homogeneity, and to much improved understanding of characteristics of interest in the population. More precisely, variables z may exist that directly or indirectly affect the distribution of a random variable X. When the value of z can be determined, this information can be used to better predict that value of X. Stress variables such as temperature may be introduced as a part of an experiment. Alternatively, z may measure some intrinsic property of a device or individual such as the factory of origin of the device or the blood chemistry of an individual. The variable z is variously called a *covariate*, a *concomitant variable*, a *regressor variable*, or an *explanatory variable*. Some covariates can be controlled or chosen by an experimenter, and other covariates are random; they may or may not be time dependent.

These models have been called *covariate models* or *regression models*. When the covariates are categorical, the term "subgroup analysis" is sometimes used, especially in a medical context. Often, a covariate model will contain a combination of continuous and categorical variables. The long history of covariate models is summarized by Kalbfleisch and Prentice (1980, p. 68).

There do not appear to be underlying principles that would lead to procedures for constructing models that take into account the observed or controlled values of covariates. Particularly when z is not time dependent, the distribution of the random variable X can usefully be regarded as a mixture, with components of the mixture indexed by z. In statistical problems, the commonly used method is to assume that the conditional survival function of X, given z has the form

$$P\{X > x \,|\, z\} = \bar{F}(x \,|\, \theta(z)), \tag{1}$$

where $\bar{F}(\cdot \,|\, \theta), \theta \in \Theta$, is a parametric family that may or may not be entirely specified. The basic procedure is rather simple, with the parameter θ treated in regression format as a function $\theta(z)$ of the covariates $z = (z_1, z_2, \ldots z_k)$.

Two most commonly used covariate functions are the linear

$$\theta(z) = \beta_1 z_1 + \beta_2 z_2 + \cdots + \beta_k z_k \tag{2}$$

and the log linear

$$\theta(z) = \exp\{\beta_1 z_1 + \beta_2 z_2 + \cdots + \beta_k z_k\} \tag{3}$$

models. For nonnegative parameters θ, the form (3) is often used. However, the β's can be positive or negative.

For parameters lying between 0 and 1, an appropriate form is

$$\theta(z) = \frac{g(z)}{1 + g(z)}, \quad g > 0. \tag{4}$$

Here, g may take the form of θ as given by (3), which insures nonnegativity.

In these models, there is no constraint on how the covariates $z_1, z_2, \ldots, z_k$ are or are not functionally related. For example, it could be that $k = 3$ and $z_3 = z_1 z_2$ or $z_3 = \log z_1 z_2$. Of course, the model that results from a function $\theta(z)$ depends on how the family $\{\bar{F}(\cdot \,|\, \theta), \theta \in \Theta\}$ is parameterized. The freedom in parameterization and in defining the covariates makes the models quite flexible. But there is an art to choosing a model for any specific application.

A. Introduction

Covariates are sometimes classified as "internal" or "external." A somewhat loose definition of these notions is given by Kalbfleisch and Prentice (2002, Section 6.3) together with a number of examples. The more precise definition suggested by Singpurwalla and Wilson (1995) is as follows. Suppose that a lifetime T has an absolutely continuous distribution. If the conditional distribution of T given the covariate is absolutely continuous, then the covariate is said to be external. Otherwise, the covariate is said to be internal.

Covariate models are often used when the relationship between variables is of particular interest. In this context, the danger of ignoring a covariate related to both another covariate and the primary variable of interest has been well documented, and can lead to what is called "Simpson's paradox."

Early recognition of this problem was made in the context of correlations. Suppose that a covariate z can take only the values 0 and 1. Then, it is possible for the correlation $Corr(X, Y)$ of two random variables to be quite high whereas $Corr(X, Y \mid z = 0)$ and $Corr(X, Y \mid z = 1)$ are low. Or, $Corr(X, Y)$ can be quite low whereas $Corr(X, Y \mid z = 0)$ and $Corr(X, Y \mid z = 1)$ are high. These facts are disturbing in a statistical analysis in which there are high correlations for men and women, but not in the combined population; similarly there may be low correlations for men and women, but high correlations in the combined population.

Example 4.C.7.a is a case in which a random variable has a decreasing hazard rate, but comes from two subpopulations each with an increasing hazard rate; ignoring a covariate that indicates the subpopulation can result in a misleading conclusion. Appleton, French and Vanderpump (1996) offer another particularly striking example. They discuss a study and follow-up that found nonsmokers dying at a higher rate than smokers. The ignored covariate was age; in the group studied, nonsmokers were significantly older than the smokers.

Regression models provide a wide scope for statistical analysis of the parameters. The aim of this analysis is to estimate regression coefficients in $\theta(z)$ as well as any other parameters from a random sample $(x_i, \theta(z_{1i}, z_{2i}, \ldots, z_{ki})), i = 1, \ldots n$. Various estimation methods such as least squares, maximum likelihood, or Bayesian methods can be used, but estimation is often complicated by data being censored. With parameter estimates, it is possible to assess the importance of the various covariates and eliminate the unimportant ones, and to obtain a better understanding of the hazard rate and the factors affecting it.

In addition to the choice of an estimation method, the choice of a model is a critical step, and the goodness-of-fit issue is important.

These issues form the bulk of research on covariate models. Further discussion of statistical issues are beyond the scope of this book, but see, for example, Andersen, Borgan, Gill, and Keiding (1993), Ansell and Phillips (1994), Cox and Oakes (1984), Crowder, Kimber, Smith and Sweeting (1991), Hosmer and Lemeshow (1999), Kalbfleisch and Prentice (2002), Lawless (1982), or Nelson (1990).

A number of quite complex models can be found in the literature, with complexity compounded in the case that time dependencies are incorporated. A number of these models are reviewed and compared by O'Quigley and Stare (2002). Other issues arise in applications. For example, the variable of interest may be the waiting time for a disease to reoccur after a treatment; but if the treatment has effected a cure, that waiting time is meaningless. For a discussion of this, see Li and Taylor (2002) or Frankel and Longmate (2002). In this chapter, only the more basic models are discussed.

B. Some Regression Models

a. Scale Parameter Regression Models: Accelerated Life Models

Normally, life testing must be completed under time constraints; practical considerations may require that the testing be carried out with stress levels greater than that would normally be encountered. In this way, relatively early failures can be observed. In these circumstances, it is a common practice to assume that a scale parameter is present that is a function of the stress levels, the idea being that time has in effect been accelerated in a specified way. Clearly, accelerated testing is subject to pitfalls, a number of which have been itemized by Meeker and Escobar (1998).

Regression models in which a scale parameter is assumed to be a function of covariates are often called *accelerated life models* regardless of the context they arise in. For a survival function $\bar{F}$ with density f and hazard rate r, the introduction of a scale parameter depending upon covariates z leads to the survival function $\bar{F}(x \mid \lambda(z)) = \bar{F}(\lambda(z)x)$, the density $f(x \mid \lambda(z)) = \lambda(z)f(\lambda(z)x)$, and the hazard rate $r(x \mid \lambda(z)) = \lambda(z)f(\lambda(z)x)/\bar{F}(\lambda(z)x)$. Most often in these models, $\theta = 1/\lambda$ is taken to be the log linear function A(3) of the covariates.

B.1. Example. Zelen (1959) studied the lifetimes of glass capacitors under different levels of temperatures and voltages. He assumed that the lifetime X of a capacitor has an exponential distribution of the form

$$\bar{F}(x\,|\,\theta) = e^{-x/\theta}, \qquad (1)$$

where $\theta = \theta(z_{i,j}) = \exp\{\beta_0 + \sum_{i,j} \beta_{ij} z_{ij}\}$. Here, the z_{ij} are indicator variables with subscripts indicating the level of temperature and voltage. The choice of parameterization is natural because θ is the expectation of X.

B.2. Example. Feigl and Zelen (1965) studied the survival times of patients with acute myelogenous leukemia using white blood count as a covariate, the exponential survival function (1), and the covariate function $\theta(z) = \beta_0 + \beta_1 z$. Other possible covariate functions mentioned are $\theta(z) = [\beta_0 + \beta_1 z]^{-1}$ and $\theta(z) = a\, e^{-bz}$.

Note that in Example B.1, the covariates were controlled by the experimenter, but in Example B.2, the covariates are observed random variables. In either case, the goal is to use the covariates to improve the prediction of X, or to understand how the covariates and X are related.

B.3. Example. Lawless (1982, p. 304) provides an analysis of survival data on advanced lung cancer patients with an assumed Weibull distribution and the reciprocal of the scale parameter being a log-linear function of seven covariates; these covariates describe the tumor type. Other examples of applications with an assumed Weibull distribution are reviewed by Smith (1991).

B.4. Example. Kalbfleisch and Prentice (2002, p. 85) consider the lognormal distribution (Section 12.B), the Pareto III (log logistic) distribution (Section 10.C), and the generalized gamma distribution (Section 9.E), and investigate the case that the reciprocal of the scale parameter is a log linear function of covariates.

b. Proportional Hazard Regression Models

The *proportional hazards model*, introduced by Cox (1972), is a semiparametric model with a frailty parameter that is taken to be a function of the covariates. To further describe this model, let $\bar{F}$ be a survival function with a corresponding hazard function $R = -\log \bar{F}$ and hazard rate $r = R'$. If a frailty parameter is now introduced, the resulting parametric family has survival function given by $\bar{F}(x\,|\,\xi) = [\bar{F}(x)]^\xi$, hazard function $R(x\,|\,\xi) = \xi R(x)$, and hazard rate $r(x\,|\,\xi) = \xi r(x)$. With the frailty parameter taken to be a function $\xi = \xi(z_1, z_2, \ldots z_k)$ of covariates, the resulting model is

$$r(x \mid \xi(z)) = r(x)\xi(z_1, z_2, \ldots z_k). \tag{2}$$

When $\xi = \xi(z_1, z_2, \ldots z_k)$ has the form A(3), the resulting model is often called the "Cox Model." Other functions of the covariates are sometimes used, but they must be positive.

The Cox model has been found by statisticians to be both flexible and tractable; consequently, it has reached a high level of importance and usage in practical applications, especially in medical research (see, e.g., Andersen, 1991). As of this writing, there are "17" papers in print discussing the Cox model, and the number is growing without an upper bound (we have borrowed this expression from William Feller, who often used "17" in his lectures to represent a large generic integer).

The model (1) is designed to take an account of observed covariates. Related models have been used to account for random heterogeneity, often due to the presence of unobserved covariates. Such a model might take the form

$$r(x \mid Z) = Zr(x), \tag{3}$$

where Z is a random variable. The models (1) and (2) are sometimes combined (see Keiding (2001); Vaupel, Manton and Stallard (1979)).

c. Proportional Odds Regression Models

In a social science context, McCullagh (1980) proposes a proportional odds model and obtains estimates using maximum likelihood. This model was further investigated in a medical context by Bennett (1983). The model is also discussed by Crowder (2001), along with accelerated life and proportional hazards models. This model takes the form

$$\frac{\bar{F}(x \mid \gamma)}{F(x \mid \gamma)} = \gamma \frac{\bar{F}(x)}{F(x)},$$

where $\gamma = \gamma(z)$ is a function of covariates z. As discussed in Section 7.F, this model leads to a survival function with a tilt parameter, that is,

$$\bar{F}(x\,|\,\gamma) = \frac{\gamma \bar{F}(x)}{1 - \bar{\gamma} \bar{F}(x)}. \tag{4}$$

Bennett (1983) proposes using the log-linear model $\gamma(z) = \exp\{\beta_1 z_1 + \beta_2 z_2 + \cdots + \beta_k z_k\}$ of A(3). As an explicit example, he considers the case that in a study of lung cancer, there are two covariates performance status as measured in a scale of 0 to 100 and the type of tumor as large, adeno, small, or squamous.

For nonnegative random variables, the hazard rate $r(\cdot\,|\,\gamma)$ of the survival function (4) is given by

$$r(x\,|\,\gamma) = \frac{1}{[1 - \bar{\gamma}\bar{F}(x)]} r(x), \quad 0 < x < \infty, \ \gamma > 0.$$

In their discussion of proportional odds regression models, Sankaran and Jayakumar (2006) point out that $\lim_{x \to \infty} r(x\,|\,\lambda) = \lim_{x \to \infty} r(x), \gamma > 0$ (see also Section 7.F.a). This is to be contrasted with the case of the proportional hazards model of Section B.b, where the relationship $r(x\,|\,\xi) = \xi r(x)$ is obtained.

In some applications, the effects of covariates diminishes with time; for such covariates, a proportional odds regression model may be more appropriate than a proportional hazards model. However, the proportional odds model has not yet been extensively used in applications.

d. Coincidence of the Models

B.5. Accelerated life and proportional hazards models. It is well known that the accelerated life model and the proportional hazards model coincide when the underlying distribution F is a Weibull distribution and the scale (frailty) parameter is a function of the covariates. In fact, this is the only distribution for which the models coincide (Kalbfleisch and Prentice, 2002); see Proposition 18.B.1). In these models, the Weibull distribution has been used extensively.

B.6. Accelerated life and proportional odds ratios models. It is shown in Proposition 18.B.3 that when the underlying random variables are nonnegative, these two models coincide if and only if the underlying distribution is a Pareto III distribution.

B.7. Proportional hazards and proportional odds ratios models. It is shown in Proposition 18.C.3 that these models coincide if and only if the underlying distribution places mass on at most two points. This means that for practical purposes, the models are distinct.

C. Regression Models for Other Parameters

Note that each of the three models discussed in Section B lead to the assumption that a particular kind of parameter is a function of covariates; for accelerated life models it is a scale parameter, for proportional hazards it is a frailty parameter, and for proportional odds it is a tilt parameter that is introduced and made a function of the covariates. Other examples exist in the literature; indeed, any of the parameters discussed in Chapter 7 can be used in conjunction with covariates.

C.1. Example. In the context of kidney transplantation, Bailey and Homer (1977) and Bailey, Homer, and Summe (1977) assume that the underlying hazard rate has the form

$$r(x) = \alpha\, e^{-\lambda x} + \delta, \quad \alpha, \lambda, \delta > 0. \tag{1}$$

With $\alpha = \xi\lambda$ and $\delta = \xi\theta\lambda$, the hazard rate (1) becomes the hazard rate 10.B(7), a relative of the Gompertz distribution that has a decreasing hazard rate. Here λ is a scale parameter and ξ is a frailty parameter. Bailey and Homer (1977) and Bailey, Homer, and Summe (1977) make use of two covariates, sex and age, and use the log-linear covariate model A(3) for their parameters.

17

Several Types of Failure: Competing Risks

> The causes of events are ever more interesting than the events themselves.
> Cisero, Epitolae ad atticun, Book IX, Section 5

When vaccination against smallpox became practical in the middle of the eighteenth century, Bernoulli (1760) became interested in the question of how life tables would be affected if deaths from smallpox were eliminated through mandatory vaccination. At the time, various life tables were available; these tables were based upon data from which it was possible to determine not just the age at death, but also whether death was due to smallpox or some other cause. Thus, Bernoulli considered individuals as facing two "competing" risks of death: death from smallpox or some other cause. His goal was to determine how the distribution of life length X of an individual would be changed if the risk of death from smallpox were eliminated by vaccination. The issue of risk removal has not received much attention in recent literature on competing risks, but see Karn (1931, 1933) or Seal (1977) for early history.

Related problems involving competing risks arise in a variety of circumstances.

Biology. Bernoulli's interest in smallpox is but one example of a large class of problems. More generally, an organism can die from a number of causes, each of which is a competing risk.

Life Testing. In life testing situations, data is often "censored"; that is, individuals are removed from a study before the event of interest can occur. The censoring event is a competing risk, and the object of interest is the life distribution with this risk eliminated.

Multiple decrement lifetimes. In certain circumstances, groups of several people are considered to survive as long as all members of the group survive. This is the case when a life insurance policy is issued that is payable at the time of first death within the group. Typically, the group may consist of husband and wife. From the insurance company's point of view, each member of the group is a competing risk.

Reliability. Mechanical, electrical, and hydraulic systems usually have a number of components. There are one or more minimal subsets of components (called *minimal cut sets*) with the property that failure of all components in the minimal subset will cause the equipment to fail. Each minimal cut set can be regarded as a mode of failure or a competing risk (see Section 5A.b for a discussion of cut sets).

Economics. Flinn and Heckman (1983) apply competing risk theory in a study of unemployed workers, where the length of the unemployment period is studied, and "risks" indicate the reason for leaving unemployment—getting a job or dropping out of the work force.

There is a substantial literature devoted to the theory of competing risks, and this chapter offers only a brief introduction. Books on the subject include those of Crowder (2001) and David and Moeschberger (1978). An excellent survey of the subject is provided by Birnbaum (1979) (see also Chiang, 1968, Chapter 11).

A. Definitions and Notation

With k competing risks, it is natural to introduce the random variables

$$X_i = \text{the hypothetical time to failure from the } i\text{th cause of failure} \\ \text{in the absence of other risks, } i = 1, 2, \ldots, k. \quad (1)$$

In this way, multivariate theory unavoidably enters even though this book is concerned with univariate models. A review of the few basics of multivariate theory that are required in the remainder of the chapter can be found in Section 20.I.

As noted below, the terminology used in this field has not been well standardized, and consequently, concepts can have several names.

a. Latent Lives and Survival Functions

The hypothetical random variable X_i defined in (1) is called the *latent life for risk i*. These hypothetical random variables are also called "net lives" or "latent variables." Without further comment, it is assumed throughout this chapter that the latent variables are positive.

Let
$$\bar{F}(x_1, \ldots, x_k) = P\{X_1 > x_1, \ldots, X_k > x_k\}$$
denote the joint survival function of $X_1, X_2, \ldots, X_k$. The corresponding distribution function is sometimes called the "multiple decrement function." The distribution function F (or the survival function $\bar{F}$) is fundamental because it determines the distribution of all the random variables of interest that arise in the theory of competing risks.

For $i = 1, 2, \ldots, k$, the marginal survival function $\bar{F}_i$ of X_i is sometimes called the "net survival function" for risk i; in the actuarial literature, the corresponding distribution function is sometimes called the ith decrement function. Of course,
$$\bar{F}_i(x_i) = \bar{F}(0, \ldots, 0, x_i, 0, \ldots, 0),$$
where x_i is in the ith place.

Let
$$U = \text{the life length of interest},$$
so that
$$U = \min(X_1, \ldots, X_k);$$
denote the survival function of U by $\bar{Q}$; that is,
$$\bar{Q}(t) = P\{\min(X_1, \ldots, X_k) > t\} = \bar{F}(t, \ldots, t).$$

Much of what follows can be regarded as directed to finding useful expressions for $\bar{Q}$.

With the assumption that
$$P\{X_i = X_j\} = 0, \quad \text{for all } i \neq j, \tag{2}$$

it is possible to define uniquely the random variable

$$J = \text{the index of the smallest of the } X_i.$$

Thus, $J = j$ means that the jth cause is the culprit, and

$$X_j < X_i, \quad \text{for all } i \neq j, \ j = 1, 2, \ldots, k.$$

Denote the probability that $J = j$ by $\pi_j, j = 1, 2, \ldots, k$. In applications, it is generally true that only U and J can be observed, by which it is meant that only the life length of the item and the cause of death or failure can be determined.

Applications are occasionally encountered in which assumption (2) does not hold because there are multiple causes. Methods for handling these applications are not discussed here, but see Tai, White, Gebski and Machin (2002).

b. Conditional and Sub-survival Functions

Three probabilistic quantities are central to the analysis of competing risks. For $j = 1, \ldots, k$, let

$$\bar{Q}(t \mid j) = P\{U > t \mid J = j\} = \frac{P\{t < X_j < \min_{i \neq j} X_i\}}{P\{X_j < \min_{i \neq j} X_i\}}, \quad t > 0, \quad (3)$$

$$\bar{Q}(t, j) = P\{U > t, J = j\} = P\{t < X_j < \min_{i \neq j} X_i\}, \quad t > 0, \quad (4)$$

$$\pi_j = \bar{Q}(0, j) = P\{J = j\} = P\{X_j < \min_{i \neq j} X_i\}. \quad (5)$$

The function $\bar{Q}(\cdot \mid j)$ is called the *conditional survival function of U, given $J = j$*. This function is also sometimes called "crude survival function" or the "cause specific survival function" for the risk j.

The function $\bar{Q}(\cdot, j)$ satisfies

$$\bar{Q}(t, j) = \pi_j \bar{Q}(t \mid j), \quad j = 1, \ldots, \ k, t \geq 0,$$

and is called the *sub-survival function* for the risk j in recognition of the fact that it is not a proper survival function. The corresponding distribution function

$$Q(t, j) = P\{U \leq t, J = j\} = P\{t \geq X_j < \min_{i \neq j} X_i\}, \quad t > 0,$$

satisfies the relationship $Q(\cdot, j) = \pi_j - \bar{Q}(\cdot, j)$. The function $\bar{Q}(\cdot, j)$ is also called the "crude survival function" or "pseudo survival function" for risk j. It follows directly from the definitions that the probability $P\{X_1 > t, X_2 > t, \ldots, X_k > t\}$ is given by

$$\bar{Q}(t) = \bar{F}(t, \ldots, t) = \sum_{j=1}^{k} \bar{Q}(t, j) = 1 - \sum_{j=1}^{k} Q(t, j)$$

$$= \sum_{j=1}^{k} \pi_j \bar{Q}(t \mid j), \quad t \geq 0. \tag{6}$$

c. Densities and Hazard Rates

The assumption (2) that $P\{X_i = X_j\} = 0$, for all $i \neq j$, has already been made, but consider the stronger assumption that the distribution function F has a density f obtained by differentiating F. Then, the distribution Q of U has the density

$$q(t) = -\frac{d}{dt}\bar{Q}(t)$$

and hazard rate

$$r(t) = q(t)/\bar{Q}(t).$$

Useful expressions for these quantities can be obtained by first observing that because F has a density, both $\bar{Q}(t, j)$ and $\bar{Q}(t \mid j)$ are differentiable.

To determine $q(\cdot)$, other quantities are useful that are also of interest in their own right:

$$q(t \mid j) = -\frac{d}{dt}\bar{Q}(t \mid j) \text{ is the } conditional \ density \ of \ U \ given \ J = j;$$

for example, $q(\cdot \mid j)$ is the density for the life length of patients who, in the face of multiple risks of death, die from the jth cause, say, kidney failure.

Although $q(\cdot \mid j)$ has been normalized to be a proper density, the quantity without normalization is also of interest:

$$q(t, j) = -\frac{d}{dt}\bar{Q}(t, j) \text{ is the } j\text{th } sub\text{-}density \ function.$$

From (6), it follows that

$$q(t) = \sum_{j=1}^{k} q(t,j) = \sum_{j=1}^{k} \pi_j q(t \mid j),$$

and

$$r(t) = \sum_{j=1}^{k} \frac{q(t,j)}{\bar{Q}(t)} = \sum_{j=1}^{k} h(t,j),$$

where

$$h(t,j) = \frac{q(t,j)}{\bar{Q}(t)}. \tag{6a}$$

The function $h(t,j)$ is sometimes called the jth *sub-hazard rate*, but it should not be confused with the ratio $q(t,j)/\bar{Q}(t,j)$.

The following proposition shows how to determine the sub-density functions from the joint survival function of the latent lives.

A.1. Proposition (Tsiatis, 1974, 1975). *If F has a density f, then for each $j = 1, \ldots, k$,*

$$q(t,j) = -\frac{\partial}{\partial x_j} \bar{F}(x_1, x_2, \ldots, x_k)\big|_{x_1 = x_2 = \cdots = x_k = t}. \tag{7}$$

Proof. It is sufficient and notationally convenient to prove the proposition for $j = 1$. Because

$$\bar{Q}(t,1) = \int_{t}^{\infty} \left[\int_{x_1}^{\infty} \cdots \int_{x_1}^{\infty} f(x_1, x_2, \ldots, x_k) \, dx_k \cdots dx_2 \right] dx_1, \quad t > 0,$$

it follows by differentiation with respect to t that

$$q(t,1) = -\int_{t}^{\infty} \cdots \int_{t}^{\infty} f(t, x_2, \ldots, x_k) \, dx_k \cdots dx_2, \quad t > 0. \tag{8}$$

Because

$$\bar{F}(x_1, \ldots, x_k) = \int_{x_1}^{\infty} \cdots \int_{x_k}^{\infty} f(z_1, \ldots, z_k) \, dz_k \cdots dz_1,$$

it follows that

$$\frac{\partial}{\partial x_1} \bar{F}(x_1, x_2, \ldots, x_k)|_{x_1=x_2=\cdots=x_k=t}$$
$$= -\int_{x_2}^{\infty} \cdots \int_{x_k}^{\infty} f(x_1, z_2, \ldots, z_k) \, dz_k \ldots dz_2|_{x_1=x_2=\cdots=x_k=t}. \quad (9)$$

Together, (8) and (9) yield (7). □

See Arnold and Brockett (1983) for a generalization of the above-mentioned proposition.

B. The Problem of Identifiability

As noted above, it is generally true that in applications, only $U = \min(X_1, \ldots, X_k)$ and the index J of the smallest of the latent lives X_i can be observed. For example, in a biological setting, only the time of death and cause of death can be observed. Is such data sufficient to estimate the joint survival function $\bar{F}$? Sometimes, the answer to this question is "yes," but in general, $\bar{F}$ cannot be estimated from such data. This is the essence of the problem of identifiability that is discussed in this section.

With observations of U and J, it is possible to estimate only $\bar{Q}(\cdot \mid j)$ and π_j, together with quantities such as $\bar{Q}(\cdot, j)$ that they determine; other quantities of interest may not be estimable from the available observations. The fact that $\bar{Q}(\cdot \mid j)$ and $\pi_j, j = 1, \ldots, k$, together do not determine the survival function $\bar{F}$ is called the *problem of identifiability*. Even when its parametric form is known, F may or may not be identifiable, that is, the parameter values may not be determined by the joint distribution of U and J; examples are given in Sections D and E.

To what extent or under what conditions do $\bar{Q}(\cdot \mid j)$ and π_j determine the joint survival function $\bar{F}$ or the marginal survival functions? This question is important because once $\bar{F}$ is known, all quantities of interest in the context of competing risks can be computed. In particular, it is possible to determine from $\bar{F}$ how the distribution of U changes when a risk is eliminated.

a. Bounds for the Joint Survival Function

The extent to which the $\bar{Q}(\cdot\,|\,j), \pi_j$ determine $\bar{F}$ can in two dimensions be answered to some extent by inequalities. The case of two dimensions is particularly important because in practice it is often the case that just one of the competing risks, say X_1, is of particular interest. Then, the other competing risks $X_2, \ldots, X_k$ can be replaced by $\min(X_2, \ldots, X_k)$, effectively reducing the problem to two dimensions.

The following proposition provides bounds for the joint survival function; these in turn yield bounds for the marginal survival functions.

B.1. Proposition (Peterson, 1976). Suppose that the random variables $X_1, \ldots, X_k$ have survival function $\bar{F}$ and $P\{X_i = X_j\} = 0, i \neq j$. For $x_1, \ldots, x_k \geq 0$,

$$\sum_{i=1}^{k} \bar{Q}(\max_{1 \leq j \leq k} x_j, i) \leq \bar{F}(x_1, \ldots, x_k) \leq \sum_{i=1}^{k} \bar{Q}(x_i, i), \quad (1)$$

$$\sum_{i=1}^{k} \bar{Q}(x_j, i) \leq \bar{F}_j(x_j) \leq \bar{Q}(x_j, j) + (1 - \pi_j), \quad j = 1, \ldots, k. \quad (2)$$

Proof. Because (2) follows from (1) by taking $x_i = 0, i \neq j$, it is only necessary to prove (1). Let

$$A = \{(z_1, \ldots, z_k) : z_i > \max_{1 \leq j \leq k} x_j, i = 1, \ldots, k\},$$

$$B = \{(z_1, \ldots, z_k) : z_i > x_i, i = 1, \ldots, k\}, \text{ and}$$

$$C = \bigcup_{i=1}^{k} \{(z_1, \ldots, z_k) : x_i < z_i \leq \min_{j \neq i} z_j\}.$$

Clearly, $A \subset B$. To see that $B \subset C$, suppose that $(z_1, \ldots, z_k) \in B$. For notational simplicity, suppose also that $z_i \leq \min_{j \neq i} z_j$. Because $(z_1, \ldots, z_k) \in B$, $z_i > x_i$, and consequently, $(z_1, \ldots, z_k) \in C$. Because of these inclusions, it follows that

$$P\{(X_1, \ldots, X_k) \in A\} \leq P\{(X_1, \ldots, X_k) \in B\} \leq P\{(X_1, \ldots, X_k) \in C\};$$

but this is (1). □

B.2. Proposition (Peterson, 1976). Suppose that $k = 2$. If nothing is known about $\bar{F}$ except the quantities $\bar{Q}(\cdot\,|\,j)$ and π_j, $j = 1, 2$, then the inequalities (1) cannot be improved.

Proof. It can be verified that the lower bound of (1) is itself a survival function with mass confined to the line $x_1 = x_2$. Random variables with this survival function achieve the lower bound of (1) with equality, but they violate A(2). To show that the lower bound of (1) cannot be improved, it is sufficient to show that the lower bound is a limit of survival functions that place no mass on the line $x_1 = x_2$ and possess the assumed values of $\bar{Q}(\cdot \mid j), \pi_j, j = 1, 2$. For $\delta \geq 0$, let $\bar{H}_\delta$ be the survival function that places mass π_1 on the line $x_2 - x_1 = \delta$ in such a way that $P\{X_1 > x_1 \mid X_1 < X_2\} = \bar{Q}(x_1, 1), x_1 \geq 0$. Similarly, $\bar{H}_\delta$ places mass π_2 on the line $x_1 - x_2 = \delta$ in such a way that $P\{X_2 > x_2 \mid X_2 < X_1\} = \bar{Q}_2(x_2, 2), x_2 \geq 0$. Then $\bar{H}_\delta$ is a survival function possessing the assumed values of $\bar{Q}(\cdot \mid j)$ and $\pi_j, j = 1, 2$, and the lower bound of (1) is the limiting survival function of the $\bar{H}_\delta$ as $\delta \to 0$.

The upper bound of (1) is itself not a survival function but a limit of the survival functions $\bar{H}_\delta$ as $\delta \to \infty$. In this limit, all probability "escapes to infinity," and this is why the upper bound is not a survival function. □

B.3. Remarks. The straight lines that comprise the support of $\bar{H}_\delta$ can be replaced by any pair of increasing lines which do not cross the line $x_1 = x_2$. Mixtures of various survival functions of this kind all satisfy the conditions of Proposition B.1. Such mixtures provide examples of various survival functions that yield the prescribed $\bar{Q}(\cdot \mid j)$ and $\pi_j, j = 1, 2$. These mixtures can have a variety of marginal distributions.

C. Assumption of Independence

In some applications, it may be possible to infer from physical considerations that the latent variables $X_1, X_2, \ldots, X_k$ are independent. The question of how assumptions of independence affect the problem of identifiability is the subject of this section.

a. The Case of Independent Latent Variables

C.1. Proposition (Berman, 1963). If the latent variables $X_i, i = 1, \ldots, k$, are mutually independent, then the sub-survival functions $\bar{Q}(\cdot, j)$ determine the distributions of the latent lives. More precisely, when densities exist, the hazard rate r_j of the jth latent life X_j is given by $r_j(t) = h(t,j)$, i.e.,

$$r_j(t) = \frac{q(t,j)}{\sum_{i=1}^{k} \bar{Q}(t,i)} = \frac{q(t,j)}{\bar{Q}(t)}, \quad j = 1, \ldots, k. \tag{1}$$

Proof. Denote the distribution of X_i by F_i. When the latent variables are independent their joint survival function is the product $\bar{F}(x_1, \ldots, x_k) = \Pi_{i=1}^{k} \bar{F}_i(x_i)$. It follows from A(4) and A(6) that

$$\bar{Q}(t,j) = \int_t^\infty \frac{\Pi_{i=1}^k \bar{F}_i(x)}{\bar{F}_j(x)} dF_j(x) = -\int_t^\infty \bar{F}(x,x,\ldots,x) \, d \log \bar{F}_j(x)$$
$$= -\int_t^\infty \bar{Q}(x) \, d \log \bar{F}_j(x) = \int_t^\infty \bar{Q}(x) \frac{f_j(x)}{\bar{F}_j(x)} \, dx. \tag{2}$$

Differentiation of (2) with respect to t yields (1). □

Alternative Proof. Again use the fact that when the latent variables are independent, their joint survival function is the product $\bar{F}(x_1, \ldots, x_k) = \Pi_{i=1}^{k} \bar{F}_i(x_i)$. Differentiation with respect to x_j yields the equation

$$\frac{\partial}{\partial x_j} \bar{F}(x_1, \ldots, x_k) = -f_j(x_j) \prod_{i=1, i \neq j}^{k} \bar{F}_i(x_i) = -r_j(x_j) \prod_{i=1}^{k} \bar{F}_i(x_i).$$

It follows from Proposition A.1 that

$$q(t,j) = -\frac{\partial}{\partial x_j} \bar{F}(x_1, x_2, \ldots, x_k)|_{x_1 = x_2 = \ldots = x_k = t}, \quad j = 1, \ldots, k,$$

and consequently,

$$q(t,j) = r_j(t) \prod_{i=1}^{k} \bar{F}_i(t),$$

and because $\Pi_{i=1}^{k} \bar{F}_i(t) = \bar{Q}(t)$, this is (1). □

Various extensions of Propositions A.1 and C.1 are known that either weaken some hypotheses about differentiability (Miller, 1977), about independence (Desu and Narula, 1977), or actually extend the model (Langberg, Proschan, and Quinzi, 1978).

C.2. Remarks (Rose, 1973). The equality (1) can hold even when the latent variables are not independent. To see this, take $k = 2, 0 < d \leq 1$, and suppose that the joint density f of the latent variables is given by

$$
\begin{aligned}
f(x_1, x_2) &= 0 \text{ if } 0 < x_1 < d/2 \text{ and } 1 - (d/2) < x_2 < 1, \\
&\quad \text{ or if } d/2 < x_1 < d \text{ and } 1 - d < x_2 < 1 - (d/2), \\
&= 2 \text{ if } 0 < x_1 < d/2 \text{ and } 1 - d < x_2 < 1 - (d/2), \\
&\quad \text{ or if } d/2 < x_1 < d \text{ and } 1 - (d/2) < x_2 < 1, \\
&= 1 \text{ elsewhere that } 0 < x_1, x_2 < 1, \\
&= 0 \text{ if } 0 < x_1, x_2 < 1 \text{ does not hold.}
\end{aligned}
\tag{3}
$$

See Figure C.1. This density has uniform marginals and satisfies (1) but is not the case of independence. □

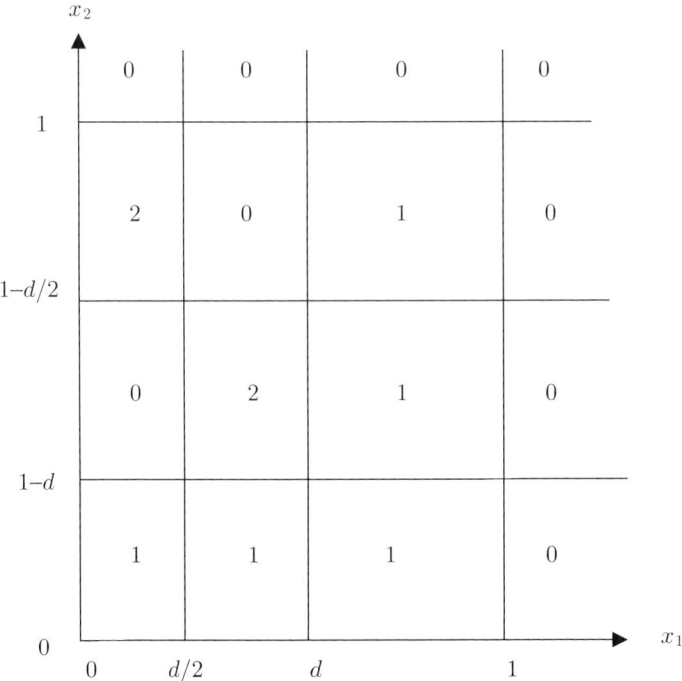

Fig. C.1. The density (3) with dependent latent variables

C.3. Example. The bivariate exponential distribution of Marshall and Olkin (1967) is the distribution of $X_1 = \min(V_1, Z), X_2 = \min(V_2, Z)$, where V_1, V_2, and Z are independent random variables with exponential distributions. In this example, $P\{X_1 = X_2\} \neq 0$ and the index J of the smallest of X_1 and X_2 is not defined. However, in this example it is possible to observe $U = \min(X_1, X_2) = \min(V_1, V_2, Z)$ and J^*, the index of the smallest of V_1, V_2, and Z. Because, V_1, V_2, and Z are independent, Proposition C.1 applies to show that the distributions of V_1, V_2, and Z are determined by the joint distribution of U and J^*. This means that the joint distribution of X_1 and X_2 can be determined. This conclusion does not depend upon the assumption that V_1, V_2, and Z have exponential distributions.

For discussions of Example C.3, see Basu and Ghosh (1978) and Arnold and Brockett (1983).

b. Independence of U and J

C.4. Proposition. Suppose that the latent variables $X_i, i = 1, \ldots, k$, are mutually independent. Then, U and J are independent if and only if there exist positive real constants $\beta_2, \ldots, \beta_k$ such that

$$\bar{F}_i(t) = [\bar{F}_1(t)]^{\beta_i}, \quad i = 2, 3, \ldots, k. \tag{4}$$

That is, the variables U and J are independent if and only if the latent variables have hazard functions that are proportional.

The condition that $X_i, i = 1, \ldots, k$, are mutually independent is not easily verified; see Section D for a discussion of this important issue.

Proposition C.4 is given by Kochar and Proschan (1991), who attribute it to Armitage (1959), Allen (1963), and Sethuraman (1965). Proposition C.4 is a direct consequence of Proposition C.5.

C.5. Proposition (Kochar and Proschan, 1991). The random variables U and J are independent if and only if there exist positive real constants $\beta_2, \ldots, \beta_k$ such that in the notation of A(6a),

$$h(t, j) = \beta_j h(t, 1), \quad j = 2, \ldots, k. \tag{5}$$

Proof. The random variables U and J are independent if and only if $Q(t, j) = Q(t)\pi_j, j = 1, \ldots, k$, i.e.,

$$q(t, j) = q(t)\pi_j, \quad j = 1, \ldots, k. \tag{6}$$

Thus, if U and J are independent, then

$$\frac{h(t,j)}{h(t,1)} = \frac{q(t,j)}{q(t,1)} = \frac{\pi_j}{\pi_1}, \quad j = 2, \ldots, k,$$

which yields (5) with $\beta_j = \pi_j/\pi_1, j = 2, \ldots, k$.

It remains to show that if (5), then U and J are independent. Multiplication of (5) by $\bar{Q}(t)$ gives

$$q(t,j) = \beta_j q(t,1), \quad j = 2, \ldots, k. \tag{7}$$

Integration of (7) on t from 0 to ∞ yields the conclusion that $\beta_j = \pi_j/\pi_1, j = 2, \ldots, k$. Finally, summation on j in (7) gives

$$q(t) = q(t,1) + \sum_{j=2}^{k} \beta_j q(t,1) = q(t,1) + \sum_{j=2}^{k} [\pi_j/\pi_1] q(t,1),$$

that is $q(t,1) = q(t)\pi_1, j = 1, \ldots, k$. Combined with (7), this yields (6). □

To obtain Proposition C.4 from Proposition C.5, combine (6) with (1) to conclude that

$$r_j(t) = \beta_j q(t,1)/\bar{Q}(t), j = 2, \ldots, k.$$

With the aid of 1.B(3), (4) follows.

C.6. Example (Kochar and Proschan, 1991). The multivariate exponential distribution of Example C.3 is not absolutely continuous, because $P\{X_1 = X_2\} > 0$. The absolutely continuous component of this distribution has been studied by Block and Basu (1974), and shown to have the density function

$$\begin{aligned}
f(x_1, x_2) \\
= [\lambda\lambda_1(\lambda_2 + \lambda_{12})/(\lambda_1 + \lambda_2)] \ \exp\{-\lambda_1 x_1 - (\lambda_2 + \lambda_{12})x_2\}, \quad x_1 < x_2, \\
= [\lambda\lambda_2(\lambda_1 + \lambda_{12})/(\lambda_1 + \lambda_2)] \ \exp\{-(\lambda_1 + \lambda_{12})x_1 - \lambda_2 x_2\}, \quad x_1 > x_2,
\end{aligned} \tag{8}$$

where $\lambda_1, \lambda_2, \lambda_{12} > 0$ and $\lambda = \lambda_1 + \lambda_2 + \lambda_{12}$. It can be shown that here the joint density of U and J is given by

$$f(u,j) = \lambda \, \exp\{-\lambda u\}[\lambda_j/(\lambda_1 + \lambda_2)], \quad j = 1, 2, \ u > 0. \tag{9}$$

This joint density is a product of the densities of U and J, showing that these variables are independent.

D. Verifiability of Independence

As noted by Cox (1959, p. 414) for the case of two risks, observations of U and J cannot yield information inconsistent with the assumption that the latent lives are independent. This important result has been studied and extended by several authors. Here, a proof is given under the assumption that $\bar{F}$ has continuous partial derivatives with respect to all of its arguments. Peterson (1975) and Miller (1977) have replaced that condition with the condition that conditional survival functions $\bar{Q}(\cdot \,|\, j)$ have no common discontinuities. More general results have been obtained by Langberg, Proschan and Quinzi (1978). Crowder (1991) has shown that even if the marginal survival functions $\bar{F}_i$ of the latent lives X_i are known, observations of U and J do not determine their joint survival function $\bar{F}$.

D.1. Proposition (Tsiatis, 1975). Let $\bar{F}$ be the joint survival function of the latent variables $X_1, \ldots, X_k$ and suppose that $P\{X_i = X_j\} = 0$, for all $i \neq j$. Let $U = \min(X_1, \ldots, X_k)$ and let J be the index of the smallest of the X_i. Suppose that the survival function $\bar{F}$ has continuous partial derivatives with respect to all of its arguments. Then, there exist independent random variables $T_1, \ldots, T_k$ with joint survival function $\bar{H}$, which has the same conditional and sub-survival distributions as $\bar{F}$. That is, if $V = \min(T_1, \ldots, T_k)$ and L is the index of the smallest of the T_i, then V and L have the same joint distribution as U and J.

Proof. To define the survival function of T_j, let $r_j(t) = q(t,j)/\bar{Q}(t)$, where $q(t,j)$ and $\bar{Q}(t)$ are determined from the joint distribution of U and J. From this hazard rate, use 1.B(3) to define

$$\bar{H}_j(x) = \exp\left\{-\int_0^x r_j(t)\, dt\right\}.$$

With $\bar{H}(x_1, \ldots, x_k) = \Pi_{j=0}^k \bar{H}_j(x_j)$, it follows from Proposition C.1 that the joint distribution of V, L is the same as that of U, J. □

E. Known Copula

Assume that $k = 2$, so there are only two competing risks. In case the net lives X_1 and X_2 are independent, the joint distribution of the net lives is identifiable. The assumption of independence is equivalent to the assumption that the copula C of X_1 and X_2 has the form

$$C(u,v) = uv, \quad 0 \leq u, \ v \leq 1.$$

(See Section 20.I.d for a discussion of copulas.) Consequently, it is natural to ask if the joint distribution is identifiable when the copula is known, but may be of some other form.

E.1. Proposition (Zheng and Klein, 1995). Suppose that the marginal distributions of X_1 and X_2 are continuous and strictly increasing on $(0, \infty)$, and the copula C of X_1 and X_2 is known. If

(a) C has a density that is strictly positive on the interior of the unit square, or more generally if

(b) as a measure on the unit square, C places positive mass on every open subset of the unit square,

then the joint distribution of X_1 and X_2 is determined by the functions $\bar{Q}(\cdot, 1)$ and $\bar{Q}(\cdot, 2)$.

The proof of this proposition is beyond the scope of this book, and is omitted.

E.2. Example. Archimedean copulas are defined in Example 20.I.23 and shown to be positive quadrant dependent. These copulas also satisfy the conditions of Proposition E.1.

A number of other examples of copulas are provided by Nelsen (2006, pp. 116–119).

a. Known Parametric Family

A word of caution is in order. Proposition E.1 requires the knowledge of the copula; knowing that the copula is a member of a parametric family of copulas may not be enough. Rose (1973) has pointed out that for indentifiability, it is also necessary that the parameters of the distribution be determined by the joint distribution of U and J.

E.3. Example (Basu and Ghosh, 1978). For the distribution of Example C.6, the joint density of U and J is given by C(9). This density depends upon λ_1, λ_2, and λ_{12} only through the quantities λ and $\lambda_j/(\lambda_1 + \lambda_2)], j = 1, 2$. But these quantities alone do not determine the parameters λ_1, λ_2, and λ_{12}. Consequently, the distribution with density C(8) is not identifiable.

E.4. Example (Basu and Ghosh, 1978; Arnold and Brockett, 1983). As noted in Example C.3, the bivariate exponential distribution of Marshall and Olkin (1967) is identifiable when the index J is suitably defined. This distribution is constructed as the joint distribution of $X = \min(V_1, Z), Y = \min(V_2, Z)$, where V_1, V_2, and Z are independent random variables with exponential distributions.

Arnold and Brockett (1983) obtain a bivariate Gompertz–Makeham distribution using a similar construction; they consider the joint distribution of

$$S = \min(X, W_1) = \min(V_1, Z, W_1), \quad T = \min(Y, W_2) = \min(V_2, Z, W_2),$$

where X, Y, V_1, V_2, and Z are independent random variables and W_1, W_2 have Gompertz distributions (W_1, W_2, V_1, V_2, and Z are mutually independent). They show that the joint distribution of S and T is identifiable. In this example, it is the structure of the underlying copula that leads to identifiability.

E.5. Example (Arnold and Brockett, 1983). Suppose that $X_1, \ldots, X_k$ have the joint survival function

$$\bar{F}(x_1, \ldots, x_k) = \int_0^\infty \exp\left\{-\sum_{i=1}^k w x_i^{a_i}\right\} dG(w). \tag{1}$$

If G has a finite moment of some positive order, then F is identifiable.

According to Proposition 11.B.1, the Pareto IV distribution is a gamma mixture of Weibull distributions. Arnold and Brockett (1983) use this fact with (1) to generate an identifiable multivariate Pareto IV distribution with survival function

$$\bar{F}(x_1, \ldots, x_k) = [1 + \sum_{i=1}^k (\lambda_i x_i)^{\alpha_i}]^{-\xi}, \quad x_i > 0, \; i = 1, \ldots, k.$$

E.6. Further examples. Additional examples of identifiable multivariate distributions are given by Arnold and Brockett (1983). Examples with exponential marginals are given by Basu and Ghosh (1978), and other examples are given by Crowder (2001).

F. Positively Dependent Latent Variables

In many applications, it is not realistic to assume that the X_i are independent, and the copula of their joint distribution is unknown. However, it may be that some kind of positive dependence is a natural assumption.

F.1. Example. As noted previously, insurance companies sometimes issue policies written on several lives, most commonly on the lives of a husband and wife. Such a policy pays a benefit to the survivor at the time of first death. From the insurance company's point of view, the time of death is the minimum of two life lengths, which represent competing risks. Husbands and wives share many risks jointly; they normally share common living conditions, ride in cars together, eat a similar diet, etc. Consequently, their life lengths are in some sense positively dependent.

F.2. Example. The components of complex equipment usually share a common environment. Electrical equipment may share voltage surges, and if packaged together, may be subject to the same hard shocks. This means that the components of the equipment have positively dependent life lengths.

F.3. Example. The bivariate exponential distribution of Example C.3 is associated because the corresponding random variables are increasing functions of independent random variables. Association is a strong sense of positive dependence that implies a positive correlation; see Section 20.I.c.

F.4. Proposition. Suppose that a device is subject to k competing risks, with respective latent variables $X_1, \ldots, X_k$. Denote the life length of the device by $U = \min(X_1, \ldots, X_k)$. If the latent variables are associated, or more generally if they are positive upper quadrant dependent, that is, if their joint survival function satisfies 20.I(14), namely, the inequality

$$\bar{F}(x_1, \ldots, x_k) \geq \prod_{i=1}^{k} \bar{F}_i(x_i),$$

then the assumption of independence leads to a pessimistic assessment of the survival function of U. That is,

$$P\{U > t\} \geq \prod_{i=1}^{k} P\{X_i > t\}. \tag{1}$$

Proof. This is a direct consequence of Proposition 20.I.12. □

F.4.a. Proposition. If the latent variables are negative upper quadrant dependent, then inequality (1) is reversed.

If the kth risk is removed, then clearly the life length of the device is increased, and consequently the probability that U exceeds t is increased. According to the following proposition, the increase is less if the latent variables are associated than if the removed latent variable is independent of the other variables. Intuitively, associated random variables tend to act in consort so that the removed variable is to some degree redundant; its removal does not affect the life length as does the removal of an independent variable.

F.5. Proposition. Suppose that $X_1, \ldots, X_k$ are associated, or more generally that $U_{k-1} = \min(X_1, \ldots, X_{k-1})$ and X_k are positively quadrant dependent. Then,

$$P\{\min(X_1, \ldots, X_{k-1}) > t\} - P\{\min(X_1, \ldots, X_k) > t\}$$
$$\leq P\{\min(X_1, \ldots, X_{k-1}) > t\}[1 - P\{X_k > t\}], \tag{2}$$

or equivalently,

$$P\{X_k > t\} \leq P\{X_k > t \mid U_{k-1} > t\}. \tag{3}$$

Proof. The fact that $U_{k-1} = \min(X_1, \ldots, X_{k-1})$ and X_k are positively quadrant dependent whenever $X_1, \ldots, X_k$ are associated is a direct consequence of Proposition 20.I.12.

The assumption of association yields directly the inequality

$$P\{X_k > t\}P\{U_{k-1} > t\} \leq P\{X_k > t, U_{k-1} > t\}; \tag{4}$$

divide through by $P\{U_{k-1} > t\}$ to obtain (3).

To obtain (2), rewrite (3) in the form

$$1 - \frac{P\{X_k > t, U_{k-1} > t\}}{P\{U_{k-1} > t\}} \leq 1 - P\{X_k > t\};$$

this is (2). □

a. Dependence from Presence of Covariates

Dependence in competing risk models is often introduced when covariates are present. Models of this kind, and the question of their identifiability have been introduced and studied, for example, by Heckman and Honoré (1989), Carling and Jacobson (1996), and Slud and Kopylev (1996).

Part VI

More About Semi-parametric Families

18

Characterizations Through Coincidences of Semiparametric Families

If a scale parameter λ is introduced in the underlying survival function $\bar{F}(x) = \exp\{-x\}$, the family $\{\bar{F}(x\,|\,\lambda) = \exp\{-\lambda x\} : \lambda > 0\}$ of exponential distributions is obtained. On the other hand, if a frailty parameter ξ is introduced in the underlying survival function $\bar{F}(x) = \exp\{-x\}$, the family $\{\bar{F}(x\,|\,\xi) = [\exp\{-x\}]^\xi = \exp\{-\xi x\} : \xi > 0\}$ is obtained. Clearly, these families coincide. Because scale parameter families and frailty parameter families do not in general coincide, it is natural to ask if there are underlying distributions other than an exponential distribution for which the introduction of scale and frailty parameters lead to the same family.

Indeed, it often happens that two semiparametric families of the form $\{\bar{H}(\cdot\,|\,\theta, \bar{F})\}$ coincide for some choice or choices of the underlying distribution F. In this chapter, pairs of the semiparametric families discussed in Chapter 7 are examined to determine the underlying distributions (assumed to be proper) for which there is coincidence. Even though location parameters are not considered here, there are still 10 families to be compared, and this means there are 45 comparisons. The underlying distributions that lead to coincidence are not known for all these pairs.

It is interesting to discover that for many of the pairs of parameters, the families of distributions generated coincide only when the underlying distribution is a member of a familiar family. In the following, parameter pairs are found with the Weibull, lognormal, gamma, Pareto, and Gompertz families as the only underlying distribution leading to coincidence. Thus, the results of this chapter provide characterizations of these distributions.

A. Introduction

It is important to remember that families can coincide but have different parameterizations. For example, it has already been noted that families obtained by introducing a frailty or scale parameter with the underlying survival function $\bar{F}(x) = \exp\{-x\}$ constitute the usual family of exponential survival functions. However, if a Laplace transform parameter is introduced, then the family $\{\bar{F}(x) = \exp\{-(s+1)x\} : s > -1\}$ is obtained; this is again the family of exponential survival functions, though it is not parameterized in the usual way.

Two families of distributions or survival functions can coincide even though they are generated by different underlying distributions. However, several of the semiparametric families introduced in Chapter 7 have an interesting property relating to this issue. Let $\bar{F}(x \mid \lambda_0)$ be a specific member of a scale parameter family, and use this as an underlying distribution to generate a new scale parameter family. Of course, the resulting family is not new, it is just the original scale family from which $F(x \mid \lambda_0)$ came. A similar result is not true if $F(x \mid \tau_0)$ is a specific member of an age parameter family. Unless $t_0 = 0$, the age parameter family generated with $F(x \mid \tau_0)$ as the underlying distribution is only a subfamily of the family from which $F(x \mid \tau_0)$ came. Those families that are generated by any member of the family as an underlying distribution include the families with scale, power, frailty, resilience, tilt, hazard exponent, and convolution. For coincidences involving one of these families, it is always possible to assume that the underlying distributions are the same. The coincidences studied involving pairs chosen from moment, Laplace transform, and age parameter families are all for identical underlying distributions, but in these cases, other coincidences may exist in which the underlying distributions are different.

In the following sections, coincidences are mostly determined by solving a functional equation, and the proofs involve a reduction of the functional equation to a functional equation with known solution. Other proofs involve solving a differential or an integral equation.

As discussed in Section 7.P, there are underlying distributions for which the introduction of a parameter of some kind does not really generate a parametric family. In particular, a distribution degenerate at 0 is not altered by the introduction of any of the parameters introduced in Chapter 7. Distributions degenerate at some point other than 0 can be given a scale parameter but not frailty or tilt parameters. One might say of the degenerate distributions not admitting parameters that the "families" they generate coincide, but this is ruled out of consideration in this chapter except in Section C.b.

a. A Method of Proof

Consider the two parametric families

$$\mathcal{F} = \{F(\cdot\,|\,\theta), \theta \in \Theta\}, \quad \mathcal{G} = \{G(\cdot\,|\,\omega), \omega \in \Omega\},$$

where

$$F(\cdot\,|\,\theta_1) \neq F(\cdot\,|\,\theta_2) \text{ whenever } \theta_1 \neq \theta_2 \text{ and } \theta_1, \theta_2 \in \Theta;$$

that is, distinct parameters lead to distinct distributions. Assume that $\mathcal{G}$ also has this property. To show that the families $\mathcal{F}$ and $\mathcal{G}$ coincide, a first step can be to show that there exists a function $\omega(\,\cdot\,)$ defined on Θ and taking values in Ω such that

$$F(\cdot\,|\,\theta) = G(\cdot\,|\,\omega(\theta)), \theta \in \Theta. \tag{1}$$

That is to say, for every $\theta \in \Theta$, there is an $\omega = \omega(\theta) \in \Omega$ such that (1) holds. This insures only that $\mathcal{F} \subset \mathcal{G}$. To complete a proof that $\mathcal{F} = \mathcal{G}$, $\mathcal{F}$ and $\mathcal{G}$ can be interchanged in the above process to show that $\mathcal{G} \subset \mathcal{F}$. But it is sufficient and usually simpler to show that for all $\omega_0 \in \Omega$, there is a $\theta \in \Theta$ such that $\omega(\theta) = \omega_0$. In this case, $\omega(\,\cdot\,)$ maps Θ onto Ω, or in other words, Ω is the *range* of $\omega(\,\cdot\,)$.

In summary, a proof that $\mathcal{F} = \mathcal{G}$ can be accomplished by showing that there exists a function $\omega(\,\cdot\,)$ with domain Θ and range Ω that satisfies (1).

Because distinct parameters identify distinct distributions in both $\mathcal{F}$ and $\mathcal{G}$, it must be that $\omega(\theta_1) \neq \omega(\theta_2)$ whenever $\theta_1 \neq \theta_2, \theta_1, \theta_2 \in \Theta$. Thus, $\omega(\,\cdot\,)$ is a one-to-one mapping of Θ onto Ω. Consequently, $\omega(\,\cdot\,)$ has an inverse; this inverse plays the role of $\omega(\,\cdot\,)$ if $\mathcal{F}$ and $\mathcal{G}$ are interchanged in the above development.

With specific choice of $\mathcal{F}$ and $\mathcal{G}$, equation (1) becomes a functional equation. In most cases, (1) can be reduced to a well-studied functional equation, one with a known solution that is discussed in Chapter 22.

b. Summary of Coincidences

Pair of parameters	Family	Proposition No.
Scale and power	degenerate at 0	C.8
Scale and frailty	Weibull or negative Weibull	B.1
Scale and resilience	Reciprocal of positive or negative Weibull	B.2
Scale and tilt	Pareto III or its negative reciprocal	B.3
Scale and hazard power	Extreme value for minimum on $(-\infty, \infty)$, possibly truncated at 0	B.4
Scale and moment	lognormal	B.5
Scale and Laplace transform	gamma	B.6
Scale and convolution	strictly stable	B.7
Scale and age	Pareto II or uniform with frailty	B.8
Power and frailty	log Weibull	B.9
Power and resilience	negative log Weibull	B.10
Power and tilt	negative log Pareto	B.11
Power and hazard power	unnamed	B.12
Power and moment	negative log gamma	B.13
Power and Laplace transform	unresolved	
Power and convolution	unresolved	
Power and age	unresolved	
Frailty and resilience	2-point distribution	C.2
Frailty and tilt	degenerate	C.3
Frailty and hazard power	degenerate	C.4
Frailty and moment	Pareto I	B.14
Frailty and Laplace transform	exponential with location parameter	B.15
Frailty and convolution	unresolved	
Frailty and age	Gompertz	B.16

A. Introduction 567

Pair of parameters	Family	Proposition No.
Resilience and tilt	2-point distribution	C.5
Resilience and hazard power	degenerate	C.9
Resilience and moment	uniform with frailty & scale	B.17
Resilience and Laplace transform	$a - X, X$ with exponential distribution	B.18
Resilience and convolution	unresolved	
Resilience and age	unresolved	
Tilt and hazard power	2-point distribution	C.6
Tilt and moment	degenerate	C.11
Tilt and Laplace transform	degenerate	C.10
Tilt and convolution	unresolved	
Tilt and age	exponential with tilt	B.19
Hazard power and moment	unresolved	
Hazard power and Laplace transform	unresolved	
Hazard power and convolution	unresolved	
Hazard power and age	unresolved	
Moment and Laplace transform	2-point distribution	C.7
Moment and convolution	gamma	B.20
Moment and age	unresolved	
Laplace transform and convolution	Poisson with scale	C.1
Laplace transform and age	unresolved	

B. Coincidences Leading to Continuous Distributions

As is apparent from the summary, coincidences of semiparametric families lead to several families of distributions that are well known, such as the Weibull, gamma, lognormal, Gompertz, and Pareto families. A few less well-known distributions also arise, such as the log Weibull and log gamma distributions.

In several of the coincidences discussed in this section, the distribution degenerate at 0 might be added to the given solutions. But, as indicated above, the distribution degenerate at 0 is disregarded in this chapter except in Section C.b.

a. Coincidences Involving Scale Parameter Families

The case of coincidence between scale and frailty families has been of particular statistical interest. For positive random variables, the following result has been given by Kalbfleisch and Prentice (2002, Section 2.4); it is the only coincidence result we are aware of that is currently in the literature. The result has appeared in the books by Lawless (1986) and Cox and Oakes (1984, p. 71).

B.1. Proposition (Scale and Frailty). With the underlying distribution F, scale and frailty parameter families coincide if and only if there exists a function $\xi(\cdot)$ with domain and range $(0, \infty)$ such that

$$\bar{F}(\lambda x) = [\bar{F}(x)]^{\xi(\lambda)} \quad \text{for all } \lambda > 0 \text{ and all } x. \tag{1}$$

From the functional equation (1), it follows that F is either a Weibull distribution, that is,

$$\begin{aligned}\bar{F}(x) &= \exp\{-(\lambda x)^\alpha\}, \quad \lambda, \alpha > 0, \quad x > 0, \\ &= 1, \quad \lambda, \alpha > 0, \quad x \leq 0,\end{aligned} \tag{2a}$$

or it is a negative Weibull distribution, that is,

$$\begin{aligned}\bar{F}(x) &= \exp\{-(-\lambda x)^{-\alpha}\}, \quad \lambda, \alpha > 0, \quad x < 0, \\ &= 0, \quad x \geq 0.\end{aligned} \tag{2b}$$

Note that both of these distributions are extreme-value distributions for minima, given, respectively, by 20.G(2) and 20.G(3), but with added scale parameter.

B. Coincidences Leading to Continuous Distributions

Proof. Because $\xi(\lambda)$ can be any positive number, it follows from (1) with $x = 0$ that either $\bar{F}(0) = 0$ or $\bar{F}(0) = 1$.

Case 1. $\bar{F}(0) = 1$ (in which case, $\bar{F}(x) = 1$ for all $x < 0$). Upon taking logarithms, (1) takes the form

$$R(\lambda x) = \xi(\lambda) R(x), \quad x, \lambda > 0. \tag{3a}$$

According to Proposition 22.B.1.a, this means that for some real constant α, $\xi(z) = z^\alpha$ and $R(x) = (\lambda x)^\alpha$. Because R is increasing and nonnegative, it follows that λ and α are positive.

Case 2. $\bar{F}(0) = 0$ (in which case, $\bar{F}(x) = 0$ for all $x > 0$). Let

$$T(x) = R(-x), \quad x > 0;$$

again take logarithms in (1) and rewrite to obtain the following analog of (3a):

$$T(\lambda x) = \xi(\lambda) T(x), \quad x, \lambda > 0. \tag{3b}$$

Again from Proposition 22.B.1.a and the fact that $\bar{F}$ is a survival function, it follows that (2b) holds.

Because $\xi(z) = z^\alpha$, it is clear that the full range of each parameter is achieved as the other parameter takes on all values in its range. It is straightforward to verify that both the Weibull distribution (2a) and the distribution given by (2b) satisfy (1). □

Kalbfleisch and Prentice (2002) prove Proposition B.1 under the assumption that $\bar{F}(0) = 1$ (case 1) and that F has a density. In this case, they are able to start with the functional equation

$$r(\lambda x) = \frac{\xi(\lambda)}{\lambda} r(x), \quad x, \lambda > 0.$$

This equation can be obtained from (3a) by differentiating; from a functional equation point of view, it has the same form as equation (3a), and its solution is also given by Proposition 22.B.1.a.

B.2. Proposition (Scale and Resilience). With the underlying distribution F, scale and resilience parameter families coincide if and only if there exists a function $\eta(\cdot)$ with domain and range $(0, \infty)$ such that

$$F(\lambda x) = [F(x)]^{\eta(\lambda)} \quad \text{for all } \lambda > 0 \text{ and all } x. \tag{4}$$

It follows that F is either of the form

$$F(x) = \exp\{-(\lambda x)^{-\alpha}\}, \quad x, \lambda, \alpha > 0,$$
$$= 0, \quad x \leq 0,$$

or it is of the form

$$F(x) = \exp\{-(-\lambda x)^{-\alpha}\}, \quad x < 0, \ \alpha, \lambda > 0,$$
$$= 1, \quad x \geq 0.$$

Note: These distribution functions are the extreme-value distributions for maxima given by 18.G(3a) and 18.G(2a), respectively.

Proof. Equation (4) can be solved by the methods analogous to those used in the proof of Proposition B.1 but with the reverse hazard rate S playing the role that R does there. As in Proposition B.1, the sign of the parameters is determined by the requirement that F be a distribution function. □

B.3. Proposition (Scale and Tilt). With the underlying distribution F, scale and tilt parameter families coincide if and only if there exists a function $\gamma(\cdot)$ with domain and range $(0, \infty)$ such that

$$\bar{F}(\lambda x) = \frac{\gamma(\lambda)\bar{F}(x)}{1 - \bar{\gamma}(\lambda)\bar{F}(x)} \quad \text{for all } \lambda > 0 \text{ and all } x \geq 0. \tag{5}$$

It follows that for some $\lambda, \alpha > 0$, either

$$\bar{F}(x) = \frac{1}{1 + (\lambda x)^\alpha}, \quad x \geq 0,$$
$$= 1, \quad x \leq 0, \tag{6}$$

or

$$\bar{F}(x) = \frac{(-\lambda x)^\alpha}{1 + (-\lambda x)^\alpha}, \quad x \leq 0,$$
$$= 0, \quad x \geq 0. \tag{7}$$

Here, the first distribution is a Pareto III distribution, defined and discussed in Section 11.B. The second distribution is the distribution of $1/(-Y)$, where Y has a Pareto III distribution.

Proof. Set $x = 0$ in (5) to obtain $\bar{\gamma}(\lambda)\{\bar{F}(0) - [\bar{F}(0)]^2\} = 0$. Because γ can take any positive value, $\bar{F}(0) - [\bar{F}(0)]^2 = 0$, and hence either $\bar{F}(0) = 1$ or $\bar{F}(0) = 0$.

Case 1. $\bar{F}(0) = 1$, so that $\bar{F}(x) = 1, x < 0$. Such a survival function satisfies (6) for all $x < 0$. For $x \geq 0$, rewrite (5) in the form

$$\frac{\bar{F}(\lambda x)}{F(\lambda x)} = \gamma(\lambda)\frac{\bar{F}(x)}{F(x)}, \quad x, \lambda > 0. \tag{8}$$

In terms of the odds ratio $\emptyset^+(x) = \bar{F}(x)/F(x)$, equation (8) becomes $\emptyset^+(\lambda x) = \gamma(\lambda)\emptyset^+(x)$; this equation has the form of 22.B(1a) and can be solved using Proposition 22.B.1.a. According to that proposition, there are constants b and c such that

$$\frac{\bar{F}(x)}{F(x)} = bx^c, \quad x > 0, \quad \text{and} \quad \gamma(\lambda) = \lambda^c.$$

This means that

$$\bar{F}(x) = \frac{bx^c}{1 + bx^c}, \quad x > 0.$$

Because $\bar{F}$ is nonnegative, $b > 0$, and because $\bar{F}$ is decreasing, $c < 0$. With the change of parameters $c = -a, b = \lambda^{-\alpha}$, the survival function (6) is obtained. Because $\gamma(\lambda) = \lambda^c$, each parameter takes on all values in $(0, \infty)$ as the other parameter ranges over the same interval.

Case 2. $\bar{F}(0) = 0$, so that $\bar{F}(x) = 0$ for all $x \geq 0$. This assures that (5) is satisfied for all $x \geq 0$. For the interval $x < 0$, let $y = -x$, and rewrite (5) in the form

$$\frac{\bar{F}(-\lambda y)}{F(-\lambda y)} = \gamma(\lambda)\frac{\bar{F}(-y)}{F(-y)}, \quad y, \lambda > 0.$$

This equation is of the form 22.B(1a) and can be solved in the manner that (8) is solved; in this way, (7) is obtained.

It can be verified directly that the distributions defined by (6) and (7) satisfy (5). □

B.4. Proposition (Scale and Hazard Power). With the underlying distribution F having hazard function R, scale and hazard power parameter families coincide if and only if there exists a function $\zeta(\cdot)$ with domain and range $(0, \infty)$ such that

$$\exp\left\{-[R(x)]^{\zeta(\lambda)}\right\} = \bar{F}(\lambda x) = \exp\left\{-R(\lambda x)\right\},$$

that is,

$$[R(x)]^{\zeta(\lambda)} = R(\lambda x) \quad \text{for all } \lambda > 0 \text{ and all } x \geq 0. \tag{9}$$

It follows that either $\bar{F}$ is of the form

$$\bar{F}(x) = \exp\left\{-\exp\left(\lambda x\right)^{\alpha}\right\}, \quad x > 0, \tag{10}$$

or it is of the form

$$\bar{F}(x) = 0, \quad x > 0, \tag{11}$$

and either

$$\bar{F}(x) = \exp\left\{-\exp\left[-(\nu\,|\,x\,|)^{\beta}\right]\right\}, \quad x < 0, \tag{12}$$

or

$$\bar{F}(x) = 1, \quad x < 0. \tag{13}$$

Here, λ, α, β, and ν are positive constants.

Proof. Note that from a functional equation point of view, (9) and (2) are the same equation. Because the scale parameter λ and the hazard power parameter ζ can take any finite positive values, it follows from (9) that

$$R(x) < \infty \text{ for all } x > 0, \quad \text{or } R(x) = \infty \text{ for all } x > 0,$$

and

$$R(x) > 0 \text{ for all } x < 0, \quad \text{or } R(x) = 0 \text{ for all } x < 0.$$

But $R(x) = \infty$ for all $x > 0$, means that $\bar{F}(x) = 0$ for all $x > 0$, and $R(x) = 0$ for all $x < 0$ means that $\bar{F}(x) = 1$ for all $x < 0$. This establishes (11) and (13).

Consider now the case that $R(x) < \infty$ all $x > 0$, If $R(x) = 0$ for some $x > 0$, then it follows from (9) that $R(x) = 0$ for all $x > 0$, and this is not possible. Consequently, $R(x) > 0$ for all $x > 0$ and (9) can be rewritten as

$$\zeta(\lambda) \log R(x) = \log R(\lambda x), \quad \lambda, x > 0. \tag{14}$$

From Proposition 22.B.1a, it follows from (14) that for some constants b and c,

$$R(x) = \exp\{bx^c\} \text{ and } \theta(\lambda) = \lambda^c.$$

This means that

$$\bar{F}(x) = \exp\{-\exp bx^c\}, \quad x > 0. \tag{15}$$

Because $\bar{F}$ is a decreasing function, either both b and c are positive or both are negative. If both are positive, then an obvious change of parameters transforms (15) to (10). If both b and c are negative, then (15) fails to be a proper survival function because then $\bar{F}(\infty) = e^{-1}$. Because $c \neq 0$, the value of $\zeta(\lambda)$ ranges over $(0, \infty)$ as λ takes on all positive values.

It remains to consider the case that $R(x) > 0$ for all $x < 0$. Again, it follows from Proposition 22.B.1.a that

$$\bar{F}(x) = \exp\{-\exp b\,|x|^c\}, \quad x < 0.$$

Because $\bar{F}$ is decreasing, and $|x|$ is decreasing in $x < 0$, it follows that b and c have opposite signs. If $b > 0, c < 0$, then $\lim_{x \to -\infty} \bar{F}(x) = e^{-1}$, which would mean that $\bar{F}$ is not a proper survival function. Consequently, $b < 0, c > 0$, and a reparameterization yields (12). □

Note that both (10) and (12) give the value $\bar{F}(0) = e^{-1}$. This means that solutions (10) and (12) can be paired up in any way to provide a solution to (9). The special case that $\lambda = \nu$ and $\alpha = \beta$ yields in this way the survival function

$$\bar{F}(x) = \exp\{-\exp[(\text{sgn } x)(\lambda\,|x|)^\alpha]\}, \quad -\infty < x < \infty,$$

where sgn x is the sign of x. If $\alpha = 1$, this is the extreme value distribution for minima with support $(-\infty, \infty)$ given by 20.G(4), but with an added scale parameter.

The following proposition involves a moment parameter, and consequently, there is a tacit assumption that the underlying distribution F satisfies $F(x) = 0, x < 0$.

B.5. Proposition (Scale and Moment). Denote by $\mathcal{B}$ the set of all β for which the βth moment μ_β of the underlying distribution F is finite. Scale and a moment parameter families coincide if and only if there exists a function $\beta(\cdot)$ with domain $(0, \infty)$ and range $\mathcal{B}$ such that

$$\frac{1}{\mu_\beta} \int_0^x z^\beta dF(z) = F(\lambda x), \quad x > 0, \ \lambda > 0. \tag{16}$$

Equation (16) is satisfied if and only if F is a lognormal distribution; then, the family is the family of lognormal distributions.

Proof. To solve (16), the first step is to show that (16) implies that F is everywhere differentiable, with density that satisfies the functional equation

$$\frac{x^{\beta(\lambda)} f(x)}{\mu_{\beta(\lambda)}} = \lambda f(\lambda x), \quad x > 0. \tag{17}$$

The second step is to convert (17) to a standard functional equation with known solution.

To show that (16) implies that F is differentiable, note that for $\beta \geq 0, \Delta > 0$,

$$x^\beta [F(x+\Delta) - F(x)] \leq \int_x^{x+\Delta} z^\beta \, dF(z)$$
$$\leq (x+\Delta)^\beta [F(x+\Delta) - F(x)]. \tag{18a}$$

According to (16),

$$\int_x^{x+\Delta} z^\beta \, dF(z) = F(\lambda(x+\Delta)) - F(\lambda x),$$

and thus, (18a) can be rewritten as

$$\frac{x^\beta \left[F(x+\Delta) - F(x)\right]}{\mu_\beta \Delta} \le \lambda \frac{F(\lambda(x+\Delta)) - F(\lambda x)}{\lambda \Delta}$$

$$\le \frac{(x+\Delta)^\beta \left[F(x+\Delta) - F(x)\right]}{\mu_\beta \Delta}. \quad (18b)$$

Because F is increasing, it is differentiable almost everywhere. If it is differentiable at x, it follows from (18b) that it is differentiable at λx (hence differentiable everywhere). By taking limits in (18b), the functional equation (17) is obtained.

Because $f(1) = 0$ implies $f(\lambda) = 0$ for all $\lambda > 0$, it follows that $f(1) \ne 0$. With $x = 1$, it follows from (17) that

$$\mu_{\beta(\lambda)} = f(1)/\lambda f(\lambda);$$

use this to rewrite (17) in the form

$$x^{\beta(\lambda)} \frac{f(x)}{f(1)} f(\lambda) = f(\lambda x), \quad x > 0. \quad (19)$$

With the notation $g(x) = f(x)/f(1)$, (19) can be rewritten in the form

$$x^{\beta(\lambda)} g(x) g(\lambda) = g(x\lambda). \quad (20)$$

It follows from Proposition 22.B.3.a that g has the form

$$g(x) = x^{bc} \exp\{c(\log x)^2/2\},$$

that is,

$$f(x) = f(1) x^{bc} \exp\{c(\log x)^2/2\}. \quad (21)$$

If this is to be a density function, then $c < 0$; set $c = -1/\sigma^2$. Next, make the change of parameters $bc = (\mu/\sigma^2) - 1$. Determine $f(1)$ by the condition that the density integrates to 1; this is possible because as is shown above, a distribution satisfying (16) has everywhere a finite derivative. According to Proposition 20.A.1, this means that the distribution cannot have a singular part.

By these means, (21) is reduced to

$$f(x) = \frac{1}{\sqrt{2\pi}\sigma x} \exp\{-[(\log x) - \mu^2]/2\sigma^2\}, \tag{22}$$

which is the density of a lognormal distribution.

Because $\beta(\lambda) = c \log \lambda$ and $\lambda(\beta) = \exp\{\beta/c\}$, it follows that the full range of each parameter is achieved as the other parameter ranges over its allowable values. Further it can be verified that the lognormal distribution satisfies (16). □

B.5.a. Remark. Motivated by a question arising in renewal theory, Vardi, Shepp and Logan (1981) consider a functional equation related to (16). They take $\beta = 1$ and identify the class of distributions for which

$$\frac{1}{\mu} \int_0^x z \, dF(z) = F(\lambda x) \tag{16a}$$

holds for some λ. The survival function on the left side of (16a) arises as the limiting distribution of the residual life in a renewal process (see Section 20.F.b), and (16a) asks for the class of underlying distributions for which the residual life distribution is a rescaled version of the underlying distribution. Another related question is also addressed by Vardi, Shepp and Logan (1981); for what underlying distributions is the limiting total lifetime distribution a rescaled version of the underlying distribution. Again, see Section 20.F.b for definitions.

B.6. Proposition (Scale and Laplace Transform). Denote by $(-c, \infty)$ the values of s for which the Laplace transform $\phi(s) = \int_{-\infty}^{\infty} e^{-sz} \, dF(z)$ of the underlying distribution F is finite. Scale and Laplace transform parameter families coincide if and only if there exists a function $s(\cdot)$ with domain $(0, \infty)$ and range $(-c, \infty)$ such that

$$\frac{1}{\phi(s(\lambda))} \int_{-\infty}^x e^{-s(\lambda)z} \, dF(z) = F(\lambda x) \text{ for all } \lambda > 0 \text{ and all } x. \tag{23}$$

Consequently, scale and Laplace transform parameter families coincide if and only if F is either a gamma distribution or the distribution of $-X$, where X has a gamma distribution.

B. Coincidences Leading to Continuous Distributions 577

Proof. From (23) it follows that for all $\Delta > 0$,

$$I = \int_x^{x+\Delta} e^{-sz}\, dF(z) = \phi(s)[F(\lambda(x+\Delta)) - F(\lambda x)]. \qquad (24)$$

Because the integrand of I is monotone, I itself admits simple bounds, which are also bounds on the right side of (24); in case $x \geq 0$,

$$e^{-s(x+\Delta)}[F(x+\Delta) - F(x)] \leq \phi(s)[F(\lambda(x+\Delta)) - F(\lambda x)]$$
$$\leq e^{-sx}[F(x+\Delta) - F(x)], \qquad (25)$$

whereas the inequalities (25) are reversed if $x + \Delta < 0$. Because F is increasing, it is differentiable almost everywhere; suppose it is differentiable at $x \geq 0$ and denote the derivative by f. Then, after dividing (25) by $\Delta > 0$, it follows that

$$e^{-sx} f(x) \leq \lambda \phi(s) \lim_{\Delta \to 0} \frac{F(\lambda(x+\Delta)) - F(\lambda x)}{\lambda \Delta} \leq e^{-sx} f(x). \qquad (26)$$

Consequently, the limit in (26) exists, F is differentiable at λx, and hence is differentiable for all $x \geq 0$ and it follows that

$$e^{-sx} f(x) = \lambda \phi(s) f(\lambda x), \quad 0 \leq x < \infty. \qquad (27)$$

The argument for $x < 0$ is similar with the same conclusion, so (27) holds for all x.

Because (27) must hold for all $\lambda > 0$, it follows that if $f(x) > 0$ for some $x > 0$, then it must be that $f(x) > 0$ for all $x > 0$. Similarly, if $f(x) > 0$ for some $x < 0$, then it must be that $f(x) > 0$ for all $x < 0$. If f is identically 0, then F is a constant and not a proper distribution function, so either $f(x) > 0$ for all $x > 0$ or $f(x) > 0$ for all $x < 0$, or both.

Case 1. $f(x) > 0$ for all $x > 0$. Set $x = 1$ in (27) to obtain

$$\phi(s(\lambda)) = e^{-s(\lambda)} f(1) / \lambda f(\lambda).$$

Substitute $\phi(s) = \phi(s(\lambda))$ from this in (27) to obtain

$$f(\lambda x) = f(x) f(\lambda)\, e^{-xs(\lambda)}\, e^{s(\lambda)} / f(1). \qquad (28)$$

578   18. Coincidences of Semiparametric Families

With the interchange of λ and x, the left-hand side of (28) is unchanged, and consequently, the same is true of the right-hand side of (28). With this interchange, it follows that $(x-1)s(\lambda) = (\lambda-1)s(x)$; this means that $s(\lambda)/(\lambda-1)$ is a constant, say c. With this notation, (28) can be rewritten as

$$e^{c(\lambda x-1)}\frac{f(\lambda x)}{f(1)} = \left[e^{c(x-1)}\frac{f(x)}{f(1)}\right]\left[e^{c(\lambda-1)}\frac{f(\lambda)}{f(1)}\right]. \tag{29}$$

Let $h(x) = e^{c(x-1)}f(x)/f(1)$ and rewrite (29) as

$$h(\lambda x) = h(\lambda)h(x), \quad x,\ \lambda > 0.$$

According to Proposition 22.A.3, this means that $h(x) = x^\alpha$ for some constant α. Thus,

$$f(x) = [f(1)\,e^c]x^\alpha\,e^{-cx}, \quad x > 0. \tag{30}$$

This function has a finite integral if $c > 0$ and $\alpha > -1$, which can be obtained from the fact that

$$\int_0^\infty x^a\,e^{-cx}\,dx = [\Gamma(a+1)]/c^{a+1}; \tag{31}$$

see Definition 23.A.1. Thus, (30) is the density of a gamma distribution. Note that $s(\lambda)$ ranges over the interval $(-c,\infty)$ as λ ranges over $(0,\infty)$; this is just the interval over which the Laplace transform of (30) is finite.

Case 2. $f(x) > 0$ for all $x < 0$. An argument similar to the above leads to the conclusion that

$$f(x) = [f(-1)\,e^\delta]|x|^\beta\,e^{-\delta x}; \tag{32}$$

If $\beta > -1$ and $\delta < 0$, then f has a finite integral over the interval $(-\infty,0)$, which again can be obtained from (31). By reparameterizing, it follows that because f satisfies (27), it must be that for some π and $\bar\pi = 1 - \pi$ in the interval $[0,1]$,

$$\begin{aligned}f(x) &= \pi\frac{\lambda^{\alpha+1}x^\alpha\,e^{-\lambda x}}{\Gamma(\alpha+1)}, & x > 0,\ \lambda > 0,\ \alpha > -1,\\ &= \bar\pi\frac{\delta^{\beta+1}x^\beta\,e^{-\delta x}}{\Gamma(\beta+1)}, & x < 0,\ \delta < 0,\ \beta > -1.\end{aligned} \tag{33}$$

With the aid of 9.A(5), it can be directly verified that the density (33) has Laplace transform

$$\phi(s) = \pi \left(\frac{\lambda}{\lambda+s}\right)^{\alpha+1} + \bar{\pi} \left(\frac{\delta}{\delta-s}\right)^{\beta+1}, \quad -\lambda < s < -\delta.$$

By using this expression for $\phi(s)$, it can be verified that (33) satisfies (27) if and only if $\pi = 0$ or 1. As in the proof of Proposition B.5, it can be shown that F can have no singular part. □

B.7. Proposition (Scale and convolution). Suppose that the underlying distribution F satisfies $F(x) = 0, x < 0$, and denote the Laplace transform of F by ϕ. With this underlying distribution, scale and convolution parameter families coincide if and only if F is strictly stable (see Definition 20.D.14).

Proof. Denote the Laplace transform of the underlying distribution F by ϕ. Scale parameter and a convolution parameter families coincide if and only if

$$\phi(s/\lambda) = [\phi(s)]^\nu \tag{34}$$

for some $\nu = \nu(\lambda), \lambda > 0$, and some $\lambda = \lambda(\nu), \nu > 0$. Because (34) must hold for positive integer values of ν, F is strictly stable.

Next, suppose that F is strictly stable. Then, for each $n = 1, 2, \ldots$, there exists $c_n > 0$ such that $\phi(c_n s) = [\phi(s)]^n$, and similarly $\phi(c_m s) = [\phi(s)]^m, m = 1, 2, \ldots$. By solving the second of these equations for ϕ and substituting in the first, it follows that

$$[\phi(s)]^{n/m} = \phi(c_m s/c_n). \tag{35}$$

Now take limits as n/m approaches ν; then, the left side of (35) has the limit $[\phi(s)]^\nu$, and so the right side of (35) must also have a limit. Because ϕ is continuous and strictly decreasing, it follows that c_m/c_n also has a limit depending on ν, say $\lambda(\nu)$. Consequently, (35) yields (34). This shows that for each ν there exists $\lambda = \lambda(\nu)$ such that (34) holds.

To show that for each $\lambda > 0$ there exists $\nu = \nu(\lambda)$ such that (34) holds, solve (34) for ν to obtain

$$\nu = [\log \phi(s/\lambda)]/[\log \phi(s)].$$

Although this value of ν need not be rational, it can be approximated to any desired degree of accuracy by a rational number n/m; because ϕ is continuous, it follows that (35) holds with $(c_m/c_n) \approx \lambda$. □

Remark. The hypothesis that F is the distribution of a nonnegative random variable is used to insure that the Laplace transform is decreasing, so the right side of (35) has a limit. There may be other ways to insure this. A similar argument can be carried out for distributions of nonpositive random variables.

Recall from Section 7.K that in this book, age parameters are introduced only when the underlying distribution is a distribution of a nonnegative random variable.

B.8. Proposition (Scale and Age). With the underlying distribution F, scale and age parameter families cannot coincide; the age parameter family is a subset of the scale parameter family if and only if there exists a function $\theta(\,\cdot\,)$ with domain $(0, \infty)$ and range contained in $(0, \infty)$ such that

$$\frac{\bar{F}(x+t)}{\bar{F}(t)} = \bar{F}(\theta(t)x) \quad \text{for all } t > 0 \text{ and all } x. \tag{36}$$

From (36), it follows that F has the form

$$\bar{F}(x) = (1 + \lambda x)^{-\xi}, \quad \xi, x > 0, \tag{37}$$

or it has the form

$$\begin{aligned}\bar{F}(x) &= (1 - \lambda x)^{\xi}, \quad \xi > 0, \ 0 \leq x \leq \lambda^{-1}, \\ &= 0, \quad x \geq \lambda^{-1}.\end{aligned} \tag{38}$$

The first of these distributions is a Pareto II distribution (Section 9.B) and the second of these distributions is a distribution uniform on $(0, \lambda^{-1})$ with frailty parameter ξ.

Proof. Set $x = 0$ to find that $\bar{F}(0) = 1$, that is, $R(0) = 0$. Upon taking logarithms, the functional equation (36) becomes

$$R(x + t) = R(t) + R(\theta(t)x). \tag{36a}$$

Because R is increasing, it is differentiable almost everywhere. Rewrite (36a) in the form

$$\frac{R(x+t) - R(t)}{x} = \frac{R(\theta(t)x) - R(0)}{\theta(t)x}\theta(t)$$

and let $x \to 0$ to conclude that R is differentiable at t if and only if it is differentiable at 0; consequently, R is differentiable everywhere. Because R is differentiable, it also follows from (36a) that θ is differentiable. Differentiate (36a) with respect to t, set $t = 0$ and use $\theta(0) = 1$ to obtain

$$r(x) = \frac{r(0)}{1 - x\theta'(0)}. \qquad (39)$$

Three cases arise:

Case 1: $\theta'(0) < 0$. Then, $r'(0) < 0$ and

$$r(x) = r(0)/[1 + \lambda x], \text{ where } \lambda = -\theta'(0) > 0. \qquad (40)$$

By integrating (40) to obtain R, (37) follows with $\xi = r(0)/\lambda$. By differentiating (36a) with respect to x and using (40), it follows that $\theta(t) = (1 + \lambda t)^{-1}$.

Note that $0 < \theta(t) \leq 1$ for $t \geq 0$, so the full range for θ is not achieved. Consequently, the age parameter yields a smaller family than the scale parameter.

Case 2: $\theta'(0) > 0$. Here, $r(x) = r(0)/[1 - \lambda x]$, where $\lambda = \theta'(0) > 0$. This together with an integration to obtain R yields (38), again with $\xi = r(0)/\lambda$. With the same procedure as in Case 1, it follows that $\theta(t) = (1 - \lambda t)^{-1}$, and consequently $\theta(t) \geq 1$. Moreover, an age parameter t for the survival function (38) is defined only for $0 < t < 1/\lambda$, a condition also necessary for $\theta(t)$ to be nonnegative. Again, the full range $(0, \infty)$ for θ is not achieved.

Case 3: $\theta'(0) = 0$. In this case, it follows from (39) that r is a constant. Differentiate (36a) with respect to x and set $x = 0$ to see that this means that θ is also a constant, and this is not possible because $\theta > 0$ is a free parameter.

By using the values of $\theta(t)$ obtained above in cases 1 and 2, it can be verified that the solutions (37) and (38) satisfy (36). □

b. Coincidences Involving Power Parameter Families

Recall that power parameters can be introduced only for nonnegative random variables, and consequently, in the following several propositions that deal with power parameters, it is tacitly assumed that $\bar{F}(0-) = 1$.

Note that in the distributions displayed below, parameters are not indicated by their usual letters. This avoids the problem of confusing parameters already in the underlying distribution and the parameters to be introduced.

B.9. Proposition (Power and Frailty). With the underlying distribution F, frailty and power parameter families coincide if and only if there exists a function $\xi(\cdot)$ with domain and range $(0, \infty)$ such that

$$[\bar{F}(x)]^{\xi(\alpha)} = \bar{F}(x^\alpha) \quad \text{for all } \alpha > 0 \text{ and all } x \geq 0. \tag{41}$$

From (41), it follows that F either has the form

$$\begin{aligned}\bar{F}(x) &= \exp\{-b(-\log x)^{-c}\}, \quad 0 \leq x \leq 1, \; b \geq 0, \; c > 0, \\ &= 0, \quad x > 1,\end{aligned} \tag{42a}$$

or

$$\begin{aligned}\bar{F}(x) &= \exp\{-b(\log x)^c\}, \quad x \geq 1, \; b > 0, \; c > 0, \\ &= 1, \quad x \leq 1,\end{aligned} \tag{42b}$$

or

$$\begin{aligned}\bar{F}(x) &= 1, \quad x < 1, \\ &= 0, \quad x \geq 1.\end{aligned} \tag{42c}$$

The survival function (42b) is the survival function of e^X, where X has Weibull distribution. This rather unfamiliar distribution is called the log Weibull distribution and is discussed in Section 15.D.d. The case $b = c = 1$ is a Pareto I survival function (see Section 11.B). The survival function (42c) is the distribution degenerate at 1; the case $b = 0$ in (42a) is the distribution degenerate at 0.

Proof. The solutions of the functional equation (41) are provided by Proposition 22.B.1.c. As indicated in the proof of that proposition, two cases must be treated separately.

Case 1: $0 < x < 1$. Either $\bar{F}(x) = 0$, $\bar{F}(x) = 1$, or

$$\bar{F}(x) = \exp\{-b(-\log x)^{-c}\} \text{ and } \xi(\alpha) = \alpha^c. \tag{43a}$$

Because $\bar{F}$ is decreasing, it must be that b and c have the same sign. To insure that $\bar{F}$ is a (proper) survival function, $b, c > 0$. Note that ξ takes all values in $(0, \infty)$ as α ranges over $(0, \infty)$.

Case 2: $x \geq 1$. In this case, either $\bar{F}(x) = 0$ or

$$\bar{F}(x) = \exp\{b'(\log x)^{c'}\} \text{ and } \xi(\alpha) = \alpha^{c'}. \tag{43b}$$

for some constants b' and c'. In order that $\bar{F}$ be decreasing, it must be that these constants have opposite signs, that is, $b = -b'$ and $c = c'$ have the same sign. To insure that $\bar{F}$ is a proper survival function, $b, c > 0$. □

B.10. Proposition (Power and Resilience). With the underlying distribution F, resilience and power parameter families coincide if and only if F either has the form

$$F(x) = \exp\{-b(\log x)^{-c}\}, \quad b \geq 0, \ c > 0, \ x \geq 1,$$
$$= 0, \quad x < 1, \tag{44a}$$

or

$$F(x) = \exp\{-b(-\log x)^c\}, \quad b, c > 0, \ 0 \leq x \leq 1,$$
$$= 1, \quad x \geq 1, \tag{44b}$$

or

$$F(x) = 0, \quad x < 1,$$
$$= 1, \quad x \geq 1. \tag{44c}$$

Proof. The distribution of X has frailty and power parameters if and only if the distribution of $1/X$ has resilience and power parameters. From Proposition B.9, it follows that the coincidence of resilience and power parameter families occurs for, and only for, the distribution of $1/X$ where X has the survival function (42a), (42b), or (42c). □

The distribution (44a) is not encountered elsewhere in this book. The distribution (44b) is a negative log Weibull distribution (15.D(5));

it is a uniform distribution with a hazard power parameter c and a parameter b that can be regarded either as a resilience or power parameter.

B.11. Proposition (Power and Tilt). With the underlying distribution F, tilt and power parameter families coincide if and only if there exists a function $\gamma(\,\cdot\,)$ with domain and range $(0, \infty)$ such that

$$\bar{F}(x^\alpha) = \frac{\gamma(\alpha)\bar{F}(x)}{1 - \bar{\gamma}(\alpha)\bar{F}(x)} \quad \text{for all } \alpha > 0 \text{ and all } x \geq 0. \tag{45}$$

It follows that tilt and power parameter families coincide if and only if F has one of the following three forms:

$$F \text{ is degenerate at 0 or 1}, \tag{46a}$$

$$\begin{aligned} F(x) &= [1 + b(-\log x)^c]^{-1}, \quad b, c > 0, \; 0 \leq x \leq 1, \tag{46b} \\ &= 1, \quad x \geq 1, \end{aligned}$$

or

$$\begin{aligned} \bar{F}(x) &= [1 + b(\log x)^c]^{-1}, \quad b, c > 0, \; x \geq 1, \tag{46c} \\ &= 1, \quad x < 1. \end{aligned}$$

The distribution (46b) is the distribution of e^{-X}, where X has a Pareto III distribution of 11.B(2). The survival function (46c) is the survival function of e^X, where X has a Pareto III distribution.

Proof. To solve (45), the cases $x < 1$ and $x \geq 1$ must be considered separately; they are not connected by (45).

Case 1: $0 < x < 1$. Either $F(x) = 0$ for all $x < 1$, or $F(x) > 0$ for all $x < 1$; in the latter case, (45) can be rewritten in terms of the odds ratio $\emptyset^+(x) = \bar{F}(x)/F(x)$ as

$$\emptyset^+(x^\alpha) = \gamma(\alpha)\emptyset^+(x), \quad 0 < x < 1, \; \alpha > 0.$$

According to Proposition 22.B.1.d, the possible solutions to this functional equation are (46a), that is,

$$\bar{F}(x) = 0, \quad 0 < x < 1,$$

or for some constants b and c,

$$\emptyset^+(x) = b(-\log x)^c \text{ and } \gamma(\alpha) = \alpha^c.$$

This leads to (46b), with $b, c > 0$ because (46b) must be a survival function.

Case 2: $x \geq 1$. Either $\bar{F}(x) = 0$ for all $x \geq 1$ (the case $\bar{F}(x) = 1$ for all $x \geq 1$ is rejected because it is an improper survival function), or it follows from Proposition 22.B.1.d and (45) with $x \geq 1$ that for some constants b and c,

$$\emptyset^+(x) = b(\log x)^c \text{ and } \gamma(\alpha) = \alpha^c.$$

This leads to (46c), with $b, c > 0$ because (46c) must be a survival function.

In all cases, the range of one parameter allows full range of the other parameter, so the same family is generated in each case. □

There are a number of underlying distributions for which the introduction of a power parameter and a hazard power parameter lead to the same family. Solutions are all related but include a degenerate distribution, a Bernoulli distribution, distributions with both an absolutely continuous and a discrete part, and absolutely continuous distributions.

B.12. Proposition (Power and Hazard Power). With the underlying distribution F, power and hazard power parameter families coincide if and only if there exists a function $\theta(\,\cdot\,)$ with domain and range $(0, \infty)$ such that

$$\bar{F}(x^\alpha) = \exp\{-[R(x)]^{\theta(\alpha)}\} \quad \text{for all } \alpha > 0 \text{ and all } x \geq 0. \quad (47a)$$

It follows that F has one of the following forms:
For $0 \leq x \leq 1$,

$$\bar{F}(x) = \exp\{-e^{-b(-\log x)^c}\}, \quad bc > 0, \qquad (48a)$$
$$\bar{F}(x) = 1, \qquad (48b)$$
$$\bar{F}(x) = \exp\{-1\}, \quad \text{or} \qquad (48c)$$
$$\bar{F}(x) = 0. \qquad (48d)$$

For $x \geq 1$,

$$\bar{F}(x) = \exp\{-e^{b'}(\log x)^{c'}\}, \quad b'c' > 0, \quad \text{or} \tag{49a}$$
$$\bar{F}(x) = 0. \tag{49b}$$

Because the introduction of a power parameter involves the assumption that F is the distribution of a nonnegative random variable,

$$\bar{F}(x) = 1, \quad x < 0.$$

The solutions (48a,b,c,d) and (49a,b) can be paired in any way that yields monotonicity.

The only absolutely continuous distribution with no discrete part is obtained with the pairing

$$(48a) \text{ with } (49a), \quad b, c, b', c' > 0;$$

this absolutely continuous distribution is concentrated on $[0, \infty)$. In particular, the Weibull distribution is obtained from (48a) and (49a) with $b = b' = \alpha$ and $c = c' = 1$.

Pairings that yield distributions degenerate at 0 include

$$(48d) \text{ with } (49b).$$

Pairings that lead to degeneracy at 1 include

$$(48b) \text{ with } (49b).$$

The pairing (48c) with (49b) leads to the Bernoulli distribution with mass e^{-1} at 1.

The pairing of

$$(48c) \text{ with } (49a), b', c' > 0$$

yields a distribution with mass at 0 and an absolutely continuous part concentrated on $[1, \infty)$.

B. Coincidences Leading to Continuous Distributions

Pairing

$$(48b) \text{ with } (49a), b, c > 0$$

yields a distribution with mass at 1 and an absolutely continuous part concentrated on $[1, \infty)$.

The pairing

$$(48a), \ b, c > 0, \text{ with } (49b)$$

yields a distribution with mass at 1 and absolutely continuous part concentrated on $(0, 1)$, whereas pairing

$$(48a), \ b, c < 0, \text{ with } (49b)$$

yields a distribution with mass at 0 and an absolutely continuous part concentrated on $[0, 1]$.

Proof. Equation (47a) can be rewritten in the form

$$R(x^\alpha) = [R(x)]^{\theta(\alpha)}, \quad x > 0. \tag{47b}$$

Because $x^\alpha \geq 1$ if and only if $x \geq 1$, (47b) provides no link between values of R on $(0, 1)$ with those on $[1, \infty)$, and it is necessary to treat the cases $x < 1$ and $x > 1$ separately.

Case 1: $x < 1$. There are three trivial solutions of (47b): If $R(x) = 1$ for all $x < 1$, then (48c) holds. If $R(x) = 0$ for all $x < 1$, then (48b) is obtained. If $R(x) = \infty$, then (48d) follows. Otherwise, it follows from Proposition 22.B.1.c that for some constants b and c,

$$R(x) = \exp\{b(-\log x)^c\} \quad \text{and} \quad \theta(\alpha) = \alpha^c. \tag{50}$$

Because R is an increasing function, b and c must have the same sign. This yields (48a).

Case 2: $x \geq 1$. Here, the trivial solutions $R(x) = 0$ and $R(x) = 1$ lead to an improper distribution, and consequently, they are rejected. However, the trivial solution $\bar{F}(x) = 0$, that is, $R(x) = \infty, x \geq 1$, is allowable and yields (49b). Otherwise, it follows from Proposition 22.B.1.c that

$$R(x) = \exp\{b'(\log x)^{c'}\} \quad \text{and} \quad \theta(\alpha) = \alpha^{c'}. \tag{51}$$

Again, the fact that R is increasing must be considered, and this again leads to the conclusion that b' and c' must be of the same sign.

The various solutions in Cases 1 and 2 can pair up in any fashion that yields a survival function, as listed in the statement of the proposition.

It can easily be verified that in both Cases 1 and 2, the range of the power parameter allows the full range of the hazard power parameter, and similarly the range of the hazard power parameter allows full range of the power parameter. Thus, the introduction of one of these parameters yields the same family as the introduction of the other parameter. □

The following proposition gives rise to both log gamma and negative log gamma distributions. These rather unfamiliar distributions are discussed in Section 15.D.

B.13. Proposition (Power and Moment). Suppose that the underlying distribution F has a twice differentiable density, and denote by $\mathcal{B}$ be the set of all β for which the βth moment μ_β of F is finite. Then, power and moment parameter families coincide if and only if there exists a function $\beta(\cdot)$ with domain $(0, \infty)$ and range $\mathcal{B}$ such that

$$x^{\beta(\alpha)} f(x)/\mu_{\beta(\alpha)} = \alpha x^{\alpha-1} f(x^\alpha), \quad x \geq 0, \quad \alpha > 0. \tag{52}$$

It follows that F is either a log gamma distribution 15.D(1) or a negative log gamma distribution 15.D(2).

Proof. Because f is a density, it is not identically zero, and in fact, $f(x_0) > 0$ for some $x_0 \neq 0$ or 1.

Case 1: $x_0 > 1$. According to (52), $f(x_0) > 0$ means that $f(x_0^\alpha) > 0$ for all $\alpha > 0$, and consequently, it follows that $f(x) > 0$ for all $x \geq 1$. Let $\phi(u) = \log f(e^u), u > 0$. Because f is twice differentiable, so ϕ is twice differentiable, and from the logarithm of (52) it follows that $\phi''(u) = \alpha^2 \phi''(\alpha u), u > 0$. Fix $u_0 > 0$, and let $\alpha u_0 = z$ to conclude that $\phi''(z) = u_0^2 \phi''(u_0)/z^2$. This equation can be solved by integration to conclude that f has the form

$$f_1(x) = a(\log x)^{\nu-1} x^{-\lambda}, \quad x > 1, \tag{53}$$

where $\lambda, \nu > 0$ and $a = \lambda/\Gamma(\nu)$ because f must be a density (see 15.D(1)).

Case 2: $x_0 < 1$. With the argument used in Case 1, it follows that $x_0 < 1$ implies $f(x) > 0$, for all x in $[0, 1]$. Set $\phi(u) = \log f(e^{-u}), u > 0$, and follow the argument of Case 1 to conclude that f has the form

$$f_2(x) = a(-\log x)^{\nu-1} x^{\lambda-1}, \quad 0 < x < 1, \tag{54}$$

where $\lambda, \nu > 0$ and $a = \lambda^{\nu-1}/\Gamma(\nu)$ so that f_2 is a density (see 15.D(2)).

Let f in (52) be the log gamma density 15.D(1) to conclude that $\beta = \lambda(1 - \alpha)$. This shows that β ranges over $(-\infty, \lambda)$ as α ranges over $(0, \infty)$; with the negative log gamma density 15.D(2) in (52), it follows that $\beta = \lambda(\alpha - 1)$ and β ranges over $(-\lambda, \infty)$ as α ranges over $(0, \infty)$. In each case, the full range of each parameter is achieved. □

c. More Coincidences Involving Frailty Parameter (Proportional Hazards) Families

Propositions B.1 and B.9 give results regarding the coincidences of scale and frailty parameter families and power and frailty parameter families. Here, other coincidences involving frailty parameter families are given.

B.14. Proposition (Frailty and moment). Suppose that the underlying distribution F has the density f, and denote by $\mathcal{B}$ the set of all β such that the βth moment μ_β of F is finite. Then, frailty and moment parameter families coincide if and only if there exists a function $\beta(\cdot)$ with domain $(0, \infty)$ and range $\mathcal{B}$ such that

$$\xi f(x)[\bar{F}(x)]^{\xi-1} = x^{\beta(\xi)} f(x)/\mu_{\beta(\xi)}, \quad x \geq 0, \ \xi > 0. \tag{55}$$

It follows from (55) that F is a Pareto I distribution.

Proof. From (55), it follows that either $f(x) = 0$ or

$$\xi[\bar{F}(x)]^{\xi-1} = x^{\beta(\xi)}/\mu_{\beta(\xi)},$$

that is,

$$\bar{F}(x) = \frac{x^{\beta(\xi)/(\xi-1)}}{(\xi\mu_{\beta(\xi)})^{1/(\xi-1)}}. \tag{56}$$

Because $\bar{F}$ is decreasing, $\beta(\xi)/(\xi - 1) < 0$, and because $\bar{F}(x) \leq 1$, (56) can hold only for $x \geq (\xi\mu_{\beta(\xi)})^{1/\beta(\xi)}$. It follows that the set where

$f(x) = 0$ is the set $\{x : x < (\xi\mu_{\beta(\xi)})^{1/\beta(\xi)}\}$. But (56) is the survival function of a Pareto I distribution.

Now, suppose that the underlying distribution is a Pareto I distribution, say with density

$$f(x) = \alpha c^\alpha / x^{\alpha+1}, \quad x \geq c > 0, \ \alpha > 0,$$
$$= 0, \quad x \leq c. \tag{57}$$

This density has βth moment $\mu_\beta = \alpha c^\beta / (\alpha - \beta), \beta < \alpha$. The introduction of a moment parameter $\beta < \alpha$ in (57) leads to the density

$$f_\beta(x) = \frac{c^{\alpha-\beta}(\alpha - \beta)}{x^{\alpha-\beta+1}}, \quad x > c,$$
$$= 0, \quad x \leq c, \tag{58}$$

whereas the introduction of a frailty parameter ξ in the density (57) leads to the density (58) with $\alpha\xi = \alpha - \beta$. This shows that β takes on all values less than α as ξ takes on all positive values. Thus, the introduction of a moment parameter and the introduction of a frailty parameter lead to the same family. □

Remark. The assumption that a density exists is essential in Proposition B.14; there are discrete distributions for which the introduction of a moment and a frailty parameter lead to the same family (including the trivial cases of degenerate distributions), but the class of such distributions has not been identified.

B.15. Proposition (Frailty and Laplace Transform). Suppose that the underlying distribution F has a density f and Laplace transform ϕ, finite on the interval $(-s_0, \infty)$. Then, frailty parameter and Laplace transform parameter families coincide if and only if there exists a function $\xi(\cdot)$ with domain $(-s_0, \infty)$ and range $(0, \infty)$ such that

$$e^{-sx}f(x)/\phi(s) = \xi(s)[\bar{F}(x)]^{\xi(s)-1}f(x) \quad \text{for all } x \text{ and } s > -s_0. \tag{59}$$

From (59), it follows that F is an exponential distribution with a location parameter.

Proof. From (59), it follows that either $f(x) = 0$, or for $\xi(s) \neq 1$,

$$\bar{F}(x) = \left[\frac{e^{-sx}}{\xi(s)\phi(s)}\right]^{1/(\xi(s)-1)} = c e^{-dx}, \tag{59a}$$

where $c = [\xi(s)\phi(s)]^{-1/[\xi(s)-1]} > 0$ and $d = s/[\xi(s) - 1]$. Because $\bar{F}(x) \leq 1$ for all x, (59a) can hold only for $x \geq (\log c)/d$. But because it has been assumed that F has a density, F is continuous, and so (59a) must hold for all $x \geq (\log c)/d$; thus, (59a) is the survival function of an exponential distribution with location parameter $(\log c)/d$.

Because the Laplace transform of the survival function (59a) is finite on the interval $(-d, \infty)$, it follows that $s_0 = d = s/[\xi(s) - 1]$, that is,

$$\xi(s) = 1 + \frac{s}{s_0}, \quad s > -s_0.$$

This function has the required domain and range. Moreover, the case excluded above that $\xi(s) = 1$ satisfies (59a) with $s = 0$. □

B.16. Proposition (Frailty and Age). There is no underlying distribution for which frailty and age parameter families coincide. The resulting families can partially coincide, with the age parameter family a subset of the frailty parameter family, if and only if

$$\frac{\bar{F}(x+t)}{\bar{F}(t)} = [\bar{F}(x)]^{\xi(t)}, \quad x, t \geq 0, \tag{60}$$

for some function ξ mapping $(0, \infty)$ into $(0, \infty)$. Equation (60) holds if and only if F is a Gompertz distribution (Proposition 10.A.3).

Proof. It is convenient to rewrite (60) in terms of the hazard function R of F by taking logarithms and changing sign to obtain

$$R(x+t) - R(t) = \xi(t)R(x), \quad x, t \geq 0. \tag{61}$$

Because ξ is not a constant function, it follows from Propositions 22.B.4 and 22.A.2 that

$$R(x) = \theta(e^{\lambda x} - 1), \quad \xi(t) = e^{\lambda t} \tag{62}$$

for some constants θ, λ, which must be positive in order that R be a hazard function. Conversely, it can be verified directly that the

Gompertz distribution satisfies (60). Inserting the Gompertz distribution $\bar{F}(x) = \exp\{-a(e^{bx} - 1)\}$ into (60) leads to the conclusion that $\xi(t) = e^{bt}$. But for positive $t, e^{bt} \geq 1$. This means that the family generated by the introduction of an age parameter is a subfamily of the one generated by the frailty parameter. □

d. Remaining Coincidences Involving Resilience Parameter Families

B.17. Proposition (Resilience and Moment). Suppose that the underlying distribution F has a density f and βth moment finite on the set $\mathcal{B}$. Then, resilience parameter and moment families coincide if and only if there exists a function $\beta(\,\cdot\,)$ with domain $(0, \infty)$ and range $\mathcal{B}$ such that

$$\eta f(x)[F(x)]^{\eta - 1} = \frac{x^{\beta(\eta)} f(x)}{\mu_{\beta(\eta)}} \quad \text{for all } x \text{ and } \eta > 0. \tag{63}$$

The functional equation (63) is satisfied if and only if F is a uniform distribution with scale and power (resilience) parameters (see 14.B.1). That is, for some positive constants c and d,

$$F(x) = (x/d)^c, \quad 0 \leq x \leq d. \tag{64}$$

Proof. Equation (63) can be solved in the same manner as equation (55). The requirement that F be a distribution function determines the sign of the constants, and the support of the distribution. Note that for the distribution (64), $\mu_\beta = cd^\beta/(c + \beta)$, where $c + \beta > 0$. From this fact, and (63), it follows that $\mathcal{B} = (-c, \infty)$ and

$$\beta(\eta) = c(\eta - 1).$$

Consequently, $\beta(\,\cdot\,)$ has the required range. □

B.18. Proposition (Resilience and Laplace Transform). Suppose that the underlying distribution F has a density f and a Laplace transform ϕ, finite on the set $\mathcal{S}$. Then, resilience parameter and Laplace transform parameter families coincide if and only if there exists a function $\eta(\,\cdot\,)$ with domain $\mathcal{S}$ and range $(0, \infty)$ such that

$$\eta(s)[F(x)]^{\eta(s)-1} f(x) = e^{-sx} f(x)/\phi(s) \quad \text{for all } x \text{ and all } s \in \mathcal{S}. \tag{65}$$

Equation (65) is satisfied if and only if F is the distribution of a random variable of the form $a - X$, where X has an exponential distribution. That is,

$$F(x) = e^{c(x-a)}, \quad -\infty < x \le a,$$
$$= 1, \quad x > a, \qquad (66)$$

for some constant a and positive constant c.

Proof. Equation (65) is satisfied for all x such that $f(x) = 0$. If $f(x) \ne 0$ and $\eta(s) = 1$, it follows from (65) that $\phi(s) = e^{-sx}$ for all x such that $f(x) \ne 0$; this means that $s = 0$. If $\eta(s) \ne 1$, then (65) can be rewritten in the form

$$F(x) = \left(\frac{e^{-sx}}{\eta(s)\phi(s)} \right)^{1/[\eta(s)-1]}. \qquad (67)$$

Because F is increasing and $F(x) \le 1$, it must be that $c = -s/(\eta(s) - 1) > 0$; thus (67) can be rewritten in the form of (66), where the constants a and c are functions of s.

Direct calculations show that the Laplace transform of (66) is finite for $s < c$, that is, $\mathcal{S} = (-\infty, c)$. Because $\eta(s) = (c - s)/c$ both when $\eta(s) = 1$ and $\eta(s) \ne 1$, it follows that $\eta(\,\cdot\,)$ has the required range. □

Remark. As in the case of frailty and Laplace transform parameters, the condition that F has a density was essential to the above proof; there are distributions F that do not have densities for which the introduction of a resilience and a Laplace transform lead to the same family. In particular, any distribution that places mass at only two points provides an example. It is not known if other examples exist.

e. Remaining Coincidences Involving Tilt Parameter Families

The solution of the functional equation in the following proposition is due to Aczél (1999). Recall that age parameters are introduced only when the underlying distribution is a distribution of a nonnegative random variable.

B.19. Proposition (Tilt and age). With the underlying distribution F, age parameter and tilt parameter families coincide if and only if there exists a function $\gamma(\,\cdot\,)$ with domain and range $(0, \infty)$

such that

$$\frac{\bar{F}(x+t)}{\bar{F}(t)} = \frac{\gamma(t)\bar{F}(x)}{1-\bar{\gamma}(t)\bar{F}(x)} \quad \text{for all } x \text{ and all } t > 0. \tag{68}$$

It follows that F is an exponential distribution with tilt parameter as described in Section 9.D.

The Pareto II distribution with unit frailty parameter also satisfies (68), but in this case the function $\gamma(\,\cdot\,)$ has a range that is a proper subset of $(0,\infty)$, and consequently, the age parameter family a proper subset of the tilt parameter family. There is no other underlying distribution for which partial coincidence occurs.

Proof. Rewrite (68) as

$$\frac{\bar{F}(x)\bar{F}(t)}{\bar{F}(x+t)} = \frac{1}{\gamma(t)} - \left(\frac{1}{\gamma(t)} - 1\right)\bar{F}(x) = \bar{F}(x) + \frac{\bar{F}(x)}{\gamma(t)}. \tag{69}$$

Because the left-hand side of (69) is symmetric in x and t, so is the right side, and consequently,

$$\bar{F}(x) + \frac{\bar{F}(x)}{\gamma(t)} = \bar{F}(t) + \frac{\bar{F}(t)}{\gamma(x)}. \tag{70}$$

The solutions $\bar{F}(x) = 1$ and $\bar{F}(x) = 0$ for all x of (70) are rejected because they are not proper survival functions. Consequently, there exists x_0 such that $\bar{F}(x_0) < 1$. With $x = x_0$, (70) yields

$$1/\gamma(t) = a\bar{F}(t) + b, \tag{71}$$

where

$$a = \frac{1 - [1/\gamma(x_0)]}{\bar{F}(x_0)} \quad \text{and} \quad b = \frac{[1/\gamma(x_0)] - \bar{F}(x_0)}{\bar{F}(x_0)}.$$

With the substitution of (71) in (70), it follows that

$$\bar{F}(x)(1 - a - b) = \bar{F}(t)(1 - a - b).$$

B. Coincidences Leading to Continuous Distributions

Because $\bar{F}$ is not a constant, it must be that $1 - a - b = 0$, and consequently, (71) yields

$$1/\gamma(t) = a\bar{F}(t) + (1-a). \tag{72}$$

In order to determine $\gamma(\cdot)$, begin by defining the function U by

$$U(x) = 1/\bar{F}(x),$$

and conclude from (69) and (72) that

$$U(x+t) = a[U(x) + U(t) - 1] + (1-a)[U(x)U(t)]. \tag{73}$$

Case 1: $a = 1$. In this case, (73) takes the form

$$U(x+t) - 1 = U(x) - 1 + U(t) - 1,$$

and in this case it follows from Proposition 22.A.1 that for some constant c, $U(x) - 1 = cx$ so that

$$\bar{F}(x) = \frac{1}{1+cx}, \quad x \geq 0.$$

Because $\bar{F}$ is a survival function, $c > 0$, and F is a Pareto II distribution with unit frailty parameter.

Here, $\gamma(t) = 1 + ct \geq 1$, so the age parameter does not generate the entire family generated by the tilt parameter.

Case 2: $a \neq 1$. In this case, it follows from Proposition 22.C.4.a that for some constants $c \neq 0, d$, and $\lambda \neq 0$,

$$U(x) = c\, e^{\lambda x} + d,$$

and consequently,

$$\bar{F}(x) = \frac{1}{c\, e^{\lambda x} + d}.$$

From (68) with $x = 0$, it follows that $\bar{F}(0) = 1$, and this means that $c + d = 1$. Because $\bar{F}$ is decreasing, c and λ are of the same sign, and in fact must be positive, for otherwise $\bar{F}$ is not a proper survival function. Consequently, $\bar{F}$ is an exponential distribution with tilt parameter

596   18. Coincidences of Semiparametric Families

given by 9.D(1) with $1/c = \gamma$. In this case, unlike Case 1, γ ranges over $(0, \infty)$ as a ranges over the same interval. □

f. Another Coincidence Involving Moment Parameter Families

B.20. Proposition (Moment and Convolution). Denote by $\mathcal{B}$ the set of all β such that the βth moment μ_β of the underlying distribution F is finite. Because of the uniqueness of Laplace transforms (Proposition 20.D.2), moment parameter and convolution parameter families coincide if and only if there exists a function $\nu(\cdot)$ with domain $\mathcal{B}$ and range $(0, \infty)$ such that

$$\frac{\int e^{-sx} x^\beta \, dF(x)}{\mu_\beta} = \left[\int e^{-sx} \, dF(x)\right]^{\nu(\beta)} \quad \text{for all } x \text{ and all } \beta \in \mathcal{B}. \quad (74)$$

Equation (74) is satisfied if and only if F is a gamma distribution. The introduction of either parameter generates the entire family of gamma distributions.

Proof. By differentiating (74) with respect to s and then setting $s = 0$, it can be determined that $\nu(\beta) = \mu_{\beta+1}/\mu_1\mu_\beta$. This means (Proposition 20.B.8), in particular, that $\nu(1) \geq 1$, with equality only for a degenerate distribution. Because degenerate distributions fail to satisfy (74), it follows that $\nu(1) > 1$. With $\beta = 1$, rewrite (74) in the form

$$-\phi'(s)/\mu_1 = [\phi(s)]^{\nu(1)}, \quad (75)$$

where ϕ is the Laplace transform of F. The differential equation (75) has the solution $\phi(s) = 1/[c + \mu_1(\nu - 1)s]^{1/(\nu-1)}$, where c is a constant and $\nu = \nu(1)$. The condition $\phi(0) = 1$ shows that $c = 1$. Consequently, this is the Laplace transform of a gamma distribution with scale parameter $\lambda = 1/[\mu_1(\nu - 1)]$ and shape parameter $1/(\nu - 1)$.

For the gamma distribution with shape parameter θ, it follows from 9.A(9) that $\nu(\beta) = \mu_{\beta+1}/\mu_1\mu_\beta = (\beta + \theta)/\theta$ and μ_β is finite for $\beta > -\theta$. Thus, ν ranges over the interval $(0, \infty)$ as the moment parameter ranges over $\mathcal{B}$. □

C. Coincidences Leading to Discrete Distributions

There are a number of coincidences that lead to distributions concentrating on one or two points. These somewhat negative results are not always easily proved, but they at least settle some questions.

C. Coincidences Leading to Discrete Distributions

Here is a more interesting result; it is a counterpart to Proposition B.20, where moment and convolution parameter families are found to coincide for the gamma distribution. Propositions B.20 and C.1 display another connection between the families of gamma distributions and Poisson distributions.

C.1. Proposition (Laplace Transform and Convolution). Denote the Laplace transform of the underlying distribution F by ϕ. Laplace transform parameter and convolution parameter families coincide if and only if

$$\int_0^\infty e^{-ux} \frac{e^{-sx}}{\phi(s)} \, dF(x) = \left[\int_0^\infty e^{-ux} \, dF(x) \right]^{\nu(s)}$$

for all $u > 0$ and all $s \in S$, (1)

where S is the set where ϕ is finite. Equation (1) is satisfied if and only if F is a Poisson distribution on some lattice $0, a, 2a, 3a, \ldots$.

Proof. Because a distribution is uniquely determined by its Laplace transform (Proposition 20.D.2), (1) is necessary and sufficient for Laplace transform parameter and convolution parameter families to coincide. Rewrite (1) in the form $\phi(s+u) = \phi(s)[\phi(u)]^{\nu(s)}$ and take logarithms to obtain

$$\psi(s+u) = \psi(s) + \nu(s)\psi(u),$$

where $\psi(s) = \log \phi(s)$. According to Proposition 22.B.4, the general solution of this functional equation is

$$\psi(s) = \alpha - \alpha \, e^{bs} \text{ and } \nu(s) = e^{bs}$$

for some constants α and b. In order for $\phi(s) = \exp\{\psi(s)\}$ to be a Laplace transform, it must have a negative first derivative and a positive second derivative (Proposition 20.D.5); from this, it follows that $\alpha < 0$ and $b < 0$. In this case, ϕ is the Laplace transform of a random variable X such that

$$P\{X = ak\} = e^{-\lambda} \lambda^k / k!, \quad k = 0, 1, 2, \ldots.$$

where $a = -b$.

Because the Laplace transform of the Poisson distribution exists for all real s, $\nu(s) = e^{bs}$ takes on all positive values, the full parameter range of a convolution parameter. Consequently, the two families coincide. □

a. Two-Point Distributions

C.2. Proposition (Frailty and Resilience). With the underlying distribution F, frailty parameter and resilience parameter families coincide if and only if

$$[\bar{F}(x)]^\xi = 1 - [1 - \bar{F}(x)]^\eta, \quad -\infty < x < \infty \tag{2}$$

for some

$$\xi = \xi(\eta) > 0 \text{ defined for all } \eta > 0 \tag{3}$$

and some

$$\eta = \eta(\xi) > 0 \text{ defined for all } \xi > 0. \tag{4}$$

It follows that F places mass on at most two points.

Proof. Rewrite (2) in the form

$$u^\xi = 1 - [1 - u]^\eta \tag{5}$$

for all u such that $u = \bar{F}(x)$ for some x, and both (3) and (4) must hold. Whatever ξ and η may be, (5) holds for $u = 0$ and $u = 1$. If $u \neq 0$ and $u \neq 1$, then from (5) it follows that

$$\xi(\eta) = \frac{-\log\left[1 - (1-u)^\eta\right]}{-\log u},$$

and

$$\eta(\xi) = \frac{-\log(1 - u^\xi)}{-\log(1-u)}.$$

Both of these quantities satisfy the nonnegativity requirement, but they can hold independently of u if and only if u is fixed. This means that $\bar{F}(x)$ can take on only one value different from 0 and 1. On the other hand, if $\bar{F}(x)$ can take on only one value between 0 and 1, then (2) to (4) are satisfied. □

C.3. Proposition (Frailty and Tilt). With the underlying distribution F, frailty parameter and tilt parameter families coincide if and

only if
$$[\bar F(x)]^\xi = \frac{\gamma \bar F(x)}{1 - \bar\gamma \bar F(x)}, \quad -\infty < x < \infty \tag{6}$$

for some
$$\xi = \xi(\gamma) > 0 \text{ defined for all } \gamma > 0 \tag{7}$$

and some
$$\gamma = \gamma(\xi) > 0 \text{ defined for all } \xi > 0. \tag{8}$$

It follows that frailty parameter and tilt parameter families coincide if and only if F places mass on at most two points.

Proof. From (6), it follows that
$$u^\xi = \frac{\gamma u}{1 - \bar\gamma u} \tag{9}$$

for all u such that $u = \bar F(x)$ for some x and both (7) and (8) must hold. Whatever ξ and γ may be, (9) holds for $u = 0$ and $u = 1$. If $u \neq 0$ and $u \neq 1$, then from (9) it follows that
$$\xi(\gamma) = \frac{\log \gamma u - \log(1 - \bar\gamma u)}{\log u}.$$

Because $\zeta(\gamma)$ cannot depend upon u, u can take on at most one value in the interval $(0, 1)$. Whenever $0 < u < 1$, $\zeta(\gamma)$ is positive, for all $\gamma > 0$, and each parameter takes on all values in $(0, \infty)$ as does the other parameter. Thus, the two families coincide whenever $\bar F(x)$ takes on only the values $0, 1$ and one intermediate value. $\square$

C.4. Proposition (Frailty and Hazard Power). With the underlying distribution F, frailty parameter and hazard exponent parameter families coincide if and only if
$$[\bar F(x)]^\xi = \exp\{-[R(x)]^\zeta\}, \quad -\infty < x < \infty \tag{10}$$

for some
$$\xi = \xi(\zeta) > 0 \text{ defined for all } \zeta > 0 \tag{11}$$

and some

$$\zeta = \zeta(\xi) > 0 \text{ defined for all } \xi > 0. \tag{12}$$

It follows that F is degenerate; in this case, the introduction of either parameter fails to change the distribution and generate a parametric family (see 7.P.3 and 7.P.6). If F is not degenerate, then it places positive mass on two points and the hazard power parameter family is a proper subfamily of the frailty parameter family.

Proof. Trivially with a degenerate underlying distribution, the introduction of a frailty and a hazard power parameter leads to the same family which consists only of the underlying distribution.

Suppose that the underlying distribution F is not degenerate. When $\bar{F}(x)$ is either 0 or 1, (10) is independent of the parameters. If $\bar{F}(x)$ is neither 0 nor 1, then $0 < R(x) < \infty$, and it follows from (10) that

$$\zeta(\xi) = \log \xi R(x)/\log R(x) = 1 + [(\log \xi)]/\log R(x)]. \tag{13}$$

Because $\zeta(\xi)$ must be independent of x, (13) can hold only for one value of x, say $x = x_0$. Let $c = R(x_0)$. If $c \leq 1$, then because $\zeta(\xi)$ is positive, it must be that $\xi \leq 1/c$; if $c \geq 1$, then for the same reason, $\xi \geq 1/c$. Consequently, the family generated by the introduction of a hazard power parameter is a proper subset of the family generated by the introduction of a frailty parameter. □

C.5. Proposition (Resilience and Tilt). With the underlying distribution F, resilience parameter and tilt parameter families coincide if and only if

$$1 - [F(x)]^\eta = \frac{\gamma \bar{F}(x)}{1 - \bar{\gamma} \bar{F}(x)}, \quad -\infty < x < \infty, \tag{14}$$

for some

$$\eta = \eta(\gamma) > 0 \text{ defined for all } \gamma > 0 \tag{15}$$

and some

$$\gamma = \gamma(\eta) > 0 \text{ defined for all } \eta > 0. \tag{16}$$

It follows that F is concentrated on at most two points.

Proof. It follows from (14) that either $\bar{F}(x) = 0, \bar{F}(x) = 1$, or

$$1 - (1-u)^\eta = \frac{\gamma u}{1 - \bar{\gamma} u} \tag{17}$$

for all u such that $u = \bar{F}(x)$ for some x, and both (15) and (16) must hold. If $u \neq 0$ and $u \neq 1$, then from (17) it follows that

$$\eta = \eta(\gamma) = 1 - \frac{\log(1 - \bar{\gamma} u)}{\log(1-u)}.$$

This value for η must be independent of $u \in (0,1)$, and therefore can hold for only one value of u, say $u = u_0$. Because $\gamma > 0$, it follows that $\eta \geq 0$. Moreover, η ranges from 0 to ∞ as γ does the same. This means that $\bar{F}(x)$ can take on the values 0, 1 and one intermediate value, that is, it can have only two "jump" points. □

C.6. Proposition (Tilt and Hazard Power). With the nondegenerate underlying distribution F, tilt and hazard power parameter families cannot coincide. The families partially coincide if and only if

$$\exp\{-(-\log u)^\zeta\} = \gamma u/(1 - \bar{\gamma} u) \tag{18}$$

for all values of u such that $u = \bar{F}(x)$ for some x, and for some

$$\gamma = \gamma(\zeta) > 0 \quad \text{defined for all } \zeta > 0,$$

and for some

$$\zeta = \zeta(\gamma) > 0 \quad \text{defined for all } \gamma > 0.$$

It follows that F places positive mass on at most two points, in which case the hazard power parameter family is a proper subfamily of the tilt parameter family.

Proof. Trivially, (18) is satisfied for $u = 0$ and $u = 1$; this means that the families coincide when the underlying distribution is degenerate. If $u \neq 0$ and $u \neq 1$, then from (18) it follows that

$$\gamma(\zeta) = \frac{\exp\{-(-\log u)^\zeta\}}{1 - \exp\{-(-\log y)^\zeta\}} \left(\frac{1-u}{u}\right) \tag{19}$$

and

$$\zeta(\gamma) = \frac{\log\{-\log[\gamma u/(1-\bar{\gamma}u)]\}}{\log(-\log u)}. \tag{20}$$

These quantities must be independent of u, and consequently can hold only for one value of $u \in (0,1)$, say $u = u_0$. This means that $\bar{F}(x)$ can take on only the values 0, u_0, and 1. If $u_0 = e^{-1}$, the introduction of a hazard power parameter ζ does not change the distribution (see Proposition 7.M.6), which corresponds to $\gamma = 1$. If $u_0 < e^{-1}$, then the family generated by the introduction of a hazard power parameter is the same family generated by the introduction of a tilt parameter restricted to the interval

$$0 \leq \gamma \leq \left(\frac{e^{-1}}{1-e^{-1}}\right)\left(\frac{1-u_0}{u_0}\right).$$

If $u_0 > e^{-1}$, then the family generated by the introduction of a hazard power parameter is the same family generated by the introduction of a tilt parameter restricted to the interval

$$\gamma \geq \left(\frac{e^{-1}}{1-e^{-1}}\right)\left(\frac{1-u_0}{u_0}\right).$$

For values of γ outside of the indicated intervals, (20) gives a negative value for ζ, and this is not possible. $\square$

C.7. Proposition (Moment and Laplace Transform). With the underlying distribution F, moment parameter and Laplace transform parameter families coincide if and only if

$$\int_t^\infty \frac{x^\beta}{\mu_\beta} \, dF(x) = \int_t^\infty \frac{e^{-sx}}{\phi(s)} \, dF(x) \tag{21}$$

for some $\beta = \beta(s)$ and some $s = s(\beta)$. It follows that the underlying distribution places mass on at most two points.

Proof. The integrands of (21) are equal, that is,

$$\frac{x^\beta}{\mu_\beta} = \frac{e^{-sx}}{\phi(s)} \tag{21a}$$

if and only if

$$\beta \log x + sx - \log \mu_\beta + \log \phi(s) = 0.$$

Because the left-hand side of this equation is a concave function of x, (21a) can have at most two solutions. Let

$$A = \{x : x^\beta/\mu_\beta > e^{-st}/\phi(s)\}, \quad B = \{x : x^\beta/\mu_\beta < e^{-st}/\phi(s)\}.$$

Because of (21),

$$\int_A \left[\frac{x^\beta}{\mu_\beta} - \frac{e^{-st}}{\phi(s)}\right] dF(x) = \int_B \left[\frac{x^\beta}{\mu_\beta} - \frac{e^{-st}}{\phi(s)}\right] dF(x) = 0,$$

and this means that the support of F is confined to the set where the integrands of (21) are equal.

Now, suppose that the support of F has at most two points; say F puts mass p on the point a and mass $\bar{p} = 1 - p$ on the point b, where $a < b$. Then, (21) holds for all $t < a$ and all $t \geq b$, so consider the case that $a \leq t < b$. Here, (21) can be rewritten in the form

$$\frac{pa^\beta}{pa^\beta + \bar{p}b^\beta} = \frac{pe^{-sa}}{pe^{-sa} + \bar{p}e^{-sb}}$$

so that $(a/b)^\beta = e^{s(b-a)}$. From this, the functions $\beta = \beta(s)$ and $s = s(\beta)$ can be obtained. Here the range of both parameters is $(-\infty, \infty)$. □

b. Degenerate Distributions

The remainder of the coincident results given here may be of less interest than those above because they lead only to degenerate distributions. Here, the "families" generated consist of nothing more the underlying distribution (see Section 7.P). Nevertheless, these results are not always easy to obtain, in part because the distributions need not have densities. They can be interpreted as saying that the families never coincide in a meaningful way.

C.8. Proposition (Scale and Power). If scale and power parameter families coincide, the underlying distribution must be degenerate at 0.

Proof. First, recall that a power parameter cannot be introduced unless the underlying random variable is nonnegative. If, with the underlying distribution F, scale and power families coincide, then $\bar{F}(\lambda x) = \bar{F}(x^{\alpha(\lambda)})$, $x, \lambda > 0$ for some $\alpha = \alpha(\lambda) > 0$. With $x = 1$, it follows that $\bar{F}(\lambda) = \bar{F}(1), \lambda > 0$, and if F is a proper distribution, this means $\bar{F}(\lambda) = \bar{F}(1) = 0, \lambda > 0$. □

C.9. Proposition (Resilience and Hazard Power). With the underlying distribution F, resilience parameter and hazard power parameter families coincide if and only if the underlying distribution is degenerate.

Proof. For notational simplicity, let $\bar{F}(x) = u$. The introduction of a resilience and a hazard power parameter leads to the same family of distributions if and only if

$$\exp\{-(-\log u)^\zeta\} = 1 - (1-u)^\eta \tag{22}$$

for some

$$\eta = \eta(\zeta) > 0 \quad \text{defined for all } \zeta > 0 \tag{23}$$

and some

$$\zeta = \zeta(\eta) > 0 \quad \text{defined for all } \eta > 0. \tag{24}$$

It follows from (22) that either $u = 0, u = 1$, or

$$\eta = \eta(\zeta) = \frac{\log\left[1 - \exp\{-(-\log u)^\zeta\}\right]}{\log(1-u)}$$

and

$$\zeta = \zeta(\eta) = \frac{\log\left[-\log\left[1 - (1-u)^\eta\right]\right]}{\log(-\log u)}.$$

There is no value of u in $(0, 1)$ for which ζ is positive for all η and consequently the range of η must be restricted (to an interval depending upon whether $u > e^{-1}$ or $u < e^{-1}$). But then, η does not take on all values in $(0, \infty)$. Consequently, neither family is a subset of the other. Although for certain values of the parameters, the generated families may have common members when the underlying distribution puts positive mass at two points, the families coincide only when the underlying distribution is degenerate. □

C.10. Proposition (Tilt and Laplace Transform). With the underlying distribution F, tilt parameter and Laplace transform parameter families coincide if and only if F is degenerate.

Proof. The introduction of these two parameters leads to the same family if and only if

$$\frac{1}{\phi(s)} \int_x^\infty e^{-sz} \, dF(z) = \frac{\gamma \bar{F}(x)}{1 - \tilde{\gamma} \bar{F}(x)} \quad \text{for all } x \tag{25}$$

and some $\gamma = \gamma(s)$. Consequently, for every $\Delta > 0$,

$$\frac{1}{\phi(s)} \int_x^{x+\Delta} e^{-sz} \, dF(z) = \frac{\gamma \bar{F}(x)}{1 - \tilde{\gamma} \bar{F}(x)} - \frac{\gamma \bar{F}(x+\Delta)}{1 - \tilde{\gamma} \bar{F}(x+\Delta)}. \tag{26}$$

Assume that $s > 0$. Because the integrand is monotone, the left hand side of (26) is bounded above by

$$\frac{e^{-sx}}{\phi(s)} \int_x^{x+\Delta} dF(z) = \frac{e^{-sx}}{\phi(s)} [\bar{F}(x) - \bar{F}(x+\Delta)],$$

and bounded below by

$$\frac{e^{-s(x+\Delta)}}{\phi(s)} \int_x^{x+\Delta} dF(z) = \frac{e^{-s(x+\Delta)}}{\phi(s)} [\bar{F}(x) - \bar{F}(x+\Delta)].$$

Consequently, with the two fractions on the right side of (26) combined, it follows from (26) that

$$\frac{e^{-s(x+\Delta)}}{\phi(s)} [\bar{F}(x) - \bar{F}(x+\Delta)] \leq \frac{\gamma [\bar{F}(x) - \bar{F}(x+\Delta)]}{[1 - \tilde{\gamma} \bar{F}(x)][1 - \tilde{\gamma} \bar{F}(x+\Delta)]}$$

$$\leq \frac{e^{-sx}}{\phi(s)} [\bar{F}(x) - \bar{F}(x+\Delta)]. \tag{27}$$

Suppose that there exists a sequence $x_1, x_2, x_3, \ldots$ tending to ∞ such that

$$[\bar{F}(x_i) - \bar{F}(x_i + \Delta)] > 0 \quad \text{for all } i. \tag{28}$$

From (28) it follows that

$$\frac{e^{-s(x_i+\Delta)}}{\phi(s)} \leq \frac{\gamma}{[1 - \bar{\gamma}\bar{F}(x_i)][1 - \bar{\gamma}\bar{F}(x_i + \Delta)]} \leq \frac{e^{-sx_i}}{\phi(s)}. \tag{29}$$

Now let $i \to \infty$ and obtain from the right hand inequality of (28) that

$$\gamma(s) \leq 0,$$

a contradiction. Consequently, for every $\Delta > 0$,

$$x_U(\Delta) = \sup\{x : [\bar{F}(x) - \bar{F}(x + \Delta)] > 0\} < \infty.$$

Next, suppose that the sequence $x_1, x_2, x_3, \ldots$ tends to $-\infty$ and satisfies (28). Take limits in (29) focusing on the left-hand inequality to find that $\infty < 1/\gamma(s)$, another contradiction. Consequently, for every $\Delta > 0$,

$$x_L(\Delta) = \inf\{x : [\bar{F}(x - \Delta) - \bar{F}(x)] > 0\} > -\infty.$$

By putting these facts together, it follows that

$$\frac{e^{-s[x_U(\Delta)+\Delta]}}{\phi(s)} \leq \frac{1}{\gamma} \leq \frac{e^{-sx_U(\Delta)}}{\phi(s)}$$

and

$$\frac{e^{-sx_L(\Delta)}}{\phi(s)} \leq \gamma \leq \frac{e^{-s[x_L(\Delta)-\Delta]}}{\phi(s)}.$$

Multiply these inequalities and take square roots to conclude that

$$\exp\left\{-\frac{s}{2}[x_L(\Delta) + x_U(\Delta) + \Delta]\right\} \leq \phi(s) \leq \exp\left\{-\frac{s}{2}[x_L(\Delta) + x_U(\Delta) - \Delta]\right\}. \tag{30}$$

Because $x_L = x_L(\Delta)$ and $x_U = x_U(\Delta)$ must be, respectively, the left and right hand endpoints of the support of F, they are independent of Δ. Consequently, it follows by letting $\Delta \to 0$ in (30) that

$$\phi(s) = \exp\left\{-s\frac{x_L + x_U}{2}\right\}$$

is the Laplace transform of a distribution degenerate at $(x_{\bar{L}} + x_{\bar{U}})/2$. In case $s < 0$, the inequalities above are reversed, but the same conclusion follows. □

C.11. Proposition (Tilt and Moment). With the underlying distribution F, tilt parameter and moment parameter families coincide if and only if F is degenerate.

Proof. The proof of this proposition follows the proof of the preceding proposition, but with

$$\frac{1}{\mu_\beta}\int_x^{x+\Delta} z^\beta\,dF(z) = \frac{\gamma \bar{F}(x)}{1 - \bar{\gamma}\bar{F}(x)} - \frac{\gamma\bar{F}(x+\Delta)}{1 - \bar{\gamma}\bar{F}(x+\Delta)}.$$

in place of (26). The conclusion is that F is degenerate at the geometric mean of x_L and x_U rather than at the arithmetic mean. □

D. Unresolved Coincidences

D.1. Power and Laplace Transform. With the underlying distribution F, power and Laplace transform parameter families coincide if and only if F satisfies

$$\bar{F}(x^\alpha) = \int_x^\infty \frac{e^{-sz}}{\phi(s)}\,dF(z), \qquad (1)$$

for some $s = s(\alpha)$ and some $\alpha = \alpha(s)$.

D.2. Power and Age. With the underlying distribution F, power and age parameter families coincide if and only if the hazard function R satisfies

$$R(x^\alpha) = R(t+x) - R(t), \qquad (2)$$

for some $\alpha = \alpha(t)$ and some $t = t(\alpha)$.

D.3. Resilience and Age. With the underlying distribution F, resilience and age parameter families coincide if and only if F satisfies

$$1 - [F(x)]^\eta = \frac{\bar{F}(x+t)}{\bar{F}(t)}, \qquad (3)$$

for some $\eta = \eta(t)$ and some $t = t(\eta)$.

D.4. Hazard Power and Moment. With the underlying distribution F, hazard power and moment parameter families coincide if and only if F satisfies

$$\exp\{-[R(x)]^\xi\} = \int_x^\infty \frac{z^\beta}{\mu_\beta}\, dF(z), \qquad (4)$$

for some $\xi = \xi(\beta)$ and some $\beta = \beta(\xi)$.

D.5. Hazard Power and Laplace Transform. With the underlying distribution F, hazard power and Laplace transform parameter families coincide if and only if F satisfies

$$\exp\{-[R(x)]^\xi\} = \int_x^\infty \frac{e^{-sz}}{\phi(s)}\, dF(z), \qquad (5)$$

for some $\xi = \xi(s)$ and some $s = s(\xi)$.

D.6. Hazard Power and Age. With the underlying distribution F, hazard power and age parameter families coincide if and only if F satisfies

$$\exp\{-[R(x)]^\xi\} = \frac{\bar{F}(x+t)}{\bar{F}(t)}, \qquad (6)$$

that is,

$$[R(x)]^\zeta = R(t+x) - R(t), \qquad (6a)$$

for some $t = t(\xi)$ and some $\xi = \xi(t)$.

Note the similarity between (6a) and (2).

D.7. Moment and Age. With the underlying distribution F, moment and age parameter families coincide if and only if F satisfies

$$\int_x^\infty \frac{z^\beta}{\mu_\beta} dF(z) = \frac{\bar{F}(t+x)}{\bar{F}(t)}, \tag{7}$$

for some $t = t(\beta)$ and some $\beta = \beta(t)$.

D.8. Laplace Transform and Age. With the underlying distribution F, Laplace transform and age parameter families coincide if and only if F satisfies

$$\int_x^\infty \frac{e^{-sz}}{\phi(s)} dF(z) = \frac{\bar{F}(t+x)}{\bar{F}(t)} \tag{8}$$

for some $t = t(s)$ and some $s = s(t)$.

19
More About Semiparametric Families

Several topics are addressed in this chapter. Criteria for semiparametric families, classification, and derivation of families are discussed. Finally, some orderings generated by semiparametric families are introduced.

A. Introduction: Stability Criteria

The semiparametric families discussed in Chapter 7 all provide methods for generating a parametric family from an underlying baseline survival function. Clearly, the underlying survival function can already have parameters.

Two important criteria for judging semiparametric families are listed in Section 7.A.c. These properties are repeated here.

If the goal is to add a parameter to enrich a family or to generate a family containing a given distribution, then it must be that the underlying distribution is a member of the generated parametric family. Thus, if $\bar{F}$ is the underlying survival function and a frailty parameter is introduced, the survival function $\bar{F}(\cdot \mid \xi) = [\bar{F}(\cdot)]^\xi$ is obtained; the underlying survival function is retrieved by taking $\xi = 1$. The following condition formalizes this requirement.

Criterion 1. The underlying distribution is a member of the parametric family. That is, for some value θ^* of the parameter θ,

$$\bar{H}(\cdot \mid \theta^*, \bar{F}) = \bar{F}(\cdot),$$

where $\bar{H}(\cdot \mid \theta, \bar{F})$ is the survival function of the semiparametric family with the real-valued parameter θ and distribution-valued parameter F

(the underlying distribution). For a note about this possibly confusing notation, see Section 7.A.b.

Criterion 2. Once the semiparametric family is used to add a parameter, its reuse may reparameterize the family, but it should fail to again add a new parameter. That is, if $\bar{H}(\cdot \mid \rho, \bar{F})$ is taken to be the underlying survival function, the result is of the form $\bar{H}(\cdot \mid \theta, \bar{F})$ but with θ replaced by some function h of ρ and θ. More formally,

$$\bar{H}(\cdot \mid \theta, \bar{H}(\cdot \mid \rho, \bar{F})) = \bar{H}(\cdot \mid h(q, \rho), \bar{F})$$

for some function h. This is a kind of *stability* property.

In this chapter, the stability property is investigated so as to gain some understanding of the kinds of parameters (other than those introduced in Chapter 7) that would have the stability property. How extensive is the set of such parameters?

B. Classification of Parameters

The parameters discussed in Chapter 7 can be classified according to their method of introduction, and these classifications are discussed in this section. The extent of the applicability of the various classifications can be determined in some cases by solving a functional equation that arises from the requirement of stability. In particular, functional equations play a role.

a. Functions of a Random Variable

Let X be a random variable with distribution function F, ψ be a function of two real variables, and

$$X_\theta = \psi(X, \theta).$$

Then, the distribution of X_θ has a parameter θ. Of course, the function ψ must have appropriate properties. First, it must be defined on a set $S \times \Theta$ where the set S includes all possible values of X, and Θ is some parameter space. Here, it is assumed that S is an interval $[c, d]$, which may be finite or infinite (in case c or d is not finite, then the endpoint of the interval must be excluded, but this technicality should not be troublesome). For the purposes of this chapter, the case $S = [0, \infty)$ is of primary interest, but [0, 1] and $(-\infty, \infty)$ are also of interest. Moreover, it is assumed that $\psi(x, \theta)$ is strictly increasing in x for each fixed θ, with inverse $\psi^{-1}(\cdot, \theta)$, denoted by $\psi_\theta^{-1}(\cdot)$. Then, in terms of the underlying survival function $\bar{F}$ of X, the survival function of X_θ can be written in the form

$$\bar{F}(x \mid \theta) = \bar{F}(\psi^{-1}(x, \theta)).$$

In order that $\bar{F}(\cdot \mid \theta)$ be a survival function, it is necessary that

$$\lim_{x \downarrow c} \bar{F}(\psi^{-1}(x, \theta)) = 1, \quad \lim_{x \uparrow d} \bar{F}(\psi^{-1}(x, \theta)) = 0.$$

This means that

$$\lim_{x \downarrow c} \psi^{-1}(x, \theta) = c, \quad \lim_{x \uparrow d} \psi^{-1}(x, \theta) = d,$$

which yields the conditions

$$\psi(c, \theta) = c, \quad \psi(d, \theta) = d. \tag{1}$$

Additional conditions are imposed by Criteria 1 and 2.

Criterion 1: In order that F belong to the parametric family of distributions of X_θ, it is sufficient to assume that for some θ_0 in Θ, $\psi(z, \theta_0) = z$.

Criterion 2: To verify the requirements of this stability property, suppose that the function ψ is reapplied to yield

$$Z = \psi(X_\theta, \rho) = \psi(\psi(X, \theta), \rho).$$

If the stability property is to hold, i.e., if Z has a distribution in the same parametric family as X_θ, then it must be that for some function h,

$$\psi(\psi(x, \theta), \rho) = \psi(x, h(\rho, \theta)). \tag{2}$$

The requirement for stability has led to this functional equation, which can now be investigated to determine the constraints it imposes upon the function ψ. With the assumption that ψ is invertible in each variable separately, the general solution to (2) is given in Proposition 22.C.2. That solution has the form

$$\psi(x,\theta) = u^{-1}(u(x) + v(\theta)), \quad h(\rho,\theta) = v^{-1}(v(\rho) + v(\theta)), \tag{3}$$

where u and v are continuous and strictly monotonic functions. Then,

$$\psi^{-1}(x,\theta) = u^{-1}(u(x) - v(\theta)). \tag{4}$$

The requirements (1) lead to the condition

$$u^{-1}(u(c) + v(\theta)) = c, \quad u^{-1}(u(d) + v(\theta)) = d,$$

that is,

$$u(c) + v(\theta) = u(c), \quad u(d) + v(\theta) = u(d).$$

Because $v(\theta)$ is not identically 0, this means that

$$u(c) = \pm\infty \quad \text{and} \quad u(d) = \pm\infty; \tag{5}$$

because u strictly is monotone, it follows from (5) that either

$$u(c) = -\infty \quad \text{and} \quad u(d) = \infty, \quad \text{or} \quad u(c) = \infty \quad \text{and} \quad u(d) = -\infty. \tag{5a}$$

It is always possible to reparameterize the family $\{\bar{F}(\cdot\,|\,\theta), \theta \in \Theta\}$ in such a way that the function h satisfies $h(\rho,\theta) = \rho\theta$. All that is necessary is to take $\theta^* = e^{v(\theta)}$, where v is the function determining h in Proposition 22.C.2. To see this, let $s(\theta) = e^{v(\theta)}$ and note from 22.C(6) that

$$h(\rho,\theta) = v^{-1}(v(\rho) + v(\theta)) = s^{-1}(e^{\log s(\rho) + \log s(\theta)}) = s^{-1}(s(\rho)s(\theta)),$$

so that $s(h(\rho,\theta)) = s(\rho)s(\theta)$. Although this does not always lead to the most natural parameterization of the family, the assumption that $h(\rho,\theta) = \rho\theta$ sometimes simplifies the theory.

In C(13) an alternative representation for $\psi(\cdot,\theta)$ is given for the case $h(\rho,\theta) = \rho\theta$, but with a different notation.

B.1. Location parameter. Let ψ be defined on $(-\infty, \infty) \times (-\infty, \infty)$ by

$$\psi(x, b) = x - b.$$

Then, the parameter b is a location parameter. This example is discussed in Section 7.B.

Here, (2) is satisfied with $h(\rho, \theta) = \rho + \theta$. In the representation (3), one can take $u(x) = -x$, $v(x) = x$, and $b = 0$. The representation (6) is obtained with $g(x) = e^x$ and $b = \log \theta$.

B.2. Scale parameter. Let ψ be defined on $[0, \infty) \times (0, \infty)$ by

$$\psi(x, \lambda) = x/\lambda.$$

Then λ is a scale parameter; this case is discussed in Section 7.C.

Here, (2) is satisfied with $h(\rho, \theta) = \rho\theta$ and in the representation (3) one can take $u(x) = -\log x$ and $v(x) = \log x$. Take $g(x) = x$ and $\lambda = 1/\theta$ to obtain the form (6).

The case that ψ is defined on $(-\infty, \infty) \times (0, \infty)$ is somewhat more complicated because it is not invertible at $x = 0$. This kind of complication is dealt with in B.3.

B.3. Power parameter. Let ψ be defined on $[0, \infty) \times (0, \infty)$ by

$$\psi(x, \alpha) = x^\alpha.$$

Then α is a power parameter; this example is discussed in Section 7.D.

In this case, (2) is satisfied with $h(\rho, \theta) = \rho\theta$, and in the representation (3), one can take $u(x) = \log(-\log x)$, for $0 < x < 1$, $u(x) = \log \log x$, for $x > 1$, and $v(x) = \log x$. The differing forms of u come about because ψ is not invertible when $x = 1$; to apply Proposition 22.C.2, it is necessary to break the problem into the two parts $0 < x < 1$ and $x > 1$.

B.4. Example. Let ψ be defined on $[0, \infty) \times [0, \infty)$ by

$$\psi(x, \theta) = e^{\theta x} - 1.$$

This example fails to satisfy $\psi(x, \theta_0) = x$ for some θ_0 in Θ. Moreover, (2) fails.

b. Functions of a Survival Function

Let $\bar{F}$ be a survival function and $\psi(x, \theta)$ be a function defined on $[0, 1] \times \Theta$. Suppose that for each fixed θ in Θ, ψ has range $[0, 1]$ and is increasing in x. Then,

$$\bar{F}(\cdot \mid \theta) = \psi(\bar{F}(\cdot), \theta)$$

is a survival function with parameter θ. If for some θ_0 in Θ, $\psi(x, \theta_0) = x$, then $\bar{F}$ is a member of the family $\{\bar{F}(\cdot \mid \theta), \theta \in \Theta\}$. In what follows, it is often convenient to use the notation $\bar{F}(\cdot \mid \theta) = \bar{F}_\theta(\cdot)$.

A connection with the uniform distribution. The conditions imposed upon the function ψ are the conditions that $\{\psi(\cdot, \theta), \theta \in \Theta\}$ is a family of distribution functions with support in $[0, 1]$. The condition that $\psi(x, \theta_0) = x$ for some θ_0 in Θ is the case that the uniform distribution belongs to the family.

Now, suppose that the function ψ is reapplied to yield

$$\bar{G} = \psi(\bar{F}_\theta, \rho) = \psi(\psi(\bar{F}, \theta), \rho)).$$

If the stability property is to hold, i.e., if $\bar{G} = \bar{F}_{h(\rho,\theta)}$ for some function h, then it must be that for some function h, (2) is again satisfied. This means that solutions are again provided by (3) and Proposition 22.C.2.

B.5. Frailty parameter. Let ψ be defined on $(0, 1) \times (0, \infty)$ by

$$\psi(x, \xi) = x^\xi.$$

Then, ξ is a frailty parameter. This parameter is discussed in Section 7.E. Note that ψ is a uniform distribution function with resilience parameter.

Here, (2) is satisfied by $h(\rho, \theta) = \rho\theta$ and the representation (3) is the same as in B.3, except that the case $x > 1$ does not arise.

B.6. Resilience parameter. Let ψ be defined on $[0, 1] \times (0, \infty)$ by

$$\psi(x, \eta) = 1 - (1 - x)^\eta.$$

Then η is a resilience parameter; this parameter is discussed in Section 7.E in conjunction with frailty parameters. Note that ψ is a uniform distribution function with frailty parameter.

Here, (2) is satisfied by $h(\rho, \theta) = \rho\theta$, and for the representation (3), $u(x) = -\log(-\log(1-x))$ and $v(x) = \log x$.

B.7. Tilt parameter. Let ψ be defined on $(0,1) \times (0, \infty)$ by

$$\psi(x, \gamma) = \frac{\gamma x}{1 - \bar{\gamma} x},$$

where $\bar{\gamma} = 1 - \gamma$ and the parameter γ is the tilt parameter discussed in Section 7.F.

Here, (2) is satisfied with $h(\rho, \theta) = \rho\theta$, and in the representation (3), $u(x) = \log(x/(1-x))$ and $v(x) = \log x$.

B.8. Hazard power parameter. Let ψ be defined on $(0,1) \times (0, \infty)$ by

$$\psi(x, \zeta) = \exp\{-(-\log x)^\zeta\}.$$

The parameter ζ is a hazard power parameter, but as cautioned in Chapter 7, take care not to confuse it with a power parameter. This parameter, discussed in Section 7.G, satisfies (2), again with $h(\rho, \theta) = \rho\theta$.

The representation of ψ from equation (3) is complicated by the fact that this ψ fails to be invertible when $\log x = 1$; it is necessary to consider separately the cases $\log x < 1$ and $\log x > 1$. For $\log x < 1$, one can take $u(x) = \log(-\log(-\log x))$, and for $\log x > 1$, $u(x) = \log\log(-\log x)$. In either case, $v(x) = \log x$.

Duality. If H is a distribution function supported by $[0, 1]$, the *dual* of H is defined by

$$H_D(x) = 1 - H(1 - x).$$

See Section 14.A.a for further discussions of duality. As noted above, semiparametric families of the form $\{\bar{F}_\theta = H(\bar{F}, \theta)\}$ are just compositions of a distribution function H supported by $[0, 1]$ with an underlying survival function. This suggests consideration also of the composition using H_D in place of H. Frailty and resilience families are connected in this way by duality. The dual of the tilt parameter family only reparameterizes the family and so gives nothing new. The dual of the hazard power family might be called a reverse hazard power family; it has received little attention in this book or elsewhere.

c. Density Modification

Suppose that F is a distribution function with support $[0, \infty)$ and density f. Let ψ be a nonnegative function defined on $(0, \infty) \times \Theta$, where Θ is a parameter space, ordinarily an interval. If

$$c(\theta) = \left[\int_0^\infty \psi(x,\theta) f(x)\, dx\right]^{-1} < \infty,$$

then $c(\theta)\psi(x,\theta)f(x)$ is a density, with parameter θ. If there exists a function h defined on $\Theta \times \Theta$ such that

$$\psi(x,\rho)\psi(x,\theta) = \psi(x, h(\rho,\theta)), \tag{6}$$

then the stability property (Criterion 2) holds; reintroduction of the parameter does not enlarge the parametric family.

It is possible to introduce density modification parameters even when F does not have a density. This is done by defining

$$c(\theta) = \left[\int_0^\infty \psi(x,\theta)\, dF(x)\right]^{-1} < \infty,$$

and by defining

$$F_\theta(x) = c(\theta) \int_0^x \psi(z,\theta)\, dF(z) < \infty.$$

Equation (6) is the functional equation 22.C(7); under reasonable regularity conditions, ψ must have the form $\psi(z,\theta) = [\phi(z)]^{g(\theta)}$, where g is monotonic. So with a change of parameterization, the only possibilities are of the form $\psi(z,\theta) = [\phi(z)]^\theta$. Two such cases are considered.

B.9. Moment parameter. Let ψ be defined on the support of $F \times \Theta$ by $\psi(z,\beta) = z^\beta$. Then, $c^{-1} = \mu_\beta$ is the βth moment of F, and the parameter space Θ (which depends upon F) is the set for which μ_β is finite. Clearly, (6) is satisfied with $h(u,v) = u+v$.

B.10. Laplace transform parameter. Let ψ be defined on the support of $F \times \Theta$ by $\psi(z,s) = e^{-sz}$. As in B.9, Θ depends upon F, but always $[0,\infty) \subset \Theta$. Here, c^{-1} is the Laplace transform of F, hence the name for this parameter.

d. Functions of a Laplace Transform

A nonnegative random variable X with distribution function F has a Laplace transform $\phi(s) = E\, e^{-sX}$ that is finite at least for nonnegative s. Under some conditions, it is possible to introduce a parameter in a manner similar to that used in B.1 and B.2 by taking a function ψ of ϕ and a parameter. However, for this procedure to be valid, the resulting functional value must again be a Laplace transform. This may depend upon F, as in the following special case.

B.11. Convolution parameter. Let ψ be defined on $(0, \infty) \times (0, \infty)$ by

$$\psi(x, \nu) = x^\nu.$$

If F is infinitely divisible (see Definition 20.D.7), then $\psi(\phi(s), \nu)$ defines a Laplace transform for all $\nu > 0$. The corresponding distribution function has the convolution parameter ν. Convolution parameters are discussed in Section 7.J. These parameters have the stability property because (1) is satisfied with $\psi(x, \nu) = x^\nu$.

A convolution parameter can be introduced without infinite divisibility, but because the parameter must be restricted to positive integer values, it is of limited interest.

B.12. Age parameters. Age parameters do not arise as an application of any of the preceding methods. Moreover, they do not seem to suggest another class of parameters.

C. Derivation of Families

In this section, a more formal approach to semiparametric families is considered using notation different from that of the previous section. Nevertheless, the notation and results are closely related to those of that section. The results are in Section D.

Suppose that S is the class of all distribution functions with support I, where typically $I = (-\infty, \infty)$ or $[0, \infty)$, and let

$$\mathcal{F} = \{\bar{H}(\cdot \mid \theta, \bar{F}) : F \in S, \theta \in \Theta\} \tag{1}$$

be a semiparametric family of survival functions with support I, with a real parameter θ and also an underlying survival function $\bar{F}$ as a parameter. For some function Φ, survival functions in $\mathcal{F}$ can be written in the form

$$\bar{H}(\cdot \mid \theta, \bar{F}) = \Phi_\theta \bar{F}(\cdot);$$

the set $\Phi = \{\Phi_\theta, \theta \in \Theta\}$ can be regarded as a set of operators mapping S to S. Several properties of families (1) or the set Φ are of interest.

In accordance with criterion 1 of Section A, the family $\{\bar{H}(\cdot \mid \theta, \bar{F}) : \theta \in \Theta\}$ is intended to represent a parametric extension of the survival function $\bar{F}$ (which may already have parameters) to a parametric family with a new parameter θ; thus it is natural to require that for some θ_0, $\Phi_{\theta_0} \bar{F} = \bar{F}$. Then, the survival function $\bar{F}$ is in the family and Criterion 1 is satisfied. To simplify notation, write $\theta_0 = \iota$. This leads to the property

$$\text{There exists } \iota \in \Theta \text{ such that } \Phi_\iota \bar{F} = \bar{F} \text{ for all } \bar{F} \in S. \qquad (2)$$

This property says that Φ has an identity operator Φ_ι. Note that this property is stronger than Criterion 1 of Section A, because here, θ_0 is independent of F.

Criterion 2 of Section A, the stability property, requires that Φ be closed under the binary operation of composition; i.e.,

$$\Phi_\rho \Phi_\theta \bar{F} = \Phi_{h(\rho,\theta)} \bar{F} \qquad (3)$$

for some function h mapping $\Theta \times \Theta$ onto Θ (that might depend upon F). Suppose further that

$$\text{for every } \theta \in \Theta, \text{ there exists a } \theta^* \text{ such that}$$
$$\Phi_{\theta^*} \Phi_\theta \bar{F} = \Phi_\iota \bar{F} \text{ for all } \bar{F} \in S. \qquad (4)$$

Here, Φ_{θ^*} is an inverse of Φ_θ; with property (4), $\Phi = \{\Phi_\theta, \theta \in \Theta\}$ forms a group with respect to the operation of composition. If in addition,

$$\Phi_\theta \Phi_\rho \bar{F} = \Phi_\rho \Phi_\theta \bar{F} \quad \text{for all} \quad \bar{F} \in S, \qquad (5)$$

then the group is commutative.

C.1. Proposition. Suppose that $\Phi = \{\Phi_\theta, \theta \in \Theta\}$ is a commutative group under the operation of composition. If for some $\bar{F} \in S, \Phi_\rho \bar{F} = \Phi_\theta \bar{F}$ implies $\rho = \theta$, then the function h of (3) is independent of F.

Proof. Let $\bar{F}_\xi = \Phi_\xi \bar{F}$, let h satisfy (3), and let h_ξ satisfy $\Phi_\rho \Phi_\theta \bar{F}_\xi = \Phi_{h_\xi(\rho,\theta)} \bar{F}_\xi$. Then,

$$\Phi_\xi \Phi_{h_\xi(\rho,\theta)} \bar{F} = \Phi_{h_\xi(\rho,\theta)} \Phi_\xi \bar{F} = \Phi_{h_\xi(\rho,\theta)} \bar{F}_\xi = \Phi_\rho \Phi_\theta \bar{F}_\xi$$
$$= \Phi_\xi \Phi_\rho \Phi_\theta \bar{F} = \Phi_\xi \Phi_{h(\rho,\theta)} \bar{F}.$$

Now, apply Φ_{ξ^*} to both sides of the equation to conclude that $h_\xi = h$. □

C.2. Proposition. If $\Phi = \{\Phi_\theta, \theta \in \Theta\}$ is a commutative group and $\bar{F}_\xi = \Phi_\xi \bar{F} = \bar{F}(\cdot \mid \xi, \bar{F})$ is any fixed member of the family generated from Φ with $\bar{F}$ as the underlying survival function, then the same family is generated from Φ with $\bar{F}_\xi$ as the underlying survival function; i.e.,

$$\{\Phi_\theta \bar{F}, \theta \in \Theta\} = \{\Phi_\theta \bar{F}_\xi, \theta \in \Theta\}.$$

Proof. Denote the inverse of Φ_ξ by Φ_{ξ^*}, so that $\Phi_{\xi^*} \Phi_\xi = \Phi_\iota$ is the identity. It follows from Proposition C.1 that $\Phi_\theta \Phi_{\xi^*} = \Phi_{h(\theta,\xi^*)}$ for some $h(\theta, \xi^*)$ in Θ. Thus,

$$\Phi_\theta \bar{F} = \Phi_\theta \Phi_{\xi^*} \Phi_\xi \bar{F} = \Phi_\theta \Phi_{\xi^*} \bar{F}_\xi = \Phi_{h(\theta,\xi^*)} \bar{F}_\xi.$$

This says that an arbitrary member of $\{\Phi_\theta \bar{F}, \theta \in \Theta\}$ is also a member of $\{\Phi_\theta \bar{F}_\xi, \theta \in \Theta\}$. □

Under the conditions of Proposition C.2, families of the form $\{\Phi_\theta \bar{F}, \theta \in \Theta\}$ generated with a single underlying distribution F form a partition of the set of all distributions for which $\Phi_\theta \bar{F}$ is defined into a set of equivalence classes.

C.3.a. Example. Of the families considered in Chapter 7, those that introduce scale, power, frailty, resilience, hazard, power, and convolution parameters all form commutative groups, with $I = [0, \infty)$. In these cases, $\Theta = (0, \infty)$ and $\iota = 1$ is the index of the identity; the inverse Φ_{θ^*} of Φ_θ is obtained with $\theta^* = 1/\theta$. Power parameter families are an exceptional case because they are defined only for underlying distributions having support contained in $[0, \infty)$.

C.3.b. Example. The families considered in Chapter 7 that introduce moment or Laplace transform parameters do not form groups, but only commutative semigroups. These families do not really fit the above format because the set $\Theta = \Theta(F)$ of θ for which $\bar{F}(\cdot \mid \theta, \bar{F})$ is defined depends upon $\bar{F}$. For moment parameter families, $0 \in \Theta$ for all F, and 0 is the identity. For Laplace transform parameter families with underlying distribution having support in $[0, \infty)$, $[0, \infty) \subset \Theta$ and again, 0 is the identity.

Any given member $\bar{F}(\cdot \mid \theta_0, \bar{F})$ of a moment parameter family, when used as an underlying distribution to generate a new moment parameter family, will generate exactly the family from which it came. Consequently, it might seem that such a family forms a group in the sense defined above, but there is a problem; the parameter of the new family which retrieves $\bar{F}$ may not be a member of $\Theta = \Theta(F)$. Similar comments apply to Laplace transform parameter families.

C.3.c. Example. For age parameter families, $\Theta = [0, \infty)$ and 0 is the identity. Here, inverses do not exist; given $\bar{F}(\cdot \mid \theta_0, \bar{F})$ for $\theta_0 > 0$, none of the distributions $\bar{F}(\cdot \mid \theta, \bar{F})$ for $\theta < \theta_0$ can be retrieved.

C.4. Remark. There are interesting examples for which the parameter θ can take on any nonnegative value including 0, and then it may be natural to take Φ_0 to be the identity operator rather than Φ_1 as in Example C.3.a, where scale, power, frailty, resilience, hazard power, and convolution parameters are discussed. A noteworthy example where it is natural to take Φ_0 to be the identity operator is the family of residual life distributions obtained by introducing an age parameter.

It is desirable for some purposes that the operators Φ_θ be monotone in the sense that

$$\bar{F}(x) \leq \bar{G}(x) \text{ for all } x \quad \Rightarrow \quad (\Phi_\theta \bar{F})(x) \leq (\Phi_\theta \bar{G})(x) \text{ for all } \theta > 0. \quad (6)$$

When F and G have support $[0, \infty)$, with the exception of Laplace transform and age parameter families, this is the case for all the parameters introduced in Chapter 7. For Laplace transform parameter families, the second inequality in (6) is reversed.

a. Stability

To formalize the notion of stability introduced at the beginning of this chapter as Criterion 2, a definition is useful.

C. Derivation of Families

C.5. Definition. A semiparametric family $\{\bar{H}(\cdot \mid \theta, \bar{F}) : \theta \in \Theta\} = \{\Phi_\theta \bar{F}(\cdot) : \theta \in \Theta\}$ is *stable* if

$$\{\Phi_\theta \bar{F}(\cdot) : \theta \in \Theta\} = \{\Phi_\rho \Phi_\theta \bar{F}(\cdot) : \theta, \rho \in \Theta\}. \tag{7}$$

The set $\{\Phi_\theta : \theta \in \Theta\}$ is stable with respect to S if (7) holds for all $F \in S$.

If $\{\Phi_\theta : \theta \in \Theta\}$ is a semigroup under composition, then for a fixed underlying survival function $\bar{F}$, the family $\{\Phi_\theta \bar{F}(\cdot) : \theta \in \Theta\}$ is stable if and only if for some function h mapping $\Theta \times \Theta$ to Θ, (3) holds, that is, $\Phi_\rho \Phi_\theta \bar{F} = \Phi_{h(\rho,\theta)} \bar{F}$.

C.6. Proposition. Suppose that $\{\Phi_\theta : \theta \in \Theta\}$ is a commutative group, (2), (3), and (4) are satisfied, and Θ is an interval (open, half-open, or closed). If $h(\rho, \theta)$ is continuous in ρ for fixed θ and continuous in θ for fixed ρ, then for some function g with domain $(-\infty, \infty)$ and range Θ,

$$h(\rho, \theta) = g(g^{-1}(\rho) + g^{-1}(\theta)). \tag{8}$$

Proof. By applying Φ_ξ to both sides of (4), it follows that

$$h(\xi, h(\rho, \theta)) = h(h(\xi, \rho), \theta). \tag{9}$$

Equation (9) is a well-studied functional equation called the "associativity equation." Because of (2), (3), and the continuity conditions, Proposition 22.C.1 can be applied to conclude that (8) holds. □

A prime example is the case $g(x) = e^x$, in which case (8) becomes

$$h(\rho, \theta) = \rho\theta. \tag{10}$$

In what follows, it is assumed that (10) holds.

The functional equation (3) with h given by (10) is solved in the two special cases described in Section B.a and B.b. Results of Section B can all be recast in the notation of this section.

b. Transformations of the Underlying Random Variable

Suppose that for some function ϕ_θ

$$\Phi_\theta \bar{F}(x) = \bar{F}(\phi_\theta(x)), \quad \theta \in \Theta, \quad F \in S. \tag{11}$$

Note that $\phi_\theta(x) = \psi^{-1}(x,\theta)$, the cumbersome notation of Section B.a. In order that $\bar{F}(\phi_\theta(\cdot))$ be a survival function, ϕ_θ must be an increasing function, with range $[0,\infty)$ when S consists of distribution functions having support $[0,\infty)$. Here, familiar examples include the scale and power parameter families, obtained, respectively, with $\phi_\theta(x) = \theta x$ and $\phi_\theta(x) = x^\theta$.

With (10) and (11), the functional equation (3) becomes

$$\phi_\rho \phi_\theta(x) = \phi_{\rho\theta}(x), \quad x > 0. \tag{12}$$

Equation (12) is a form of a well-known functional equation called the "translation equation" discussed in Proposition 22.C.2, but standard techniques can be applied to directly solve (12).

C.7. Proposition. Suppose that (12) holds for all $\rho, \theta > 0$, and that ϕ_θ has range $[0,\infty)$. Suppose also that $\phi_\theta(x)$ is strictly increasing in θ for some $x = x_0 > 0$. Then, there is a strictly increasing function g defined on some interval I with range $(0,\infty)$ such that

$$\phi_\theta(x) = g(\theta g^{-1}(x)), \quad \theta, x > 0. \tag{13}$$

Proof. Let $g(\theta) = \phi_\theta(x_0)$. With $x = x_0$ in (12), it follows that $\phi_\rho(g(\theta)) = g(\rho\theta)$; but g is strictly increasing, so it has an inverse and (13) follows. □

The function g of this proof is defined on $(0,\infty)$, but functions defined on other intervals can also be used in (13). The representation (13) is not unique; different functions g can lead to the same ϕ_θ. In fact, $g_1(\theta g_1^{-1}(x)) = g_2(\theta g_2^{-1}(x)), x, \theta > 0$, if and only if the composition $g_2^{-1} g_1$ is homogeneous, i.e.,

$$g_2^{-1} g_1(\theta x) = \theta g_2^{-1} g_1(x). \tag{14}$$

Condition (14) insures that the families generated through (13) with g_1 and g_2 are the same. Perhaps more interesting is the requirement that the families $(\Phi_\theta \bar{F})(\cdot), \theta > 0$, generated by g_1 and g_2 be the same. This requirement leads to a condition weaker than (14), namely,

$$g_2^{-1} g_1(\theta x) = a(\theta) g_2^{-1} g_1(x) \tag{15}$$

for some monotone transformation a of the parameter space.

The lack of uniqueness in the representation (13) makes it difficult to solve (13) for g when ϕ_θ is given. However, the lack of uniqueness makes it easier to find a g that satisfies the equation.

C.8.a. Example (Scale parameter families). Let $g(x) = x, x > 0$. Here $\phi_\theta(x) = g(\theta g^{-1}(x)) = \theta x$, and the parameter introduced is a scale parameter. Any function of the form $g(x) = cx, x > 0, c > 0$, would work just as well. But more generally, (15) is satisfied if g_1 and g_2 are powers of x.

C.8.b. Example (Power parameter families). Let $g(x) = e^x$, $-\infty < x < \infty$. Then, $\phi_\theta(x) = g(\theta g^{-1}(x)) = x^\theta$, and the parameter introduced is a power parameter.

C.8.c. Example. If $g(x) = e^x - 1, x > 0$, then $\phi_\theta(x) = (x+1)^\theta - 1$, a variant of Example C.8.b.

C.8.d. Example. With the rather simple function $g(x) = \log(1+x)$, the relatively complicated function $\phi_\theta(x) = \log[\theta(e^x - 1) + 1]$ is obtained. With the underlying baseline survival function $\bar{F}(x) = e^{-\lambda x}, x > 0$, this example leads to the survival function

$$\bar{F}_\theta(x) = [\theta e^x - \bar\theta]^{-\lambda}, \quad x > 0.$$

This distribution is an exponential distribution with tilt parameter $\theta = 1/\gamma$, added frailty parameter λ, and unit scale parameter. The distribution has a monotone hazard rate, increasing from $\lambda\theta$ to λ if $\theta < 1$, and decreasing from $\lambda\theta$ to λ if $\theta > 1$ (see Section 9.D, where the case with unit frailty parameter is discussed in more detail).

Note that a different distribution is obtained if, starting with an exponential distribution, a frailty parameter is added before the tilt parameter.

c. Transformations of the Survival Function

As in Section B.b, suppose that with the notation $\psi_\theta(\cdot)$ in place of $\psi(\cdot, \theta)$

$$\Phi_\theta \bar{F}(x) = \psi_\theta(\bar{F}(x)), \quad \theta, x > 0. \tag{16}$$

Here the functions ψ_θ map $[0, 1]$ onto $[0, 1]$. The most familiar example of this form of transformation is $\psi_\theta(x) = x^\theta, \theta > 0, 0 \leq x \leq 1$, which leads to a frailty parameter, and the family obtained is a proportional hazards family.

In this case, the functional equation (3) again becomes (12), but with the modification that the domain and range of ψ_θ are both $[0, 1]$.

C.9. Proposition. Suppose that (12) holds for all $p, \theta > 0$, and x in the interval $(0, 1)$. Suppose also that $\psi_\theta(x)$ is strictly increasing in θ for some $x = x_0 > 0$. Then, there is a strictly increasing function g defined either on $(-\infty, \infty), (-\infty, 0)$ or $[0, \infty)$ with corresponding range $(0, 1)$, $(0, 1]$, or $[0, 1)$ such that

$$\psi_\theta(x) = g(\theta g^{-1}(x)), \quad \theta > 0, \ 0 \leq x \leq 1. \tag{17}$$

The proof of Proposition C.9 is essentially the same as the proof of Proposition C.7. Note that (17) looks the same as (13), but the range of x is different.

C.10.a. Example. Let $g(x) = e^x, -\infty < x \leq 0$. Then, $\psi_\theta(x) = g(\theta g^{-1}(x)) = x^\theta$, so that $\bar{F}_\theta(x) = [\bar{F}(x)]^\theta$ and the parameter introduced is a frailty parameter.

C.10.b. Example. If $g(x) = 1 - e^{-x}, x \geq 0$, it follows that $\psi_\theta(x) = 1 - (1-x)^\theta$, so that $F_\theta(x) = [F(x)]^\theta$ and θ is a resilience parameter.

C.10.c. Example. If $g(x) = x/(1+x), x \geq 0$, it follows that $\psi_\theta(x) = \theta x/[1 - \bar{\theta} x]$ and $\bar{F}_\theta(x) = \theta \bar{F}(x)/[1 - \bar{\theta}\bar{F}(x)]$. This is the geometric-extreme stable extension of F, i.e., the family obtained from F by introducing a tilt parameter.

C.10.d. Example. If $g(x) = \exp\{-e^{-x}\}, -\infty < x < \infty$, then $\psi_\theta(x) = \exp\{-(-\log x)^\theta\}, \bar{F}_\theta(x) = \exp\{-[R(x)]^\theta\}$, and a hazard exponent parameter has been introduced.

D. Orderings Generated by Semiparametric Families

There are a number of properties of real functions ψ of a real variable that can be defined in terms of crossings of a "grid." In its most general form, a grid is a collection $\mathcal{G}$ of functions with the same domain as ψ. The common domain ordinarily is an interval, and the property is defined by conditions on the manner in which ψ crosses functions in $\mathcal{G}$. Here are some familiar examples.

D.1.a. Example. Suppose that $\mathcal{G}$ consists of all the constant functions. Then, ψ crosses functions in $\mathcal{G}$ only from below (above) if and only if ψ is increasing (decreasing).

D.1.b. Example. Suppose that $\mathcal{G}$ consists of all linear functions. Then, the following conditions are equivalent:

(i) ψ crosses functions in $\mathcal{G}$ at most twice. If there are two crossings, ψ crosses first from above and then from below and

(ii) ψ is convex.

D.1.c. Example. Suppose that $\mathcal{G}$ consists of the survival functions of exponential distributions, and ψ is a survival function. Then, the condition that ψ crosses functions in $\mathcal{G}$ only from above is the condition that ψ has an increasing hazard rate average (see Section 5.B).

Semiparametric families can be used to define grids, and under appropriate conditions, the properties they define can be extended to define orderings of distributions. For example, suppose that the grid consists of the survival functions $\{\Phi_\theta \bar{H} : \theta \in \Theta\}$ and that $\bar{G}$ crosses such survival functions only from above. Now, consider a second grid, one consisting of the survival functions in $\{\Phi_\theta \bar{G} : \theta \in \Theta\}$, and suppose that $\bar{F}$ crosses such survival functions only from above. Under appropriate conditions, this implies that $\bar{F}$ crosses functions in $\{\Phi_\theta \bar{H} : \theta \in \Theta\}$ only from above. Then there is a basis for defining an ordering based upon the semiparametric family $\{\Phi_\theta(\cdot) : \theta \in \Theta\}$; say $\bar{F} \leq \bar{G}$ if $\bar{F}$ crosses functions in $\{\Phi_\theta \bar{G} : \theta \in \Theta\}$ only from above. Of course, a key ingredient is transitivity.

It is apparent from (iv) of Proposition 2.C.11 that the the star ordering (Definition 2.C.10) is a grid ordering. Similarly, the convex ordering (Definition 2.C.7) can be regarded as a grid ordering as is apparent from (iii) of Proposition 2.C.8, though this involves not just a single crossing of the grid, but two crossings, as in Example D.1.b.

The discussion here is limited to orderings based upon a single crossing; most of these orderings have not been previously considered.

A function g is said to *cross the function h at most once, and only from above* if either $g - h$ is of one sign (they do not cross at all) or if there exists x_0 such that

$$g(x) \geq h(x), \quad x \leq x_0 \quad \text{and} \quad g(x) \leq h(x), \quad x \geq x_0. \tag{1}$$

A convenient shorthand notation, that takes the form

$$g(x) - h(x) : +, -, \tag{2}$$

is discussed in Section 21.B.g.

D.2. Definition. For a fixed set $\{\Phi_\theta, \theta \in \Theta\}$ of operators defining a semiparametric family as in Section C, write $\bar{F} \leq \bar{G}$ if for all $\theta \in \Theta, \bar{F}$ crosses $\Phi_\theta \bar{G}$ at most once, and only from above.

D.3. Proposition. Suppose that $\{\Phi_\theta, \theta \in \Theta\}$ forms a group under composition and that $\Theta \subset (0, \infty)$. Let $I \subset (0, \infty)$ be an interval and suppose further that for all distributions F and G with support I,

$$\theta < \rho \implies \Phi_\theta \bar{F}(x) \leq \Phi_\rho \bar{F}(x) \quad \text{for all } x \in I, \tag{3}$$

$$\bar{F}(x) - \bar{G}(x) : +, - \text{ implies } \Phi_\theta \bar{F}(x) - \Phi_\theta \bar{G}(x) : +, - \text{ for all } \theta \in \Theta. \tag{4}$$

Suppose further that for all p in $(0, 1)$, all continuous distribution functions F with support I and all $x_0 \in I$, there exists θ_0 such that

$$\Phi_{\theta_0} \bar{F}(x_0) = p. \tag{5}$$

Then, restricted to survival functions $\bar{F}$ and $\bar{G}$ that are continuous and have support I, the ordering of Definition D.2 is reflexive and transitive, i.e., $\bar{F} \leq \bar{F}$ and $\bar{F} \leq \bar{G} \leq \bar{H}$ implies $\bar{F} \leq \bar{H}$.

Proof. First, consider reflexivity. Let Φ_ι be the identity of the group $\{\Phi_\theta, \theta \in \Theta\}$. Because of (3), $\theta > \iota$ implies $\bar{F}(x) \leq \Phi_\theta \bar{F}(x), x \in I$, and the inequality is reversed for $\theta < \iota$. Thus, the survival functions do not cross, so almost vacuously, $\bar{F} \leq \bar{F}$.

To show reflexivity, suppose that for some $x_0 \in I, \bar{F}(x_0) = \Phi_{\theta_0} \bar{H}(x_0)$. Then because of (5), there exists θ_0 such that $\bar{F}(x_0) = \Phi_{\theta_0} \bar{G}(x_0)$. Because $\bar{F} \leq \bar{G}$ and $\bar{G} \leq \bar{H}$, it follows with the aid of (5) that

$$\bar{F}(x) \geq \Phi_{\theta_0} \bar{G}(x), \quad x < x_0; \quad \bar{F}(x) \leq \Phi_{\theta_0} \bar{G}(x), \quad x > x_0, \tag{6}$$

and

$$\Phi_{\theta_0} \bar{G}(x) \geq \Phi_\theta \bar{H}(x), \quad x < x_0; \quad \Phi_{\theta_0} \bar{G}(x) \leq \Phi_\theta \bar{H}(x), \quad x > x_0. \tag{7}$$

But (6) and (7) together show that $\bar{F}$ can cross $\Phi_\theta \bar{H}$ only from above. □

a. Transformations of the Underlying Random Variable

Suppose that C(11) holds and that ψ_θ is an increasing function with range I. Moreover, because of (3), it follows that $\chi_x(\theta) = \psi_\theta(x)$ is increasing in θ and has an increasing inverse.

Suppose that $\bar{F} \leq \bar{G}$ that is, $\bar{F}(x) - \bar{G}(\psi_\theta(x)) : +, -$ for all $\theta \in \Theta$. Thus, $\bar{G}^{-1}\bar{F}(x) - \chi_x(\theta) : -, +$ for all $\theta \in \Theta$, or $\chi_x^{-1}\bar{G}^{-1}\bar{F}(x) - \theta : -, +$ for all $\theta \in \Theta$. It follows that in this ordering,

$$\bar{F} \leq \bar{G} \text{ if and only if } \chi_x^{-1}\bar{G}^{-1}\bar{F}(x) \text{ is increasing in } x. \tag{8}$$

D.3.a. Example. If $\phi_\theta(x) = \theta x$ as in Example C.8.a and $I = [0, \infty)$, then the ordering (8) is the star ordering

$$\bar{G}^{-1}\bar{F}(x)/x \text{ is increasing in } x > 0. \tag{9}$$

D.3.b. Example. If $\psi_\theta(x) = x^\theta$ as in Example C.8.b, then the conditions for (8) fail, in particular, (4) fails, and (8) does not yield an ordering.

b. Transformations of the Survival Function

Suppose that C(16) holds, i.e., $\Phi_\theta \bar{F}(x) = \psi_\theta(\bar{F}(x)), \theta, x > 0$, in which case the functions ψ_θ map $[0, 1]$ onto $[0, 1]$. As for transformations of random variables (Section B.a), $\chi_x(\theta) = \psi_\theta(x)$ is increasing in θ and has an increasing inverse.

Suppose that $\bar{F} \leq \bar{G}$, that is, $\bar{F}(x) - \psi_\theta(\bar{G}(x)) : +, -$ for all $\theta \in \Theta$. It follows in Section B.a that in this ordering,

$$\bar{F} \leq \bar{G} \text{ if and only if } \chi_x^{-1}\bar{G}^{-1}\bar{F}(x) \text{ is increasing in } x, 0 < x < 1. \tag{10}$$

D.4.a. Example. Suppose that $\psi_\theta(x) = x^\theta$ as in Example C.8.b, the case of scale parameter families, and suppose that $I = [0, \infty)$. Then the ordering (10) is the ordering

$$[\log \bar{F}(x)]/[\log \bar{G}(x)] \text{ is increasing in } x > 0. \tag{11}$$

D.4.b. Example. If $\psi_\theta(x) = 1 - (1-x)^\theta$ as in Example C.10.b, the case of power parameter families, and if $I = [0, \infty)$, then the ordering

(10) is the ordering

$$[\log F(x)]/[\log G(x)] \text{ is increasing in } x > 0. \tag{12}$$

D.4.c. Example. If $\psi_\theta(x) = \theta x/[1 - \bar\theta x]$ and $I = [0, \infty)$, then the ordering (10) is the ordering

$$\frac{(1-x)\bar F \bar G^{-1}(x)}{x(1 - \bar F \bar G^{-1}(x))} \text{ is increasing in } x > 0. \tag{13}$$

E. Related Stronger Orders

The orderings of Section D are all of the form $\bar F \leq \bar G$ if for all $\theta \in \Theta$, $\bar F$ crosses $\Phi_\theta \bar G$ at most once, and only from above. For notational simplicity, let $\Phi_\theta \bar G = \bar G_\theta$ and rewrite the condition for $\bar F \leq \bar G$ as

$$\bar F(x) - \bar G_\theta(x) : +, - \text{ for all } \theta \in \Theta. \tag{1}$$

Let R be the hazard function of F and let $-\log \bar G_\theta(x) = R_\theta(x)$ be the hazard function of G_θ. Then (1) is equivalent to

$$R(x) - R_\theta(x) : -, + \text{ for all } \theta \in \Theta. \tag{2}$$

Now consider a new relationship.

E.1. Definition. Write $\bar F \leq_r \bar G$ to mean that

$$r(x) - r_\theta(x) : -, + \text{ for all } \theta \in \Theta \tag{3}$$

where r and r_θ are the hazard rates, respectively, of F and G_θ.

E.2. Proposition. If $\bar F \leq_r \bar G$, then $\bar F \leq \bar G$ in the sense of Definition D.2.

Proof. The indicator function $K(x, y) = 0$ if $x < y$, and $K(x, y) = 1$ if $x \geq y$, $-\infty < x, y < \infty$, is totally positive of order ∞ (see 22.B.5). Consequently, by the variation diminishing property of such functions,

$$\int_0^\infty [r(x) - r_\theta(x)]K(x, y) \, dx = \int_y^\infty [r(x) - r_\theta(x)] \, dx = R(x) - R_\theta(x)$$

has at most one sign change, in the order $-,+$ if there is a sign change. But this is another way of saying (2) holds, i.e., $\bar{F} \leq \bar{G}$. □

E.3. Example. If $r_\theta(x) = \theta$ for all $x, \theta > 0$, then condition (3) is the condition that r is increasing. The corresponding weaker condition (2) is the condition that the corresponding survival function has an increasing hazard rate average.

Conditions of the form (3) are sometimes reflexive and transitive, and so they can be used to define orderings; but the distributions that can be so ordered may be somewhat restricted.

E.4. Example. Let R_F and R_G be the hazard functions, respectively, of the distribution functions F and G. The ordering of Example D.4.a is the ordering (2) with $R_\theta(x) = \theta R_G(x), x, \theta > 0$. With the respective hazard rates of F and G denoted by r_F and r_G, condition (3) becomes the condition that

$$r_F(x)/r_G(x) \text{ is increasing in } x > 0. \tag{4}$$

This condition clearly defines a reflexive and transitive ordering.

If F and G are Weibull distributions with respective parameters λ, α_1 and λ, α_2, then (4) becomes

$$r_F(x)/r_G(x) = [\alpha_1 \lambda (\lambda x)^{\alpha_1 - 1}]/[\alpha_2 \lambda (\lambda x)^{\alpha_2 - 1}] = \alpha_1 (\lambda x)^{\alpha_1 - \alpha_2}/\alpha_2,$$

which is increasing for $\alpha_1 \geq \alpha_2$. Thus, the ordering of (4) orders the family of Weibull distributions for fixed scale parameter λ.

Part VII

Complementary Topics

20

Some Topics from Probability Theory

> The theory of probability, as a mathematical discipline, can and should be developed from axioms in exactly the same way as Geometry and Algebra.
> Andreii Nikolaevich Kolmogorov, *Foundations of the Theory Probability*

A. Foundations

In order to describe the behavior of a random quantity such as a life length, certain probabilistic descriptions have become standard. Although these are well known, a brief elementary outline is offered, with an emphasis on motivation. This will help set notation and terminology; for those not familiar with these foundations, it may add some depth of understanding.

a. Probability Spaces

Start with a specific random experiment in mind and denote the set of possible outcomes by Ω. For the particularly simple experiment of tossing a coin once, one might, for example, take Ω to be $\{0,1\}$ or $\{H,T\}$. Or one might want to include points in Ω to represent the remote possibilities that the coin will stand on edge, or roll away, or whatever. Another simple experiment is to spin a balanced "spinner," and record the direction of the pointer. Here, it is natural to take Ω to be $[0, 2\pi)$ or $(-\pi, \pi)$, but $[0, 1]$ would be just as good if angles were properly normalized.

It is often desirable to define the "probability" of various "events" (sets of possible outcomes), i.e., to define a function P on the set $\mathcal{F}$ of events (represented as subsets of Ω) with the following properties:

(a) $P\{A\} \geq 0$, for all $A \in \mathcal{F}$.
(b) $P\{\Omega\} = 1$.
(c) If $A_1, A_2, A_3, \ldots$ is a sequence of pairwise disjoint sets in $\mathcal{F}$, then
$$P\{\cup A_i\} = \sum_i A_i.$$

In the case of the balanced spinner, the probability of an interval should be proportional to the length of the interval. With this proviso, it is well known that the probability function cannot be extended to all subsets of Ω while preserving properties (a) to (c). Fortunately, the "nonmeasurable" sets are obscure, difficult to find, and of little or no practical interest. But this problem means that the domain $\mathcal{F}$ of the probability function must be somehow restricted. Only the subsets of Ω to be assigned a probability are called "events." The collection $\mathcal{F}$ of events is required to satisfy the following properties:

(d) $\Omega \in \mathcal{F}$.
(e) If $A \in \mathcal{F}$ then the complement $A^c \in \mathcal{F}$.
(f) If $A_1, A_2, A_3, \ldots \in \mathcal{F}$, then the union $\cup A_i \in \mathcal{F}$.

These properties insure that $\mathcal{F}$ is a natural domain for a function P satisfying (a) to (c). The function P is said to be a *probability measure* and the triple $(\Omega, \mathcal{F}, P)$ is called a *probability space*.

b. Random Variables

Almost everyone is familiar with the concept of a "random variable" even though they may not know the name. Common examples include the number of "heads" in n tosses of a coin, measurement errors, or the life length of a light bulb. But to treat these variables mathematically, one needs a mathematical definition of a random variable.

Now, a real-valued random variable, say X, is determined by a specific experimental outcome (a point ω in Ω), and this means that X must be a real-valued function defined on Ω. But X cannot be just any function on Ω because expressions of the form $P\{\omega : X(\omega) \in E\} = P\{X \in E\}$ need to be defined for some reasonably large set of subsets E of the real line. Of course, this means that sets $\{\omega : X(\omega) \in E\}$ must be in $\mathcal{F}$. To start with, the set $\mathcal{B}$ of subsets E of the real line for which this is to hold ought to include all intervals of the real line. Moreover, it should be closed under the taking of complements, as well as under the formation of countable unions and intersections. This is just a way of saying that $\mathcal{B}$ should include the *Borel subsets* of the real line. Functions

X defined on Ω with the property that $\{\omega : X(\omega) \in E\}$ is in $\mathcal{F}$ for all Borel subsets E of the real line are said to be *Borel measurable*, and so X is a *random variable* if X is a Borel measurable function defined on Ω.

A pleasant fact is that to work with such random variables, it is usually unnecessary to define them as functions on Ω or even to identify Ω.

c. Distribution Functions

Probabilities of the form $P\{X \in E\}$ would be difficult to specify completely for all Borel sets E; fortunately, these probabilities are determined once they are specified for a reasonably simple collection of Borel sets, the intervals of the form $(-\infty, x)$, where x is a real number. Thus, the *distribution function*

$$F(x) = P\{X \in (-\infty, x]\} = P\{X \leq x\}$$

plays a fundamental role. More specifically, it can be shown that a function F defined on $(-\infty, \infty)$ is a distribution function of some random variable if and only if F is

(i) nondecreasing,
(ii) right continuous (i.e., $\lim_{z \downarrow x} F(z) = F(x)$), and satisfies
(iii) $\lim_{z \to -\infty} F(z) \geq 0, \lim_{z \to \infty} F(z) \leq 1$.

Random variables $X_1, X_2, \ldots, X_n$ are said to be *independent* if

$$P\{X_1 \in E_1, X_2 \in E_2, \ldots, X_n \in E_n\} = \prod_{i=1}^{n} P\{X_i \in E_i\}$$

for all Borel sets $E_1, E_2, \ldots, E_n$.

d. Classification of Distribution Functions

A distribution function F is said to be a *proper distribution* if equality holds in (iii), i.e., if

$$\lim_{z \to -\infty} F(z) = 0, \quad \lim_{z \to \infty} F(z) = 1.$$

If one of these limits fails to hold, that is, if $\lim_{z \to -\infty} F(z) > 0$ or $\lim_{z \to \infty} F(z) < 1$, F is said to be an *improper distribution*.

Most distributions encountered in this book are proper. An improper distribution can be thought of as the distribution of some random variable that takes on either the value $-\infty$ or ∞ with positive probability. Because this book is concerned primarily with distributions of nonnegative random variables, improper distributions which are encountered have the property that $\lim_{z \to -\infty} F(z) < 1$. Such improper distributions are encountered as waiting times for events that may never occur. These distributions sometimes arise in the context of competing risks (Chapter 17) (see also 10.A(6)).

Discrete and absolutely continuous distribution functions are defined in Chapter 1 (Definitions 1.B.2.a and 1.B.2.b), but there is one more basic class. Though they are at most rare in applications, distribution functions exist which are everywhere continuous and have a zero derivative except at an insignificant number of points (i.e., at points in a set of Lebesgue measure 0). These continuous distribution functions do not have densities, and are said to be *singular*. Every distribution function F can be written as a linear combination

$$F = aF_D + bF_{AC} + cF_S,$$

where $a, b, c \geq 0, a + b + c = 1, F_D$ is a discrete distribution function, F_{AC} is an absolutely continuous distribution function, and F_S is a singular distribution function.

A discrete distribution is called a *lattice distribution* if the corresponding random variable takes on only values $0, a, 2a, 3a, \ldots$ that are nonnegative integer multiples of a positive number a. The examples of discrete distributions given in Section E are all lattice distributions; the concept arises also in Section F.b.

Most of the distribution functions considered in the book are absolutely continuous, but discrete and mixtures of discrete and absolutely continuous distributions are sometimes encountered. Singular distributions play little or no practical role in one dimension, but they are important in more than one dimension.

A distribution function that is differentiable except on a set of Lebesgue measure 0 may or may not have a singular part. The following rather technical proposition gives conditions under which the distribution function has no singular part; this proposition is used in the proof of 18.B.5.

A.1. Proposition. If the distribution function F is differentiable everywhere and the derivative is either bounded or finite and integrable, then F has no singular part.

For a proof of this result, see Natanson (1955, p. 133, 266).

e. Unimodality

The concept of unimodality is defined in Definition 1.B.4. Here, some additional results are given, the first proposition being due to Khintchine (see Feller, 1971, p. 158).

A.2. Proposition. A distribution function F is unimodal with mode at 0 if and only if it is the distribution of a product

$$X = UZ$$

of independent random variables, where the random variable U is uniformly distributed on $[0, 1]$.

Proposition A.2 can be expressed in terms of Laplace transforms as follows.

A.3. Proposition. A distribution function F is unimodal if and only if it has a Laplace transform of the form

$$E\, e^{-sX} = \int_0^1 E\, e^{-sZ/x}\, dx = \int_0^1 \int e^{-sz/x}\, dG(z)\, dx$$

for some random variable Z and corresponding distribution G.

Proposition A.2 is sometimes useful for demonstrating unimodality, and for constructing unimodal distributions. The proposition can be modified for distributions concentrated on $[0, \infty)$ that have their mode at the origin. When such distributions have densities, they have densities decreasing on $[0, \infty)$.

A.4. Proposition. A distribution function F such that $F(0) = 0$ is concave on $[0, \infty)$ if and only if it is the distribution of a product

$$X = UZ$$

of independent random variables, where the random variable U is uniformly distributed on $[0, 1]$ and Z is a nonnegative random variable.

For more about unimodality, see Dharmadhikari and Joag-dev (1988).

f. Inverse Distribution Functions

Inverse distribution functions are introduced in Section 1.I, where it is indicated that there is a certain arbitrariness to the definition. The arbitrariness to the definition of the inverse of F comes about when

F is "flat" over some interval I. If $F(x) = p$ for all x in I, then the knowledge that $F(x) = p$ does not identify x. The definition used in this book, repeated here, takes x to be the right-hand endpoint of the interval I (see A.6 for a further discussion).

A.5. Definition. For a distribution function F, the *inverse* F^{-1} *of* F is the function defined by

$$F^{-1}(p) = \sup\{z : F(z) \leq p\}, \quad 0 \leq p < 1,$$
$$= \sup\{z : F(z) < 1\}, \quad p = 1. \qquad (1)$$

Similarly, the inverse $\bar{F}^{-1}$ of the survival function $\bar{F}$ is defined by

$$\bar{F}^{-1}(p) = \sup\{z : \bar{F}(z) \geq p\}, \quad 0 \leq p < 1,$$
$$= \sup\{z : \bar{F}(z) > 0\}, \quad p = 1.$$

With these definitions,

$$F^{-1}(1-p) = \bar{F}^{-1}(p). \qquad (1a)$$

A definition equivalent to (1) is

$$F^{-1}(p) = \inf\{z : F(z) > p\}, \quad 0 \leq p < 1,$$
$$= \inf\{z : F(z) = 1\}, \quad p = 1. \qquad (1b)$$

Here, it is to be understood that if the set $\inf\{z : F(z) = 1\}$ is empty, $F^{-1}(1) = \infty$.

Note that if $F(x) < 1$,

$$F^{-1}(F(x)) = \sup\{z : F(z) \leq F(x)\} = \sup\{z : F(z) = F(x)\}. \qquad (2)$$

If $F(z) = p$ for all z in an interval I of positive length containing x, then F^{-1} is discontinuous at p and $F^{-1}(F(x))$ is the right-hand endpoint of the interval I. It follows that

$$F^{-1}(F(x)) \geq x, \qquad (3a)$$

but

$$F^{-1}(F(x)) = x \text{ if } F \text{ is strictly increasing at } x. \qquad (3b)$$

See Figure A.1.

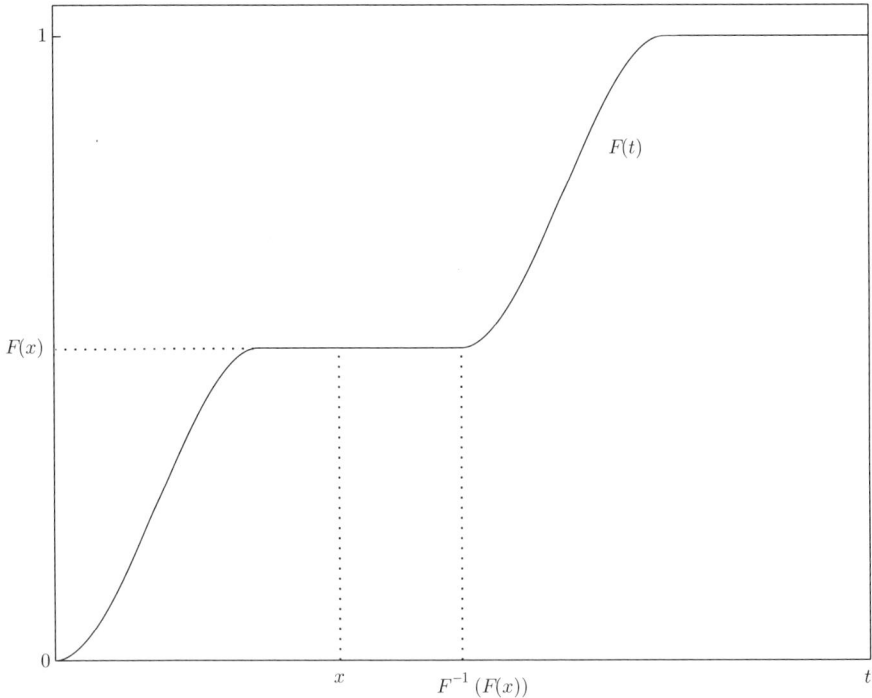

Fig. A.1. An example where $F^{-1}(F(x)) > x$

Because $F(\inf\{z : F(z) > p\}) \geq p$, it follows that

$$F(F^{-1}(p)) \geq p, \tag{4a}$$

but if $F(x) = p$ for a unique x, then

$$F(F^{-1}(p)) = p. \tag{4b}$$

See Figure A.2.

Since

$$\lim_{u \downarrow p} F^{-1}(u) = \lim_{u \downarrow p} \sup\{z : F(z) \leq u\}$$
$$= \sup\{z : F(z) \leq p\} = F^{-1}(p),$$

it follows that F^{-1} is right continuous.

A.6. Alternative definition. Some authors define $F^{-1}(p)$ to be

$$\inf\{z : F(z) \geq p\} = \sup\{z : F(z) < p\}, \quad 0 < p \leq 1,$$

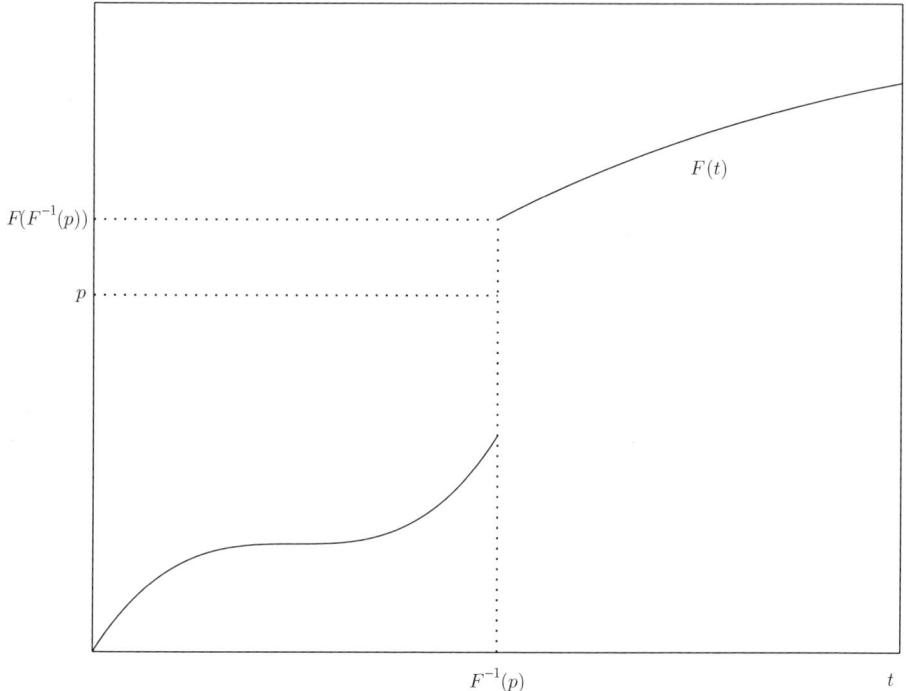

Fig. A.2. An example where $F(F^{-1}(p)) > p$

but this is a definition different from Definition A.5; if $F(z) = p$ for all z in an interval I of positive length, then this alternative definition takes $F^{-1}(p)$ to be the *left-hand* endpoint of I. In this book, conditions (1) or (1a) are used to define the inverse because right continuity is used in Chapters 2 and 19. The distinction is important, e.g., when considering the inverse of an empirical distribution.

It is convenient to record the inverse of a finitely discrete distribution.

A.7. Example. Suppose that F is the discrete distribution with jump points x_i and that $F(x_i) = p_i$, $i = 1, 2, \ldots, n$. Let $p_0 = 0$. Then one can determine directly from Definition A.5 that $F^{-1}(0) = x_1$ and

$$F^{-1}(p) = x_i, \quad p_{i-1} \leq p < p_i,$$

for $i = 1, 2, \ldots, n$, and

$$F^{-1}(1) = x_n.$$

A.8. Proposition. If $G(x) = F(\lambda x)$ for some $\lambda > 0$ and all $x \geq 0$, then

$$F^{-1}(p) = \lambda G^{-1}(p), \quad 0 \leq p \leq 1.$$

This result can be obtained directly from Definition A.5.

A.9. Proposition. If X is a random variable with a continuous distribution function F which is strictly increasing on the interval where it is concentrated, then $Y = F(X)$ has a distribution uniform on $[0, 1]$, i.e.,

$$P\{Y \leq y\} = y, \quad 0 \leq y \leq 1.$$

Here, Y is called the *probability integral transform* of X.

Proof. Because F is strictly increasing on its interval of concentration, equality holds in (4a), and

$$P\{F(X) \leq u\} = P\{X \leq F^{-1}(u)\} = FF^{-1}(u) = u. \quad \square$$

Because of (3b) and in spite of (3a) it follows that

$$\int_a^b x^r \, dF(x) = \int_a^b [F^{-1}F(x)]^r \, dF(x) = \int_{F(a)}^{F(b)} [F^{-1}(p)]^r \, dp. \quad (5)$$

This result does not depend upon the form of Definition A.5; the same result holds for the Alternative Definition A.6.

In Chapters 2 and 5, the quantity $G^{-1}F$ are encountered. It is sometimes convenient to make use of the fact that

$$G^{-1}F = \bar{G}^{-1}\bar{F}, \quad (6)$$

that is,

$$G^{-1}F(x) = \sup\{z : G(z) \leq F(x)\}$$
$$= \sup\{z : \bar{G}(z) \geq \bar{F}(x)\} = \bar{G}^{-1}\bar{F}(x).$$

Similarly, it can be determined from (1b) that

$$FG^{-1}(p) = 1 - \bar{F}\bar{G}^{-1}(1-p). \quad (7)$$

For the derivative of these and related expressions, see 22.A.4.b.

B. Moments

Suppose that $Y = \psi(X)$ and that the random variables X and Y have respective distribution functions F and G. Then, according to Definition 1.B.4,

$$EY = \int y \, dG(y).$$

In cases where F is known but not G, there is a standard way to compute EY directly without first obtaining G.

B.1. Proposition. If $Y = \psi(X)$, then

$$EY = \int \psi(x) \, dF(x).$$

The choices $\psi(x) = x^r$ and $\psi(x) = e^{-sx}$ are particularly important.

B.2. Definition. If

$$\int |x|^r \, dF(x) < \infty,$$

then

$$\mu_r = \int x^r \, dF(x) < \infty, \tag{1}$$

and μ_r is called the rth *moment* of F.

Because of Proposition B.1, the expressions $\mu_r = EX^r$ can be used interchangeably.

Caution about notation. Traditionally, the letter "k" is used for the kth moment where k is an arbitrary integer. Because the letter "k" carries with it the implication that it is an integer, the letter "r" is often used in place of "k," as in Definition B.2. This traditional use of the letter "r" goes back at least to the classic book of Hardy, Littlewood and Pólya (1934), and subsequent authors have continued to use that notation. But in this book, the letter "r" is also used to denote a hazard rate. Of course, hazard rates are functions, but when the notation "μ_r" is used, r is some real number. The usage of "r" should be clear from the context.

B.3. Proposition. Suppose that $F(x) = 0$ for all $x < 0$. For any $r > 0$ such that μ_r exists,

$$EX^r = r \int_0^\infty \bar{F}(x) x^{r-1} \, dx. \tag{2}$$

This formula can be established from (1) through an integration by parts; alternatively, some readers may prefer to write $x^r = \int_0^x r z^{r-1} \, dz$ in (1) and use Fubini's theorem (24.B.1). The equation also appears as 1.C(4). The formula (2) can be quite useful when there is a simple expression for $\bar{F}$.

B.4. Proposition. The rth moment of the distribution function F is given by

$$EX^r = \int_0^1 [F^{-1}(p)]^r \, dp. \tag{3}$$

B.5. Proposition. If the rth moment μ_r of F exists (finitely), and $0 < s < r$, then $\mu_s < \infty$.

Proof. Let X be a random variable with distribution function F. Then,

$$E|X|^s = \int |x|^s \, dF(x) = \int_{\{|x|<1\}} |x|^s \, dF(x) + \int_{\{|x|\geq 1\}} |x|^s \, dF(x)$$

$$\leq P\{|X| < 1\} + \int_{\{|x|\geq 1\}} |x|^r \, dF(x) < \infty. \qquad \square$$

The existence or nonexistence of moments can sometimes be determined quite easily from the hazard rate, as indicated by the following proposition.

B.6. Proposition (Barlow, Marshall and Proschan, 1963). Suppose that F is a distribution with support $[0, \infty)$ and hazard rate r, and that $s > 0$. If

$$s < \liminf_{x \to \infty} x \, r(x),$$

then $\mu_s < \infty$. If

$$s > \limsup_{x \to \infty} x \, r(x),$$

then $\mu_s = \infty$.

Proof. To prove the first part, let $L = \liminf xr(x)$. For every $\varepsilon > 0$, there exists x_0 such that $x \geq x_0$ implies $x\, r(x) \geq (L - \varepsilon)$. Thus, for $x \geq x_0$,

$$R(x) = \int_0^{x_0} r(z)\, dz + \int_{x_0}^x r(z)\, dz \geq R(x_0) + \int_{x_0}^x \frac{L - \varepsilon}{z}\, dz$$
$$= R(x_0) + (L - e)(\log x - \log x_0).$$

From this and the fact that $\bar{F}(x) = e^{-R(x)}$, it follows that

$$\bar{F}(x) \leq \bar{F}(x_0)[x_0/x]^{L-\varepsilon}, \quad x > x_0.$$

The result now follows from (2). The proof of the second part is similar. □

B.6.a. Example. Suppose that $\bar{F}(x) = x^{-c}, x \geq 1$, for some constant $c > 0$. Then, $r(x) = c/x, x > 1$, and $x\, r(x) = c, x > 1$. Consequently, F has finite moments of all positive orders less than c, but moments of order greater than c are not finite.

B.7. Lyapunov's inequality. Denote the rth absolute moment of X by ν_r. That is,

$$\nu_r = E|X|^r.$$

If $a \geq b \geq c$, then

$$\nu_b^{a-c} \leq \nu_c^{a-b}\, \nu_a^{b-c}. \tag{4}$$

For a proof of this result, see Marshall and Olkin (1979, p. 459) or Gut (2005, p. 129). Lyapunov's inequality can be restated in the form

$$\log \nu_b \leq \bar{\alpha} \log \nu_a + \alpha\, \log \nu_c, \tag{5}$$

a statement that $\log \nu_r$ is convex. For a related result, see 21.B.11.a.

Of course, for nonnegative random variables, the absolute moments and the ordinary moments coincide. With $c = 0$ in (4), it follows that for nonnegative random variables,

$$\mu_r^{1/r} \text{ is decreasing in } r > 0. \tag{6}$$

A well-known consequence of (6) is that $(EX^2)^{1/2} \geq EX$, or in other words, the variance of X is nonnegative.

a. Moment Problems

If the random variable X has finite moments, then these moments can be computed from the distribution function F of X. A related question with a long history is: For a given sequence $\{\mu_n,\ n = 0, 1, \ldots\}$, when does there exist a random variable X with some distribution F for which

$$\mu_k = EX^k = \int x^k\, dF(x), \quad k = 0, 1, 2, \ldots?$$

This question is known as the problem of moments. Three classical moment problems mentioned here are (i) the *Hamburger moment problem*, where F is concentrated of $(-\infty, \infty)$, the *Stieltjes moment problem*, where F is consentrated on $[0, \infty)$, and the *Hausdorff moment problem*, where F is concentrated on $[0, 1]$. A standard reference for moment problems is the book of Shohat and Tamarkin (1943), where proofs of the following propositions can be found (see also Karlin and Shapley, 1953).

B.8. Proposition. The Hamburger moment problem has a solution if and only if $\mu_0 = 1$ and the determinants

$$\det[\mu_{i+j}]_{i,j=0}^{n} \geq 0, \quad n = 1, 2, \ldots. \tag{7}$$

Note that $n = 2$ in (7) is the inequality

$$\det \begin{pmatrix} \mu_0 & \mu_1 \\ \mu_1 & \mu_2 \end{pmatrix} = \mu_2 - \mu_1^2 \geq 0,$$

which says that the variance is nonnegative.

See also Example 21.B.11.a.

B.9. Proposition. The Stieltjes moment problem has a solution if and only if $\mu_0 = 1$, (7) is satisfied, and in addition,

$$\det[\mu_{i+j+1}]_{i,j=0}^{n} \geq 0. \tag{8}$$

B.10. Proposition. The Hausdorff moment problem has a solution if and only if $\mu_0 = 1$, and

$$\Delta^k \mu_n \geq 0, \quad k, n = 0, 1, 2, \ldots, \tag{9}$$

where Δ is the signed difference operator

$$\Delta^0 \mu_n = \mu_n,$$
$$\Delta^1 \mu_n = \mu_n - \mu_{n+1},$$
$$\Delta^k \mu_n = \mu_n - \binom{k}{1}\mu_{n+1} + \binom{k}{2}\mu_{n+2} - \binom{k}{3}\mu_{n+3}$$
$$+ \cdots + (-1)^k \mu_{n+k}. \tag{10}$$

With $k=1$ and $n=2$, (10) states that $\mu_2 \le \mu_1$, an inequality obvious from the fact that $0 \le X \le 1$ means that $X^2 \le X$. Consequently, $P\{0 \le X \le 1\} = 1$ implies

$$0 \le \mathrm{Var}(X) = \mu_2 - \mu_1^2 \le \mu_1 - \mu_1^2 = \mu_1(1-\mu_1) \le 1/4.$$

Additional moment inequalities are given in Proposition 21.B.17.

b. Fisher's Information

Fisher's information involves probability density functions, which are defined and discussed in Section 1.B.b. Fisher's information I_θ for a parameter θ in a family of density functions $f(\cdot|\theta)$ differentiable with respect to θ is defined by

$$I_\theta = \int_{-\infty}^{\infty} \left[\frac{\partial}{\partial \theta} \log f(t\,|\,\theta)\right]^2 f(t\,|\,\theta)\,dt = E\left[\frac{\partial}{\partial \theta} \log f(t\,|\,\theta)\right]^2.$$

If f is twice differentiable with respect to θ, then

$$I_\theta = -E\left[\frac{\partial^2}{\partial \theta^2} \log f(t\,|\,\theta)\right].$$

To see this, twice differentiate both sides of the equality $\int f(t\,|\,\theta)dt = 1$ with respect to θ, to first obtain $\int [\partial \log f(t\,|\,\theta)/\partial \theta] f(t) dt = 0$, and then

$$\int \left[\frac{\partial^2}{\partial \theta^2} \log f(t\,|\,\theta) + \left(\frac{\partial}{\partial \theta} \log f(t\,|\,\theta)\right)^2\right] f(t)\,dt = 0.$$

Fisher's information also has a representation in terms of the hazard rate.

B.11. Proposition (Efron and Johnstone, 1990).

$$I_\theta = \int_{-\infty}^{\infty} \left[\frac{\partial \log r(t \mid \theta)}{\partial \theta}\right]^2 f(t \mid \theta)\, dt.$$

c. Equilibrium Distributions

For any distribution function F with finite mean μ such that $F(x) = 0$ for $x < 0$, the corresponding equilibrium distribution (stationary renewal distribution) has density

$$f_{(1)}(x) = \frac{\bar{F}(x)}{\mu} = \frac{\bar{F}(x)}{\int_0^\infty \bar{F}(z)\, dz}, \quad x \geq 0$$
$$= 0, \quad x < 0, \tag{11}$$

is briefly introduced in Section 1.B.h. This distribution is encountered, for example, in Propositions 4.C.6, 4.C.15, 5.D.2, 5.E.6, and also in Section F.b of this chapter.

Sometimes the equilibrium distribution of the equilibrium distribution of F is considered. The density of this distribution is denoted as $f_{(2)}$. This process can be continued to define $f_{(s)}$ for any positive integer s.

By using a technique sometimes known in analysis as "fractional integration," Barlow, Marshall and Proschan (1963) give a definition of $f_{(s)}$, which allows s to take on any nonnegative value, integer or not, just so long as the sth moment of F is finite. For $s > 0$, let

$$\gamma^{(s)}(t) = (-t)^{s-1}/\Gamma(s), \quad t \leq 0,$$
$$= 0, \quad t > 0. \tag{12}$$

It can be verified by using 23.B(2) that

$$\gamma^{(r+s)}(t) = \int_{-\infty}^{\infty} \gamma^r(x)\gamma^{(s)}(t-x)\, dx, \quad r, s > 0.$$

Let

$$f_{(s)}(x) = \int_{-\infty}^{\infty} \left[\frac{\gamma^{(s)}(t)}{\lambda_s}\right] f(x-t)\, dt, \quad s > 0,$$
$$= f(x), \quad s = 0, \tag{13}$$

where f is a density on the positive axis and $\lambda_s = \mu_s/\Gamma(s+1)$ is the normalized moment of 1.C(3). When λ_s exists finitely, $f_{(s)}$ is a density on $[0, \infty)$ and $f_{(s+1)}$ is the equilibrium distribution of $f_{(s)}$.

Denote the rth moment of $f_{(s)}$ by $\mu_r^{(s)}$, and the rth normalized moment by $\lambda_r^{(s)} = \mu_r^{(s)}/\Gamma(r+1)$. Barlow, Marshall and Proschan (1963) show that $\lambda_r^{(s)} = \lambda_{r+s}/\lambda_s$; this relates the normalized moments of $f_{(s)}$ to those of f.

For $s = 1$, (13) becomes

$$f_{(1)}(x) = \int_{-\infty}^{0} \frac{f(x-t)}{\mu} \, dt = \frac{\bar{F}(x)}{\mu}.$$

This equilibrium distribution has hazard rate

$$r_{(1)}(t) = \frac{\bar{F}(t)/\mu}{\int_{t}^{\infty} [\bar{F}(x)/\mu] \, dx} \qquad (14)$$

from which it follows that

$$r_{(1)}(t) = 1/m(t), \qquad (15)$$

where m is the mean residual life 1.B(11) of F. The relationship (15) was obtained by Brown (1981) and by Deshpande, Kochar and Singh (1986). It follows that F has a decreasing mean residual life if and only if $F_{(1)}$ has an increasing hazard rate. Deshpande, Kochar and Singh (1986) also introduce several other classes of distributions (not discussed in this book) that they characterize in terms of properties of $F_{(1)}$.

C. Convergence

C.1. Definition. A sequence $\{F_n\}$ of probability distributions *converges in distribution* to a probability distribution F if $\lim_{n\to\infty} F_n(x) = F(x)$ at all points x, where F is continuous. If X_i has the distribution $F_i, i = 1, 2, \ldots$, then the sequence $\{X_n\}$ is said to *converge in distribution* to X, where X has the distribution function F.

Convergence in distribution is also called *weak convergence*. Note that convergence in distribution does not provide information about convergence of random variables having the distributions. However, if the limiting distribution F is degenerate, say at m, then it must be that corresponding random variables converge in probability to m (Definition C.5).

C.2. Proposition. The following conditions are equivalent:

(i) $\{X_n\}$ *converges in distribution* to X.
(ii) $\lim_{n\to\infty} Eg(X_n) = Eg(X)$ for all real, bounded uniformly continuous functions g.
(iii) $\limsup_{n\to\infty} P\{X_n \in A\} \leq P\{X \in A\}$ for all closed sets A.
(iv) $\liminf_{n\to\infty} P\{X_n \in B\} \geq P\{X \in B\}$ for all open sets B.

This result is discussed by Billingsley (1968, p. 24) as a part of a much more complete treatment. See also Billingsley (1995, p. 335 and p. 378) or Gut (2005, Chapter 5, particularly p. 226).

C.3. Definition. Distributions G and H are said to be of the *same type* if there exist $a > 0$ and b such that

$$G(ax + b) = H(x) \quad \text{for all } x. \tag{1}$$

In this case, a is often called a "scale" parameter and b is called a "location" or "centering" parameter.

The terminology of Definition C.3 is used by Feller (1971, p. 45), but because (1) can be rewritten as

$$G(y) = H\left(\frac{y-b}{a}\right) \quad \text{for all } y. \tag{1a}$$

the definition can lead to some confusion: both a and $1/a$ are scale parameters. See the comments after Definition 7.C.1.

C.4. Proposition. Suppose that $F_n(a_n x + b_n)$ converges to $G(x)$ at all points x, where G is continuous, and that $F_n(a_n x + b_n)$ converges to $H(x)$ at all points x, where H is continuous. Then, G and H are of the same type.

For a proof of this result, see Feller (1971, p. 253).

C.5. Definition. The sequence $\{X_n\}$ *converges in probability* to X if for every $\varepsilon > 0$,

$$\lim_{n\to\infty} P\{|X_n - X| > \varepsilon\} = 0;$$

the sequence $\{X_n\}$ *converges almost surely* to X if for every $\varepsilon > 0$,

$$\lim_{m \to \infty} P\{|X_n - X| > \varepsilon \text{ for some } n \geq m\} = 0.$$

Clearly, almost sure convergence implies convergence in probability. This explains the nomenclature for the following two laws of large numbers.

C.6.a. Strong law of large numbers. Let $\{X_n\}$ be a sequence of independent identically distributed random variables such that $E|X_n| = \mu$ is finite, and let $S_n = X_1 + \cdots + X_n$. Then $\{S_n/n\}$ converges almost surely to μ.

C.6.b. Weak law of large numbers. Let $\{X_n\}$ be a sequence of independent identically distributed random variables such that $E|X_n| = \mu$ is finite, and let $S_n = X_1 + \cdots + X_n$. Then $\{S_n/n\}$ converges in probability to μ.

When variances are finite, the weak law of large numbers can be proved using Chebyshev's inequality; a proof of the strong law can be similarly obtained using Kolmogorov's inequality. See, e.g., Feller (1971, Chapter 7) or Gut (2005, Chapter 6) for this proof as well as various generalizations and applications.

C.7. Proposition. Suppose that $\{X_n\}$ is a sequence of random variables converging in distribution to X and $\{Z_n\}$ is a sequence of random variables that converges in probability to the constant c. Let

$$U_n = X_n + Z_n, \quad V_n = X_n Z_n, \quad W_n = X_n/Z_n.$$

Then, $\{U_n\}$ converges in distribution to $X + c$ and $\{V_n\}$ converges in distribution to cX. When $c \neq 0$, $\{W_n\}$ converges in distribution to X/c.

A classic reference for this important theorem is Cramér (1946, p. 254) and the result is given by Gut (2005, p. 249). A more abstract treatment is offered by Billingsley (1968).

C.8. Central limit theorem. Let $\{X_n\}$ be a sequence of independent identically distributed random variables with mean $EX_n = \mu$ and finite variance σ^2. Let $S_n = X_1 + \cdots + X_n$. Then,

$$\frac{S_n - n\mu}{\sigma\sqrt{n}}$$

converges in distribution to a random variable with a standard normal distribution. That is,

$$\lim_{n \to \infty} P \left\{ \frac{S_n - n\mu}{\sigma \sqrt{n}} \leq x \right\} = \Phi(x),$$

where

$$\Phi(x) = \int_{-\infty}^{x} \frac{1}{\sqrt{2\pi}} e^{-t^2/2} \, dt.$$

D. Laplace Transforms and Infinite Divisibility

In Chapters 4 and 5, a number of classes of distributions are defined in terms of their distribution or survival functions. Other important classes of distributions, most particularly the infinitely divisible distributions, are more easily defined in terms of the corresponding Laplace transforms, which uniquely define the distributions.

The Laplace transform of a distribution function F (or random variable X with distribution F) is defined in 1.C.4. Here, the definition is repeated more formally.

D.1. Definition. The *Laplace transform* of a distribution F on $(-\infty, \infty)$ is defined by

$$\phi(s) = \int_{-\infty}^{\infty} e^{-sx} \, dF(x) \tag{1}$$

for values of s such that the integral exists. If X is a random variable with distribution function F, then ϕ is also called the Laplace transform of X. The function $\text{mgf}(s) = \phi(-s)$ is called the *moment generating function* of F or of X.

Clearly, it is always the case that $\phi(0) = 1$. If the support of F is contained in $[0, \infty)$, the Laplace transform exists finitely for all $s \geq 0$ because then, it is decreasing in s. For the purposes of this book, it is sufficient to consider only the restriction of ϕ to the set $[0, \infty)$.

D.2. Proposition (Uniqueness). A probability distribution is uniquely determined by its Laplace transform.

For a proof on this important result, see Feller (1971, p. 430), Billingsley (1995, p. 286), or Gut (2005, p. 189).

D.2.a. Remark. Equation (1) shows how to obtain the Laplace transform of a given distribution F. As might be expected from Proposition D.2, inversion formulas exist, which allow the recovery of F from a given Laplace transform. Although inversion formulas are beyond the scope of this book, it should be noted that tables of Laplace transforms have been published that can substitute for the inversion formulas and for (1) in many cases. See, for example, Abramowitz and Stegun (1964) or for more complete tables, the classic book of Erdélyi, Magnus, Oberhettinger, and Tricomi (1954).

D.3. Proposition (Continuity). Let $\{F_n\}$ be a sequence of probability distributions with respective Laplace transforms $\{\phi_n\}$. If $\{F_n\}$ converges in distribution to the probability distribution F, then $\{\phi_n\}$ converges pointwise on $(0, \infty)$ to the Laplace transform of F. Conversely, if $\{\phi_n\}$ converges pointwise on $(0, \infty)$ to a function ϕ such that $\lim_{s \to 0} \phi(s) = 1$, then ϕ is the Laplace transform of a distribution function F and $\{F_n\}$ converges in distribution to F.

This proposition is proved by Feller (1971, p. 431) and Gut (2005, p. 242) in terms of the moment generating function.

If the support of F is contained in $[0, \infty)$, the Laplace transform ϕ can itself be regarded as a survival function (although as such, it need not be proper, and may have mass at ∞); indeed, Laplace transforms are mixtures of exponential survival functions. From this observation, Propositions 4.B.6 and 4.B.7, it follows that Laplace transforms are logarithmically convex. However, not all logarithmically convex functions are Laplace transforms. To exhibit necessary and sufficient conditions for a survival function to be a Laplace transform, a definition is required.

D.4. Definition. A function ϕ defined on $[0, \infty)$ is *completely monotone* if it possesses derivatives of all orders and if the nth derivative $\phi^{(n)}$ satisfies

$$(-1)^n \phi^{(n)}(s) \geq 0, \quad s > 0.$$

D.5. Proposition (Bernstein, 1928). A function ϕ defined on $[0, \infty)$ is the Laplace transform of a probability distribution function F for which $F(x) = 0, x < 0$, if and only if it is completely monotone and $\phi(0) = 1$.

For a proof of this proposition, see Feller (1971, p. 439). For a more detailed discussion of this result, see Donoghue (1974).

D.5.a. Proposition. If the function ϕ defined on $[0, \infty)$ is the Laplace transform of a probability distribution function F for which $F(x) = 0, x < 0$, then $\log \phi$ is convex.

Because ϕ can be regarded as a mixture of exponential survival functions, this fact follows from Corollary 3.D.4.a.

a. Convolutions and Infinite Divisibility

If X and Y are independent nonnegative random variables with respective distribution functions F and G, then the distribution $H = F * G$ of $X + Y$ is called the *convolution* of F and G, and is given by

$$H(x) = F * G(x) = \int_0^x G(x - z) \, dF(z).$$

For a proof of this result, see Gut (2005, pp. 67–68).

Because $H = F * G$ is the distribution of $X + Y = Y + X$, it follows that

$$F * G = G * F;$$

moreover,

$$(F_1 * F_2) * F_3 = F_1 * (F_2 * F_3).$$

D.6. Proposition. For any real-valued function ψ with domain that includes the support of H and for which the integrals exist,

$$E\psi(X + Y) = \int \psi(z) \, dH(z) = \iint \psi(x + y) \, dF(x) \, dG(y).$$

If ϕ_F and ϕ_G, respectively, are the Laplace transforms of F and G, then the Laplace transform of the convolution H is the product $\phi_F \phi_G$. In particular, if $F = G$, then the convolution of F with itself, denoted F^{2*}, has the Laplace transform $(\phi_F)^2$. More generally, the n-fold convolution of F with itself, denoted F^{n*}, has Laplace transform $(\phi_F)^n$. This raises a question: When does a Laplace transform have, for all positive integers n, the form $(\phi_F)^n$? In other words, if ϕ is a Laplace transform, when is $\phi^{1/n}$ a Laplace transform for all positive integers n? The answer, stated formally in Proposition D.9, is that ϕ must be the Laplace transform of an infinitely divisible distribution.

D.7. Definition. A random variable X or its distribution is said to be *infinitely divisible* if for all positive integers n, X can be written in the form

$$X = X_1 + X_2 + \cdots + X_n,$$

where $X_1, X_2, \ldots, X_n$ are identically distributed and independent.

D.8. Proposition. A function ϕ is the Laplace transform of an infinitely divisible distribution if and only if $\phi = e^{-\psi}$, where ψ has a completely monotone derivative and $\psi(0) = 0$.

For a proof of this result, see Feller (1971, p. 450). The following result is an immediate consequence of Proposition D.8.

D.9. Proposition. The function ϕ^θ is a Laplace transform for all $\theta > 0$ if and only if ϕ is the Laplace transform of an infinitely divisible distribution.

D.9.a. Remark. For a family of distributions with a convolution parameter $\nu > 0$, the Laplace transform must have the form ϕ^ν, where of course, ϕ is the Laplace transform of an infinitely divisible distribution.

The following proposition is particularly important in the context of this book, because of the central role played by the exponential distribution.

D.10. Proposition (Steutel, 1967, 1969). Every mixture of exponential distributions is infinitely divisible.

Feller (1971, p. 452) quotes this result and calls it a "surprising observation"; it is also a very useful result. See also Steutel (1973) for further discussion. The following proposition is equivalent to Proposition D.10, but may be more appealing.

D.10.a. Proposition (Goldie, 1967). If X has an exponential distribution and Y is a nonnegative random variable independent of X, then $Z = XY$ has an infinitely divisible distribution.

D.11. Proposition (Bondesson, 1979; 1992, p. 73). If f is a density having the form

$$f(x) = cx^{\beta-1}g(x), \quad \beta > 0, \quad x > 0,$$

where $g(0) = 1$ and g is the Laplace transform of a generalized gamma convolution (see Section 9.I), then f is the density of a generalized gamma convolution, and hence is infinitely divisible.

D.12. Definition. Denote the n-fold convolution of F with itself by F^{n*}. A distribution of the form

$$G(x) = \sum_{n=0}^{\infty} e^{-\lambda} \frac{\lambda^n}{n!} F^{n*}(x) \qquad (2)$$

is called a *compound Poisson distribution*.

Note that the distribution function G of (2) is the distribution of $\sum_{i=1}^{N} X_i$, where N has a Poisson distribution and the X_i are independent and have the distribution F.

D.13. Proposition. The following classes of distributions coincide:

(i) Infinitely divisible distributions.

(ii) Limits of sequences of compound Poisson distributions.

(iii) Limits of sequences of infinitely divisible distributions.

For a proof of this proposition and further details, see Feller (1971, pp. 303, 557).

b. Stable Distributions

Stable distributions form a particularly important class of infinitely divisible distributions. It can be seen from the definition below that these distributions can be regarded as a natural extension of the normal distribution. As with the normal distribution, stable distributions can be defined in any number of dimensions, but here only the case of one dimension is considered. For an introduction to stable distributions, see Feller (1971, p. 170).

D.14. Definition. Let $X_1, X_2, \ldots$ be a sequence of independent random variables all having the same distribution F as X. Then F (or X) is said to be *stable* if for each n, there exist norming constants $a_n > 0$ and b_n such that $X_1 + X_2 + \cdots + X_n$ has the same distribution as $a_n X + b_n$. The common distribution F of the X_i is said to be *strictly stable* if $b_n = 0$ for all n, i.e., if for each $n = 1, 2, \ldots$, the nth convolution F^{n*} of F with itself is, after a change of scale, F.

It is known that the only possible norming constants are $a_n = n^{1/\alpha}$, and that $0 < \alpha \leq 2$. The number α is called the *characteristic exponent*, or simply the *exponent* of the distribution. Clearly, all stable distributions are infinitely divisible.

Familiar examples of stable distributions are the Cauchy distribution as well as the degenerate distributions ($\alpha = 1$) and the normal distribution ($\alpha = 2$). It can be shown that if F is stable with exponent $\alpha \neq 1$, then there exists a constant b such that $F(x + b)$ is strictly stable. The normal and Cauchy distributions centered at the origin are strictly stable. A further example with $\alpha = 1/2$ is given in Section 15.H.

E. Some Discrete Distributions

The several discrete distributions that appear in earlier chapters of this book are briefly discussed here. For a comprehensive discussion of discrete distributions, see Johnson, Kotz and Kemp (1992).

a. Bernoulli Distributions

If X is a random variable such that

$$P\{X = 0\} = 1 - p, \quad P\{X = 1\} = p, \tag{1}$$

where $0 \leq p \leq 1$, then the distribution of X is called a *Bernoulli distribution*.

If X has the Bernoulli distribution (1), then $EX^r = p$, for all $r > 0$, Var $X = p(1-p)$, and X has the Laplace transform $\phi(s) = E e^{-sX} = (1 - p) + p e^{-s}, s \geq 0$. In most applications, X can be interpreted as indicating the outcome of a trial that can result only in success ($X = 1$) or failure ($X = 0$). Because it is natural to think of X as indicating whether or not some specified event occurs, X is often called an *indicator random variable*. Indicator random variables are often denoted by I.

b. Binomial Distributions

If X represents the number of successes in n independent trials, each with success probability p, then X takes on the values $0, 1, \ldots, n$, and

$$P\{X = k\} = \binom{n}{k} p^k (1-p)^{n-k}, \quad k = 0, 1, \ldots, n. \tag{2}$$

This is the probability mass function of the *binomial distribution* with parameters n and p. If X_i is 1 or 0 according to whether the ith trial results in success or failure, then because the trials are independent, the X_i are independent and $X = X_1 + X_2 + \cdots + X_n$. Thus, (2) is the n-fold convolution of (1) with itself, and consequently,

$$EX = np, \quad \text{Var } X = np(1-p),$$

and X has the Laplace transform

$$\phi(s) = Ee^{-sX} = [(1-p) + pe^{-s}]^n, \quad s \geq 0.$$

Of course, these quantities can be computed directly from (2).

The upper tail of the binomial distribution is given by formula 23.B(8).

c. Poisson Distributions

Suppose that X_n has a binomial distribution with parameters n and $p_n = \lambda/n$. Then,

$$\lim_{n \to \infty} P\{X_n = k\} = \lim_{n \to \infty} \frac{n!}{k!(n-k)!} \left(\frac{\lambda}{n}\right)^k \left(1 - \frac{\lambda}{n}\right)^{n-k} = e^{-\lambda} \frac{\lambda^k}{k!}.$$

If

$$P\{Y = k\} = e^{-\lambda} \frac{\lambda^k}{k!}, \quad k = 0, 1, \ldots, \tag{3}$$

then Y is said to have a *Poisson distribution*. As the derivation demonstrates, Poisson distributions are limits in distribution of sequences of Bernoulli distributions.

If Y has the probability mass function (3), then $EY = \text{Var } Y = \lambda$, and Y has the Laplace transform given by

$$\lim_{n \to \infty} \left[\left(1 - \frac{\lambda}{n}\right) + \frac{\lambda}{n} e^{-s}\right]^n = \lim_{n \to \infty} \left[1 + \frac{\lambda}{n}(e^{-s} - 1)\right]^n = \exp\{\lambda(e^{-s} - 1)\}.$$

When this Laplace transform is regarded as a survival function, it is the survival function of a negative Gompertz distribution 10.A(6).

The upper tail of the Poisson distribution is given by formula 23.A(8).

d. Geometric Distributions

Let $X_i, i = 1, 2, \ldots,$ be a sequence of independent random variables with the common Bernoulli distribution having parameter p. Think of the X_i as indicating the outcomes (success or failure) in a sequence of Bernoulli trials. Two random variables often of interest are

$Y =$ the number of trials required to achieve a success, and
$Z =$ the number of failures before the first success.

Of course, $Y = Z + 1$, but it is important to distinguish between these random variables. Because the trials are independent,

$$P\{Y = k\} = P\{\text{first } k-1 \text{ trials fail}\} \, P\{k\text{th trial succeeds}\},$$

it follows from (2) that

$$P\{Y = k\} = (1-p)^{k-1} p, \quad k = 1, 2, \ldots. \tag{4}$$

The distribution (4) is called the *geometric (p) distribution* in this book. From (4) or by a similar argument, it follows that,

$$P\{Z = k\} = (1-p)^k p, \quad k = 0, 1, \ldots; \tag{5}$$

on occasion, this is also called a geometric distribution.

From direct computations, it follows that $EY = 1/p$ and $EZ = (1/p) - 1$; $\text{Var } Y = \text{Var } Z = (1-p)/p^2$. It can also be computed that Y and Z have the Laplace transforms

$$E e^{-sY} = \frac{p e^{-s}}{1 - e^{-s}(1-p)}, \tag{6a}$$

$$E e^{-sZ} = \frac{p}{1 - e^{-s}(1-p)}. \tag{6b}$$

Note that the Laplace transform (6a) can be regarded as an exponential survival function with a tilt parameter. The distribution (4) has a remarkable reproductive property given in the following proposition.

E.1. Proposition. If $X_1, X_2, X_3, \ldots$ are independent random variables all with a geometric (p) distribution, and if N has a geometric

(q) distribution, then $W = X_1 + X_2 + \cdots + X_N$ has a geometric (pq) distribution.

Proof. The Laplace transform of W is

$$E e^{-sW} = \sum_{n=1}^{\infty} \left(\frac{p\,e^{-s}}{1 - e^{-s}(1-p)}\right)^n (1-q)^{n-1} q = \frac{pq\,e^{-s}}{1 - e^{-s}(1-pq)}.$$

The result follows from the uniqueness of the Laplace transform (see Proposition D.2.) □

e. Negative Binomial Distributions

Negative binomial distributions are more general than the geometric distributions and they involve the same sequence of independent Bernoulli random variables with common parameter p. However, here the random variable considered is

$$Z_r = \text{the number of failures before the } r\text{th success.}$$

Note that for $k = 0, 1, \ldots$.

$$P\{Z_r = k\} = P\{\text{Exactly } r - 1 \text{ successes in the first } r - 1 + k \text{ trials}\}$$
$$\times P\{\text{success on the } r + k\text{th trial}\}$$
$$= \binom{k+r-1}{r-1} p^{r-1}(1-p)^k p = \binom{k+r-1}{r-1} p^r (1-p)^k, \tag{7}$$

The expectation, variance, and Laplace transform of Z_r can all be formally computed, but there is a more direct way. Note that

$$Z_r = W_1 + W_2 + \cdots + W_r,$$

where W_i is the number of failures between the $(i-1)$ and ith success. The W_i are independent and all have the geometric distribution (5). Thus, (7) is the r-fold convolution of (5) with itself, and results about (5) can be applied to conclude that

$$EZ_r = r\frac{1-p}{p}, \quad \text{Var } Z_r = r\frac{1-p}{p^2}, \quad E e^{-sZ_r} = \left(\frac{p}{1 - e^{-s}(1-p)}\right)^r. \tag{8}$$

The negative binomial distribution can be generalized by allowing r to take noninteger values (in which case, the derivation involving Z_r no longer applies). Note that the coefficient in (7) can be rewritten in the form

$$\binom{k+r-1}{r-1} = \frac{\Gamma(k+r)}{\Gamma(r)\Gamma(k+1)},$$

where Γ is the gamma function defined in Section 21.A. Using this expression, the *generalized negative binomial distribution* is defined by

$$P\{Z_r = k\} = \frac{\Gamma(k+r)}{\Gamma(r)\Gamma(k+1)} p^r (1-p)^k, \quad k = 0, 1, \ldots, \quad r > 0. \quad (9)$$

When r is an integer, (9) reduces to (7); for this generalization, the formulas (8) still apply.

f. Relationships Between the Poisson and Negative Binomial Distributions

The Poisson distribution is derived above in the standard way as a limit of binomial distributions. It is less well-known that the Poisson distribution can also be obtained as a limit of negative binomial distributions. In particular, if $r(1-p)/p = \lambda$ so that $p = r/(\lambda + r)$, then with the aid of 23.A(6) and 24.A.3 it follows that

$$\lim_{r \to \infty} \frac{\Gamma(k+r)}{\Gamma(r)\Gamma(k+1)} p^r (1-p)^k = \lim_{r \to \infty} \frac{r^k}{k!} \left(\frac{r}{\lambda+r}\right)^r \left(\frac{\lambda}{\lambda+r}\right)^k = \frac{\lambda^k}{k!} e^{-\lambda}.$$

Next, suppose that Z has a Poisson distribution with parameter δ, where δ is random and has the gamma distribution with density 9.A(1). Then

$$P\{Z = k\} = \int_0^\infty e^{-\delta} \frac{\delta^k}{k!} \frac{\lambda(\lambda\delta)^{\nu-1}}{\Gamma(\nu)} e^{-\lambda\delta} \, d\delta,$$

$$= \frac{\Gamma(\nu+k)}{k!\Gamma(\nu)} \left(\frac{\lambda}{\lambda+1}\right)^\nu \left(\frac{\lambda}{\lambda+1}\right)^k, \quad k = 0, 1, \ldots. \quad (10)$$

If $\nu = r$ is an integer and if $p = \lambda/(\lambda+1)$, then (10) reduces to (9). This derivation of (9) provides a simple way to show that (9) does in fact define a probability mass function (see also 23.B(7)).

Note that if X has a binomial distribution, then Var $X < EX$; if X has a Poisson distribution, then Var $X = EX$. But if X has a negative binomial distribution, then Var $X > EX$. Consequently, the comparison of the sample mean and variance is sometimes used to distinguish between these distributions.

F. Poisson and Pólya Processes: Renewal Theory

Poisson and Pólya processes are both *counting processes*, i.e., they are processes $N(t), t \geq 0$, which record the number of "events" that occur in the time interval $[0, t]$. These processes can be defined directly as indicated below, but it may be more instructive to discuss their origins in terms of sequences of Bernoulli trials.

a. Poisson Processes

Suppose that an infinite sequence of independent Bernoulli trials are made at times $1/n, 2/n, \ldots$, where the success probability is $p_n = \lambda/n$. Then, in the interval of time $[0, t]$, $[nt]$ trials are performed (here, $[x]$ is the integer part of x, the greatest integer less than or equal to x), and the expected number of "events" or "successes" in the interval is $[nt]p_n \approx \lambda t$. As shown in the discussion of the Poisson distribution, this means that as $n \to \infty$, the random variable $N(t)$ has a limiting Poisson distribution with parameter λt. Now because the Bernoulli trials are independent, the number of successes in disjoint intervals of time are independent. Consequently, the successes counted in disjoint intervals by the limiting process $N(t)$ are independent. Thus, $N(t), t \geq 0$, is said to have *independent increments*. The limiting process $N(t), t \geq 0$, is said to be a *Poisson process with rate* λ.

In a Poisson process with rate λ, the waiting time T to the first event is greater than t if and only if $N(t) = 0$. That is,

$$P\{T > t\} = e^{-\lambda t}.$$

More generally, the waiting time T_ν to the νth event in the Poisson process is greater than t if and only if $N(t)$ is less than ν. Thus,

$$P\{T_\nu > t\} = \sum_{k=0}^{\nu-1} \frac{e^{-\lambda t}(\lambda t)^k}{k!}.$$

This is the survival function of a gamma distribution, introduced and discussed in Section 9.A.

The Poisson process with rate $\lambda > 0$ can be defined as a counting process such that

(i) $N(0) = 0$,

(ii) the increments are independent, and

(iii) $N(t+s) - N(s)$ has a Poisson distribution with parameter λt.

Alternatively, the Poisson process with rate $\lambda > 0$ can be defined as a renewal process with interarrival times having an exponential distribution with parameter λ.

b. Renewal Processes

Suppose that when a device that is in service fails, say at time X_1, it is immediately replaced; the replacement is a new device of the same kind as the failed device and has a life length X_2. If this process of replacement upon failure continues indefinitely, a renewal process is generated. Such a process is defined in terms of a sequence $\{X_i, i = 0, 1, \ldots\}$ of interarrival times that are independent and identically distributed nonnegative random variables. "Renewals" are deemed to occur at times $S_1 = X_1, S_2 = X_1 + X_2, S_3 = X_1 + X_2 + X_3, \ldots$ and $N(t)$ counts the number of renewals in the time interval $[0, t]$. With the convention that $S_0 = 0$, $N(t) = \sup\{n : S_n \leq t\}$. The process $\{N(t), t \geq 0\}$ is called a *renewal process* with *underlying distribution* F. The distribution F is also called the *general lifetime*. If F is an exponential distribution, the process is a Poisson process, described in Section F.a.

At t, the waiting time Y_t for the next renewal is called the "excess" or "residual life' at time t; and the time Z_t since the last renewal is called the age at time t. More formally,

$$Y_t = S_{N(t)+1} - t, \quad Z_t = t - S_{N(t)}.$$

If the common distribution F of the X_i has the mean μ and is not a lattice distribution, then as $t \to \infty$, the limiting distributions of both Y_t and Z_t are the equilibrium distribution (Section B.c), that is, the distribution with density

$$f_{(1)}(x) = \frac{\bar{F}(x)}{\mu} = \frac{\bar{F}(x)}{\int_0^\infty \bar{F}(z)\, dz}, \quad x \geq 0. \qquad (1)$$

The quantity $W_t = Y_t + Z_t$ is sometimes called the *total lifetime*. Although W_t is the life length of the device in service at time t, it does not have the distribution F. The information that a particular device is in service at time t indicates that the device is likely to have a longer than normal life length because longer time spans are more likely to "catch" t. Indeed, there are underlying distributions F for which Y_t alone has a distribution that is stochastically larger than F.

A *stationary renewal process* is a process for which the first life length X_1 has the equilibrium distribution, and then subsequent X_i all have the distribution F. For a stationary renewal process, the distributions of Y_t and Z_t are independent of t, and Y_t has the density (1). Moreover, for a stationary renewal process, the distribution of $W_t = Y_t + Z_t$ is independent of t and is given by

$$F_W(x) = \frac{1}{\mu} \int_0^x z \, dF(z), \quad x \geq 0. \tag{2}$$

This distribution is a member of a family introduced in Section 7.H. As noted above, it is to be expected that W_t is in some sense larger than X, where X is a random variable with distribution F; according to Proposition 7.H.3, W is larger than X in the likelihood ratio order.

In the context of reliability theory, the "rate of occurrences of failures" (that is, of renewals) is of interest. For a Poisson process, this rate is λ, but in general it may be time dependent.

Renewal theory, the study of renewal processes, includes a very substantial body of results that are beyond the scope of this book. See, for example, Smith (1958), Cox (1962), Feller (1971, Chapter XI), Barlow and Proschan (1975, pp. 161–178), or Asmussen (2003).

c. Urn Models and Pólya Processes

The Pólya processes, like Poisson processes, can be derived from sequences of Bernoulli trials, but the trials are neither independent nor identically distributed. Instead they are the results of sampling according to a Pólya urn model.

From an urn containing a red and b black balls, a ball is removed; its color is noted, and the ball is returned to the urn together with s additional balls of the same color. The experiment is then repeated using the newly constituted urn, and in this way, a sequence of trials is performed. The case $s > 0$ describes the *Pólya urn model*.

The results of this section are taken from Marshall and Olkin (1993), who also discuss a bivariate version.

d. The Case $s > 0$

Let X be the number of red balls drawn when sampling according to Pólya's urn model. For $s > 0$, the probability mass function h^+ of X is given by

$$h^+(x \mid a, b; s, n) = P\{X = x\}$$
$$= \binom{n}{x} \frac{\Gamma\left(\frac{a}{s} + x\right) \Gamma\left(\frac{b}{s} + n - x\right)}{\Gamma\left(\frac{b}{s}\right)} \frac{\Gamma\left(\frac{a+b}{s}\right)}{\Gamma\left(\frac{a+b}{s} + n\right)}, \quad x = 0, 1, \ldots n. \quad (3)$$

Here, X is said to have a *Pólya–Eggenberger distribution*.

If drawings occur at times $1/m, 2/m, \ldots$, then $h^+(x \mid a, b, s, mt)$ gives the probability of exactly x red balls by time t, $t = 1/m, 2/m, \ldots$. The number of red balls by time t is a discrete time process, which has a continuous time limit $N(t)$ as $m \to \infty$ called a *Pólya process*. More explicitly, in (3), set $n = mt, a/(a+b) = \theta/\beta m, s/(a+b) = 1/\beta m$ so that $a/s = \theta$, $b/s = \beta_m - \theta$. With the use of 23.A(6), it follows that $\Gamma(c+d)/\Gamma(c) \approx c^d$ for large c, from which it can be verified that

$$g^+(x \mid \theta, \beta, t) = P\{N(t) = x\} = \lim_{m \to \infty} h^+(x \mid \theta s, (m - \theta \beta)s, s, mt)$$
$$= \frac{\Gamma(\theta + x)}{x!\,\Gamma(\theta)} \left(\frac{t}{t+\beta}\right)^x \left(\frac{\beta}{t+\beta}\right)^\theta, \quad x = 0, 1, \ldots, \quad \text{and} \quad t \geq 0. \quad (4)$$

According to E(10), the density g^+ is a mixture of Poisson distributions obtained by regarding the parameter of the Poisson distribution as a random variable with a gamma distribution:

$$g^+(x \mid \theta, \beta, t) = \int_0^\infty \frac{(\lambda t)^x}{x!} e^{-\lambda t} a(\lambda \mid \theta, \beta) \, d\lambda, \quad (5)$$

where

$$a(\lambda \mid \theta, \beta) = \frac{\beta^\theta \lambda^{\theta - 1}}{\Gamma(\theta)} e^{-\beta \lambda} \quad (6)$$

is a gamma density with scale parameter β and shape parameter θ. With this representation it is apparent that the increments of the Pólya process, conditioned on the value of the Poisson parameter, are independent. But unconditionally, their joint distributions are gamma mixtures of independent Poisson distributions.

Because of (5), the Pólya process can be considered to be a Poisson process with a random parameter having a gamma distribution.

The waiting time T_k^+ for the kth jump in the Pólya process with $s > 0$ has survival function $\bar{F}_k^+$ given by $\bar{F}_k^+(t) = \sum_{x=0}^{k-1} P\{N(t) = x\}$. Using (4), this can be given more explicitly in the form

$$\bar{F}_k^+(t) = \sum_{x=0}^{k-1} g^+(x \,|\, \theta, \beta, t) = \int_0^\infty \Psi(t \,|\, \mu, k) \, a(\mu \,|\, \theta, \beta) \, d\mu$$

$$= \int_0^\infty \int_t^\infty \frac{(\mu z)^{k-1} \mu \exp\{-\mu z\}}{(k-1)!} a(\mu \,|\, \theta, \beta) \, dz \, d\mu$$

$$= \frac{1}{B(k, \theta)} \int_t^\infty \frac{\beta^k z^{k-1}}{(1 + \beta z)^{k+\theta}} \, dz, \tag{7}$$

where $\Psi(\cdot \,|\, \mu, k)$ is the survival function of a gamma distribution with scale parameters μ and β, and shape parameters k and θ, a is given by (6), and B is the beta function defined in Section 23.B. The density of this survival function is given in 11.D(1).

The mixture (5) was introduced by Greenwood and Yule (1920) as a model originally thought to contradict the presence of "contagion"; subsequently, it was found that the same distribution arises from the Pólya urn model, which was designed to reflect the presence of "contagion." Thus, the equivalence, perhaps first noticed by Lundberg (1940, 1964), is rather remarkable and has received considerable attention in the literature. See, for example, Feller (1943) and Bates and Neyman (1952) for illuminating discussions, and Lundberg (1940, 1964) for more about the Pólya process.

e. The Case $s < 0$

For $r = -s > 0$, drawing from the urn must cease when all balls have been removed. Except when a/r and b/r are integers, an additional complication arises: As the trials progress and the urn empties, a time comes when the urn discipline breaks down because there are less than r balls in the urn of the required color to be removed. In such a case, assume that all of the remaining balls of that color are removed, and then sampling continues unless or until the urn is empty.

Again, let X be the number of red balls drawn in the first n trials. The distribution h^- of X satisfies

$$h^-(x \mid a, b; r, n) = \binom{n}{x} \frac{\Gamma\left(\frac{a}{r}+1\right)\Gamma\left(\frac{b}{r}+1\right)\Gamma\left(\frac{a+b}{r}-n+1\right)}{\Gamma\left(\frac{a}{r}-x+1\right)\Gamma\left(\frac{b}{r}-n+x+1\right)\Gamma\left(\frac{a+b}{r}+1\right)} \quad (8)$$

for x an integer such that

$$\max\,[0, n-(b/r)] \leq x \leq \min\,[(a/r), n]. \quad (9)$$

Unless a/r and b/r are both integers this interval need not include all the possible values of X. Indeed, if $n - (b/r) > 0$ and b/r is not an integer, then $n - [b/r] - 1$ is a possible value. If $a/r > n$ and a/r is not an integer, then $[a/r] + 1$ is a possible value. The fact that (8) does not completely define the probability mass function of X will not cause problems in taking the limit.

Now, suppose that drawings occur at times $1/m, 2/m, \ldots$, so that $h^-(x \mid a, b, r, mt)$ is the probability of exactly x red balls by time $t = 1/m, 2/m, \ldots$. As in the case $s > 0$, set $n = mt$, $a/(a+b) = \theta/\beta m$, and $r/(a+b) = 1/\beta m$, so that $a/r = \theta$ and $b/r = \beta m - \theta$. Then, the conditions of (9) are satisfied for large m if and only if $0 \leq x \leq \theta$ and $0 \leq t \leq 1/\beta$. Thus,

$$g^-(x \mid \theta, \beta, t) = P\{N(t) = x\} = \lim_{m \to \infty} h^-(x \mid \theta r, (\beta m - \theta)r; r, mt)$$

$$= \frac{\Gamma(\theta+1)(t/\beta)^x [1-(t/\beta)]^{\theta-x}}{x!\,\Gamma(\theta-x+1)}, \quad (10)$$

$x = 0, 1, 2, \ldots$, $x \leq 1/\beta$, and $0 \leq t \leq \beta$.

Again, when θ is not an integer, (10) is not a complete definition of the probability mass function of $N(t)$ because $P\{N(t) = [\theta] + 1\} > 0$. This probability can be obtained from (10) by subtracting from 1 the sum of the values given in (10).

The waiting time T_k^- for the kth jump in the process has survival function $\bar{F}_k^-$ given by

$$\bar{F}_k^-(t) = \sum_{x=0}^{k-1} g^-(x \mid \theta, \beta, t)$$

$$= \int_t^{1/\beta} \frac{(1-\beta z)^{\theta-k} \beta^k z^{k-1}}{B(k, \theta - k + 1)}\, dz, \quad \theta \geq k, \quad 0 \leq t \leq 1/\beta. \quad (11)$$

This is the survival function of a beta distribution with scale parameter β, having the form of 14.E(2).

In both (4) and (10), with $\theta/\beta = \lambda$,

$$\lim_{\beta \to \infty} P\{N(t) = x\} = \frac{(\lambda t)^x}{x!} e^{-\lambda t}, \tag{12}$$

which shows the Poisson process emerging as a limit of the Pólya process.

G. Extreme-Value Distributions

Extreme value distributions are limiting distributions for normalized minima and maxima of independent identically distributed random variables. To be more precise, let $X_1, X_2, X_3, \ldots$ be independent and identically distributed, and for $n = 1, 2, \ldots$ let

$$U_n = \min(X_1, X_2, \ldots, X_n), \quad V_n = \max(X_1, X_2, \ldots, X_n). \tag{1}$$

If there exist a_n and b_n such that $a_n U_n + b_n$ converges in distribution, then what might the limiting distribution be? Similarly, if $a_n V_n + b_n$ converges in distribution, what might its limit be? For the maximum, one possible limiting distribution, W_2^* of (3a), was found by Fréchet (1927). Complete answers to these questions were found by Fisher and Tippett (1928) and again by Gnedenko (1943); for a more detailed history, see Johnson, Kotz and Balakrishnan (1995, p. 1). See also Kotz and Nadarajah (2000) and Castillo, Hadi, Balakrishnan and Sarabia (2005) for additional treatments of extreme-value theory. It is perhaps surprising that the only possible limiting distributions for minima have survival functions of the same type as

$$\bar{W}_1(x) = \exp\{-x^\alpha\}, \quad x \geq 0, \ \alpha > 0, \tag{2}$$
$$\bar{W}_2(x) = \exp\{-(-x)^{-\alpha}\}, \quad x \leq 0, \ \alpha > 0, \tag{3}$$

or

$$\bar{\Lambda}(x) = \exp\{-e^x\}, \quad -\infty < x < \infty. \tag{4}$$

The distribution W_1 has come to be known as the Weibull distribution, and W_2 is the distribution of $-1/X$, where X has a Weibull distribution. The third of these distributions, Λ, is the distribution of log X, where X has an exponential distribution.

Because $V_n = \max(X_1, X_2, \ldots, X_n) = -\min(-X_1, -X_2, \ldots, -X_n)$, the limiting distributions for V_n can be obtained directly from the limiting distributions for U_n; these are the same type as

$$W_1^*(x) = \exp\{-(-x)^\alpha\}, \quad x \leq 0, \; \alpha > 0, \tag{2a}$$

$$W_2^*(x) = \exp\{-x^{-\alpha}\}, \quad x \geq 0, \; \alpha > 0, \tag{3a}$$

or

$$\Lambda^*(x) = \exp\{-e^{-x}\}, \quad -\infty < x < \infty. \tag{4a}$$

For a derivation of these results, see Barlow and Proschan (1975, pp. 231–236) or Kotz and Nadarajah (2000).

a. Domains of Attraction

Retain the notation of (1). If $a_n U_n + b_n$ converges in distribution to a random variable with the distribution G, then there is no choice of the norming constants a_n and b_n for which the sequence converges in distribution to a distribution of a type different from G. A similar statement can be made about the convergence of $a_n V_n + b_n$. Consequently, the following definition can be made.

G.1. Definition. If $a_n U_n + b_n$ converges in distribution to a random variable with the distribution G, the common distribution F of the X_i is said to belong to the *minimum domain of attraction* of G. If $a_n V_n + b_n$ converges in distribution to a random variable with the distribution H, the common distribution F of the X_i is said to belong to the *maximum domain of attraction* of H.

The domains of attraction W_1, W_2, W_1^*, and W_2^* were characterized by Gnedenko (1943). To describe these results, it is convenient to define

$$x_0 = \inf\{x : F(x) > 0\}, \quad x^0 = \sup\{x : F(x) < 1\}.$$

G.2. Proposition (Gnedenko, 1943). *The distribution function F belongs to the*

G. Extreme-Value Distributions

(i) minimum domain of attraction of W_1 if and only if

$$\lim_{t \downarrow 0} \frac{F(x_0 + tx)}{F(x_0 + t)} = x^\alpha, \quad x > 0;$$

(ii) minimum domain of attraction of W_2 if and only if

$$\lim_{t \to -\infty} \frac{F(t)}{F(tx)} = x^\alpha, \quad x > 0;$$

(iv) maximum domain of attraction of W_1^* if and only if

$$\lim_{t \downarrow 0} \frac{\bar{F}(x^0 - tx)}{\bar{F}(x^0 - t)} = x^\alpha, \quad x > 0;$$

(v) maximum domain of attraction of W_2^* if and only if

$$\lim_{t \to -\infty} \frac{\bar{F}(t)}{\bar{F}(tx)} = x^\alpha, \quad x > 0.$$

A proof of these results has been given by Barlow and Proschan (1975) and Kotz and Nadarajah (2000).

Sufficient conditions for F to belong to the domains of attraction of Λ and Λ^* were obtained by von Mises (1936). For these results, it is assumed that the distribution F has a density f, hazard rate $r(t) = f(t)/\bar{F}(t)$ and reverse hazard rate $s(t) = f(t)F(t)$.

G.3. Proposition (von Mises, 1936). If $\lim_{t \downarrow x_0} \frac{d}{dt} \frac{1}{s(t)} = 0$, then F is in the minimum domain of attraction of Λ. If $\lim_{t \uparrow x^0} \frac{d}{dt} \frac{1}{r(t)} = 0$, then F is in the maximum domain of attraction of Λ^*.

A proof of this proposition has been given by Markus and Pinsky (1969). A detailed treatment that includes necessary and sufficient conditions for distributions to be in the domain of attraction of Λ or Λ^* has been given by Resnick (1987).

b. Residual Life Distributions and Maximum Domains of Attraction

According to a definition of Pickands (1975), a distribution function G is a *generalized Pareto distribution* if it has, for positive λ and ξ, one of the following forms:

$$\bar{G}(x \mid \lambda) = (1 + \lambda x)^{-\xi}, \quad x \geq 0, \tag{5}$$

$$\bar{G}(x \mid \lambda) = e^{-\lambda x}, \quad x \geq 0, \tag{6}$$

or

$$\bar{G}(x \mid \lambda) = (1 - \lambda x)^{\xi}, \quad 0 \leq x \leq 1/\lambda. \tag{7}$$

See also Smith (1989). Here, (5) is a Pareto II distribution, (6) is an exponential distribution, and (7) is a uniform distribution with scale and frailty parameters. (See Section 7.E for a general discussion of frailty parameters.)

G.4. Proposition (Pickands, 1975). A continuous distribution F with $\bar{F}(u) = 0, \bar{F}(x) > 0, x < u$, is in the maximum domain of attraction of one of the extreme value distributions W_1^*, W_2^*, or Λ^* if and only if

$$\lim_{t \to u} \inf_{0 < \lambda < \infty} \sup_{0 \leq x < \infty} \left| \frac{\bar{F}(x+t)}{\bar{F}(t)} - \bar{G}(x \mid \lambda) \right| = 0,$$

where G is given by either (5), (6), or (7).

Most distributions ordinarily encountered lie in the domain of attraction of one of the extreme value distributions, and this proposition says that for large t, the residual life distributions can be approximated by one of the distributions (5), (6), or (7). This is not the same as saying that residual life distributions converge weakly to one of these distributions. Indeed, that weak limit can be degenerate at 0 or it may be an improper distribution (see also Proposition 1.B.13). The Pareto IV distribution is an example in which the residual life distribution diverges to an improper distribution with all mass at ∞, but can be approximated by a Pareto II distribution (4); see 11.B(9). The Gompertz–Makeham distribution is an example in which the residual life distribution converges weakly to the distribution degenerate at 0, but can be approximated by the exponential distribution (5); see Section 10.C.d. And the beta distribution is an example with a residual life survival function approximated by (7); see 14.C(5).

H. Chebyshev's Covariance Inequality

To statisticians and probabilists, the name "Chebyshev's inequality" usually refers to the inequality that bounds the probability of deviating from the expectation in terms of the variance. However, there is another less well-known inequality also called "Chebyshev's inequality" that is the subject of this section.

Real-valued functions f and g defined on the same domain D are said to be *similarly ordered* if

$$[f(x) - f(y)][g(x) - g(y)] \geq 0 \tag{1}$$

for all x, y in D.

The notion of correlation and covariance are defined in Definition I.5. The intuitive meaning of a positive correlation (or equivalently, a positive covariance) between two random variables is that if one variable is large, then the other tends to be large. Thus, it is to be expected that similarly ordered functions of a single random variable are positively correlated.

H.1. Proposition. If the real-valued functions f and g are defined on the same domain D and are similarly ordered, if X is a random variable taking values in D, and if the expectations exist, then

$$Ef(X)g(X) \geq Ef(X)Eg(X). \tag{2}$$

Proof. Introduce a random variable Y with the same distribution as X but independent of X. From (1), it follows directly that $[f(X) - f(Y)][g(X) - g(Y)] \geq 0$; consequently,

$$E[f(X) - f(Y)][g(X) - g(Y)] \geq 0. \tag{3}$$

Because X and Y are identically distributed and independent, (3) reduces to (2). □

This Chebyshev's inequality states that when f and g are similarly ordered, the covariance (or correlation) of $f(X)$ and $g(X)$ is nonnegative.

A discrete version of Proposition H.1 is given in Hardy, Littlewood and Pólya (1952, Theorem 43, p. 43), who label the result "Chebyshev's inequality." A more general result is given by Hardy, Littlewood and

Pólya (1952, Theorem 236, p. 168), where Proposition H.1 for monotonic functions is attributed to Chebyshev.

I. Multivariate Basics

The purpose of this section is to outline a few basics of multivariate theory that are required particularly in Chapter 17.

I.1. Definition. The function F defined on $\mathcal{R}^k = (-\infty, \infty)^k$ by

$$F(x_1, \ldots, x_k) = P\{X_1 \leq x_1, \ldots, X_k \leq x_k\} \tag{1}$$

is called the *joint distribution function* of the random variables $X_1, \ldots, X_k$. The function $\bar{F}$ defined on $\mathcal{R}^k$ by

$$\bar{F}(x_1, \ldots, x_k) = P\{X_1 > x_1, \ldots, X_k > x_k\} \tag{2}$$

is called the *joint survival function* of $X_1, \ldots, X_k$.

From the joint distribution function of $X_1, \ldots, X_k$, it is possible to determine the joint distribution function of any subset A of these variables because $P\{X_i \leq x_i, i \in A\}$ can be obtained from the function $F(x_1, \ldots, x_k)$ by setting $x_j = \infty, j \notin A$. In particular, the *marginal distribution function* F_i of X_i is given by

$$F_i(x_i) = F(\infty, \ldots, \infty, x_i, \infty, \ldots, \infty), \; i = 1, 2, \ldots, k,$$

where x_i is the ith component.

Note that unlike the univariate case in which $\bar{F}(x) = 1 - F(x)$, in the multivariate case $\bar{F}(x_1, \ldots, x_k) \neq 1 - F(x_1, \ldots, x_k)$ for all $x_1, \ldots, x_k$. For example, with $k = 2$,

$$\bar{F}(x_1, x_2) = 1 - F_1(x_1) - F_2(x_2) + F(x_1, x_2).$$

When $k = 2$, necessary and sufficient conditions for a function to be a joint distribution function take a simple form. These conditions are

$$F(x_2, y_2) - F(x_1, y_2) - F(x_2, y_1) + F(x_1, y_1) \geq 0$$
$$\text{for all } x_1 < x_2, \quad y_1 < y_2,$$

and
$$\lim_{x\to\infty, y\to\infty} F(x,y) = 1, \quad \lim_{x\to-\infty, y\to-\infty} F(x,y) = 0.$$

Similarly, necessary and sufficient conditions for a distribution to be a joint survival function are

$$\bar{F}(x_1, y_1) - \bar{F}(x_1, y_2) - \bar{F}(x_2, y_1) + \bar{F}(x_2, y_2) \geq 0$$
$$\text{for all } x_1 < x_2, \quad y_1 < y_2,$$

and

$$\lim_{x\to\infty, y\to\infty} \bar{F}(x,y) = 0, \quad \lim_{x\to-\infty, y\to-\infty} \bar{F}(x,y) = 1.$$

I.2. Definition. A nonnegative function f of k real variables for which

$$F(x_1, \ldots, x_k) = \int_{-\infty}^{x_1} \cdots \int_{-\infty}^{x_k} f(z_1, \ldots, z_k) \prod_{i=1}^{k} dz_i \tag{3}$$

is called a *joint probability density* of $X_1, \ldots, X_k$ of F.

I.3. Definition. The random variables $X_1, \ldots, X_k$ are said to be *mutually independent* if

$$F(x_1, \ldots, x_k) = \prod_{i=1}^{k} F_i(x_i) \quad \text{for all } x_1, \ldots, x_k. \tag{4}$$

An equivalent condition for mutual independence is that the joint survival function is a product of the marginal survival functions, i.e.,

$$\bar{F}(x_1, \ldots, x_k) = \prod_{i=1}^{k} \bar{F}_i(x_i) \quad \text{for all } x_1, \ldots, x_k. \tag{5}$$

If a joint probability density exists, then the random variables $X_1, \ldots, X_k$ are mutually independent if and only if the joint probability density is the product of the marginal probability densities.

I.4. Remark. Even though the random variables $X_1, \ldots, X_k$ are pairwise independent, that is, every pair of these variables is independent, it may be that the variables are not mutually independent. For simple examples, see Feller (1968, p. 220), Billingsley (1995, p. 54), or Gut (2005, p. 71).

a. Covariances and Correlations

I.5. Definition. Suppose that $Var\, X_i < \infty, i = 1, 2$. Then,

$$Cov(X_1, X_2) = EX_1X_2 - EX_1 EX_2 \tag{6}$$

is finite and is called the *covariance* of X_1 and X_2;

$$Corr(X_1, X_2) = Cov(X_1, X_2) / [Var\, X_1\, Var\, X_2]^{1/2} \tag{7}$$

is called the *correlation* of X_1 and X_2.

The standard result,

$$-1 \leq Corr(X_1, X_2) \leq 1,$$

can be proved using the Cauchy–Schwarz inequality.

With straightforward computations it can be verified that

$$Cov(a_1 X_1 + b_1, a_2 X_2 + b_2) = a_1 a_2\, Cov(X_1, X_2),$$

and if $a_1 a_2 > 0$,

$$Corr(a_1 X_1 + b_1, a_2 X_2 + b_2) = Corr(X_1, X_2).$$

For nonnegative random variables, a formula often convenient for computations is

$$Cov(X_1, X_2) = \int_0^\infty \int_0^\infty [F(x_1, x_2) - F_1(x_1) F_2(x_2)]\, dx_1\, dx_2. \tag{8}$$

Note also that

$$F(x_1, x_2) - F_1(x_1) F_2(x_2) = \bar{F}(x_1, x_2) - \bar{F}_1(x_1) \bar{F}_2(x_2) \tag{9}$$

and this provides an alternative integrand.

If X_1 and X_2 are independent, then their covariance (hence, correlation) is 0, but a correlation of 0 does not imply that the variables are independent. For example, if (X_1, X_2) takes on the values $(0, 1), (0, -1), (1, 0), (-1, 0)$ each with probability $1/4$, then $EX_1 = EX_2 = EX_1 X_2 = Corr(X_1, X_2) = 0$, but these two random variables are not independent. On the other hand, if the variables X_1 and X_2 have a bivariate normal distribution with zero correlation, then they are independent.

I.6. Proposition (James, 1986). If X is a random variable with distribution function F, then

$$Cov\{\gamma(X), \phi(X)\} = \int_{-\infty}^{\infty} [\gamma(t) - E\{\gamma(X) \mid X > t\}][\{\phi(t) - E\{\phi(X) \mid X > t\}] \, dF(t).$$

James states that "this lemma can be proved by expanding the integrand, writing the conditional expectations as integrals and interchanging the order of integration."

As noted above, correlations always fall into the interval $[-1, 1]$. The following proposition gives the conditions on the marginal distributions under which it is possible to attain this full range of correlations.

I.7. Proposition (Moran, 1967). Suppose that the marginal distributions F_i of $X_i, i = 1, 2$, are fixed and given and that both variables have finite variances. For all $\rho \in [-1, 1]$, there exists a joint distribution function F with marginal distributions F_1 and F_2 and correlation ρ if and only if

(i) there exist constants a and b such that $aX_2 + b$ has the distribution F_1, and

(ii) the distribution F_1 is symmetric about its mean.

Moran advances the following argument. Rescale X_2 to have the same mean and variance as X_1. If there exists a joint distribution with marginals F_1 and F_2 and correlation $\rho = 1$, then for that joint distribution, $E(X_1 - X_2)^2 = 0$ so $X_1 = X_2$ with probability 1. If there exists a joint distribution for which the correlation $\rho = -1$, then $E(X_1 + X_2)^2 = 0$ so $X_1 = -X_2$ with probability 1. If both these conditions are possible, then the distribution of X_1 must be symmetric. Thus, Proposition I.7 gives the conditions under which the correlation can achieve both extremes 1 and -1. If this is possible, then convex combinations of the joint distributions having these extreme correlations can achieve any intermediate values of the correlation.

b. The Hoeffding–Fréchet Bounds

Clearly, the marginal distributions do not determine the joint distribution, but bounds can be given for the joint distribution if the marginals are known. These bounds have generated considerable interest, and several proofs have been given.

I.8. Proposition (Fréchet, 1951; Hoeffding, 1940). If the joint distribution has marginal distributions F_1 and F_2, then

$$\max\left[F_1(x_1) + F_2(x_2) - 1, 0\right] \leq F(x_1, x_2) \leq \min\left[F_1(x_1), F_2(x_2)\right]. \quad (10)$$

Note that the upper and lower bounds of (10) are both joint distribution functions with marginal distributions F_1 and F_2. It follows from (8) that the maximum and minimum correlations possible with marginal distributions F_1 and F_2 are attained, respectively, by the upper and lower bounds of (10). In particular, bivariate distributions with exponential marginal distributions can have correlations only in the range $[1 - \pi^2/6, 1]$; this result was first obtained by Moran (1967).

The extension of Proposition I.8 to the multivariate case is not entirely trivial; see Joe (1997, Chapter 3).

c. Notions of Dependence

A number of qualitative concepts of positive and negative dependence can be found in the literature, and only a few are introduced here. For more extensive discussions of the subject, see Joe (1997), Spizzichino (2001), or Müller and Stoyan (2002).

I.9. Definition (Lehmann, 1966). Random variables X_1, X_2 are said to be *positively quadrant dependent* if their joint distribution function F satisfies

$$F(x_1, x_2) \geq F_1(x_1) F_2(x_2), \quad (11)$$

or equivalently (because of (9)), if

$$\bar{F}(x_1, x_2) \geq \bar{F}_1(x_1) \bar{F}_2(x_2). \quad (12)$$

If nonnegative random variables X_1 and X_2 are positively quadrant dependent, then they are positively correlated. This fact follows directly from (8).

In higher dimensions, the appropriate definition of "positive quadrant dependence" is not entirely clear, because the natural extension of (11) to higher dimensions,

$$F(x_1, \ldots, x_k) \geq \prod_{i=1}^{k} F_i(x_i), \quad \text{for all } x_1, \ldots, x_k \quad (13)$$

is not equivalent to the corresponding extension

$$\bar{F}(x_1, \ldots, x_k) \geq \prod_{i=1}^{k} \bar{F}_i(x_i) \quad (14)$$

of (12). Condition (13) is called *positive lower orthant dependence*, and (14) is called *positive upper orthant dependence*.

I.10. Example (Frailty models). Suppose that $X_1, \ldots, X_k$ are conditionally independent, given the event $\Theta = \theta$, where Θ is a scalar random variable with distribution function H. When θ acts as a frailty parameter, the conditional joint distribution of $X_1, \ldots, X_k$ given $\Theta = \theta$ has the form

$$\bar{F}(x_1, \ldots, x_k \mid \Theta = \theta) = \exp\left\{-\theta \sum_{i=1}^{k} R_i(x_i)\right\},$$

and the unconditional survival function of the X_i is given by

$$\bar{F}(x_1, \ldots, x_k) = \int \exp\left\{-\theta \sum_{i=1}^{k} R_i(x_i)\right\} dH(\theta).$$

Then $\bar{F}$ satisfies (14) so that $X_1, \ldots, X_k$ are positive upper orthant dependent. To see this, let $a = \sum_{i=1}^{m} R_i(x_i)$, $b = \sum_{i=m+1}^{k} R_i(x_i)$; it follows from Chebyshev's inequality 18.H.1 that

$$\bar{F}(x_1, \ldots, x_k) = E\left[\exp\left\{-\Theta(a+b)\right\}\right] \geq E\left[\exp\left\{-\Theta a\right\}\right] E\left[\exp\left\{-\Theta b\right\}\right]$$

$$= \bar{F}(x_1, \ldots, x_m, 0, \ldots, 0) \bar{F}(0, \ldots, 0, x_{m+1}, \ldots, x_k).$$

Iteration of this inequality yields (14). The proof of lower orthant dependence is similar.

A stronger notion of dependence that implies both (13) and (14) is that of association. To introduce this notion, first note Proposition H.1. If f and g are similarly ordered and ϕ is an increasing function, then the compositions ϕf and ϕg are similarly ordered. Consequently, $\phi f(X)$ and $\phi g(X)$ have a nonnegative covariance for any increasing function ϕ. This idea can be carried over to random variables that do not arise as functions of a single random variable X.

I.11. Definition (Esary, Proschan and Walkup, 1967). Random variables $X_1, \ldots, X_k$ are said to be *associated* if

$$\operatorname{Cov}(\phi(X_1, \ldots, X_k), \psi(X_1, \ldots, X_k)) \geq 0 \tag{15}$$

for all increasing functions ϕ, ψ for which the covariance exists.

I.12. Proposition. If $X_1, \ldots, X_k$ are associated, then for $j = 1, \ldots, k-1$,

$$P\{X_1 > x_1, \ldots, X_k > x_k\}$$
$$\geq P\{X_1 > x_1, \ldots, X_j > x_j\} P\{X_{j+1} > x_{j+1}, \ldots, X_k > x_k\} \tag{16a}$$

and

$$P\{X_1 \leq x_1, \ldots, X_k \leq x_k\}$$
$$\geq P\{X_1 \leq x_1, \ldots, X_j \leq x_j\} P\{X_{j+1} \leq x_{j+1}, \ldots, X_k \leq x_k\}. \tag{16b}$$

Proof. To prove (16a), let

$$I_1(u_1, \ldots, u_j) = 1 \quad \text{if } u_i > x_i, i = 1, \ldots, j,$$
$$= 0 \quad \text{otherwise},$$

$$I_2(u_{j+1}, \ldots, u_k) = 1 \quad \text{if } u_i > x_i, i = j+1, \ldots, k,$$
$$= 0 \quad \text{otherwise}.$$

Because these indicator functions are increasing and $X_1, \ldots, X_k$ are associated, the covariance of $I_1(X_1, \ldots, X_j)$ and $I_2(X_{j+1}, X_k)$ is nonnegative. This statement is (16a). The proof of (16b) is similar. □

The following proposition follows directly from Proposition I.12.

I.12.a. Proposition. If $X_1, \ldots, X_k$ are associated, then with the notation of Definition I.9,

$$\bar{F}(x_1, \ldots, x_k) = P\{X_1 > x_1, \ldots, X_k > x_k\}$$
$$\geq \prod_{i=1}^{k} P\{X_i > x_i\} = \prod_{i=1}^{k} \bar{F}_i(x_i) \tag{17a}$$

and

$$F(x_1,\ldots,x_k) = P\{X_1 \leq x_1,\ldots,X_k \leq x_k\}$$
$$\geq \prod_{i=1}^{k} P\{X_i \leq x_i\} = \prod_{i=1}^{k} F_i(x_i). \tag{17b}$$

The conditions for association are clearly too cumbersome to check directly. Fortunately, sufficient conditions for association are known. First, it is sufficient that (15) hold only for functions ϕ, ψ that are binary (i.e., take only the values 0 and 1), or bounded and continuous. Even these are too cumbersome to check directly, and most known cases of association come from the following proposition, which says that increasing functions of independent random variable are associated.

I.13. Proposition (Esary, Proschan and Walkup, 1967). If

$$X_i = \phi_i(Y_1,\ldots,Y_n), \quad i = 1,\ldots,k,$$

where the ϕ_i are increasing functions and $Y_1,\ldots,Y_n$ are independent, then $X_1,\ldots,X_k$ are associated.

The concept of association has the following properties; see Esary, Proschan and Walkup (1967) for proofs.

(i) Members of any subset of associated random variables are associated.

(ii) If two sets of associated random variables are independent, then their union is associated.

(iii) A set consisting of a single random variable is associated.

(iv) Increasing functions of associated random variables are associated.

(v) Limits in distribution of associated random variables are associated.

(vi) If random variables $X_1,\ldots,X_k$ are associated, then both (13) and (14) hold.

I.14. Example (Order statistics). Suppose that $X_1,\ldots,X_k$ are associated, and denote by $X_{(1)},\ldots,X_{(k)}$ their values ordered increasingly, so that $X_{(1)} \leq \cdots \leq X_{(k)}$. Then, the $X_{(i)}$ are increasing functions of the unordered X_i, and by property (iv), they are associated. An important special case of this example arises when $X_1,\ldots,X_k$ are independent and identically distributed random variables, in which case the variables

$X_{(1)}, \ldots, X_{(k)}$ are known as *order statistics*. As a consequence of the fact that order statistics are associated, $\text{Cov}(X_{(i)}, X_{(j)}) \geq 0$.

I.15. Example (Minimal cut sets). Consider a coherent system (Section 5.A) with k minimal cut sets, say $C_1, \ldots, C_k$. If the components of the system have independent life lengths and $X_1, \ldots, X_k$ are the life lengths of the cut sets, then the system life length $U = \min(X_1, \ldots, X_k)$ and the X_i are associated.

I.16. Example. Suppose that U, V, and W are independent and either

$$X = U + V, \quad Y = U + W, \tag{18}$$

or

$$X = \min(U, V), \quad Y = \min(U, W), \tag{19}$$

or U, V, and W are nonnegative and

$$X = UW, \quad Y = VW. \tag{20}$$

Then, by (ii) and (iv) of Proposition I.13, X and Y are associated. Note that the structure of (18) can be used to generate the bivariate normal distribution with positive correlation. The models (18), (19), and (20) are often extended to higher dimensions.

I.16.a. Example (Bivariate gamma distribution). If U, V, and Z are independent random variables with gamma distributions, then $X = U + Z, Y = V + Z$ have a bivariate gamma distribution, and X and Y are associated.

I.16.b. Example (Bivariate F distribution). If U, V, and Z are independent random variables with gamma distributions, then apart form the usual normalization by degrees of freedom, the joint distribution of $X = U/Z$ and $Y = V/Z$ is a bivariate F distribution. Because X and Y are increasing functions of the independent variables U, V, and $1/Z$, they are associated. The fact that they are positive quadrant dependent is due to Kimball (1951) and is important in the analysis of variance.

I.16.c. Example (Bivariate beta distribution). If U, V, and Z are independent random variables with gamma distributions having a common scale parameter, then $X = U/(U+Z), Y = V/(V+Z)$ have a bivariate beta distribution. The argument similar to that used in Example I.16.a shows that these variables are associated.

I.16.d. Example (A multivariate exponential distribution). For each subset A of $\{1, 2, \ldots, k\}$, suppose that T_A has an exponential distribution. Further suppose that all of these random variables are independent, and let $X_i = \min_{\{A \text{ such that } i \in A\}} T_A$. Then, $X_1, \ldots, X_k$ have the multivariate exponential distribution of Marshall and Olkin (1967), and by Proposition I.13, $X_1, \ldots, X_k$ are associated.

I.17. Example. Univariate extreme value distributions are discussed in Section G. The natural multivariate extension of these distributions is reviewed by Marshall and Olkin (1983); they show that if $X_1, \ldots, X_k$ have a multivariate extreme value distribution, then these variables are associated.

I.18. Example (Jogdeo, 1977). Suppose that $X_1, \ldots, X_k$ are conditionally independent, given the event $\Theta = \theta$, where Θ is a scalar random variable. That is, the conditional joint distribution of $X_1, \ldots, X_k$ has the form

$$F(x_1, \ldots, x_k \mid \theta) = \prod_{i=1}^{k} G(x_i \mid \theta).$$

Denote the distribution of Θ by H. Then, $X_1, \ldots, X_k$ have the joint distribution

$$F(x_1, \ldots, x_k) = \int \prod_{i=1}^{k} G(x_i \mid \theta) \, dH(\theta).$$

If each $G(x_i \mid \theta)$ is stochastically increasing in θ, then $X_1, \ldots, X_k$ are associated. This example can be extended to the case that θ is an associated vector; for related results, see Spizzichino (2001, p. 123). This example is to be compared with Example I.10.

In cases where survival and distribution functions do not have a simple form and association is not apparent, a following still stronger condition is sometimes easy to check.

I.19. Definition. Suppose that the random variables $X_1, \ldots, X_k$ have the joint density f. These random variables or their joint distribution are said to be *multivariate totally positive of order* 2 if

$$f(x \vee y) f(x \wedge y) \geq f(x) f(y),$$

where

$$x \vee y = (\max[x_1, y_1], \ldots, \max[x_k, y_k]),$$
$$x \wedge y = (\min[x_1, y_1], \ldots, \min[x_k, y_k]).$$

The following proposition provides a key connection between total positivity and association.

I.20. Proposition (Karlin and Rinott, 1980). If $X_1, \ldots, X_k$ are multivariate totally positive of order 2, then they are associated.

I.21. Proposition (Karlin and Rinott, 1980). Suppose that X_0, $X_1, \ldots, X_k$ are independent random variables and that X_i has a logarithmically concave density $f_i, i = 1, \ldots, k$. Suppose further that X_0 has a density. Then, $X_0 + X_1, \ldots, X_0 + X_k$ have a joint density that is multivariate totally positive of order 2.

I.21.a. Example. If $X_0, X_1, \ldots, X_k$ are independent random variables, each having a gamma distribution, then $X_0 + X_1, \ldots, X_0 + X_k$ have a multivatiate gamma distribution that is multivariate totally positive of order 2. This multivariate gamma distribution has been studied by various authors.

I.22. Proposition (Karlin and Rinott, 1980). Suppose that $X_1, \ldots, X_k$ are independent random variables and X_i has the density $f_i, i = 1, \ldots, k$. Let X_0 be a positive random variable. If for $i = 1, 2, \ldots, k$, either $f_i(u/v)$ or $f_i(uv)$ is totally positive of order 2 in $-\infty < u < \infty, v > 0$, then $X_0 X_1, \ldots, X_0 X_k$ and $X_1/X_0, \ldots, X_k/X_0$ each have joint densities that are multivariate totally positive of order 2.

d. Copulas

A multivariate distribution C with all marginal distributions uniform on $[0, 1]$ is called a *copula*. If C is the joint distribution of k random variables, and if $F_1, \ldots, F_k$ are univariate distribution functions, then $C(F_1, \ldots, F_k)$ is a k-variate distribution with marginal distributions $F_1, \ldots, F_k$. Thus, the function C can be thought of as a function which "couples" the marginals; as such it is also sometimes called a "dependence function." All k-variate distributions arise in this fashion from at least one copula. If F is a continuous k-variate distribution with marginal distributions $F_1, \ldots, F_k$, then the copula giving rise to F is unique and is given by

$$C(u_1,\ldots,u_k) = F(F_1^{-1}(u_1), F_2^{-1}(u_2),\ldots, F_k^{-1}(u_k)).$$

A number of copulas have been proposed and studied in the literature; for an extensive discussion, see Nelsen (2006).

In two dimensions, the most important copulas may be the copula for independence,

$$C(u,v) = uv, \quad 0 \le u,\ v \le 1,$$

the copula for the upper Fréchet bound (10)

$$C(u,v) = \min(u,v), \quad 0 \le u,\ v \le 1,$$

and the copula for the lower Fréchet bound (10)

$$C(u,v) = \max(u+v-1, 0), \quad 0 \le u,\ v \le 1.$$

I.23. Example. Let ϕ be a Laplace transform of a nonnegative random variable Z (not degenerate at 0). Then, ϕ is defined on $(0,\infty)$, is decreasing, and

$$C(u,v) = \phi(\phi^{-1}(u) + \phi^{-1}(v)), \quad 0 \le u,\ v \le 1,$$

is a copula. This special kind of copula is said to be an *Archimedean copula*, a term due to Genest and McKay (1986). Such copulas are positive quadrant dependent. To see this, rewrite the condition

$$C(u,v) = \phi(\phi^{-1}(u) + \phi^{-1}(v)) \le uv \qquad (21)$$

as

$$E\exp\{-[\phi^{-1}(u) + \phi^{-1}(v)]Z\} \ge uv;$$

let $\phi^{-1}(u) = s, \phi^{-1}(v) = t$ to conclude that (21) is equivalent to the condition that

$$E\,e^{-(s+t)Z} \ge E\,e^{-sZ} E\,e^{-tZ}, \quad s,t \ge 0.$$

This follows directly from Chebyshev's inequality H.1. Most of the copulas that are found in the literature turn out to be Archimedean. For a listing of these copulas and a more detailed discussion of the relation between copulas and multivariate distributions, see Marshall and Olkin (1988), Hutchinson and Lai (1990), or Nelsen (2006, pp. 116–119). The use of copulas in finance is discussed by Cherubini, Luciano, and Vecchiato (2005).

21
Convexity and Total Positivity

A. Convex Functions

Convexity is a fundamental analytic tool and the concept appears in many contexts. For more extensive discussions, see, for example, Roberts and Varberg (1973), Rockafeller (1970), van Tiel (1984), or Webster (1994).

In the following, the notation $\bar{\alpha} = 1 - \alpha$ is used. The natural domain of a convex function, defined below, is a convex set.

A.1. Definition. A subset $\mathcal{A}$ of $\mathcal{R}^n$ is said to be *convex* if $x, y \in \mathcal{A}$ implies $\alpha x + \bar{\alpha} y \in \mathcal{A}$ for all $\alpha \in [0, 1]$.

A.2. Definition. A real-valued function ϕ defined on a convex subset $\mathcal{A}$ of $\mathcal{R}^n$ is said to be *convex* if

$$\phi(\alpha x + \bar{\alpha} y) \leq \alpha \phi(x) + \bar{\alpha} \phi(y) \tag{1}$$

for all $x, y \in \mathcal{A}$ and all $\alpha \in [0, 1]$. If the inequality (1) is reversed, then ϕ is said to be *concave*. In other words, a function ϕ is concave if and only if $-\phi$ is convex. In the following, conditions are given only for convexity because corresponding conditions for concavity can be obtained with a sign change.

A.3. Proposition.
(i) A function ϕ defined on $\mathcal{A}$ is convex if and only if

$g(\alpha) = \phi(\alpha x + \bar{\alpha} y)$ is a convex function of $\alpha \in [0, 1]$
for all $x, y \in \mathcal{A}$.

(ii) In case $n = 1$ and $\mathcal{A} = I$ is an interval, ϕ is convex if and only if

$$\frac{\phi(y_1) - \phi(x_1)}{y_1 - x_1} \leq \frac{\phi(y_2) - \phi(x_2)}{y_2 - x_2}$$

whenever $x_1 < y_1 \leq y_2$, $x_1 \leq x_2 < y_2$.

(iii) In case $n = 1$ and $\mathcal{A} = I$ is an interval, ϕ is convex if and only if

$$\phi(x + \Delta) - \phi(x) \text{ is increasing in } x \text{ for all } \Delta > 0; x, x + \Delta \in I.$$

(iv) If $n = 1, \mathcal{A} = (a, b)$ is an open interval and ϕ is differentiable on (a, b), then ϕ is convex if and only if the derivative ϕ' of ϕ is increasing on (a, b).

(v) If $n = 1, \mathcal{A} = (a, b)$ is an open interval and ϕ is twice differentiable on (a, b), then ϕ is convex if and only if the second derivative ϕ'' of ϕ is nonnegative.

(vi) The matrix $H = H(x_1, \ldots, x_n) = (h_{ij}(x_1, \ldots, x_n))$, where

$$h_{ij}(x_1, \ldots, x_n) = \frac{\partial^2 \phi(x_1, \ldots, x_n)}{\partial x_i \partial x_j},$$

is called the *Hessian matrix*. If $\mathcal{A} \in \mathcal{R}^n$ is an open convex set and ϕ is twice differentiable, then ϕ is convex on $\mathcal{A}$ if and only if the Hessian matrix H is positive semidefinite on $\mathcal{A}$.

Proof of (iii). Add the inequality (1) to the inequality obtained from (1) by interchanging α and $\bar{\alpha}$ to obtain the inequality

$$\phi(\alpha x + \bar{\alpha} y) - \phi(x) \leq \phi(y) - \phi(\bar{\alpha} x + \alpha y).$$

Assume that $x < y$ and take $\Delta = \bar{\alpha}(y - x)$ to obtain (iii). □

For a proof of the remaining and related results, see, for example, Marshall and Olkin (1979, Chapter 16), Roberts and Varberg (1973), or Rockafeller (1970).

A.3.a. Example. The function $h(x, \theta) = x^\alpha / \theta^\beta$ has the Hessian matrix H given by

$$H(x, \theta) = \begin{pmatrix} \dfrac{\alpha(\alpha - 1)t^{\alpha-2}}{\theta^\beta} & -\dfrac{\alpha\beta t^{\alpha-1}}{\theta^{\beta+1}} \\ -\dfrac{\alpha\beta t^{\alpha-1}}{\theta^{\beta+1}} & \dfrac{\beta(\alpha + 1)t^\alpha}{\theta^{\beta+2}} \end{pmatrix}.$$

Thus, h is convex in $(x,\theta) \in (0,\infty) \times (0,\infty)$ when $\alpha(\alpha-1) \geq 0$, $\beta(\beta+1) \geq 0$, and $\alpha\beta(\alpha-\beta-1) \geq 0$. In particular, h is convex in $x, \theta > 0$ when $\beta = \alpha - 1 \geq 0$. A direct proof of these results, without computing the Hessian matrix, would be possible but considerably more intricate.

a. Differentiability of Convex Functions

In (iii) of Proposition A.3, the function ϕ is differentiable by assumption. In some applications, it is desirable to avoid this assumption.

A.4. Proposition. Let ϕ be a finite convex function defined on an open interval I of the real line. Then, ϕ is differentiable except possibly on a countable subset of I. Moreover, ϕ' is continuous and increasing relative to the dense subset D of I where ϕ is differentiable.

For a proof of this proposition, see Rockafellar (1970, p. 244).

b. Compositions of Convex Functions

A.5. Proposition. If $\phi_1, \phi_2, \ldots, \phi_k$ are convex functions defined on the convex set $\mathcal{A} \in \mathcal{R}^n$ and h is an increasing convex function defined on $\mathcal{R}^k$, then $\psi(x) = h(\phi_1(x), \phi_2(x), \ldots, \phi_k(x))$ is convex on $\mathcal{A}$. Similarly, if $\phi_1, \phi_2, \ldots, \phi_k$ are concave functions defined in the convex set $\mathcal{A} \in \mathcal{R}^n$ and h is an increasing concave function defined on $\mathcal{R}^k$, then $\psi(x) = h(\phi_1(x), \phi_2(x), \ldots, \phi_k(x))$ is concave on $\mathcal{A}$.

In particular, when $k = n = 1$ and $\mathcal{A} = \mathcal{R}$ it follows that

(i) if ϕ is convex and h is both increasing and convex, then $h(\phi(\,\cdot\,))$ is convex;

(ii) if ϕ is concave and h is both increasing and concave, then $h(\phi(\,\cdot\,))$ is concave;

(iii) if ϕ is convex and h is both decreasing and concave, then $h(\phi(\,\cdot\,))$ is concave;

(iv) if ϕ is concave and h is both decreasing and convex, then $h(\phi(\,\cdot\,))$ is convex.

A.5.a. Observation. If in (i) to (iv), the hypotheses concerning convexity (concavity) of h are replaced by log convexity (log concavity), then the composition $h\phi$ is log convex (concave).

A.5.b. The sum of convex functions is convex and the sum of concave functions is concave. The product of log convex functions is log convex and the product of log concave functions is log concave.

A.6. Proposition. (i) If ϕ is a positive function and $\log \phi$ is convex, then ϕ is convex. (ii) If ϕ is a positive concave function, then $\log \phi$ is concave.

It is not unusual to encounter functions ϕ for which $\log \phi$ takes a simpler form than ϕ itself. Then, (i) may be useful for proving that ϕ is convex.

c. Inverse of a Monotone Convex Function

A.7. Proposition. If ϕ is a strictly increasing convex function defined on a possibly infinite interval (a, b), then the inverse ϕ^{-1} of ϕ is strictly increasing and concave.

Proof. Because ϕ is increasing and convex,

$$\alpha x + \bar{\alpha} y = \phi^{-1}\phi(\alpha x + \bar{\alpha} y) \leq \phi^{-1}(\alpha \phi(x) + \bar{\alpha}\phi(y)). \qquad (2)$$

Let $u = \phi(x), v = \phi(y)$, so that $x = \phi^{-1}(u), y = \phi^{-1}(v)$ and substitute in (2) to obtain

$$\alpha \phi^{-1}(u) + \bar{\alpha} \phi^{-1}(v) \leq \phi^{-1}(\alpha u + \bar{\alpha} v).$$

Thus, ϕ^{-1} is concave. The monotonicity of ϕ^{-1} is well known. ☐

If ϕ is differentiable, an alternative proof of Proposition A.7. can be given using derivatives. From 24.A.4.a, it follows that

$$\frac{d\phi^{-1}(x)}{dx} = \frac{1}{\phi'\phi(x)};$$

By hypothesis, ϕ' and ϕ are both increasing, and consequently, their composition is increasing. This means that $1/\phi'\phi(x)$ is decreasing, and consequently, ϕ^{-1} is concave.

d. Starshaped and Superadditive Functions

A.8. Definition. A real-valued function ϕ defined on $[0, \infty)^n$ is said to be *starshaped* if

$$\phi(\alpha x) \leq \alpha \phi(x) \text{ for all } \alpha \in [0, 1] \text{ and all } x \in [0, \infty)^n.$$

Starshaped functions are sometimes called "$[0, 1]$—subhomogeneous" functions. See, for example, Burai and Száz (2005).

A. Convex Functions   691

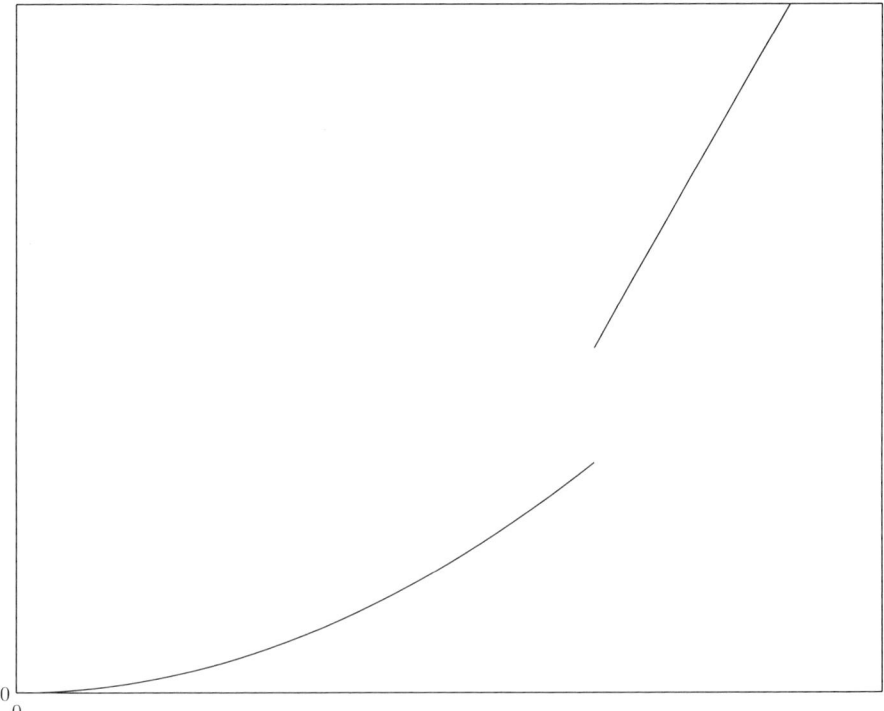

Fig. A.1. Example of a starshaped function

A.9. Proposition. A real-valued function ϕ defined on $[0, \infty)$ is starshaped if and only if $\phi(0) \leq 0$ and $\phi(x)/x$ is increasing in $x > 0$.

To understand the geometric meaning of starshapedness, picture an observer located at the origin; the function ϕ is starshaped if and only if none of the area above the graph of ϕ is hidden from the observer by the graph itself. Starshapedness is weaker than convexity, which requires all area above the graph to be visible to an observer situated anywhere on the graph.

See Figure A.1 for the graph of a star-shaped finction.

A.10. Definition. A real-valued function ϕ defined on a subset $\mathcal{A}$ of $\mathcal{R}^n$ is said to be *superadditive* if $x, y \in \mathcal{A}$ implies $x + y \in \mathcal{A}$ and

$$\phi(x+y) \geq \phi(x) + \phi(y).$$

If $-\phi$ is superadditive, then ϕ is said to be *subadditive*.

A.11. Proposition. Let ϕ be a real-valued function defined on $[0, \infty)$. (i) If $\phi(0) \leq 0$ and ϕ is convex, then ϕ is starshaped. (ii) If ϕ is starshaped, then it is superadditive.

Proof. (i) If $\phi(0) \leq 0$ and ϕ is convex, then by taking $y = 0$ in (1) it follows that ϕ is starshaped. (ii) If ϕ is starshaped, then

$$\phi(\alpha(x+y)) \leq \alpha\phi(x+y); \tag{3}$$

by successively taking $\alpha = x/(x+y)$, then $\alpha = y/(x+y)$ in (3) and adding the results, it follows that ϕ is superadditive. □

e. Jensen's Inequality

A.12. Proposition. Suppose that ϕ is a convex function defined on the open convex subset $\mathcal{A}$ of $\mathcal{R}^n$, and suppose that X is a random variable taking values in $\mathcal{A}$ with a finite expectation. Then,

$$E\phi(X) \geq \phi(EX). \tag{4}$$

This fundamental inequality of Jensen has a version for conditional expectations, which can be properly stated only with measure–theoretic language (see, for example, Gut (2005, p. 477)).

A.13. Proposition. Suppose that ϕ is a convex function defined on the open convex subset $\mathcal{A}$ of $\mathcal{R}^n$ and X is a random variable defined on the probability space $(\Omega, \mathcal{B}, P)$ taking values in $\mathcal{A}$. Let $\mathcal{F} \subset \mathcal{B}$ be a σ-algebra such that $E(X \,|\, \mathcal{F})$ exists finitely almost everywhere (P), that is, except possibly on a set of probability 0. Then with probability 1,

$$E[\phi(X) \,|\, \mathcal{F}] \geq \phi(E[X \,|\, \mathcal{F}]). \tag{5}$$

A.13.a. Example. When $\mathcal{B}$ is the smallest σ-algebra for which the random variable Y is measurable, then (5) takes the more familiar form

$$E[\phi(X) \,|\, Y] \geq \phi(E[X \,|\, Y]) \quad \text{with probability 1.}$$

f. Prékopa's Theorem

A.14. Theorem (Prékopa, 1971). Suppose that ϕ is a nonnegative function defined and log concave on $\mathcal{R}^m \times \mathcal{R}^n$. If the function

$$h(x) = \int_{R^n} \phi(x, z) \, dz$$

is finite for all x, then h is log concave on $\mathcal{R}^m$. When $m = 1$ and mild regularity conditions are satisfied, this can be written in the form

$$\int_{R^n} \phi(x,z) \, dz \int_{R^n} \frac{\partial^2 \phi(x,z)}{\partial x^2} \, dz \leq \left[\int_{R^n} \frac{\partial \phi(x,z)}{\partial x} dz \right]^2.$$

For a proof of Prékopa's theorem, see Brascamp and Lieb (1975).

A.14.a. Example. If f and g are log-concave densities, then according to Proposition 4.B.3, their convolution is log concave. To prove this using Theorem A.14, observe that because f and g are log concave it follows that $\phi(x,z) = f(x-z)g(z)$ is log concave. It follows from Theorem A.14 that the convolution $h(x) = \int f(x-z)g(z) \, dz$ is log concave.

For another application of Theorem A.14, see Proposition 4.C.1.i.

g. Preservation of Convexity, Starshapedness, and Superadditivity

The following propositions are sometimes useful in proving convexity or starshapedness, but first some notation is required. For any function ϕ mapping $[0, \infty)$ into $[0, \infty]$ such that $\phi(0) = 0$, let

$$g_\alpha(x) = [\phi(x^{1/\alpha})]^\alpha, \quad \alpha \neq 0.$$

A.15. Proposition (Marshall and Proschan, 1972b). Let $\mathcal{C}$ be the class of increasing convex functions ϕ defined on $[0, \infty)$ with $\phi(0) = 0$. Then, g_α is in $\mathcal{C}$ whenever ϕ is in $\mathcal{C}$ if and only if $\alpha \geq 1$.

The proof of this result is rather lengthy, and is not reproduced here.

A.16. Proposition (Marshall and Proschan, 1972b). Let $\mathcal{C}$ be the class of increasing starshaped functions ϕ defined on $[0, \infty)$ with $\phi(0) = 0$. Then, g_α is in $\mathcal{C}$ whenever $\alpha \neq 0$.

Proof. As a consequence of the definition of starshapedness, g_α is in C if and only if $\phi(0) = 0$ and, on $(0, \infty)$, $\phi(x) = x\zeta(x)$ for some increasing nonnegative function ζ. Thus, $g_\alpha(0) = 0$, and $g_\alpha(x) = x[\zeta(x^{1/\alpha})]^\alpha$, $x > 0$. Because $[\zeta(x^{1/\alpha})]^\alpha$ is increasing and nonnegative on $(0, \infty)$, it follows that g_α is in C. □

A.17. Proposition. The pointwise limit of convex functions, all defined on the same set, is a convex function. Similar statements can be made about concavity, starshapedness, and superadditivity.

This proposition follows directly from the fact that the inequalities defining the various properties are preserved under pointwise limits.

A.18. Proposition. The set of convex functions forms a convex cone; that is, the sum of convex functions is convex, and every positive multiple of a convex function is convex. The same statement can be made about the sets of concave, starshaped, and superadditive functions.

A.19. Proposition (Hardy, Littlewood and Pólya, 1929). A convex function ϕ defined on a finite interval can be approximated uniformly by finite linear combinations of linear functions and functions of the form $\phi_{(a)}(x)$, where $\phi_{(a)}(x) = 0, x \leq a, \phi_{(a)}(x) = x - a, x \geq a$.

See also Marshall and Olkin (1979, p. 448) for further details.

B. Total Positivity

Total positivity is an indispensable tool in several places in this book. Although the theory of total positivity was already well advanced in the early 1950s, it is not as well known as it deserves to be. A brief outline is offered here (digested from a brief survey of Marshall and Olkin (1979)). The most complete discussion of the field currently available is that of Karlin (1968), who did much to advance the field (see also Gantmakher and Krein, 1961). A more recent collection of papers on the subject has been edited by Garsca and Micchelli (1996). A survey of the subject is offered by Barlow and Proschan (1965). Both Barlow and Proschan (1965) and Marshall and Olkin (1979) offer brief comments about the history of the subject, and interesting historical details not well known are given by Pinkus (1996).

B.1. Definition. Let A and B be subsets of the real line. A function K defined on $A \times B$ is said to be *totally positive of order k*, denoted TP_k, if for all $x_1 < \cdots < x_m$, $y_1 < \cdots < y_m$, $(x_i \in A, y_i \in B)$, and for all $m, 1 \leq m \leq k$

$$K\begin{pmatrix} x_1, \ldots, x_m \\ y_1, \ldots, y_m \end{pmatrix} \equiv \det \begin{bmatrix} K(x_1, y_1) & \ldots & K(x_1, y_m) \\ \vdots & & \vdots \\ K(x_m, y_1) & \ldots & K(x_m, y_m) \end{bmatrix} \geq 0. \quad (1)$$

When the inequalities (1) are all strict, K is said to be *strictly totally positive of order k* (STP$_k$). If K is TP$_k$ (STP$_k$) for all $k = 1, 2, \ldots$, then it is said to be *totally positive (strictly totally positive) of order ∞*, written TP$_\infty$ (STP$_\infty$).

Some consequences of the definition are as follows.

B.2. If g and h are nonnegative functions defined on A and B, respectively, and if K is TP$_k$ on $A \times B$, then $g(x)h(y)K(x,y)$ is TP$_k$ on $A \times B$.

B.3. If g and h are increasing functions defined, respectively, on A and B, and if K is TP$_k$ on $g(A) \times h(B)$, then $K(g(x), h(y))$ is TP$_k$ on $A \times B$.

B.4. If K is TP$_k$ on $A \times B$ and $A_0 \subset A, B_0 \subset B$, then K is TP$_k$ on $A_0 \times B_0$.

a. Examples

With the aid of B.2 and B.3, many examples of totally positive functions are obtainable from a few relatively basic examples.

B.5. Example. The function

$$K(x,y) = e^{xy}, \quad -\infty < x, \ y < \infty, \quad \text{is STP}_\infty. \quad (2)$$

In this case, the positivity of the relevant determinants is well known because they are generalized Vandermonde determinants. See, e.g., Pólya and Szegö (1972, p. 46, Problem 76) for a discussion of these determinants.

B.5.a. Example. The function

$$K(x,y) = x^y, \quad 0 \leq x < \infty, \ -\infty < y < \infty, \quad \text{is STP}_\infty.$$

B.5.b. Example. The function

$$K(x,y) = e^{-(x-y)^2}, \quad -\infty < x, \ y < \infty, \quad (3)$$

is STP$_\infty$.

B.6. Example. The indicator function

$$K(x,y) = 1 \quad \text{if } x \leq y, \qquad (4)$$
$$= 0 \quad \text{if } x > y, \ -\infty < x, \ y < \infty,$$

is TP_∞.

B.6.a. Example. The indicator function

$$K(x,y) = 0 \quad \text{if } x < y,$$
$$= 1 \quad \text{if } x \geq y, \ -\infty < x, \ y < \infty, \qquad (5)$$

is TP_∞.

B.7. Definition. A real-valued function f defined on $(-\infty, \infty)$ is said to be a *Pólya frequency function of order k* (PF_k) if the function $K(x,y) = f(y-x), -\infty < x, y < \infty$, is totally positive of order k.

Note that the term *frequency function* in this context does not require that f be a density, though with $k = 1$, total positivity requires that f be nonnegative.

B.8. Proposition (Schoenberg, 1951). The function

$$K(x,y) = f(y-x), \quad -\infty < x, \ y < \infty,$$

is TP_2 if and only if f is nonnegative and $\log f$ is concave on $(-\infty, \infty)$. Thus, f is log concave on $(-\infty, \infty)$ if and only if f is PF_2.

This important fact is often useful.

B.9. Proposition. The function

$$K(x,y) = f(y+x), \quad 0 \leq x, \ y < \infty,$$

is TP_2 in $x, y \geq 0$ if and only if f is nonnegative and $\log f$ is convex on $[0, \infty)$.

b. The Basic Composition Formula

The following lemma is called the *basic composition formula*; it will be recognized as a generalization of the well-known *Binet–Cauchy formula* of matrix theory.

B.10. Lemma (Andréief, 1883; Pólya and Szegö, 1925, 1972, p. 61, Problem 68). If σ is a σ–finite measure and the integral $M(x,y) = \int K(x,z)L(z,y) \, d\sigma(z)$ converges absolutely, then with the notation introduced in (1),

$$M\begin{pmatrix}x_1,\ldots,x_m \\ y_1,\ldots,y_m\end{pmatrix} = \int\cdots\int_{z_1<\ldots<z_m} K\begin{pmatrix}x_1,\ldots,x_m \\ z_1,\ldots,z_m\end{pmatrix} L\begin{pmatrix}z_1,\ldots,z_m \\ y_1,\ldots,y_m\end{pmatrix} d\sigma(z_1)\ldots d\sigma(z_m).$$

A proof of this result is outlined by Karlin (1968, p. 17).

A consequence of the basic composition formula is the following theorem, one of the most basic and useful properties of totally positive functions.

B.11. Theorem. If K is TP_m on $A \times B$, L is TP_n on $B \times C$, and σ is a sigma finite measure, then

$$M(x,y) = \int K(x,z) L(z,y)\, d\sigma(z)$$

is $\text{TP}_{\min(m,n)}$ on $A \times C$.

B.11.a. Example. Here is an application of Theorem B.11. According to Example B.5.a, the function $K(x,r) = x^r$ is totally positive in x and r. From Theorem B.11, it follows that

$$\mu_{r+s} = \int_0^\infty x^r x^s\, dF(x)$$

is totally positive in r and s for all r,s such that the moments exist. This result is to be compared with Proposition 20.B.8.

c. The Variation Diminishing Property

To a considerable degree, the interest of totally positive functions is due to their variation diminishing property investigated early by Schoenberg (1930). To describe this property, the following definition is useful.

B.12. Definition. If $f\colon B \to (-\infty, \infty)$, where B is a subset of $(-\infty, \infty)$, then the *number of sign changes* of f on B is the supremum of the numbers of sign changes in sequences of the form $f(x_1),\ldots,f(x_m)$ where m is finite, $x_1,\ldots,x_m$ are points in B, $x_1 < \cdots < x_m$, and zero values in the sequence are discarded.

B.13. Theorem. Let $K\colon A \times B \to (-\infty, \infty)$ be Borel-measurable and TP_k. Let σ be a regular σ-finite measure on B, and $f : B \to (-\infty, \infty)$

be a bounded measurable function such that the integral

$$g(x) = \int_B K(x,y) f(y) \, d\sigma(y)$$

converges absolutely. If f changes sign at most $j \leq k-1$ times on B, then g changes sign at most j times on A. Moreover, if g changes sign j times, then it must have the same arrangement of signs as does f.

d. Log Concavity

B.14. Proof of Proposition 4.B.3. Proposition 4.B.3 states that if f and g are log-concave densities, then their convolution

$$h(x) = \int f(x-z) g(z) \, dz$$

is log concave. To prove this result, first use B.7 to restate the log-concavity in terms of total positivity of order 2, and then use B.11. □

B.15. Proof of Proposition 4.B.8. To show that $\bar{F}$ is log concave whenever f is log concave, write

$$\bar{F}(x-z) = \int_0^\infty K(x,y) \, f(y-z) \, dy,$$

where K, given by (4), is the indicator function of the set $\{x \leq y\}$. By B.7, $f(y-z)$ is TP_2 and because K is TP_∞, it follows from B.11 that $\bar{F}$ is TP_2. Again from B.7, it follows that $\bar{F}$ is log concave. To show that F is log concave, use K as defined in (5).

To show that $\bar{F}$ is log convex whenever f is log convex, replace $f(y-z)$ by $f(y+z)$ in the above proof. □

e. Crossings of Distributions and Moments

Let F and G be distributions with corresponding rth moments $\mu_r(F)$ and $\mu_r(G)$. As noted in 1.C(4), if $F(x) = 0$, $x < 0$, then

$$\mu_r(F) = r \int_0^\infty \bar{F}(x) x^{r-1} \, dx,$$

and a similar formula holds for $\mu_r(G)$. It can be seen directly from this formula that if $\mu_r(F) = \mu_r(G)$, then it cannot be that $F(x) \leq G(x)$ for all x unless equality holds for all x. That is to say, F and G must cross at least once. The following result makes a stronger statement.

B.16. Moment crossings. Suppose that F and G are distribution functions with support $[0, \infty)$ and corresponding moments $\mu_r(F)$ and $\mu_r(G)$. If F crosses G k times, then $\mu_r(F)$ crosses $\mu_r(G)$ at most k times (as functions of r).

Proof. Use equation 1.B(4) to write

$$\mu_r(F) - \mu_r(G) = r \int_0^\infty x^{r-1}[\bar{F}(x) - \bar{G}(x)] \, dx.$$

Now, the results follow from Theorem B.13 and the fact (Example B.5.a) that x^{r-1} is totally positive in x and r. □

The noncentral distributions discussed in the first three sections of Chapter 15 are represented as mixtures; the densities are infinite series. The following proposition, a direct application of Theorem B.13, is useful in determining the shape of these densities.

B.17. Proposition. If the sequence $\{a_j\}_{j=1}^\infty$ has at most $k-1$ sign changes in j, and if $K(j, x)$ is TP_k in $x \geq 0$ and $j = 0, 1, \ldots$, then

$$g(x) = \sum_{j=0}^\infty a_j K(j, x)$$

has at most $k-1$ sign changes in $x \geq 0$. If g has $k-1$ sign changes, then they must be in the same order as the sign changes of a_j.

f. Crossing Shorthand Notation

Sign changes are defined in Definition B.12; the term "crossing" is often used in reference to a sign change. A function g is said to *cross the function h at most once, and only from above* if either $g - h$ is of one sign (they do not cross at all), or if there exists x_0 such that

$$g(x) \geq h(x), x < x_0, \quad g(x) \leq h(x), x > x_0. \tag{6}$$

It is sometimes convenient to indicate this property with the shorthand notation

$$g(x) - h(x) : +, -, \tag{7}$$

which can be thought of as saying that $g(x) - h(x)$ can go only from $+$ to $-$ as x increases, with the possibility that no sign change occurs.

There is kind of a calculus that goes with the shorthand notation (7) that depends upon the fact that if ζ is a function strictly increasing and continuous on the range of g and h, then (6) is equivalent to

$$\zeta(g(x)) - \zeta(h(x)) : +, -. \qquad (8)$$

As a consequence of (8), it follows that (7) can be easily transformed to various equivalent conditions such as

$cg(x) - ch(x) : +, -,$ where c is a positive constant,
$(g(x) + a) - (h(x) + a) : +, -,$ where a is a constant,
$\log g(x) - \log h(x) : +, -.$

In fact, if h itself is strictly increasing and continuous, then the inverse h^{-1} is a possible choice for ζ so that (7) is equivalent to

$$h^{-1}g(x) - x : +, -.$$

22
Some Functional Equations

A number of functional equations are encountered in the earlier chapters of this book, especially in Chapter 18. The solutions to these equations are not as widely known as they might be, and are reviewed here.

An extensive discussion of functional equations has been provided by Aczél (1966), who traces the subject to papers of 1747–1750 by D'Alembert, papers of 1755–1770 by Euler, and especially to a paper of 1821 by Cauchy. A more recent book on the subject with a number of applications in probability, statistics, and economics has been written by Castillo and Ruiz-Cobo (1992). Functional equations have been used extensively to characterize distributions by Lukacs and Laha (1964), Kagan, Linnik and Rao (1973), and Azlarov and Volodin (1986). Functional equations and applications to information theory are discussed by Aczél and Daróczy (1975).

A. Cauchy's Equations

The following closely related equations bear the name of Cauchy:

$$f(x+y) = f(x) + f(y), \tag{1}$$
$$f(x+y) = f(x)f(y), \tag{2}$$
$$f(xy) = f(x)f(y), \tag{3}$$
$$f(xy) = f(x) + f(y). \tag{4}$$

Several of these equations are encountered in this book. Also, variants arise, which can be reduced to one of these equations; these are discussed later. Propositions A.1, A.2, A.3, and A.4 relate, respectively, to equations (1), (2), (3), and (4).

A.1. Proposition. If the function f is continuous at a point or bounded in some interval, and if (1) holds for all real x, y, then there exists a constant c such that

$$f(x) = cx \quad \text{for all } x. \tag{5}$$

If (1) holds for all nonnegative (positive) x and y, then (5) holds for all nonnegative (positive) x.

See Aczél (1966, Chapter 2) or Castillo and Ruiz-Cobo (1992, Chapter 1) for a discussion of equation (1). Proposition A.1 can also be obtained from Proposition A.2 by taking logarithms; a proof of Proposition A.2 is offered after Proposition A.5.

A.1.a. Proposition. If the function f satisfies the equation

$$f\left(\frac{x+y}{2}\right) = \frac{f(x)+f(y)}{2} \quad \text{for all real } x, y, \tag{1a}$$

then $f(x) + f(0)$ satisfies (1). Hence, if f satisfies the conditions of Proposition A.1, then f is a linear function.

Proof. Following Aczél (1966, p. 43), let $y = 0$ in (1a) to obtain

$$f\left(\frac{x}{2}\right) = \frac{f(x)+f(0)}{2}. \tag{1b}$$

Replace x by $x + y$ in (1b) to obtain

$$\frac{f(x)+f(y)}{2} = f\left(\frac{x+y}{2}\right) = \frac{f(x+y)f(0)}{2}.$$

This equation gives (1) with $f(x) + f(0)$ in place of $f(x)$. □

A.2. Proposition. Suppose that the function f is continuous at a point or bounded in some interval. If f satisfies the functional equation (2) for all x, y, then either $f(x) = 0$ for all x, or there exists a constant a such that

$$f(x) = \exp\{ax\}. \tag{6}$$

If (2) holds for all nonnegative (positive) x, y, then (6) holds for all nonnegative (positive) x.

A.3. Proposition. Suppose that the function f is continuous at a point or bounded in some interval. If f is a positive function and (3) holds for all $x, y > 0$, that is, if

$$f(xy) = f(x)f(y) \quad \text{for all } x, y > 0,$$

then for some constant c,

$$f(x) = x^c, \quad x > 0. \tag{7}$$

If (3) holds for all real x, y, then there are several possibilities. For some constant c, and for all x,

(i) $f(x) = |x|^c$, (ii) $f(x) = |x|^c \operatorname{sgn} x$,
(iii) $f(x) = \operatorname{sgn} x$, (iv) $f(x) = 0$, (v) $f(x) = 1$. (8)

Proof. For the solution when $x, y > 0$, let $g(z) = f(e^z)$ and apply Proposition A.2. For the case that (3) holds for all real x, y, see Aczél (1966, Theorem 3, p. 41) or Castillo and Ruiz-Cobo (1992, Theorem 2.4.5, p. 30). □

A.4. Proposition. Suppose that the function f is continuous at a point or bounded in some interval. If f is a positive function and (4) holds for all $x, y > 0$, then for some constant c,

$$f(x) = c \log x, \quad x > 0. \tag{9}$$

If (4) holds for all real x, y, then $f(x) = 0$ for all x.

If (2) holds, then by iteration, it follows that $f(x_1 + x_2 + \cdots + x_k) = \Pi_{i=1}^k f(x_i)$, and if all of the x_i are equal, it follows that

$$[f(x)]^k = f(kx), \quad k = 1, 2, \ldots. \tag{10}$$

Thus, (10) is a weaker equation than (2), and as such, it possibly could have more solutions.

A.5. Proposition. Suppose that f satisfies the regularity conditions of Proposition A.1. If (10) holds for all x and for all positive integers

k, then there exist constants a,b such that

$$f(x) = e^{ax}, \quad x \geq 0, \quad f(x) = e^{bx}, \quad x \leq 0. \tag{11}$$

If (10) holds for all nonnegative (positive) x, then (11) holds for all nonnegative (positive) x.

Proof Assuming Continuity. Let $x \geq 0$ and $y = kx$ to obtain from (10) the equation

$$[f(y)]^{1/k} = f(y/k), \quad y > 0. \tag{12}$$

This means that (10) holds for $k = 1, 1/2, 1/3, \ldots$. With $mx = 1$ in (10), it follows that

$$[f(1/m)]^k = f(k/m) \quad \text{and} \quad f(1) = [f(1/m)]^m;$$

thus,

$$f(k/m) = [f(1/m)]^k = [f(1)]^{k/m}. \tag{13}$$

By using the assumption of continuity and letting $k/m \to x \geq 0$ in (13), (11) follows, where $f(1) = e^a$. For the case that $x \leq 0$, the argument is similar, and $f(-1) = e^b$. □

To prove Proposition A.2 under the assumption of continuity, use Proposition A.5 and (2) with $x = -y$. A simple proof of Proposition A.2 under the assumption that f is bounded in some interval is given by Feller (1968, p. 459). Because continuity at some point implies boundedness in some interval, Feller's proof covers both regularity conditions stated in Proposition A.2. The ideas of his proof can easily be adapted to prove Proposition A.5. A thorough discussion of Proposition A.2, its various relatives, and its history are given by Aczél (1966, Chapter 2).

B. Variants of Cauchy's Equations

A number of variations and generalizations of Cauchy's functional equations have been discussed in the literature, and in particular by Aczél (1966). Some of these functional equations are encountered in this book and are briefly discussed here.

a. Transformations

The functional equation

$$f((x^\alpha + y^\alpha)^{1/\alpha}) = f(x)f(y)$$

can be rewritten in the form

$$g(x^\alpha + y^\alpha) = g(x^\alpha)g(y^\alpha),$$

where $g(x) = f(x^{1/\alpha})$. Thus, g satisfies the functional equation A(2), and with continuity, it follows that $f(x) = \exp\{cx^\alpha\}$ for some real constant c and all x. The essential property here is that the function $\phi(x) = x^\alpha$ is continuous and strictly monotone in $x \geq 0$, so that it has a continuous inverse. Variants of Cauchy's equation can be found with any such function.

b. Pexider Equations

Pexider equations are similar to the basic equations of Cauchy, but they involve more than one function. Two forms of these equations are discussed here together with some special cases. These forms are

$$f(xy) = g(x)h(y), \tag{1}$$
$$f(x+y) = g(x)h(y). \tag{2}$$

The particularly simple special case that $f = h$ in (1) arises repeatedly in Chapter 18, and is spelled out in detail in Proposition B.1a. But first, the general case is considered.

B.1. Proposition. Suppose that f, g, and h are real functions defined on $(0, \infty)$ and that f is continuous at some point. If (1) holds for all $x, y > 0$, then for some constants a, b, c, either

$$f(x) = abx^c, \quad g(x) = ax^c, \quad h(x) = bx^c, \quad x > 0,$$

or

$$f(x) = g(x) = 0 \text{ and } h \text{ is arbitrary,}$$

or

$$f(x) = h(x) = 0 \text{ and } g \text{ is arbitrary.}$$

For a proof of this result, see Aczél (1966, Theorem 4, p. 144).

B.1.a. Proposition. Suppose that f and g are real functions defined on $(0, \infty)$ and that f is continuous at some point. If

$$f(xy) = g(x)f(y) \quad \text{for all } x, y > 0, \tag{1a}$$

then either

$$f(x) = bx^c, \quad g(x) = x^c, \quad x > 0,$$

for some constants b and c, or

$$f(x) = 0, \quad \text{for all } x > 0, \quad \text{and } g \text{ is arbitrary.}$$

Equation (1a) is easily transformed to Cauchy's equation A(3) as follows: Set $y = 1$ in (1a) to conclude that either $f(x) = 0$ for all x or $g(x) = f(x)/f(1)$. With this, (1a) takes the form

$$\frac{f(xy)}{f(1)} = \frac{f(x)}{f(1)} \frac{f(y)}{f(1)}.$$

Another variant of Proposition B.1.a that arises several times in Chapter 18 is the following.

B.1.b. Proposition. Suppose that ϕ is a positive function and g is a real function, both defined on $(0, \infty)$. If

$$\phi(xy) = [\phi(x)]^{g(y)}, \quad x, y > 0, \tag{1b}$$

then $g(y) = y^c$ for some real c and $\phi(x) = \exp\{bx^c\}$ for some real b.

Proof. Take logarithms in (1b) and let $f(x) = \log \phi(x)$ to transform (1b) to (1a). □

B.1.c. Proposition. Suppose that ϕ and g are positive functions defined on $(0, \infty)$ that satisfy the functional equation

$$\phi(x^y) = [\phi(x)]^{g(y)}, \quad x, y > 0. \tag{1c}$$

Because $x^y \geq 1$ if and only if $x \geq 1$, (1c) provides no link between the values of ϕ in $(0, 1)$ and on $[1, \infty)$. Consequently, these cases must be treated separately.

Case 1: $x < 1$. Either

$$\phi(x) = \exp\{b(-\log x)^c\} \quad \text{and} \quad g(y) = y^c$$

for some constants b and c, or

$$\phi(x) = 0 \quad \text{for all } x < 1, \quad \text{and } g \text{ is arbitrary,}$$

or

$$\phi(x) = 0 \quad \text{for all } x < 1, \quad \text{and } g \text{ is arbitrary.}$$

Case 2: $x \geq 1$. Either

$$\phi(x) = \exp\{b'(\log x)^{c'}\} \quad \text{and} \quad g(y) = y^{c'}$$

for some constants b' and c', or

$$\phi(z) = 0 \quad \text{for all } x \geq 1, \quad \text{and } g \text{ is arbitrary,}$$

or

$$\phi(z) = 1 \quad \text{for all } x \geq 1, \quad \text{and } g \text{ is arbitrary.}$$

Proof. Let $x = e^u$, that is, $u = \log x$, and let $f(u) = \log \phi(e^u)$. Then, (1c) becomes (1a) and the results follows from Proposition B.1.a. □

B.1.d. Proposition. Let ψ be a real-valued function and g be a positive function, each defined on $(0, \infty)$. If

$$\psi(x^y) = g(y)\psi(x), \quad x, y > 0, \tag{1d}$$

then for $x < 1$, either

$$\psi(x) = b(-\log x)^c \quad \text{and} \quad g(y) = y^c$$

for some constants b and c, or
$$\psi(x) = 0.$$

For $x \geq 1$, either
$$\psi(x) = b'(\log x)^{c'} \quad \text{and} \quad g(y) = y^{c'},$$
for some constants b' and c', or
$$\psi(x) = 0.$$

Proof. This result follows immediately from Proposition B.1.c. □

B.2. Proposition. Suppose that f, g, and h are real functions defined on $(-\infty, \infty)$ and f is continuous at some point. If
$$f(x+y) = g(x)h(y) \quad \text{for all real } x, y, \tag{2}$$
then either
$$f(z) = ab\, e^{cz}, \quad g(z) = a\, e^{cz}, \quad h(z) = b\, e^{cz}, \quad -\infty < z < \infty$$
for some constants a, b, and c, or
$$f(z) = g(z) = 0 \text{ and } h \text{ is arbitrary},$$
or
$$f(z) = g(z) = 0 \text{ and } g \text{ is arbitrary}.$$

Proof. In (2), let $x = e^u, y = e^v$ to obtain this from Proposition B.1. □

B.2.a. Proposition. Suppose that f and g are real functions defined on $(-\infty, \infty)$ and f is continuous at some point. If
$$f(x+y) = g(x)f(y) \quad \text{for all real } x, y, \tag{2a}$$
then either
$$f(x) = b\, e^{cx}, \quad g(x) = e^{cx}, \quad -\infty < x < \infty$$

for some constants b and c, or

$$f(x) = 0 \text{ and } g \text{ is arbitrary.}$$

Proof. This is an immediate consequence of Proposition B.2. □

The following proposition is similar to Proposition B.2.a; only the domain of the functions has been changed. This modified proposition does not follow directly from Proposition B.2.

B.2.b. Proposition. Suppose that f and g are real functions defined on $[0, \infty)$ and f is continuous at some point. If

$$f(x+y) = g(x)f(y) \quad \text{for all } x, y \geq 0, \tag{2b}$$

then either

$$f(x) = b\, e^{cx}, \quad g(x) = e^{cx}, \quad 0 \leq x < \infty$$

for some constants b, c, or

$$f(x) = 0 \text{ and } g \text{ is arbitrary.}$$

Proof. In (2b), set $y = 0$ to conclude that $f(x) = g(x)f(0)$. If $f(0) = 0$, then $f(x) = 0$ for all $x > 0$. If $f(0) \neq 0$, it follows that $g(x) = f(x)/b$, where $b = f(0)$. Then, (2b) can be rewritten in the form

$$f(x+y) = \frac{f(x)}{b} f(y) \quad \text{for all } x, y \geq 0,$$

that is,

$$\frac{f(x+y)}{b} = \frac{f(x)}{b} \frac{f(y)}{b} \quad \text{for all } x, y \geq 0.$$

It follows from Proposition A.2 that

$$\frac{f(x)}{b} = e^{cx}, \quad x \geq 0,$$

for some constant c, and the result follows. □

c. Sincov's Equation

The following proposition is discussed by Aczél (1966, p. 64) and attributed to D.M. Sincov.

B.3. Proposition. If $a > 0$ and for all x, y,

$$f(x+y) = a^{xy} f(x) f(y), \tag{3}$$

then $f(x) = a^{x^2/2} g(x)$, where g is an arbitrary solution to the functional equation A(2). Thus, under reasonable regularity conditions, solutions of (3) have the form $f(x) = a^{(x^2/2)+cx}$.

B.3.a. Proposition. If ϕ and h are nonnegative functions defined on $[0, \infty)$ such that

$$\phi(xy) = x^{h(y)} \phi(x) \phi(y), \tag{4}$$

then under regularity conditions such as continuity at a point, for some constants b, c,

$$\phi(x) = x^{bc} \exp\{c(\log x)^2/2\}, \quad h(x) = c \log x. \tag{5}$$

Proof. With x and y interchanged in (4), it follows that $x^{h(y)} = y^{h(x)}$. This means that $h(y) \log x = h(x) \log y$, and so $(\log z)/h(z)$ is a constant. Consequently, for some constant b, $h(z) = b \log z$. It follows that (4) can be written in the form

$$\exp\{b (\log x)(\log y)\} \phi(x) \phi(y) = \phi(xy). \tag{6}$$

Let $u = \log x$, $v = \log y$, and $f(u) = \phi(e^u)$ to obtain from (6) the functional equation

$$f(u+v) = a^{uv} f(u) f(v), \quad -\infty < u, v < \infty, \tag{7}$$

where $a = e^b$. The solution of (7), given by Proposition B.3, has the form

$$f(x) = cx a^{x^2/2} = cx \exp\{bx^2/2\}, \quad x > 0.$$

In terms of ϕ, this solution becomes

$$\phi(x) = x^{bc} \exp\{c(\log x)^2/2\}.$$

□

B.4. Proposition. Suppose that f and g are real functions defined on $[0, \infty)$ and that

$$f(x+y) = f(x)g(y) + f(y) \quad \text{for all } x, y \geq 0. \tag{8}$$

Let f_0 be an arbitrary solution of A(2) and let f_1 be an arbitrary solution of A(1). The most general solutions of (8) have the form

(i) $f(x) = \alpha[1 - f_0(x)], \quad g(x) = f_0(x), \quad x \geq 0,$
(ii) $f(x) = f_1(x), \quad g(x) = 1, \quad x \geq 0,$
(iii) $f(x) = a \neq 0, \quad g(x) = 0, \quad x \geq 0,$
or
(iv) $f(x) = 0, \quad x \geq 0, \quad g$ arbitrary.

This proposition is a special case of Theorem 1 of Aczél (1966, p. 150); see also Castillo and Ruiz-Cobo (1992, p. 49).

Proof Assuming f is Differentiable. Differentiate (8) with respect to x to obtain

$$f'(x+y) = f'(x)g(y). \tag{9}$$

If $f'(0) = 0$, then it follows from (9) that $f'(z) = 0$ for all $z \geq 0$ and either (iii) or (iv) hold. Suppose that $f'(y) \neq 0$. From (9) it follows that $g(y) = f'(y)/f'(0)$ and (9) can be written in the form

$$g(x+y) = g(x)g(y). \tag{10}$$

From Proposition A.2, it follows that $g(x) = e^{ax}$ for some constant a. If $a = 0$, then (ii) holds; if $a \neq 0$, then because $g(y) = f'(y)/f'(0) = e^{ay}$, f has the form $f(x) = \xi e^{ax} + c$ for some constant c, which must be $-\xi$ to satisfy (8). □

B.4.a. Proposition. The function f satisfies the equation

$$f(x+y) = f(x)g(y) + h(y) \tag{11}$$

if and only if f is a constant, or (8) holds with $f(x) - f(0)$ in place of $f(x)$ and $h(y) = f(y) - f(0)g(y)$.

Proof. With $y = 0$ in (11), it follows that either f is a constant or $g(0) = 1$ and $h(0) = 0$. Suppose that f is not a constant. In (11), set $x = 0$ to obtain $h(y) = f(y) - f(0)g(y)$. With this value inserted, (11) becomes

$$f(x+y) - f(0) = [f(x) - f(0)]g(y) + [f(y) - f(0)]. \qquad \square$$

C. Some Additional Functional Equations

a. The Associativity Equation

The equation

$$F[F(x,y)z] = F[x, F(y,z)] \tag{1}$$

is called the *associativity equation*. This equation has been studied extensively in quite general contexts; see Aczél (1966, Section 6.2) and the references therein or Castillo and Ruiz-Cobo (1992, p. 94). But in this book the interest is in the case that the F is a real-valued function of real variables.

C.1. Proposition. Let I be an interval of the real line, and suppose that F maps $I \times I$ to I. If F satisfies (1), is continuous in each argument the other being fixed, and if there exist numbers e and $x^{-1} \in I$ such that

$$F(e, x) = x, \tag{2}$$

$$F(x^{-1}, x) = e, \tag{3}$$

then and only then, there exists a continuous and strictly monotonic function f defined on $(-\infty, \infty)$ with range I such that

$$F(x, y) = f[f^{-1}(x) + f^{-1}(y)]. \tag{4}$$

Equation (1) can hold for functions F continuous in each argument and satisfying (2), (3) only if the interval I is open. A function g satisfies (4) if and only if for some constant $a, g(x) = f(ax)$.

The proof of this result is beyond the scope of this book, but see Aczél (1966, p. 254), or Castillo and Ruiz-Cobo (1992, p. 94).

b. The Transformation Equation and a Relative

The transformation equation arises in equation 7.F(24), where a particular solution is found.

For the following, see Aczél (1966, p. 316) or Castillo and Ruiz-Cobo (1992, p. 101).

C.2. Proposition. Let ψ be a continuous function defined on a real rectangle. Suppose that ψ is invertible in each variable, i.e., the equation $\psi(x, y) = c$ can be solved uniquely for x when y is fixed, and uniquely for y when x is fixed. The general solution to the functional equation

$$\psi(\psi(x, y), z) = \psi(x, h(y, z)) \tag{5}$$

can be written in the form

$$\psi(x, y) = u^{-1}(u(x) + v(y)), \quad h(x, y) = v^{-1}(v(x) + v(y)), \tag{6}$$

where u and v are continuous and strictly monotonic functions. The representation (6) is not unique: $u(x)$ and $v(x)$ can be replaced by $u^*(x) = u(ax + b), v^*(x) = av(x)$, where a and b are constants.

If $h(x, y) = xy$, then by taking $g(x) = e^{u(x)}$, ψ can be written in the form

$$\psi(x, y) = g^{-1}(yg(x)),$$

where g is strictly monotone.

If ψ is defined on $(0, 1) \times (0, 1)$, $u(x) = \log[x/(1-x)]$, $v(x) = \log x$, then the particular solution to equation (5) is $h(x, y) = xy, \psi(x, y) = xy/[1 - x(1-y)]$. This is the solution found in the proof of Proposition 7.F.12 where the additional requirement of ψ is that it be a probability generating function for some random variable taking on positive integer values. Whether other probability generating functions satisfy (5) is not known.

Equation (5) can be regarded as a special case of the more general functional equation

$$F(G(x, y), z) = H(M(x, z), N(y, z)).$$

Another special case is given in the following proposition.

C.3. Proposition. Let F be a real function defined on $A \times B$ and G be a function that maps $A \times B \to A$, where A and B are intervals of the real line. Suppose that

(i) $F(x, y)$ has continuous partial derivatives F_1, F_2;
(ii) for some constant c, $F_2(c, y) \neq 0$;
(iii) for some constant a, $F(a, y)$ is constant;
(iv) $F(c, y) = z$ has a unique solution $(c \neq 0)$;
(v) $F \neq G, F(x, y) \neq xy$.

Then every solution of the equation

$$F(w, G(x, y)) = F(w, x) F(w, y) \qquad (7)$$

has the form

$$F(x, y) = f(x)^{g(y)}, \qquad (8)$$

where f and g are strictly monotonic and differentiable.

Equation (8) is discussed by Castillo and Ruiz-Cobo (1992, p. 132).

c. Addition Equations

Equations of the form

$$f(x + y) = \Phi(f(x), f(y)) \qquad (9)$$

are called *addition equations*.

C.4. Proposition. If Φ is a polynomial, the general solutions of (9) that are continuous at a point have the form

$$f(x) = ax + b \quad \text{or} \quad f(x) = ae^{cx} + b.$$

See Aczél (1966, p. 61) or Castillo and Ruiz-Cobo (1992, p. 33) for further discussion of this result.

C. Some Additional Functional Equations

C.4.a. Proposition. The most general solution of the polynomial addition equation

$$f(x+y) = af(x)f(y) + bf(x) + bf(y) + c, \quad x, y > 0, \tag{10}$$

where $a > 0$ and $c = (b^2 - b)/a$ has the form

$$f(x) = \frac{e^{\theta x} - b}{a}. \tag{11}$$

Proof. (Aczél, 1966, p. 60). Rewrite (10) as

$$af(x+y) + b = [af(x) + b][af(y) + b], \quad x, y > 0,$$

which with $h(x) = af(x) + b$ can be rewritten as

$$h(x+y) = h(x)h(y), \quad x, y > 0.$$

It follows from Proposition A.2 that for some θ, $h(x) = e^{\theta x}$, and thus (11) holds. □

23

Gamma and Beta Functions

A. The Gamma Function

The gamma function Γ is variously known as "Euler's integral of the second kind," "Euler's integral," or as the "factorial function" because $\Gamma(n) = (n-1)!$ for every positive integer n. The gamma function can be viewed as a continuous extension of the factorial function.

A.1. Definition. The integral

$$\Gamma(z) = \int_0^\infty t^{z-1} e^{-t} \, dt, \tag{1}$$

defined for $z > 0$, is called the *gamma function*.

The gamma function can also be defined and is finite on much of the complex plane, including noninteger negative values, but apart from Proposition A.2.b, the restrictive definition given here is adequate for the purposes of this book. See Erdélyi, Magnus, Oberhettinger, and Tricomi (1953) or Abramowitz and Stegun (1964); for a more detailed account of the gamma function, see Artin (1931, 1964).

For a graph of the gamma function see Figure A.1.

A.2.a. Proposition. $\Gamma(z+1) = z\Gamma(z), \quad 0 < z < \infty.$

Proof. This result can be proved through an integration by parts. Alternatively, it can be shown using Fubini's theorem 24.B.1 concerning the interchange of order of integration:

$$\Gamma(z+1) = \int_0^\infty t^z e^{-t} \, dt = \int_{t=0}^\infty e^{-t} \int_{s=0}^t z s^{z-1} \, ds \, dt$$
$$= z \int_{s=0}^\infty s^{z-1} \int_{t=s}^\infty e^{-t} \, dt \, ds = z \int_{s=0}^\infty s^{z-1} e^{-s} \, ds = z\Gamma(z). \quad \square$$

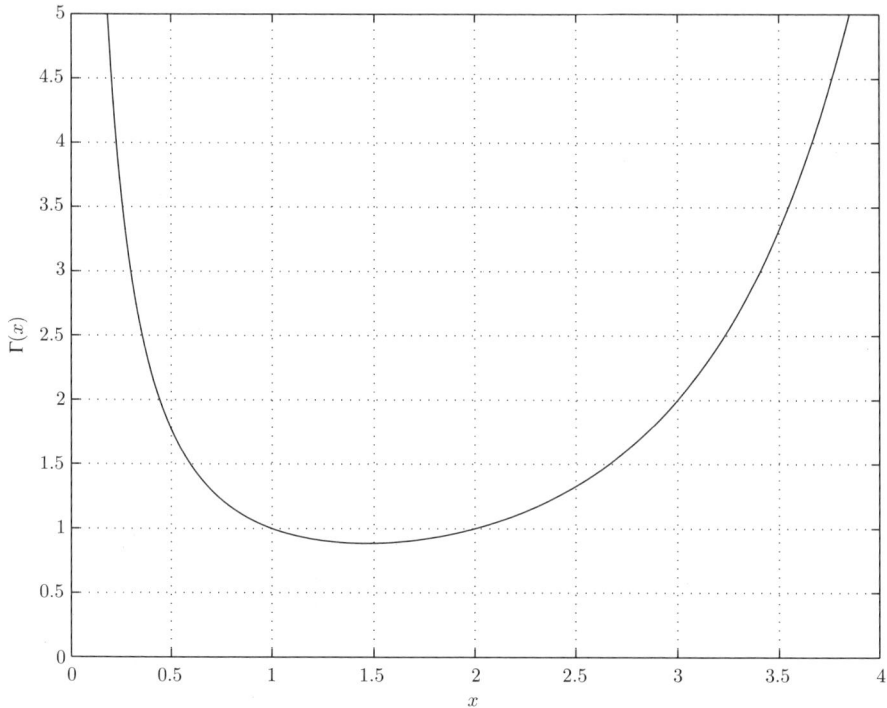

Fig. A.1. Graph of the gamma function for positive argument

A.2.b. Proposition (Artin, 1931, 1964; Bohr and Mollerup, 1922). Let f be a function with domain that includes $(0, \infty)$. If

(i) $f(x+1) = xf(x)$,
(ii) f is logarithmically convex on $(0, \infty)$, and
(iii) $f(1) = 1$,

then, on its domain, f is identical with the gamma function.

A.2.c. Proposition. $\Gamma(n+1) = n!$, $n = 0, 1, 2, \ldots$.

Proof. By definition, $0! = 1$, and a straightforward integration yields that $\Gamma(1) = 1$. Consequently, this result follows from Proposition A.2 via an induction argument. □

By making the change of variables $u = \sqrt{t}$ in (1), it follows that

$$\Gamma(z) = 2 \int_0^\infty u^{2z-1} e^{-u^2} \, du. \qquad (2)$$

This form of the function is sometimes more convenient than (1).

A.3. Proposition. $\Gamma(1/2) = \sqrt{\pi}$.

Proof. From (2), with $z = 1/2$, it follows that

$$\Gamma(1/2) = 2\int_0^\infty e^{-u^2}\,du = \sqrt{\pi}.$$

The last equality here is familiar because it states that the density of the normal distribution with mean 0 and variance $1/2$ integrates to unity. It is proved by writing

$$[\Gamma(1/2)]^2 = 4\int_0^\infty\int_0^\infty e^{-u^2-v^2}\,du\,dv,$$

and integrating after a change to polar coordinates. □

A.4. Proposition.

$$\Gamma(2z) = \frac{2^{2z-\frac{1}{2}}}{\sqrt{2\pi}}\Gamma(z)\Gamma\left(z+\frac{1}{2}\right). \tag{3}$$

More generally,

$$\Gamma(nz) = (2\pi)^{(1-n)/2}\, n^{nz-\frac{1}{2}}\prod_{k=0}^{n-1}\Gamma\left(z+\frac{k}{n}\right). \tag{4}$$

Although the formula (3) is known as the *Gauss duplication formula* (see Abramowitz and Stegun, 1964, p. 256), it is in fact due to Legendre. The more general result (4) is due to Gauss (see Artin, 1931, 1964).

Proof of (3). (See Webster, 1994). Make use of the fact (A.8.a) that $\log \Gamma(x)$ is convex in $x > 0$. Let

$$f(x) = \frac{2^{x-1}}{\sqrt{\pi}}\Gamma\left(\frac{x}{2}\right)\Gamma\left(\frac{x+1}{2}\right);$$

because f is the product of log-convex functions, it is log convex. Furthermore, (i) and (iii) of Proposition A.2.b are satisfied, and consequently it follows from that proposition that $f(x) = \Gamma(x)$. □

a. Stirling's Formula

A.5. Proposition. $\lim_{t\to\infty} \dfrac{\Gamma(t+1)}{e^{-t}t^{t+(1/2)}\sqrt{2\pi}} = 1$.

The approximation

$$\Gamma(t+1) \sim e^{-t} t^{t+(1/2)} \sqrt{2\pi}$$

for large t is known as *Stirling's formula*. It is given here for reference without proof.

From Stirling's formula, it can be deduced that for large z,

$$\Gamma(az+b) \sim \sqrt{2\pi}\, e^{-az} (az)^{az+b-\frac{1}{2}}. \tag{5}$$

From (5), it follows that for large z,

$$\frac{\Gamma(az+b+c)}{\Gamma(az+b)} \sim (az)^c \tag{6}$$

(see Abramowitz and Stegun, 1964, p. 257).

A.5.a. $\lim_{z \to 0} \Gamma(z) = \infty$, $\lim_{z \to 0} z\Gamma(z) = 0$. These results can be obtained using (5).

b. Total Positivity

A.6. Proposition. The function $\Gamma(x+y)$ is totally positive in x and $y > 0$.

First Proof. Because $\Gamma(r)$ is the $(r-1)$th moment of an exponential distribution, this result is a special case of Example 21.B.11.a. □

Second Proof. According to Example 21.B.5.a, the function t^x is totally positive in $t > 0$ and all x. Write $\Gamma(x+y)$ in the form $\Gamma(x+y) = \int_0^\infty t^x t^{y-1} e^{-t}\, dt$ and apply Theorem 21.B.11 to complete the proof. □

c. The Incomplete Gamma Function: Tail of a Poisson Distribution

Introduce the notation

$$\Gamma_x(z) = \int_0^x t^{z-1} e^{-t}\, dt, \quad x \geq 0. \tag{7}$$

The notation $\Gamma_x(z) = \gamma(x, z)$ is also found in the literature.

A.7. Definition. The function

$$I_x(z) = \Gamma_x(z)/\Gamma(z) \tag{8}$$

is called the *incomplete gamma function*.

The term "incomplete gamma function" is also sometimes applied to Γ_x itself, without the normalizing denominator. The name was also applied by Pearson (1934, 1968) to the ratio $I(u, z-1) = \Gamma_{u\sqrt{z}}(z)/\Gamma(z)$, which he tabulated extensively.

By expanding the exponential factor of the integrand of (7) in a power series and integrating term by term, it can be shown that

$$\Gamma_x(z) = \sum_{j=0}^{\infty} \frac{(-1)^j}{j!} \frac{x^{z+j}}{(z+j)}.$$

Other expansions of Γ_x can be given in terms of hypergeometric functions. In particular,

$$I_x(z) = x^z \, e^{-x} \sum_{j=0}^{\infty} \frac{x^j}{\Gamma(z+1+j)} \tag{9}$$

$$= \frac{x^z \, e^{-x}}{\Gamma(z+1)} \, {}_1F_1(1; z+1; x),$$

where

$${}_1F_1(a;b;x) = \sum_{j=0}^{\infty} \frac{\Gamma(a+j)}{\Gamma(a)} \frac{\Gamma(b)}{\Gamma(b+j)} \frac{x^j}{j!}$$

is the confluent hypergeometric function; see Abramowitz and Stegun (1964, Chapters 6, 13); these authors use the notation ${}_1F_1(a;b;x) = M(a,b,x)$.

When z is an integer, (9) can be rewritten as

$$\sum_{j=z}^{\infty} \frac{x^j}{j!} e^{-x} = \frac{1}{\Gamma(z)} \int_0^x t^{z-1} \, e^{-t} \, dt = I_x(z). \tag{10}$$

This is a well-known formula for the upper tail of a Poisson distribution.

d. Convexity and Concavity Results

A.8. Proposition. For fixed x, the incomplete gamma function $\Gamma_x(z)$ is logarithmically convex in $z > 0$, hence convex. The reciprocal $1/\Gamma_x(z)$ of the incomplete gamma function is logarithmically concave in $z > 0$.

Proof. With the aid of Hölders inequality (Proposition 24.B.5), it follows that

$$\Gamma_x(\alpha u + \bar{\alpha} v) = \int_0^x t^{\alpha u + \bar{\alpha} v - 1} e^{-t} \, dt = \int_0^x (t^{u-1} e^{-t})^\alpha (t^{v-1} e^{-t})^{\bar{\alpha}} \, dt$$

$$\leq \left[\int_0^x t^{u-1} e^{-t} \, dt \right]^\alpha \left[\int_0^x t^{v-1} e^{-t} \, dt \right]^{\bar{\alpha}} = [\Gamma_x(u)]^\alpha [\Gamma_x(v)]^{\bar{\alpha}}.$$

Logarithmic convexity implies convexity (Proposition 21.A.5), and it also implies that $-\log \Gamma_x(z) = \log[1/\Gamma_x(z)]$ is concave. □

A.8.a. Because $\Gamma(z) = \lim_{x \to \infty} \Gamma_x(z)$, it follows that the gamma function has the same convexity and concavity properties as does the incomplete gamma function. The log convexity of the gamma function can also be directly obtained from the representation (Abramowitz and Stegun (1964, p. 258)

$$\log \Gamma(z) = \int_0^\infty \left[(z-1) e^{-t} - \frac{e^{-t} - e^{-zt}}{1 - e^{-t}} \right] \frac{1}{t} \, dt,$$

because the integrand is a convex function of z.

Log convexity and log concavity are total positivity statements (see 21.B.7 and 21.B.9), and this fact can be used to obtain inequalities for the gamma and incomplete gamma function.

For additional results related to Proposition A.8.a, see Marshall and Olkin (1979, pp. 73–75).

B. The Beta Function

B.1. Definition. The function

$$B(u, v) = \int_0^1 t^{u-1} (1-t)^{v-1} dt \tag{1}$$

is called the *beta function*.

B. The Beta Function

It can be shown that the integral (1) is finite for $0 < u, v < \infty$. Although the domain of the beta function can be extended, it is this domain that is relevant for the applications in this book.

B.2. Proposition. $B(u, v) = B(v, u)$.

Replacement of t by $1 - t$ in (1) is all that is required to prove this result. It also follows directly from Proposition B.5.

B.3. Proposition. The beta function (1) has the representation

$$B(u, v) = \int_0^\infty \frac{z^{u-1}}{(1+z)^{u+v}} dz, \quad 0 < u, v < \infty.$$

To prove this result, make the change of variables $t = z/(1+z)$ in (1).

B.4. Proposition. $B(u+1, v) = \dfrac{u}{u+v} B(u, v), \quad 0 < u, v < \infty.$

Proof. Use Proposition B.3 and Fubini's theorem 24.B.1 to write

$$B(u+1, v) = \int_0^\infty \frac{z^u}{(1+z)^{u+v+1}} dz = \int_{t=0}^\infty \int_{z=0}^t u z^{u-1} dz \frac{1}{(1+t)^{u+v+1}} dt$$

$$= \int_{z=0}^\infty \int_{t=z}^\infty u z^{u-1} \frac{1}{(1+t)^{u+v+1}} dt \, dz$$

$$= \int_{z=0}^\infty \frac{u}{u+v} \frac{z^{u-1}}{(1+t)^{u+v}} dz = \frac{u}{u+v} B(u, v). \quad \square$$

B.5. Proposition. $B(u, v) = \dfrac{\Gamma(u)\Gamma(v)}{\Gamma(u+v)}, \quad 0 < u, v < \infty.$

Proof (See Webster, 1994). Define

$$f(u, v) = \frac{\Gamma(u+v)}{\Gamma(v)} B(u, v);$$

for fixed v, it follows from A.8.a and B.7 that $f(u, v)$ is a product of functions log convex in $u > 0$, and consequently, $f(u, v)$ is log convex in $u > 0$. Additionally,

$$f(1, v) = \frac{\Gamma(1+v)}{\Gamma(v)} B(1, v) = v B(1, v) = v \int_0^1 (1-z)^{v-1} dz = 1,$$

and (using Proposition B.4)

$$f(u+1,v) = \frac{\Gamma(u+1+v)}{\Gamma(v)} B(u+1,v) = \frac{(u+v)\Gamma(u+v)}{\Gamma(v)} \frac{u}{u+v} B(u,v)$$

$$= u \frac{\Gamma(u+v)}{\Gamma(v)} B(u,v) = u f(u,v).$$

It follows from Proposition A.2.b that $f(u,v) = \Gamma(u)$. □

Alternatively, Proposition B.5 can be proved by using A(2) to rewrite the numerator $\Gamma(u)\Gamma(v)$, and then changing to polar coordinates.

a. The Incomplete Beta Function

B.6. Definition. Let

$$B_x(u,v) = \int_0^x t^{u-1}(1-t)^{v-1} dt, \quad 0 < u, v < \infty, \quad 0 \le x \le 1. \quad (2)$$

As with Proposition B.3, B_x can be rewritten in the form

$$B_x(u,v) = \int_0^{x/(1-x)} \frac{z^{u-1}}{(1+z)^{u+v}} dz. \quad (3)$$

The ratio

$$I_x(u,v) = \frac{B_x(u,v)}{B(u,v)} \quad (4)$$

is called the *incomplete beta function*. Some caution is required here because the same term is sometimes applied to the function B_x.

The incomplete beta function is tabulated by Pearson (1968). It has a series representation in terms of the Gauss hypergeometric function F defined by the series

$$F(a;b;c;x) = \sum_{j=0}^{\infty} \frac{\Gamma(a+j)}{\Gamma(a)} \frac{\Gamma(b+j)}{\Gamma(b)} \frac{\Gamma(c)}{\Gamma(c+j)} \frac{x^j}{j!}, \quad (5)$$

another commonly used notation is $F(a,b;c;x) = {}_2F_1(a,b;c;x)$. This function is discussed by Abramowitz and Stegun (1964, Chapter 15).

According to Abramowitz and Stegun (1964, Formulas 26.5.23 and 15.3.3),

$$B_x(u,v) = \frac{x^u}{u} F(u, 1-v; u+1; x) = \frac{x^u(1-x)^v}{u} F(1, u+v; u+1; x). \tag{6}$$

Considerable simplification occurs in (5) when $b = c$ and $0 < x < 1$. Then, the series is independent of b, and can be summed to yield

$$F(a,b;b;x) = \sum_{j=0}^{\infty} \frac{\Gamma(a+j)}{\Gamma(a)} \frac{x^j}{j!} = \frac{1}{(1-x)^a}. \tag{7}$$

See, e.g., Abramowitz and Stegun, (1964, Formula 15.1.8).

B.7. Proposition. For each fixed x and v, $\log B_x(u,v)$ is convex in $u > 0$. In particular, $\log B(u,v)$ is convex in $u > 0$ for fixed u.

Proof. By Hölders inequality (24.B.5) it follows that

$$B_x(\alpha s + \bar{\alpha} t, v) = \int_0^x [z^{s-1}(1-z)^{v-1}]^{\alpha} [z^{t-1}(1-z)^{v-1}]^{\bar{\alpha}} dz$$

$$\leq \left[\int_0^x z^{s-1}(1-z)^{v-1} dz \right]^{\alpha} \left[\int_0^x z^{t-1}(1-z)^{v-1} dz \right]^{\bar{\alpha}}$$

$$= [B_x(s,v)]^{\alpha} [B_x(t,v)]^{\bar{\alpha}}. \qquad \square$$

b. Tail of a Binomial Distribution

The formula

$$\sum_{j=k}^{n} \binom{n}{j} p^j (1-p)^{n-j} = \frac{\int_0^p t^{k-1}(1-t)^{n-k} dt}{B(k, n-k+1)} = I_p(k, n-k+1) \tag{8}$$

gives a well-known and often useful connection between the upper tail of a binomial distribution and the incomplete beta function. See, e.g., Abramowitz and Stegun (1964, Formula 26.5.24).

Formula (8) is equivalent to

$$\sum_{j=0}^{k-1} \binom{n}{j} p^j (1-p)^{n-j} = \frac{\int_p^1 t^{k-1}(1-t)^{n-k} dt}{B(k, n-k+1)}, \tag{9}$$

and this formula can be extended to the case of noninteger n;

$$\sum_{j=0}^{k-1} \frac{\Gamma(\eta+1)}{j!\Gamma(\eta-j+1)} p^j (1-p)^{\eta-j} = \frac{\int_0^p t^{k-1}(1-t)^{\eta-k} dt}{B(k, \eta-k+1)}, \quad (10)$$

where $\eta - k + 1 > 0$. This formula can be proved by induction on k using an integration by parts (see Marshall and Olkin, 1993).

Equation (8) gives the tail of a binomial distribution in terms of an incomplete beta function. Just as the Poisson distribution can be obtained from the binomial distribution, the tail of a Poisson distribution can be obtained from (8) by setting $p = \lambda/n$ and then letting $n \to \infty$.

c. Tail of a Negative Binomial Distribution

The following formula, a counterpart to (8), gives the tail of a negative binomial distribution in terms of the incomplete beta function:

$$\sum_{j=s}^{k-1} \frac{\Gamma(r+j)}{\Gamma(r)j!} p^j (1-p)^r = \frac{1}{B(r,s)} \int_0^{p/(1-p)} \frac{t^{s-1}}{(1+t)^{r+s}} dt$$

$$= \frac{1}{B(r,s)} \int_0^p t^{s-1}(1-t)^{r-1} dt. \quad (11)$$

This relationship is discussed by Rider (1962).

From (8) and (11), Morris (1963) noted that

$$\sum_{j=0}^{n-k} \binom{k+j-1}{j} p^k (1-p)^j = \sum_{j=k}^{n} \binom{j-1}{k-1} p^k (1-p)^{j-k}. \quad (12)$$

d. The Liouville–Dirichlet Integral

Although this book does not deal with multivariate distributions, multivariate extension of the beta integral (1) is given here for completeness. Dirichlet's extension of the beta integral is

$$\int \cdots \int_{0 \leq x_i, \Sigma x_i < 1} \prod_1^k x_i^{a_i-1} \left(1 - \sum_1^k x_i\right)^{a_0-1} \prod_1^k dx_i$$

$$= \frac{\prod_0^k \Gamma(a_i)}{\Gamma(\sum_0^k a_i)}, \quad a_i > 0, \quad i = 0, 1, \ldots, k, \quad (13)$$

which is often denoted $B(a_0, a_1, \ldots, a_k)$. When $k = 1$, this is the beta function of Definition B.1. Its usefulness stems in part from the fact that tails of the multinomial distribution are equal to incomplete Dirichlet distributions (see Olkin and Sobel, 1965; Olkin, 1972).

Liouville generalized (13) by substituting $f(\sum x_i)$ for $(1 - \sum x_i)^{a_0-1}$ to obtain

$$\int_0^\infty \cdots \int_0^\infty \frac{\prod x_i^{a_i-1}}{\prod \Gamma(a_i)} f\left(\sum x_i\right) \prod dx_i = \int_0^\infty \frac{z^{a_0-1}}{\Gamma(\sum a_i)} f(z)\, dz, \quad (14)$$

where $a_0 = \sum a_i$.

For further discussions of the history of the Dirichlet distribution and its applications, see Sobel, Uppuluri and Frankowski (1980).

24

Some Topics from Analysis

The facts outlined in this chapter are included only for reference and are stated without proof. Primes are used to indicate derivatives.

A. Basic Results from Calculus

A.1. Integration by parts. If f and g are absolutely continuous functions defined on $[a, b]$, then

$$\int_a^b f(t)g'(t)\,dt = f(t)g(t)\big|_a^b - \int_a^b f'(t)g(t)\,dt. \tag{1}$$

A.2. L'Hospital's rule. If f and g have continuous derivatives and $f(a) = g(a) = 0$, then

$$\lim_{x \to a} \frac{f(x)}{g(x)} = \lim_{x \to a} \frac{f'(x)}{g'(x)}, \tag{2}$$

whether or not that limit is finite.

A.3. Exponential function as a limit.

$$\lim_{z \to \infty} \left(1 + \frac{x}{z}\right)^z = e^x. \tag{3}$$

A.4. Chain rule for differentiation. Let f be a differentiable function defined on the interval (a, b) and g be a differentiable function taking values in (a, b). Then, the composition $h = f(g)$ is a differentiable function with derivative $h'(x) = f'(g(x))g'(x)$.

A.4.a. Derivative of an inverse. If F is a strictly increasing function with derivative f, inverse F^{-1}, and corresponding survival function $\bar{F} = 1 - F$, then

$$\frac{d}{dz}f^{-1}(z) = \frac{1}{fF^{-1}(z)} \quad \text{and} \quad \frac{d}{dz}\bar{F}^{-1}(z) = -\frac{1}{f\bar{F}^{-1}(z)}.$$

Proof. Write $FF^{-1}(z) = z$; differentiate using the chain rule. The derivative of $\bar{F}^{-1}$ is similarly obtained. □

A.4.b. If F and G are strictly increasing distribution functions with derivatives f and g, then

$$\frac{d}{dz}G^{-1}F(z) = \frac{f(z)}{gG^{-1}F(z)}, \quad \frac{d}{dz}\bar{G}^{-1}\bar{F}(z) = \frac{f(z)}{g\bar{G}^{-1}\bar{F}(z)},$$

$$\frac{d}{dp}FG^{-1}(p) = \frac{fG^{-1}(p)}{gG^{-1}(p)}, \quad \frac{d}{dp}\bar{F}\bar{G}^{-1}(p) = \frac{f\bar{G}^{-1}(p)}{g\bar{G}^{-1}(p)}.$$

A.5. Differentiation of integrals. If

$$\Phi(x) = \int_{\psi_1(x)}^{\psi_2(x)} \phi(x, y)\, dy,$$

where ψ_1 and ψ_2 have derivatives continuous in a closed interval $[x_0, x_1]$, $\phi(x, y)$ and $\partial \phi(x,y)/\partial x$ are continuous in the region $x_0 \leq x \leq x_1$, $\psi_1(x) \leq y \leq \psi_2(x)$, then the derivative Φ' of Φ with respect to x is given by

$$\Phi'(x) = \int_{\psi_1(x)}^{\psi_2(x)} \frac{\partial}{\partial x}\phi(x,y)\, dy - \psi'_1(x)\phi(x, \psi(x)) + \psi'_2(x)\phi(x, \psi_2(x)).$$

A.6. Let h be a differentiable function of two real variables, and let

$$h_1(u, v) = \frac{\partial h(u, v)}{\partial u}, \quad h_2(u, v) = \frac{\partial h(u, v)}{\partial v}.$$

If ϕ and ψ are differentiable real-valued functions of a real variable, then

$$\frac{d}{dx}h(\phi(x), \psi(x)) = h_1(\phi(x), \psi(x))\phi'(x) + h_2(\phi(x), \psi(x))\psi'(x).$$

B. Some Results Concerning Lebesgue Integrals

B.1. Interchange of order of integration: Fubini's theorem.
Interchanging the order of integration in multiple integrals has been carried out with little comment in this book. Such interchanges are justified by the following simplified version of Fubini's theorem.

If either of the integrals

$$\int_{y \in B} \int_{x \in A} |\phi(x,y)|\, dF(x)\, dG(y) \quad \text{or} \quad \int_{x \in A} \int_{y \in B} |\phi(x,y)|\, dG(y)\, dF(x)$$

exist, then both of these integrals of ϕ exist and

$$\int_{y \in B} \int_{x \in A} \phi(x,y)\, dF(x)\, dG(y) = \int_{x \in A} \int_{y \in B} \phi(x,y)\, dG(y)\, dF(x).$$

For a further discussion of Fubini's theorem, see for example, Billingsley (1995, p. 233) or Gut (2005, p. 65).

B.2. Interchange of a limit and an integral: Lebesgue monotone convergence theorem. The interchange of a limit and an integration can be justified in several ways, and a paticularly useful way is by using the Lebesgue monotone convergence theorem. The following somewhat simplified version of that theorem is sufficient for the purposes of this book.

Let $\phi_n, n = 1, 2, \ldots$, be a nondecreasing sequence of nonnegative functions such that $\lim_{n \to \infty} \phi_n = \phi_0$. If $\int_A |\phi_n(x)|\, dF(x) < \infty, n = 1, 2, \ldots$, and $\lim_{n \to \infty} \int_A \phi_n(x)\, dF(x) < \infty$, then $\int_A |\phi(x)|\, dF(x) < \infty$ and $\lim_{n \to \infty} \int_A \phi_n(x)\, dF(x) = \int_A \phi_0(x)\, dF(x)$.

See Billingsley (1995, p. 208) or Gut (2005, p. 55).

B.3. Interchange of a limit and an integral: Lebesgue dominated convergence theorem. Another classical theorem allowing for interchange of a limit and an integration is the Lebesgue dominated convergence theorem; this theorem is discussed, for example, by Billingsley (1995), p. 209) and Gut (2005, p. 57). Again, a somewhat simplified version is stated here that serves the purpose of this book.

Let $\phi_n, n = 1, 2 \ldots$, be a sequence of measurable functions such that $\lim_{n \to \infty} \phi_n = \phi_0$ except possibly on a set of probability 0. Suppose there exists a measurable function g such that

(i) $\int |g(x)|\, dF(x) < \infty$, and
(ii) $|\phi| \leq |g|$ except possibly on a set of probability 0.

Then, $\lim_{n\to\infty} \int \phi_n(x)\,dF(x)$ exists and $\lim_{n\to\infty} \int \phi_n(x)\,dF(x) = \int \lim_{n\to\infty} \phi_n(x)\,dF(x)$.

B.4. Modified Bessel function K_θ of the third kind. The modified Bessel function K_θ of the third kind (also called *Macdonald's function*, and even the "modified Bessel function of the second kind" in the statistical literature) can be defined in several equivalent forms by

$$\begin{aligned} K_\theta(z) &= \frac{1}{2} \int_0^\infty y^{\theta-1} \exp\left\{-\frac{z}{2}\left(y + \frac{1}{y}\right)\right\} dy \\ &= \frac{1}{2} \int_0^\infty \frac{1}{y^{\theta+1}} \exp\left\{-\frac{z}{2}\left(y + \frac{1}{y}\right)\right\} dy \\ &= \frac{1}{2} \left(\frac{z}{2}\right)^\theta \int_0^\infty \frac{1}{t^{\theta+1}} \exp\left\{-t - \frac{z^2}{4t}\right\} dt. \end{aligned}$$

The second form here is obtained from the first when y is replaced by $1/y$; the equivalence of these forms shows that $K_\theta(z) = K_{-\theta}(z)$. The third form is obtained by letting $t = yz/2$.

For $\theta > -1/2$,

$$K_\theta(z) = \left(\frac{z}{2}\right)^\theta \frac{\sqrt{\pi}}{\Gamma(\theta + (1/2))} \int_1^\infty e^{-zt}(t^2 - 1)^{\theta-(1/2)}\,dt.$$

Modified Bessel functions arise in Section 13.B.a.

B.5. Hölder's inequality. Let f_i be nonnegative functions, and let $q_i \geq 0$, $i = 1, \ldots, n$, be numbers such that $\sum_{i=1}^n q_i = 1$. With the assumption that the integrals on the right side of the following inequality exist finitely,

$$\int \prod_{i=1}^n f_i^{q_i}\,d\mu \leq \prod_{i=1}^n \left(\int f_i\,d\mu\right)^{q_i}.$$

For a proof of this result, see, for example, Marshall and Olkin (1979, p. 457).

References

Aalen, Odd O. and Håkon K. Gjessing (2001). Understanding the shape of the hazard rate: A process point of view. *Statistical Science* **16**, 1–22.

Abramowitz, Milton and Irene A. Stegun (eds.) (1964). *Handbook of Mathematical Functions*. National Bureau of Standards, Applied Mathematics Series **55**, U.S. Government Printing Office, Washington D.C. (reprinted by Dover Publications, 1970).

Aczél, J. (1966). *Lectures on Functional Equations and Their Applications*. New York: Academic Press.

Aczél, J. and Z. Daróczy (1975). *On Measures of Information and Their Characterizations*. New York: Academic Press.

Adamidis, K. and S. Loukas (1998). A lifetime distribution with decreasing failure rate. *Statistics and Probability Letters* **39**, 35–42.

A-Hameed, M.S. and F. Proschan (1973). Nonstationary shock models. *Stochastic Processes and their Applications* **1**, 383–404.

A-Hameed, M.S. and F. Proschan (1975). Shock models with underlying birth process. *Journal of Applied Probability* **12**, 18–28.

Ahuja, J.C. and Stanley W. Nash (1967). The generalized Gompertz-Verhulst family of distributions. *Sankhyā* **29**, 141–156.

Aitchison, J. and J.A.C. Brown (1957). *The Lognormal Distribution*. Cambridge, U.K.: Cambridge Univ. Press.

Allen, W.R. (1963). A note on conditional probability of failure when hazards are proportional. *Operations Research* **11**, 658–659.

Amoroso, Luigi (1925). Ricerche intorno alla curva dei redditi. *Annali di Mathematica*, Series IV, **2**, 123–159.

An, Mark Yuying (1998). Logconcavity versus logconvexity: A complete characterization. *Journal of Economic Theory* **80**, 350–369.

Andersen, Per Kragh (1991). Survival analysis 1982–1991: The second decade of the proportional hazards regression model. *Statistics in Medicine* **10**, 1931–1941.

Andersen, Per Kragh, Ørnulf Borgan, Richard D. Gill and Niels Keiding (1993). *Statistical Models Based on Counting Processes.* New York: Springer-Verlag.

Ando, T. (1989). Majorization, doubly stochastic matrices, and comparison of eigenvalues. *Linear Algebra and its Applications* **118**, 163–248.

Andréief, C. (1883). Note sur une relation les intégrals définies des produits des functions. *Mémoires Societé des Sciences Physiques et Naturelles de Bordeaux* **2**, 1–14.

Ansell, J.I. and M.J. Phillips (1994). *Practical Methods for Reliability Data Analysis.* Oxford, U.K.: Clarendon Press.

Appleton, David R., Joyce M. French and Mark P.J. Vanderpump (1996). Ignoring a covariate: An example of Simpson's paradox. *The American Statistician* **50**, 340–341.

Armitage, P. (1959). The comparison of survival curves. *Journal of the Royal Statistical Society* **A122**, 279–300.

Arnold, Barry C. (1975). Multivariate exponential distributions based on hierarchical successive damage. *Journal of Applied Probability* **12**, 142–147.

Arnold, Barry C. (1983). *Pareto Distributions.* Fairland, Maryland: International Co-operative Publishing House.

Arnold, B.C. (1987). *Majorization and the Lorenz Order: a Brief Introduction.* New York: Springer-Verlag.

Arnold, B.C. (1989). A logistic process constructed using geometric minimization. *Statistics and Probability Letters* **7**, 253–257.

Arnold, Barry C. and Patrick L. Brockett (1983). Identifiability for dependent multiple decrement/competing risk models. *Scandinavian Actuarial Journal*, **10**, 117–127.

Arnold, Barry C. and Richard A. Groeneveld (1995). Measuring skewness with respect to the mode. *The American Statistician* **49**, 34–38.

Arnold, Barry C. and J.S. Huang (1975). Characterizations. *The Exponential Distribution. Theory, Methods and Applications*, N. Balakrishnan, and J.S. Huang (eds.), pp. 185–203. Amsterdam, The Netherlands: Gordon and Breach.

Arnold, Barry C. and Dean Isaacson (1976). On solutions to $\min(X,Y) \stackrel{d}{=} aX$ and $\min(X,Y) \stackrel{d}{=} aX \stackrel{d}{=} bY$. *Zeitschrift für Warscheinlichkeitstheorie und verwandte Gebiete* **35**, 115–119.

Artin, E. (1931, 1964). *Einführung in die Theorie der Gammafunktion*, Hamburger Mathematische Einzelschriften, Heft II, Verlag B.G. Teubner, Leipzig. [Translated by M. Butler, *The Gamma Function.* 1964, New York: Holt, Rinehart & Winston.]

Asmussen, Søren (2003). *Applied Probability and Queues*, 2nd edn. New York: Springer-Verlag.

Azlarov, T.A., and N.A. Volodin (1986). *Characterization Problems Associated with the Exponential Distribution.* New York: Springer-Verlag.

Badia, F.G., M.D. Berrade, C.A. Campos, and M.A. Navascués (2001). On the behavior of aging characteristics in mixed populations. *Probability in the Engineering and Informational Sciences* **15**, 83–94.

Bailey, R. Clifton and Louis D. Homer (1977). Computations for a best match strategy for kidney transplantation. *Transplantation* **23**, 329–336.

Bailey, R. Clifton, Louis D. Homer, and J.P. Summe (1977). A proposal for the analysis of kidney graft survival. *Transplantation* **24**, 309–315.

Balakrishnan, N. and A. Clifford Cohen (1991). *Order Statistics and Inference.* Boston, Massachusetts: Academic Press.

Balakrishnan, N. and Basu, A. P. (eds.) (1995). *The Exponential Distribution. Theory, Methods and Applications.* Amsterdam, The Netherlands: Gordon and Breach.

Barlow, Richard E. (1979). Geometry of the total time on test transform. *Naval Research Logistics Quarterly* **26**, 393–402.

Barlow, Richard E. (1985). A Bayes explanation of an apparent failure rate paradox. *IEEE Transactions on Reliability* **34**, 107–108.

Barlow, R.E., D.J. Bartholomew, J.M. Bremner, and H.D. Brunk (1972). *Statistical Inference Under Order Restrictions.* London, U.K.: John Wiley & Sons.

Barlow, Richard E. and Raphael Campo (1975). Total time on test processes and applications to failure data analyses. *Reliability and Fault Tree Analyses*, R.E. Barlow, J.B. Fussell, and N.D. Singpurwalla, (eds.), pp. 451–481. Philadelphia, Pennsylvania: Society for Industrial and Applied Mathematics.

Barlow, Richard E. and Albert W. Marshall (1964). Bounds for distributions with monotone hazard rate, I and II. *The Annals of Mathematical Statistics* **35**, 1234–1257 and 1258–1274.

Barlow, Richard E. and Albert W. Marshall (1965). Tables of bounds for distributions with monotone hazard rate. *Journal of the American Statistical Association.* **60**, 872–890.

Barlow, R.E. and A.W. Marshall (1967). Bounds on interval probabilities for restricted families of distributions. *Proceedings of the Fifth Berkeley Symposium on Mathematical Statistics and Probability, Vol. III*, L. LeCam and J. Neyman, (eds.), pp. 229–257. Berkeley and Los Angeles: University of California Press.

Barlow, Richard E., Albert W. Marshall, and Frank Proschan (1963). Properties of probability distributions with monotone hazard rate. *The Annals of Mathematical Statistics* **34**, 375–389.

Barlow, Richard E. and Frank Proschan (1965, 1996). *Mathematical Theory of Reliability.* New York: John Wiley & Sons. (Reprinted Philadelphia, Pennsylvania: Society for Industrial and Applied Mathematics.)

Barlow, Richard E. and Frank Proschan (1975). *Statistical Theory of Reliability and Life Testing.* New York: Holt, Rinehart and Winston.

Barndorff-Nielsen, O.E. (1978). Hyperbolic distributions and distributions on hyperbole. *Scandinavian Journal of Statistics* **5**, 151–157.

Barndorff-Nielsen, O.E. and C. Halgreen (1977). Infinite divisibility of the hyperbolic and generalized inverse Gaussian distributions. *Zeitschrift für Wahrscheinlichkeitstheorie und verwandte Gebiete* **38**, 309–311.

Bartoszewicz, Jaroslaw (1986). Dispersive ordering and the total time on test transformation. *Statistics and Probability Letters* **4**, 285–288.

Bartoszewicz, Jaroslaw (1995). Stochastic order relations and the total time on test transform. *Statistics and Probability Letters* **22**, 103–110.

Bassan, Bruno, Yosef Rinott and Yehuda Vardi (2002). On stochastic comparisons of excess times. Discussion paper number 302. Center for the Study of Rationality, Hebrew University, Jerusalem.

Basu, A.P. (1965). On characterizing the exponential distribution by order statistics. *Annals of the Institute of Statistical Mathematics* **17**, 93–96.

Basu, A.P. and J.K. Ghosh (1978). Identifiability of the multinormal and other distributions under competing risks model. *Journal of Multivariate Analysis* **8**, 413–429.

Bates, G. and J. Neyman (1952). Contributions to the theory of accident proneness. *University of California Publications in Statistics* **1**, 215–275.

Beckenbach, Edwin F. and Richard Bellman (1961). *Inequalities.* Berlin, Germany: Springer-Verlag.

Bencala, Kenneth and John H. Seinfeld (1976). On frequency distributions of air pollutant concentrations. *Atmospheric Environment* **10**, 941–950.

Bennett, Steve (1983). Analysis of survival data by the proportional odds model. *Statistics in Medicine* **2**, 273–277.

Bergman, Bo (1979). On age replacement and the total time on test concept. *Scandinavian Journal of Statistics* **6**, 161–168.

Berman, S.M. (1963). Note on extreme values, competing risks and semi-Markov processes. *The Annals of Mathematical Statistics* **34**, 1104–1106.

Bernoulli, Daniel (1760). Essai d'une novelle analyse de la mortalité causée par la petite verole & des avantages de l'inoculation pour la prévenir. *Mémoires de Mathematique et de Physique, Tirés des Registres de l'Académie Royale des Sciences, Paris.* pp. 1–45.

Bernstein, Serge (1928). Sur les fonctions absolument monotones. *Acta Mathematica* **52**, 1–66.

Bertholon, Henri, Nicolas Bousquet, and Gilles Celeux (2004). An alternative competing risk model to the Weibull distribution in lifetime data analysis. *Institut National de Recherche en Informatique et en Automatique*, Rapport de Recherche No. 5265. (Also *Lifetime Data Analysis*, 2006, **12**, 481–504).

Bhattacharyya, G.K. and Arthur Fries (1982). Fatigue failure models—Birnbaum-Saunders vs. Inverse Gaussian. *IEEE Transactions on Reliability* **31**, 439–440.

Bickel, P.J. and E.L. Lehmann (1975). Descriptive statistics for nonparametric models I. Introduction. *The Annals of Statistics* **3**, 1038–1044.

Billingsley, Patrick (1968). *Convergence of Probability Measures*. New York: John Wiley & Sons.

Billingsley, Patrick (1995). *Probability and Measure*, 3rd edn. New York: John Wiley & Sons.

Birnbaum, Z.W. (1942). An inequality for Mills' ratio. *The Annals of Mathematical Statistics* **21**, 272–279.

Birnbaum, Z.W. (1948). On random variables with comparable peakedness. *The Annals of Mathematical Statistics* **19**, 76–81.

Birnbaum, Z.W. (1979). *On the mathematics of competing risks*. U.S. Department of Public Health, Education, and Welfare. Publication No. (PHS) 79-1351.

Birnbaum, Z.W., J.D. Esary, and A.W. Marshall (1966). A stochastic characterization of wearout for components and systems. *The Annals of Mathematical Statistics* **37**, 816–825.

Birnbaum, Z.W., J.D. Esary, and S.C. Saunders (1961). Multicomponent systems and structures, and their reliability. *Technometrics* **12**, 55–77.

Birnbaum, Z.W. and S.C. Saunders (1969). A new family of life distributions. *Journal of Applied Probability* **6**, 319–327.

Blanda, Keven P. and H.L. MacGillivray (1988). Kurtosis: A critical review. *The American Statistician* **42**, 111–119.

Blanda, Keven P. and H.L. MacGillivray (1990). Kurtosis and spread. *Canadian Journal of Statistics* **18**, 17–30.

Block, Henry and A.P. Basu (1974). A continuous bivariate exponential extension. *Journal of the American Statistical Association* **69**, 1031–1037.

Block, Henry and Harry Joe (1997). Tail behavior of the failure rate functions of mixtures. *Lifetime Data Analysis* **3**, 269–288.

Block, Henry W, Yulin Li, and Thomas H. Savits (2001). Behavior of failure rates of mixtures and systems. Technical report. Department of Statistics, University of Pittsburgh, Pittsburgh, Pennsylvania.

Block, Henry W., Yulin Li, and Thomas H. Savits (2003a). Preservation of properties under mixture. *Probability in the Engineering and Informational Sciences* **17**, 205–212.

Block, Henry W., Yulin Li, and Thomas H. Savits (2003b). Initial and final behaviour of failure rate functions for mixtures and systems. *Journal of Applied Probability* **40**, 721–740.

Block, Henry W., Jei Mi, and Thomas H. Savits (1993). Burn-in and mixed populations. *Journal of Applied Probability* **30**, 692–702.

Block, Henry W. and Thomas H. Savits (1976). The IFRA closure problem. *The Annals of Probability* **4**, 1030–1032.

Block, Henry W. and Thomas H. Savits (1997). Burn-in. *Statistical Science* **12**, 1–19.

Block, Henry W., Thomas H. Savits, and Harshinder Singh (1998). The reversed hazard rate function. *Probability in the Engineering and Informational Sciences* **12**, 69–90.

Block, Henry W., Thomas H. Savits, and Eshetu T. Wondmagegnehu (2003). Mixtures of distributions with increasing linear failure rates. *Journal of Applied Probability* **40**, 485–504.

Bohr, Harald and Mollerup, Johannes (1922). *Lærebog i Mathematisk Analyse III*, Kobenhaven, Denmark.

Boland, Philip J. and Emad El-Neweihi (1995). Component redundancy vs system redundancy in the hazard rate ordering. *IEEE Transactions on Reliability* **44**, 614–619.

Boland, Philip J., Emad El-Neweihi, and Frank Proschan (1994). Applications of the hazard rate ordering in reliability and order statistics. *Journal of Applied Probability* **31**, 180–192.

Bondesson, Lennart (1978). On infinite divisibility of powers of a gamma variable. *Scandinavian Actuarial Journal* 48–61.

Bondesson, Lennart (1979). A general result on infinite divisibility. *The Annals of Probability* **7**, 965–979.

Bondesson, Lennart (1992). *Generalized Gamma Convolutions and Related Classes of Distributions and Densities*. New York: Springer-Verlag.

Bosch, A.J. (1977). Eine Characterisierung der Exponentialverteilungen. *Zeitschrift für Angewandte Mathematik und Mechanik* **57**, 609–610.

Box, G.E.P. and D.R. Cox (1964). An analysis of transformations revisited. *Journal of the Royal Statistical Society* **B26**, 211–252.

Bradley, Dorothy H., Edwin L. Bradley, and David C. Naftel (1984). A generalized Gompertz-Rayleigh model as a survival distribution. *Mathematical Biosciences* **70**, 195–202.

Brascamp, H.J. and E.H. Lieb (1975). Some inequalities for Gaussian measures and long-range order of the one-dimensional plasma. *Functional Integration and Its Application*, A.M. Arthurs (ed.). Oxford, U.K.: Clarendon Press.

Brockmeyer, E., H.L. Halstrøm, and A. Jensen (1948, 1960). The Life and Works of A.K. Erlang. *Acta Polytechnica Scandinavica. (Applied Mathematics and Computing Machinery Series No.6)*. Copenhagen, Denmark: The Danish Academy of Technical Sciences.

Brown, Mark (1981). Further monotonicity properties for specialized renewal processes. *The Annals of Probability* **9**, 891–895.

Brown, Mark (1983). Approximating IMRL distributions by exponential distributions with applications to first passage times. *The Annals of Probability* **11**, 419–427.

Brown, Mark (1984). Proximity between distributions: An inequality and its applications. *Reliability Theory and Models*, Mohamen S. Abdel-Hameed, Erhan Çinlar, and Joseph Quinn (eds.), pp. 257–266. Orlando, Florida: Academic Press.

Brown, Mark (1987). Inequalities for distributions with increasing failure rate. *Contributions to the Theory and Applications of Statistics. A Volume in Honor of Herbert Solomon*, A.E. Gelfand (ed.), pp. 3–17. New York: Academic Press.

Brown, Mark (1990). Error bounds for exponential approximations of geometric convolutions. *The Annals of Probability* **18**, 1388–1402.
Brown, Mark (2001). Exploiting the waiting time paradox: Applications of the renewal length transformation. City College, CUNY report, New York.
Brown, Mark and Guangping Ge (1984). Exponential approximations for two classes of aging distributions. *The Annals of Probability* **12**, 869–875.
Burai, Pál and Száz, Árpád (2005). Relationships between homogeneity, subadditivity and convexity properties. Univerzitet u Beogradu. *Publikacije Elektrotehničkog Fakulteta. Serija Matematika i Fizika* **16**, 77–87.
Burr, I.W. (1942). Cumulative frequency functions. *The Annals of Mathematical Statistics* **13**, 215–232.
Camp, B.H. (1922). A new generalization of Tschebyscheff's statistical inequality. *Bulletin of the American Mathematical Society* **28**, 427–432.
Canfield, Ronald V. and Leon E. Borgman (1975). Some distributions of time to failure for reliability applications. *Technometrics* **17**, 263–268.
Carling, Kenneth and Tor Jacobson (1996). Identification of dependent competing risks models. *Lifetime Data: Models in Reliability and Survival Analysis*, N.P. Jewell, A.C. Kimber, M-L. T. Lee, and G.A. Whitmore (eds.), pp. 59–63. Dordrecht, The Netherlands: Kluwer.
Castillo, Enrique, Ali S. Hadi, N. Balakrishnan, and José M. Sarabia (2005). *Extreme Value and Related Models with Applications in Engineering and Science*. Hoboken, New Jersey: John Wiley & Sons.
Castillo, Enrique and Maria Reyes Ruiz-Cobo (1992). *Functional Equations and Modelling in Science and Engineering*. New York: Marcel Dekker.
Champernowne, D.G. (1937). The theory of income distribution. *Econometrica* **5**, 379–381.
Champernowne, D.G. (1952). The graduation of income distributions. *Econometrica* **20**, 591–615.
Chandra, Mahesh and Nozer D. Singpurwalla (1981). Relationships between some notions which are common to reliability theory and economics. *Mathematics of Operations Research* **6** 113–121.
Charlier, C.V.L. (1906). Über das Fehlergesetz. *Arkiv für Mathematik, Astronomi och Fysik* **2**(8). 9 pp.
Chebyshev, P.L. (1874). Sur les valeurs limités des intégrals. *Journal de Mathematiques Pures et Appliquées* **19**(2), 157–160.
Chen, Y.Y., M. Hollander, and N.A. Langberg (1983). Tests for monotone mean residual life, using randomly censored data. *Biometrics* **39**, 119–127.
Chen, Zhenmin (2000). A new two-parameter lifetime distribution with bathtub shape or increasing failure rate. *Statistics and Probability Letters* **49**, 155–161.
Cherubini, Umberto, Elisa Luciano, and Walter Vecchiato (2005). *Copula Methods in Finance*. New York: John Wiley & Sons.
Chhikara, R.S. and J.L. Folks (1977). The inverse Gaussian distribution as a lifetime model. *Technometrics* **19**, 461–468.

Chhikara, Raj S. and J. Leroy Folks (1989). *The Inverse Gaussian Distribution*. New York: Marcel Dekker.

Chiang, Chin Long (1968). *Introduction to Stochastic Processes in Biostatistics*. New York: John Wiley & Sons.

Cirillo, R. (1979). *The Economics of Vilfredo Pareto*. London: Frank Case.

Clayton, D.G. (1974). Some odds ratio statistics for the analysis of ordered categorical data. *Biometrika* **61**, 525–531.

Cox, D.R. (1959). The analysis of exponentially distributed life-times with two types of failure. *Journal of the Royal Statistical Society* **B21**, 411–422.

Cox, D.R. (1962). *Renewal Theory*. London, U.K.: Methuen.

Cox, D.R. (1972). Regression models and life tables (with discussion). *Journal of the Royal Statistical Society* **B39**, 86–94.

Cox, D.R. and D. Oakes (1984). *Analysis of Survival Data*. London, U.K.: Chapman and Hall.

Cramér, Harald (1946). *Mathematical Methods of Statistics*. Princeton, New Jersey: Princeton Univ. Press.

Cramér, Harald (1972). On the history of certain expansions used in mathematical statistics. *Biometrika* **59**, 205–207.

Crawford, Gordon B. (1966). Characterization of geometric and exponential distributions. *The Annals of Mathematical Statistics* **37**, 1790–1795.

Crow, Edwin L. and Kunio Shimizu (eds.) (1988). *Lognormal Distributions: Theory and Applications*. New York: Marcel Dekker.

Crowder, Martin (1991). On the identifiability crisis in computing risk analysis. *Scandinavian Journal of Statistics* **18**, 223–233.

Crowder, Martin J. (2001). *Classical Competing Risks*. Boca Raton, Florida: Chapman and Hall/CRC.

Crowder, M.J., A.C. Kimber, R.L. Smith, and T.J. Sweeting (1991). *Statistical Analysis of Reliability Data*. London, U.K.: Chapman and Hall.

Dabrowska, Dorota M. and Kjell Doksum (1988). Estimation and testing in a two-sample generalized odds-rate model. *Journal of the American Statistical Association* **83**, 744–749.

Dahiya, Ram C. and Syed A. Hossain (1996). A modification of Goel-Okomoto model. *Lifetime Data: Models in Reliability and Survival Analysis*, Nicholas P. Jewell, Alan C. Kimber, Mei-Ling Ping Lee, and G.A. Whitmore (eds.), pp. 77–84. Dordrecht, The Netherlands: Kluwer.

David, H.A. and M.L. Moeschberger (1978). *The Theory of Competing Risks*. London, U.K.: Charles Griffin.

David, H.A. and H.N. Nagaraja (2003). *Order Statistics*, 3rd edn. Hoboken, New Jersey: John Wiley & Sons.

Davis, Henry T. and Michael Feldstein (1979). The generalized Pareto law as a model for progressively censored data. *Biometrika* **66**, 299–306.

De Morgan, Augustus (1860). On a property of Mr. Gompertz's law of mortality. *The Assurance Magazine and Journal of the Institute of Actuaries* (London) **8**, 181–184.

De Morgan, Augustus (1861a). On the unfair suppression of due acknowledgement to the writings of Mr. Benjamin Gompertz. *The Assurance Magazine and Journal of the Institute of Actuaries* (London) **9**, 86–89.

De Morgan, Augustus (1861b). On Gompertz's law of mortality. *The Assurance Magazine and Journal of the Institute of Actuaries* (London) **9**, 214–215.

Deshpande, Jayant V., Subhash C. Kochar, and Harshinder Singh (1986). Aspects of positive ageing. *Journal of Applied Probability* **23**, 748–758.

Desmond, Anthony (1985). Stochastic models of failure in random environments. *Canadian Journal of Statistics* **13**, 171–183.

Desmond, A.F. (1986). On the relationship between two fatigue-life models. *IEEE Transactions on Reliability* **25**, 167–169.

Desu, M.M. (1971). A characterization of the exponential distribution by order statistics. *The Annals of Mathematical Statistics* **42**, 837–838.

Desu, M.M. and Subhash C. Narula (1977). Reliability estimation under competing causes of failure. *The Theory and Applications of Reliability*, Vol. II, Chris P. Tsokos and I.N. Shimi (eds.), pp. 471–481. New York: Academic Press.

Dharmadhikari, Sudhakar and Kumar Joag-dev (1988). *Unimodality, Convexity, and Applications*. Boston, Massachusetts: Academic Press.

Dhillon, B.S. (1981). Life Distributions. *IEEE Transactions on Reliability* **30**, 457–459.

Doksum, K. (1969). Star-shaped transformations and the power of rank tests. *The Annals of Mathematical Statistics* **40**, 1167–1176.

Donoghue, W.F., Jr., (1974). *Monotone Matrix functions and Analytic Continuation*. New York: Springer-Verlag.

Dubey, S.D. (1968). A compound Weibull distribution. *Naval Research Logistics Quarterly* **15**, 179–188.

Dugué, D. (1941). Sur un nouveau type de courbe de fréquence. *Comptes Rendus de l'Academie des Sciences, Paris* **213**, 634–635.

Edgeworth, F.Y. (1904). The law of error. *Transactions of the Cambridge Philosophical Society* **20**, 36–65 and 113–114.

Edmonds, T.R. (1832). *Life Tables Founded Upon the Discovery of a Numerical Law*. London, U.K.: J. Moyes.

Edmonds, T.R. (1861a). On the discovery of the law of human mortality, and on the antecedent discoveries of Dr. Price and Mr. Gompertz. *The Assurance Magazine and Journal of the Institute of Actuaries* (London) **9**, 170–184.

Edmonds, T.R. (1861b). On the law of human mortality and Mr. Gompertz's new exposition of his law of mortality. *The Assurance Magazine and Journal of the Institute of Actuaries* (London) **9**, 327–341.

Efron, Bradley and Iain Johnstone (1990). Fisher's information in terms of the hazard rate. *The Annals of Statistics* **18**, 38–62.

Eggenberger, F. and G. Pólya (1923). Über die Statistik verketer Vorgänge. *Zeitschrift für Angewandte Mathematik und Mechanik* **1**, 279–289.

Elderton, W. Palin (1906). *Frequency Curves and Correlation.* London, U.K.: Layton.

Elderton, W.P. (1934). An approximate law of survivorship and other notes on the use of frequency curves in actuarial statistics. *Journal of the Institute of Actuaries* **65**, 1–37.

Elderton, William Palin and Norman Lloyd Johnson (1969). *Systems of Frequency Curves.* London: Cambridge Univ. Press.

Epstein, B. and M. Sobel (1953). Life testing. *Journal of the American Statistical Association* **48**, 486–502.

Erdélyi, A., W. Magnus, F. Oberhettinger, and F.G. Tricomi (1953). *Higher Transcendental Functions.* New York: McGraw-Hill.

Erdélyi, A., W. Magnus, F. Oberhettinger, and F.G. Tricomi (1954). *Tables of Integral Transforms,* Vol. 1. New York: McGraw-Hill.

Erlang, A.K. (1920). The application of the theory of probability in telephone administration. *Proceedings of the Scandinavian H.C. Örsted Congress, Copenhagen.* (Reprinted in E. Brockmeyer, J. Halstrøm, and A. Jensen. The Life and Works of A.K. Erlang, *Transactions of the Danish Academy of Technical Sciences,* 1948, No. 2 pp. 172–200, Copenhagen, Denmark).

Esary, J.D. and A.W. Marshall (1964). System structure and the existence of a system life. *Technometrics* **6**, 459–462.

Esary, J.D. and A.W. Marshall (1970). Coherent life functions. *SIAM Journal of Applied Mathematics* **18**, 810–860.

Esary, J.D. and A.W. Marshall (1973). Multivariate geometric distributions generated by a cumulative damage process. Technical report 55#Y73041A. Naval Postgraduate School, Monterey, California.

Esary, J.D. and A.W. Marshall (1974). Families of components, and systems, exposed to a compound Poisson damage process. *Reliability and Biometry,* Frank Proschan and R.J. Serfling (eds.), pp. 31–46. Philadelphia, Pennsylvania: Society for Industrial and Applied Mathematics.

Esary, J.D., A.W. Marshall, and F. Proschan (1970). Some reliability applications of the hazard transform. *SIAM Journal of Applied Mathematics* **18**, 849–860.

Esary, J.D., A.W. Marshall, and F. Proschan (1973). Shock models and wear processes. *The Annals of Probability* **1**, 627–649.

Esary, J.D., F. Proschan, and D.W. Walkup (1967). Association of random variables, with applications. *The Annals of Mathematical Statistics* **38**, 1466–1474.

Evans, Ralph A. (1990). Weibullitis. *IEEE Transactions on Reliability* **39**, 513.

Evans, Ralph A. (2000). Editorial: Populations and hazard rates. *IEEE Transactions on Reliability* **49**, 250.

Feigl, Polly and Marven Zelen (1965). Estimation of exponential survival probabilities with concomitant information. *Biometrics* **21**, 826–838.

Feller, W. (1943). On a general class of "contagious" distributions. *The Annals of Mathematical Statistics* **14**, 389–400.

Feller, William (1968). *An Introduction to Probability Theory*, Vol. I, 3rd edn. New York: John Wiley & Sons.

Feller, William (1971). *An Introduction to Probability Theory*, Vol. II, 2nd edn. New York: John Wiley & Sons.

Ferguson, Thomas S. (1965). A characterization of the exponential distribution. *The American Mathematical Monthly* **72**, 256–260.

Findeisen, Peter (1978). A simple proof of a classical theorem which characterizes the gamma distribution. *The Annals of Statistics* **6**, 1165–1167.

Fisher, R.A. (1925, 1958). *Statistical Methods for Research Workers*, 13th edn. 1958. Edinburgh, UK: Oliver and Boyd. (1st edn. published 1925.)

Fisher, R.A. and L.H.C. Tippett (1928). Limiting forms of the frequency distribution of the largest or smallest member of a sample. *Proceedings of the Cambridge Philosophical Society* **24**, 180–190.

Fisk, Peter R. (1961). The graduation of income distributions. *Econometrica* **29**, 171–184.

Fletcher, Harvey (1911). A verification of the theory of Brownian movements and a direct determination of the value of NE for gaseous ionization. *The Physical Review* **33**, 81–110.

Flinn, C.J. and J.J. Heckman (1983). Are unemployment and out of the labor force behaviorally distinct labor force states? *Journal of Labor Economics* **1**, 28–42.

Folks, J.L. and R.S. Chhikara (1978). The inverse Gaussian distribution and its statistical applications—A review. *Journal of the Royal Statistical Society* **B40**, 263–289.

Frankel, Paul and Jeffrey Longmate (2002). Parametric models for accelerated and long-term survival: A comment on proportional hazards. *Statistics in Medicine* **21**, 3279–3289.

Fréchet, Maurice (1927). Sur la loi de probabilité de l'écart maximum. *Annales de la Société Polonaise de Mathématique, Cracovie* **6**, 93–116.

Fréchet, Maurice (1950). *Généralités sur les Probabilités. Éléments Aléatories*, 2nd. edn. Borel series, Traité du calcul des probabilités et du ses applications, Div. I, Pt. III Gauthier-Villars, Paris.

Fréchet, Maurice (1951). Sur les tableaux de corrélation dont les marges sont donnés. *Annales de l'Université de Lyon*. Section A, Series 3, **14**, 53–77.

Freudenthal, A.M. and M. Shinozuka (1961). Structural safety under conditions of ultimate load failure and fatigue. Technical report WADD-TR-61-77. Wright Air Development Division.

Galambos, Janos and Samuel Kotz (1978). *Characterization of Probability Distributions*. Lecture Notes in Mathematics, Vol. 675. Berlin, Germany: Springer-Verlag.

Galton, F. (1879). The geometric mean in vital and social statistics. *Proceedings of the Royal Society of London* **29**, 365–367.

Gantmakher, Feliks Ruvinovich and M.G. Kre'in (1961). *Oscillation Matrices and Kernels and Small Vibrations of Mechanical Systems*. English translation from the Russian edition of 1950. U.S. Atomic Energy Commission, Office of Technical Information Extension.

Gasca, Mariano and Charles A. Micchelli (1996). *Total Positivity and Its Applications*. Dordrecht, The Netherlands: Kluwer.

Gastwirth, J.L. (1971). A general definition of the Lorenz curve. *Econometrica* **42**, 191–196.

Gastwirth, J.L. (1972). The estimation of the Lorenz curve and Gini index. *Review of Econonomic Statistics* **54**, 306–316.

Gauss, Carl Friedrich (1823). Theoria combinationis observationum erroribus minimis obnoxiae. *Commentationes Societatis Regiae Scientiarum Gottingensis* **5**, 33–90. (Supplementum (1828) **6**, 57–98.). *Werk* (1880) **4**, 10–11 (Goettingen).

Gaver, D.P. and M. Acar (1979). Analytical hazard representations for use in reliability, mortality, and simulation studies. *Communications in Statistics: Simulation and Computation* **B8**, 91–111.

Gawronski, C.R. (1984). On the bell-shape of stable densities. *The Annals of Probability* **12**, 230–242.

Genest, C. and J. McKay (1986). Copules archimédiennes et familles de lois bidimensionnelles dont les marges sont donnés. *Canadian Journal of Statistics* **14**, 145–159.

Gertsbakh, I.B. (1984). Asymptotic methods in reliability theory: A review. *Advances in Applied Probability* **16**, 147–175.

Gertsbakh, I.B. (1989). *Statistical Reliability Theory*. New York: Marcel Dekker.

Gertsbakh, I.B. and Kh.B. Kordonsky (1969). *Models of Failure*. Berlin, Germany: Springer-Verlag.

Ghai, Gauri L. and Jie Mi (1999). Mean residual life and its association with failure rate. *IEEE Transactions on Reliability* **48**, 262–265.

Ghitany, M.E., F.A. Al-Awadhi, and L. Khalfan (2006). Marshall-Olkin extended Lomax distribution and its application to censored data. Unpublished manuscript.

Ghosh, B.K. (1973). Some monotonicity theorems for χ^2, F and t distributions with applications. *Journal of the Royal Statistical Society* **B35**, 480–492.

Gibrat, R. (1930). Une loi des réparations économiques: L'effet proportionelle. *Bulletin de Statistique Géneral, France*, **19**, 469–ff.

Gibrat, R. (1931). *Les Inegalités Economiques*. Libraire du Recueil Sirey, Paris.

Gini, C. (1912). Variabilitá e Mutabilitá. Contributo allo studio delle distribuzioni e relazioni statistiche. *Studi Economico-Giuridici della Università de Cagliari*, Anno 3, Part 2, 80 pp.

Glaser, Ronald E. (1980). Bathtub and related failure rate characterizations. *Journal of the American Statistical Association* **75**, 667–672.

Gleser, Leon Jay (1989). The gamma distribution as a mixture of exponential distributions. *The American Statistician* **43**, 115–117.

Gnedenko, B.V. (1943). Sur la distribution limite du terme maximum d'une série aléatoire. *Annals of Mathematics* **44**, 423–453.

Goldie, Charles (1967). A class of infinitely divisible random variables. *Proceedings of the Cambridge Philosophical Society* **63**, 1141–1143.

Gompertz, B. (1825). On the nature of the function expressive on the law of human mortality. *Philosophical Transactions of the Royal Society of London* **115**, 513–583.

Gompertz, Benjamin (1861). Letter from Benjamin Gompertz. *The Assurance Magazine and Journal of the Institute of Actuaries* (London) **9**, 296–298.

Good, I.J. (1955). On the weighted combination of significance tests. *Journal of the Royal Statistical Society* **B17**, 264–265.

Gorski, Andrew C. (1968). Beware of Weibull euphoria. *IEEE Transactions on Reliability* **17**, 202–203.

Gottfried, Paul (1990). Comment on: "On the hazard rate of the lognormal distribution". *IEEE Transactions on Reliability* **39**, 519.

Govan, John (1899). On Gompertz's law of mortality. *The Assurance Magazine and Journal of the Institute of Actuaries* (London) **34**, 147–150.

Gradshteyn, I.S. and I.M. Ryzhik (1994). *Tables of Integrals, Series, and Products*, 5th edn. Boston, Massachusetts: Academic Press.

Gray, Peter (1858). On Mr. Gompertz's method for the adjustment of tables of mortality. *The Assurance Magazine and Journal of the Institute of Actuaries* (London) **7**, 121–130.

Greenwood, Major and G. Udney Yule (1920). An inquiry into the nature of frequency distributions representative of multiple happenings with particular reference to the occurrence of multiple attacks of disease or of repeated accidents. *Journal of the Royal Statistical Society* **83**, 255–279.

Grosswald, E. (1976). The Student t-distribution of any degree of freedom is infinitely divisible. *Zeitschrift für Wahrscheinlichkeitstheorie und verwandte Gebiete* **36**, 103–109.

Guess, Frank and Frank Proschan (1988). Mean Residual Life: Theory and Applications. *Handbook of Statistics*, Vol. 7: *Quality Control and Reliability*, pp. 215–224, P.R. Krishnaiah and C.R. Rao, (eds.). Amsterdam, The Netherlands: North Holland.

Gupta, P.L. and R.C. Gupta (1996). Ageing characteristics of the Weibull mixtures. *Probability in the Engineering and Informational Sciences* **10**, 591–600.

Gupta, Ramesh C. (1979). On the characterization of survival distributions in reliability by properties of their renewal densities. *Communications in Statistics—Theory and Methods* **A8**, 685–697.

Gupta, Ramesh C. and Robin Warren (2001). Determination of change points of non-monotonic failure rates. *Communications in Statistics—Theory and Methods* **30**, 1903–1920.

Gurland, John and Jayaram Sethuraman (1994). Reversal of increasing failure rates when pooling failure data. *Technometrics* **36**, 416–418.

Gurland, John and Jayaram Sethuraman (1995). How pooling failure data may reverse increasing failure rates. *Journal of the American Statistical Association* **90**, 1416–1423.

Gut, Allan (2005). *Probability: A Graduate Course.* New York: Springer Science-Business Media.

Hadwiger, H. (1940). Eine analytische reproductions funktion für biologische gesamtheiten. *Skandinavisk Aktuarietidskrift* **23**, 101–113.

Haines, Andrew and Nozer Singpurwalla (1974). Some contributions to the stochastic characterization of wear. *Reliability and Biometry*, pp. 47–80, Frank Proschan and R.J. Serfling (eds.). Philadelphia, Pennsylvania: Society for Industrial and Applied Mathematics.

Hall, W.J. and Hongyue Wang (2006). Use of life tables to extrapolate survival from clinical trial data. Unpublished manuscript, University of Rochester Medical Center.

Hall, W.J. and J.A. Wellner (1981). Mean residual life. *Statistics and Related Topics*, pp. 169–184, M. Csörgo, D.A. Dawson, J.N.K. Rao, and A.K. Saleh, (eds.). Amsterdam, The Netherlands: North-Holland.

Hallinan, Arthur J., Jr. (1993). A review of the Weibull distribution. *Journal of Quality Technology* **25**, 85–93.

Hardy, G.H., J.E. Littlewood, and G. Pólya (1929). Some simple inequalities satisfied by convex functions. *Messenger of Mathematics* **58**, 145–152.

Hardy, G.H., J.E. Littlewood, and G. Pólya (1952). *Inequalities*, 2nd edn. Cambridge, UK: Cambridge Univ. Press. (1st edn, published 1934.)

Harris, Carl M. and Nozer D. Singpurwalla (1968). Life distributions derived from stochastic hazard functions. *IEEE Transactions on Reliability* **17**, 70–79.

Heckman, James J. and Bo E. Honoré (1989). The identifiability of the competing risks model. *Biometrika* **76**, 325–330.

Hjorth, Urban (1980). A reliability distribution with increasing, decreasing, constant and bathtub-shaped failure rates. *Technometrics* **22**, 99–107.

Hoeffding, W. (1940). Maszstabinvariante Korrelationstheorie. *Schriften des Mathematischen Instituts und des Instituts für Angewandte Mathematik der Universität Berlin* **5**, 181–233.

Hollander, Myles and Frank Proschan (1984). Nonparametric concepts and methods in reliability. *Handbook of Statistics*, Vol. 4, P.R. Krishnaiah and P.K. Sen (eds.), pp. 613–655. Amsterdam, The Netherlands: Elsevier Science Publisher.

Hong, C.S. and L.F. Herk (1996). Incremental risk aversion and diversification preference. *Journal of Economic Theory* **70**, 180–200.

Hosmer, David W., Jr., and Stanley Lemeshow (1999). *Applied Survival Analysis: Regression Modeling of Time to Event Data*. New York: John Wiley & Sons.

Hougaard, Philip (1984). Life table methods for heterogeneous populations: Distributions describing the heterogeneity. *Biometrika* **71**, 75–83.

Hougaard, Philip (1986a). Survival models for heterogeneous populations derived from stable distributions. *Biometrika* **73**, 387–396.

Hougaard, Philip (1986b). A class of multivariate failure time distributions. *Biometrika* **73**, 671–178.

Hu, Taizhong, Asok K. Nanda, Huiliang Xie, and Zegang Zhu (2004). Properties of some stochastic orders: A unified study. *Naval Research Logistics Quarterly* **51**, 193–216.

Hutchinson, T.P. and C.D. Lai (1990). *Continuous Bivariate Distributions, Emphasising Applications*. Adelaide, Australia: Rumsby Scientific Publishing.

Ibragimov, I.A. (1956). On the composition of unimodal distributions. *Theory of Probability and its Applications* **1**, 255–260.

James, Ian R. (1986). On estimating equations with censored data. *Biometrika* **73**, 35–42.

Jensen, Finn and Niels Erik Petersen (1982). *Burn-in*. Chichester, U.K.: John Wiley & Sons.

Jiang, R. and D.N.P. Murthy (1998). Mixture of Weibull distributions—Parametric characterization of failure rate function. *Stochastic Models and Data Analysis* **14**, 47–65.

Jiang, R., D.N.P. Murthy, and P. Ji (2001). n-fold Weibull multiplicative model. *Reliability Engineering and System Safety* **74**, 211–219.

Joe, Harry (1997). *Multivariate Models and Dependence Concepts*. London, U.K.: Chapman & Hall/CRC.

Jogdeo, K. (1977). Association and probability inequalities. *The Annals of Statistics* **3**, 495–504.

Johnson, Norman L., Samuel Kotz, and Adrienne W. Kemp (1992). *Discrete Distributions*. New York: John Wiley & Sons.

Johnson, Norman L. and Samuel Kotz (1970a). *Continuous Univariate Distributions-1*. New York: Houghton Mifflin.

Johnson, Norman L. and Samuel Kotz (1970b). *Continuous Univariate Distributions-2*. New York: Houghton Mifflin.

Johnson, Norman L., Samuel Kotz and N. Balakrishnan (1994). *Continuous Univariate Distributions*. Vol. 1, 2nd edn. New York: John Wiley & Sons.

Johnson, Norman L., Samuel Kotz, and N. Balakrishnan (1995). *Continuous Univariate Distributions*, Vol. 2, 2nd edn. New York: John Wiley & Sons.

Jørgensen, Bent (1982). *Statistical Properties of the Generalized Inverse Gaussian distribution*. Lecture Notes in Statistics, Vol. 9. New York: Springer-Verlag.

Jørgensen, Bent, V. Seshadri, and G.A. Whitmore (1991). On the mixture of the inverse Gaussian distribution with its complementary reciprocal. *Scandinavian Journal of Statistics* **18**, 77–89.

Joyce, W.B. and P.J. Anthony (1988). Failure rate of a cold—or hot-spared component with a lognormal lifetime. *IEEE Transactions on Reliability* **37**, 299–307.

Kagan, A.M., Yu. V. Linnik, and C. Radhakrishna Rao (1973). *Characterization problems in mathematical statistics*. (Translated from Russian by B. Ramachandran), New York: John Wiley & Sons.

Kakwani, Nanak C. (1980). *Income Inequality and Poverty*. New York: Oxford Univ. Press.

Kalbfleisch, John D. and Ross L. Prentice (1980). *The Statistical Analysis of Failure Time Data*. New York: John Wiley & Sons.

Kaminsky, Kenneth S. (1983). An aging property of the Gompertz survival function and a discrete analog. Technical report no. 1983-8. Department of Mathematical Statistics, University of Umeå, Umeå, Sweden.

Karamata, J. (1932). Sur une inégalité rélative aux fonctions convexes. *Publications Mathématique de l'Université de Belgrade* **1**, 145–148.

Karlin, Samuel (1968). *Total Positivity*, Vol. I. Stanford, California: Stanford Univ. Press.

Karlin, Samuel, Frank Proschan, and Richard E. Barlow (1961). Moment inequalities of Pólya frequency functions. *Pacific Journal of Mathematics* **11**, 1023–1033.

Karlin, Samuel and Yosef Rinott (1980). Classes of orderings of measures and related correlation inequalities. I. Multivariate totally positive distributions. *Journal of Multivariate Analysis* **10**, 467–498.

Karlin, Samuel and Herman Rubin (1956). The theory of decision procedures for distributions with monotone likelihood ratio. *The Annals of Mathematical Statistics* **27**, 272–299.

Karlin, S. and L.S. Shapley (1953). *Geometry of Moment Spaces*. (Memoirs of the American Mathematical Society, No. 12). Providence, Rhode Island: American Mathematical Society.

Karn, M. Noel (1931). An inquiry into the various death-rates and the comparative influence of certain diseases on the duration of life. *Annals of Eugenics* **4**, 279–326.

Karn, M. Noel (1933). A further study of methods of constructing life tables when certain causes of death are eliminated. *Biometrika* **25**, 329–337.

Keiding, Niels (2001). Comment on a paper by Aalen and Gjessing. *Statistical Science* **16**, 19–20.

Keilson, J. and F.W. Steutel (1974). Mixtures of distributions, moment inequalities and measures of exponentiality and normality. *The Annals of Probability* **2**, 112–130.

Keilson, Julian and Ushio Sumita (1982). Uniform stochastic ordering and related inequalities. *Canadian Journal of Statistics* **10**, 181–198.

Kendall, Maurice G. and Alan Stuart (1963). *The Advanced Theory of Statistics*, Vol. 1, 2nd edn. New York: Hafner Publishing Co.

Kent, John T. (1981). Convolution mixtures of infinitely divisible distributions. *Mathematical Proceedings of the Cambridge Philosophical Society* **90**, 141–153.

Kimball, A.W. (1951). On tests of significance in analysis of variance. *The Annals of Mathematical Statistics* **22**, 600–602.

Kingman, J.F.C. (1978). Discussion of "The inverse Gaussion distribution and its statistical application—A review", by J.L. Folks and R.S. Chhikara. *Journal of the Royal Statistical Society* **B40**, 263–289.

Kirmani, S.N.U.A and Ramesh C. Gupta (2001). On the proportional odds model in survival analysis. *Annals of the Institute of Statistical Mathematics* **53**, 203–216.

Klefsjö, Bengt (1981). HNBU survival under some shock models. *Scandinavian Journal of Statistics* **8**, 39–47.

Klefsjö, Bengt (1982a). On aging properties and total time on test transform. *Scandinavian Journal of Statistics* **9**, 37–41.

Klefsjö, Bengt (1982b). The HNBUE and HNWUE classes of life distributions. *Naval Research Logistics Quarterly* **29**, 331–344.

Kleiber, Christian and Samuel Kotz (2002). A characterization of income distributions in terms of generalized Gini coefficients. *Social Choice and Welfare* **19**, 789–794.

Kleiber, Christian and Samuel Kotz (2003). *Statistical Size Distributions in Economics and Actuarial Sciences*. New York: John Wiley & Sons.

Klutke, Georgia-Ann, Peter C. Keissler, and M.A. Wortman (2003). A critical look at the bathtub curve. *IEEE Transactions on Reliability* **52**, 125–129.

Kochar, Subhash C. and Frank Proschan (1991). Independence of the time and cause of failure in the multiple dependent competing risks model. *Statistica Sinica* **1**, 295–299.

Kochar, Subhash, Hari Mukerjee, and Francisco J. Samaniego (1999). The "signature" of a coherent system and its applications in comparisons among systems. *Naval Research Logistics Quarterly* **46**, 507–523.

Konstantinowsky, D. (1914). Elektrische Ladungen und Brown'sche Bewegung sehr kleiner Metallteilchen in Gasen. *Sitzungsberichte der Kaiserlichen Akademie der Wissenschaften* **123**, 1697–1752.

Kotz, Samuel and Saralees Nadarajah (2000). *Extreme Value Distributions: Theory and Applications*. London, U.K.: Imperial College Press.

Kotz, Samuel and Nozer D. Singpurwalla (1999). On a bivariate distribution with exponential marginals. *Scandinavian Journal of Statistics* **26**, 451–464.

Kunitz, Harald (1989). A new class of bathtub-shaped hazard rates and its application in a comparison of two test-statistics. *IEEE Transactions on Reliability* **38**, 351–354.

Kupka, Joseph and Sonny Loo (1989). The hazard and vitality measures of ageing. *Journal of Applied Probability* **26**, 532–542.

Lai, C.D., M. Xie, and D.N.P. Murthy (2001). Bathtub-shaped failure rate life distributions. *Handbook of Statistics*, Vol. 20, pp. 69–104, N. Balakrishnan and C.R. Rao. (eds.). Amsterdam, The Netherlands: Elsevier Science Publisher.

Lancaster, H.O. (1966). Forerunners of the Pearson χ^2. *The Australian Journal of Statistics* **8**, 117–126.

Lancaster, H.O. (1969). *The Chi-Squared Distribution*. New York: John Wiley & Sons.

Lancaster, H.O. (1982). Chi-square distribution. *Encyclopedia of Statistical Sciences*, Vol. 1, pp. 439–442, S. Kotz and N.L. Johnson (eds.). New York: John Wiley & Sons.

Langberg, N., F. Proschan, and A.J. Quinzi (1978). Converting dependent models into independent ones. *The Annals of Probability* **6**, 174–181.

Lawless, J.F. (1982). *Statistical Models and Methods for Lifetime Data*. New York: John Wiley & Sons.

Lawless, J.F. (1986). A note on lifetime regression models. *Biometrika* **73**, 509–512.

Lawrance, A.J. and P.A.W. Lewis (1977). An exponential moving-average sequence and point process (EMA1). *Journal of Applied Probability* **14**, 98–113.

Lawrance, A.J. and P.A.W. Lewis (1980). The exponential autoregresssive-moving average EARMA (p,q) process. *Journal of the Royal Statistical Society* **B42**, 150–161.

Lawrance, A.J. and P.A.W. Lewis (1983). Simple dependent pairs of exponential and uniform random variables. *Operations Research* **31**, 1179–1197.

Lee, Eliza T. (1992). *Statistical Methods for Survival Data Analysis*, 2nd edn. New York: John Wiley & Sons.

Leemis, Lawrence M. (1986). Lifetime distribution identities. *IEEE Transactions on Reliability* **35**, 170–174.

Lehmann, E.L. (1955). Ordered families of distributions. *The Annals of Mathematical Statistics* **26**, 399–419.

Lehmann, E.L. (1966). Some concepts of dependence. *The Annals of Mathematical Statistics* **37**, 1137–1153.

Lehmann, E.L. and J. Rojo (1992). Invariant directional orderings. *The Annals of Statistics* **29**, 2100–2110.

Li, Chin-Shang and Jeremy M.G. Taylor (2002). A semi-parametric accelerated failure time cure model. *Statistics in Medicine* **21**, 3235–3247.

Likeš, J. (1967). Distributions of some statistics in samples from exponential and power-function populations. *Journal of the American Statistical Association* **62**, 259–271.

Lindsay, Bruce G. (1995). *Mixture models: theory, geometry and applications.* NSF-CBMS Regional Conference Series in Probability and Statistics, Vol. 5.

Institute of Mathematical Statistics, Hayward, California; American Statistical Association, Alexandria, Virginia.
Ljubo, M. (1965). Curves and concentration indices for certain generalized Pareto distributions. *Statistical Review* **15**, 257–260. (In Serbo-Croatian, English summary.)
Lomax, K.S. (1954). Business failures: Another example of the analysis of failure data. *Journal of the American Statistical Association* **49**, 847–852.
Lorenz, M.O. (1905). Methods of measuring concentration of wealth. *Journal of the American Statistical Association* **9**, 209–219.
Losinger, Willard C. (1997). The Lorenz curve applied to livestock populations. *Chance* **10**, 19–22.
Lukacs, Eugene (1955). A characterization of the gamma distribution. *The Annals of Mathematical Statistics* **26**, 319–324.
Lukacs, E. and R.G. Laha (1964). *Applications of characteristic functions.* New York: Hafner Publishing Company.
Lundberg, O. (1964). *On Random Processes and their Application to Sickness and Accident Statistics*, 2nd edn. Stockholm, Sweden: Almqvist and Wiksell. (1st edn. published 1940.)
Lynch, James D. (1999). On conditions for mixtures of increasing failure rate distributions to have an increasing failure rate. *Probability in the Engineering and Informational Sciences* **13**, 33–36.
Lynn, Nicholas J. and Nozer D. Singpurwalla (1997). Comment: "Burn-in" makes us feel good. *Statistical Science* **12**, 13–19.
Ma, C. (1999). Uniform stochastic ordering on a system of components with dependent lifetimes induced by a common environment. *Sankhyā* **A61**, 218–228.
MacGillivray, H.L. (1992). Shape properties of the g- and h- and Johnson families. *Communications in Statistics: Theory and Methods* **21**, 1233–1250.
MacGillivray, H.L. and K.P. Blanda (1988). The relationships between skewness and kurtosis. *Australian Journal of Statistics* **30**, 319–337.
Makeham, William Matthew (1860). On the law of mortality and the construction of annuity tables. *The Assurance Magazine and Journal of the Institute of Actuaries* (London) **8**, 301–310.
Makeham, William Matthew (1867). On the law of mortality. *The Assurance Magazine and Journal of the Institute of Actuaries* (London) **13**, 325–358.
Makeham, William Matthew (1890). On further development of Gompertz's law. *The Assurance Magazine and Journal of the Institute of Actuaries* (London) **28**, 152–159, 185–192, 316–332.
Malik, Henrick John (1967). Exact distribution of the quotient of independent generalized gamma variables. *Canadian Mathematical Bulletin* **10**, 463–465.
Mallows, C.L. (1963). A generalization of Chebyshev inequalities. *Proceedings of the London Mathematical Society* **13**(3), 385–412.

Mann, H.B. and D.R. Whitney (1947). On a test of whether one of two random variables is stochastically larger than the other. *The Annals of Mathematical Statistics* **18**, 50–60.

Mann, N.R., R.E. Schafer, and N.D. Singpurwalla (1974). *Methods for Statistical Analysis of Reliability and Life Data*. New York: John Wiley & Sons.

Markov, A.A. (1898). On the limiting values of integrals in connection with interpolation. *Zapiski Imperatorskoi Akademii Nauk. Fiziko-Matematicheskoe Otdelenie* (8) 6. *Selected Papers on Continued Fractions and the Theory of Functions Deviating Least from Zero*, No. 5, pp. 146–230, OGIZ, Moscow-Leningrad, 1948, (in Russian).

Markus, M. and M. Pinsky (1969). On the domain of attraction of $e^{-e^{-x}}$. *Journal of Analysis and Applications* **28**, 440–449.

Marsaglia, George (1974). Extensions and applications of Lukacs' characterization of the gamma distribution. *Proceedings of the Symposium on Statistics and Related Topics*. Carleton University, Ottawa, Canada.

Marsaglia, George (1989). The $X+Y$, X/Y characterization of the gamma distribution. *Contributions to Probability and Statistics: Essays in Honor of Ingram Olkin*, L.J. Gleser, M.D. Perlman, S.J. Press, and A.R. Sampson (eds.), pp. 91–98. New York: Springer-Verlag.

Marshall, Albert W. (1991). Multivariate stochastic orderings and generating cones of functions. *Stochastic Orders and Decisions Under Risk*, K. Mosler and M. Scarsini (eds.), pp. 231–247. (Institute of Mathematical Statistics Lecture Notes—Monograph Series, Vol. 19). Hayward, California: Institute of Mathematical Statistics.

Marshall, Albert W. (1994). A system model for reliability studies. *Statistica Sinica* **4**, 549–565.

Marshall, A.W., J.C. Meza, and I. Olkin (2001). Can data recognize its parent distribution? *Journal of Computational and Graphical Statistics* **10**, 555–580.

Marshall, Albert W. and Ingram Olkin (1967). A multivariate exponential distribution. *Journal of the American Statistical Association* **62**, 30–44.

Marshall, Albert W. and Ingram Olkin (1979). *Inequalities: Theory of Majorization and Its Applications*. New York: Academic Press.

Marshall, Albert W. and Ingram Olkin (1983). Domains of attraction of multivariate extreme value distributions. *The Annals of Probability* **11**, 168–177.

Marshall, Albert W. and Ingram Olkin (1985). A family of bivariate distributions generated by the bivariate Bernoulli distribution. *Journal of the American Statistical Association* **80**, 332–338.

Marshall, Albert W. and Ingram Olkin (1988). Families of multivariate distributions. *Journal of the American Statistical Association* **83**, 834–841.

Marshall, Albert W. and Ingram Olkin (1990). Multivariate distributions generated from mixtures of convolution and product families. *Topics in Statis-*

tical Dependence, H.W. Block, A.R. Sampson, and T.H. Savits (eds.), pp. 371–393. (Institute of Mathematical Statistics Lecture Notes-Monograph Series Vol. 16). Hayward, California: Institute of Mathematical Statistics.

Marshall, Albert W. and Ingram Olkin (1993). Bivariate distributions from Pólya's urn model for contagion. *Journal of Applied Probability* **30**, 497–508.

Marshall, Albert W. and Ingram Olkin (1997). A new method of adding a parameter to a family of distributions with applications to the exponential and Weibull families. *Biometrika* **84**, 641–652.

Marshall, Albert W. and Frank Proschan (1972a). Classes of distributions applicable in replacement policies, with renewal theory implications. *Proceedings of the 6th Berkeley Symposium on Mathematical Statistics and Probability*, Vol. I, ed. by L. LeCam, J. Neyman and E.L. Scott, pp. 395–415. Berkeley: University of California Press.

Marshall, Albert W. and Frank Proschan (1972b). Convexity preserving scale transforms. *Inequalities*, Vol. III, pp. 225–234, O. Shisha (ed.). New York: Academic Press.

McAlister, D. (1879). The law of the geometric mean. *Proceedings of the Royal Society of London* **29**, 367–375.

McCullagh, P. (1980). Regression models for ordinal data, with discussion. *Journal of the Royal Statistical Society* **B42** 109–142.

McDonald, James B. and Dale O. Richards (1987). Hazard rates and generalized beta distributions. *IEEE Transactions on Reliability* **36**, 463–466.

McEwen, R.P. and B.R. Parresol (1991). Moment expressions and summary statistics for the complete and truncated Weibull distribution. *Communications in Statistics: Theory and Methods* **20**, 1361–1372.

McGill, W.J. and J. Gibbon (1965). The general-gamma distribution and reaction time. *Journal of Mathematical Psychology* **2**, 1–18.

Meeker, William Q. and Luis A. Escobar (1998). Pitfalls of accelerated testing. *IEEE Transactions on Reliability* **47**, 114–118.

Meidell, B. (1922). Sur une problème du calcul des probabilités et les statistiques mathematiques. *Comptes Rendus de l'Academie des Sciences, Paris* **175**, 806–808.

Meyer, Paul A. (1966). *Probability and Potentials*. Waltham, Massachusetts: Blaisdell Publishing Co.

Mi, Jie (1995). Bathtub failure rate and upside-down mean residual life. *IEEE Transactions on Reliability* **44**, 388–391.

Mi, Jie (1998). A new explanation of decreasing failure rate of a mixture of exponentials. *IEEE Transactions on Reliability* **47**, 460–462.

Miller, D.R. (1977). A note on independence of multivariate lifetimes in competing risk models. *The Annals of Statistics* **5**, 576–579.

Miller, Rupert G. Jr. (1981). *Survival Analysis*. New York: John Wiley & Sons.

von Mises, R. (1936). La distribution de la plus grande de n valeurs. *Revue Mathématique de l'Union Interbalkanique* **1**, 141–160. Selected Papers of Richard von Mises, 1964, Vol. 2, pp. 271–294. Providence, Rhode Island: American Mathematical Society.

de Moivre, A. (1724). *Treatise of Annuities Upon Lives.* London.

Moran, P.A.P. (1967). Testing for correlation between non-negative variates. *Biometrika* **54**, 385–394.

Morris, K.W. (1963). A note on direct and inverse binomial sampling. *Biometrika* **50**, 544–545.

Mudholkar, Govind S. and Alan D. Hutson (1996). The exponentiated Weibull family: Some properties and flood data application. *Communications in Statistics: Theory and Methods* **25**, 3059–3083.

Mudholkar, Govind S. and Georgia D. Kollia (1994). Generalized Weibull family: A structural analysis. *Communications in Statistics: Theory and Methods* **A23**, 1149–1171.

Mudholkar, Govind S. and Ila C. Sarkar (1999). A proportional hazards modelling of multisample reliability data. *Communications in Statistics: Theory and Methods* **28**, 2079–2101.

Mudholkar, Govind S. and Deo Kumar Srivastava (1993). Exponentiated Weibull family for analyzing bathtub failure-rate data. *IEEE Transactions on Reliability* **42**, 299–302.

Mudholkar, Govind S., Deo Kumar Srivastava, and Marshall Freimer (1995). The exponentiated Weibull family: A reanalysis of the bus-motor-failure data. *Technometrics* **37**, 436–445.

Mudholkar, Govind S., Deo Kumar Srivastava, and Georgia D. Kollia (1996). A generalization of the Weibull distribution with application to the analysis of survival data. *Journal of the American Statistical Association* **91**, 1575–1563.

Müller, Alfred and Dietrich Stoyan (2002). *Comparison Methods for Stochastic Models and Risks.* Chichester, U.K.: John Wiley & Sons.

Murthy, D.N. Prabhakar, Min Xie, and Renyan Jiang (2004). *Weibull Models.* Hoboken, New Jersey: John Wiley & Sons.

Murthy, V.K., G. Swartz, and K. Yuen (1973). Realistic models for mortality rates and estimation, I and II. Technical reports. Department of Biostatistics, University of California, Los Angeles, California.

Muth, Eginhard J. (1977). Reliability models with positive memory derived from the mean residual life function. *The Theory and Applications of Reliability*, Vol. II, pp. 401–435, C.P. Tsokos and I.N. Shimi (eds.). New York: Academic Press.

Nanda, Asok K. and Moshe Shaked (2001). The hazard rate and the reversed hazard rate orders with applications to order statistics. *Annals of the Institute of Statistical Mathematics* **53**, 853–864.

Natanson, I.P. (1955). *Theory of Functions of a Real Variable.* (Translated by Leo Boron.) New York: Frederick Ungar.

Navarro, Jorge, Felix Belzunce, and José M. Ruiz (1997). New stochastic orders based on double truncation. *Probability in the Engineering and Informational Sciences* **11**, 395–402.

Nelsen, Roger B. (1999). *An Introduction to Copulas*, 2nd edn. (Lecture Notes in Statistics Vol. 139). New York: Springer-Verlag.

Nelson, Wayne (1990). *Accelerated Testing: Statistical Models, Test Plans, and Data Analyses*. New York: John Wiley & Sons.

Nelson, Wayne (2004). *Applied Life Data Analysis*. New York: John Wiley & Sons.

Nguyen, Truc T., John T. Chen, and Arjun K. Gupta (2003). A proof of the conjecture on positive skewness of generalized inverse Gaussian distributions. *Biometrika* **90**, 245–250.

Oakes, D. (1989). Bivariate survival models induced by frailties. *Journal of the American Statistical Association* **84**, 487–493.

O'Cinneide, Colm A. and Adrian E. Raftery (1989). A continuous multivariate exponential distribution that is multivariate phase type. *Statistics and Probability Letters* **7**, 323–325.

Ogborn, M.E. (1953). On the nature of the function expressive of the law of human mortality. *Journal of the Institute of Actuaries* **79**, 170–212.

Oja, H. (1981). On the location, scale skewness and kurtosis of univariate distributions. *Scandinavian Journal of Statistics* **8**, 154–168.

Olkin, Ingram (1972). Monotonicity properties of Dirichlet integrals with applications to the multinomial distribution and the analysis of variance. *Biometrika* **59**, 303–307.

Olkin, Ingram and Milton Sobel (1965). Integral expressions for tail probabilities of the multinomial and negative multinomial distributions. *Biometrika* **52**, 167–179.

O'Quigley, John and Janez Stare (2002). Proportional hazards models with frailties and random effects. *Statistics in Medicine* **21**, 3219–3233.

Ord, J.K. (1972). *Families of Frequency Distributions*. London, U.K.: Charles Griffin.

Palm, C. (1946). Specialnummer för Teletrafikteknik. Tekn. Meddelanden från Kungl. Telegrafstyrelsen.

Pareto, V. (1897). *Cours d'economie Politique*, Vol. II. Lausanne, Switzerland: F. Rouge.

Parzen, Emanuel (2004). Quantile probability and statistical data modeling. *Statistical Science* **19**, 652–662.

Pearson, Karl (1895). Contributions to the mathematical theory of evolution. II. Skew variations in homogeneous material. *Philosophical Transactions of the Royal Society of London* **A186**, 343–414.

Pearson, Karl (ed.) (1968). *Tables of the Incomplete Beta Distribution*, 2nd edn. London, U.K.: Cambridge Univ. Press. (1st edn. published 1934.)

Perks, Wilfred (1932). On some experiments in the graduation of mortality statistics. *Journal of the Institute of Actuaries* **43**, 12–57.

Peterson, Arthur V., Jr. (1975). *Nonparametric estimation on the competing risk problem.* Doctoral dissertation, Stanford University, Stanford, California.

Peterson, Arthur V. (1976). Bounds for a joint distribution function with fixed sub-distribution functions: Application to competing risks. *Proceedings of the National Academy of Sciences of the United States of America* **73**, 11–13.

Pham, T.G. and N. Turkkan (1994). The Lorenz and the scaled total-time-on-test transform curves: a unified approach. *IEEE Transactions on Reliability* **43**, 76–83.

Phelps, Robert R. (1966). *Lectures on Choquet's Theorem.* Princeton: Van Nostrand.

Phillips, R.S. (1954). An inversion formula for Laplace transforms and semigroups of linear operators. *Annals of Mathematics* **59**, 325–356.

Pickands, J. (1975). Statistical inference using extreme order statistics. *The Annals of Statistics* **3**, 119–131.

Pinkus, Allan (1996). Spectral properties of totally positive kernels and matrices. *Total Positivity and Its Applications,* pp. 477–511, M. Gasca and C.A. Micchelli (eds.). Dordrecht, The Netherlands: Kluwer.

Pitman, E.J.G. (1937). The "closest" estimates of statistical parameters. *Proceedings of the Cambridge Philosophical Society* **33**, 212–222.

Pólya, G. and G. Szegö (1972). *Problems and Theorems in Analysis,* Vol. I. (Translated by D. Aeppli), Springer-Verlag, Berlin and New York (1st ed. in German, 1925).

Prékopa, A. (1971) Logarithmic concave measures with application to stochastic programming. *Acta Scientiarum Mathematicarum. (Szeged.)* **32**, 301–315.

Prentice, R.L. (1975). Discrimination among some parametric models. *Biometrika* **62**, 607–614.

Press, S. James (2003). *Subjective and Objective Bayesian Statistics,* 2nd edn. Hoboken, New Jersey: John Wiley & Sons.

Price, R. (1771). *Observations on reversionary payments.* London, U.K.: R. Cadell. Read before the Royal Society in April 1769.

Proschan, Frank (1963). Theoretical explanation of observed decreasing failure rate. *Technometrics* **5**, 375–384.

Proschan, Frank and Myles Hollander (1984). Nonparametric concepts and methods in reliability. *Handbook of Statistics,* Vol. 4, P. R. Krishnaiah and P.K. Sen (eds.), pp. 613–655. Amsterdam, The Netherlands: Elsevier.

Rachev, S.T. and S. Resnick (1991). Max-geometric infinite divisibility and stability. *Communications in Statistics. Stochastic Models* **7**, 191–218.

Raftery, Adrian E. (1984). A continuous multivariate exponential distribution. *Communications in Statistics: Theory and Methods* **13**, 947–965.

Rajarshi, Sujata and M.B. Rajarshi (1988). Bathtub distributions: A review. *Communications in Statistics: Theory and Methods* **17**, 2597–2621.

Rayleigh, Lord J.W.S. (1880). On the resultant of a large number of vibrations of the same pitch and of arbitrary phase. *The London, Edinburgh, and Dublin Philosophical Magazine and Journal of Science*, 5th series, **10**, 73–798.

Rayleigh, Lord J.W.S. (1919). On the problem of random vibrations, and of random flights in one, two, or three dimensions. *The London, Edinburgh, and Dublin Philosophical Magazine and Journal of Science*, 6th series, **37**, 321–347.

Resnick, Sidney I. (1987). *Extreme Values, Regular Variation, and Point Processes.* New York: Springer-Verlag.

Richards, Dale O. and James McDonald (1987). A general methodology for determining distributional forms with applications in reliability. *Journal of Statistical Planning and Inference* **16**, 365–376.

Rider, Paul R. (1962). The negative binomial distribution and the incomplete beta function. *The American Mathematical Monthly* **69**, 302–304.

Rieck, James R. (1999). A moment-generating function with applications to the Birnbaum-Saunders distribution. *Communications in Statistics: Theory and Methods* **29**, 2213–2222.

Roberts, A.W. and D.E. Varberg (1973). *Convex Functions.* Princeton, New Jersey: Academic Press.

Rockafeller, R. Tyrrell (1970). *Convex Analysis.* Princeton Math. Series, No. 28. Princeton, New Jersey: Princeton Univ. Press.

Rojo, Javier (1995). Characterization of some concepts of aging. *IEEE Transactions on Reliability* **44**, 285–290.

Rojo, Javier (1996). On tail categorization of probability laws. *Journal of the American Statistical Association* **91**, 378–384.

Rolski, Tomasz (1975). Mean residual life. *Bulletin of the International Statistical Institute* **46**, 266–270.

Rose, David Melvin (1973). *An Investigation of Dependent Competing Risks.* Doctoral dissertation, University of Washington, Seattle, Washington.

Rosin, P. and E. Rammler (1933). The laws governing the fineness of coal. *Journal of the Institute of Fuels* **6**, 29–36.

Ross, S.M. (1979). Multivalued state component systems. *The Annals of Probability* **7**, 379–383.

Royden, H.L. (1953). Bounds on a distribution function when its first n moments are given. *The Annals of Mathematical Statistics* **24**, 361–376.

Ruben, Harold (1974). Non-central chi-square and gamma revisited. *Communications in Statistics* **3**, 607–633.

Samaniego, Francisco J. (1985). On closure of the IFR class under formation of coherent systems. *IEEE Transactions on Reliability* **34**, 69–72.

Sankaran, P.G. and K. Jayakumar (2006). On proportional odds models. (Unpublished).

Särndal, Carl-Erik (1971). The hypothesis of elementary errors and the Scandinavian school of statistical theory. *Biometrika* **58**, 375–207.

Saunders, Sam C. (2007). *The Statistical Theory of Reliability and Life Testing for Engineers and Scientists.* New York: Springer-Verlag.

Savits, Thomas H. (1985). A multivariate IFR class. *Journal of Applied Probability* **22**, 197–204.

Schoenberg, I.J. (1930). Über variationsvermindernde lineare Transformationen. *Mathematische Zeitschrift* **32**, 321–328.

Schoenberg, I.J. (1951). On Pólya frequency functions, I. The totally positive functions and their Laplace transforms. *Journal d'Analyse Mathematique* **59**, 199–230.

Schrödinger, E. (1915). Zür Theorie der Fall-und Steigversuche an Teilchenn mit Brownscher Bewegung. *Physikalische Zeitschrift* **16**, 289–295.

Seal, Hilary L. (1969). *Stochastic Theory of Risk in Business.* New York: John Wiley & Sons.

Seal, Hilary L. (1977). Studies in the history of probability and statistics. XXXV. Multiple decrements or competing risks. *Biometrika* **64**, 429–439.

Sengupta, Debasis and Asok K. Nanda (1999). Log-concave and concave distributions in reliability. *Naval Research Logistics* **46**, 419–433.

Seshadri, V. (1993). *The Inverse Gaussian Distribution.* Oxford, U.K.: Clarendon Press.

Seshadri, V. (1999). *The Inverse Gaussian Distribution. Statistical Theory and Applications.* New York: Springer-Verlag.

Sethuraman, J. (1965). On the characterization of the three limiting types of the extreme. *Sankhyā* **27**, 357–364.

Shaked, Moshe and J. George Shanthikumar (1994). *Stochastic Orders and Their Applications.* San Diego: California: Academic Press.

Shaked, Moshe and J. George Shantikumar (2007). *Stochastic orders.* New York: Springer Science-Business Media.

Sherwin, David J. (1997). Concerning bathtubs, maintained systems and human frailty. *IEEE Transactions on Reliability* **46**, 162.

Shimizu, Kunio and Edwin L. Crow (1988). History, genesis and properties. *Lognormal Distributions*, E. L. Crow and K. Shimizu (eds.), pp. 1–25. New York: Marcel Dekker.

Shohat, J.A. and J.D. Tamarkin (1943). *The Problem of Moments.* Providence, Rhode Island: American Mathematical Society.

Shuster, Jonathan (1968). On the inverse Gaussian distribution function. *Journal of the American Statistical Association* **63**, 1514–1516.

Singpurwalla, Nozer D. (2003). The hazard potential. (Unpublished).

Singpurwalla, Nozer D. and Simon P. Wilson (1993). The warranty problem: Its statistical and game theoretic aspects. *SIAM Review* **35**, 17–42.

Singpurwalla, Nozer D. and Simon P. Wilson (1995). The exponentiation formula of reliability and survival: Does it always hold? *Lifetime Data Analysis* **1**, 187–194.

Slud, Eric V. and Leonid Kopylev (1996). Dependent competing risks with time-dependent covariates. *Lifetime Data: Models in Reliability and Survival Analysis*, Nicholas P. Jewell, Alan C. Kimber, Mei-Ling Ping Lee, and G.A. Whitmore (eds.), pp. 323–330. Dordrecht, The Netherlands: Kluwer.

Slymen, D.J. and P.A. Lachenbruch (1984). Survival functions arising from two families and generated by transformations. *Communications in Statistics: Theory and Methods* **13**, 1179–1201.

Smith, Richard L. (1989). Extreme value analysis of environmental time series: An application to trend detection in ground-level ozone. *Statistical Science* **4**, 367–393.

Smith, Richard L. (1991). Weibull regression models for reliability data. *Reliability and System Safety* **34**, 55–77.

Smith, W.L. (1958). Renewal theory and its ramifications. *Journal of the Royal Statistical Society* **B20**, 243–302.

Smoluchowski, M. von (1915). Notiz über die Berechnung der Brownsche Molekular-bewegung bei der Ehrenhaft-millikanschen Versuchsanordnung. *Physikalische Zeitschrift* **16**, 318–321.

Snedecor, G.W. (1934). *Calculation and Interpretation of the Analysis of Variance*. Ames, Iowa: Collegiate Press.

Sobel, Milton, V.R.R. Uppuluri and K. Frankowski (1980). Dirichlet distributions—Type I. *Selected tables in Mathematical Statistics*, Vol. IV. Providence, Rhode Island: American Mathematical Society.

Spizzichino, Fabio (2001). *Subjective Probability Models for Lifetimes*. Boca Raton, Florida: Chapman & Hall/CRC.

Sprague, T.B. (1861). On Mr. Gompertz's law of human mortality, and Mr. Edmonds's claims to its independent discovery and extension. *The Assurance Magazine and Journal of the Institute of Actuaries* (London) **9**, 288–295.

Spurgeon, E.F. (1932). *Life Contingencies*, 3rd edn. Cambridge, UK.: Cambridge Univ. Press.

Stacy, E.W. (1962). A generalization of the gamma distribution. *The Annals of Mathematical Statistics* **33**, 1187–1192.

Steutel, F.W. (1967). Note on the infinite divisibility of exponential mixtures. *The Annals of Mathematical Statistics* **38**, 1303–1305.

Steutel, F.W. (1969). Note on completely monotone densities. *The Annals of Mathematical Statistics* **40**, 1130–1131.

Steutel, F.W. (1973). Some recent results in infinite divisibility. *Stochastic Processes and their Applications* **1**, 125–143.

Stoyan, Dietrich (1977). *Qualitative Eigenschaften und Abschätzungen stochastischer Modelle*. Berlin, Germany: Akademie-Verlag. [*Comparison Methods for Queues and Other Stochastic Models*. 1983. (Translated by Daryl J. Daley). Chichester, U.K.: John Wiley & Sons.]

Sweet, Arnold L. (1990). On the hazard rate of the lognormal distribution. *IEEE Transactions on Reliability* **39**, 325–328.

Szekli, R. (1995). *Stochastic Ordering and Dependence in Applied Probability.* (Lecture Notes in Statistics, Vol. 97). New York: Springer-Verlag.

Tai, Bee-Choo, Ian R. White, Val Gebski, and David Machin (2002). On the issue of 'multiple' first failures in competing risk analysis. *Statistics in Medicine* **21**, 2243–2255.

Taillie, C. (1981). Lorenz ordering within the generalized gamma family of income distributions. *Statistical Distributions in Scientific Work*, Vol. 6, C. Taillie, G.P. Patil, and B. Baldessari (eds.), pp. 181–192. Dordrecht, The Netherlands: Reidel.

Teicher, Henry (1960). On the mixture of distributions. *The Annals of Mathematical Statistics* **31**, 55–73.

Teicher, Henry (1962). Identifiability of mixtures. *The Annals of Mathematical Statistics* **32**, 244–248.

Thiele, T.N. (1872). On a mathematical formula to express the rate of mortality throughout the whole of life, tested by a series of observations made use of by the Danish Life Insurance Company of 1871. *Journal of the Institute of Actuaries* **16**, 313–329.

Thorin, Olof (1977a). On the infinite divisibility of the Pareto distribution. *Scandinavian Actuarial Journal* 31–40.

Thorin, Olof (1977b). On the infinite divisibility of the lognormal distribution. *Scandinavian Actuarial Journal* 121–148.

Thorin, Olof (1978). Proof of a conjecture of L. Bondesson concerning infinite divisibility of powers of the gamma variable. *Scandinavian Actuarial Journal* 151–164.

Thyrion, P. (1964). Les lois exponentielles composées. *Association Royale des Actuaires Belges Bulletin* **62**, 35–44.

van Tiel, J. (1984). *Convex Analysis.* Chichester, U.K.: John Wiley & Sons.

Titterington, D.M., A.F.M. Smith, and U.E. Makov (1985). *Statistical Analysis of Finite Mixture Distributions.* Chichester, U.K.: John Wiley & Sons.

Tsiatis, Anastasios Athansios (1974). *Evaluation of Competing Risks.* Doctoral dissertation, University of California, Berkeley, California.

Tsiatis, Anastasios (1975). A nonidentifiability aspect of the problem of competing risks. *Proceedings of the National Academy of Science of the United States of America* **72**, 20–22.

Tweedie, M.C.K. (1941). A mathematical investigation of some electrophoretic measurements of colloids. Unpublished M. Sc. thesis, University of Reading, U.K.

Tweedie, M.C.K. (1945). Inverse statistical variates. *Nature* **155**, 453.

Tweedie, M.C.K. (1956). Some statistical properties of the inverse Gaussian distribution. *Virginia Journal of Science* **7**(3), 160–165.

Vardi, Y., L.A. Shepp, and B.F. Logan (1981). Distribution functions invariant under residual-lifetime and length-biased sampling. *Zeitschrift für Wahrscheinlichkeitstheorie und verwandte Gebiete* **56**, 415–426.

Vaupel, J.W., K.G. Manton, and E. Stallard (1979). The impact of heterogeneity in individual frailty on the dynamics of mortality. *Demography* **16**, 439–454.

Vaupel, James W. and Anatoli I. Yashin (1985). Heterogeneity ruses: Some surprising effects of selection on population dynamics. *The American Statistician* **39**, 176–185.

Verhulst, P.F. (1838). Notice sur la loi que la population suit dans accroissement. *Correspondance mathématique et physique, publiée par L.A.J. Quetelet*, **10**, 113–121.

Verhulst, P.F. (1845). Recherches mathématiques sur la loi d'accroissement de la population. *Nouvelles Mémoires de l'Académie Royale des Sciences et Beaux-Arts de Belgique, Series* 2, **20**, 32 pp.

Voda, Viorel Gh. (1989). New models in durability tool-testing: Pseudo-Weibull distribution. *Kybernetika* **25**, 209–215.

Wald, A. (1947). *Sequential Analysis*. New York: John Wiley & Sons.

Walker, Stephen G. and David A. Stephens (1999). A multivariate family of distributions on $(0, \infty)^p$. *Biometrika* **86**, 703–709.

Wang, Jane-Ling, Hans-Georg Müller, and William B. Capra (1998). Analysis of oldst-old mortality lifetables revisited. *The Annals of Statistics* **26**, 126–163.

Wang, Y.H. (1976). A functional equation and its application to the characterization of Weibull and stable distributions. *Journal of Applied Probability* **13**, 385–391.

Wang, Yao, Anwar M. Hossain, and William J. Zimmer (2003). Monotone log–odds rate distributions in reliability analysis. *Communications in Statistics: Theory and Methods* **32**, 227–244.

Webster, Roger (1994). *Convexity*. Oxford, U.K.: Oxford Univ. Press.

Weibull, W. (1939a). A statistical theory of the strength of materials. *Ingeniörs Vetenskaps Akademiens Handligar*, No. 151, Stockholm, Sweden.

Weibull, W. (1939b). The phenomenon of rupture in solids. *Ingeniörs Vetenskaps Akademiens Handligar*, No. 153, Stockholm, Sweden.

Weibull, W. (1951). A statistical distribution function of wide applicability. *Journal of Applied Mechanics* **18**, 293–297.

Weibull, W. (1952). Discussion: A statistical distribution function of wide applicability. *Journal of Applied Mechanics* **19**, 233–234.

Whitmore, G.A. and M. Yalovky (1978). A normalizing logarithmic transformation for inverse Gaussian random variables. *Technometrics* **20**, 207–208.

Wittstein, Theodor (1883). The mathematical law of mortality. (Translated by D.A. Bumsted.) *The Assurance Magazine and Journal of the Institute of Actuaries* (London) **24**, 153–173.

Woolhouse, W.S.B. (1863). Observations on Gompertz's law of mortality and the dependence between it and Simpson's rule for finding the value of an annuity on three lives. *The Assurance Magazine and Journal of the Institute of Actuaries* (London) **10**, 121–130.

Xie, M., T.N. Goh, and Y. Tang (2002). A modified Weibull extension with bathtub-shaped failure rate function. *Reliability Engineering and Systems Safety* **76**, 279–285.

Yashin, Anatoli (2004). Semiparametric models in the studies of aging and longevity. *Parametric and Semiparametric Models with Applications to Reliability, Survival Analysis and Quality of Life*, M.S. Nikulin, N. Balakrishnan, M. Mesbah, and N. Limnios (eds.), pp. 149–164. Boston, Massachusetts: Birkhäuser.

Yitzhaki, Shlomo (2003). Gini's mean difference: A superior measure of variability for non-normal distributions. *Metron* **LXI** 285–316.

Zelen, Marvin (1959). Factorial experiments in life testing. *Technometrics* **1**, 269–288.

Zheng, Ming and John P. Klein (1995). Estimates of marginal survival for dependent competing risks based on an assumed copula. *Biometrika* **82**, 127–138.

Zigangirov, K.S. (1962). Expression for the Wald distribution in terms of normal distribution. *Radio Engineering and Electronic Physics* **7**, 164–166.

van Zwet, W.R. (1964). *Convex Transformations of Random Variables*. Amsterdam, The Netherlands: Mathematisch Centrum.

Author Index

Aalen, Odd O. 133, 183, *733*
Abdel-Hameed, Mohamen S. *738*
Abramowitz, Milton 325, 444, 654, 717, 719, 720, 721, 722, 724, 725, *733*
Acar, M. 527, *744*
Aczél, Janos 370, 371, 593, 701, 702, 703, 704, 706, 710, 711, 712, 713, 714, 715, *733*
Adamidis, K. 338, *733*
Aeppli, D. *756*
A-Hameed, M.S. 183, *733*
Ahuja, J.C. 441, *733*
Aitchison, J. 431, 432, 438, *733*
Al-Awadhi, F.A. 411, *744*
Allen, W.R. 552, *733*
Amoroso, Luigi 348, *733*
An, Mark Yuying 98, *733*
Andersen, Per Kragh 536, 538, *733*, *734*
Ando, T. 64, *734*
Andréief, C. 696, *734*
Ansell, J.I. 536, *734*
Anthony, P.J. 438, *748*
Appleton, David R. 535, *734*
Armitage, P. 552, *734*
Arnold, Barry C. 42, 68, 72, 73, 246, 273, 296, 305, 330, 399, 412, 424, 547, 552, 556, 557, *734*
Arthurs, A.M. *738*
Artin, E. 717, 718, 719, *734*

Asmussen, Søren 665, *734*
Azlarov, T.A. 291, 296, 701, *734*

Badia, F.G. 81, 161, *735*
Bailey, R. Clifton 382, 540, *735*
Balakrishnan, N. 98, 177, 238, 273, 291, 313, 317, 323, 345, 364, 401, 431, 432, 433, 441, 467, 669, *734*, *735*, *739*, *747*, *750*, *762*
Baldessari, B. *760*
Barlow, Richard E. 8, 13, 36, 40, 80, 107, 109, 110, 111, 117, 119, 143, 150, 155, 157, 161, 167, 179, 195, 199, 202, 206, 208, 225, 315, 645, 649, 650, 665, 671, 694, *735*, *748*
Barndorff-Nielsen, O.E. 451, 460, 465, 466, *736*
Bartholomew, D.J. 36, 40, *735*
Bartoszewicz, Jaroslaw 67, *736*
Bassan, Bruno 170, *736*
Basu, A.P. 219, 302, 552, 553, 556, 557, *735*, *736*, *737*
Bates, G. 667, *736*
Beckenbach, Edwin F. 337, *736*
Bellman, Richard 337, *736*
Belzunce, Felix 58, 61, *755*
Bencala, Kenneth 433, *736*
Bennett, Steve 242, 244, 538, *736*
Bergman, Bo 174, *736*
Berman, S.M. 549, *736*
Bernoulli, Daniel 541, *736*
Bernstein, Serge 654, *736*

Berrade, M.D. 81, 161, *735*
Bertholon, Henri 528, *736*
Bhattacharyya, G.K. 467, *736*
Bickel, P.J. 47, *736*
Billingsley, Patrick 40, 651, 652, 652, 676, 731, *737*
Birnbaum, Z.W. 47, 137, 151, 156, 438, 467, 542, *737*
Blanda, Keven P. 48, 67, *737*, *751*
Block, Henry W. 13, 81, 87, 88, 114, 115, 128, 133, 158, 159, 162, 167, 173, 179, 232, 553, *737*, *738*, *753*
Boethius, Anicius Manlius Severinus 97
Bohr, Harald 718, *738*
Boland, Philip J. 59, 145, 146, 151, *738*
Bondesson, Lennart 314, 348, 353, 359, 360, 656, *738*
Borgan, Ørnulf 536, *734*
Borgman, Leon E. 408, 528, *734*, *739*
Boron, Leo *754*
Bosch, A. J. 305, *738*
Bousquet, Nicolas 528, *736*
Box, G.E.P. 449, *738*
Bradley, Dorothy H. 257, 394, 395, 396, *738*
Bradley, Edwin L. 257, 394, 395, 396, *738*
Brascamp, H.J. 693, *738*
Bremner, J.M. 36, 40, *735*
Brockett, Patrick L. 547, 552, 556, 557, *734*
Brockmeyer, E. 313, *738*, *742*
Brown, J.A.C. 431, 432, 433, *733*
Brown, Mark 53, 167, 169, 203, 214, 245, 260, 650, *738*, *739*
Brunk, H.D. 36, 40, *735*
Burai, Pál 690, *739*
Burr, I.W. 353, *739*
Butler, M. *734*

Camp, B.H. 199, *739*
Campo, Raphael 36, 107, 155, *735*
Campos, C.A. 81, 161, *735*
Canfield, Ronald V. 408, 528, *739*
Capra, William B. 364, 366, *761*
Carling, Kenneth 559, *739*

Carroll, Lewis 291
Castillo, Enrique 238, 669, 701, 702, 703, 711, 712, 713, 714, *739*
Cauchy, A.L. 701
Celeux, Gilles 528, *736*
Champernowne, D.G. 425, *739*
Chandra, Mahesh 42, 328, *739*
Charlier, C.V.L. 73, *739*
Chebyshev, P.L. 208, *739*
Chen, John T. 465, *755*
Chen, Y.Y. 169, *739*
Chen, Zhenmin 393, *739*
Cherubini, Umberto 686, *739*
Chhikara, Raj S. 451, 452, 456, 458, *739*, *740*, *743*, *749*
Chiang, Chin Long 542, *740*
Chung, Kai Lai 99
Cicero, Marcus Tullius *541*
Çinlar, Erhan *738*
Cirillo, R. 400, *740*
Clayton, D.G. 242, *740*
Cohen, A. Clifford 238, *735*
Cox, D.R. 17, 449, 536, 537, 554, 568, 665, *738*, *740*
Cramér, Harald 6, 432, 652, *740*
Crawford, Gordon B. 302, *740*
Crow, Edwin L. 431, 432, *740*, *758*
Crowder, Martin J. 536, 538, 542, 554, 557, *740*
Csörgo, M. *746*

Dabrowska, Dorota M. 255, 285, *740*
Dahiya, Ram C. 390, *740*
D'Alembert, Jean le Rond 701
Daley, Daryl J. *759*
Daróczy, Z. 701, *733*
David, H.A. 238, 542, *740*
Davis, Henry T. 525, *740*
De Morgan, Augustus 363, 369, 370, 371, 384, *740*, *741*
Deshpande, Jayant V. 650, *741*
Desmond, Anthony 469, 471, *741*
Desu, M.M. 301, 551, *741*
Dharmadhikari, Sudhakar 639, *741*
Dhillon, B.S. 393, 409, *741*
Doksum, Kjell 65, 255, 284, *740*, *741*
Donoghue, W.F. Jr. 654, *741*

Dubey, S.D. 241, 407, 441, *741*
Dugué, D. 451, 459, *741*

Edgeworth, F.Y. 73, *741*
Edmonds, T. R. 363, *741*
Efron, Bradley 649, *741*
Eggenberger, F. 422, *742*
Elderton, William Palin 5, 6, 313, *742*
El-Neweihi, Emad 59, 145, 146, 151, *738*
Epstein, B. 291, *742*
Erdélyi, A. 462, 466, 654, 717, *742*
Erlang, A.K. 313, *742*
Esary, J.D. 90, 137, 143, 144, 145, 147, 148, 151, 156, 165, 183, 184, 186, 306, 307, 680, 681, *737*, *742*
Escobar, Luis A. 536, *753*
Euler, Leonhard 701
Evans, Ralph A. 297, 323, *742*

Feigl, Polly 537, *743*
Feldstein, Michael 525, *740*
Feller, William 3, 99, 100, 184, 187, 245, 261, 281, 345, 440, 651, 652, 653, 656, 657, 665, 667, 676, 704, *743*
Ferguson, Thomas S. 302, *743*
Findeisen, Peter 318, *743*
Fisher, R.A. 323, 669, *743*
Fisk, Peter R. 401, 425, *743*
Fletcher, Harvey 466, *743*
Flinn, C.J. 542, *743*
Folks, J. Leroy 451, 452, 456, 458, *739*, *740*, *743*, *749*
Frankel, Paul 536, *743*
Frankowski, K. 727, *759*
Fréchet, Maurice 199, 669, 678, *743*
Freimer, Marshall 353, *754*
French, Joyce M. 535, *734*
Fries, Arthur 467, *736*
Freudenthal, A.M. 467, *743*
Fussell, J.B. *735*

Galambos, Janos 291, 296, 302, *743*
Galton, F. 72, 431, 433, *744*
Gantmakher, Feliks Ruvinovich 694, *744*

Gasca, Mariano 694, *744*, *756*
Gastwirth, J.L. 42, 45, *744*
Gauss, Carl Friedrich 198, 199, *744*
Gaver, D.P. 527, *744*
Gawronski, C. R. 530, *744*
Ge, Guangping 214, *739*
Gebski, Val 544, *760*
Genest, C. 685, *744*
Gertsbakh, I.B. 167, 304, *744*
Ghai, Gauri L. 18, *744*
Ghitany, M.E. 411, *744*
Ghosh, B.K. 501, *744*
Ghosh, J.K. 552, 556, 557, *736*
Gibbon, J. 320, *753*
Gibrat, R. 432, *744*
Gide, André 195
Gill, Richard D. 536, *734*
Gini, C. 44, *744*
Gjessing, Håkon K. 133, 183, *733*
Glaser, Ronald E. 128, 135, 350, 492, *745*
Gleser, Leon Jay 312, *745*, *752*
Gnedenko, B.V. 323, 669, *745*
Goh, T.N. 393, *762*
Goldie, Charles 656, *745*
Gompertz, Benjamin 5, 363, 365, 375, 380, *745*
Good, I.J. 320, *745*
Gorski, Andrew C. 323, *745*
Gottfried, Paul 439, *745*
Govan, John 386, *745*
Gradshteyn, I.S. 462, *745*
Gray, Peter 363, *745*
Greenwood, Major 667, *745*
Groeneveld, Richard A. 72, 73, 424, *734*
Grosswald, E. 466, *745*
Guess, Frank 18, *745*
Gupta, Arjun K. 465, *755*
Gupta, P.L. 68, *745*
Gupta, Ramesh C. 17, 20, 88, 136, 162, 165, 169, 242, 244, 250, 251, *746*, *749*
Gurland, John 113, 114, 133, 407, *745*, *746*
Gut, Allan 261, 646, 651, 652, 654, 655, 676, 692, 731, *746*

Hadi, Ali S. 669, *739*
Hadwiger, H. 451, *746*
Haines, Andrew 172, *746*
Halgreen, C. 451, 460, 466, *736*
Hall, W.J. 18, 529, *746*
Hallinan, Arthur J. Jr. 321, 323, *746*
Halphen, Étienne 451, 459
Halstrøm, H.L. 313, *738*, *742*
Hardy, G.H. 231, 333, 644, 673, 694, *746*
Harris, Carl M. 80, 407, *746*
Harvey, Paul 137
Heckman, James J. 542, 559, *743*, *746*
Herk, L.F. 77, *747*
Heyde, C.C. 440
Hjorth, Urban 325, 527, 528, *746*
Hoeffding, W. 678, *746*
Hollander, Myles 98, 169, *739*, *746*, *756*
Homer, Louis D. 382, 540, *735*
Hong, C.S. 77, *747*
Honoré, Bo E. 559, *746*
Hosmer, David W., Jr. 536, *747*
Hossain, Anwar M. 192, *761*
Hossain, Syed A. 390, *740*
Hougaard, Philip 233, 530, *747*
Hu, Taizhong 227, *747*
Huang, J.S. 296, *734*
Hutchinson, T.P. 305, 686, *747*
Hutson, Alan D. 353, 354, *754*

Ibragimov, I.A. 100, *747*
Isaacson, Dean 330, *734*

Jacobson, Tor 559, *739*
James, Ian R. 677, *747*
Jayakumar, K. 256, 539, *757*
Jensen, A. 313, *738*, *742*
Jensen, Finn 121, *747*
Jewell, Nicholas P. *739*, *740*, *759*
Ji, P. 332, 354, *747*
Jiang, Renyan 126, 131, 285, 323, 331, 332, 349, 354, 393, 528, *747*, *754*
Joag-dev, Kumar (Jogdeo, K.) 639, 683, *741*, *747*

Joe, Harry 81, 87, 128, 133, 678, *737*, *741*
Johnson, Norman Lloyd 6, 98, 177, 273, 291, 313, 317, 323, 345, 364, 401, 432, 433, 441, 442, 658, 669, *747*, *750*
Johnstone, Iain 649, *741*
Jørgensen, Bent 452, 465, *747*, *748*
Joyce, W.B. 433, *748*

Kagan, A. M. 701, *748*
Kakwani, Nanak C. 68, *748*
Kalbfleisch, J.D. 225, 401, 412, 534, 535, 536, 537, 539, 568, 569, *748*
Kaminsky, Kenneth S. 372, *748*
Karamata, J. 63, *748*
Karlin, Samuel 98, 647, 684, 694, 697, *748*
Karn, M. Noel 541, *748*
Keiding, Niels 536, 538, *734*, *748*
Keilson, Julian 54, 281, *748*
Keissler, Peter C. 121, *749*
Kemp, Adrienne W. 658, *747*
Kendall, Maurice G. 313, *749*
Kent, John T. 281, *749*
Khalfan, L. 411, *744*
Kimball, A.W. 682, *749*
Kimber, A.C. 536, *739*, *740*, *759*
Kingman, J.F.C. 6, *749*
Kirmani, S.N.U.A 20, 162, 165, 242, 244, 250, 251, *749*
Klefsjö, Bengt 171, 174, 177, *749*
Kleiber, Christian 42, 44, *749*
Klein, John P. 555, *762*
Klutke, Georgia-Ann 121
Kochar, Subhash C. 147, 552, 553, 650, *741*, *749*
Kollia, Georgia D. 521, *754*
Kolmogorov, Andreii Nikolaevich 635
Konstantinowsky, D. 467, *749*
Kopylev, Leonid 559, *759*
Kordonsky, Kh.B. 304, *744*
Kotz, Samuel 42, 44, 98, 177, 183, 273, 291, 296, 302, 313, 317, 323, 345, 364, 401, 431, 432, 433, 441, 442, 658, 669, 670, 671, *743*, *747*, *749*, *750*
Kre'in, M.G. 694, *744*, *747*

Krishnaih, P.R. 745, 746, 756
Kunitz, Harald 393, 749
Kupka, Joseph 18, 750

Lachenbruch, P.A. 445, 446, 448, 759
Laha, R. G. 701, 751
Lai, C.D. 121, 305, 686, 747, 750, 751
Lancaster, H.O. 317, 750
Langberg, N. 169, 551, 554, 739, 750
Lawless, J.F. 536, 537, 568, 750
Lawrance, A.J. 305, 750
LeCam, L. 735, 753
Lee, Eliza T. 382, 750
Lee, Mei-Ling Ting, 739, 740, 759
Leemis, Lawrence M. 393, 750
Lehmann, E.L. 47, 53, 57, 61, 678, 736, 750
Lemeshow, Stanley 536, 747
Lewis, P.A.W. 305, 750
Li, Chin-Shang 536, 750
Li, Yulin 87, 88, 114, 115, 159, 162, 167, 173, 232, 737
Lieb, E.H. 693, 738
Likeš, J. 320, 750
Limnios, N. 762
Lindsay, Bruce G. 79, 750
Linnik, Yu. V. 701, 748
Littlewood, J.E. 321, 333, 644, 673, 694, 746
Ljubo, M. 525, 751
Lochbaum, David 121
Logan, B.F. 576, 760
Lomax, K.S. 369, 401, 751
Longmate, Jeffrey 536, 743
Loo, Sonny 18, 750
Lorenz, M.O. 42, 751
Losinger, Willard C. 45, 751
Loukas, S. 338, 733
Luciano, Elisa 686, 739
Lukacs, E. 318, 701, 751
Lundberg, O. 667, 751
Lynch, James D. 114, 115, 751
Lynn, Nicholas J. 117, 751

Ma, C. 226, 751
MacGillivray, H.L. 48, 67, 449, 737, 751
Machin, David 544, 760

Magnus, W. 462, 466, 654, 717, 742
Makeham, William Matthew 322, 363, 375, 376, 380, 387, 751
Makov, U.E. 79, 80, 760
Malik, Henrick John 417, 751
Mallows, C.L. 201, 751
Mann, H.B. 47, 50, 752
Mann, N.R. 291, 752
Manton, K. G. 233, 538, 761
Markov, A.A. 198, 752
Markus, M. 671, 752
Marsaglia, George 318, 752
Marshall, Albert W. 13, 52, 63, 64, 77, 90, 100, 109, 110, 111, 119, 142, 143, 144, 145, 147, 148, 151, 156, 165, 175, 176, 177, 179, 183, 184, 186, 187, 188, 195, 199, 202, 206, 208, 242, 261, 263, 282, 302, 305, 306, 307, 338, 347, 552, 556, 645, 646, 649, 650, 665, 683, 686, 688, 693, 694, 722, 726, 732, 735, 737, 742, 752, 753
McAlister, D. 431, 433, 753
McCullagh, P. 244, 538, 753
McDonald, James B. 350, 492, 753, 757
McEwen, R.P. 325, 753
McGill, W.J. 320, 753
McKay, J. 685, 744
Meeker, William Q. 536, 753
Meidell, B. 199, 753
Mesbah, M. 762
Meyer, Paul A. 62, 65, 753
Meza, J.C. 347, 752
Mi, Jie 18, 87, 131, 737, 744, 753
Micchelli, Charles A. 694, 744, 756
Miller, D.R. 550, 554, 753
Miller, Rupert G. Jr. 97, 753
von Mises, Richard 13, 671, 754
Moeschberger, M.L. 542, 740
de Moivre, Abraham 363, 754
Mollerup, J. 718, 738
Moran, P.A.P. 677, 678, 754
Morris, K.W. 726, 754
Mosler, K. 752
Mudholkar, Govind S. 353, 354, 521, 754
Mukerjee, Hari 147, 749

Müller, Alfred 48, 52, *754*
Müller, Hans-Georg 364, 366, 678, *761*
Murthy, D.N. Prabhakar 126, 131, 285, 323, 331, 332, 349, 354, 393, 528, *747*, *750*, *754*
Murthy, V.K. 121, 525, *754*
Muth, Eginhard J. 17, 18, *754*

Nadarajah, Saralees 323, 669, 670, 671, *749*
Naftel, David C. 257, 394, 395, 396, *738*
Nagaraja, H. N. 238, *740*
Nanda, Asok K. 53, 103, 151, 227, *747*, *754*, *758*
Narula, Subhash C. 551, *741*
Nash, Ogden 271
Nash, Stanley W. 441, *733*
Natanson, I.P. 638, *754*
Navarro, Jorge 58, 61, *755*
Navascués, M.A. 81, 161, *735*
Nelsen, Roger B. 291, 555, 685, 686, *755*
Nelson, Wayne 536, *755*
Neyman, J. 667, *735*, *736*, *753*
Nguyen, Truc T. 465, *755*
Nikulin, M.S. *762*

Oakes, D. 233, 536, 568, *740*, *755*
Oberhettinger, F. 462, 466, 654, 717, *742*
O'Cinneide, Colm A. 305, *755*
Ogborn, M.E. 363, *755*
Oja, H. 73, *755*
Olkin, Ingram 52, 63, 64, 100, 119, 142, 154, 195, 242, 261, 263, 282, 302, 305, 307, 338, 347, 552, 556, 646, 665, 683, 686, 688, 694, 722, 726, 727, 732, *752*, *753*, *755*
O'Quigley, John 536, *755*
Ord, J.K. 79, *755*
Ovid, Publius Ovidius Naso 533

Palm, C. 320, *755*
Pareto, V. 399, *755*
Parresol, B.R. 325, *753*
Parzen, Emanuel 35, *755*

Patil, G.P. *760*
Pearson, Karl 5, 47, 73, 313, 721, 724, *755*
Perks, Wilfred 392, *755*
Perlman, M.D. *752*
Petersen, Niels Erik 121, *747*
Peterson, Arthur V., Jr. 548, 554, *756*
Pham, T.G. 45, *756*
Phelps, Robert R. 65, *756*
Phillips, M.J. 536, *734*
Phillips, R.S. 203, *756*
Pickands, J. 401, 477, 671, 672, *756*
Pinkus, Allan 694, *752*, *756*
Pinsky, M. 671, *752*
Pitman, E.J.G. 50, *756*
Pólya, G. 231, 333, 422, 644, 673, 694, 695, 696, *742*, *746*, *756*
Pope, Alexander 47
Prékopa, A. 693, *756*
Prentice, R.L. 225, 401, 412, 521, 534, 535, 536, 537, 539, 568, 569, *748*, *756*
Press, S. James 80, *752*, *756*
Price, R. 120, *756*
Proschan, Frank 8, 13, 18, 59, 90, 98, 109, 110, 111, 118, 119, 143, 147, 148, 150, 151, 158, 161, 165, 167, 175, 176, 177, 179, 183, 184, 186, 187, 188, 195, 206, 225, 315, 552, 553, 554, 645, 649, 650, 665, 670, 671, 680, 681, 693, 694, *733*, *735*, *738*, *742*, *745*, *746*, *748*, *749*, *750*, *753*, *756*

Quinn, J. *738*
Quinzi, A.J. 551, 554, *750*

Rachev, S.T. 245, 246, *756*
Raftery, Adrian E. 305, *755*, *756*
Rajarshi, M.B. 121, 131, *757*
Rajarshi, Sujata 121, 131, *757*
Ramachandran, B. *748*
Rammler, E. 323, *757*
Rao, C. R. 701, *745*, *748*, *750*
Rao, J.N.K. *746*
Rayleigh, Lord J.W.S. 329, *757*

Resnick, Sidney I. 245, 246, 671, *756, 757*
Richards, Dale O. 350, 492, *753, 757*
Rider, Paul R. 726, *757*
Rieck, James R. 470, *757*
Rinott, Yosef 170, 684, *736, 748*
Roberts, A.W. 687, 688, *757*
Rockafeller, T. 687, 688, 689, *757*
Rojo, Javier 15, 53, 57, 61, 154, 163, 172, *750, 757*
Rolski, Tomasz 169, *757*
Rose, David Melvin 551, 555, *757*
Rosin, P. 323, *757*
Ross, S.M. 156, *757*
Royden, H.L. 208, *757*
Ruben, Harold 501, *757*
Rubin, Herman 98, *748*
Ruiz, José M. 58, 61, *755*
Ruiz-Cobo, Maria Reyes 238, 701, 702, 703, 711, 712, 713, 714, *739*
Ryzhik, I.M. 462, *745*

Sakh, A.K. *746*
Samaniego, Francisco J. 147, 150, *749, 757*
Sampson, A.R. *752, 753*
Sankaran, P.G. 256, 539, *757*
Sarabia, José M. 669, *739*
Sarkar, Ila C. 421, *754*
Särndal, Carl-Erik 6, *758*
Saunders, Sam C. 187, 451, 467, 470, *737, 758*
Savits, Thomas H. 13, 87, 88, 108, 114, 115, 158, 159, 162, 167, 173, 179, 232, 331, *737, 738, 753, 758*
Schafer, R.E. 291, *752*
Schoenberg, I.J. 696, 697, *758*
Schrödinger, E. 451, 467, *758*
Scott, E.L. *753*
Seal, Hilary L. 383, 541, *758*
Seinfeld, John H. 433, *736*
Sen, P.K. *746, 756*
Sengupta, Debasis 103, *758*
Serfling, R.J. *742, 746*
Seshadri, V. 451, 452, 455, 458, *748, 758*
Sethuraman, Jayaram 113, 114, 133, 407, 552, *746, 758*

Shaked, Moshe 13, 48, 49, 52, 53, 55, 64, 66, 67, 68, 75, 76, 151, 177, 179, *754, 758*
Shanthikumar, J. George 13, 48, 49, 52, 55, 64, 66, 67, 68, 75, 76, 177, 179, *758*
Shapley, L.S. 647, *748*
Shepp, L.A. 576, *760*
Sherwin, David J. 11, *758*
Shimi, I.N. *741, 754*
Shimizu, Kunio 431, 432, *740, 758*
Shinozuka, M. 467, *743*
Shisha, O. *753*
Shohat, J.A. 647, *758*
Shuster, Jonathan 455, 459, *758*
Simcov, D.M. 710
Singh, Harshinder 13, 179, 650, *737, 741*
Singpurwalla, Nozer D. 11, 12, 42, 80, 117, 172, 183, 291, 293, 328, 407, 535, *735, 739, 746, 749, 751, 752, 758*
Slud, Eric V. 559, *759*
Slymen, D.J. 445, 446, 448, *759*
Smith, A.F.M. 79, 80, *760*
Smith, Richard L. 536, 537, 672, *740, 759*
Smith, W.L. 665, *759*
Smoluchowski, M. von 451, *759*
Snedecor, G.W. 400, *759*
Sobel, Milton 291, 727, *742, 755, 759*
Spizzichino, Fabio 678, 683, *759*
Sprague, T.B. 363, *759*
Spurgeon, E.F. 5, *759*
Srivastava, Deo Kumar 353, 521, *754*
Stacy, E. W. 348, *759*
Stallard, E. 233, 538, *761*
Stare, Janez 536, *755*
Stegun, Irene A. 325, 444, 654, 717, 719, 720, 721, 722, 724, 725, *733*
Stephens, David A. 330, *761*
Steutel, F.W. 101, 281, 408, 656, *748, 759*
Stoyan, Dietrich 48, 52, 678, *754, 759*
Stuart, Alan 313, *749*
Sumita, Ushio 54, *748*
Summe, J.P. 382, 540, *735*

Swartz, G. 525, *754*
Sweet, Arnold L. 436, *759*
Sweeting, T.J. 536, *740*
Száz, Árpád 690, *739*
Szegö, G. 695, 696, *756*
Szekli, R. 48, *760*

Tai, Bee-Choo 544, *760*
Taillie, C. 319, *760*
Tamarkin, J.D. 647, *758*
Tang, Y. 393, *762*
Taylor, Jeremy M.G. 536, *750*
Teicher, Henry 79, *760*
Thiele, T.N. 321, *760*
Thorin, Olof 353, 359, 408, 440, *760*
Thyrion, P. 407, *760*
van Tiel, J. 687, *760*
Tippett, L.H. C. 323, 669, *743*
Titterington, D. M. 79, 80, *760*
Tricomi, F.G. 462, 466, 654, 717, *742*
Tsiatis, Anastasios Athanasios 546, 554, *760*
Turkkan, N. 45, *756*
Tweedie, M.C.K. 451, 452, *760*

Uppuluri, V.R.R. 727, *759*

Vanderpump, Mark P.J. 535, *734*
Varberg, D.E. 687, 688, *757*
Vardi, Yehuda 170, 576, *736*, *760*
Vaupel, James W. 87, 233, 538, *761*
Vecchiato, Walter 686, *739*
Verhulst, P. F. 333, *761*
Voda, Viorel Gh. 349, *761*
Volodin, N.A. 291, 296, 701, *734*

Wald, A. 451, *761*
Walker, Stephen G. 330, *761*

Walkup, D.W. 680, 681, *742*
Wang, Hongyue 529, *746*, *746*
Wang, Jane-Ling 364, 366, *761*
Wang, Y.H. 330, *761*
Wang, Yao 192, *761*
Warren, Robin 136, *746*
Webster, Roger 687, 719, 723, *761*
Weibull, W. 309, 323, 324, *761*
Wellner, J.A. 18, *746*
White, Ian R. 544, *760*
Whitmore, G.A. 458, 470, *739*, *740*, *748*, *759*, *761*
Whitney, D.R. 47, 50, *752*
Wilson, Simon P. 12, 535, *758*
Wittstein, Theodor 121, *761*
Wondmagegnehu, Eshetu T. 331, *738*
Woolhouse, W.S.B. 369, 371, *761*
Wortman, M.A. 121, *749*

Xie, Huiliang 227, *747*
Xie, Min 121, 285, 323, 331, 349, 393, 528, *750*, *754*, *762*

Yalovky, M. 458, *761*
Yashin, Anatoli 87, 375, *761*, *762*
Yitzhaki, Shlomo 44, *762*
Yuen, K. 525, *754*
Yule, G.U. 667, *745*

Zelen, Marvin 536, 537, *743*, *762*
Zheng, Ming 555, *762*
Zhu, Huiliang 227, *747*
Zigangirov, K. S. 455, *762*
Zimmer, William J. 192, *761*
van Zwet, W.R. 47, 71, 73, 319, *762*

Subject Index

abbreviations
  for distribution classes, 181
accelerated life models, 536
age parameter families, 264–265
  and stability, 619
  inverse distributions of, 264
  ordering, 265
  TTT transforms of, 265
Archimedean copula, 685
associated random variables, 558, 680–683
associativity equation, 712

balayage, 62
basic composition formula, 696
bathtub hazard rates, 120–133
  delayed, 131
  from minima, 128
  from mixtures, 122
  inverted, 131
Bernoulli distribution, 658
Bernoulli process, 304
Bessel functions, modified, 461, 732
beta distribution, 479–489
  and log concavity, 484
  bivariate, 682
  density shape, 480
  derivation of, 486
  generalized, 488
    limits of, 488
  hazard rate of, 484
  moments of, 485

  ordering of, 487
  residual life of, 486
  transformed, 521
beta function, 722
  incomplete, 724
Binet-Cauchy formula, 696
binomial distribution, 658
  tail of, 725
Birnbaum-Saunders
  distribution, 466–471
    derivation of, 470
bivariate
  beta distribution, 682
  distribution, bounds for, 678
  exponential distribution, 552, 553, 556
  F distribution, 682
  gamma distribution, 682
  Gompertz-Makeham
    distribution, 556
Box-Cox transformation, 449
Brownian motion and
  inverse Gaussian, 451
burn-in, 121
Burr distribution, 401

Cauchy equations, 701
  variants of, 704
cause specific survival function, 544
central limit theorem, 652
characteristic exponent
  of stable distribution, 657

Subject Index

characterization by coincidence, 563–609
  summary of, 566
Chebyshev covariance inequality, 673
Chebyshev inequality
  application of, 652
  for nonnegetive variables, 208
chi-square distributions, 317–318
  and inverse Gaussian, 459
  noncentral, 497
Choquet's theorem, 65
closure properties of nonparametric families, summary of, 182
coefficient of variation, 24, 69
  for exponential distributions, 295
  for NBUE, NWUE, 197
coherent life functions, 144
coherent structure functions, 138
coherent systems, 137–151
  ordering, 151
  with IHRA components, 156
  with NBU components, 165
coincidence of
  regression models, 539
  semiparametric families, 563, 609
    summary of, 566
    unresolved, 607
commutative semigroups, 621
competing risks, 541–559
completely monotone densities, 100
completely monotone functions, 654
component distribution of a mixture, 91
composite distributions, 523
composition formula, basic, 696
compound distributions, 27, 80
compound Poisson distribution, 657
concave distributions,
  bounds for, 208
  preservation properties of, 178
concave functions, 687
concomitant variables, 533
conditional
  mixing distributions, 84
  survival functions, 544
cone orders, 76
contagion, 422, 667

convergence
  almost surely, 652
  in distribution, 650
  in probability, 651
  weak, 650
convex cone, 76
convex functions, 687–694
  composition of, 689
  equivalent conditions for, 689
  inverse of, 690
  preservation properties, 693
convex order, 62
  equivalent conditions for, 63
convex transform order, 71
  equivalent conditions for, 71
convolution, 35, 655
  of IHR, 110
  of IHRA, 158
  of NBU, 165
  of NBUE, 175
convolution parameter families, 261–263
  and infinite divisibility, 262, 281
  mixtures of, 268, 280
  ordering, 262
  stability in, 619
copulas, 684
  and competing risks, 555
  and Fréchet bounds, 685
  and Laplace transforms, 685
  Archimedean, 685
correlation, 676
counting process, 663
covariance, 676
  formulas for, 676
covariate models, 533
crossings of distributions, 627, 698
  notation for, 699
crude survival function, 544
cumulative hazard functions, 11
cut set, 140

damage threshold models, 185
decreasing hazard rate average
  (see DHRA)
decreasing mean residual life
  (see DMRL)

Subject Index 773

decreasing reverse hazard rate
    (see DRHR)
decrement function, 543
densities
    completely monotone, 100
    decreasing, 178
    log concave, 31
    log convex, 31
DHR distributions, 31, 116–120
    bounds for, 202
    conditions for, 105
    equilibrium distributions of, 119
    mixtures of, 119
    moment inequalities for, 195
    properties of, 117, 118
DHRA distributions, 151, 160
    bounds for, 201
    equivalent conditions for, 160
    mixtures of, 160
    moment inequalities for, 196
differentiation
    chain rule for, 729
    of an inverse, 730
    of integrals, 730
Dirichlet integral, 726
dispersive order, 65
    equivalent conditions for, 66
distributions
    absolutely continuous, 9
    algebraic structure of, 494
    classification of, 637
    composite, 523
    compound, 80
    concave, 178
    conditional mixing, 84
    crossings of, 699
    equilibrium, 170, 649
        DHR for, 119
    function, 6, 7, 637
    inverse, 20
    lattice, 638
    log-concave, 101
    mixing, 80
    model, 80
    multivariate, 674
    not admitting parameters, 285
    of same type, 25, 651
    prior, 80
    predictive, 80
    product family, 237
    shape of, 67
    stable, 529, 657
        characterization of, 579
    support of, 9
    TTT transform of, 172
    underlying, 217
    with bounded support
        moments of, 474
DMRL distributions, 169–173
    and hazard rate order, 170
    mixtures of, 173
domain of attraction, 670
DRHR distributions, 178–180
    bounds for, 202
    closure properties of, 179
    mixtures of, 179
    residual life distribution of, 179
duality
    of frailty and resilience, 235
    of distributions, 474
duplication formula, 719

empirical distributions, 39
equilibrium distributions, 18, 284, 649
    and DHR, 119
    and IHR, 111
    hazard rate of, 650
Erlang distribution, 313
excess function, 17
expected value, 15
exponential distributions, 28, 291–307
    and DHR, DHRA, IHR, IHRA, NBU, NBUE, NWU, NWUE, 300
    and mixtures, 306
    as a limit, 300, 304
    as approximation, 214
    bivariate, 552, 553, 556
    characterization of, 296–298, 300, 301, 302
    closure under minima, 302
    coefficient of variation of, 295
    convergence of, 299
    density of, 292
    functional equation for, 301

exponential distributions (*cont.*)
  geometric distribution relationship, 304
  Gini Index of, 293
  hazard rate of, 292
  infinite divisibility of, 302, 656
  Laplace transform of, 295, 303
  limits for, 299
  location parameter added characterization of, 590
  Lorenz curve of, 293
  mean residual life of, 298
  mixtures of, 399, 656
  moments of, 28, 294
  multivariate, 683
  odds ratio, 293
  order statistics of, 302
  ordering for, 304
  Poisson processes connection, 663
  random sums for, 305
  residual life, 299
  resilience parameter added, 333
  scaled, 320
  thresholds, 306
  tilt parameter added, 113, 338
    characterization of, 594
    ordering, 345
  TTT transform of, 293
extreme stable family, 243
extreme value distributions, 323, 669
  transformed, 445
  characterization of, 570
  domain of attraction for, 672
  power parameter added, characterization of, 572

F distribution
  bivariate, 682
  density of, 422
  moments of, 419
  generalized, 411–418
    density of, 412
    from mixtures, 417
    from ratios, 417
    hazard rates of, 417
    limits of, 418
    ordering of, 423

failure rate (see hazard rate)
fatigue-life distribution, 467
Feller–Pareto distribution, 412
first-order stochastic dominance, 77
Fisher information, 648
Fisk distribution, 401
force of mortality, 11
frailty and resilience in consort, 238
frailty models, multivariate positive quadrant dependence of, 679
frailty parameter families, 233–242
  Gini index of, 240
  hazard rates of, 234
  inverse distribution of, 239
  majorization in, 242
  mixing, 241
  mixture of, 268
  order preservation in, 240
  ordering, 239
  stability of, 616
  TTT transform of, 239
Fréchet-Hoeffding bounds, 678
  and copulas, 685
Fubini's theorem, 731
functional equations, 701–715
  addition equations, 714
  associativity equation, 712
  Cauchy, 701
  Pexider, 705
  Simcov, 710
  transformation equation, 713

gamma distribution, 29, 310–321
  bivariate, 682
  characterizations of, 576, 596
  convolution parameter in, 312
  density properties, 31, 314
  derivation of, 311
  generalized, 348, 417
    hazard rates of, 350
    infinite divisibility of, 656
  hazard rate of, 315
  Laplace transform of, 311
  limits of, 319
  mixture representation, 312
  ordering, 318

Poisson processes connection, 663
residual life of, 316
TTT transform of, 317
gamma function, 717
characterization of, 718
incomplete, 720
total positivity of, 720
gamma-Weibull distribution
infinite divisibility of, 353
Gauss duplication formula, 719
generalized beta distribution, 488
limits of, 488
generalized gamma convolutions
infinite divisibility of, 359
limits of, 359
mixtures of, 359
generalized gamma distributions
(see gamma distributions, generalized
generalized negative binomial
distribution, 662
generalized Pareto distribution, 671
generalized Weibull distribution,
extended, 523
geometric distribution, 660
and tilt parameter families, 252
geometric-extreme stable, 245
Gibrat distribution
law of proportional effects, 432
Gini index, 69
Glivenko-Cantelli theorem, 40
Gompertz distribution, 364–375
coincidence characterization, 591
differential equation for, 364
functional equations for, 369–373
generalized exponential form, 373
hazard power parameter extension, 394
mixtures of, 113, 374
modified negative, 390
moments of, 367
negative, 368
negative-positive, 390
odds ratio derivation, 373
ordering, 374
Perks extension, 392
power parameter extension, 393
summary of extensions, 396

Gompertz-Makeham distribution, 375–389
bivariate, 556
complete monotonicity of, 383
differential equation for, 380
extended parameter range, 381
functional equations for, 384–386
moments of, 380
ordering, 386
random minimun derivation, 383
residual life distribution of, 386
second Makeham extension
differential equation for, 387
extended parameter range, 388
power parameter extension, 395
Gompertz-Rayleigh distribution, 396

Hamburger moment problem, 647
harmonic new better than used
in expectation (see HNBUE)
harmonic new worse than used in
expectation (see HNWUE)
Hausdorff moment problem, 647
hazard functions, 10
reverse, 13
hazard potential, 11, 183
hazard power parameter family, 256–258
hazard rates of, 257
inverse distribution, 258
ordering, 258
properties of, 257
TTT transform of, 258
hazard rates,
bathtub, 10, 97, 120–133
for location parameters, 221
for scale parameter family, 225
limiting, 87
monotone, 31, 104–120
of equilibrium distribution, 650
of mixtures, 27, 81, 125
reverse, 13
sum of, 524
unimodal, 131
hazard rate averages
monotone, 151–161

hazard rate order, 52–55
  and DMRL, 170
  equivalent conditions for, 52
hazard rate shape
  determination of, 133
  of coherent systems, 147
  of mixtures, 89
hazard transforms
  of coherent system, 147
  of mixtures, 89
heavy tail distribution, 399
Hessian matrix
  and convexity, 688
HNBUE, 177
HNWUE, 177
Hoeffding-Fréchet bound, 678
hypergeometric function, 721
Hölder's inequality, 732

IFR (see IHR)
IHR distributions, 31, 104–116
  bounds for, 212
  conditions for, 105
  equivalent conditions for, 107
  hazard function convexity, 105
  mixtures of, 111
  moment inequalities for, 195
  moments of, 109
  odds ratio characterization, 107
  PF_2 characterization, 106
  preservation theorems for, 109
  residual life distribution of, 111
  TTT transform of, 107
  unimodality for, 116
IHRA distributions, 151–161
  and crossings, 64
  bounds for, 201, 206, 211
  characterization of, 152
  closure under coherent systems, 156
  convolutions of, 158
  equivalent conditions for, 152
  mixtures of, 159
  moment inequalities for, 196
  moments of, 154
  properties of, 154
  residual life distributions of, 158

IMRL distributions, 169–173
  mixtures of, 172
incomplete beta function, 724
incomplete gamma function, 720
  and concavity, 722
increasing hazard rate average
  (see IHRA)
increasing mean residual life
  (see IMRL)
infinite divisibility, 261, 656
  and convolution families, 281
  exponential distribution, 656
  inverse Gaussian, 454
  Weibull distribution, 328
integration by parts, 729
inverse,
  derivative of, 730
  distribution of reciprocal random variable, 32
inverse distribution functions, 20, 35
inverse Gaussian distributions, 451–466
  and Brownian motion, 451
  chi-square distribution relationship, 459
  coefficient of variation, 457
  density of, 452
  generalized, 459
    convolutions of, 465
    hazard rate of, 463
    infinite divisibility of, 466
    Laplace transform of, 464
    moments of, 464
    ordering, 465
  hazard rate of, 456
  infinite divisibility of, 454
  Laplace transform of, 464
  moments of, 457
  normal limit, 458
  norming constants and Bessel functions, 461
  ordering, 458
  reciprocal of, 459
inverse survival function, 36
inverted bathtub hazard rate, 131

Jensen's inequality, 65, 692
joint distribution, conditional, 683

Subject Index   777

joint distribution function, 674
  bounds for, 548

k-out-of-n system, 139
Kolmogorov inequality, 652

lack of memory property, 222, 230, 296
Laplace transforms, 25, 653–658
  continuity of, 654
  convolutions, 655
  function of, 619
  uniqueness of, 653
Laplace transform order, 76
Laplace transform parameter families, 260, 618
  inverse distribution of, 261
  ordering, 261
  TTT transform of, 261
latent lives, 543, 549
  positive dependence of, 557
lattice distribution, 638
law of large numbers, 652
Lebesgue dominated convergence, 731
Lebesgue monotone convergence, 731
L'Hospital's rule, 729
life functions, coherent, 144
likelihood ratio order, 56–59
Liouville-Dirichlet integral, 726
location, measures of, 24
location parameter families, 220–224
  hazard rate of, 221
  inverse distribution of, 223
  ordering, 224
  stability of, 615
  TTT transform, 223
log Cauchy distribution
  hazard rate of, 443
  moments of, 444
log concave densities, 31, 98–103
  convolution of, 698
  hazard rate of, 102
  moment inequalities for, 195
  residual life of, 103
  reverse hazard rate of, 102
  unimodality of, 99

log concave distributions, 101, 103–116
  mixture of, 114, 115
log concave functions, 689
log convex densities, 31, 98
log convex functions, 689
log extreme value distribution, 442
log gamma distributions, 509–514
  characterization of, 588
  hazard rate of, 513
log Gompertz distribution, 518
log logistic distribution, 441
log Student's t distribution, 445
  negative, 514
log Weibull distribution, 442, 514–517
  characterization of, 582
  hazard rate of, 514
  negative, 514
logarithmic distributions, 427–449
  monotone hazard rate of, 430
  negative, 430
  properties of, 428
logistic distribution, 345
  differential equation for, 345
  stability for, 246
logistic law of growth, 345
lognormal distribution, 30, 427, 431–441
  characterization of, 574
  coefficient of variation of, 440
  density of, 434
  derivation of, 433
  growth models, 432
  hazard rate of, 135, 436
  infinite divisibility of, 440
  median of, 440
  mode of, 440
  moments of, 439
  ordering, 441
  physical models for, 431
  unimodality of, 436
Lomax distribution, 401
Lorenz curve, 42
Lorenz order, 68
Lyapunov inequality, 646

Macdonald's function, 732
maintenance policies, 187
majorization for proportional
    hazards, 242
marginal distribution function, 674
Markov's inequality, 198
    improvements of, 200
    reversals of, 204
maxima, distribution of, 33
Maxwell-Boltzman distribution, 329
mean residual life function, 15
Mellin transform, 25
Mills' ratio, 11
minimal cut sets, 682
minimum, distribution of, 32
mixing distribution, 80
mixtures, 79–94, 267–283
    and minima, 92
    closure under, 61
    hazard rates for, 81
    introduction of, 26
    of nonparametric families
        DHRA distributions, 160
        DMRL distributions, 173
        DRHR distributions, 179
        IHRA distributions, 159
        IMRL distributions, 172
        NBU distributions, 168
        NWU distributions, 166
        NWUE distributions, 176
    of semiparametric families,
        267–283
        convolution parameter families,
            280
        power parameter families, 232
        product and frailty
            parameter famailies, 268
        resilience parameter families,
            277
        scale parameter families, 267
    preservation of orders under, 94
model distribution, 80
moments, 22, 644–648
    existence of, 646
    formulas for, 645
    inequalities for, 195–198
    normalized, 23, 195, 294
    of IHR distributions, 109

of IHRA distributions, 196
of NBU distributions, 196
of NBUE distributions, 197
moment crossings, 699
moment generating function, 23, 653
moment parameter family, 258–260
    and stability, 618
    inverse distribution, 259
    ordering, 259
    TTT transform, 259
moment problems
    Hamburger, 647
    Hausdorff, 647
    Stieltjes, 647
mortality law and graduation, 5
multiple correlation distribution
    noncentral, 505
    unimodality of, 506
multiple decrement function, 543
mutual independence, 675

NBU distributions, 161–168
    bounds for, 203
    coherent systems of, 165
    convolution of, 165
    equivalent conditions for, 162
    moments of, 163, 196
    normalized moments of, 196
    odds ratio for, 165
    properties of, 163
    random sums of, 167
    replacement policy
        characterizations, 187
    residual life of, 166
NBU ordering, 386
NBUE distributions, 173–177
    bounds for, 203, 206
    coefficient of variation, 197
    convolutions of, 175
    inequalities for, 177
    normalized moments, 197
    preservation of, 174
    TTT transform of, 174
negative binomial distribution
    tail of, 726
negative log gamma distribution,
    513
    characterization of, 588

Subject Index   779

negative Weibull distribution
  characterization of, 568
net survival function, 543
new better than used distributions
  (see NBU)
new better than used in expectation
  (see NBUE)
new worse than used distributions
  (see NWU)
new worse than used in expectation
  (see NWUE)
noncentral F distribution, 501
  doubly, 503
  moments of, 501
  ordering of, 503
  unimodality of, 502
noncentral beta distribution, 504
  doubly, 508
  ordering of, 509
noncentral chi-square distribution, 497
  log-concavity of, 498
  unimodality of, 498
noncentral gamma distribution, 498
  Laplace transform of, 498
  ordering, 500
noncentral multiple correlation, 509
normal distribution as limit, 653
normalized moments, 23, 294
NWU distributions, 161–168
  bounds for, 203
  equivalent conditions for, 162
  moment inequalities, 196
  mixtures of, 166
NWUE distributions, 173–177
  bounds for, 204
  coefficient of variation, 197
  mixture of, 176
  normalized moments, 197
  replacement policy
    characterizations, 187
  TTT transform of, 174

odds ratio, 19
  and IHR, 107
  and NBU, 165
  and nonparametric families, 192
order statistics, 682

ordering coherent systems, 151
ordering semiparametric families
  age parameter, 265
  convolution parameter, 262
  frailty parameter, 239
  hazard power parameter, 258
  Laplace transform parameter, 261
  location parameter, 224
  moment parameter, 259
  power parameter, 230
  resilience parameter, 239
  scale parameter, 226
  tilt parameter, 250
orderings
  and crossings, 630
  generated by semiparametric
    families, 626
orders
  cone, 76
  equivalences for classes
    of distributions, 182
  of distributions, 47–77
    convex, 62
    convex transform, 70
    dispersive, 65
    hazard rate, 52
    Laplace transform, 76
    likelihood ratio, 56
    Lorenz, 68
    preservation properties of, 60
    star, 73
    stochastic, 47
    superadditive, 75
    summary of relationships, 75, 77
  partial, 49
  preservation
    under mixtures, 94
    with frailty parameters, 230
    with power parameters, 230
    with resilience parameters, 230
    with scale parameters, 227
    with tilt parameters, 250
  properties for semiparametric
    families, summary of, 283
  reflexivity of, 48
  transitivity of, 48

parallel system, 139
parameters
   classification of, 612
   successive addition of, 265
Pareto distributions, 399–411
   descendants of, 402
   from exponential mixtures, 411
   ordering of, 423
Pareto I distribution, 400
   characterization of, 589
Pareto II distribution, 400
   characterization of, 580, 594
   density, 401
   infinite divisibility of, 408
   with tilt parameter, 411
Pareto III distribution, 401
   and exponential with tilt, 410
   and log logistic distribution, 441
   characterization of, 570, 584
   density, 401
   generalization of, 424
   with resilience parameter, 402
Pareto IV distribution, 401
   as a limit, 408
   density, 401
   mixture representation, 407
   ordering, 424
   residual life distribution of, 407
partial orders, 49
path set, 140
peakedness, 47
percentiles, bounds for, 211
Pexider equation, 299, 371
planned replacements, 187
Poisson distributions, 659
   characterization of, 597
   compound, 657
   tail of, 721
Poisson mixture, 497
Poisson process, 303, 304, 663
Pólya frequency function, 696
   of order 2 (see PF2), 99
Pólya process, 663, 666
   waiting time in, 422, 487
Pólya urn models, 422
Pólya-Eggenberger distribution, 666
positive quadrant dependence, 678

power parameter families, 228–232
   and stability, 615
   hazard rates of, 230
   mixtures of, 232
   ordering, 230
   TTT transform of, 229
predictive distribution, 80
preorder, 49
prior distribution, 80
probability foundations, 635–643
probability density, 8
probability integral transform, 643
probability mass functions, 8
product families, 237
   mixtures of, 268
   survival function, 7
proportional hazards families, 233–242
   and majorization, 242
   Gini index of, 240
   hazard rates of, 234
   inverse distribution of, 239
   mixtures of, 268
   ordering, 239
   TTT transform of, 239
proportional hazards regression models, 537
   proportional odds family (see tilt parameter families)
proportional odds regression models, 538
proportional reverse hazards family, 234
   hazard rates of, 235
Prékopa's theorem, 693
pseudo survival function, 545
pseudo-Weibull distribution, 349

quantile, 35

random variables, 636
   functions of, 612
   independence of, 637
   indicator, 658
   transformations of, 629
rates of mortality, 5
Rayleigh distribution, 328

regression models, 533–540
    accelerated life, 536
    proportional hazards, 537
    proportional odds, 538
regressor variables, 533
reliability function, 142
renewal distribution
    stationary, 18, 111
renewal process, 664
    stationary, 665
renewal theory, 187–192, 663
replacement policies, 187–192
    age and block, 187
    and monotone hazard rate, 188
    and NBU, 187
    and NWU, 187
    and stochastic orders, 191
residual life distributions, 14
    and IHRA, 158
    convergence of, 15
    of a mixture, 84
    order closure under, 60
residual life family
        (see age parameter families)
resilience parameter families, 234
    frailty parameters and, 238
    Gini index for, 240
    mixing, 241, 278
    order preservation with, 240
    ordering, 239
    stability of, 616
reverse hazard function, 13
reverse hazard power parameter, 257
reverse hazard rate, 13
reverse hazard rate order, 53, 58

scale parameter families, 26, 224
    hazard rates of, 225
    inverse distributions of, 225
    mixing, 267
    order perservation with, 227
    ordering, 226
    stability of, 615
    TTT transaform of, 225
second-order stochastic dominance, 77
semiparametric families, 217–287, 611–631

coincidences of, 563–609
criteria for, 219, 611, 622
derivation of, 619
summary of, 218
summary of order properties, 283
series system, 139
shock models, 182
similarly ordered, 673
Simpson's paradox, 535
skewness, 70
    measures of, 72
spread
    measures of, 24
stability of semiparametric families, 219, 611, 622
stable distributions, 529, 657
    characterization of, 579
    index of, 529
Stacy's distribution, 348
standard deviation, 24
star order, 73
    equivalent conditions for, 74
starshaped functions, 690
    hazard transforms, 147
starshapedness, preservation of, 693
stationary renewal distribution
        (see equilibrium distribution)
stationary renewal process, 665
Stieltjes moment problem, 647
Stirling's formula, 719
stochastic dominance
    first-order, 77
    second-order, 77
stochastic order, 47, 50
    equivalent conditions for, 50
stochastic comparisions
    summary of, 191
structure function, 138
subadditive function, 691
sub-density function, 545
sub-survival function, 544
summaries for nonparametric families
    of abbreviations, 181
    of closure properties, 182
    of relationships, 181
    of stochastic comparisons, 191
sums, distribution of, 34

superadditive function, 691
superadditive order, 75
superadditive order
 and NBU, NWU, 162
superadditivity, preservation of, 693
survival functions, 7
 alternatives for determination, 21
 inequalities for, 198–214
 log-concavity of, 101
 products of, 523
 transformations of, 616, 625, 629
survivor function, 8

threshold models
 cumulative damage, 185
 random, 186
tilt parameter families, 242–256
 derivation from
  geometric extreme stability, 245
  mixtures, 248
  proportional odds, 243
 generalizations of, 255
 inverse distributions of, 249
 order preservation with
  introduction of, 250
 ordering, 250
 properties of, 251, 252
 TTT transform of, 250
 with underlying exponential
  distributions, 113
total lifetime, 665
total positivity, 694–700
 and logconcavity, 698
 multivariate, 683
totally positive functions
 composition of, 697
TTT statistic, 39
TTT transforms, 36
 and hazard rates, 40
 convergence of, 40
 for location parameter families, 223
 for power parameter families, 229

uniform distribution
 and IHR, 116
 and log-concavity, 99
 and stability, 616
 dual of, 493
 with added parameters, 489
 with frailty parameter,
  characterization of, 580
 with one added parameter, 475–479
unimodality, 10, 639
 strong, 99
unimodal hazard rates, 131
upper quadrant dependence, 558
urn models, 665

variation diminishing property, 697
Verhulst distribution, 332

wearout, 148
Weibull distributions, 29, 321–333
 characterization of, 330, 568
 coefficient of variation, 327
 density of, and concavity, 31
 extended family, 331
 generalized, 331, 521
  hazard rate for, 522
 Gini index of, 327
 hazard rate properties, 324
 infinite divisibility of, 328
 inverse, 331
 mixture representation of, 330
 mixtures of, 331
 moments of, 325
 ordering, 327
 residual life of, 332, 355
 TTT transform of, 326
 with resilience parameter, 353
 with tilt parameter, 355

Springer Series in Statistics *(Continued from page ii)*

Knottnerus: Sample Survey Theory: Some Pythagorean Perspectives
Konishi: Information Criteria and Statistical Modeling
Küchler/Sørensen: Exponential Families of Stochastic Processes
Kutoyants: Statistical Inference for Ergodic Diffusion Processes
Lahiri: Resampling Methods for Dependent Data
Lavallée: Indirect Sampling
Le Cam: Asymptotic Methods in Statistical Decision Theory
Le Cam/Yang: Asymptotics in Statistics: Some Basic Concepts, 2nd edition
Le/Zidek: Statistical Analysis of Environmental Space-Time Processes
Liu: Monte Carlo Strategies in Scientific Computing
Manski: Partial Identification of Probability Distributions
Mielke/Berry: Permutation Methods: A Distance Function Approach, 2nd edition
Molenberghs/Verbeke: Models for Discrete Longitudinal Data
Mukerjee/Wu: A Modern Theory of Factorial Designs
Nelsen: An Introduction to Copulas, 2nd edition
Pan/Fang: Growth Curve Models and Statistical Diagnostics
Politis/Romano/Wolf: Subsampling
Ramsay/Silverman: Applied Functional Data Analysis: Methods and Case Studies
Ramsay/Silverman: Functional Data Analysis, 2nd edition
Reinsel: Elements of Multivariate Time Series Analysis, 2nd edition
Rosenbaum: Observational Studies, 2nd edition
Rosenblatt: Gaussian and Non-Gaussian Linear Time Series and Random Fields
Särndal/Swensson/Wretman: Model Assisted Survey Sampling
Santner/Williams/Notz: The Design and Analysis of Computer Experiments
Schervish: Theory of Statistics
Shao/Tu: The Jackknife and Bootstrap
Simonoff: Smoothing Methods in Statistics
Song: Correlated Data Analysis: Modeling, Analytics, and Applications
Sprott: Statistical Inference in Science
Stein: Interpolation of Spatial Data: Some Theory for Kriging
Taniguchi/Kakizawa: Asymptotic Theory for Statistical Inference for Time Series
Tanner: Tools for Statistical Inference: Methods for the Exploration of Posterior
    Distributions and Likelihood Functions, 3rd edition
Tillé: Sampling Algorithms
Tsaitis: Semiparametric Theory and Missing Data
van der Laan/Robins: Unified Methods for Censored Longitudinal Data and Causality
van der Vaart/Wellner: Weak Convergence and Empirical Processes: With Applications
    to Statistics
Verbeke/Molenberghs: Linear Mixed Models for Longitudinal Data
Weerahandi: Exact Statistical Methods for Data Analysis

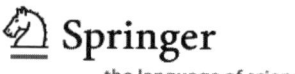

springer.com

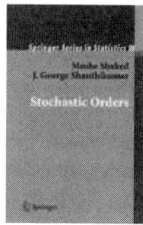

Stochastic Orders
Moshe Shaked and J. George Shanthikumar

This reference text presents comprehensive coverage of the various notions of stochastic orderings, their closure properties, and their applications. Some of these orderings are routinely used in many applications in economics, finance, insurance, management science, operations research, statistics, and various other fields. And the value of the other notions of stochastic orderings needs further exploration. This book is an ideal reference for those interested in decision making under uncertainty and interested in the analysis of complex stochastic systems. It is suitable as a text for advanced graduate course on stochastic ordering and applications.

2007. 475 pp. (Springer Series in Statistics) Hardcover
ISBN 978-0-387-32915-4

Reliability, Life Testing and the Prediction of Service Lives
Sam C. Saunders

This book is intended for students and practitioners who have had a calculus-based statistics course and who have an interest in safety considerations such as reliability, strength, and duration-of-load or service life. This book unifies the study of cumulative-damage distributions, (i.e., inverse-Gaussian and its reciprocal) with "fatigue-life."

2006. 340 p. (Springer Series in Statistics) Hardcover ISBN 0-387-32522-0

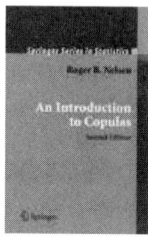

An Introduction to Copulas
Second Edition
Roger B. Nelsen

In this book the student or practitioner of statistics and probability will find discussions of the fundamental properties of copulas and some of their primary applications. The applications include the study of dependence and measures of association, and the construction of families of bivariate distributions. With nearly a hundred examples and over 150 exercises, this book is suitable as a text or for self-study.

2006. 270 pp. (Springer Series in Statistics) Hardcover
ISBN 978-0-387-28659-4

Easy Ways to Order▶ Call: Toll-Free 1-800-SPRINGER • E-mail: orders-ny@springer.com • Write: Springer, Dept. S8113, PO Box 2485, Secaucus, NJ 07096-2485 • Visit: Your local scientific bookstore or urge your librarian to order.